Using AutoCAD® 2005: Advanced

Using AutoCAD® 2005: Advanced

RALPH GRABOWSKI

autodesk® Press

THOMSON
DELMAR LEARNING™ Australia • Canada • Mexico • Singapore • Spain • United Kingdom • United States

autodesk® Press

Using AutoCAD® 2005: Advanced

Ralph Grabowski

Vice President, Technology and Trades SBU:
Alar Elken

Editorial Director:
Sandy Clark

Senior Acquisitions Editor:
James DeVoe

Senior Development Editor:
John Fisher

Marketing Director:
Dave Garza

Channel Manager:
Dennis Williams

Marketing Coordinator:
Casey Bruno

Production Director:
Mary Ellen Black

Production Manager:
Andrew Crouth

Production Editor:
Thomas Stover

Art & Design Specialist:
Mary Beth Vought

Cover Image:
Photonica

Printed in the United States
2 3 4 5 XXX 07 06 05

For more information contact Delmar Learning
Executive Woods
5 Maxwell Drive, PO Box 8007,
Clifton Park, NY 12065-8007
Or find us on the World Wide Web at
www.delmarlearning.com

Library of Congress Cataloging-in-Publication Data:
Card Number: [Number]

ISBN: 1-4018-8386-9

NOTICE TO THE READER

BRIEF CONTENTS

APPENDICES

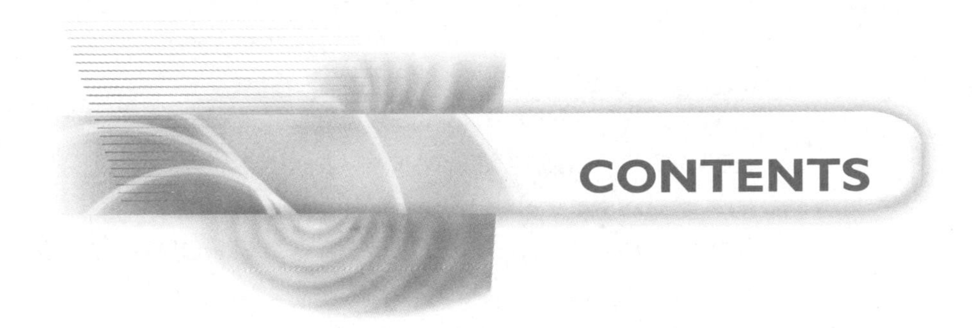

CONTENTS

This icon indicates command or options new to AutoCAD 2005.

UNIT II — CONNECTING WITH OTHER PROGRAMS

Chapter 5 — Importing and Exporting Files 141

UNIT III — THREE DIMENSIONAL DESIGN

Chapter 10 — Basic 3D Drafting and Editing 281

UNIT V — CUSTOMIZING AND PROGRAMMING

APPENDICES

With more than 4,000,000 users around the world, AutoCAD offers engineers, architects, drafters, interior designers, and many others, a fast, accurate, and extremely versatile drawing tool.

Now in its 13th edition, *Using AutoCAD 2005* makes using AutoCAD a snap by presenting easy-to-master, step-by-step tutorials covering AutoCAD's commands and methods. *Using AutoCAD 2005* consists of two volumes: *Basics* gets you started with AutoCAD, teaching commands for basic 2D drafting; *Advanced* explores sophisticated 2D features, 3D design, rendering, and customizing.

USING THIS BOOK

Using AutoCAD 2005: Advanced describes advanced AutoCAD features in these areas:

- Advanced 2D Drafting
- Connecting with Other Programs
- Three Dimensional Design
- Rendering and Imaging
- Customization and Programming

The basic use for each command is given, followed by one or more tutorials that help you understand how the command works. This is followed by a comprehensive look at the command and its many variations.

Problems at the end of every chapter are designed to use the specific commands covered in the book to that point. Review questions reinforce the concepts taught in the chapter. This method allows you to pace your learning; not everyone grasps each command in the same amount of time.

CONVENTIONS

This book uses the following conventions.

Keys

Several references are made to keyboard keystrokes in this book, such as **ENTER**, **CTRL**, **ALT**, and function keys (**F1**, **F2**, and so on). Note that these keys might be found in different locations on different keyboards.

Control and Alternate Keys

Some commands are executed by holding down one key, while pressing a second key. The control key is labeled **CTRL**, and is always used in conjunction with another key or a mouse button.

To access menu commands from the keyboard, hold down the **ALT** key while pressing the underlined letter. The **ALT** key is also used in conjunction with other keys.

Flip Screen and Online Help

The text and graphics windows are alternately displayed by using the "Flip Screen" key. The **F2** key is used for this function.

To access Autodesk's online help, press **F1**. You can press this key in the middle of a command for help on the command.

Command Nomenclature

When a command sequence is shown, the following notations are used:

Command: **mline** *(Press* **ENTER**.*)*

Current settings: Justification = Top, Scale = 1.00, Style = STANDARD

Specify start point or [Justification/Scale/STyle]: *(Pick a point, or enter coordinates.)*

Specify next point or [Undo]: *(Pick point 1.)*

Specify next point or [Close/Undo]: *(Press* **ENTER**.*)*

Boldface text designates user input. This is what you type at the keyboard.

(Pick a point.) means that you enter a point in the drawing to show AutoCAD where to place the object. To pick the point, you can either click on the point, or type the x, y, z coordinates.

(Pick point 1.) The book often illustrates designated points in drawings. The points are labelled as "Point 1," and so on.

(Press **ENTER**.*)* means you press the **ENTER** key; you do not type "enter."

[Close/Undo] is how AutoCAD lists command *options*. The words are surrounded by square brackets. At least one letter is always capitalized. This means you enter the capital letter as the response to select the option. For example, if the option is **Undo**, simply entering **U** chooses the option. Additional options are separated by the slash mark. When two options start with the same letter, then two letters are capitalized, and you must enter both.

<Default> is how AutoCAD indicates *default* values. The numbers or text are enclosed in angle brackets. The default value is executed when you press **ENTER**.

If the command and its options can be entered by other methods, these are described in the book. This includes commands that can be selected from menus and toolbars, or entered as keyboard shortcuts:

- From the **Draw** menu, choose **Spline**.
- From the **Draw** toolbar, choose the **Spline** button.
- At the 'Command:' prompt, enter the **spline** command.
- Alternatively, enter the alias **spl** at the keyboard.

Note: Tips and warnings are highlighted as notes in the text. Look for the distinctive icon.

WHAT'S NEW IN AUTOCAD 2005

Fully updated to AutoCAD 2005, this book includes new information specific to advanced AutoCAD use:

- The new **SHEETSET** command and related commands create sets of drawings.
- The **XATTACH** and **IMAGEATTACH** commands have new options.
- The toolbar on the **3DCLIP** command's window has added buttons.
- The dialog boxes of several other commands are changed.
- The **TBCUSTOMIZE** command determines whether toolbars can be changed.
- There are added aliases, keyboard shortcuts, and system variables.

The *New In 2005* icon alerts you to new and changed commands throughout the text.

ONLINE COMPANION

The Online Companion™ is your link to AutoCAD on the Internet. We've compiled supporting resources with links to a variety of sites. You find out not only about training and education, industry sites, and the online community, but also about valuable archives compiled for AutoCAD users from various Web sites.

In addition, there is information of special interest to readers of *Using AutoCAD*. This includes updates, information about the author, and a page where you can send your comments. You can find the Online Companion at www.autodeskpress.com/resources/olcs/index.asp.

E.RESOURCE

e.Resource™ is an educational resource that creates a truly electronic classroom. It is a CD-ROM containing tools and instructional resources that enrich the classroom and make preparation time shorter. The elements of e.Resource link directly to the text and combine to provide a unified instructional system. With e.Resource you can spend your time teaching, not preparing to teach.

Features contained in e.Resource include:

- **Syllabus** — lesson plans created by chapter. You have the option of using these lesson plans with your own course information.
- **Chapter Hints** — objectives and teaching hints that provide the basis for a lecture outline that helps you present concepts and material.
- **Answers to Review Questions** — solutions that enable you to grade and evaluate end-of-chapter tests.
- **PowerPoint® Presentation** — slides that provide the basis for a lecture outline that helps you to present concepts and material. Key points and concepts can be graphically highlighted for student retention.
- **Exam View Computerized Test Bank** — over 800 questions of varying levels of difficulty are provided in true/false and multiple-choice formats, so you can assess student comprehension.

- **AVI Files** — video files, listed by topic, that allow you to view quick videos illustrating and explaining key concepts.
- **DWG Files** — a list of *.dwg* files that match many of the figures in the textbook. These files can be used to stylize the PowerPoint presentations.

You can learn more about the e.Resource (ISBN 1401884016) from www.autodeskpress.com/resources/eresource/index.asp

WE WANT TO HEAR FROM YOU!

Many of the changes to the look and feel of this new edition came by way of requests from and reviews by users of our previous editions. We'd like to hear from you as well! If you have any questions or comments, please contact:

The CADD Team
c/o Autodesk Press
5 Maxwell Drive
Clifton Park NY 12065-8007

or visit our Web site at www.autodeskpress.com

ACKNOWLEDGMENTS

We would like to thank and acknowledge the professionals who reviewed the manuscript to help us publish this AutoCAD text:

James Freygang, Ivy Tech State College, South Bend, Indiana

Alex Lepeska, Renton Technical College, Renton, Washington

David Pitzer, Pocono Lake, Pennsylvania

Dr. Jerry Pat Spicer, Western Illinois University, Macomb, Illinois

Technical Editor: Bill Fane, British Columbia Institute of Technology, Burnaby BC, Canada

Copy Editor: Stephen Dunning, Douglas College, Coquitlam BC, Canada

Additional Material: Kevin Standiford prepared much of the material for Chapter 7, "Introducing Database Links."

ABOUT THE AUTHOR

Ralph H. Grabowski has been writing about AutoCAD since 1985, and is the author of over sixty books on computer-aided design. He received his B.A.Sc. degree in Civil Engineering from the University of British Columbia.

Mr. Grabowski publishes *upFront.eZine*, the weekly email newsletter about the business of CAD, as well as a series of CAD e-books under the eBooks.onLine imprint. He is the former Senior Editor of *CADalyst* magazine, the original magazine for AutoCAD users. His Web site is at www.upfrontezine.com, and his Weblog is at worldcadaccess.typepad.com.

UNIT I

Advanced Drafting

Drawing with Advanced Objects

Drafting in AutoCAD mostly uses basic objects, such as lines, circles, arcs, and polylines. AutoCAD includes some objects that are used for specialized purposes, such as multilines for floor plans of buildings, and splines for ship hulls.

In this chapter, you learn to draw and edit multilines and splines using these commands:

MLINE draws up to sixteen parallel (multiple) lines.

MLSTYLE creates and sets styles for mlines.

MLEDIT edits mline intersections and segments.

SPLINE draws spline curves based on NURBS mathematics.

SPLINEDIT edits splines.

FINDING THE COMMANDS

On the **DRAW** and **MODIFY II** toolbars:

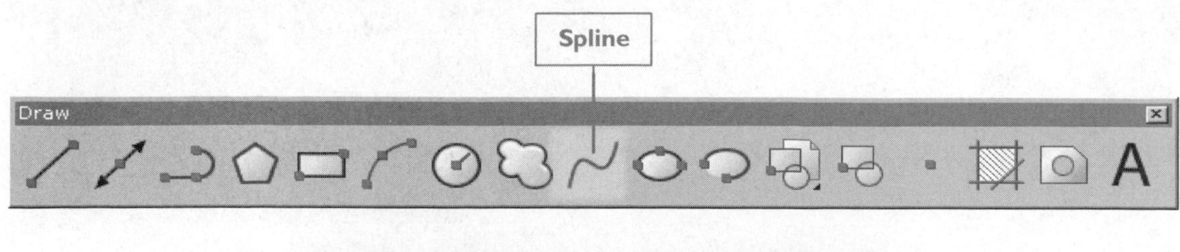

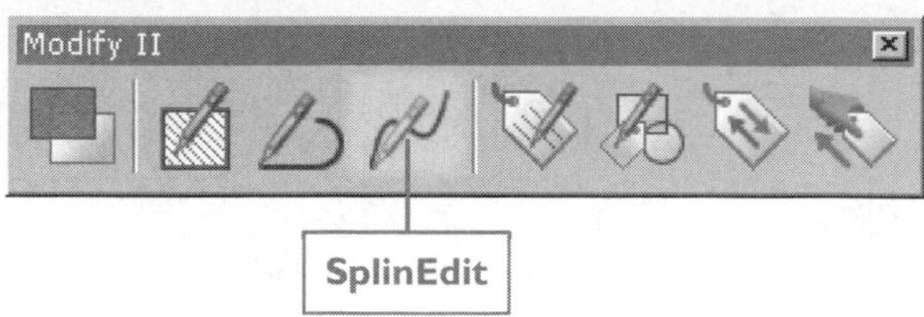

On the **DRAW**, **FORMAT**, and **MODIFY** menus:

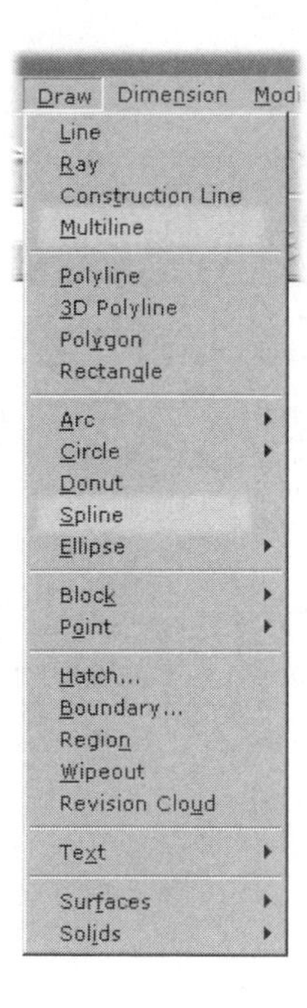

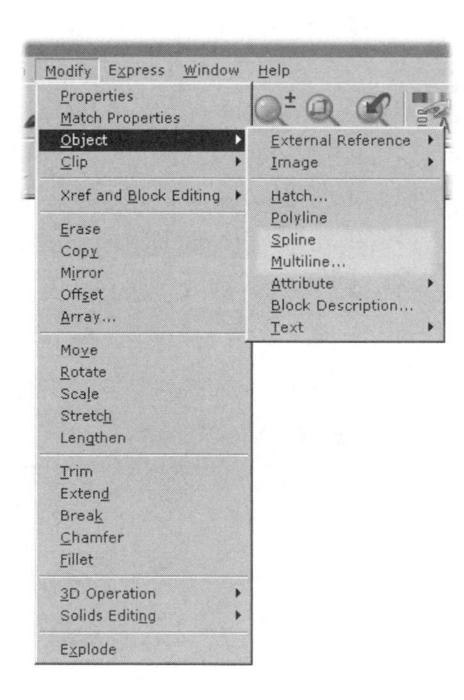

MLINE

The **MLINE** command draws up to sixteen parallel (multiple) lines.

This command is useful for drawing walls, alternating colored lines, and anything else that involves multiple lines. Multiple lines (or "mlines") can be filled with color, and can have a variety of end caps.

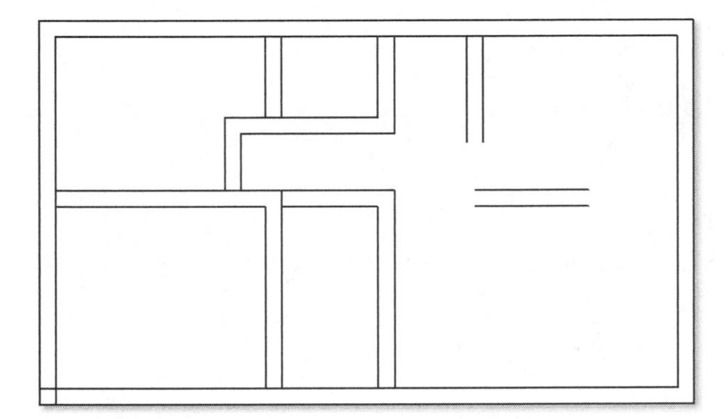

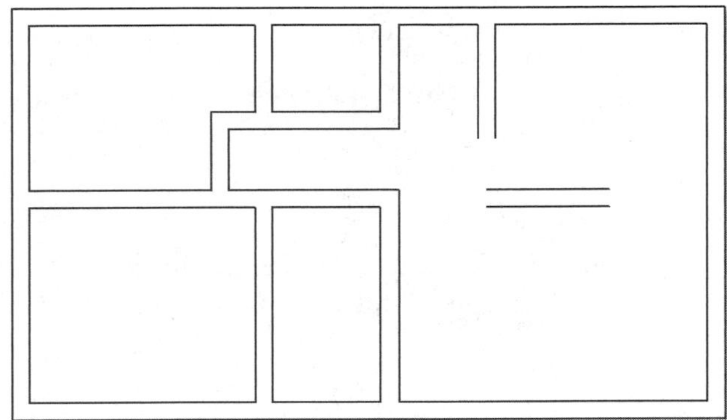

Left: *Floor plan drawn with multilines using MLine.*
Right: *Same floor plan after editing with MlEdit.*

The **MLINE** command operates like the **LINE** command, asking for a start point, and then the next points. If you don't specify a style, AutoCAD draws the mline as a pair of lines 1.0 units apart.

BASIC TUTORIAL: DRAWING MULTIPLE LINES

1. To draw with mlines, start the **MLINE** command:
   - From the **Draw** menu, choose **Multiline**.
   - At the 'Command:' prompt, enter the **mline** command.
   - Alternatively, enter the alias **ml** at the keyboard.

   Command: **mline** *(Press* ENTER.*)*
2. In all cases, AutoCAD displays the current settings, and then prompts you to specify a start point.

   Current settings: Justification = Top, Scale = 1.00, Style = STANDARD

   Specify start point or [Justification/Scale/STyle]: *(Pick a point, or enter coordinates.)*
3. Pick additional points:

   Specify next point: *(Pick another point.)*

   Specify next point or [Undo]: *(Pick another point.)*

   Specify next point or [Close/Undo]: *(Pick another point.)*
4. Press **ENTER** to exit the command:

   Specify next point or [Close/Undo]: *(Press* ENTER.*)*

Note: AutoCAD LT has a different command for drawing parallel lines. Although **DLINE** is limited to drawing just two parallel lines, it is more "intelligent" than the **MLINE** command, because it automatically cleans up intersections.

DRAWING MLINES: ADDITIONAL METHODS

The **MLINE** command has a number of options that control the look and size of mlines being drawn. In addition, system variables store values associated with the options.

- **Justification** option determines how mlines are drawn relative to the cursor.
- **Scale** option changes the width of mlines.
- **STyle** option selects an mline style.
- **CMLJUST** system variable stores the current justification.
- **CMLSCALE** system variable stores the current scale.

Let's look at them all.

Justification

The **Justification** option specifies how the mlines are drawn relative to the cursor pick points. The default is top, which means the pick points define the top of the multiline.

Enter justification type [Top/Zero/Bottom] <top>: *(Enter an option.)*

Top is the line with the most positive offset.

Zero is the center of the mline, specifically with an offset of 0.

Bottom is the line with the most negative offset.

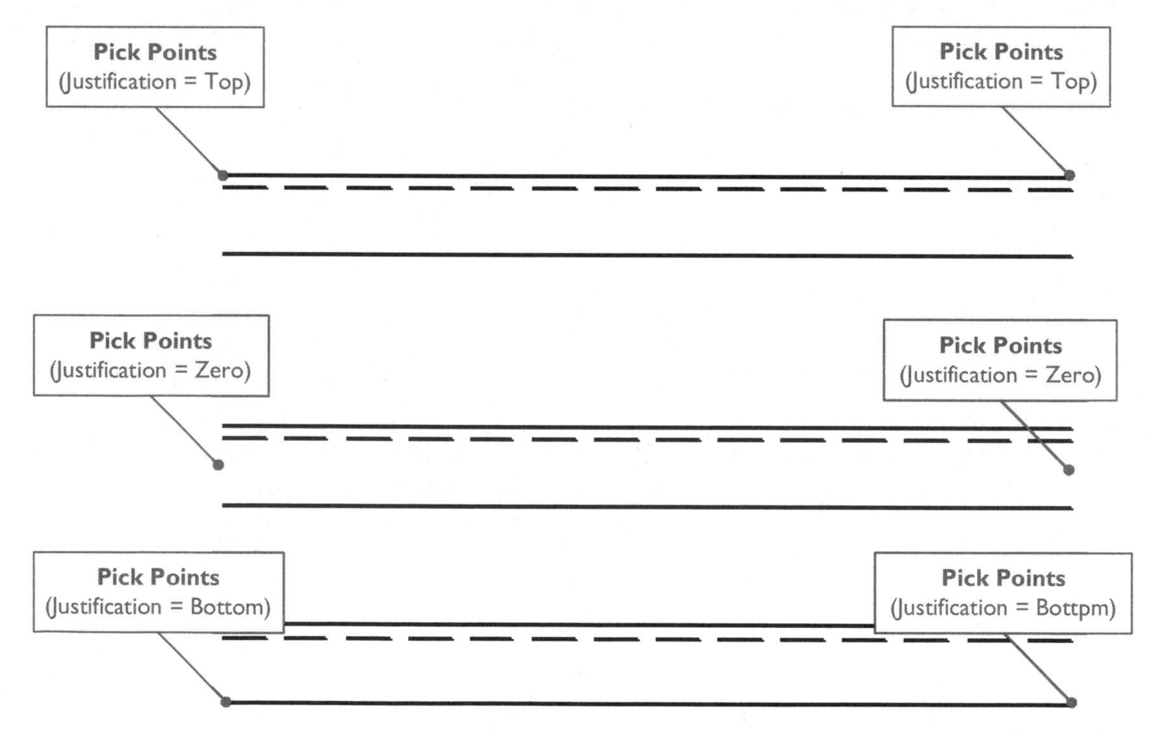

Mline justification: top, zero, or bottom.

Offsets are described in detail later under the **MLSTYLE** command.

Scale

The **Scale** option changes the overall width of the multiline.

Enter mline scale <1.0>: *(Enter a scale factor.)*

Enter a value of 2 to double the width of the mline, or 0.5 to make the mline half as wide. You can enter a negative scale factor, such as -1, which "flips" the mline about its zero-offset point. This also affects the justification: the mline is drawn above the pick points if justification is set to Top.

Enter a factor of 0 to collapse the mline to a single line.

The length of mlines is unaffected by the scale factor. If the mline has linetypes, their scaling is also unaffected by this option.

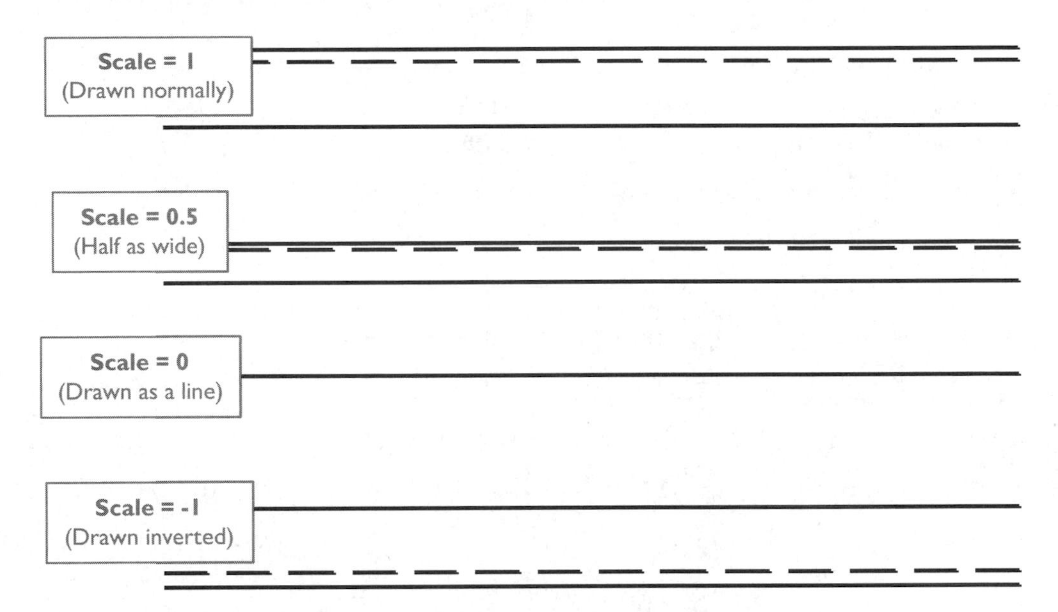

Changing the scale changes only the width of the mline.

STyle

The **STyle** option selects a multiline style. See the **MLSTYLE** command.

CMlJust

The **CMLJUST** system variable (short for "current multiline justification") holds the current mline justification. You can use this variable to change the justification outside of the **MLINE** command.

CMlJust	*Meaning*
0	Top (default)
1	Middle
2	Bottom

CMlScale

The **CMLSCALE** system variable (short for "current multiline scale") holds the current scale factor for mlines. Default value is 1 for imperial drawings and 20 for metric drawings.

Notes: You cannot find the area of a room drawn with multilines using the **AREA** command's **Object** option. Instead, you must pick points at the corners of the room; use the INTersection object snap for greater accuracy.

You cannot use the **TRIM, EXTEND, BREAK,** or **LENGTHEN** commands on mlines. Instead, use the **MLEDIT** command; for breaking mlines, use **MLEDIT Cut All**.

You cannot use the **FILLET, CHAMFER, OFFSET,** or **MATCHPROP** commands on mlines. For filleting with a radius of 0, use **MLEDIT Corner Joint**. You can, however, use the **COPY, MOVE, ERASE, EXPLODE, ROTATE, SCALE, MIRROR,** and **STRETCH** commands on mlines.

MLSTYLE

The **MLSTYLE** command creates and sets styles for mlines.

This command displays a dialog box that performs three functions: (1) specifies the elements that make up the mline; (2) specifies the mline properties; and (3) saves and loads mline style files.

BASIC TUTORIAL: CREATING MLINE STYLES

1. To create styles for multilines, start the **MLSTYLE** command:
   - From the **Format** menu bar, choose **Multiline**.
   - At the 'Command:' prompt, enter the **mlstyle** command.

   Command: **mlstyle** *(Press* **ENTER**.*)*

   In all cases, AutoCAD displays the Multiline Styles dialog box.

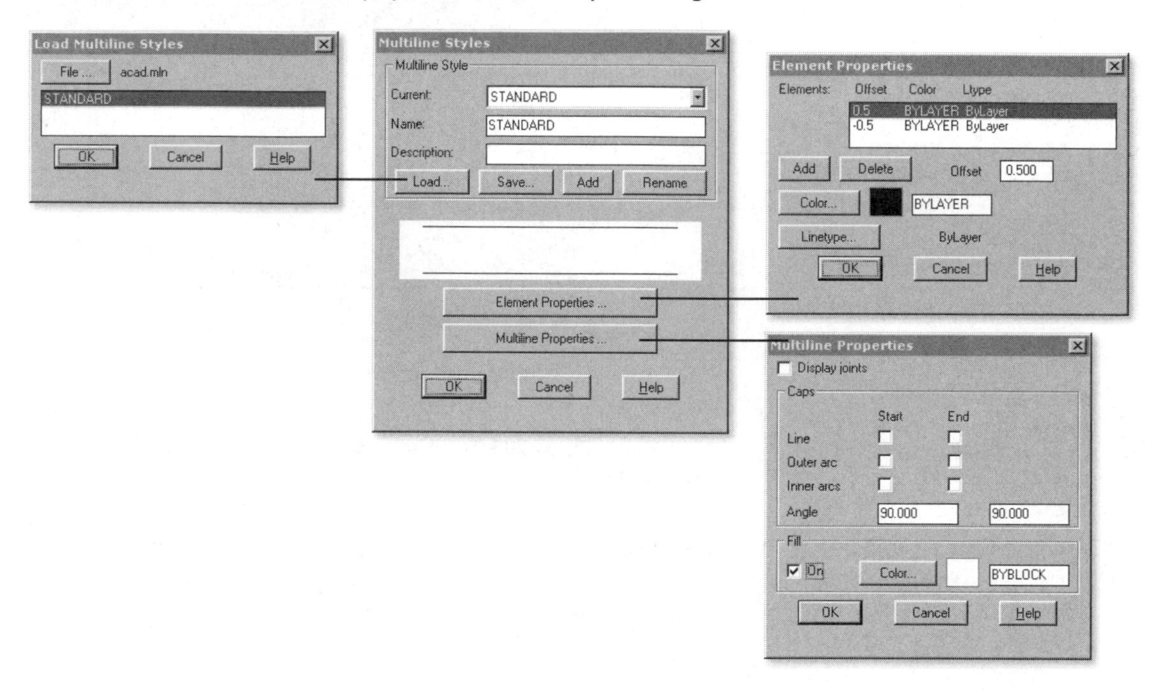

Mline styles consist of elements and properties.

   Every new AutoCAD drawing holds a simple mline style called "Standard," which consists of two lines, one unit apart.

   In this tutorial, you create a new style that mimics the insulated exterior wall of houses: two lines 6" apart, with the insulation linetype between them.

2. The first step is to create a new mline style. In the **Name** text box, type a descriptive name:

   Name: **insulated**

   and a description:

   Description: **6" walls with pink insulation**

   And then click **Add**.

3. Click the **Elements Properties** button to define the lines that make up the mline.

   The walls are each 3" from the center line, so redefine the two existing lines.

   a. In the Elements list, select the first element.
   b. In the **Offset** text box, change 0.5 to 3.

      Offset: **3**

c. Change the offset of the second element from -0.5 to -3.
   Offset: **-3**

4. Between the two walls lines is the insulation. Its offset is zero.
   a. Click **Add** to add the new element. (You might find that two elements with offset 0 are added; if so, select one, and then click **Delete**.)
   b. Change the linetype by clicking **Linetype**; if necessary, load the "Batting" linetype.
      Click **OK** to exit the dialog box.

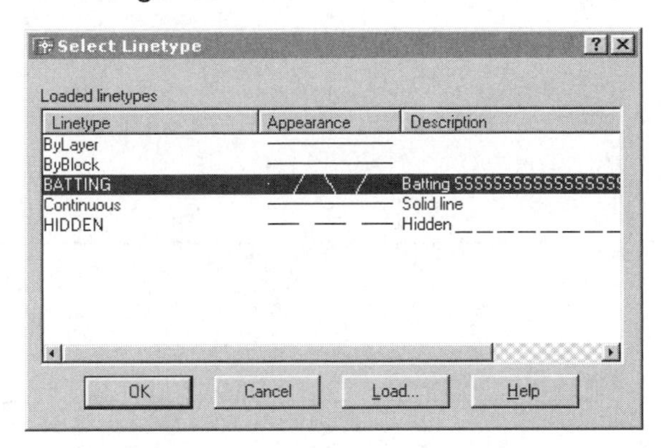

Select the linetype; if necessary, click Load to load it into the drawing.

   c. Change the color of the insulation to pink: click **Color**, and then select magenta (color 4).
      Click **OK** to exit the dialog box.
   d. The Element Properties dialog box should look similar to the figure below.

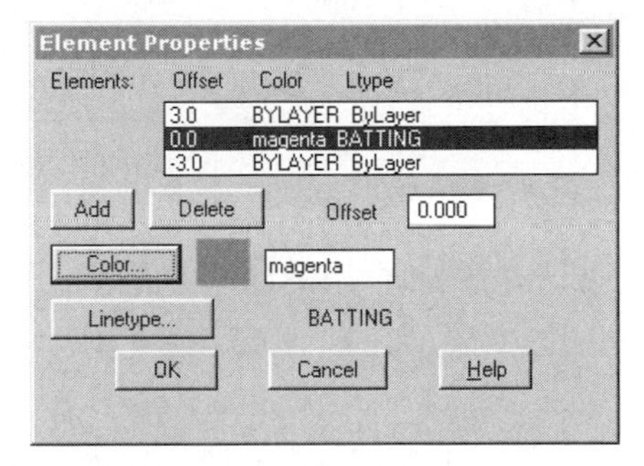

The elements for a 6" wall with pink insulation.

   Click **OK** to exit the dialog box.

5. Back in the Multiline Styles dialog box, notice the preview of the mline.

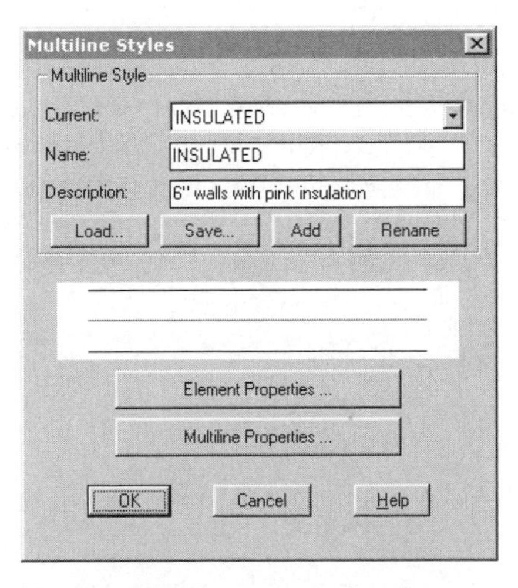

The dialog box previews the mline.

You should save the mline style for later use.

a. Click **Save.**

b. AutoCAD displays the Save Multiline Style dialog box. Enter a file name:

   File name: **insulate**

c. Click **Save**, and then **OK.**

6. Use the **MLINE** command to test the new multiline. Notice that the linetype is not drawn until you end the **MLINE** command.
7. If the linetype is too small (or too large), change its size with the **LTSCALE** command. For this linetype, I find the scale needs to be 4x larger:

   Command: **ltscale**

   Enter new linetype scale factor <1.0000>: **4**

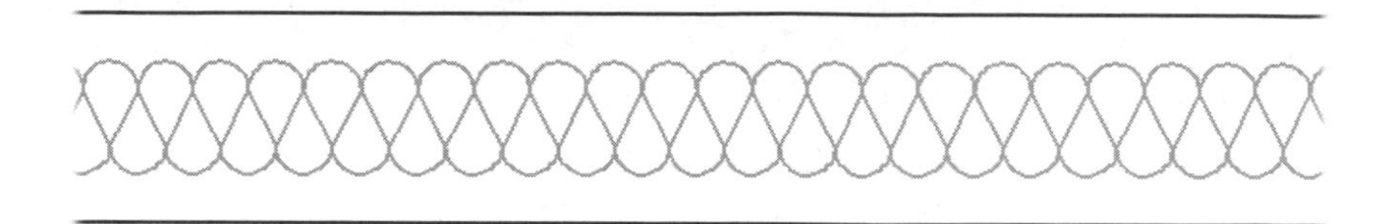

The multiline shows two walls with insulation.

Technical editor Bill Fane notes that setting the linetype scale large enough for Batting may affect the scale of other linetypes in the drawing. His solution is to not use Batting, but to fill the two wall lines with pink color, using the **Fill** option. Also, he thinks that color #241 looks better than magenta as the pink used by insulation.

You can use MLSTYLE to create a single mline with alternating colors using the following style properties:

Element	***Offset***	***Color***	***Linetype***
1	0	Black	Continuous
2	0	Yellow	Hidden

Using mlines to create a single line with alternating colors.

You may need to adjust the linetype scale to a larger value, such as 10 or 100. Experiment with other linetypes.

Notes: You cannot modify existing mline styles. After creating an mline style, you can edit the style, but when you attempt to save the changes, AutoCAD complains mysteriously, "Invalid style," and does not save them. The work-arounds are: (1) delete the style, and then create a new style with the changed properties; or (2) save the changed style to an *.mln* file, and then open it in a new drawing; or (3) if the style is unused in the drawing, save the changes to an *.mln* file, use the **PURGE** command to remove the unused mline style, and then load the *.mln* file with **MLSTYLE.**

AutoCAD prevents you from loading mline styles already in use in drawings. (AutoCAD complains, "Style already in use.") This means you cannot override one mline style with another.

You can not delete or rename the "Standard" multiline style.

MULTILINE STYLES: ADDITIONAL METHODS

The Multiline Properties dialog box controls the fill color and endcap style of mlines. In addition, the mline styles can be saved.

- ***.mln*** files hold mline style definitions.
- **CMLSTYLE** system variable stores the name of the current mline style.

Let's look at each.

Multiline Properties Dialog Box

The Multiline Properties dialog box controls the fill color and endcap style of multilines.

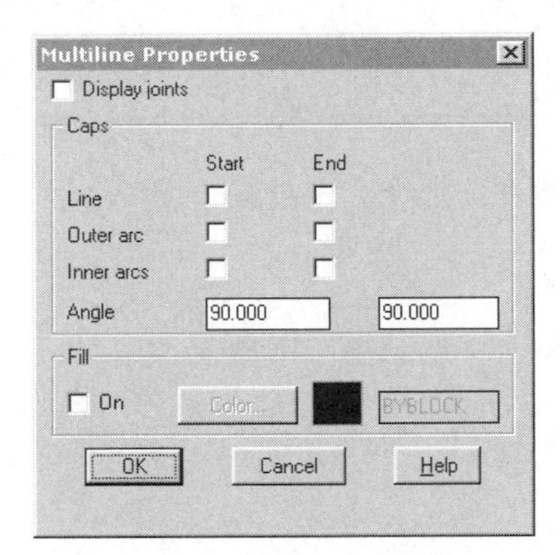

The Multiline Properties dialog box.

Display Joints

The **Display Joints** check box toggles whether the joints ("miters") at vertices are displayed.

Caps

The **Caps** options determine the look of mline endcaps. You can set the caps for the start and end independently:

Line draws a straight line at the ends of the mline.

Outer Arc draws an arc between the outermost mline elements.

Inner Arcs draws an arc between all inner mline elements; if there is an uneven number of elements, the center element is unconnected.

Angle specifies the angle of the endcaps. Permitted range of angles is from 10 to 170 degrees.

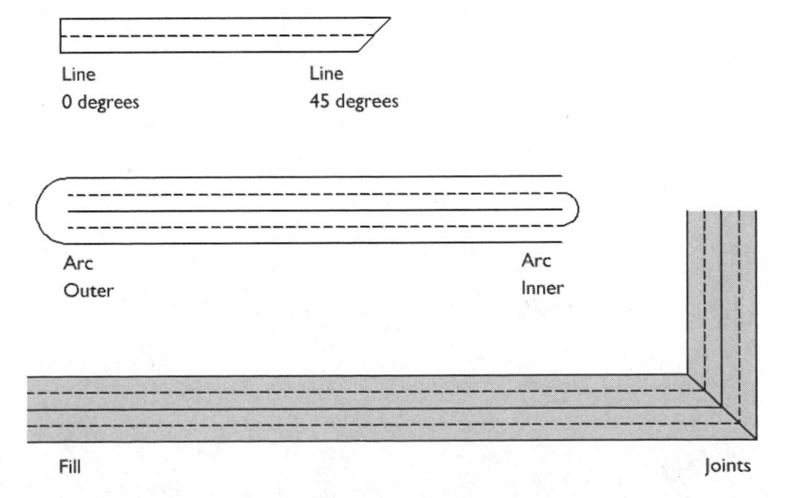

Multilines with end caps, miters, and fills.

Fill

The **On** option toggles the fill color for the mline.

Color selects the fill color.

You can apply none, some, or all of the properties to the mline style. The mline illustrated below has all options turned on, including an end angle of 10 degrees.

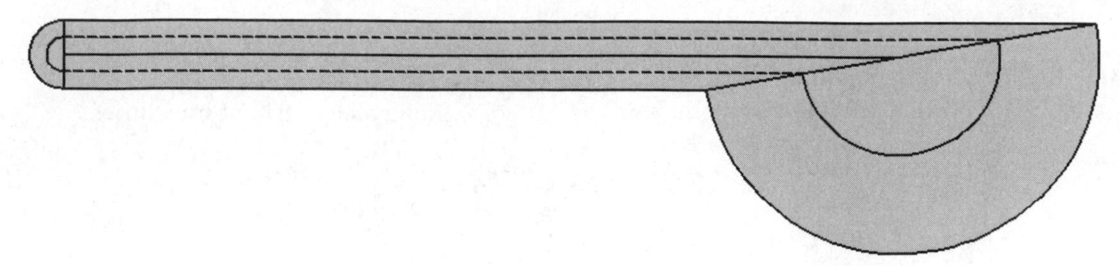

All properties applied to an mline.

.mln

The *.mln* file stores multilines styles in a DXF-like format. The "Standard" multiline file looks like this:

```
MLSTYLE
2
 STANDARD
70
 0
3

62
 0
51
 90.00000000000000
52
 90.00000000000000
71
 2
49
 0.50000000000000
62
 256
6
 BYLAYER
49
 -0.50000000000000
62
 256
6
 BYLAYER
0
```

CMlStyle

The **CMLTYLE** system variable stores the name of the current mline style.

MLEDIT

The **MLEDIT** command edits multiline intersections and segments.

This command works through a dialog box: select an option from the dialog box, and then apply it to a portion of the mline. Once you select an option, AutoCAD repeats the prompts so that you can apply the same editing action to other mlines.

The **-MLINE** command provides access to the same editing options through the command-line.

MlEdit	*-MlEdit*	*Action*	*Meaning*
Intersections:			
	CC	Closed Cross	First mline: *all* lines are trimmed. Crossing mline: *no* lines trimmed, creating the overlap look.
	OC	Open Cross	First mline: *all* lines are trimmed. Crossing mline: *outside* lines are trimmed.
	MC	Merged Cross	Both mlines: *outside* lines are trimmed.
T-intersections:			
	CT	Closed Tee	First mline is trimmed to second mline. Intersecting mline: *no* lines trimmed, equivalent to TRIM.
	OT	Open Tee	First mline is trimmed to second mline. Intersecting mline: *outside* line next to first mline is trimmed.
	MT	Merged Tee	First mline is trimmed to second mline Intersecting mline: *all* lines next to first mline are trimmed.
Corners:			
	CJ	Corner Joint	Intersecting mlines: all lines trimmed, equivalent to FILLET command.
Vertices:			
	AV	Add Vertex	Adds a vertex to an mline.
	DV	Delete Vertex	Removes a vertex from an mline.
Walls:			
	CS	Cut Single	Removes a portion of a single line between two pick points.
	CA	Cut All	Removes a portion of all lines between two pick points; useful for creating gaps for door and window blocks; equivalent to the BREAK command.
	WA	Weld All	Joins two mlines; useful for filling in gaps created with the Cut All option.

Notes: The edits performed by **MLEDIT** are not associative. After cleaning up the intersection of two mlines, for example, if you then move one mline, the intersection stays where it was, creating gaps in the mlines. You need to clean up the new intersection; holes can be fixed with the **Weld All** option.

To move the vertex (intersection or endpoints) of an mline, use grips editing.

When gaps are cut in mlines with **Cut All**, AutoCAD does not apply endcaps to the gaping wounds.

Use the **STRETCH** command to move door and window blocks within multiline "walls."

BASIC TUTORIAL: EDITING MLINES

1. To edit mlines, start the **MLEDIT** command:
   - From the **Modify** menu, choose **Objects**, and then **Multilines**.
   - Double-click the mline.
   - At the 'Command:' prompt, enter the **mledit** command.

     Command: **mledit** *(Press* **ENTER**.*)*

   In all cases, AutoCAD displays the Multiline Edit Tools dialog box.

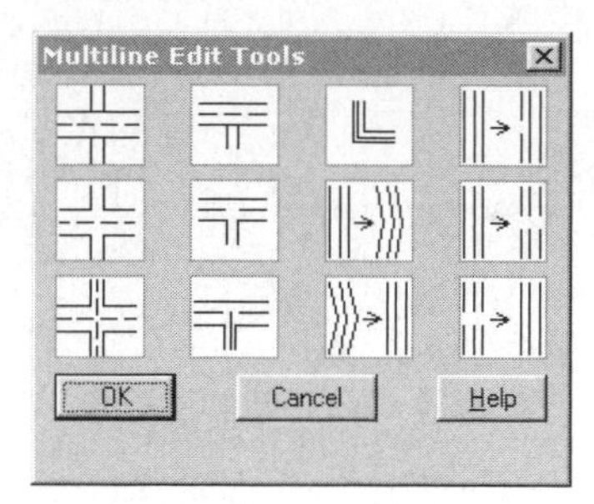

Multiline Edit Tools dialog box.

2. Select an mline editing option (icon), and then click **OK**.
3. AutoCAD prompts you to select one or two mlines, depending on the editing action.

     Select mline: *(Pick an mline segment.)*
4. AutoCAD performs the editing action, and then repeats the prompt to select an mline. Press **ENTER** to exit the command.

     Select mline: *(Press* **ENTER**.*)*

SPLINE

The **SPLINE** command draws spline curves based on NURBS mathematics.

Splines defined the shape of boat hulls, aircraft surfaces, car bodies, and interpolating elevation points to create contour maps. Before CAD, drafters used long, thin strips of wood or plastic to help draw the curve (also known as "lofting"). Heavy weights, called "ducks" because of their shape, held the strips in place.

For drawing splines in AutoCAD, Autodesk implemented NURBS, short for "nonuniform rational B-splines." (There are many kinds of splines: cubic, Bezier, nonuniform, rational, and so on.) The advantages of NURBS over the other kinds of splines are that algorithms (mathematical formulae) generate them reasonably quickly, and they are generalizations of other kinds of splines, including non-rational B-splines, non-rational Bezier, and rational Bezier curves.

NURBS are excellent for drawing smooth flowing curves, but poor for curves with sharp corners. NURBS can define any kind of curve, including simple ones like circles and arcs, and even straight lines; the amount of computation involved, however, is so high that simple formulae are used for the lines, arcs, and circles.

When you draw a spline with the **SPLINE** command, you specify the location of the control points and the tangency of the endpoints. AutoCAD draws the spline, as well as the hidden spline frame (which you can reveal by turning on the **SPLFRAME** system variable).

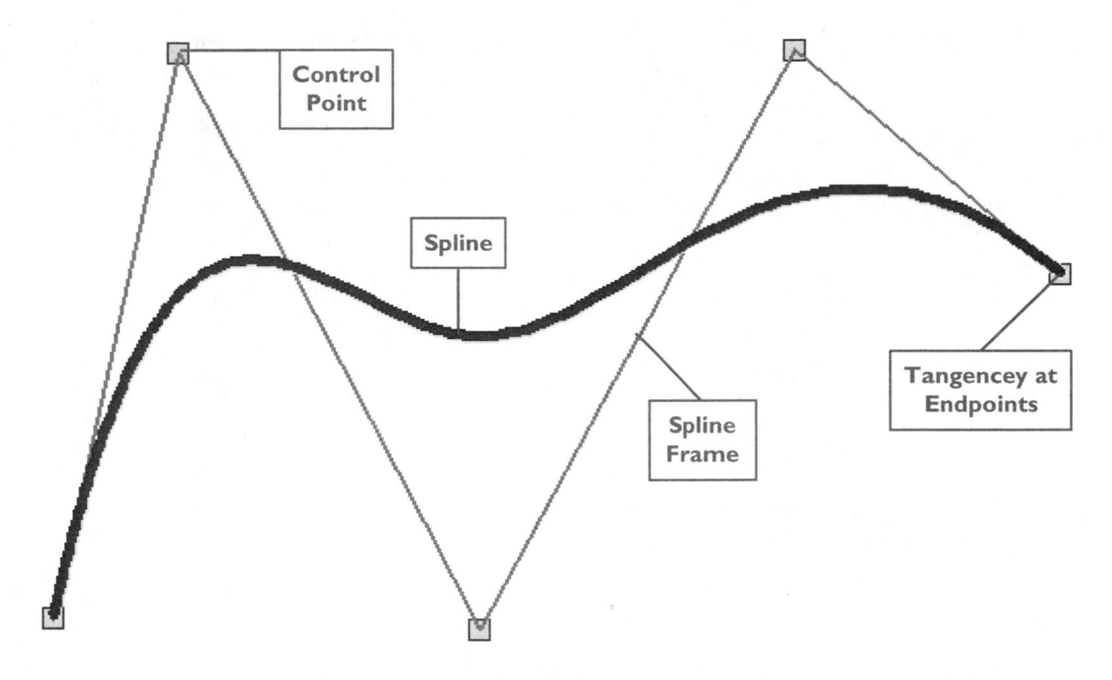

Elements of AutoCAD splines.

The **SPLINE** command in AutoCAD needs to know these things: (1) an initial starting point; (2) additional points; (3) start tangency; and (4) end tangency.

BASIC TUTORIAL: DRAWING SPLINES

1. To draw spline curves, start the **SPLINE** command:
   - From the **Draw** menu, choose **Spline**.
   - From the **Draw** toolbar, choose the **Spline** button.
   - At the 'Command:' prompt, enter the **spline** command.
   - Alternatively, enter the alias **spl** at the keyboard.

   Command: **spline** *(Press* ENTER.*)*
2. In all cases, AutoCAD prompts you to pick the starting point for the spline:

   Specify first point or [Object]: *(Pick a point.)*

   The curve is not drawn until the next point is picked.
3. AutoCAD prompts you to pick additional points.

   Specify next point: *(Pick a point.)*

   Specify next point or [Close/Fit tolerance] <start tangent>: *(Pick a point.)*
4. **Undo** is a "hidden" option. You can enter **u** at the "Specify next point" to undo the last curve segment, and pick another point.
5. Press **ENTER** to end entering points, after which you must provide tangency information.

   Specify next point or [Close/Fit tolerance] <start tangent>: *(Press* ENTER.*)*
6. AutoCAD prompts you to define the start and end tangents.

   If you are not sure about tangency, then press **ENTER** for both prompts. AutoCAD constrains the start and end of the spline at the first and last pick points. In most cases, this is what you want.

   Specify start tangent: *(Press* ENTER.*)*

   Specify end tangent: *(Press* ENTER.*)*

If you pick points for tangency, they affect the entire spline, as the figure below illustrates.

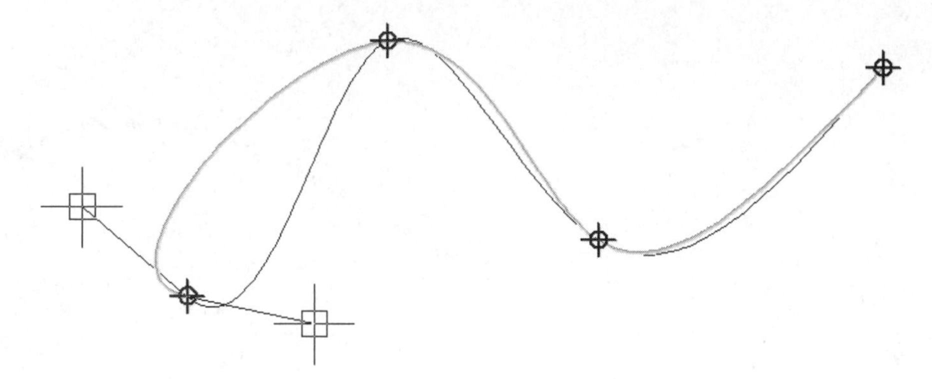

Tangency points affect the shape of the entire spline.

Meaning of NURBS

As mentioned, NURBS (singular, not a plural word) is short for "***n***on-***u***niform ***r***ational ***B***-***s***pline." The three parts of the name mean the following:

Non-uniform means the spline uses *knots* of varying spacing; in contrast, uniform splines use knots of constant spacing.

Rational means the spline's control points can have varying weights; in contrast, non-rational splines don't use weights.

B-spline means the splines use the *basis* spline algorithm to interpolate the position of the curve between control points; it is a more generalized version of the Bezier spline. ("B" is short for *basis*.)

The shape of the spline is determined by control points, weights, knot points, and degrees.

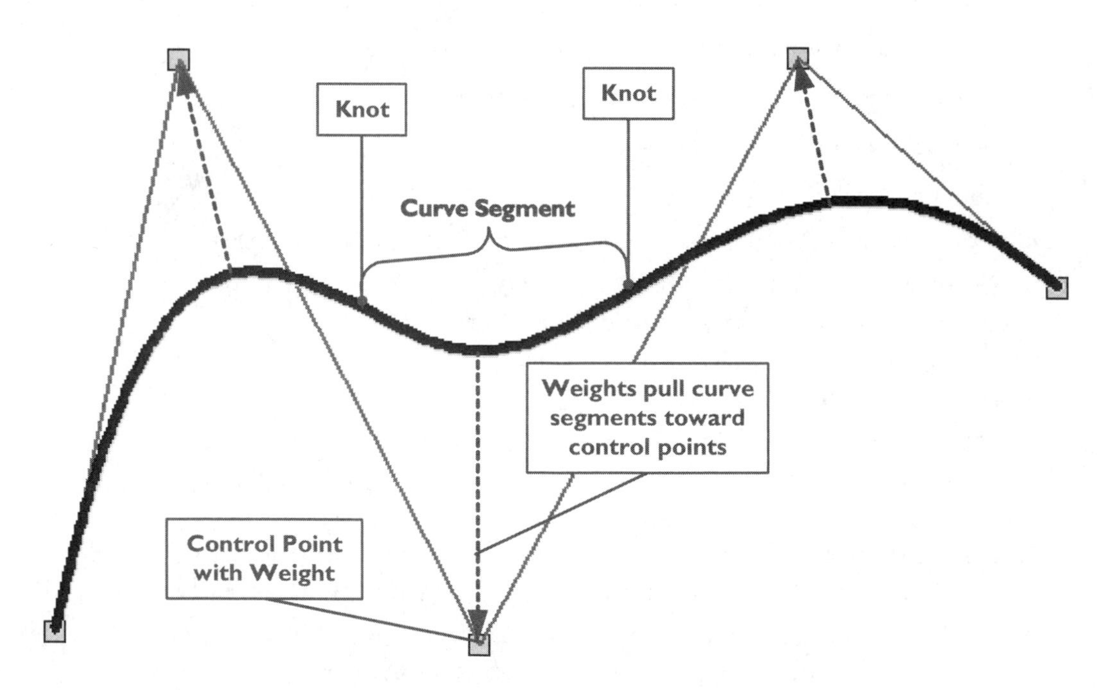

The stronger the weight, the closer the curve segment gets to the control point.

Control points "influence" the shape of the curve. The curve does not go usually through the control points, but goes near them.

Each control point has a *weight*; the stronger the weight, the more the curve is pulled toward the control point. It's not until the weight reaches infinity that the curve passes through the control point; when the weight is 0, the control point has no effect on the curve (which is how NURBS defines a straight line).

Knot points divide the curve into curve segments; they are located at the curve inflection points (where the curve segments change direction) and at the two endpoints. Change the knot, and the length of the segment changes.

The *degree* defines the accuracy of the spline in mimicing specific shapes; more degrees, more accuracy, but also more computation. The degree is related to segments and control points:

Segments = Degrees +1

Knots = Segments + 1

Control points = Knots + 1

Thus, a 2-degree curve shown below has three segments, four knots, and five control points. There is one control point for each segment, plus one at each end of the spline.

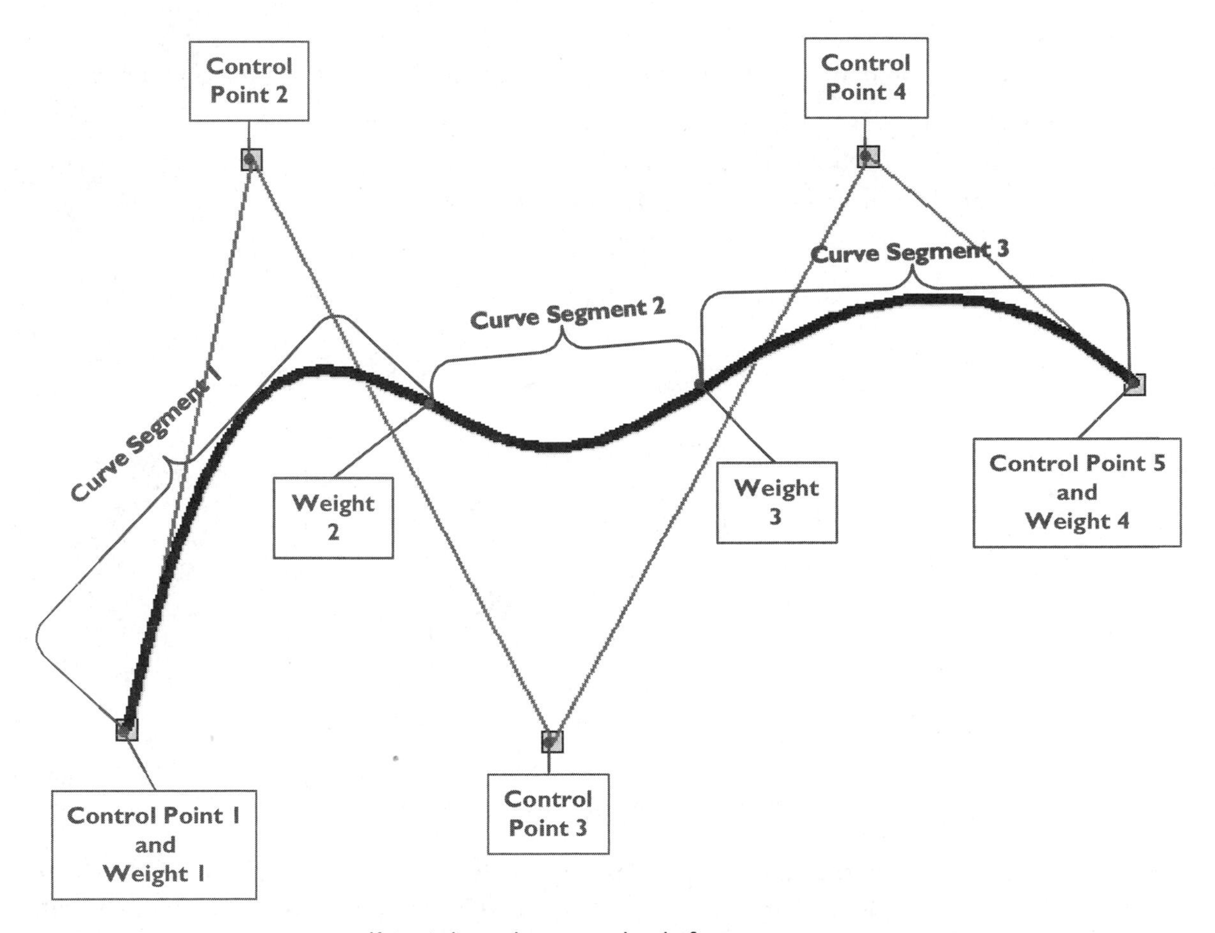

Knots indicate the start and end of curve segments.

For a detailed mathematical description of splines, see "An Interactive Introduction to Splines" at www.people.nnov.ru/fractal/Splines/Intro.htm, which includes interactive (Java) illustrations, and "Unit 6: B-spline Curves" at www.cs.mtu.edu/~shene/COURSES/cs3621/NOTES.

Note: Instead of *weight*, Autodesk uses the term "fit tolerance," and inverts the definition. A fit tolerance of 0 causes the curve segment to go through the control point (equivalent to a weight of infinity). The value of zero is AutoCAD's default.

DRAWING SPLINES: ADDITIONAL METHODS

The **SPLINE** command has several options for creating splines.

- **Object** option converts polylines to splines.
- **Fit Tolerance** option determines how close the curve segments are drawn to control points.
- **Close** option closes the spline.
- **DELOBJ** system variable.

Let's look at each.

Object

The **Object** option converts spline-fitted polylines to NURB spline objects.

Select objects: *(Pick one or more polyline splines.)*

If you attempt to convert any other kind of polyline or object, AutoCAD complains, "Only spline fitted polylines can be converted to splines. Unable to convert the selected object."

Fit Tolerance

The **Fit Tolerance** option determines how close the curve segments are drawn to the pick points.

Specify fit tolerance <0.0000>: *(Enter a value between 0 and 1.)*

The range is from 0 to 1; the default value of 0 forces the spline through the pick points. You can apply a different fit tolerance to each pick point, except at the endpoints.

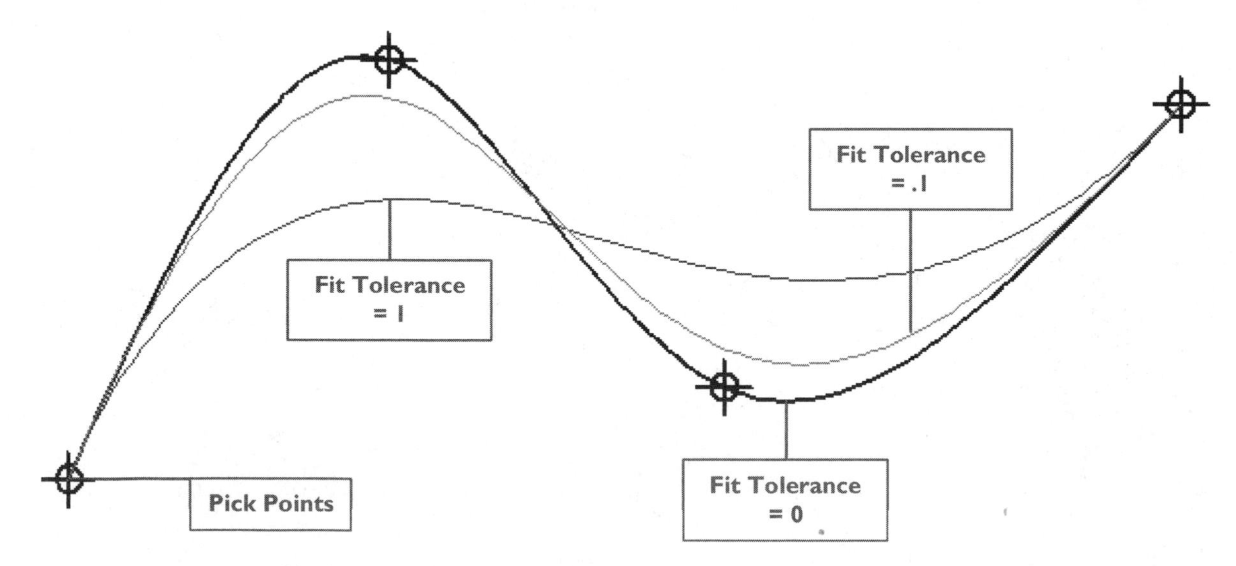

Fit tolerance determines how closely curve segments approach pick points.

The technical editor notes that any positive value works for the fit tolerance, but the effect may not be noticeable when the control points are close together. Fit tolerance affects *all* pick points, except the start and end.

Close

The **Close** option creates a loop; the start point is joined with the end point.

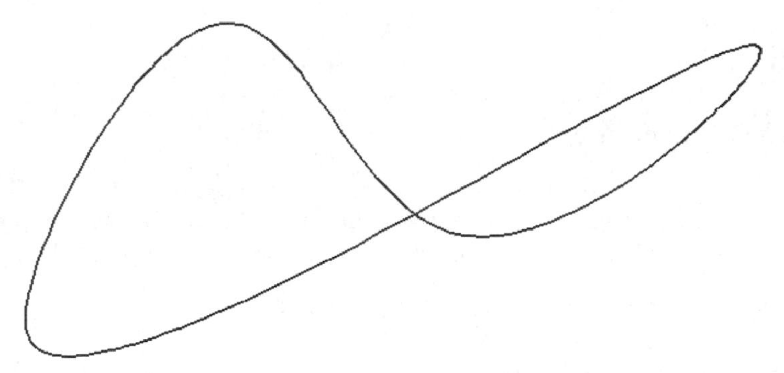

Close option joins the spline's endpoints to create a loop.

DelObj

The **DELOBJ** system variable determines whether the polyline is erased by the **Object** option.

When you use the **SPLINE** command's **Object** option to convert a splined polyline into a NURB spline, AutoCAD normally erases the polyline. If you wish to keep the polyline, change the value of **DELOBJ** to 0. This system variable is also used by other commands that convert objects, such as **BOUNDARY** and **REGION**.

Command: **delobj**

Enter new value for DELOBJ <1>: **0**

DelObj	*Meaning*
0	Retains polyline object.
1	Deletes polyline object.

NURB Splines vs. PLINE/PEDIT Splines

In addition to the **SPLINE** command, AutoCAD has one other method for drawing spline curves: the **PEDIT** command's **Spline** option. It converts polylines to uniform quadratic and cubic B-splines made of straight-line approximations. The **SPLINETYPE** system variable determines whether the polyline spline is a quadratic (set to 5) or a cubic (6) spline. The **SPLINESEGS** system variable specifies how many straight-line segments (default = 8) approximate the curve between vertices.

Splines created with **SPLINE** are more accurate and eighteen times more efficient than polyline splines. (AutoCAD needs less RAM and drawing files use less disk space.) You can convert polyline splines into NURBS with the **SPLINE** command's **Object** option, or with the **SPLINEDIT** command.

Caution: Selecting the polyline with **SPLINEDIT** immediately converts it to a spline object, even if you immediately exit the command.

There are, however, drawbacks to using the NURB splines created by the **SPLINE** command, drawbacks that are handled by polyline splines. For instance, other programs may not understand Autodesk's implementation of NURBS, and so splines must be converted to polylines. (To convert spline objects to polyline splines, save the drawing in R12 DXF format.)

NURBS loses its fit data when you use the **SPLINEDIT** command's **Purge** option while editing fit data, elevate its order; add control points, change the weight of control point, change the fit tolerance, move a control point, trim, break, stretch, or lengthen the spline.

And NURB splines have other drawbacks: they cannot be joined; cannot be given a thickness; don't work with the **CIRCLE** command's **TTR** (tangent, tangent, radius) option; and parallel copies created with the **OFFSET** command are noncontinuous. (Reinaldo Togores of the Universidad de Cantabria details the differences not documented by Autodesk personales.unican.es/togoresr/Splines-en.html.)

SPLINEDIT

The **SPLINEDIT** command edits splines.

This command allows you to add, move, and delete fit points, change tangent points, add control points, change weights of control points, change the order of the spline, switch between open and closed splines, and change the tolerance. In addition, it converts splined polylines to NURB splines.

As an alternative, you can use grips editing to move control points.

BASIC TUTORIAL: EDITING SPLINES

Before editing a spline, it is helpful to turn on the display of its frame.

1. To turn on the display of the spline's frame, set the **SPLFRAME** system variable to 1 (on), followed by the **REGEN** command:

   Command: **splframe**

   Enter new value for SPLFRAME <0>: **1**

   Command: **regen**

   Regenerating model.

2. With the cursor, pick a spline.
   Notice that AutoCAD highlights the spline with grips located at the pick points along the spline and at the vertices (control points) of the frame.

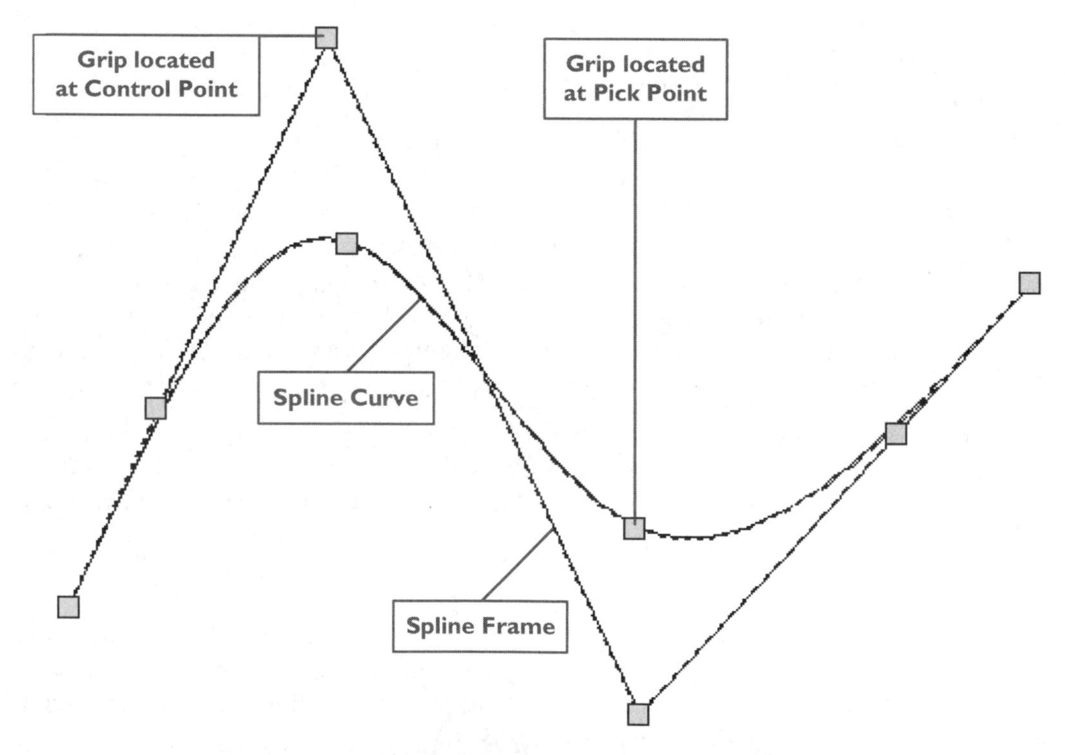

AutoCAD places grips at control and pick points.

3. Select a grip, and then use the grips editing commands to move (stretch) the grip, copy and move the spline, rotate the spline about the selected grip, resize (scale) the spline, and mirror the spline.
4. When done, press **ESC** to exit grips editing mode.

Autodesk calls the pick points "fit points." You can remove these points with the **Purge** option of the **SPLINEDIT** command's **Fit data** option. If you edit a frame grip, you can no longer edit a pick-point grip.

EDITING SPLINES: ADDITIONAL METHODS

1. To edit splines, start the **SPLINEDIT** command:
   - From the **Modify** menu, choose **Object,** and then **Spline**.
   - From the **Modify II** toolbar, choose the **Edit Spline** button.
   - At the 'Command:' prompt, enter the **splinedit** command.
   - Alternatively, enter the alias **spe** at the keyboard.

   Command: **splinedit** *(Press* ENTER*.)*
2. In all cases, AutoCAD prompts you to select a single spline.

   Select spline: *(Select a single spline.)*

   If you select a splined polyline, it is converted to a NURB spline.
3. Specify an option:

   Enter an option [Fit data/Close/Move vertex/Refine/rEverse/Undo]:

Let's look at each option.

Fit data

The **Fit data** option allows you to edit "fit points," Autodesk's name for the points you picked originally to place the spline. (This option does not appear if you previously grip-edited the spline.)

Enter a fit data option

[Add/Close/Delete/Move/Purge/Tangents/toLerance/eXit] <eXit>: *(Enter an option.)*

Fit data has the following options:

Add

The **Add** option adds fit points (misnamed "control point" in the prompt) to the spline. This is done in two steps: (1) select an existing fit point, and AutoCAD highlights it and the adjacent fit point; and (2) pick a point to position the new fit point.

Specify control point <exit>: *(Select an existing fit point.)*

Specify new point <exit>: *(Pick a point, or press* ENTER *to exit the option.)*

AutoCAD adds a fit point at the point you pick, and redraws the spline to take into account the added points.

Picking the first or last fit points allows you to extend the spline by adding fit points beyond the ends of the spline.

Close/Open

The **Close** option joins the two ends of the spline to form a loop. When closed, the **Open** option opens up the spline by removing the segment between the start and end points.

Delete

The **Delete** option prompts you to select fit points, which are then removed.

Specify control point <exit>: *(Select an existing fit point.)*

Press ENTER when done removing fit points to exit this option. To remove all fit points at once, use the **Purge** option.

Move
The **Move** option moves fit points. It is easier to use grips editing to move fit points.

Purge
The **Purge** option removes the fit data from the spline.

Tangents
The **Tangents** option edits the tangency at the spline's start and end points. The option prompts you to relocate the tangents, as follows:

> Specify start tangent or [System default]: *(Pick a point, or press* ENTER *to skip changing this tangent.)*
>
> Specify end tangent or [System default]: *(Pick a point, or press* ENTER *to skip changing this tangent.)*

The **System Default** option calculates the default tangents at the ends. Autodesk suggests using the TANgent and PERpendicular object snap modes to snap the spline's endpoint to geometry on other objects.

Tolerance
The **Tolerance** option changes the tolerance (inverse weight) of control points.

> Enter fit tolerance <0>: *(Enter a value between 0 and 1.)*

Close/Open
The **Close** option joins the two ends of the spline to form a loop. When closed, the **Open** option opens up the spline by removing the segment between the original start and end points.

Move vertex
The **Move vertex** option moves "vertices," Autodesk's name for the control points, located at the ends of the frame segments. It is easier to use grips editing to move vertices.

Refine
The **Refine** option is the most powerful, allowing you to change the order, add control points, and change the weight — concepts described earlier in this chapter.

> Enter a refine option [Add control point/Elevate order/Weight/eXit] <eXit>:

Refine has the following options:

Add Control Point
The **Add control point** option adds control points to the spline.

> Specify a point on the spline <exit>: *(Pick a point.)*

The point is not necessarily added at the point you pick; AutoCAD adds the control point between the two closest existing control points.

Elevate Order

The **Elevate order** option increases the number of control points on the spline. The current order is shown below as "n."

Enter new order <*n*>: *(Enter a value between n and 26.)*

The frame receives more control points; the spline curve does not change.

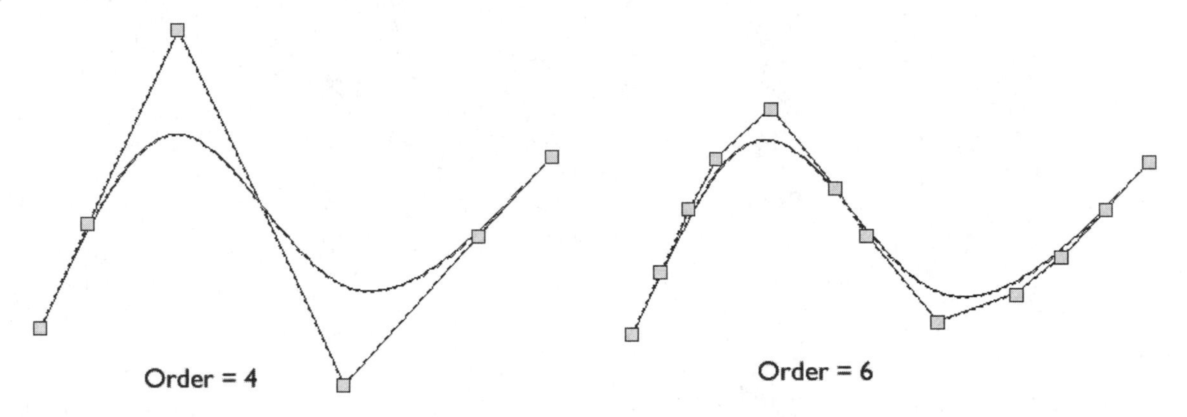

Increasing spline's order from 4 to 6.

Weight

The **Weight** option changes the weight at individual control points. The larger the weight, the closer the spline curve is pulled to the control point — the opposite of tolerance.

Enter new weight (current = 1) or [Next/Previous/Select point/eXit] <N>: *(Enter a new value, type an option, or press* ENTER *to exit the option.)*

Use the **Select point** option to select the control point to change. A larger value, such as 10, tugs the spline curve toward the control point; a smaller value, such as 0.1 pushes the curve away from the control point.

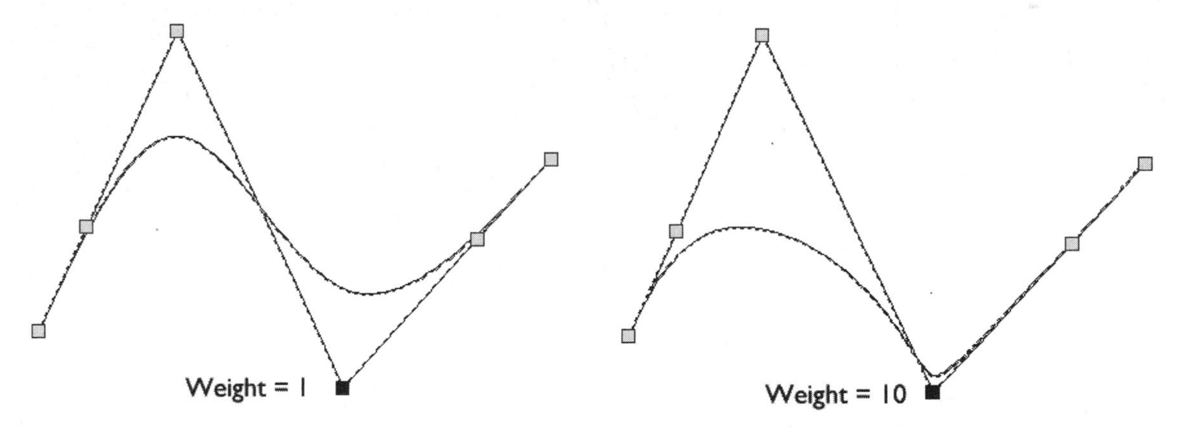

Larger weight tugs the spline curve toward the control point.

You can enter any value larger than 0, including extremely large numbers, such as 1E+17, but the resulting spline may become weird.

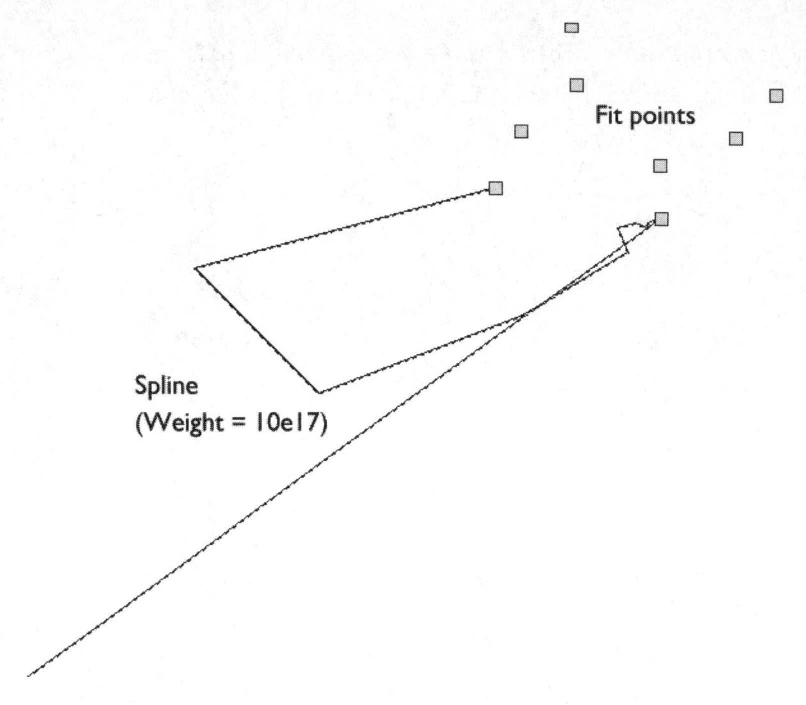

Too large a value for weight creates unexpected results.

rEverse

The **rEverse** option reverses the direction of the spline: the end point becomes the start point, and vice vera. AutoCAD reports, "Spline has been reversed."

Undo

The **Undo** option undoes the last operation, without needing to exit the command.

(AutoCAD does not have an option for editing knots.)

EXERCISES

1. From the CD-ROM, open the *Ch02MLine.dwg* file, a drawing of a floor plan. Use the **MLEDIT** command to clean up the intersections.

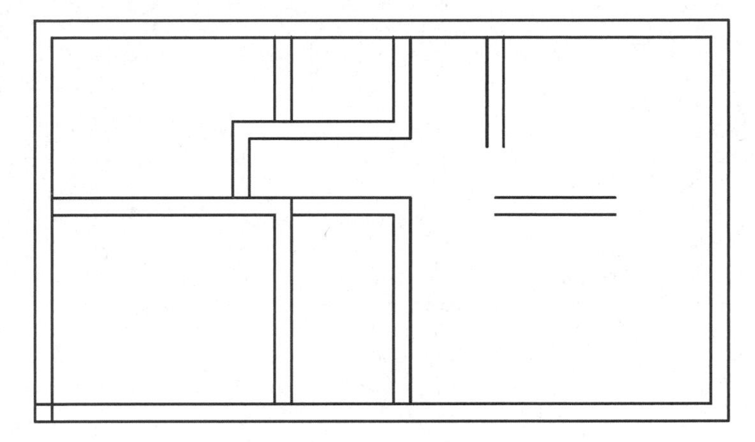

Ch02MLine.dwg

2. Use the **MLSTYLE** command to create multiline styles with the following properties:

Style Name	*Description*	*Element 1*	*Element 2*	*Element 3*
4-WALL	4" interior walls	Offset 2 Continuous	Offset -2 Continuous	...
6-WALL	6" exterior walls	Offset 3 Continuous	Offset 0 Batting	Offset -3 Continuous

3. Use the **MLINE** command to draw a floor plan, such as of your school, your home, or of another building. If necessary, clean up with the **MLEDIT** command.
4. Add doors and windows to the floor plan, which you can find in the *House Designer.dwg* file on the CD-ROM. (Add them with DesignCenter.)

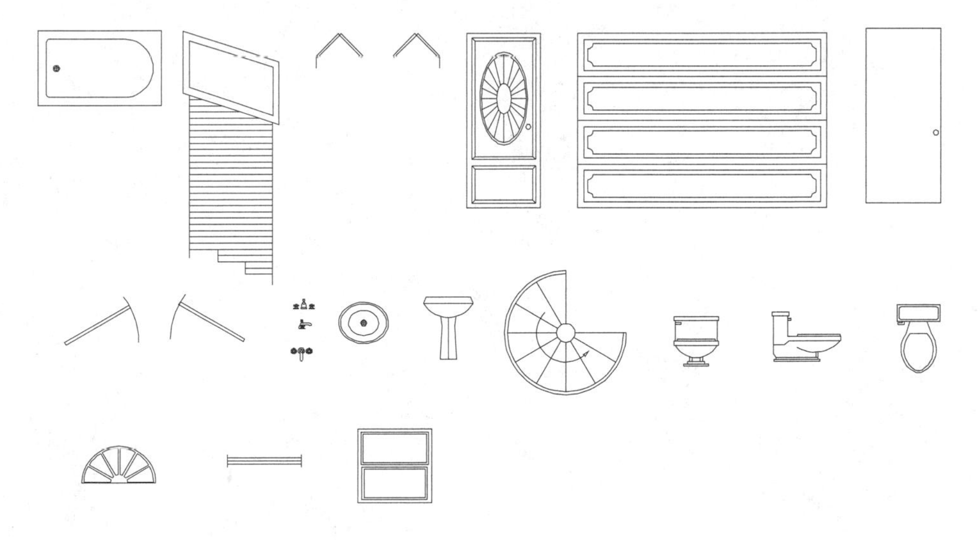

House Designer.dwg

5. From the CD-ROM, open the *Ch02SalisburyHill.dwg* file, a drawing of contour points for Salisbury Hill.

   Use the **SPLINE** command to recreate the contour map of the hill.

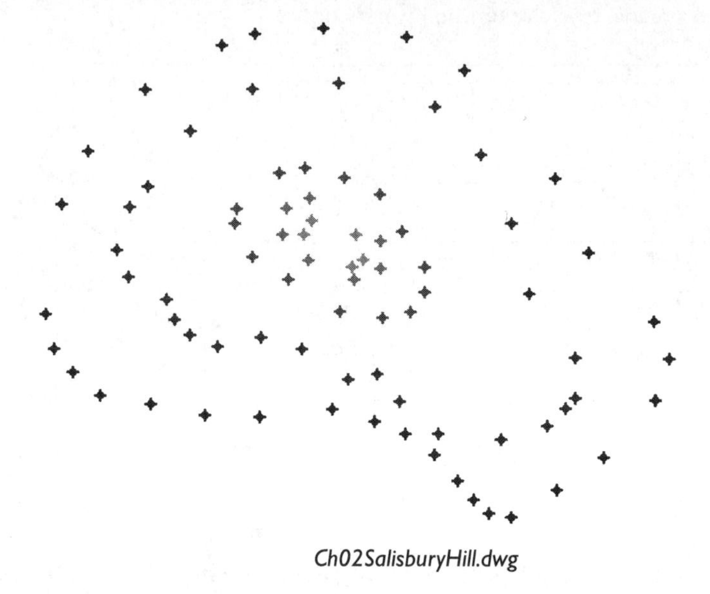

Ch02SalisburyHill.dwg

Tips: Set object snap to NODe, and turn off all others.

The elevations (0' through 200') are on separate layers.

When complete, use the **View | 3D Views | NE Isometric** menu to view the hill in 3D.

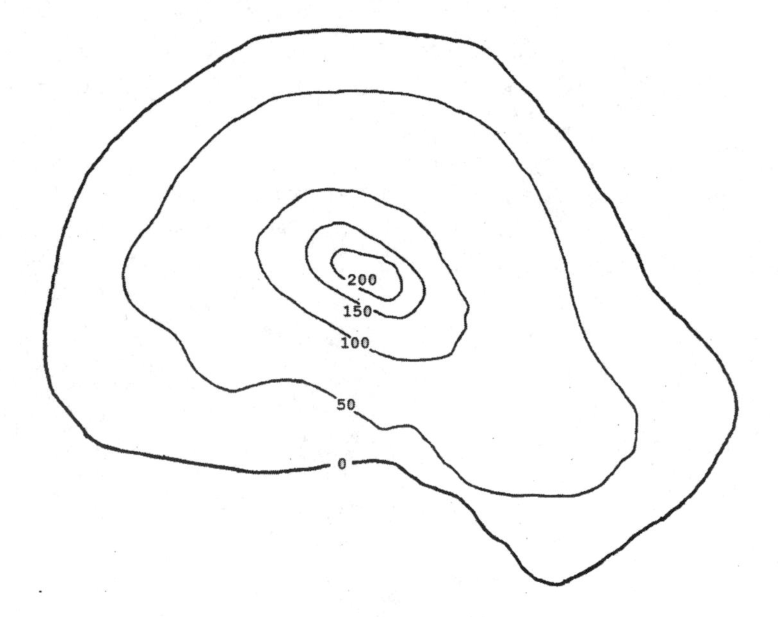

6. From the CD-ROM, open the *Ch02WoodLake.dwg* file, a drawing of contour points for Wood Lake, North Dakota.

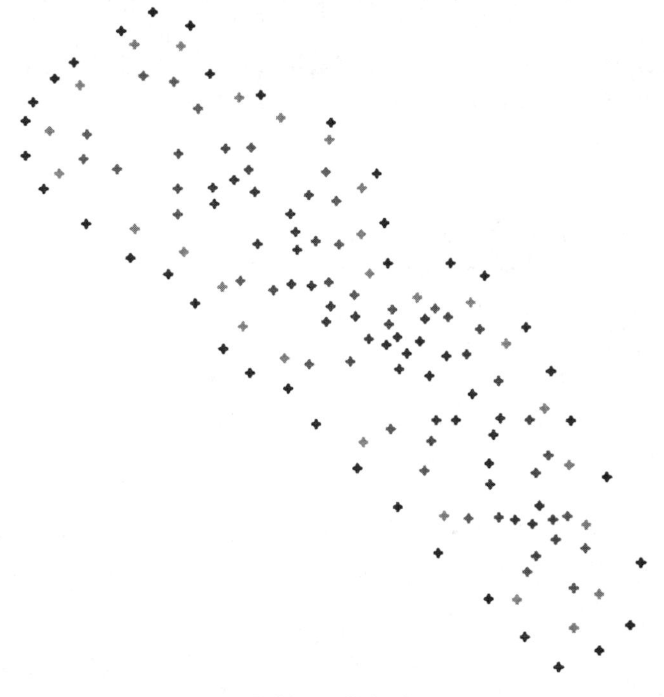

Ch02WoodLake.dwg

Use the **SPLINE** command to recreate the contour map of the lake bottom. The result should look similar to the original map illustrated below.

www.state.nd.us/gnf//fishing/lakedata.html

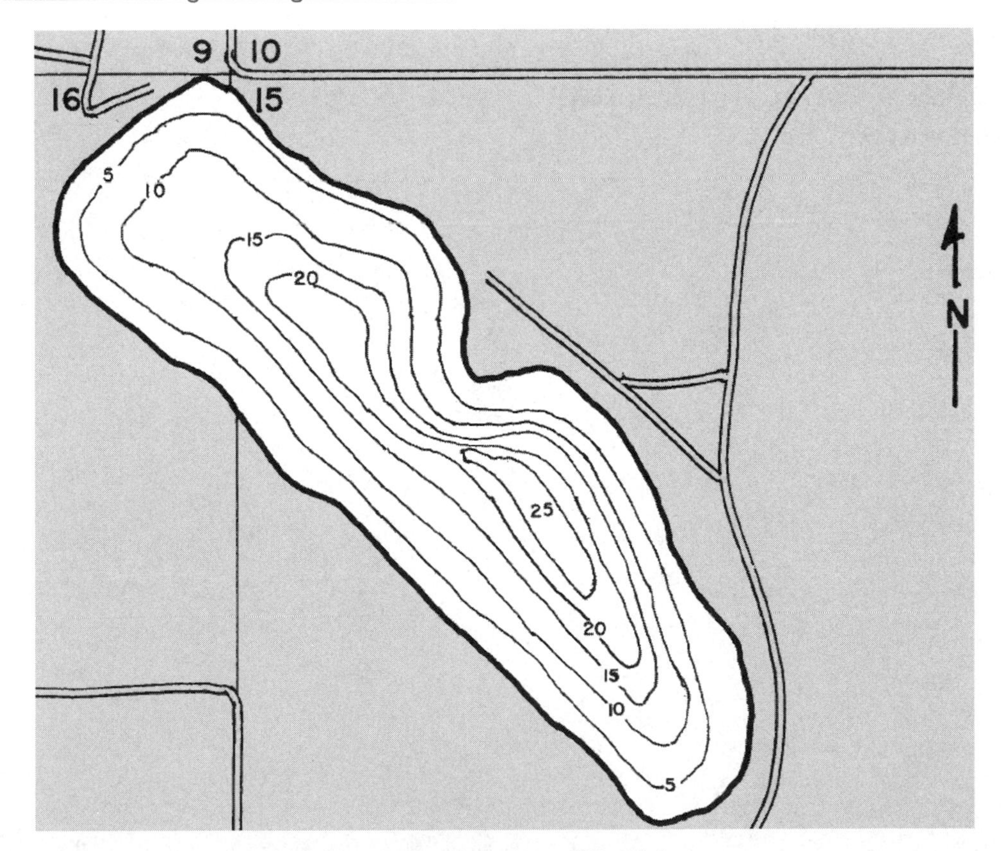

Contours for lake bottom of Wood Lake.

CHAPTER REVIEW

1. What are multilines good for drawing?
   Splines?
2. Explain the effect of the **MLINE** command's **Scale** option on multilines.
3. What happens to multilines when the scale is:
   0
   -1
   10
4. What command creates new multiline styles?
5. How do you share mlines between drawings?
6. How would you clean up an mline intersection?
   Fill a gap in a multiline?
7. Can you modify a multiline style?
8. Describe how to add a door symbol in a wall made of multilines.
9. What command lets you move doors and windows in mline walls without changing the multiline?
10. List some of the differences between NURB splines and polyline splines.
11. How many lines can MLine draw at a time?
12. Which command edits:
    NURBS splines?
    Multilines?
    Polyline splines?
13. What command replaces **MLINE** in AutoCAD LT?
14. Explain the acronym "NURBS."
15. How does weight affect a spline?
    What does AutoCAD call weights?
16. How would you increase the accuracy of the spline?
17. Describe the purpose of the **SPLINEDIT** command's **Object** option.
18. Can you use grips editing on multilines?
    On splines?

Working with Multiple Drawings

In this chapter, you learn about two methods AutoCAD uses to work with more than one drawing a time. The first method opens additional drawings, each in its own window; you can copy and paste objects between the drawings. The second method opens additional drawings in the same window; the additional drawings are said to be "externally referenced," because they are external to the current drawing, and used primarily for reference purposes. (Externally-referenced drawings are called "xrefs," hence the many commands that begin with "X.")

In this chapter, you learn both methods of working with multiple drawings:

OPEN opens multiple drawings at once, while **SYSWINDOWS** controls the windows in which each drawing is displayed.

COPYCLIP and **PASTECLIP** copy and paste elements between drawings.

XATTACH attaches external drawings for referencing (modified in AutoCAD 2005).

XREF controls referenced drawings, while **PROJECTPATH** controls paths to xrefs.

XREFNOTIFY alerts you to changes in xrefs.

XCLIP hides parts of xrefs.

XREF Bind and **XBIND** insert all (or parts) of xrefs as blocks in the current drawing.

XOPEN opens xrefs for editing, while **REFEDIT** edits xrefs and blocks in-place.

FINDING THE COMMANDS

On the **REFERENCE** and **REFEDIT** toolbars:

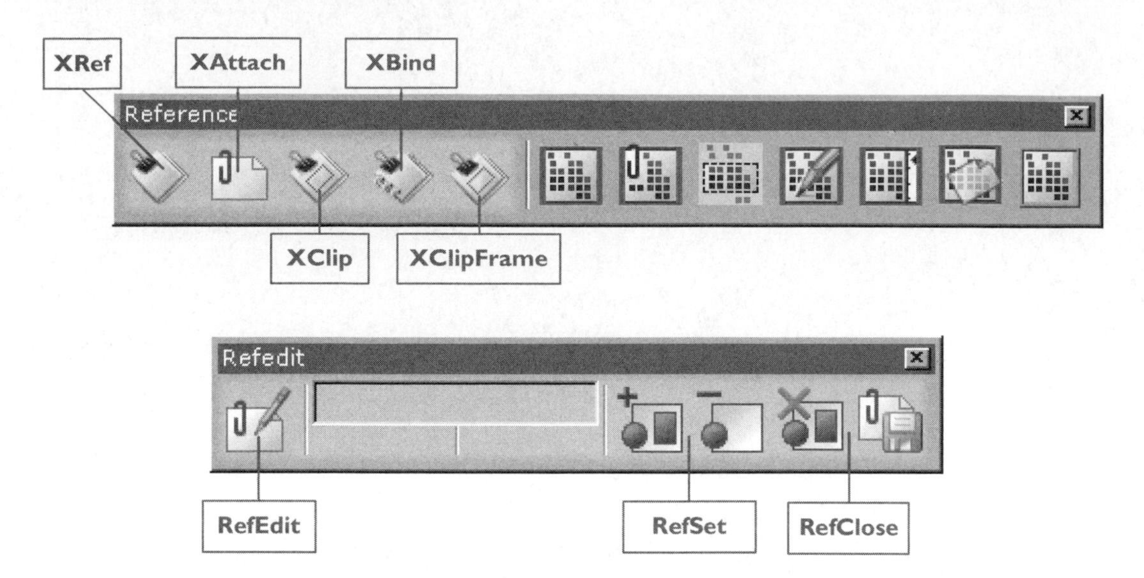

On the **EDIT**, **INSERT**, and **MODIFY** menus:

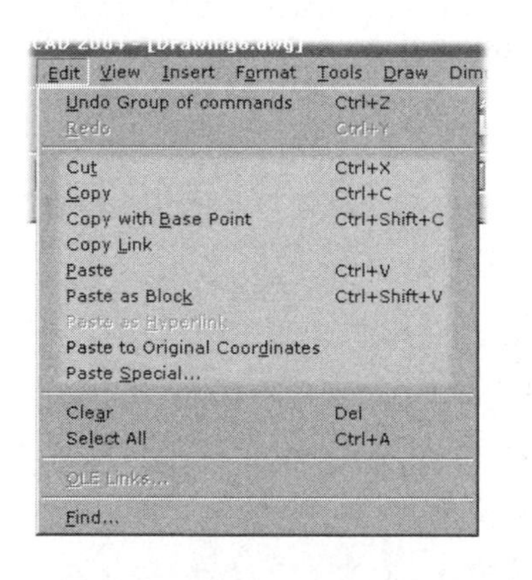

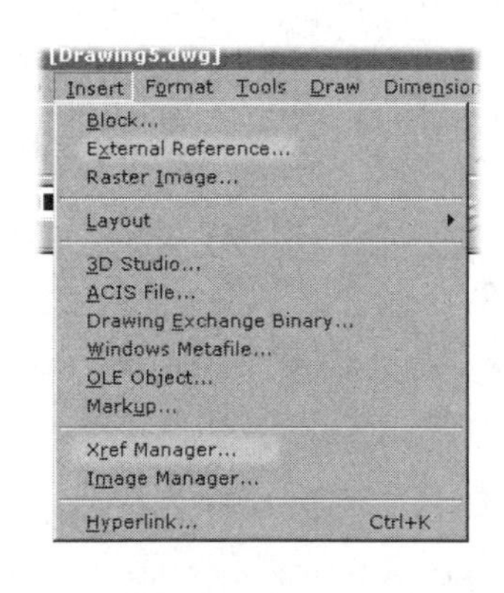

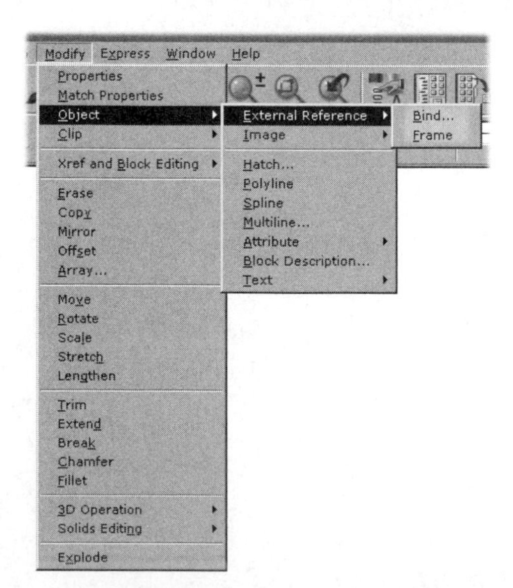

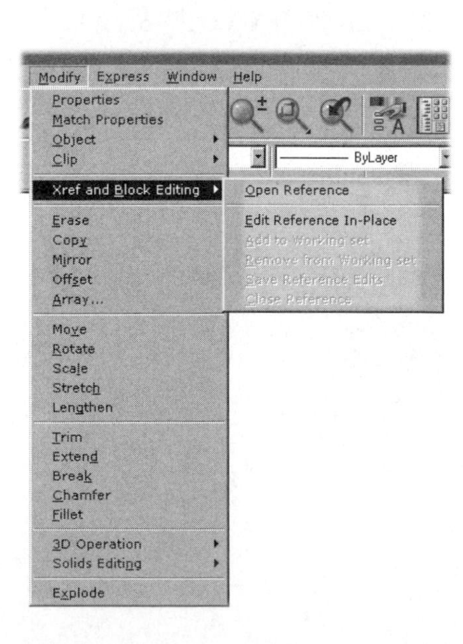

OPEN

The **OPEN** command opens one or more drawings at a time.

AutoCAD provides commands for controlling the windows, each containing a separately-opened drawing: minimizing and maximizing them, switching between them, and closing them.

(Historically, software programs written for Windows at first handled just one drawing or document at a time. In the early 1990s, Microsoft added MDI, short for "multiple document interface," a fancy term for opening two or more documents. Autodesk waited until AutoCAD 2000 before adding MDI, because of AutoLISP's difficulties in handling multiple drawings. The **SDI** system variable is still available for turning AutoCAD back into a "single document interface" program for compatibility reasons.)

TUTORIAL: OPENING MULTIPLE DRAWINGS

1. To open more than one drawing at a time, start the **OPEN** command:
   - From the menu bar, choose **File**, and then **Open**.
   - From the **Standard** toolbar, choose the **Open** button.
   - At the 'Command:' prompt, enter the **open** command.
   - Alternatively, press the **CTRL+O** shortcut keystroke.

   Command: **open** *(Press* **ENTER**.*)*

   In most cases, AutoCAD displays the Select File dialog box. (If the **FILEDIA** system variable is set to 0, prompts are instead displayed on the command line.)
2. Browse to a folder containing drawings.
3. To open more than one drawing at a time, hold down the **CTRL** key, and then select each drawing you wish to open.

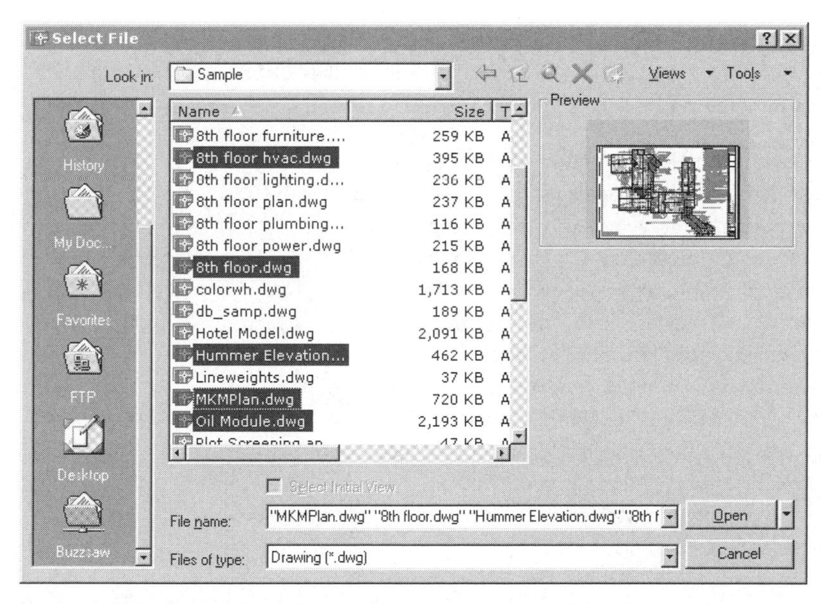

Hold down the CTRL *key to select more than one drawing to open.*

4. Click **Open**. AutoCAD opens each drawing in its own window. (It may take longer than you expect to open many drawings.)

5. To confirm that all drawings have opened, select the **Window** menu. Notice the list of drawing names. The check mark next to a drawing name indicates the drawing currently displayed "on top."

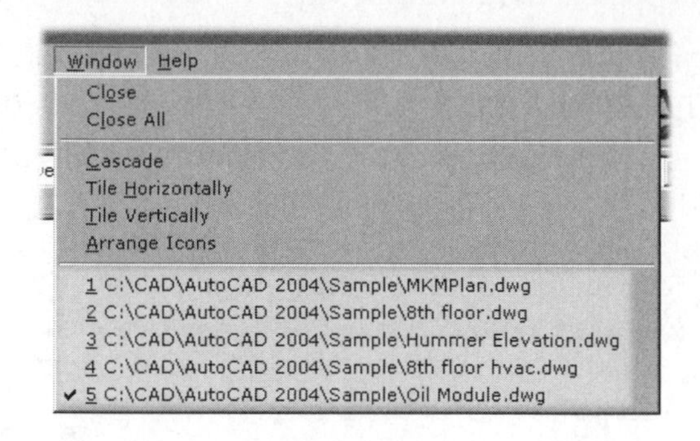

The Windows menu lists the names of all opened drawings.

6. To switch to another drawing, select its name from the list in the **Window** menu.

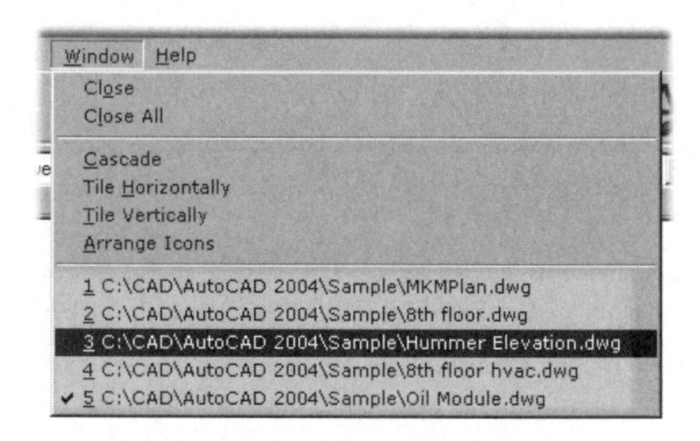

To switch to another drawing, select its file name.

Note: The **OPEN** command only opens *.dwg* and *.dxf* files. Other commands provide access to other file formats understood by AutoCAD (see Chapter 5):

ACISIN opens *.sat* files created by ACIS-compatible CAD programs.
DXBIN opens *.dxb* (drawing exchange binary) files created by CAD\camera.
DXFIN opens *.dxf* (drawing interchange format) files created by other CAD programs.
FILEOPEN opens drawings without a dialog box; useful for scripts and macros, but limits AutoCAD to running in SDI mode (single drawing interface).
IMAGEATTACH attaches raster (bitmap) images into drawings.
INSERT inserts other drawings as blocks into the current drawing.
PARTIALOAD loads additional portions of partially-loaded drawings.
REPLAY displays images in *.bmp*, *.tif*, and *.tga* formats.
RMLIN inserts *.rml* (redline markup language) files created by Autodesk's Volo View software.
WMFIN inserts *.wmf* (Windows meta format) files.
XOPEN opens externally-referenced drawings in their own windows.
3DSIN opens *.3ds* models created by 3D Studio.

CONTROLLING THE DISPLAY OF MULTIPLE DRAWINGS

AutoCAD provides several options for controlling the display of multiple drawings, as do most Windows applications. The **SYSWINDOWS** command controls windows from the keyboard, and is useful for macros and programs. The **Windows** menu lists several commands; several more are "hidden" in the little-used **System** menu. In addition, Microsoft sanctions several keyboard shortcuts for quickly switching between documents.

SysWindows Command

The **SYSWINDOWS** command arranges windows within AutoCAD (short for "system windows").

> Command: **syswindows**
>
> Enter an option [Cascade/tile Horizontal/tile Vertical/Arrange icons]: *(Enter an option.)*

The options are identical to those of the **Windows** menu, described next.

Windows Menu

The **Windows** menu includes the following commands for controlling windows:

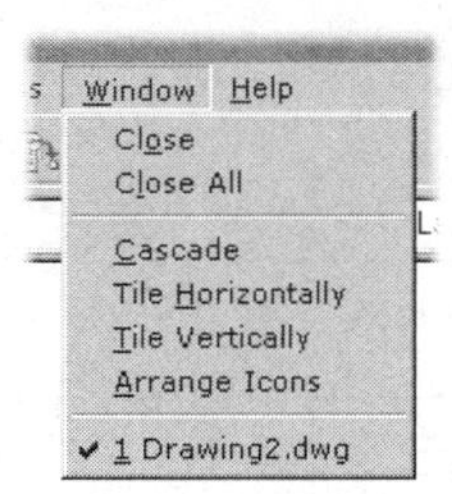

Windows menu controls windows.

Close closes the current window (or press **ALT+F4**); if necessary, AutoCAD prompts you to save the drawing first, as with the **CLOSE** command.

Close All closes all windows; AutoCAD prompts you to save the drawings, as with the **CLOSEALL** command.

Cascade displays the windows in a staggered, overlapped fashion.

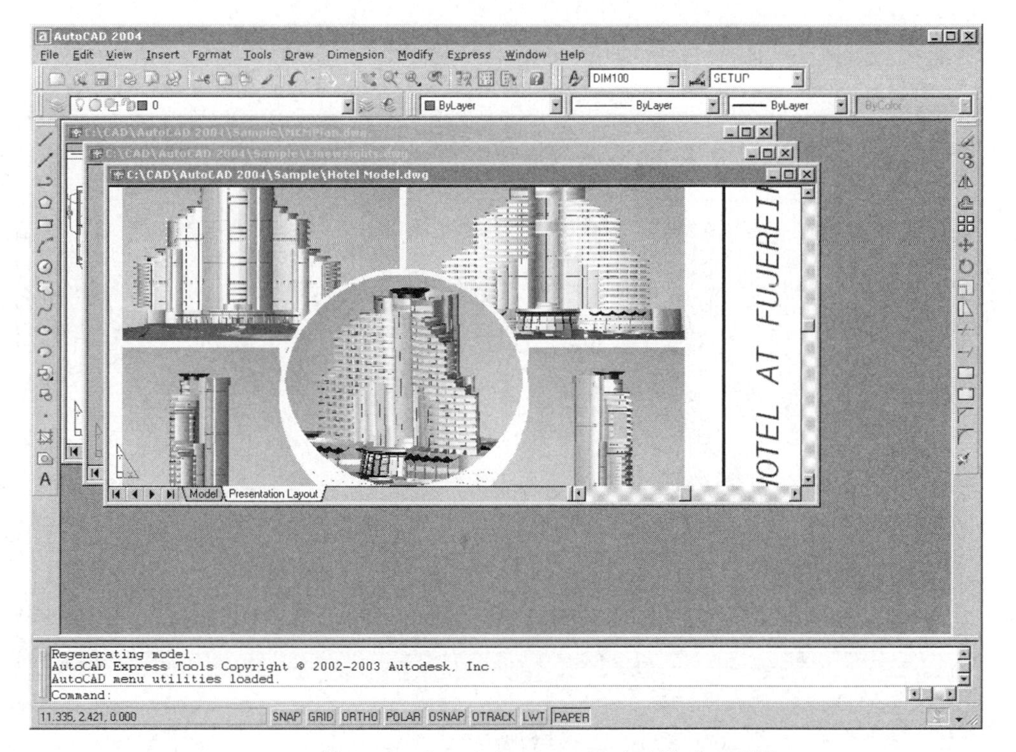

Three drawing windows cascaded within AutoCAD.

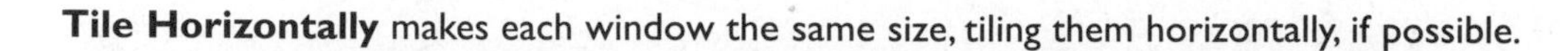

Tile Horizontally makes each window the same size, tiling them horizontally, if possible.

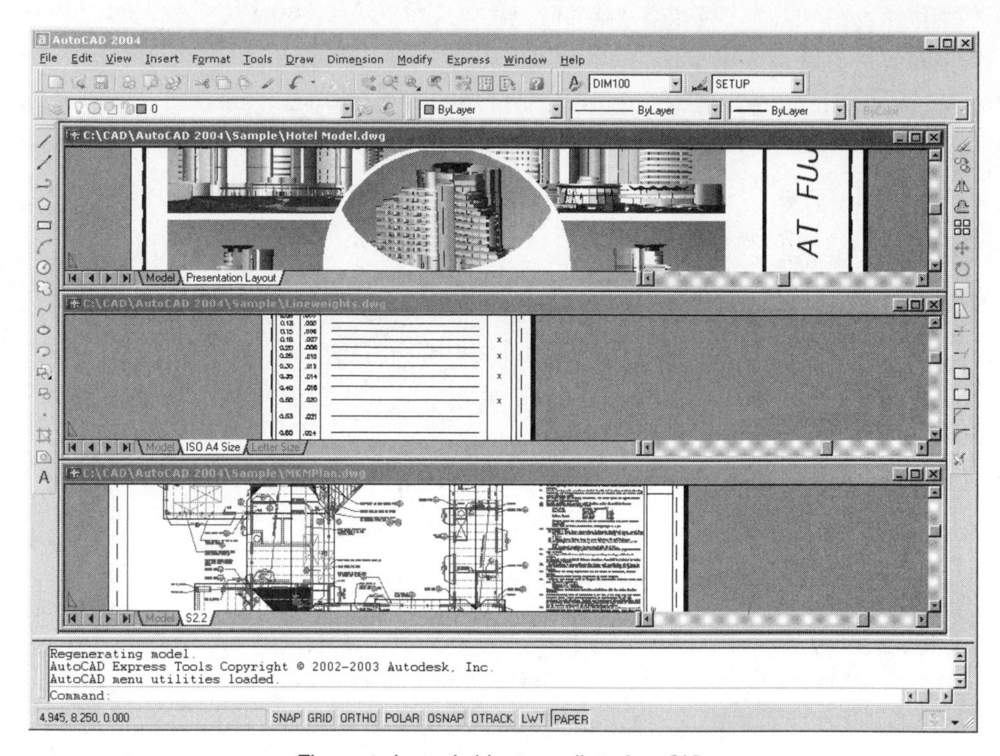

Three windows tiled horizontally in AutoCAD.

Tile Vertically also makes each window the same size, but tiles them vertically, if possible.

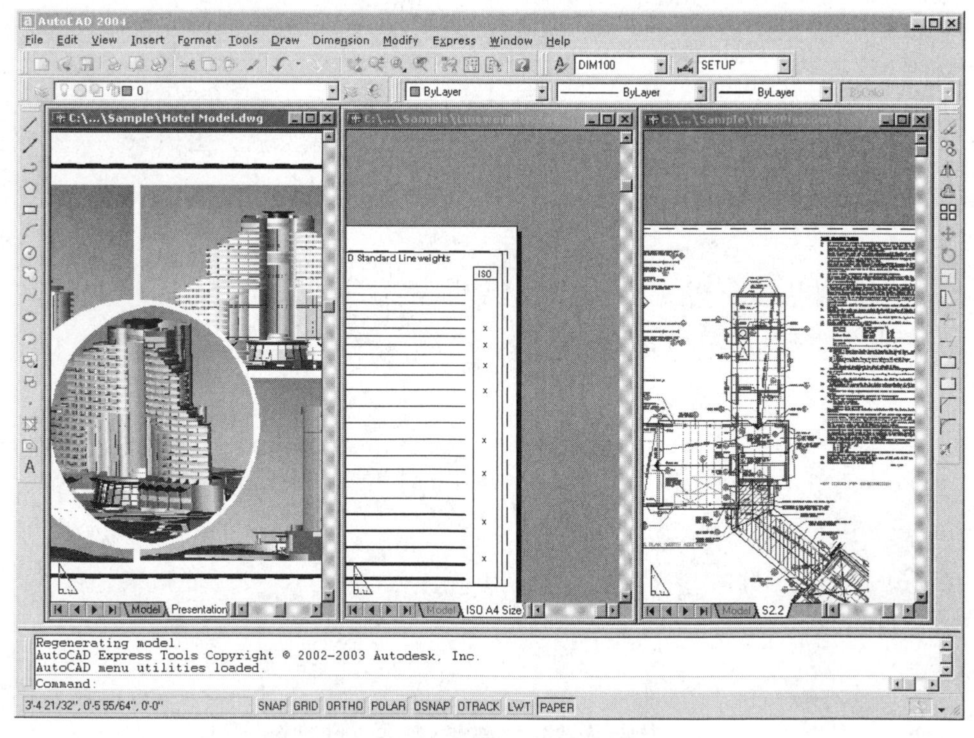

Three drawing windows tiled vertically in AutoCAD.

When there are one, four, or eight drawings open in AutoCAD, the **Tile Horizontally** and **Tile Vertically** commands operate identically.

Arrange Icons lines up scattered icons neatly along the bottom of the AutoCAD window; this option works only when windows are minimized. (See the next section.)

(Historically, the first release of Windows displayed software programs only in *tiled windows*. This is where Windows got its name. Subsequent releases allowed for overlapping windows.)

System Menu

The **System** menu is an icon that appears in one of two locations. When a window is maximized, it appears at the left end of the menu bar; when a window is tiled or minimized, the menu appears at the left end of the window's title bar.

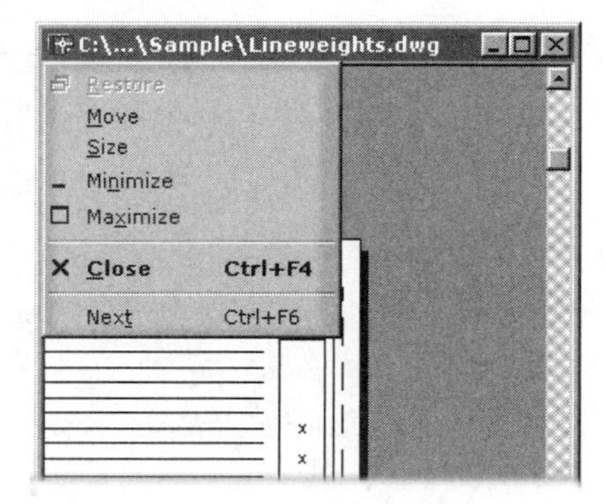

The System menu appears at the left end of title bars or the menu bar.

Click the menu to reveal the following windows-related commands:

Restore *tiles* windows.

Move allows you to move the window using the cursor keys on the keyboard. Generally, it is more convenient to move windows by dragging their title bars with the cursor.

Size allows you to change the size of the window using cursor keys, after placing the cursor over the edge of the window to be resized. As with Move, it's usually easier to resize the window by dragging one of the four edges with the cursor. (Grab one of the four corners to resize two edges at the same time.)

Minimize reduces windows to icons.

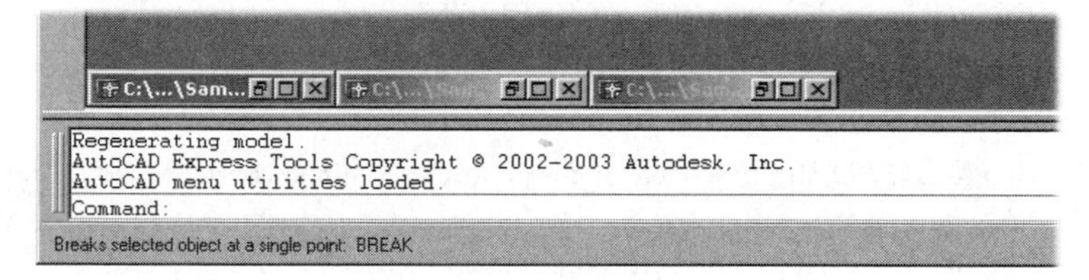

Drawing windows minimized as icons.

Maximize maximizes the size of windows within the confines of AutoCAD.

Close (**CTRL+F4**) closes the current window.

Next (**CTRL+F6**) switches to the next window. As an alternative, you can press **CTRL+TAB**, or else select another drawing from the **Window** menu. Or, left-click any window to make it current.

COPYCLIP AND PASTECLIP

The **COPYCLIP** command (short for "copy to Clipboard") copies selected objects to the Windows Clipboard, while the **PASTECLIP** command (short for "paste from Clipboard") pastes objects from Clipboard to the drawing — if the objects are valid.

These two commands are particularly useful for copying objects between drawings. (Use the **COPY** command to copy objects within a single drawing, because it is more efficient.)

TUTORIAL: COPYING AND PASTING OBJECTS

1. To copy objects between drawings, start AutoCAD with two drawings.
2. In one drawing, start the **COPYCLIP** command:
   - From the menu bar, choose **Edit**, and then **Copy**.
   - From the **Standard** toolbar, choose the **Copy to Clipboard** button.
   - At the 'Command:' prompt, enter the **copyclip** command.
   - Alternatively, enter the **CTRL+C** shortcut keystroke at the 'Command:' prompt.

   Command: **copyclip** *(Press* ENTER*.)*
3. In all cases, AutoCAD prompts you to select the objects to copy:

   Select objects: *(Select one or more objects.)*

   Press **CTRL+A** to select all objects in the current model layout space.
4. Press **ENTER** to end object selection.

   Select objects: *(Press* ENTER *to end object selection.)*

   AutoCAD invisibly copies the selected objects to Clipboard.

5. Switch to the second drawing using **CTRL+TAB**.
6. To paste the object from Clipboard, start the **PASTECLIP** command:
   - From the menu bar, choose **Edit**, and then **Paste**.
   - From the **Standard** toolbar, choose the **Paste from Clipboard** button.
   - At the 'Command:' prompt, enter the **pasteclip** command.
   - Alternatively, enter the **CTRL+V** shortcut keystroke at the 'Command:' prompt.

   Command: **pasteclip** *(Press* ENTER*.)*
7. In all cases, AutoCAD prompts you to select the insertion point:

   Specify insertion point: *(Pick a point in the drawing.)*

   AutoCAD places the objects at the point you specify.

Notes: Within AutoCAD, the **COPYCLIP** and **PASTECLIP** commands are limited to copying and pasting objects. To copy named tables (block definitions, layers, text and dimension styles, layouts, linetypes, and xrefs) between drawings, use the **ADCENTER** command to access DesignCenter.

AutoCAD is not limited to pasting its own objects in drawings. You can copy an object from just about any other application, and then paste it into AutoCAD — documents, spreadsheets, audio clips, and movies. When the object in the Clipboard is not an AutoCAD object, AutoCAD pastes it in the upper left corner of the current viewport.

AutoCAD provides additional commands specific to the copy'n paste process:

CUTCLIP copies selected objects to Clipboard, and then erases them from the drawing.

COPYBASE copies selected objects to Clipboard, after prompting for a base point; objects are then pasted as a block with the **PASTEBLOCK** command.

COPYLINK copies all objects in the current viewport to Clipboard; objects are pasted with a link back to the original drawing.

PASTEBLOCK (Paste as block) pastes objects into the drawing as a block; available only when Clipboard contains an AutoCAD object. When Clipboard contains a block, it is pasted as a nested block.

PASTEASHYPERLINK adds hyperlinks to selected objects; this command is available only when Clipboard contains hyperlink data.

PASTEORIG (Paste to Original Coordinates) pastes blocks in the drawing using its original insertion coordinates; available only when the Clipboard contains an AutoCAD block, and when a different drawing is open.

PASTESPEC allows you to select the format for pasted objects; also allows OLE operations (object linking and embedding).

There is a sixth paste command: you can paste text from Clipboard into the command prompt area using the **CTRL+V** shortcut. As an alternative, right-click the command area and select **Paste**. If AutoCAD understands the text, it acts upon it. For example, you can paste commands, scripts, or AutoLISP code, and AutoCAD executes them.

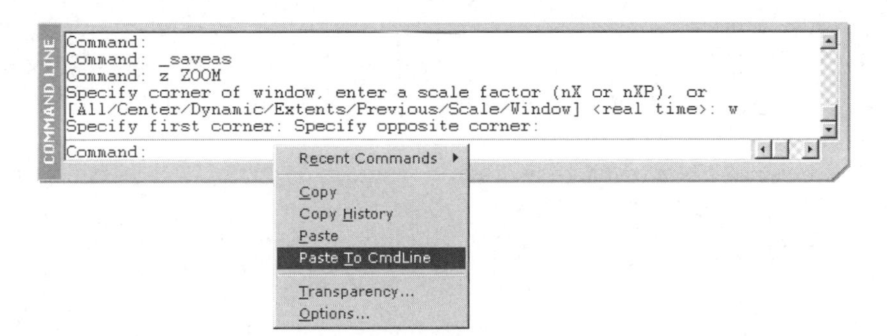

Pasting text in command line.

XATTACH

The **XATTACH** command displays the Select Reference File dialog box, so that you can attach *externally-referenced drawings*, or "xrefs," for short.

One of the strengths of CAD is its ability to draw components or symbols, and then assemble them into complete drawings. In Chapter 6 of *Using AutoCAD: Basics*, you learned how to do this with the **BLOCK** and **INSERT** commands. The drawback to inserting blocks, however, is in updating them. Drafting offices commonly store symbols in a central location (often on a server computer), and not on individual computers, so that the symbols can be updated easily. Each time symbols are updated in the central computer, the symbols should be updated in all drawings that use the symbols. In the old days, this meant using the **insert name=** command repeatedly for every updated symbol in every drawing!

The solution instead is to use externally referenced drawings, where AutoCAD maintains a link to the original *.dwg* file. Each time you open drawings containing xrefs, AutoCAD automatically loads the updated xrefs; your drawings are always up to date! As of AutoCAD 2004, an informative message alerts you that xrefs have been updated by another drafter.

(There is a place for inserting xrefs as blocks: when you want to freeze a drawing, so that it does not change. This can happen during stages of a project, or at project completion. Combine all xrefs into a single drawing with the **XREF** command's **Bind** option.)

Inserting blocks is best when you *don't* want the drawing to change; attaching xrefs is best when you expect change or the same blocks are reused many times. In summary, the differences are:

	Inserting Blocks	***Xrefing Drawings***
Stored in the drawing:	Yes	No, only links are stored
Increases file size of drawing:	Yes	Slightly
Contains attributes:	Yes	No
Updates manually:	Yes	Yes
Updates automatically:	No	Yes, when drawing is opened
Clips easily:	No	Yes
Toggles display through:	Layers	Layers and temporary unloading
Removed easily from drawing:	No	Yes

USING XREFS

Xrefs are commonly used in two areas: drawing borders and base plans.

Drawing Borders

Drawing borders and title blocks have a common look, often determined by the design firm or the client's requirements. The two figures below show a drawing border on its own, and then xref'ed into a drawing. When changes are made to the drawing border, the revision is updated.

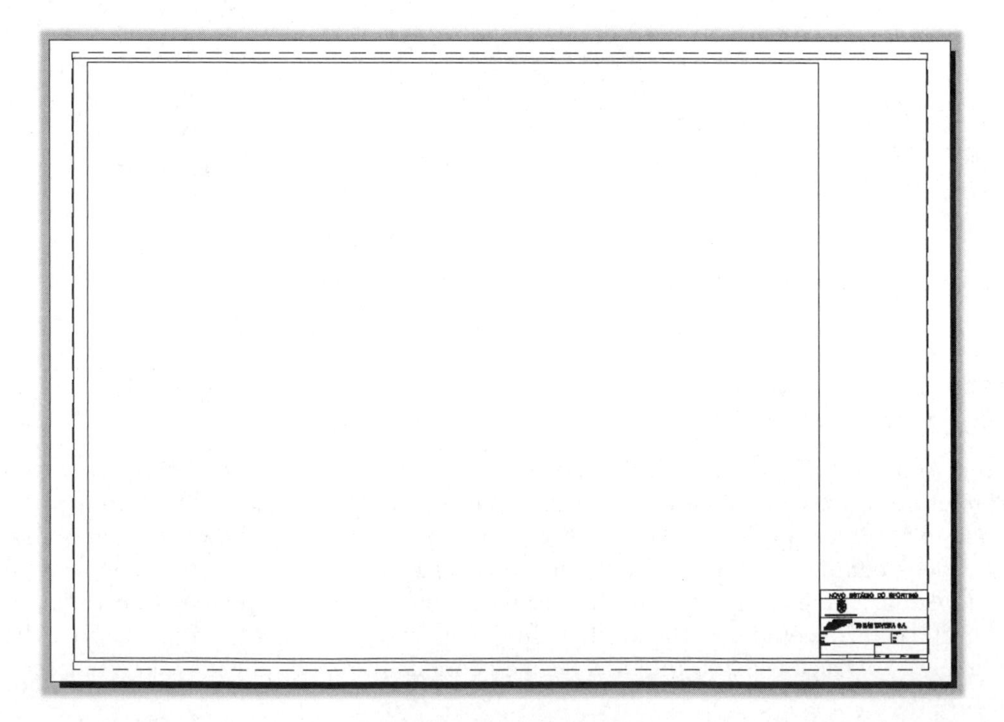

Drawing of border and title block.

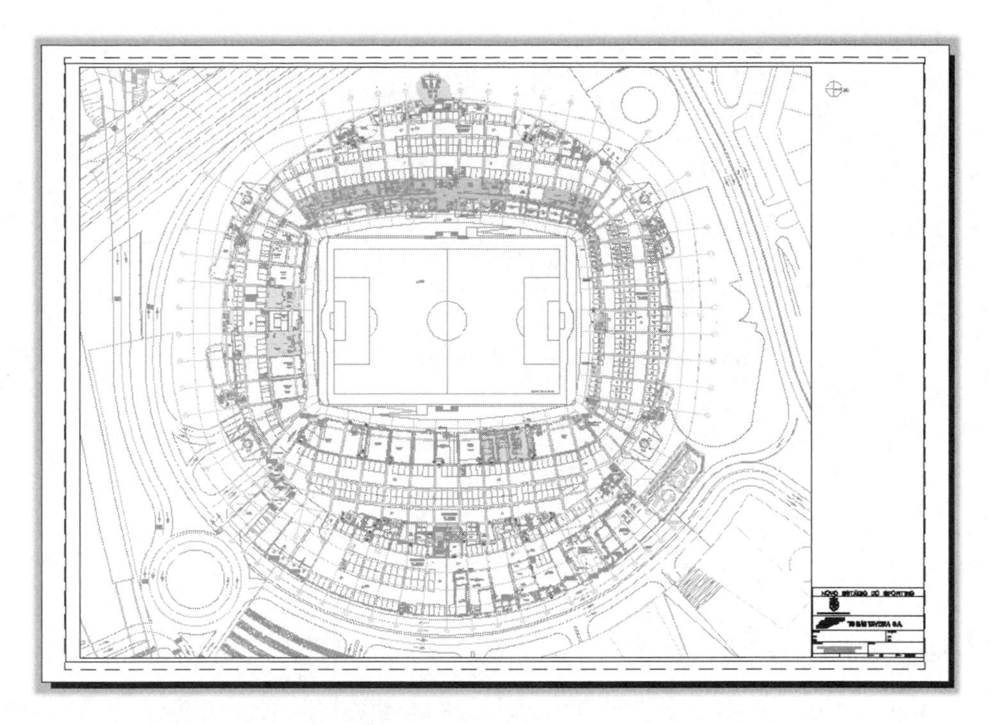

Title block and border xref'ed into the drawing of a stadium.

Base Plans

Base plans are common to all project drawings. The base plan could be the site plan, which shows elevations, roads, and other geographical elements. It could be the building's outer walls, core elements (stairways, elevators), and structural columns, as shown below.

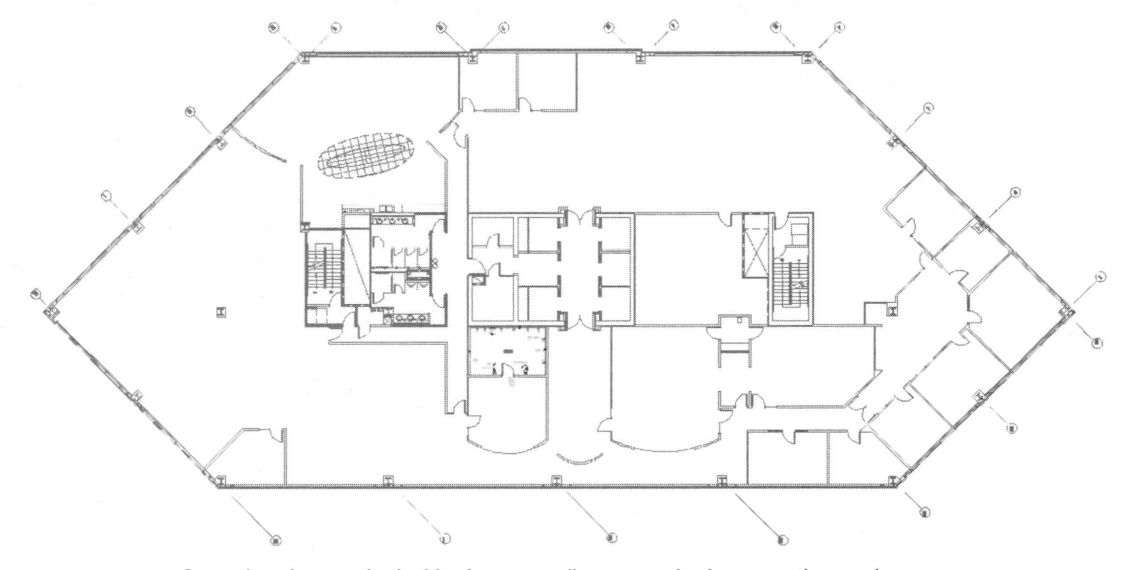

Base plan showing the building's outer walls, structural columns, and core elements.

The base plan is used by other disciplines for their design work. In the figure below, the facilities manager has placed the furniture in a drawing that xrefs the base plan.

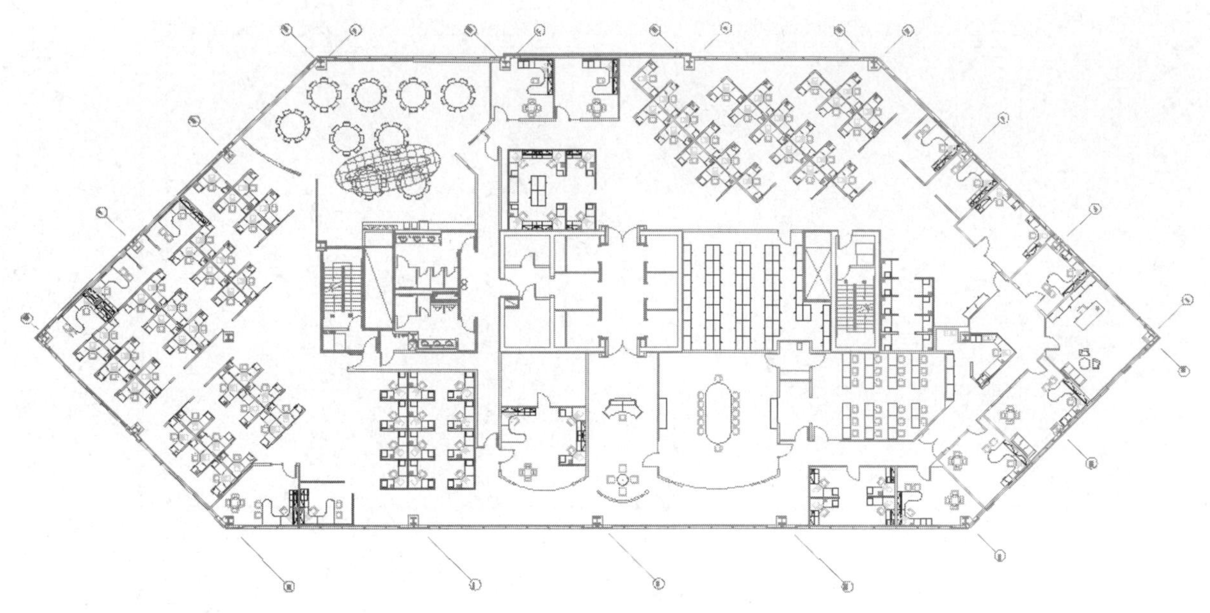

Furniture placed in a drawing, using the base plan as an xref.

An xref drawing can reference another drawing; this is called a "nested xref." In the figure below, the electrical designer has designed the electrical components to meet the needs of the furniture and the constraints of the base plan. Both the facilities manager and the electrical designer use the same base plan. Because the base plan is an xref, both CAD users are confident they are working with the latest revision of the base plan.

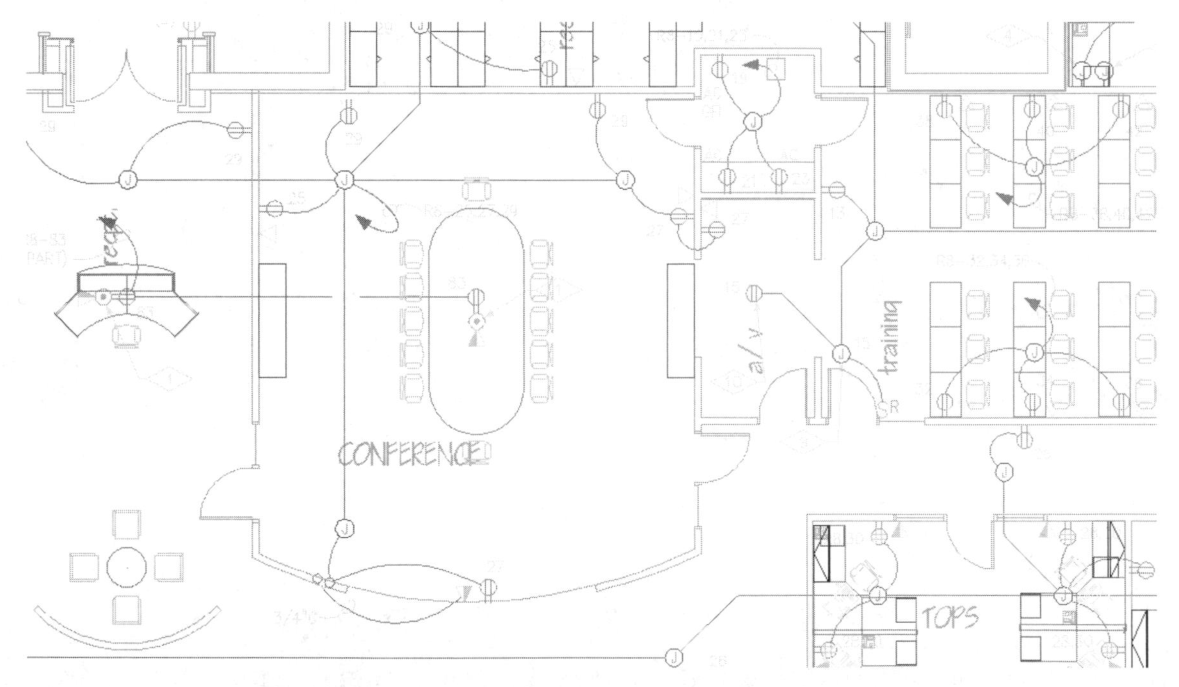

Electrical design based on the furniture plan and constrained by the base plan.

A drawing can have many xrefs; I think one number quoted is 32,000. Too many, of course, leaves a muddled mess, as shown below.

USING XREFS IN AUTOCAD

AutoCAD provides commands and system variables to work with externally-referenced drawings. In general, you use them in this order:

1. **XATTACH** attaches *.dwg* files as externally-referenced drawings (xrefs).
2. **XREF** controls the display and paths of xrefs; also detaches xrefs.
3. **XNOTIFYTIME** specifies how often AutoCAD checks for changes to xrefed drawings.
4. **XCLIP** hides portions of xrefed drawings.
5. **XOPEN** opens a selected xref in another window for editing; **REFEDIT** allows *in situ* editing of xrefs.

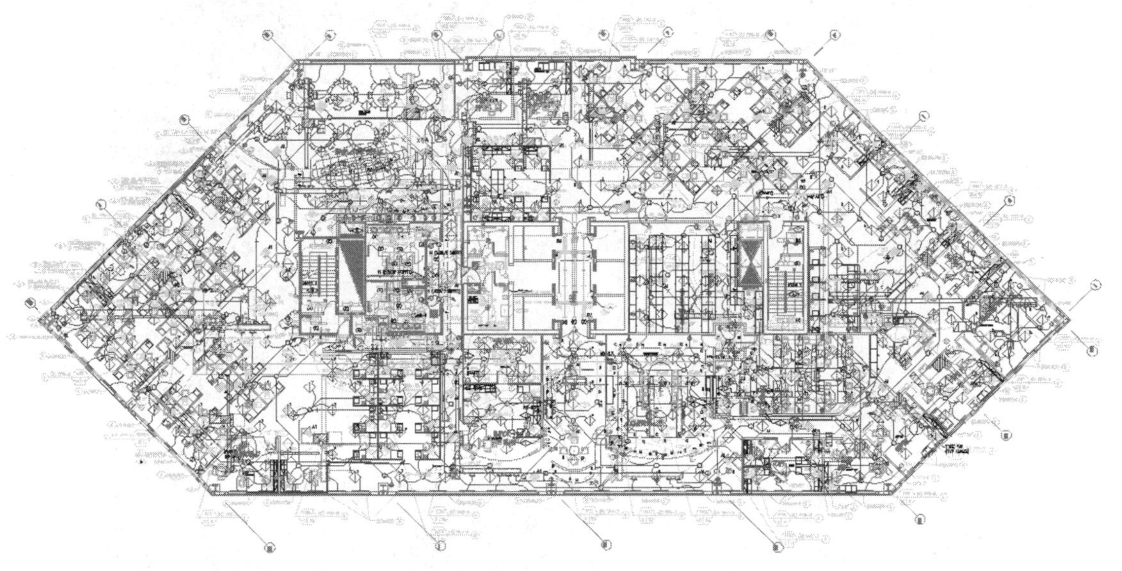

Too many drawings xrefed at one time making plans difficult to read.

Changes made in xrefed drawings are reflected in the master drawing, either: (1) the next time the master drawing is opened; or (2) the next time the xref is updated while the master drawing is open. The updates occur until you bind the externally-referenced drawing.

You can *bind* the externally-referenced drawing to the master drawing with the **Bind** option of the **XREF** command. This has the effect of adding all the contents of the externally-referenced drawing as if it were inserted like a block.

Alternatively, you can selectively bind parts of xrefs to the master drawing: blocks, text styles, dimension styles, layers, and linetypes. This is done with the **XBIND** command, and is an alternative to DesignCenter for importing named objects from other drawings.

Note: When you attach drawing as xrefs, other users can still open the original drawings for editing. AutoCAD does not "lock" *.dwg* files used as xrefs.

AutoCAD uses two terms for the drawing in which xrefs are placed: "master drawing" and "host drawing." Other writers also use the term "parent drawing." In this book, we use the term "master drawing."

TUTORIAL: ATTACHING DRAWINGS

1. To attach drawings to the current drawing, start the **XATTACH** command:
   - From the menu bar, choose **Insert**, and then **External Reference**.
   - From the **Reference** toolbar, choose the **External Reference Attach** button.
   - At the 'Command:' prompt, enter the **xattach** command.
   - Alternatively, enter the **xa** alias at the 'Command:' prompt.

   Command: **xattach** *(Press* ENTER.*)*

   In all cases, AutoCAD displays the Select Reference File dialog box, which looks and acts like the Select File dialog box presented by the **OPEN** command:

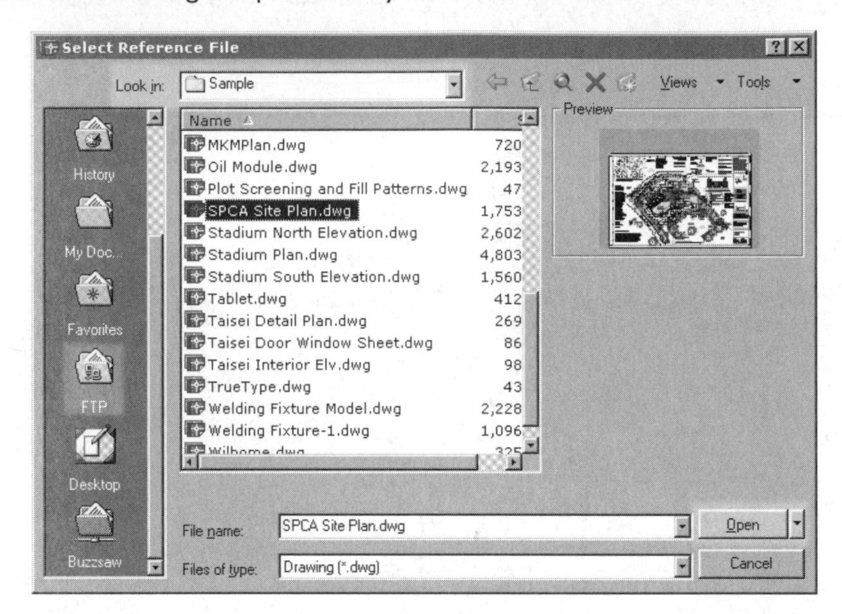

Select Reference File dialog box.

2. Select one or more *.dwg* drawing files, and then click **Open**.

   You can reference drawings located on your computer, or from computers connected to yours through the local network or the Internet:
   - To reference drawings located on the network, click the **Look in** drop list, and then select a network drive, or "My Network Places" (or similar name). Navigate to the computer, drive, and folder holding the drawings, select the drawings, and then click **Open**.
   - To select drawings located on the Internet, click **FTP** in the Places list. Double-click an FTP location, and then wait for the connection to be made. Select the drawing file, and then click **Open**. (AutoCAD stores a copy of the drawing on your computer in the *c:\documents and settings\username\local settings\temporary internet files* folder.)

   In all cases, notice that AutoCAD displays the External Reference dialog box, which looks like the Insert dialog box.

CHANGES IN AUTOCAD 2005: XATTACH COMMAND

AutoCAD 2005 adds an option to the External Reference dialog box: the **Uniform Scale** option ensures that the Y and Z scale factors are the same as the X scale factor.

In addition, AutoCAD 2005 adds the **XREFTYPE** system variable to determine the default for loading external references as attachments or overlays.

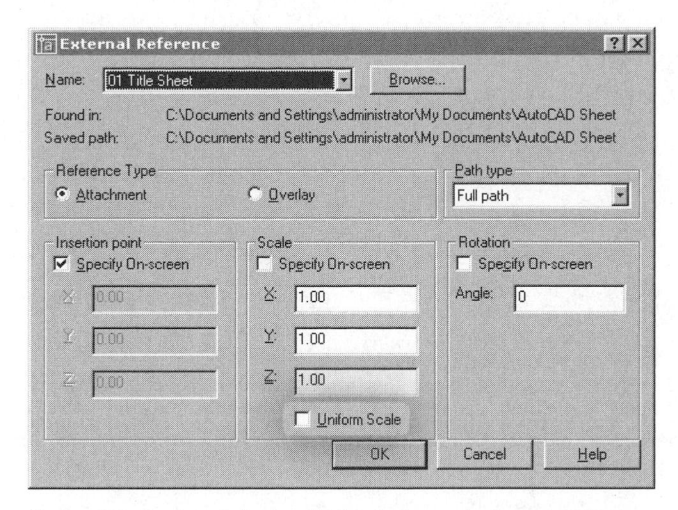

External Reference dialog box looks similar to the Insert dialog box.

3. Click **OK**. Notice that the referenced drawing appears.

   On the status line, AutoCAD makes a report similar to the following:

   ```
   Attach Xref "Filename": C:\CAD\AutoCAD 2005\Sample\filename.dwg
   "Filename" loaded.
   ```

Notes: Only model space *.dwg* drawing files can be attached as references; you cannot attach paper space views, *.dxf* files, or other kinds of CAD files. The exception are images or raster files, which are attached with the **IMAGEATTACH** command.

ATTACHING DRAWINGS: ADDITIONAL METHODS

The External Reference dialog box contains options for attaching drawings as references. In addition, several system variables control the loading of xrefs.

- **Browse** option selects a different *.dwg* drawing file to attach.
- **Reference Type** option determines how nested xrefs are displayed.
- **Path Type** option specifies how the path is stored.
- **Insertion Point** option specifies where the xref is placed in the drawing.
- **Scale** option specifies the xref scale factor.
- **Rotation** option specifies the xref rotation angle.
- **XLOADCTL** system variable determines the default for attachments or overlays.
- **XREFTYPE** system variable toggles demand loading (new to AutoCAD 2005).
- **XLOADPATH** specifies the path for storing temporary copies of demand-loaded xrefs.

Let's look at each option and variable.

Browse

The **Browse** option displays a file dialog box, so that you can select different *.dwg* drawing files to attach.

The **Found In** and **Saved Path** items are usually the same, listing the drive and folder names where the xref is located. Check the **Found In** path to ensure that the correct xref has been loaded. (The full path consists of the drive name, such as C:, and one or more folder names, such as \dwg\project.)

In some cases, your computer may hold two or more drawings of the same name. In other cases, AutoCAD cannot immediately find the xref, and so hunts for any file of the same name, using the search order listed below in the note.

The **Saved Path** item describes how the path to the xref is stored in AutoCAD. Depending on the option you select under Path Type, AutoCAD stores either no path, part of a path, or the full path.

Note: When the xref is no longer located in the specified folder (or if no path is saved for the xref), AutoCAD searches for it in this order:

1. Path to the folder of the master drawing.
2. Paths defined by the **Projects** item in the **Files** tab of the **OPTIONS** command, and **PROJECTNAME**.
3. Paths defined by the **Search** item in the **Files** tab of the **OPTIONS** command.
4. Path to the folder specified by the Windows application shortcut's **Start-in** option.

Reference Type

The **Reference Type** option determines how nested xrefs are displayed, whether attached or overlaid.

Recall that nested xrefs reference one or more additional references. This option answers the question, "Should nested xrefs be visible?" When the answer is "Yes," choose **Attach**. This ensures that all nested xrefs are displayed.

When the answer is "No," choose **Overlay**. This ensures nested xrefs are *not* displayed; only the first level of the xref drawing is shown.

The difference between attach and overlay can be hard to understand. Robyn Stables of EduCAD (New Zealand) taught me to think of nested xrefs as a stack of drawings. When you *attach* the nested xrefs, it's as if you staple together the drawings — you get all of them. When you *overlay*, you don't staple; you get just the topmost drawing.

In summary:

Reference Type	*Displays Nested Xrefs*
Attachment	Yes
Overlay	No

Path Type

The **Path Type** option specifies how the path is stored.

Recall that the *path* tells AutoCAD where to locate *.dwg* files. It consists of the drive and folder names, such as:

c:\cad\project1

The c: is the name of the C: drive, while \cad\project1 are folder names. As noted earlier, xref files accessed through FTP are copied to your computer. Their path looks the same.

If the path leads to another computer, then the networked computer's name is added:

\\downstairs\c:\cad\project1

There are three options for storing the path to the xref's *.dwg* file: full, relative, and none.

Full Path is sometimes called "hard coded," because there is no flexibility: the xref's *.dwg* file must be located where the path says it is. You use this option when it is important that AutoCAD not access any other drawing file by the same name. The path saved by AutoCAD looks similar to this:

c:\cad\project1\filename.dwg

Relative Path is relative to the folder structure of the master drawing. The master drawing can be moved to a different drive or computer; provided the folder structure on the new host match that of the old one, AutoCAD finds and loads the xrefs.

If the xrefs cannot be found, then AutoCAD searches for the xrefs along paths defined by the **Files** tab of the **OPTION** command.

```
project1\filename.dwg
```

No Path eliminates the path altogether. AutoCAD expects the xrefs in the same folder as the master drawing. The "path" saved by AutoCAD is just the file name:

```
filename.dwg
```

Again, if xrefs cannot be found, then AutoCAD searches for them along specified paths.

The figure illustrates the three forms of path saved by AutoCAD. Look carefully at **Saved Path:**

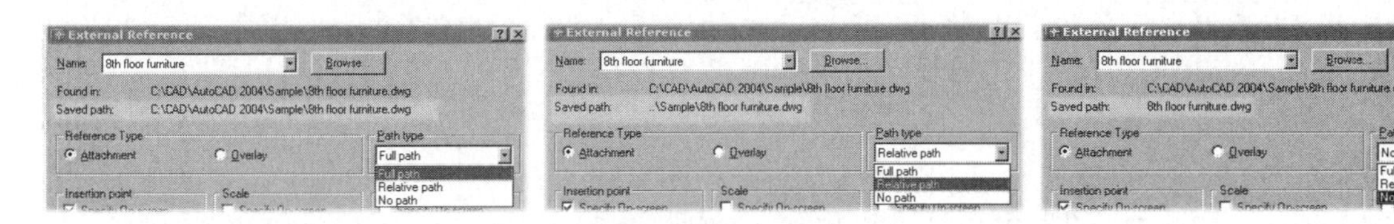

Three types of paths stored by AutoCAD.

In summary:

Path Type	***Xref is Found...***
Full path	... unless *.dwg* file is moved, or path name is changed.
Relative path	... when its folder is below the master drawings's folder.
No path	... when it is in the same folder as the master drawing.

Autodesk provides the following warnings related to relative paths: "You must save the current drawing before you can set the path type to Relative Path. For a nested xref, a relative path always references the location of its immediate host and not necessarily the currently open drawing. The Relative Path option is not available if the referenced drawing is located on a different local disk drive or on a network server."

Insertion Point

The **Insertion Point** option specifies where the xref is placed in the drawing.

You can specify the x, y, and z coordinates in the dialog box, or check the **Specify on-screen** option to locate the xref relative to other objects in the drawing:

Specify insertion point or [Scale/X/Y/Z/Rotate/PScale/PX/PY/PZ/PRotate]: *(Enter coordinates, or an option.)*

The insertion point is relative to the point defined in the **BASE** command in the xref, which in most cases is the origin (0,0,0).

Scale

The **Scale** option specifies the xref's scale factor relative to the insertion point. You can specify the x, y, and z scale factors in the dialog box, or check the **Specify on-screen** option to size the xref relative to other objects in the drawing:

Enter X scale factor, specify opposite corner, or [Corner/XYZ] <1>: *(Enter a value, or an option.)*

Enter Y scale factor <use X scale factor>: *(Enter a value, or press* **ENTER**.*)*

The scale factor is relative to the original drawing. For instance, enter 0.5 to make the xref half as large, 1.0 for the same size, and 2.0 for twice as large.

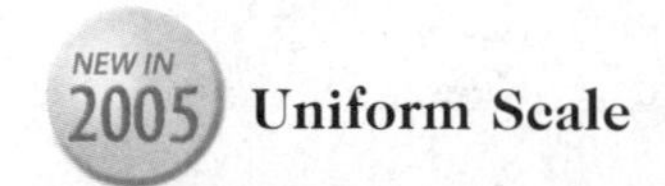

Uniform Scale

AutoCAD 2005 adds an option to the External Reference dialog box: when turned on (check mark shows) the **Uniform Scale** option ensures that the Y and Z scale factors are the same as the X scale factor.

Rotation

The **Rotation** option specifies the xref's rotation angle about the insertion point. You can specify the angle in the dialog box, or check the **Specify on-screen** option to align the xref with objects in the drawing:

Specify rotation angle <0.00>: *(Enter a value, or pick two points in the drawing.)*

XLoadCtl , XrefType, and XLoadPath

The XLOADCTL system variable controls *demand-loading,* which determines whether originals or copies of xref files are opened by the XATTACH command. The XREFTYPE system variable toggles demand loading. The XLOADPATH system variable specifies the path for storing copies of demand-loaded xrefs.

XLoadCtl

To improve performance when working with large xrefs that are clipped and/or have many frozen layers, AutoCAD uses demand loading, which means that AutoCAD loads into memory only the xref data needed to display the xref. When the clipping boundary is changed, or layers thawed, AutoCAD automatically reads more xref data.

XLoadCtl	*Meaning*
0	Entire xref is loaded (demand loading is off).
1	Needed portion of xref is opened (demand loading is on).
2	(*Default value*) Copy of xref is created (demand loading is on).

To turn off demand loading, change the value of XLOADCTL to 0; to load the original xref instead of a copy, change the value to 1.

XrefType

The XREFTYPE system variable determines which **Reference Type** is selected as the default in the External Reference dialog box: **Attachment** or **Overlay**.

XrefType	*Meaning*
0	Attachment is displayed as the default.
1	Overlay is displayed as the default.

XLoadPath

To improve performance further when xrefs are accessed across networks, AutoCAD stores a copy of the xrefs in a folder on your computer.

To specify the location of the copies, change the path stored by the XLOADPATH system variable:

Command: **xloadpath**

Enter new value for XLOADPATH, or . for none
<"C:\DOCUME~1\ADMINI~1\LOCALS~1\Temp\">: *(Enter new path, and press* ENTER.*)*

XRefCtl

The **XREFCTL** system variable determines whether AutoCAD creates *.xlg* log files that record all activity related to xrefs.

The log file is given the same name as the master drawing, such as *8th floor.xlg*, and is stored in the same folder as the master drawing. The log file records all actions by the **XATTACH** and **XREF** commands. AutoCAD normally does not create the log file. To turn on the logging feature, change the value of **XREFCTL** to 1.

XRefCtl	*Meaning*
0	Log files *not* written.
1	Log files written.

An example of the log file content:

```
==============================
Drawing: C:\CAD\AutoCAD 2005\Sample\8th floor.dwg
Date/Time: 09/04/06 16:41:35
Operation: Attach Xref(s)
==============================
Drawing: C:\CAD\AutoCAD 2005\Sample\8th floor.dwg
Date/Time: 09/04/06 16:41:35
Operation: Unload Xref
==============================
```

ALTERNATIVE METHODS: ATTACHING XREFS

AutoCAD provides alternative methods for attaching xrefs. One alternative is the **XREF** command's **Attach** option, which operates identically to the **XATTACH** command.

Another alternative is DesignCenter, which you may find more useful than **XATTACH**. It previews xrefs, and quickly attaches them when you simply drag them into drawings. For greater control, right-click the xref, and then select **Attach as Xref** from the shortcut menu; this displays the same External Reference dialog box as does **XATTACH**.

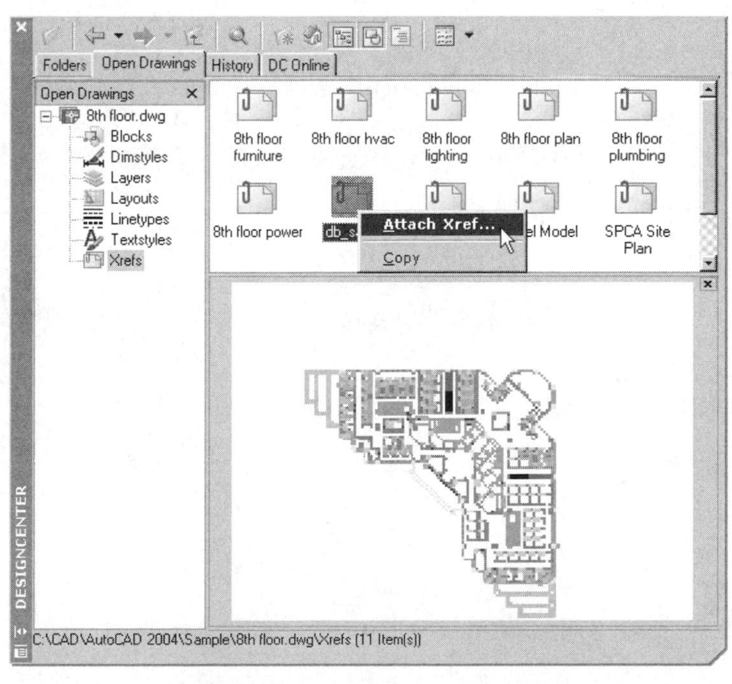

DesignCenter previews and attaches xrefs.

Named Objects

To help identify the layers, blocks, dimension and text styles, and linetypes that belong to xref drawings, AutoCAD adds a prefix to their names. For example, suppose an xref drawing is called *widget.dwg*, and has a layer named "Details." When you attach *widget* to a master drawing, the layer is displayed as:

widget | details

Notice that AutoCAD prefixes the layer name with the xref drawing name and a vertical bar (|). This allows layers of the same name to coexist among referenced drawings. As shown by the dialog box below, some layer names are:

8th floor hvac | m-hvac-fpbs

8th floor lighting | e-lite-circ-neww

8th floor plumbing | m-isom-ftng

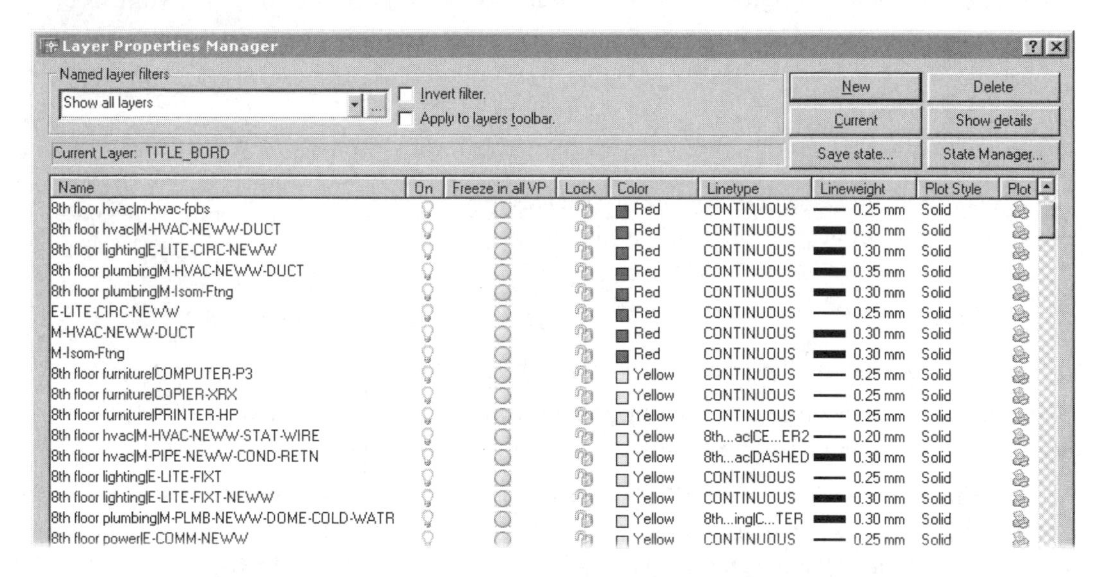

Dialog box showing xref'ed layer names prefixed by the xref drawing filename.

The **Named layer filters** droplist (located in the upper left corner of the dialog box) toggles the display of xref layers. To view the names of layers associated with a specific xref drawing, select the name of an xref drawing, such as "8th floor furniture."

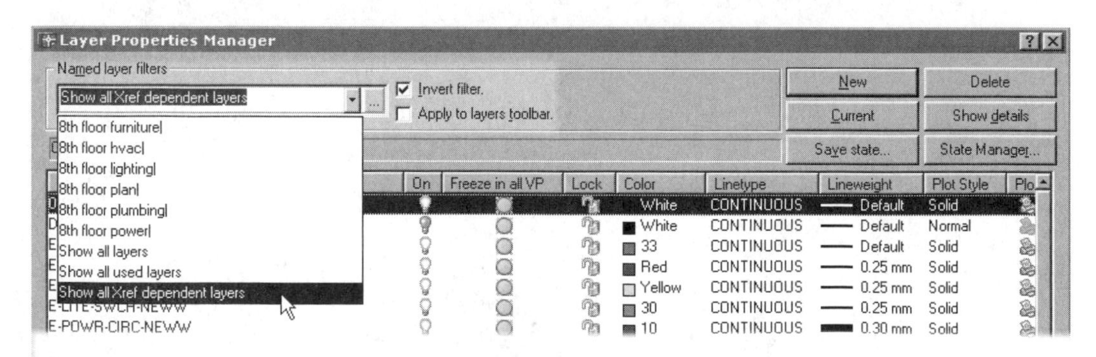

Dialog box filters the display of xref'ed layer names.

To turn off xref layer names:

1. Select **Show all Xref dependent layer names.**
2. Click **Invert filter.**

AutoCAD displays the names of all layers *not* part of xref'ed drawings.

This naming convention holds true for blocks, dimension and text styles, and linetypes. (If xrefs contain variable block attributes, they are ignored.) Note that the names of xref'ed layers cannot be changed until after they are bound.

When you bind drawing *widget* or layer "Details," AutoCAD changes the vertical bar to 0, as follows:

```
widget$0$details
```

You can use the **RENAME** command to change the layer name to something more meaningful.

(Historically, the $ symbol represented a temporary file or name, one that could be changed or erased without much worry.)

WORKING WITH XREFS IN PROJECTS

Xrefs are fine until you send the master drawing and its xrefs to a different computer, perhaps to another office or to your client — or even take the files home to work on. The Achilles heel of xrefs is their path names: paths depend on the drive and folder names being correct. It is unlikely that any two computers have the identical folder names.

Autodesk has provided a solution through the **PROJECTNAME** system variable, which stores a *project name* in the master drawing: the project name points to search paths. Project names and associated search paths are specified in the **Files** tab of the **OPTIONS** command.

When AutoCAD cannot find the xref's *.dwg* files, it searches the paths associated with the project name. If it still cannot find it, then AutoCAD search searches its own paths.

TUTORIAL: CREATING PROJECT NAMES

1. To create project names for master drawings, start the **OPTIONS** command. (From the **Tools** menu, select **Options.**)
2. In the Options dialog box, click the **Files** tab.
3. In the Search Paths list, look for "Project Files Search Path," and then click the **+** sign. The entry probably reads "Empty."

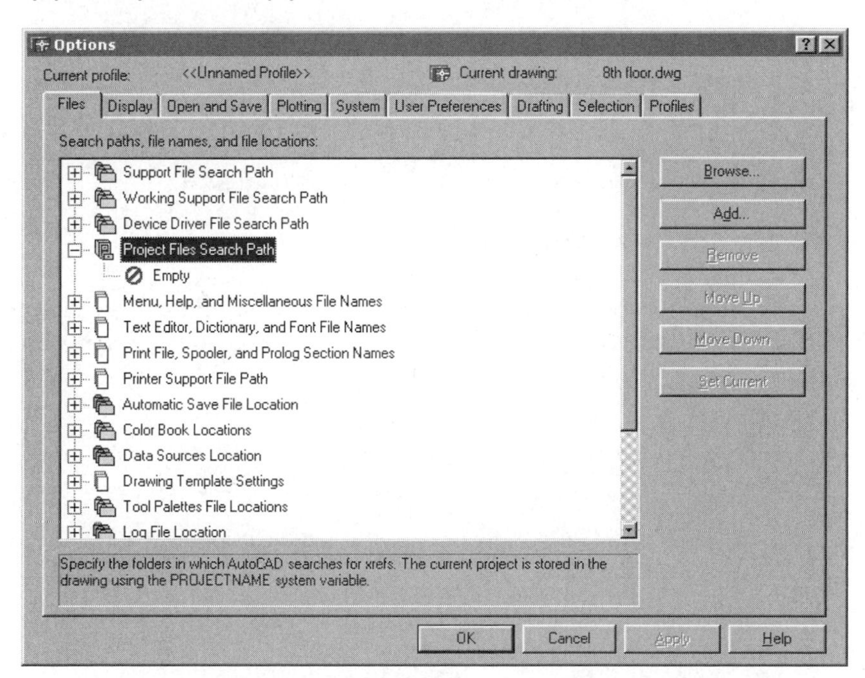

Options dialog box controls project names and related search paths.

4. To add a project name, click **Add**.

   Notice that AutoCAD adds a generic project name, "Project1." If you wish, change the name to something meaningful.

Add a project name...

5. Click the + next to Project1. Notice that the search paths are empty.
6. To add a search path, select "Empty," and then click **Browse**. Notice the Browse for Folder dialog box.
7. Select a drive and folder(s), and then click **OK**. Notice that AutoCAD adds the path to the project name.

... and a search path.

8. You can add more project names and search paths, following the steps listed above.

Two project names with associated search paths.

   When there is more than one project name, you need to make one current. Select one, and then click **Current**. This is the value stored in the **PROJECTNAME** system variable.

9. Click **OK** to exit the dialog box.

Notes: Create two project names: one for your computer, and a second for the other computer. When the master and xref drawings are on the other computer, select the other project name.

To ensure that all xrefs travel with the master drawing, use the **ETRANSMIT** command to collect all support files required by the master drawing.

XREF

The XREF command controls the attachment, display, and paths of externally-referenced drawings.

TUTORIAL: CONTROLLING XREFS

To control the status of referenced drawings, start the **XREF** command:

- From the **Insert** menu, choose **Xref Manager**.
- From the **Reference** toolbar, choose the **External Reference** button.
- At the 'Command:' prompt, enter the **xref** command.
- Alternatively, enter the **xr** alias at the 'Command:' prompt.

Command: **xref** *(Press* ENTER.*)*

In all cases, AutoCAD displays the Xref Manager dialog box, which lists the names of all attached drawings, and indicates their status.

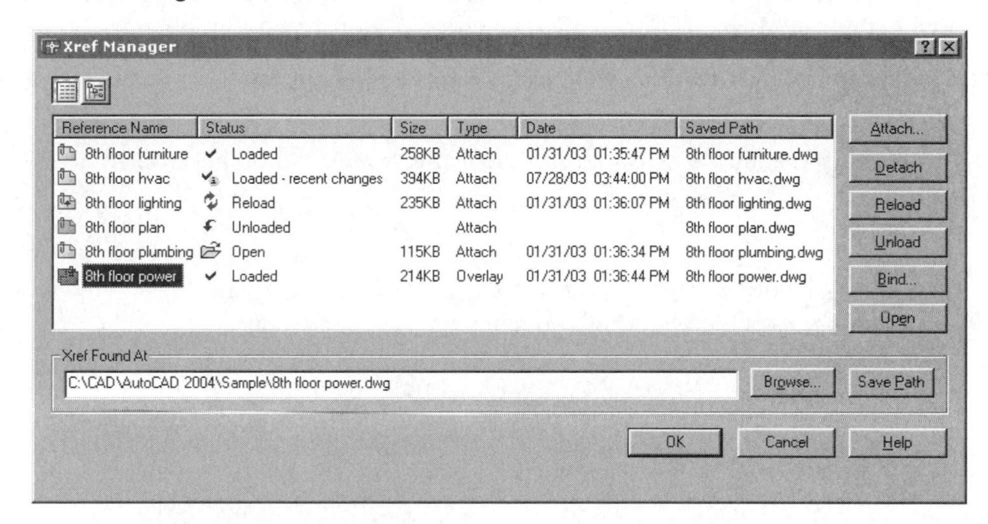

Xref Manager dialog box lists the names of attached drawings.

The names of attached drawings are displayed in two views: list and tree. Press **F3** and **F4** to switch between list and tree view, or click the icons.

List View

The **List** view displays a list of all attached xrefs, as illustrated by the figure above. The view shows information about each xref: reference name, status, type, file date, file size, and the saved path. (Click a header to sort the list in alphabetical order; click a second time to sort in reverse order.) The meaning of the headers is:

Reference Name lists the names of the xrefs stored in the drawing. The name is usually the same as the filename. To change the name, click it twice, and then enter a new name; nested xrefs cannot be renamed.

Status specifies whether the xrefs are:

Status	*Meaning*
Loaded	(*Default*.) Attached to the drawing.
Loaded - recent changes	Attached to drawing, but the xref's *.dwg* file has been saved with changes that are not reflected in the loaded copy.
Reload	Will be reloaded after the Xref Manager is closed.
Unloaded	Will be unloaded after the Xref Manager is closed.
Unreferenced	Attached to the drawing, but erased.
Not Found	No longer found in the search paths.
Unresolved	Cannot be read by AutoCAD.
Orphaned	Attached to an unreferenced, unresolved, or not found xref.

Size indicates the file size in KB (kilobytes); not displayed when the xref is unloaded, not found, or unresolved.

Type indicates whether the xref is attached or overlaid. To change the xref between the two types, click the word **Attach** or **Overlay** until it changes; nested xrefs cannot be changed.

Date indicates the date and time the xref was last saved; not displayed when the xref is unloaded, not found, or unresolved.

Saved Path indicates the name of the xref's *.dwg* file, which is not necessarily the same as the reference name.

Tree View

The **Tree** view displays how xrefs are attached to each other. (Click the **Tree View** button.) It is useful for determining nested xrefs, which are shown by the gray icons.

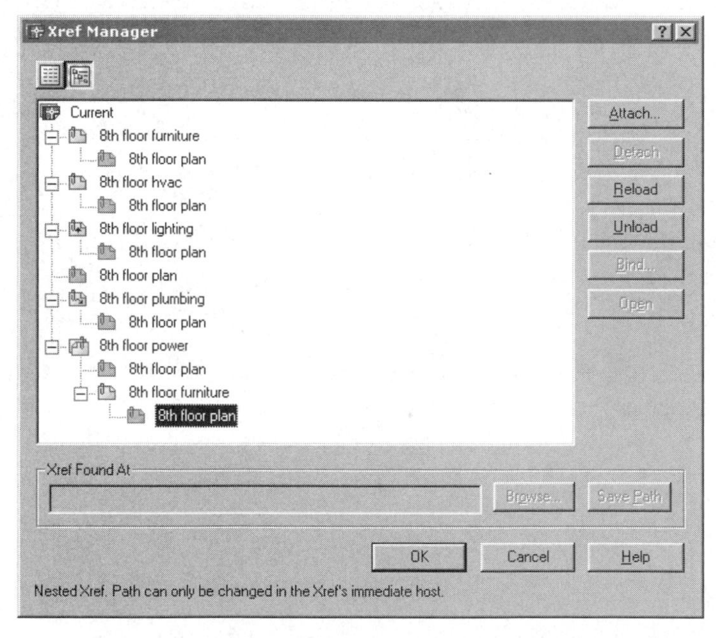

The Tree view illustrates the xref hierarchy.

Controlling Xrefs

The **Attach** button displays the External Reference dialog box when an xref is selected; when no xref is selected, it displays the Select Reference File dialog box. See the XATTACH command.

The **Detach** button removes the selected xrefs from the master drawing, after exiting this dialog box. Use the **Attach** button to return the xref to the drawing. Nested xrefs cannot be detached. Under some conditions, xrefs cannot be detached, and so the button is grayed out. Nested xrefs cannot be detached, nor can xrefs referenced by other xrefs or blocks.

The **Reload** button reloads the selected xrefs, after exiting this dialog box. Use this button to update xrefs that have been edited by other users.

The **Unload** button temporarily removes selected xrefs. A better name for this button would be "Freeze," because it hides the xref so that it is not displayed or regenerated.

The **Bind** button displays the Bind Xrefs dialog box, as described later in this chapter. Nested xrefs cannot be bound.

The **Open** button opens the selected xref for editing in a new window. See XOPEN command.

Xref Found At Options

The **Xref Found At** text box reports the full path of the selected xref. This is the actual location of the xref's *.dwg* file, and might be different from the saved path described by the XATTACH command.

The **Browse** button allows you to select different paths and file names for the selected xref.

The **Save Path** button saves the path shown in the **Xref Found At** text box for the selected xref. This allows you to change the path to the xref's *.dwg* file.

CONTROLLING XREFS: ADDITIONAL METHODS

A number of system variables provide additional control over xrefs:

- **XREFNOTIFY** system variable and **XREFNOTIFYTIME** registry variable control notifications of changed and missing xrefs.
- **TRAYICONS, TRAYNOTIFY,** and **TRAYTIMEOUT** system variables control the display of notifications.
- **VISRETAIN** system variable determines whether changes to xref layers are retained.

Let's look at them.

XRefNotify and XRefNotifyTime

The XREFNOTIFY system variable and XREFNOTIFYTIME registry variable control notifications about changed and missing xrefs.

When you open drawings, AutoCAD checks for xrefs. If the drawing has xrefs, AutoCAD loads them, and then displays the xref icon in the tray at the end of the status bar. (When drawings have no xrefs, no icon is displayed.)

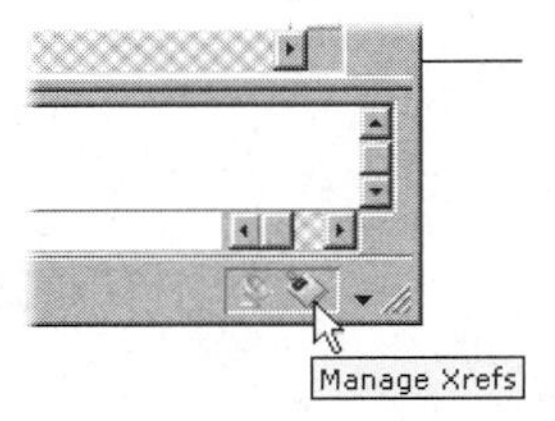

Xref icon notifies that drawings have xrefs attached.

In addition, AutoCAD checks every five minutes to determine whether xrefs have changed. If so,

AutoCAD displays a yellow balloon near the tray, alerting you to the change.

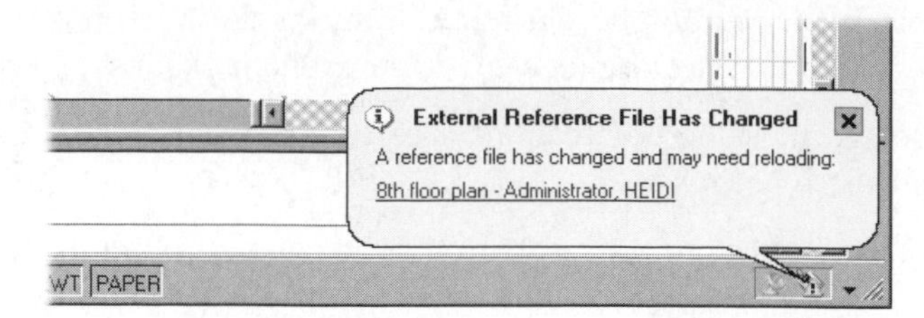

Balloon notifies you of an xref that was edited and saved.

The balloon reports the file name of the modified drawing (*8th floor plan* in the example above), as well as the login name and computer network name on which the drawing resides. In the figure above, "Administrator" is the login name (or user name), while "HEIDI" is the network name of the computer.

When "8th floor plan - Administrator, HEIDI" is clicked, AutoCAD dismisses the balloon and displays the Xref Manager dialog box. Notice that the dialog box lists the names of xrefs needing reloading. Select the reference name, and then click **Reload** — assuming you actually want the xref updated in the master drawing.

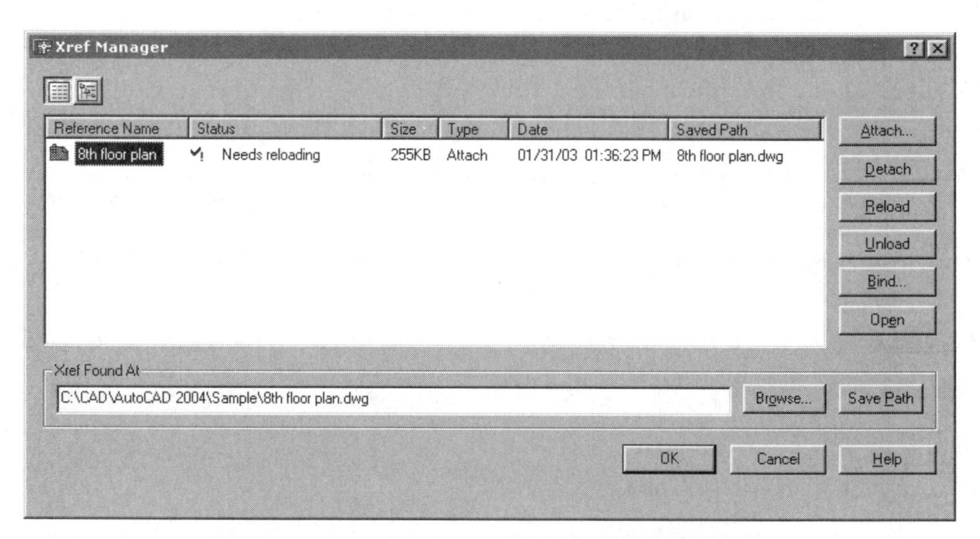

Xref Manager dialog box displayed by xref notification balloon.

The tray's xref icon has two commands "hidden" in it. Right-click the icon to access the commands:

- **Xref Manager** displays the dialog box of the same name.
- **Reload Xrefs** reloads all xrefs, without using the Xref Manager dialog box.

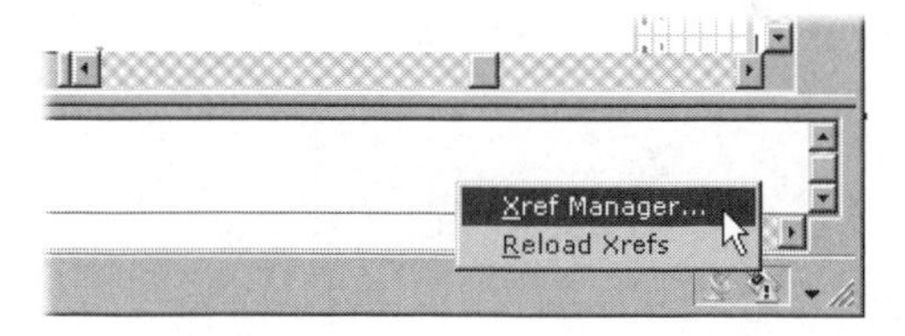

Right-clicking xref icon displays shortcut menu.

The **XREFNOTIFY** system variable controls the xref notification process:

XRefNotify	***Meaning***
0	Xref notification turned off.
1	Displays xref icon when xrefs are attached, and displays yellow alert icon when xrefs are missing.
2	(*Default.*) Also displays balloon messages when xrefs are modified.

To change the interval between checks, such as to ten minutes, change the value of the **XNOTIFYTIME** registry variable, as follows:

Command: **(setenv "XNOTIFYTIME" "10")**

You can enter any value between "1" and "10080" minutes (seven days). It is important that you type "XNOTIFYTIME" all in uppercase letters, and surround it with quotation marks.

TrayIcons, TrayNotify, and TrayTimeout

The **TRAYICONS**, **TRAYNOTIFY**, and **TRAYTIMEOUT** system variables control the display of notifications.

System Variable	***Meaning***
TrayIcons	Toggles the display of the tray at the right end of the status bar.
TrayNotify	Toggles whether notifications are displayed in the tray.
TrayTimeout	Determines the duration in seconds that notifications are displayed.

VisRetain

The **VISRETAIN** system variable determines whether changes to the on/off and freeze/thaw visibility, color, linetype, lineweight, and plot styles of xref layers are saved in the master drawings.

When you change xref layers, AutoCAD saves the changes in the master drawing. (The original xref is unaffected. For instance, you might change the color of one layer, and freeze another layer.) The next time you open the master drawing, AutoCAD opens the xref drawing, and changes the settings of the two layers.

If you prefer that the master drawing *not* remember the changes you make to the xref layers, then change the value of **VISRETAIN** to 0.

This variable has two settings, **0** (forget xref layer changes) and **1** (remember changes):

VisRetain	***Meaning***
0	Changes made to xref layers are *not* saved with the master drawing. The next time the master drawing is opened, the xref drawing's layer table is reloaded.
1	(*Default value.*) Changes to xref layers are saved with the master drawing's layer table. The next time the master drawing is opened, xref layer settings stored in the master drawing take precedence over the settings stored in the xref file.

XCLIP

The XCLIP command hides portions of xref drawings and blocks.

Sometimes you may want to hide part of an xref'ed drawing. To do this, you *clip* the xref by drawing a border around the area to show; the area outside the border is hidden from view. This command works on xrefs and blocks, creating both 2D clipping boundaries and 3D clipping planes. It does not clip non-xref parts of drawings, i.e., elements of the master drawing.

The clipping boundary is made of line segments and polylines; open polylines are treated as closed. (Curved polylines are decurved, but splined polylines retain their curves.) The clip can be toggled off to display the entire xref.

TUTORIAL: CLIPPING XREFS

1. To clip the display of xrefs and blocks, start the XREF command:
   - Right-click an xref, and select **Clip Xref** from the shortcut menu.
   - From the **Reference** toolbar, choose the **External Reference Clip** button.
   - At the 'Command:' prompt, enter the **xclip** command.
   - Alternatively, enter the **xc** alias at the 'Command:' prompt.

   Command: **xclip** *(Press* ENTER.*)*
2. In all cases, AutoCAD prompts you to select one or more xrefs:

   Select objects: *(Pick one or more xrefs.)*

   Select objects: *(Press* ENTER *to end object selection.)*
3. Define a new rectangular clipping boundary, as follows:

   Enter clipping option

   [ON/OFF/Clipdepth/Delete/generate Polyline/New boundary] <New>: **n**

   Specify clipping boundary:

   [Select polyline/Polygonal/Rectangular] <Rectangular>: **r**

   Specify first corner: *(Pick a point.)*

   Specify opposite corner: *(Pick another point.)*

   Notice that all of the xref outside of the boundary disappears from view. Whether you see the boundary itself depends on the setting of the XCLIPFRAME system variable, which toggles the boundary display.

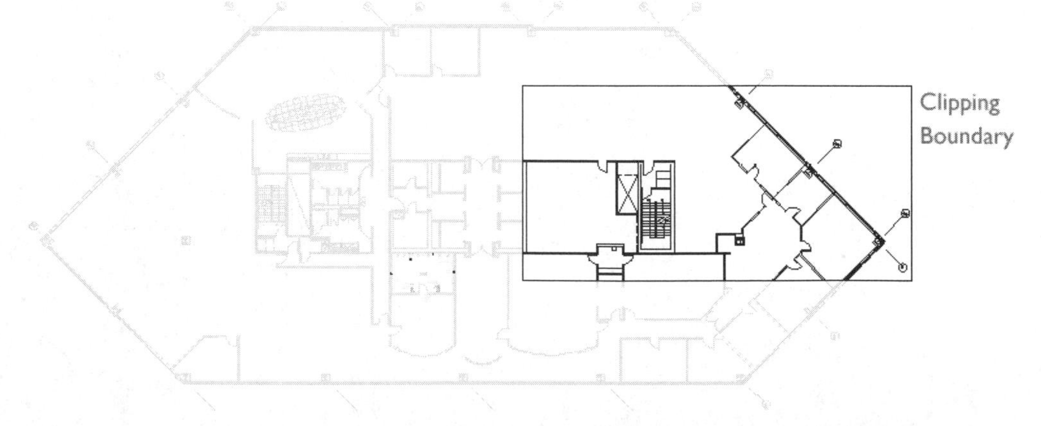

Xref outside clipping boundary disappears (shown in light gray for illustrative purposes).

CLIPPING XREFS: ADDITIONAL METHODS

A number of options and a system variable provide additional control over xrefs:

- **On** and **Off** options toggle the display of the clipped xref.
- **Clipdepth** option sets the front and back clipping planes on 3D xrefs.
- **Delete** option erases the clipping boundary from the drawing.
- **Generate polyline** option copies clipping boundary as a polyline.
- **XCLIPFRAME** system variable toggles the display of xref clipping boundaries.

Let's look at them.

On and Off

The **ON** and **OFF** options toggle the display of the clipped xref. The meaning of the options is the reverse of what you might expect. ON means "off": when you enter ON, AutoCAD turns off the display of the clipped portion of the xref. Another way to think of this is: ON means clipping is turned on. These options apply to the selected xref only.

ClipDepth

The **Clipdepth** option sets the front and back clipping planes. This option works together with the clipping boundary to create a *clipping volume*. The clipping planes are perpendicular to the boundary. This option provides the following prompts:

Specify front clip point or [Distance/Remove]: *(Pick a point.)*

Specify back clip point or [Distance/Remove]: *(Pick another point.)*

The **Distance** option specifies that the clipping plane is placed a distance from the clipping boundary:

Specify distance from boundary: *(Enter a distance.)*

The **Remove** option erases both front and back clipping planes.

Delete

The **Delete** option erases the clipping boundary and planes from the drawing. (You cannot use the ERASE command.)

Generate Polyline

The **Generate polyline** option copies the clipping boundary as a polyline. The polyline takes on the current settings for color, layer, linetype, and lineweight. The purpose of this option is to be able to edit the boundary with the PEDIT command, which allows you to create complex clipping boundaries.

XClipFrame

The XCLIPFRAME system variable toggles the display of all xref clipping boundaries. The boundary is displayed when this variable is set to 1.

Command: **xclipframe**

Enter new value for XCLIPFRAME <0>: **1**

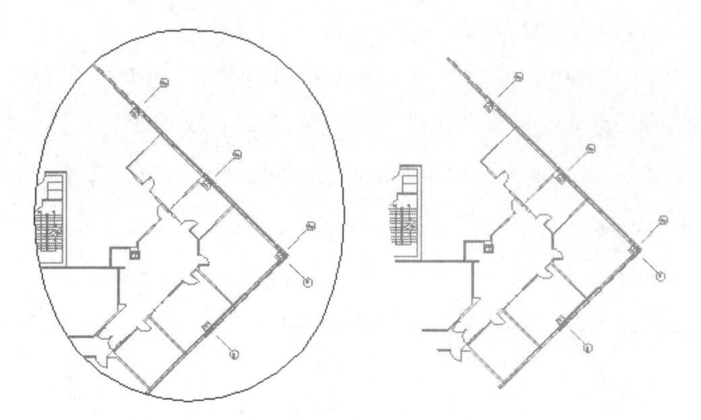

***Left**: Clipping frame (boundary) turned on.*
***Right**: Clipping frame turned off.*

XREF BIND AND  XBIND

The XREF command's **Bind** option adds entire xrefs to master drawings. The entire xref is inserted; you cannot select portions of the drawing. If the xref (now a block) is clipped, the clipping boundary is retained until the block is exploded.

This command combines (or merges) xrefs with the master drawing, creating a single drawing. (The links to xrefs are removed; the xrefs become blocks in the master drawing.) This is useful at the end of a project, when you want to "freeze" a set of drawings, so that future changes to referenced drawings no longer affect the master drawing.

There are two options for binding xrefs: binding and inserting. The sole difference between the two options is how the names of objects (blocks, text styles, layer names, and so on) are handled:

- **Bind**: names change from *dwgname|objname* to *dwgname*0*objname*.
- **Insert**: names change from *dwgname|objname* to just plain *objname*.

In contrast, the XBIND command adds selected named objects only, rather than the entire xref.

TUTORIAL: BINDING XREFS

1. To combine xrefs with master drawings, start the XREF command.
2. In the Xref Manager dialog box, select the xref drawings you want to bind.
   Some xrefs cannot be bound, in which case the **Bind** button grays out. Examples include drawings needing reloading (because they have changed), and drawings containing proxy objects.
3. Click the **Bind** button. Notice the Bind Xrefs dialog box.

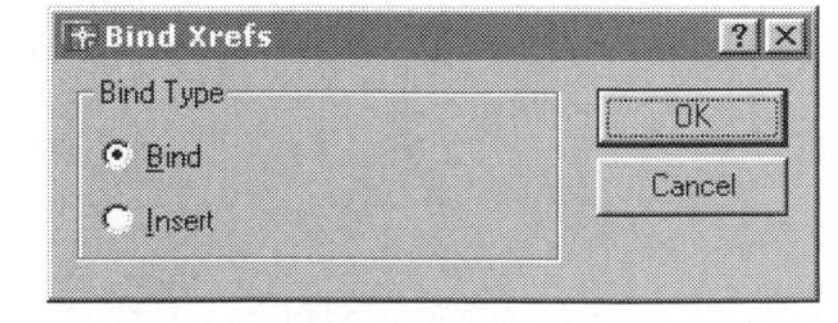

Xrefs can be bound or inserted into master drawings.

4. Choose **Bind** or **Insert**.
5. Click **OK.** The xrefs are inserted as blocks, with the same names as the xrefs. The drawing looks no different, until you inspect the layer listing.

If you wish, use the **EXPLODE** command to convert the blocks into individual elements.

The **XREF Bind** command binds the entire xref drawing to the master drawing. There are times, however, when you may want to bind only a linetype or layer to the drawing, without binding the rest of the xef. Do this with the **XBIND** or **ADCENTER** commands.

When you select **Modify | Object | External References | Bind** from the menu bar or enter the **XBIND** command, AutoCAD displays the Xbind dialog box.

The dialog box initially displays a list of xrefs on the left side. Notice the + symbol next to each drawing name. Click it to display a list of named objects in that drawing.

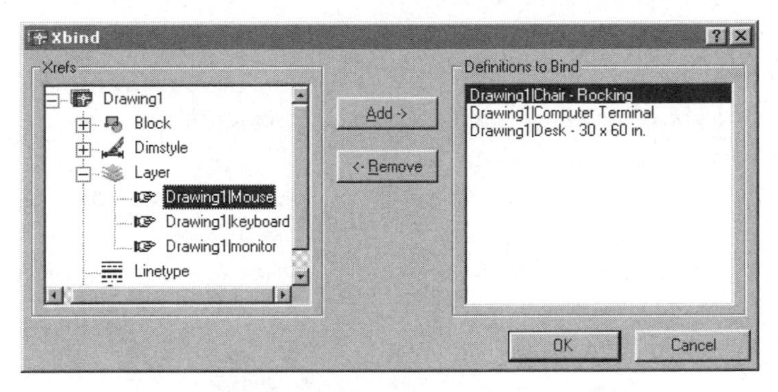

XBind dialog box selects which named copies to bind.

To bind a named object to the current drawing, select its name (such as a block name), then choose the **Add** button. You read earlier how AutoCAD replaces the vertical bar in object names with 0. If there is already a named object of the same name in the master drawing, AutoCAD increments the 0 to create dwgname1objname, and so on.

Click **OK**, and the named objects are added to the master drawing. (They are not "erased" from the xref.)

Note: The **BINDTYPE** system variable affects only the command-line oriented **-XREF** command's **Bind** option. The system variable determines ahead of time how object names are converted:

BindType Value	*Object Name Format*
0	Bind style: *dwgname0objname*
1	(Default.) Insert style: *objname*

XOPEN AND REFEDIT

The **XOPEN** command opens xrefs in their own windows for editing, while the **REFEDIT** command edits xrefs and blocks *in situ*.

Historically, xrefs were meant to be seen, and not edited. With subsequent releases of AutoCAD, Autodesk added additional editing capabilities: first the **REFEDIT** command, and then the **XOPEN** command with AutoCAD 2004.

In general, you would use the **XOPEN** command, because it is easier to understand. You would use the **REFEDIT** command only when you need to: (1) edit an xref in-place, so that you see it in relation to the master drawing and other xrefs; or (2) edit a block without exploding and redefining it.

(If someone else is editing the drawing, you cannot edit it with **XOPEN** or **REFEDIT**)

TUTORIAL: EDITING XREFS

1. To edit xrefs, start the **XOPEN** command:
   - At the 'Command:' prompt, enter the **xopen** command.
   - From the **Modify** menu, choose **Xref and Block Editing** button, and then **Open Reference**.
   - As an alternative, select the xref name in the **XREF** command's dialog box, and then click the **Open** button.

   Command: **xopen** *(Press* ENTER.*)*
2. AutoCAD prompts you to select an xref:

   Select xref: *(Pick a single xref.)*

   Notice that AutoCAD opens the xref in a new window. (If you are working in the **XREF** command's dialog box, the window appears after you exit the dialog box.)
3. Make editing changes to the drawing.
4. Close the window with the **CLOSE** command, and then click **Yes** when asked if you want to save changes to the drawing.
5. Reload the xref to update the master drawing: right-click the xref icon on the tray, and select **Reload Xrefs**.

In-Place Editing

AutoCAD allows you to edit xrefs and blocks in-place using the **REFEDIT** command. This means you can edit xrefs while seeing them in context (unlike the **XOPEN** command), and can edit blocks without exploding them. (**XOPEN** does not edit blocks at all.) This is called "reference editing" or "in-place editing."

In-place editing can be difficult to figure out. On top of the large number of restrictions listed later, the process involves three commands in a specific sequence of steps. Here is an overview:

Step 1. Start the **REFEDIT** command.

Step 2. Select a single xref or block; if nested, you must select a nested item. If you need to edit additional xrefs and blocks, add them with the **REFSET** command.

Step 3. Edit the objects making up the xrefs and/or blocks.

Step 4. When done editing, save or reject the changes with the **REFCLOSE** command.

Not every xref or block can be edited with **REFEDIT**. Non-uniformly scaled blocks cannot be referenced edited, nor can blocks inserted with the **MINSERT** command. Associative dimensions and associative hatches, both a form of block, cannot be edited. Any level of nested xrefs and blocks can be edited. In addition, automatic saves are disabled during reference editing.

TUTORIAL: IN-PLACE EDITING

1. To edit xrefs and blocks, start the **REFEDIT** command:
   - Right-click an xref, and select **Edit Xref in-place** from the shortcut menu.
   - From the **Modify** toolbar, choose **Xref and Block Editing**, and then **Edit Reference In-Place**.
   - From the **RefEdit** toolbar, choose the **Edit xref or block** button.
   - At the 'Command:' prompt, enter the **refedit** command.

   Command: **refedit** *(Press* ENTER.*)*
2. In all cases, AutoCAD prompts you to select one xref or block:

   Select reference: *(Pick an xref or a block.)*

   AutoCAD displays the Reference Edit dialog box. Its primary purpose is to alert you to nested objects. If the xref or block contains no nestings, click **OK** and carry on.

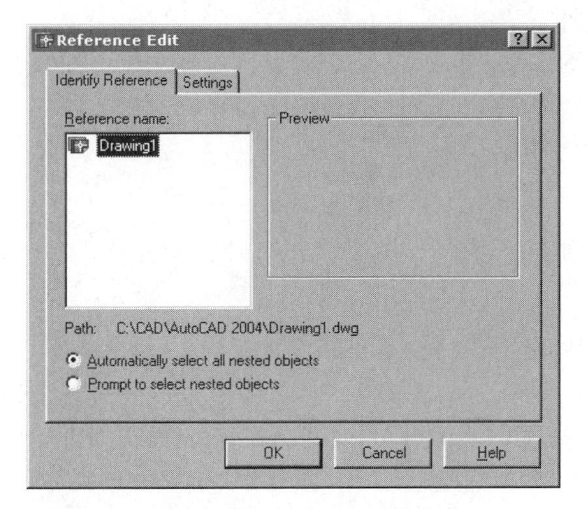

Reference Edit dialog box with no nested xrefs or blocks.

Other times, the list of nested xrefs and blocks can get quite hairy, as displayed by the dialog box below, which lists only the names of *nested* xrefs and blocks.

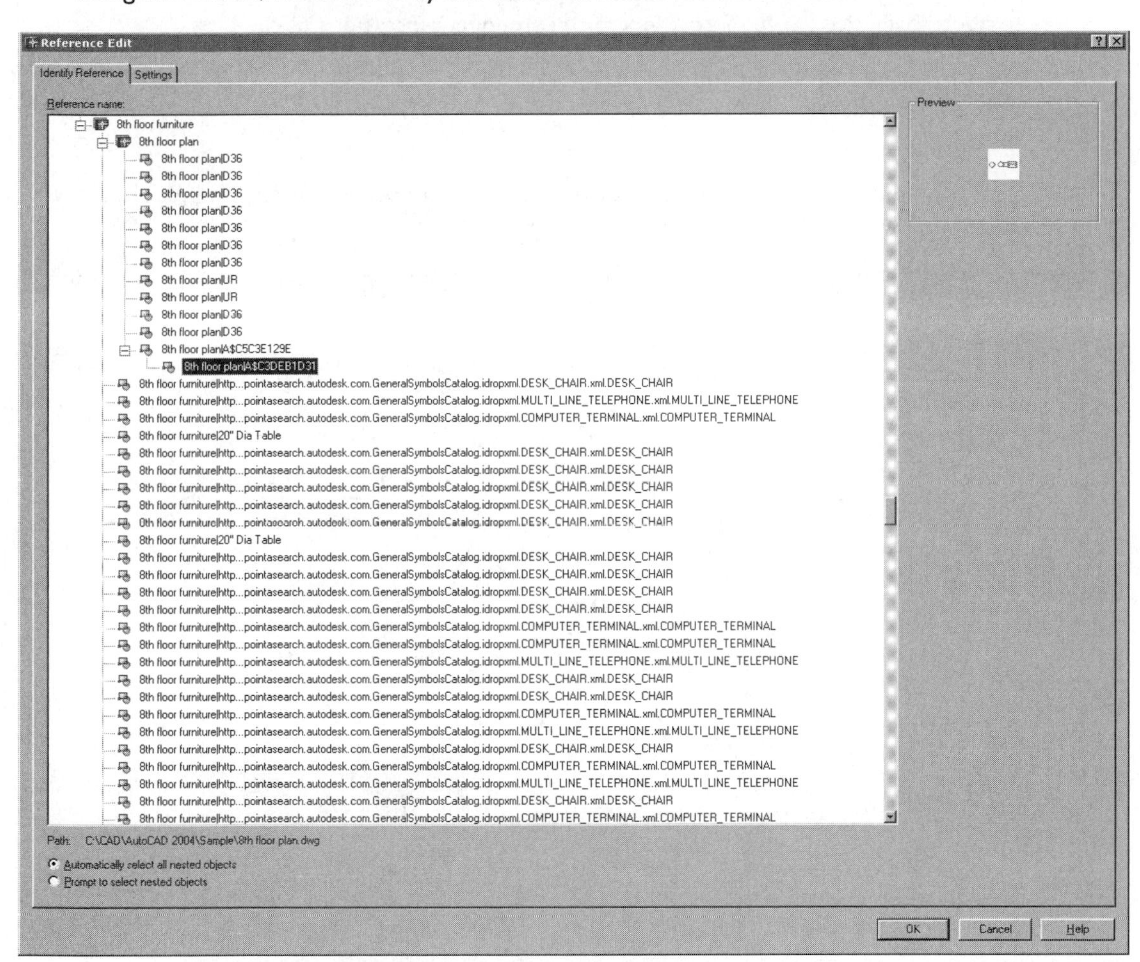

Drawing with many nested xrefs and blocks.

3. To help you deal with nested objects, the dialog box has two options:
   - **Automatically Select All Nested Objects** incudes all nested objects automatically in the reference editing session.

- **Prompt to Select Nested Objects** prompts you to select specific objects for editing, after you close this dialog box:

  Select nested objects: *(Select one or more objects to be edited.)*

  Select nested objects: *(Press* ENTER *to exit selection.)*

4. Click the **Settings** tabs for additional options:

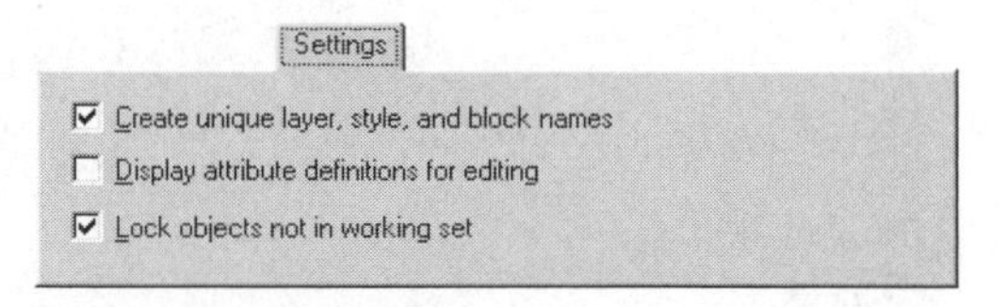

Settings tab of RefEdit command's dialog box.

- **Create unique layer, style, and block names** changes names in a manner similar to bound xrefs: the "|" changes to "0". When off, names are unchanged.
- **Display Attribute Definitions for Editing** hides attribute data, and reveals attribute definitions. (This setting is available only when blocks with attributes are selected for editing; constant attributes cannot be edited.) Note that when changes are saved at the end of the reference editing session, the attributes of the original blocks are unchanged; the edited attribute definitions come into effect with subsequent insertions of the block.
- **Lock objects not in working set** prevents editing of objects not selected, much like placing them on locked layers.

5. Click **OK**.

   Notice that AutoCAD "fades" all the drawing but that part being reference edited, and opens the RefEdit toolbar.

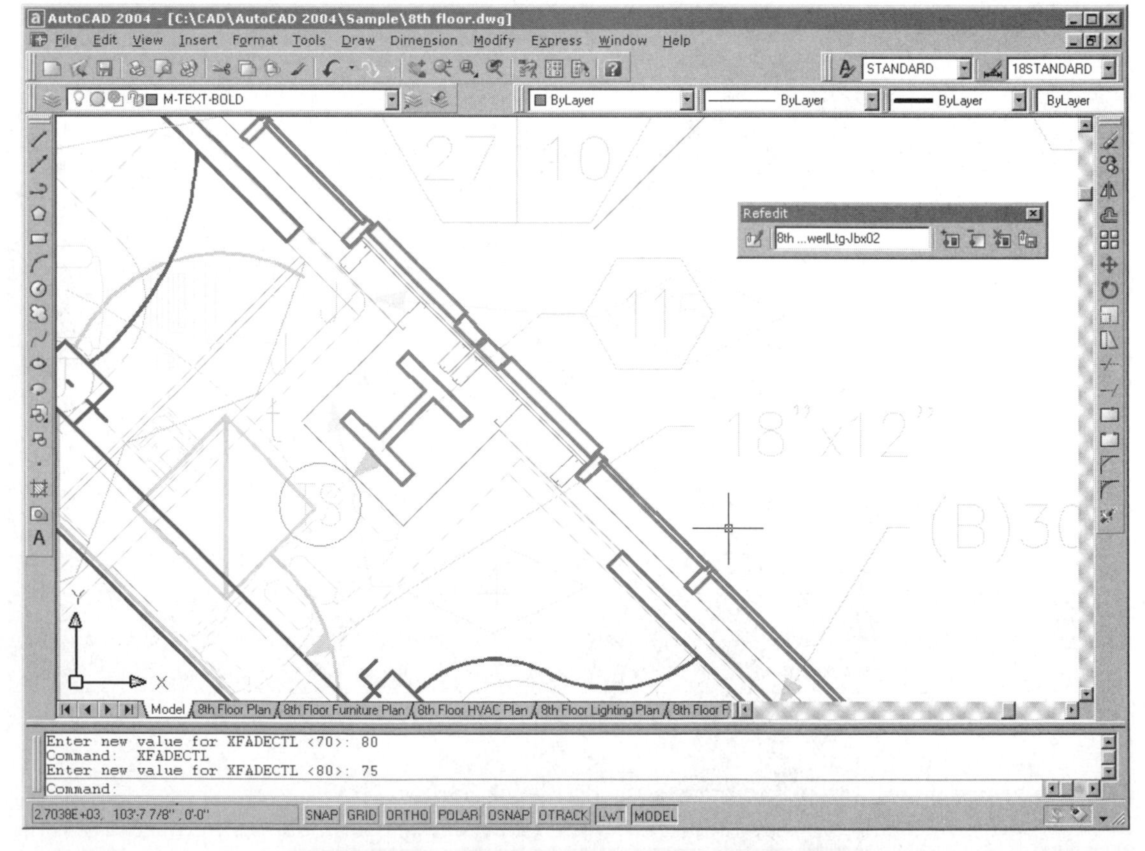

RefEdit displays a toolbar, and fades objects not being edited.

6. You can now apply most of AutoCAD's editing commands to the selected objects. For example, you can change colors, lineweights, and linetypes; you can move, erase, and copy objects; you can lengthen, fillet, and chamfer objects; and you can add new objects.
7. Curiously, **REFEDIT** initially limits you to selecting a single xref or block. You can, however, add and remove objects from the set being edited through the **REFSET** command, or more easily through buttons on the RefEdit toolbar:

Add and Remove buttons.

- **Add objects to working set** prompts you to select one or more xrefs and blocks.

  Command: _refset

  Transfer objects between the RefEdit working set and host drawing...

  Enter an option [Add/Remove] <Add>: _add

  Select objects: *(Select one or more xrefs and blocks.)*

  Select objects: *(Press* **ENTER** *to end the selection process.)*

- **Remove objects from working set** prompts you to remove xrefs and blocks.

8. When done editing, you need to save (or discard) the changes, and return the drawing to its "normal" state. You can do this through the **REFCLOSE** command, or more easily through buttons on the RefEdit toolbar:

Discard and Save buttons.

- **Discard changes to reference** negates all editing changes you made to the xref or block, and exits reference editing mode. AutoCAD displays a warning dialog box, to help save you from making mistakes.

- **Save back changes to reference** saves editing changes you made to the xref or block, and exits in-place editing mode. AutoCAD displays a warning box to make sure you make the right decision.

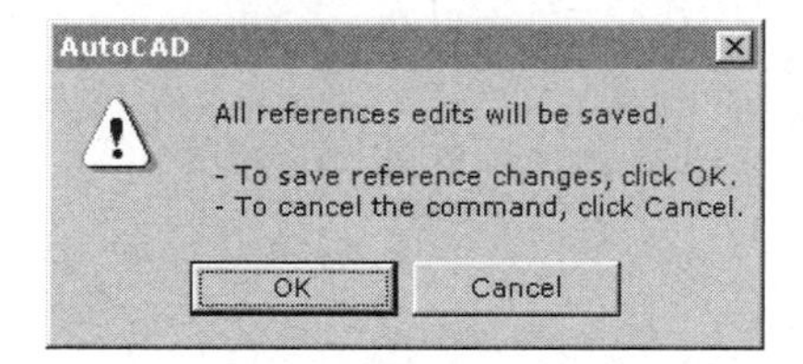

AutoCAD checks to make sure you want to save the change.

You can use the UNDO command to reverse all changes made by REFEDIT and related commands.

CLIPPING XREFS: ADDITIONAL METHODS

A number of system variable provide additional control over reference editing:

- **REFEDITNAME** system variable reports the name of the reference being edited.
- **XEDIT** system variable determines whether the current drawing can be edited in-place when being referenced by another drawing.
- **XFADECTL** system variable controls the fading of objects not edited in-place.

Let's look at the last two items in greater detail.

XEdit

The **XEDIT** system variable determines whether an xref can be edited in-place. If you want to prevent in-place editing, set this system variable to 0, which provides you with greater control over who can and cannot edit drawings.

XEdit	*Meaning*
0	Cannot be in-place edited.
1	Can be in-place edited.

The drawback to **XEDIT** is that its value is stored in every drawing, so its setting must be changed in every drawing, one by one.

XFadeCtl

The **XFADECTL** system variable sets the amount of fading of objects not edited in-place (short for "reference fade control"). The variable is initially set to 50; change it towards 90 for a lighter image, and toward 0 for a darker image.

Command: **xfadectl**

Enter new value for XFADECTL <50>: **70**

I find a value of 60 or 70 is best, helping the refedit set to stand out from the rest of the drawing.

EXERCISES

1. At the same time, open the following three drawings from the CD-ROM:
   - *17_35.dwg*
   - *desktop.dwg*
   - *office.dwg*

   a. Use three different methods to switch between the drawings: the **Windows** menu, **CTRL+TAB**, and **CTRL+F6**. Which do you find the easiest?
   b. Close the *17_25.dwg* drawing. How many drawings remain open?

2. Continue from the previous exercise, and switch to *desktop.dwg*.
   a. Copy the entire drawing of the computer to Clipboard. (Hint: use **CTRL+C** and **CTRL+A**.)
   b. Switch to *office.dwg*, and paste the computer onto the desk.
   c. Close the drawings, saving your work.

3. Start a new drawing with the *ANSI E-named plot styles.dwt* template, and then attach *desktop.dwg* as an xref.
   a. Use **XOPEN** to open *desktop.dwg* in another window, and then move the mouse symbol to the other side of the keyboard.
   b. Save the changes to the *desktop.dwg* drawing, and then switch back to the master drawing.
   c. Update the master drawing to show the changes.
   d. What is the file name of the master drawing? Of the xref?
   e. Close *desktop.dwg*.

4. Continue from the previous exercise:
   a. Use the **REFEDIT** command to change the color of all objects in *desktop.dwg* to yellow.
   b. Save the master drawing.
   c. Open *desktop.dwg*. Did the drawing change its color? Why, or why not?
   d. Close *desktop.dwg*.
   e. Switch back to the master drawing, and use the **U** command until the xref returns to its original colors.
   f. Open *desktop.dwg* again. What color is it?

5. Continue from the previous exercise:
   a. Use **XCLIP** to hide the keyboard and mouse; only the monitor should be visible.
   b. Toggle the display of the clipping boundary. Which command did you use?

6. Continue from the previous exercise:
   a. Use the **Bind** option of the **XREF** command to bind *desktop.dwg* into the master drawing. Does the clipping remain?
   b. Explode the Desktop block. Does the clipping remain?

CHAPTER REVIEW

1. Which keys do you hold down to open more than one drawing at a time with the **OPEN** command?
2. Describe two ways to switch between two or more open drawings in AutoCAD?
3. How do you exit a drawing without exiting AutoCAD?
4. Explain how you might copy objects from one drawing to another.
5. What is "xref" short for?
   What is the "master drawing"?
   What is a "nested xref"?
6. List three advantages to attaching drawings as xrefs over inserting them as blocks:
   a.
   b.
   c.
7. List three advantages to inserting drawings as blocks over attaching them as xrefs:
   a.
   b.
   c.
8. Can external references be edited?
   If so, how?
9. Name an advantage and an disadvantage of **REFEDIT** over **XOPEN**:
   Advantage:
   Disadvantage:
10. How can you access xref *.dwg* files from the Internet?
11. When your master drawing contains an xref, can that xref be edited by another user?
12. An xref called *hospital.dwg* has a layer named "Emergency." How would the layer name appear in a master drawing?
13. Explain the purpose of "clipping."
14. Which commands place externally-referenced drawings?
15. How can you obtain a list of the external references attached to a master drawing?
16. How do you convert an external reference to a block?
17. Under what condition would you want to convert the external references to blocks?
18. How do you remove an external reference from a drawing?
19. Must externally-referenced drawing remain in its original drive and folder?
20. List two ways to update xrefs in master drawings:
    a.
    b.
21. How do you get AutoCAD to create a log file of xref activity? *turn on xrefctl system var*
22. Explain the difference between the two types of xref:
    Attach
    Overlay
23. List several types of objects that the **PASTECLIP** command can paste into AutoCAD drawings.
24. When does the **PASTEASHYPERLINK** command become active?
25. Does the **PASTEORIG** command allow you to paste a block back into the same drawing it came from?

Managing Sheet Sets

AutoCAD 2005 introduces an entirely new way of working with more than one drawing: sheet sets. These are collections of drawings made from model space, paper space, and named views. Sheet sets are the opposite of layouts, which place one or more views of a drawing on a single sheet. In contrast, sheet sets collect two or more drawings into a single set, which makes them ideal for controlling projects. Sheets sets are handled through a control panel, called the "Sheet Set Manager" — a form of project management software.

In this chapter, you learn how to set up and use sheet sets with these commands:

NEWSHEETSET runs the Create Sheet Set wizard (new to AutoCAD 2005).

OPENSHEETSET opens sheet set data files (new to AutoCAD 2005).

SHEETSET opens the Sheet Set Manager window (new to AutoCAD 2005).

SHEETSETHIDE closes the Sheet Set Manager window (new to AutoCAD 2005).

SSFOUND, **SSLOCATE**, **SSMAUTOOPEN**, and **SSMSTATE** control the Sheet Set Manager window (new to AutoCAD 2005).

UPDATETHUMBSNOW forces the preview images to be updated in the Sheet Set Manager window (new to AutoCAD 2005).

ARCHIVE packages sheets sets for archival purposes (new to AutoCAD 2005).

FINDING THE COMMANDS

On the **STANDARD** toolbar:

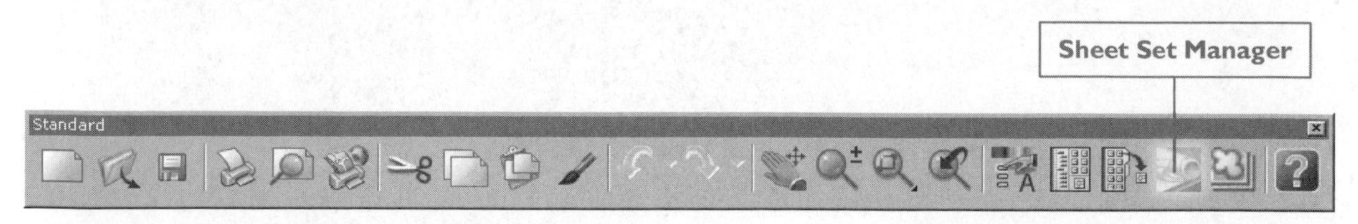

On the **FILE** and **TOOLS** menu:

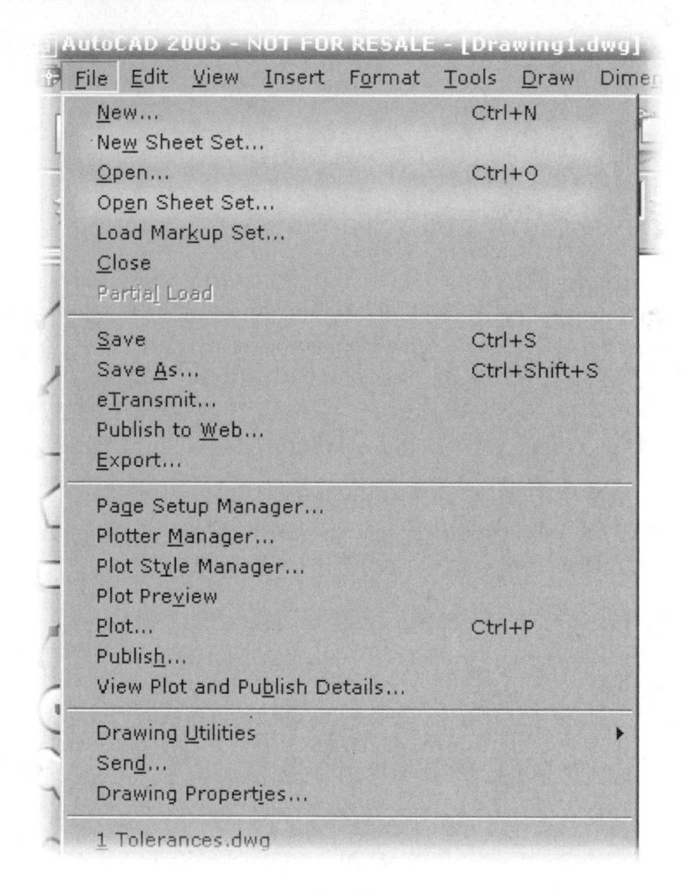

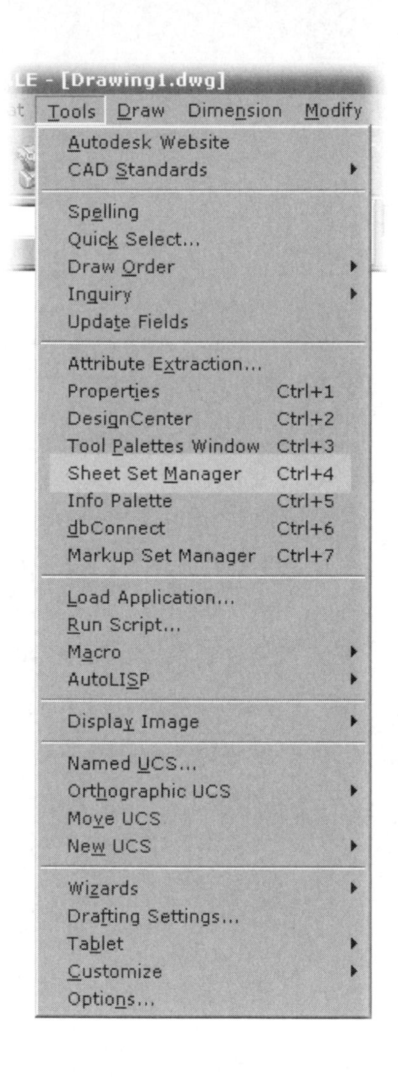

ABOUT SHEET SETS

Projects consist of dozens and even hundreds of drawings. In the past, AutoCAD provided no assistance in organizing large numbers of drawings. Instead, you used work-arounds, such as purchasing drawing management software from third parties, or creating a structure of folders in which to store and catalog drawings. AutoCAD 2004 was helpful, in a small way, by allowing Windows Explorer to display thumbnail images of *.dwg* files.

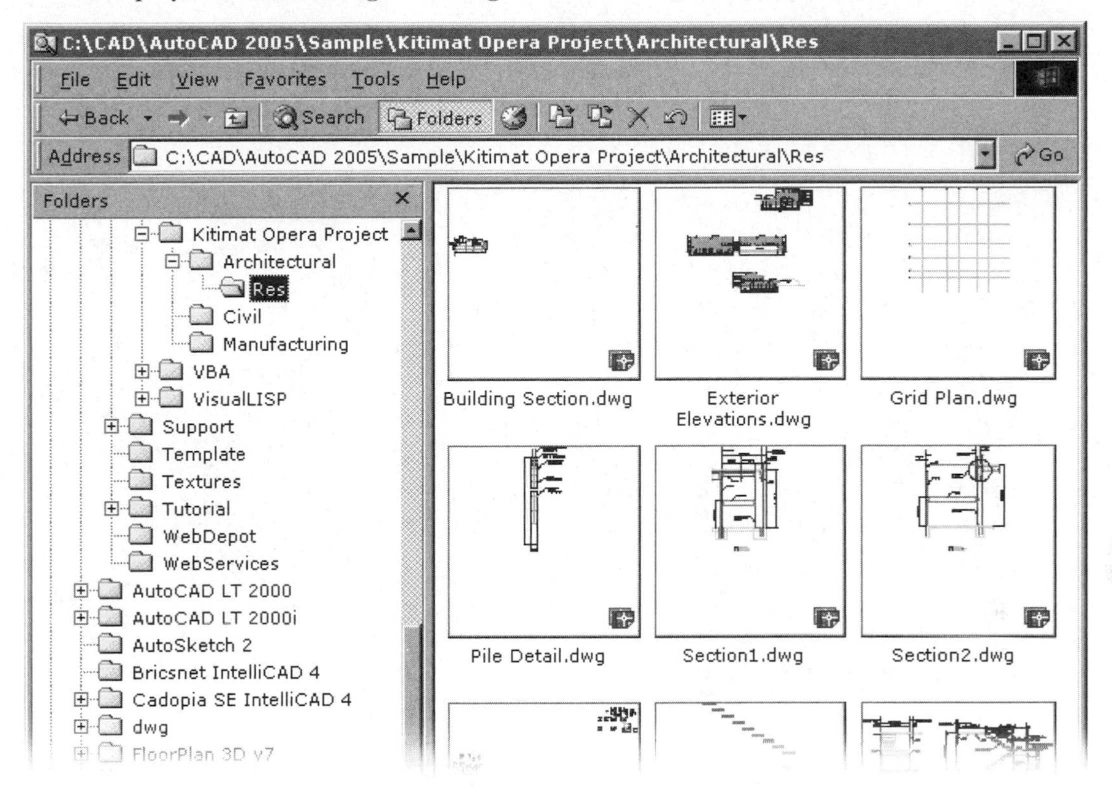

Folders and thumbnails turn Windows Explorer into a simple drawing manager.

SHEET SETS: THE TECHNICAL VIEW

The Sheet Set Manager does not store drawings. Instead, it holds indices that point to the names of layouts and views in drawing files. The indices also store the names of *.dwg* files and the folders in which they reside.

All this data is stored in *.dst* files — short for "Data Sheet seT" — which are coded in binary so that you cannot monkey with them. When you open a sheet set with the **OPENSHEETSET** command, you select a *.dst* file. When you erase a sheet, you erase only its pointer; the source drawing is unaffected.

Warning! There is a lot of data stored in *.dst* files, and Autodesk recommends backing them up with the same diligence that you do drawing files. Each time a sheet set data file is opened, AutoCAD creates a backup copy (*.ds$*). Use this file for recovering from corrupted or mismanaged data files.

Every drawing knows which sheet set it belongs to. When you open such a drawing, it's quite likely that the Sheet Set Manager also opens. That can become annoying, so turn off the **SSMAUTOOPEN** system variable to prevent the window from automatically obstructing your view.

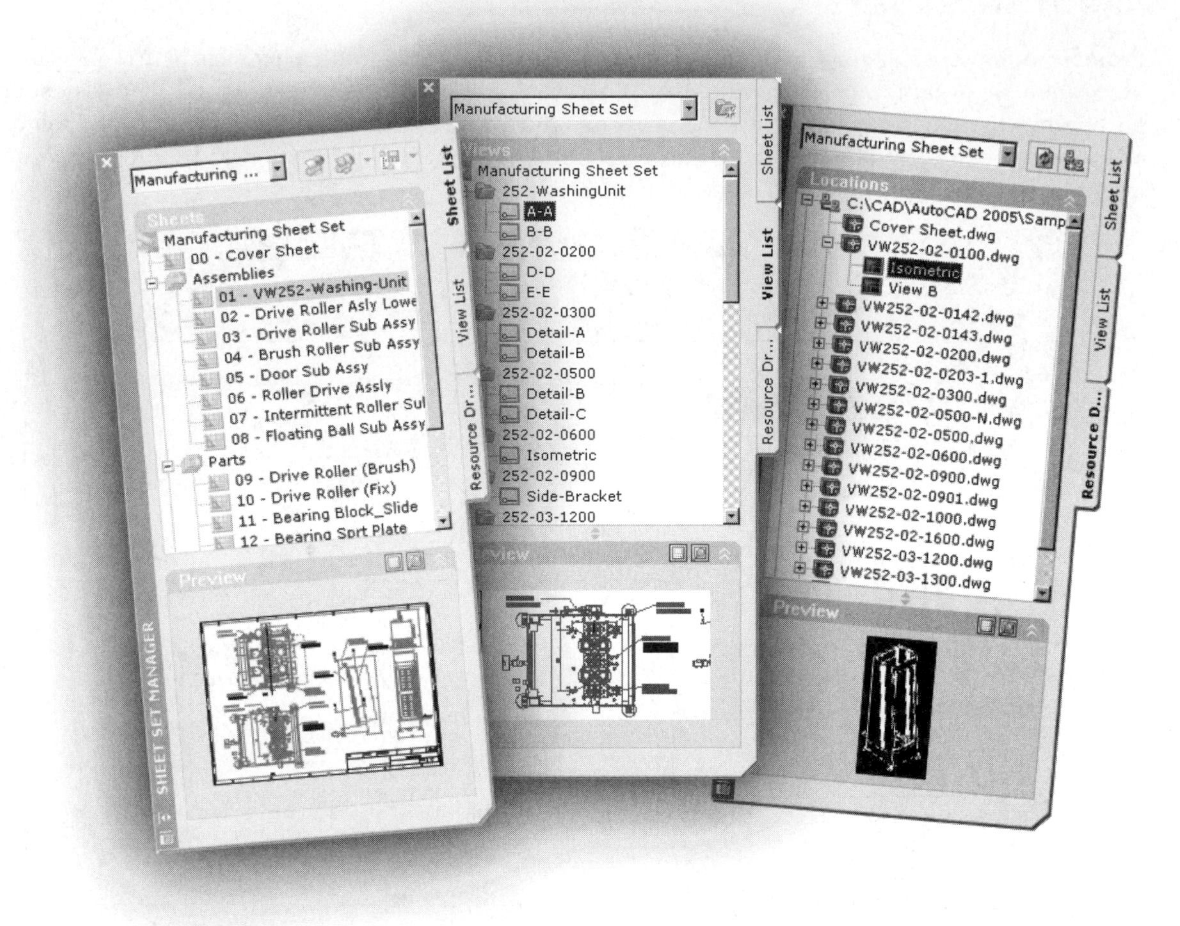

Examples of the Sheet Set Manager window.

With AutoCAD 2005, Autodesk introduces the Sheet Set Manager to organize drawings by project. Project drawings are divided into multiple categories, such as by discipline or by function. The Sheet Set Manager can be thought of as a project-oriented version of AutoCAD's DesignCenter.

The power and complexity of the Sheet Set Manager lie in its many tools. Some of the tasks it performs include:

- Open drawings by double-clicking sheet names.
- Preview drawings and views as thumbnails.
- Generate *title sheets*, tables that list the names of sheets in the set.
- Attach cross-reference labels and callout blocks to views.
- Add custom properties.
- Publish sets in DWF format and to plotters.
- Archive all drawings in the set.

Technical editor Bill Fane provides this example of a large architectural project. The project manager defines sheet sets that include references to any drawing in any drive/folder, and jumps directly to the desired sheet. The architect works on the general drawings in one folder, while the structural, electrical, and HVAC engineers have their own folders — perhaps on different network drives.

The **SHEETSET** command can automatically open and plot all of the sheets in a given set, in the order you specify. Plotted sheet sets come out already collated. The project manager can define an "HVAC third floor" set of plans, an "Electrical fourth floor" set, and so on. The engineers are then easily able to plot the desired sets without having to open each drawing themselves.

Individual drawings can also contain hyperlink references to sheets. Double-click an annotation, such as "Sheet A2," and that sheet opens for you. When sheets are added and removed within a set then all hyperlink references update, pointing to the correct sheets.

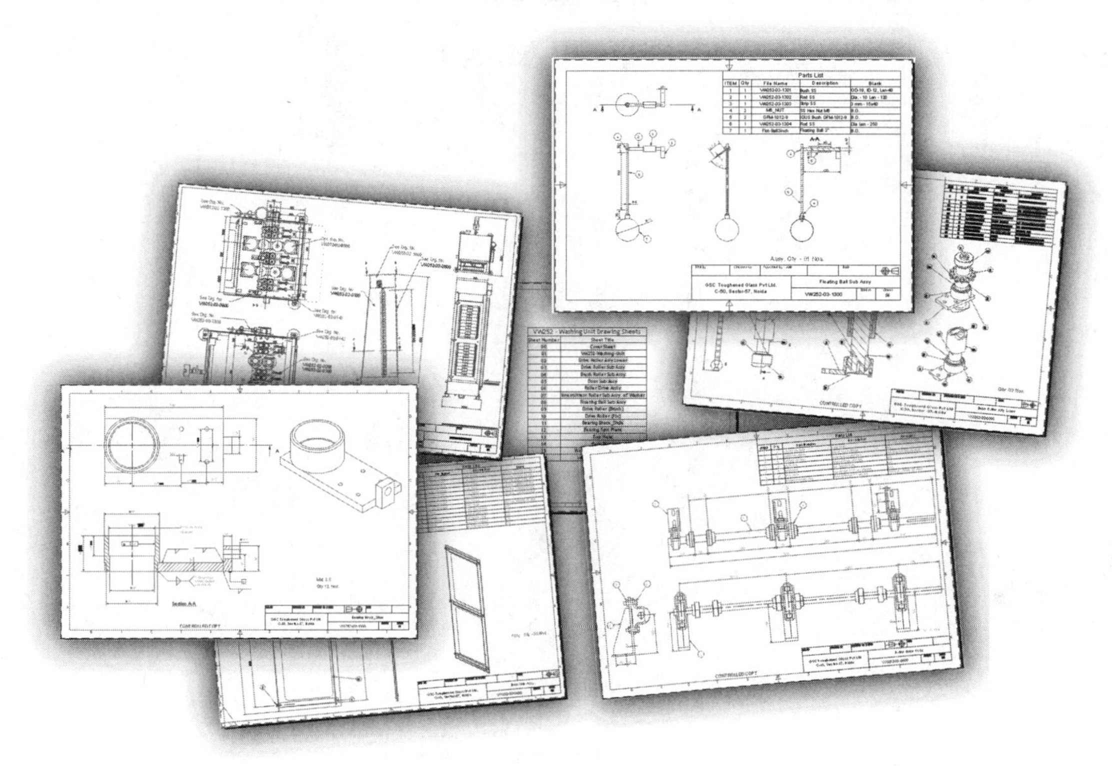

Sheet sets collect all related drawings for projects.

SHEET SETS

Sheet sets are collections of two or more drawings. Use sheet sets when you need a method of handling all the drawings that make up a project. But you can do more with sheet sets than just that. You can also use them to place labeled views in drawings — this is an alternative to viewports in paperspace layouts.

Given that Sheet Set Manager is a CAD management tool, it makes sense for the firm's CAD manager to control it. Implementing it is complex enough that drafters probably should not spend their valuable time figuring it out. The responsibilities of the CAD manager include:

- Creating the folders in which to store drawings.
- Defining the template files for each discipline.
- Producing blocks for titles and detail bubbles.
- Defining custom properties and plot styles,.
- Standardizing title sheets.

These are discussed on the pages following.

TUTORIAL: SHEET SET MANAGER OVERVIEW

The Sheet Set Manager is central to managing sheet sets.

1. Start AutoCAD with a new (blank) drawing.
2. To see the Sheet Set Manager, start the **SHEETSET** command. From the **Tools** menu, choose **Sheet Set Manager**.

   AutoCAD displays the Sheet Set Manager window, empty of any sheets.

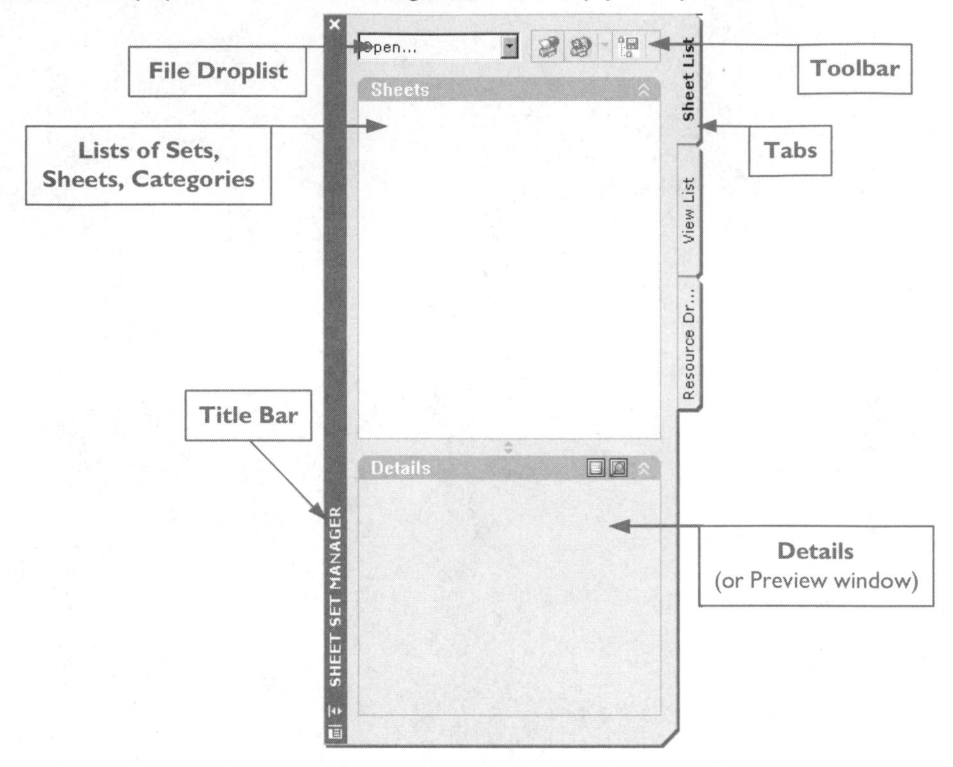

Elements of the Sheet Set Manager window.

The Sheet Set Manager window consists of user interface elements that look similar to the DesignCenter and the Properties windows.

File Droplist — accesses *.dst* files.

List of Sets, Sheets, Categories — lists the names of drawings and views that make up sheets. Also listed here are categories and subsets for organizing sheets.

Title Bar — identifies the sheet set, and drags the window around the screen. (If your computer has two monitors, you can drag the window to the second monitor.)

Toolbar — displays icons for plotting sets, or publishing them as *.dwf* files. (The toolbar for the View and Resource tabs have other icons.)

Tabs — change the view of sheets:

Sheet List lists the names of sheets, and organizes them in subsets.

View List lists names of views in sheets.

Drawing Resources lists the folders, *.dwg* files, and view names from which sheets are constructed.

Details and Previews — describes the selected sheet by its details or through a thumbnail preview image.

Other than the toolbar, there don't seem to be many commands for controlling sheets and sets. It turns out that many of the Sheet Set Manager's commands and options are "hidden" in shortcut menus, as described later in this chapter.

3. To view an example of a sheet set, click the **File** droplist, and then select **Open**.
4. In the Open Sheet Set dialog box, select "IRD Addition.dst," and then click **Open**. (This file is found in AutoCAD's *\sample\sheet set\architectural* folder.)
   The Sheet Set Manager fills with text and icons.
   Notice that no drawings are opened.

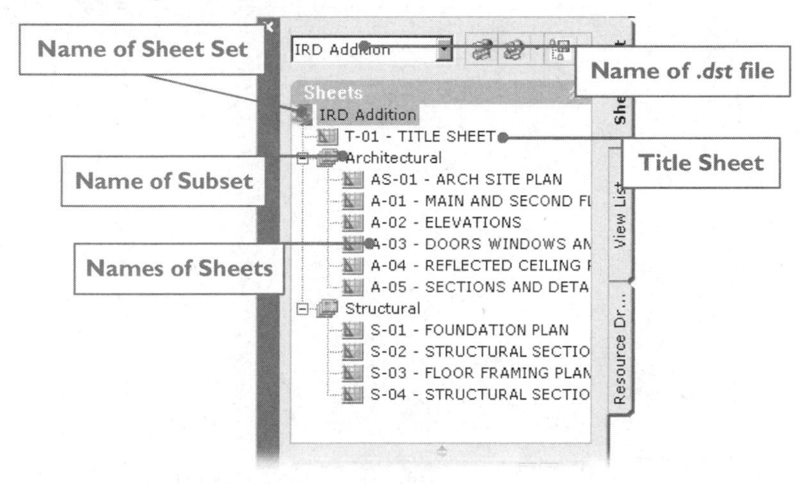

Elements of the Sheet List tab, listing names of the sheet, subsets, and set sheets.

Let's first look at the top half of the window.

SHEETS SECTION

The Sheets section lists the names of the sheets. But it also lists the name for the entire sheet set ("IRD Addition"), and the names of subsets ("Architectural" and "Structural"). AutoCAD uses icons to help distinguish between the different types of names, but I find them too similar in appearance to be helpful.

(To access online help for the Sheet Set Manager, click its title bar, and then press **F1**.)

 Sheet Set

The Sheet Set is the name of the sheet set, "IRD Addition," which is based on the name of the project — International Road Dynamics Office Addition. (This also happens to be the name of the sheet set data file, *IRD Addition.dst*.) There is one master sheet set; only one can be open in AutoCAD at a time. Underneath the sheet set name are the names of sheets and subsets.

How are new sheet sets created? You use the **NEWSHEETSET** command to create them from scratch or to copy them from existing sheet sets. As an alternative to creating sheets, you can drag the drawings and views onto the current drawing.

Sheets

Sheets are the names of drawing layouts and views. They are listed in alphabetical order.

To open the drawing associated with a sheet name, simply double-click it. Wasn't that the easiest way you've every opened a drawing? If you plan to modify the drawing, always do so when the Sheet Set Manger is open; otherwise the associated data is not updated.

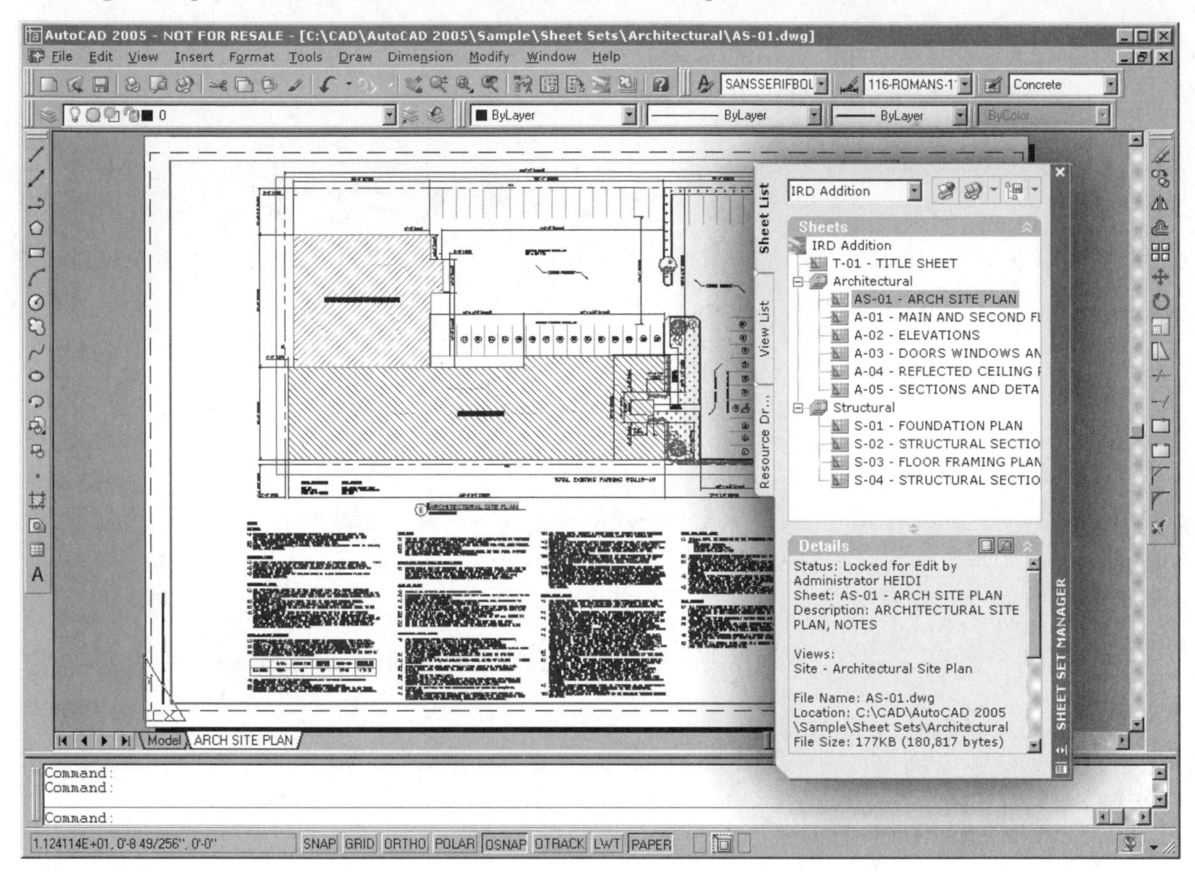

Double-click sheet names to see the related resource drawings.

How are sheets created? You right-click the sheet set name, and then select the commands to import layouts as sheets, or to create new sheets.

You can erase sheets; don't worry, you're not erasing the resource drawing upon which the sheet is based. If you change your mind, use the **U** command immediately; otherwise, you will need to reconstruct the sheet and its data.

Title Sheet

The title sheet is often used as a table of contents, listing the numbers and names of the drawings in the set. Autodesk calls this the "sheet list table." To see the title sheet, double-click "T-01 TITLE SHEET." Notice that AutoCAD opens the drawing (illustrated below). Look for the sheet index in the lower right corner.

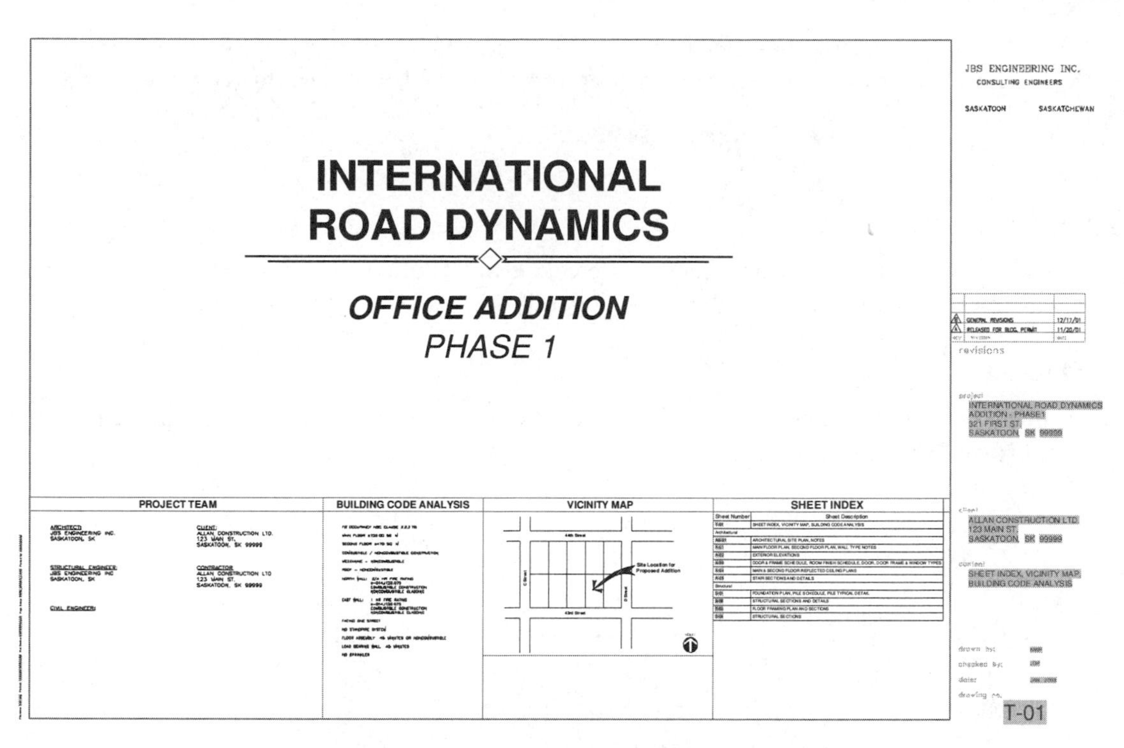

The title sheet contains the sheet index and other reference data.

The text with gray background is *field text*, which updates itself automatically. When the T-01 in the drawing above (lower right corner) changes, the T-01 in the sheet index updates itself to reflect the change.

How are title pages created? Right-click the sheet set name, and then select **Insert Sheet List Table**. AutoCAD automatically generates the content of the title sheet, placing field text in a table.

SHEET INDEX	
Sheet Number	Sheet Description
T-01	SHEET INDEX, VICINITY MAP, BUILDING CODE ANALYSIS
Architectural	
AS-01	ARCHITECTURAL SITE PLAN, NOTES
A-01	MAIN FLOOR PLAN, SECOND FLOOR PLAN, WALL TYPE NOTES
A-02	EXTERIOR ELEVATIONS
A-03	DOOR & FRAME SCHEDULE, ROOM FINISH SCHEDULE, DOOR, DOOR FRAME & WINDOW TYPES
A-04	MAIN & SECOND FLOOR REFLECTED CEILING PLANS
A-05	STAIR SECTIONS AND DETAILS
Structural	
S-01	FOUNDATION PLAN, PILE SCHEDULE, PILE TYPICAL DETAIL
S-02	STRUCTURAL SECTIONS AND DETAILS
S-03	FLOOR FRAMING PLAN AND SECTIONS
S-04	STRUCTURAL SECTIONS

The sheet index acts as a table of contents for the set.

Subsets

Subsets help organize sheets into categories. In this sheet set, the categories are "Architectural" and "Structural." Civil engineering projects might have subsets named "Site Utility Plans" and "Profiles." Subsets are listed in alphabetical order, and can be nested within subsets.

Subsets are optional; you don't need to use them in sheet sets. When a set has too many sheets, however, you probably should start adding categories to help segregate sheets.

How are subsets created? Right-click the sheet set name, and then select **New Subset**. After creating new subsets, you can move sheets around by dragging the names to new locations. Sheets do not need to part of any subset. Notice that the title sheet is not part of one.

Note: To close the Sheet Set Manager window, click the **x** on its title bar. Or, from the **Tools** menu, select **✓Sheet Set Manager**. Or, press **CTRL+4**. Alternatively, enter the **sheetsethide** command:

Command: **sheetsethide**

DETAILS/PREVIEW

Switch your attention to the lower half of the Sheet Set Manager window, which shows you details or preview images of the selected sheets — depending on which button you click.

Details

Click the **Details** button to see information about the selected item. The information displayed varies, according to the sheet set, sheet, or subset.

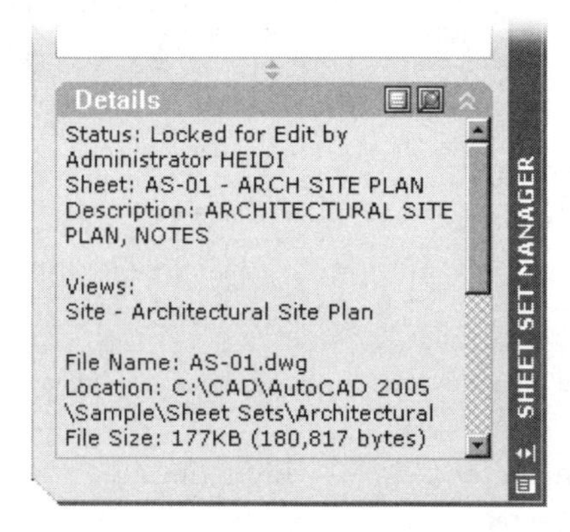

Details list information about sheet sets, sheets, and subsets.

Preview

Click the **Preview** button to see a thumbnail image of the selected sheet. (Sheet set and subset names lack previews.) The **UPDATETHUMBSNOW** command forces AutoCAD to update all of the preview images displayed by the Sheet Set Manager.

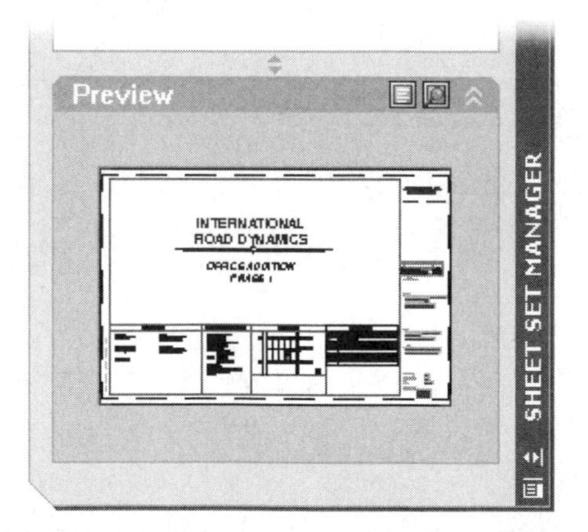

Preview shows thumbnail images of the selected sheet.

VIEW LIST

The Sheet Set Manager displays two more groups of information, which are accessed by tabs. Click on the **View List** tab to see a list of category and view names.

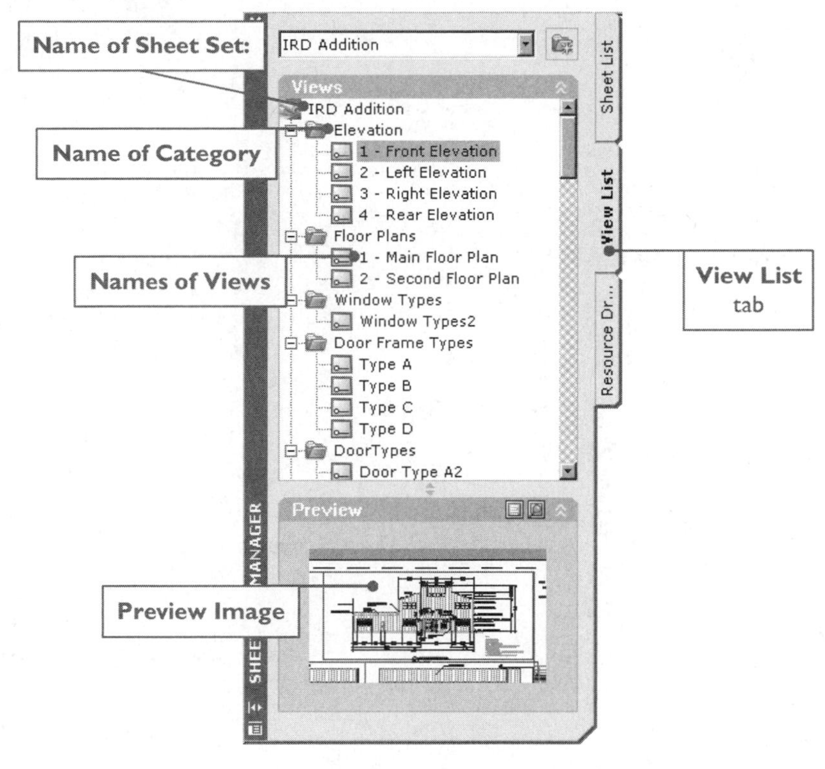

Elements of the View List tab, listing names of views in categories.

Categories

The view names are segregated by *categories*, such as "Elevation" and "Floor Plans." Like subsets, *categories* are optional, but useful when the drawing contains many views. Category names are listed in alphabetical order. You move views between categories by dragging them to new locations.

How are categories created? Right-click the sheet set name, and then select **New View Category**.

View Names

Views are parts of sheets; one analogy is to think of views like viewports in paperspace.

As you click on view names, the Preview window shows different parts of drawings. Double-click a view name to bring up the drawing and the related view.

How are views created? From named views generated by the VIEW command, and only those created in AutoCAD 2005.

Note: Only views created in AutoCAD 2005 are listed on the View List tab.

RESOURCE DRAWINGS

The third tab lists "resources," which are the drawing files used by the sheet set. Click the Resource Drawings tab.

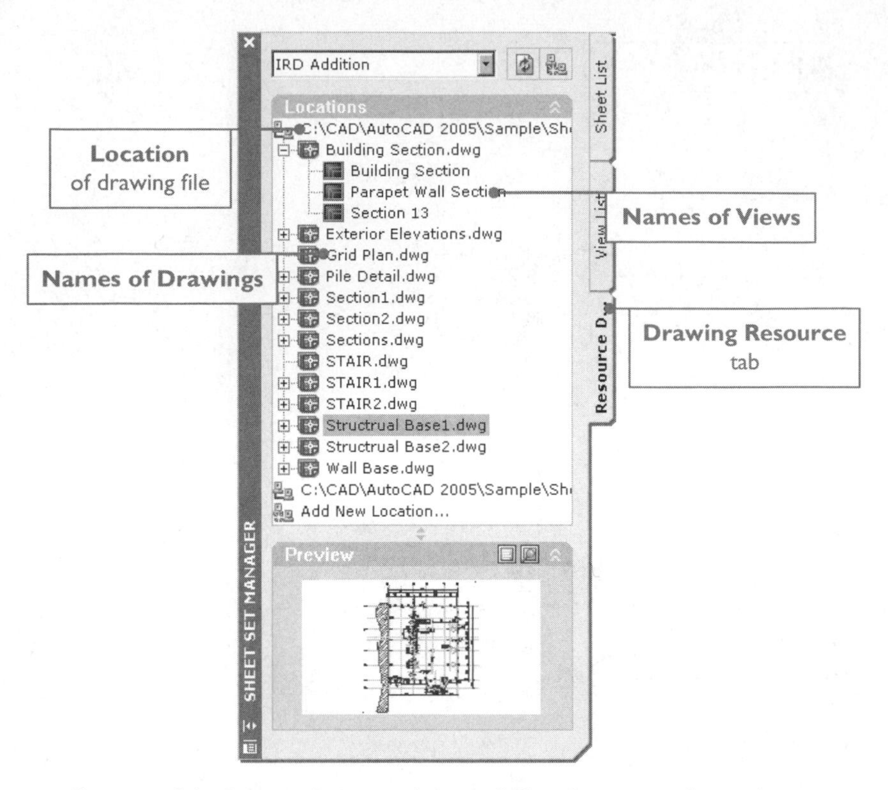

Elements of the Resource Drawing tab, listing folders, drawings, and named views.

Each folder (a.k.a. "location") contains one or more AutoCAD drawing files; drawings optionally contain named views. Click the **+** signs to see the names of drawings in each folder and the names of views in each drawing.

Location

Locations are the names of the drive and folders that hold the AutoCAD drawings used by the sheet set; you must have at least one location. Each location holds one or more *.dwg* drawing and *.dwt* template files. The folders can be located on your computer or on any networked computer.

Locations can contain folders, which show up in the list as folder icons. Locations are listed alphabetically.

How are locations created? On the Resource Drawing tab's toolbar, click the **Add New Location** button, and then select a folder. If the folder does not exist, click the **Create New Folder** button on the Browse for Folder dialog box's toolbar.

Drawings

Drawings are the basis of sheets, holding the layouts and views needed to create a sheet. Any drawing file found in the location folder is automatically included on the list. Drawing files are listed in alphabetical order.

How are drawings added? Use Windows Explorer to drag drawing files from other folders; alternatively, in AutoCAD, use the **SAVEAS** command to save the drawing to the folder.

Views

The *views* listed here are those created in *model space* only. In the Preview window, views are shown with a black background. View names are listed alphabetically.

How are views added to drawings? Click the VIEW command's **New** button, as illustrated below.

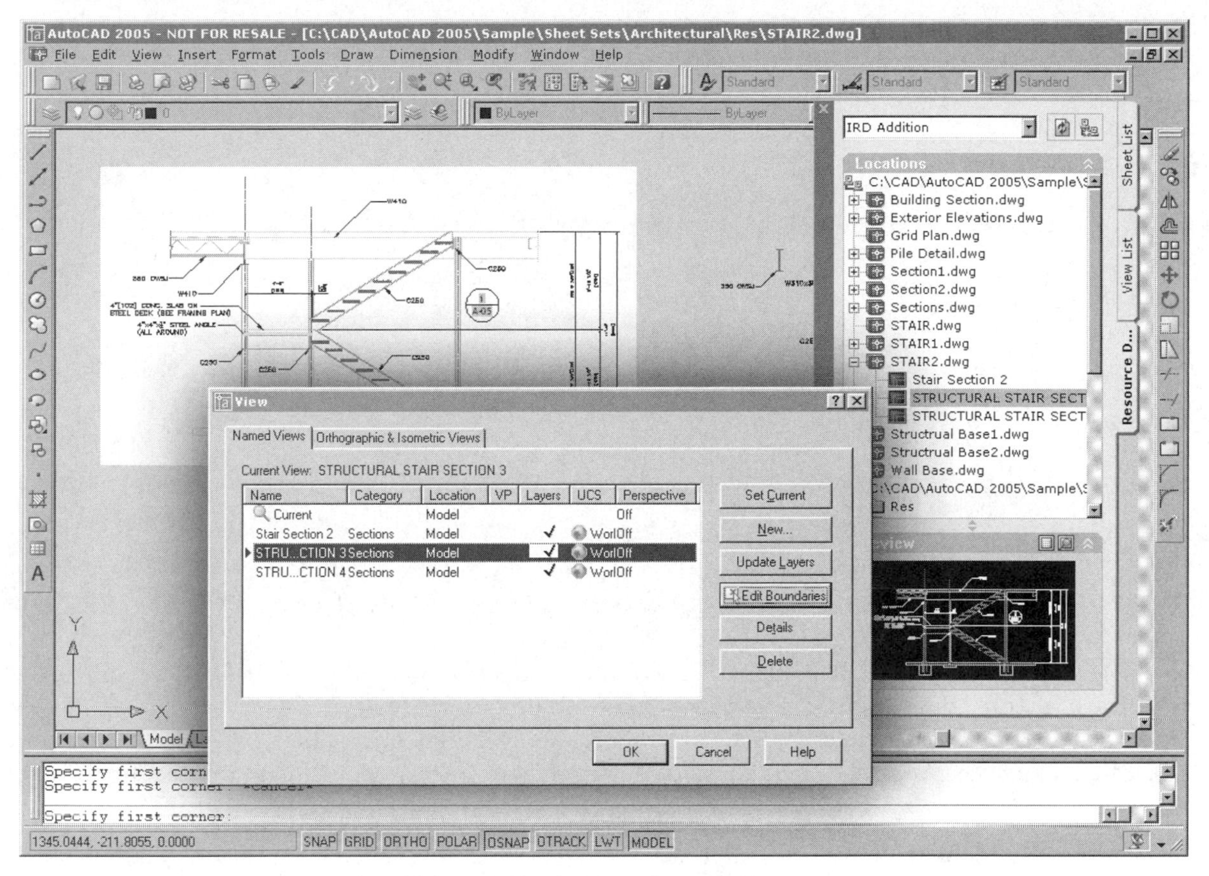

Views are created by the View command's New option.

PREPARING FOR SHEET SETS

Before creating sheet sets from existing drawings, the CAD manager should prepare by performing these steps.

Move Drawings to Organized Folders

Using Windows Explorer, move drawing files to folders named for projects. Large projects may require you further to segregate drawings into subfolders named after one of the classification systems discussed below.

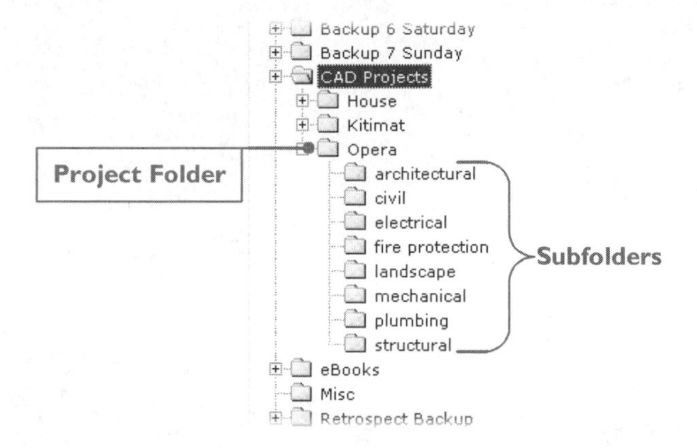

Using folders to segregate drawings by project and discipline.

Usually, it's up to the CAD manager to design a classification system for your firm; in some cases, clients will dictate the system. For example, some architectural design firms use the layer guidelines from the American Institute of Architects to segregate drawings into folders based on the names of disciplines, such as architecture and structural engineering. (The guidelines can be viewed online at www.slcc.edu/tech/techsp/arch/courses/ARCH2510/CAD_Stds/layerindex.htm.)

Internationally, some firms use the International Organization of Standards system, which is similar to that of the AIA. Construction firms use the MasterFormat guidelines from the Construction Specifications Institute <www.csinet.org>, which is based on the activities and materials used for construction, such as pouring concrete and installing electrical devices.

There are many other classification systems to choose from, including standards from the California Department of Transportation, the Swedish building element classification system, and the US Coast Guard.

If you need a structure to work with, you may want to consider one of the groupings listed on the following page.

Implementing this classification system can take hours; remember also to take into account the time it may take to obtain agreement from all stakeholders.

Create Master Templates

The next step is to create *.dwt* drawing template files, which are used to create new sheets. The templates contain:

- Standard drawing border and title block.
- Default units of linear and angle measurement
- Standard layer names, colors, linetypes, and lineweights.
- Plot styles.
- Standard dimension, text, and table styles.

AIA Major Categories
Architectural
Civil
Electrical
Fire Protection
Landscape
Mechanical
Plumbing
Structural

CSI MasterFormat	
00	Requirements
01	General
02	Site
03	Concrete
04	Masonry
05	Metals
06	Wood and Plastics
07	Thermal and Moisture
08	Finishes
09	Equipment
10	Specialties
11	Equipment
12	Furnishings
13	Special Construction
14	Conveying Systems
15	Mechanical
16	Electrical

AutoCAD includes template drawings for many standard sizes of drawings, ranging from A-size (A4 in metric) to E-size (A0). The templates are drawn to the standards stipulated by ANSI, DIN, ISO, JIS, and Chinese (Gb). In addition, a generic architectural and D-size template are available.

To access these predrawn templates, use the **OPEN** command, and then change **Files of Type** to "Drawing Template (*.dwt)". Modify the templates to meet your needs, for example by adding your firm's name.

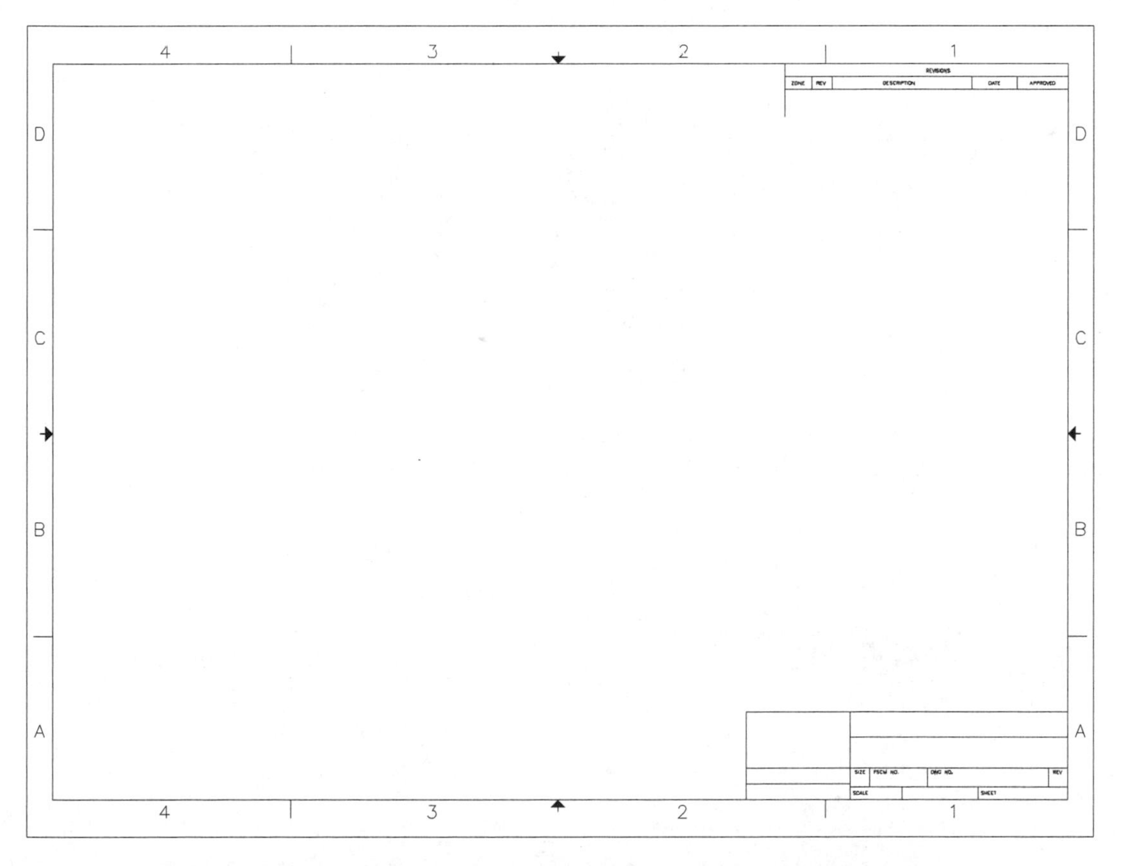

Template drawings contain a border and title block, as well as predetermined settings.

Draw Callout and View Label Blocks

It is common for drawings to refer to detail views found in other drawings. To identify the cross-references, *callouts* are used, also known as *reference tags*, *detail keys*, and *building section keys*. The three callouts illustrated below (left) use a split circle to identify the detail view: the upper half contains the detail number, while the lower half contains the drawing number.

Left: *Examples of callout blocks.*
Right: *Example of a view label block.*

In addition, it is common to label the views with a title, view number, and scale. Often, the scale of the detail differs from the rest of the drawing, as illustrated above (right).

The Sheet Set Manager accommodates both of these practices by allowing you to specify callout and view label blocks. When you insert them in sheets, AutoCAD automatically fills in the data for you.

These blocks use field text stored as attributes to update their values automatically. Field text is identified by the rectangular gray background. (Attributes are covered in Chapter 6 of this book, while field text is discussed in Chapter 16 of *Using AutoCAD: Basics*.) Create your own blocks, or borrow from the samples included in AutoCAD *\sheetset* folder. The blocks illustrated here are provided by the "IRD Addition" drawings found in the *\architectural* folder.

The figure below illustrates the attribute and field text names used to create the callout blocks. The attributes are named **V#** and **S#**. (Set the attribute definitions to **Preset** to prevent AutoCAD from issuing prompts when inserting the blocks.)

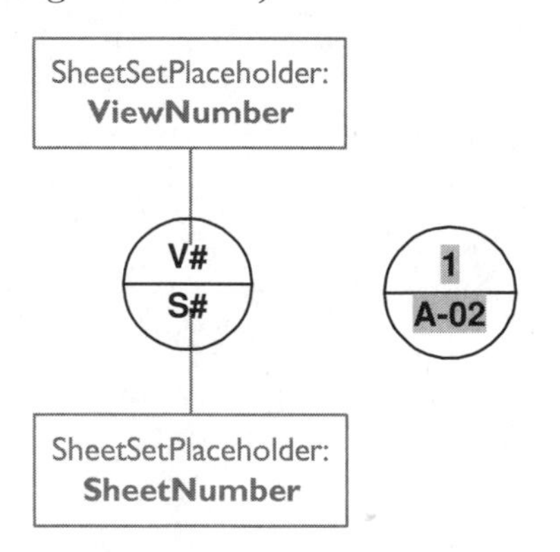

Attributes and field text used in callout blocks.

Illustrated below are the attribute and field text names used for the view label blocks. The attributes are named **#**, **VIEWNAME**, and **VPSCALE**. The view name is underlined by the **%%u** metacharacters.

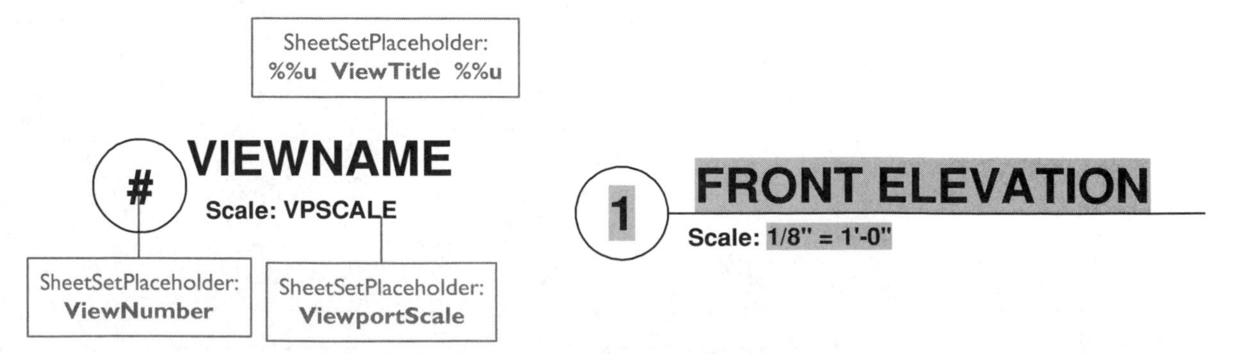

Attributes and field text used in view label blocks.

Sheet Names

It is good CAD management practice to assign standardized sheets names that assist AutoCAD in sorting them correctly.

Give each drawing a prefix that determines its subset. For example, architectural sheets could start with A. Follow this by a number that specifies the order of the sheet in the set, such as A-01. Refer to a layer standard, such as AIA or CSI, to help name drawings.

It's important to prefix single-digit numbers with a zero; otherwise AutoCAD sorts the sheets incorrectly, as illustrated by the table below. For example, if your project has more than nine sheets, use the 0 prefix for sheets 0 through 9; if you think you might have more than 99 sheets, use additional zero padding, such as 001 and 010.

Sorted Incorrectly	***Sorted Correctly***
A-10	A-01
A-1	A-02
A-2	...
...	A-09
A-9	A-10

Create Page Setup Overrides

The master template needs to include a page setup for plotting and publishing sheet sets. This allows you to plot all sheets the same way, overriding page setups applied to the source drawings. Use the PAGESETUP command to select the plotter and paper size. You may also want to specify the plot area (Layout, usually), plot scale, plot style, and orientation.

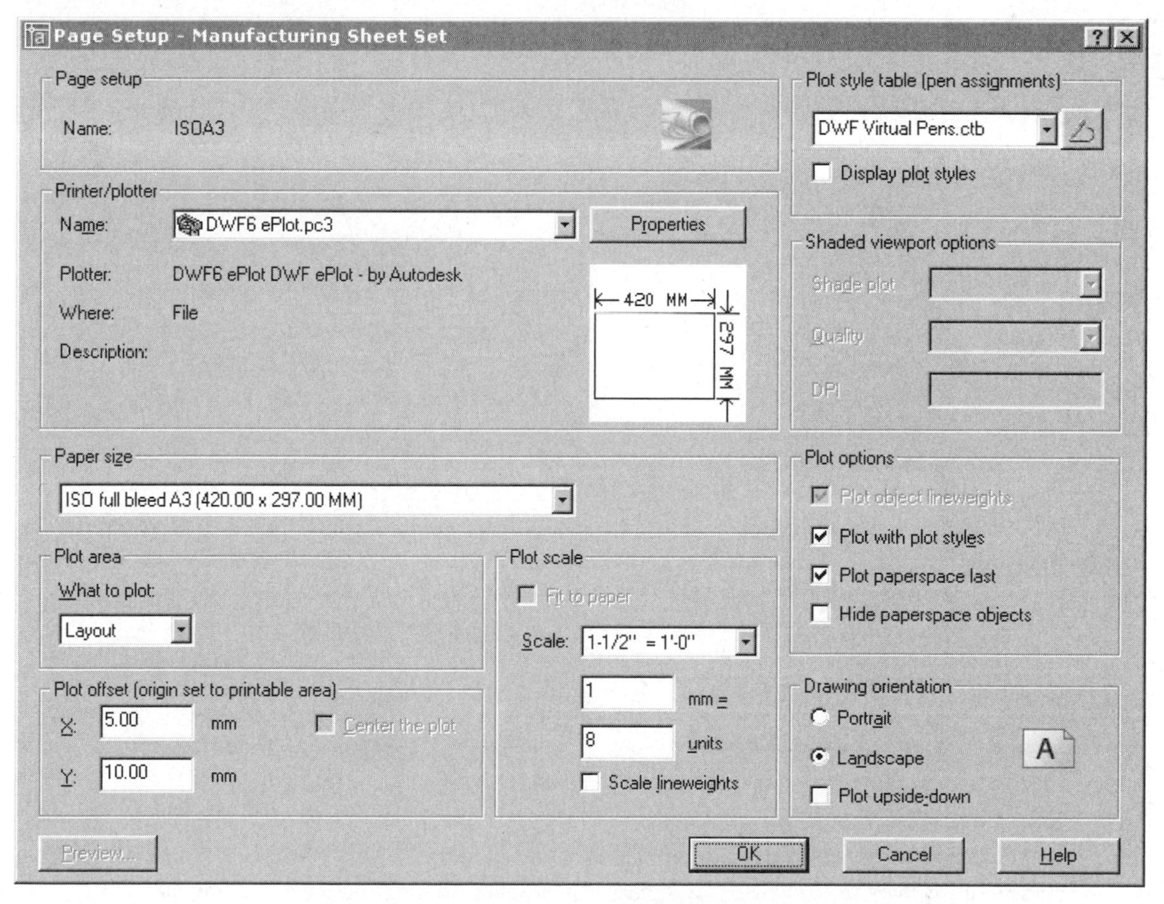

Page setups ensure consistent plotting of all sheets in the set.

Save the Templates Correctly

Save the drawings as templates using the **SAVEAS** command. You can store them in AutoCAD's *templates* folder, or in the folders you created for the projects.

Customize Sheet Set Properties

Each sheet set has *properties* that define its name, location on the computer, and so on. You can use the **NEWSHEETSET** command to run a wizard that helps you define the properties, or you can enter them yourself: right click the name of the sheet set, and then from the shortcut menu, select **Properties**.

The dialog box has two primary sections. The upper half consists of Sheet Set and Sheet Creation items; these are found in every Sheet Set Properties dialog box. The lower half consists of *custom properties*, which are created by the user.

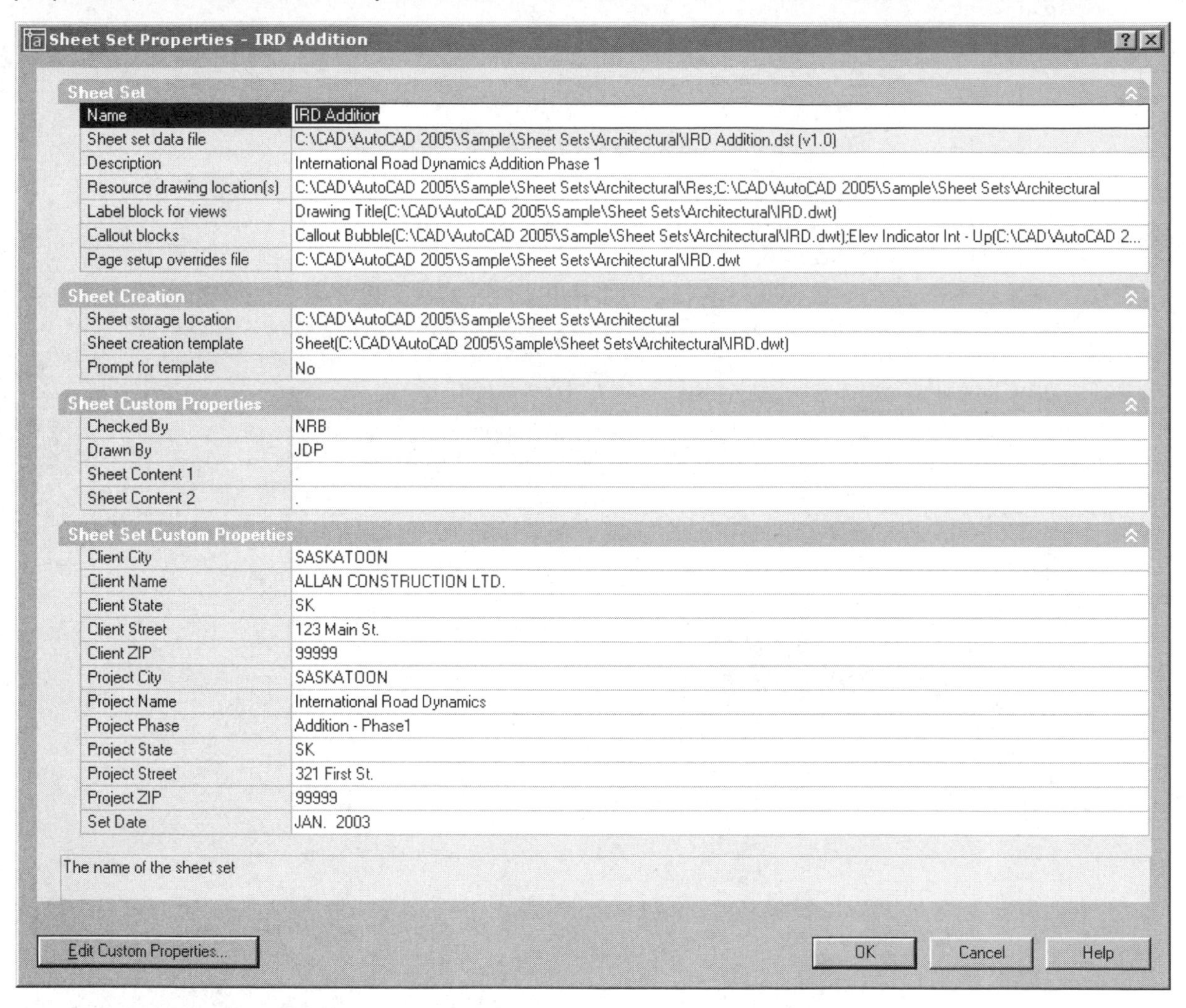

Properties of the sheet set are created and edited in this dialog box.

Properties in white can be edited; those in gray cannot. All properties are accessible through field text.

Sheet Set

Name — name of the sheet set displayed in the Sheet Set Manager.

Sheet set data file — names of drives, folders, and *.dst* files.

Description — user description of the project.

Resource drawing location(s) — names of folders holding the *.dwg* drawing files used by sheets in this set.

Label block for views — names of blocks (folders and names of the *.dwt* or *.dwg* files holding the view label blocks).

Callout blocks — names of blocks (folders and names of the *.dwt* or *.dwg* files holding the callout blocks).

Page setup overrides file — names of the drives, folders, and the *.dwt* template files holding the page (plot) setup parameters.

Sheet Creation

Sheet Storage Location — names of drive and folders in which sheets are stored.

Sheet Creation Template — names of drive and folders of the *.dwt* or *.dwg* file used as the master template for new sheets.

Prompt for Template — determination of whether users can select their own template file: **Yes** means the master template can be overridden; **No** means users must use the file specified by the Sheet Creation Template field.

NEWSHEETSET

The NEWSHEETSET command starts the Create Sheet Set wizard, which guides you through the process required to create new sheet sets. In the following pages, you perform these tutorials: (1) copy drawings into a project folder; (2) create the new sheet set; (3) copy label and callout blocks; (4) create and label views; and (5) generate a title sheet.

TUTORIAL 1: CREATING THE PROJECT FOLDER

Before creating a new sheet set, you need to designate a folder for holding all drawings and related files for your project.

1. Use Windows Explorer to create a folder on your computer called *\Using AutoCAD Sheet Sets.*
2. From AutoCAD's *\sample* folder, hold down the **CTRL** key, and then drag a copy of the *MKMPlan.dwg* file into the newly created folder.

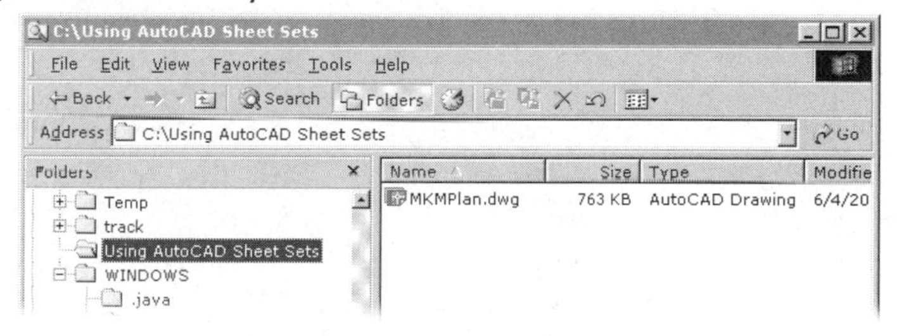

New folder with the AutoCAD drawing file.

TUTORIAL 2: CREATING NEW SHEET SETS

1. To create a new sheet set, start the **NEWSHEETSET** command:
   - From the **File** menu, choose **New Sheet Set**.
   - From the **Tools** menu, choose **Wizards**, and then choose **New Sheet Set**.
   - In the Sheet Set Manager, click the **Open** droplist, and then choose **New Sheet Set**.
   - At the 'Command:' prompt, enter the **newsheetset** command.

   Command: **newsheetset** *(Press* ENTER.*)*

   AutoCAD displays the Create Sheet Set wizard.

2. New sheet sets are generated either from an existing sheet set, or else "from scratch" by specifying the names of folders holding the drawings.

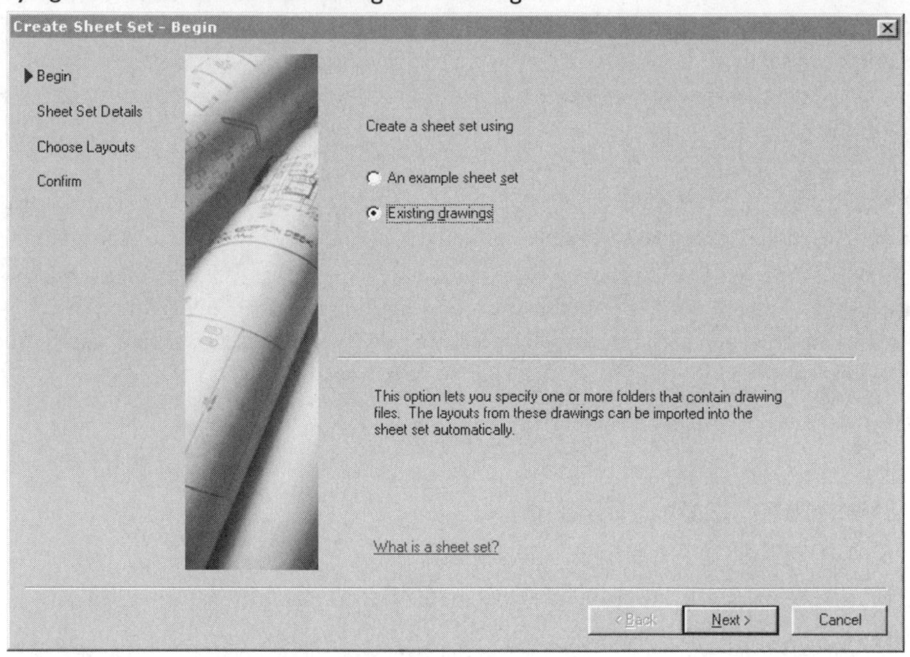

Begin stage of the Newsheetset wizard.

- **An example sheet set** — use this option when your drawings are not yet organized, or when you are required to use a client's sheet set structure.
- **Existing drawings** — use this option when you already have drawings logically organized in folders.

Note: When you select the **An example sheet set** option, AutoCAD next displays a list of six sample sheet sets, namely metric and imperial versions for each of architectural, civil, and mechanical. Their default categories are listed below. You can later add to and remove from the categories, as required.

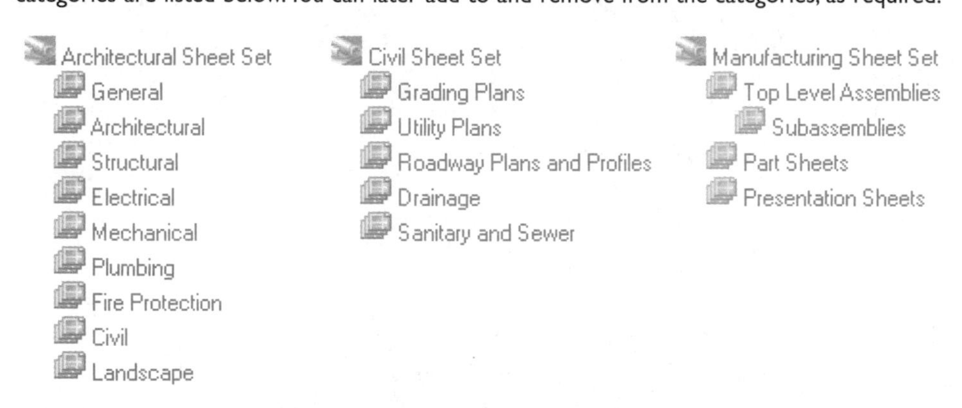

For this tutorial, select **Existing drawings**, and then click **Next**.

3. Enter a name for the sheet set. The usual choice would be the name of the project. For this tutorial, enter the following:

Name of the new sheet set:	**MKM Expansion Project**
Description (optional):	**Custom residence for Santa Rosa CA**
Store sheet set data file here:	**C:\Using AutoCAD Sheet Sets**

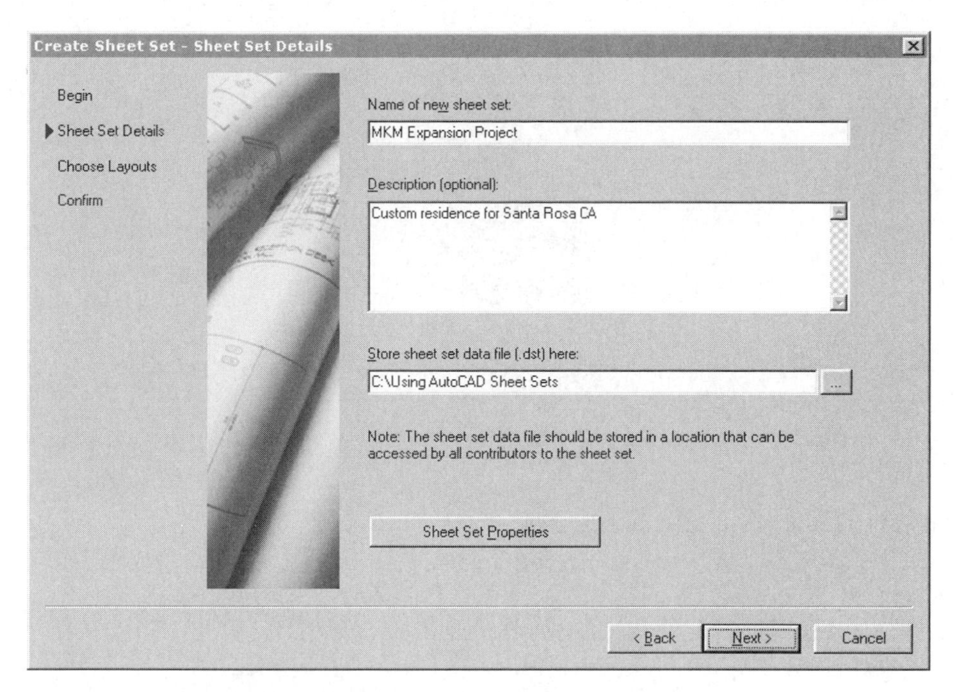

The Sheet Set Details stage of the Newsheetset wizard.

There is no need to click the **Sheet Set Properties** button now; the properties can be edited later. Click **Next**.

Notes: Store sheet set data (*.dst*) files on a network drive, because then they can be accessed by all users. These important files are also more likely to be backed-up when stored in a central location.

The AutoCAD Design Center does not operate while the Sheet Set Manager is open.

4. Select the drawings to include with this sheet set by clicking **Browse**.
   For this tutorial, select the *C:\Using AutoCAD Sheet Sets* folder.
   Notice that AutoCAD lists the names of folders and drawings in the folder, as well as the names of layouts found in each drawing (if any).

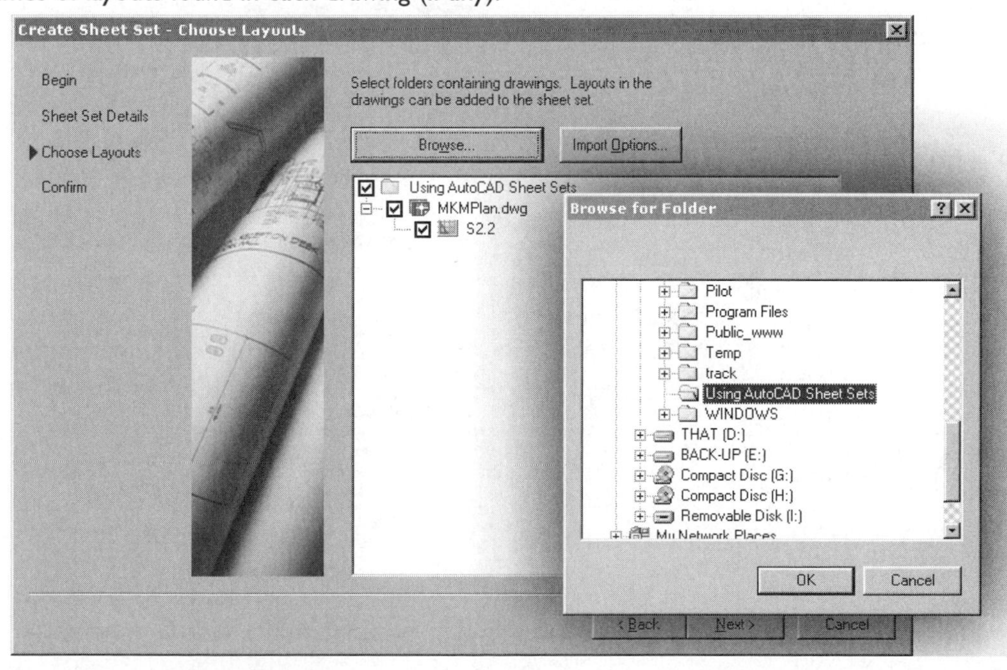

For this tutorial, select *mkmplan.dwg* and its layout.

(When you click **Import Options**, the dialog box allows you to specify whether sheet titles are fixed by the drawing's file name. Leave this option turned on. The other options can be left off, because we are not working with subsets at this point.)

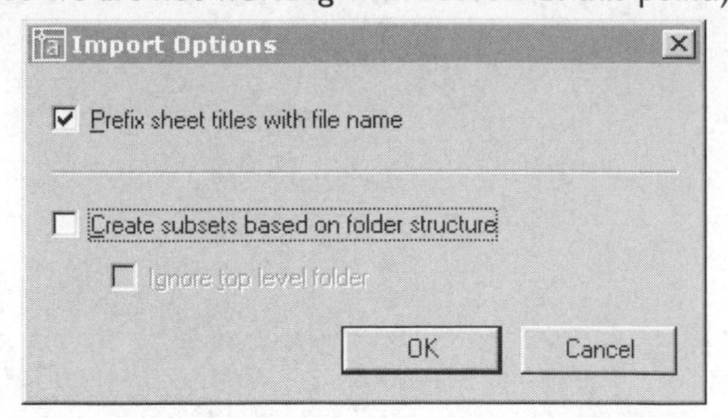

Selecting the drawing import options.

5. Click **Next** to carry on to the final step.

   AutoCAD provides an overview of your choices, and warns you, "Note: Creation of the sheet set is complete only after each sheet in the sheet set has been opened and saved." (You can change any options by clicking the **Back** button.)

6. Click **Finish**.

   AutoCAD opens the Sheet Set Manager, and lists the name of the (single, solitary) layout from the *mkmplan.dwg* drawing.

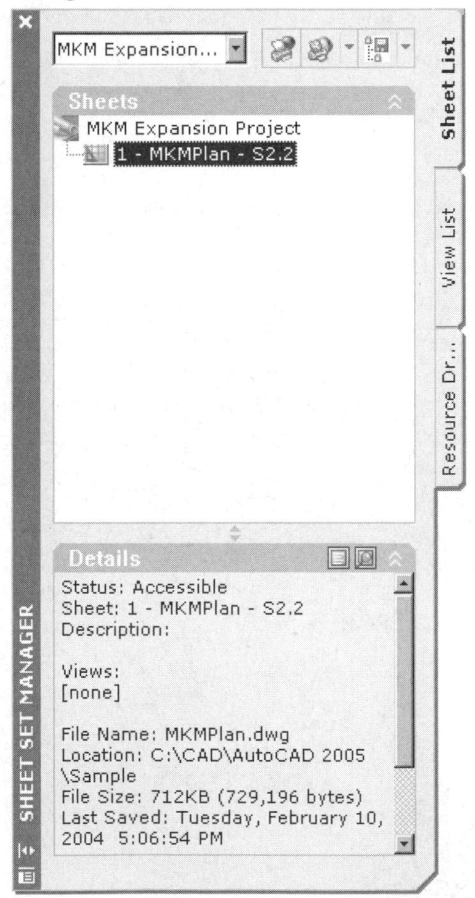

The Sheet Set Manager populated with its first sheet..

Note: To populate the sheet set, right-click "MKM Expansion Project", and then select **New Sheet** to create a blank sheet. Alternatively, select **Import Layout as Sheet** to import sheets from the layouts of other drawings.

7. Follow AutoCAD's warning, and save the sheets, as follows:
   a. In the Sheet Set Manager, double-click "1-mkmplan-s2.2". The drawing opens.
   b. In AutoCAD, use the **QSAVE** command to save the drawing.
   c. Close the drawing with the **CLOSE** command.

TUTORIAL 3: ADDING LABEL AND CALLOUT BLOCKS

You can create custom callout and label blocks in the master template file. Alternatively, you can add them to the current sheet set by copying the blocks from another drawing. The latter process takes two steps: (1) find the drawing containing the blocks; and then (2) copy them to the Sheet Set Manager.

1. In the Sheet Set Manager, right-click the sheet set name, "MKM Expansion Project."
   From the shortcut menu, select **Properties**. Notice that the Sheet Set Properties dialog box has fields that list the callout and label blocks names, but which are currently blank.
2. In the **Label Block for Views** item, click the **...** button.
3. In the Select Block dialog box, click the **...** button.
   In the Select Drawing dialog box, change the **Files of Type** to template (*.*dwt*). Go to AutoCAD's *\sample\sheet sets\architectural* folder.
4. Select the *IRD.dwt* template file, and then click **Open**. (This template file contains the label and callout blocks we need to borrow for this sheet set tutorial.)

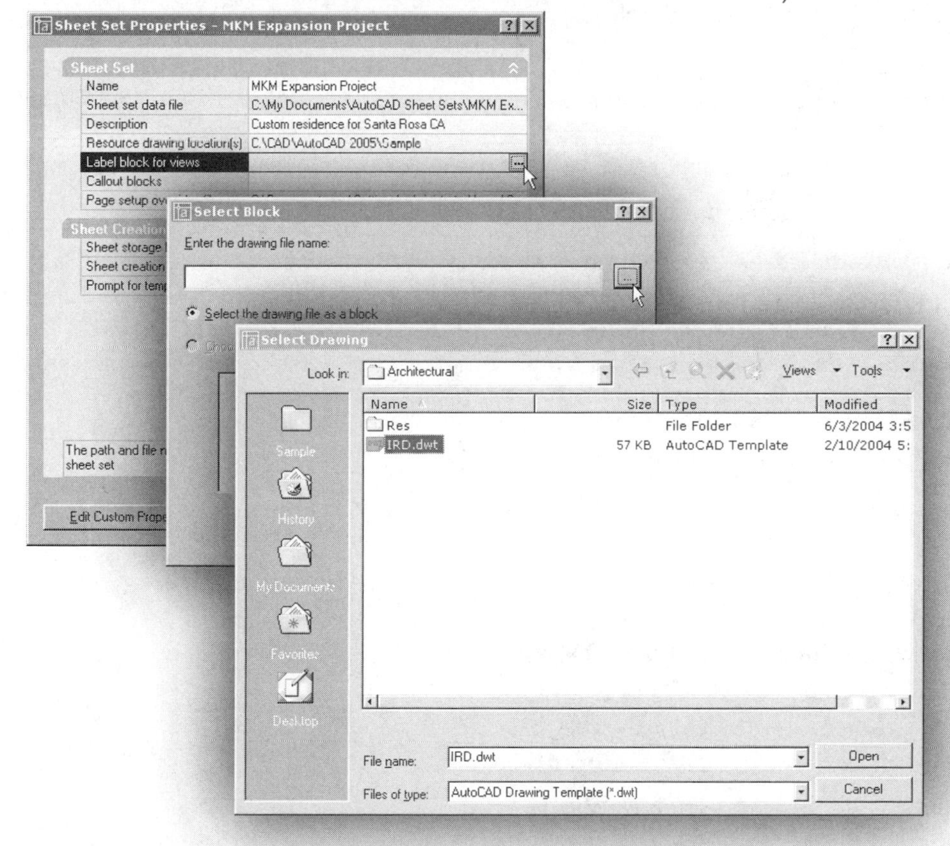

Selecting drawings containing blocks for the sheet set.

5. Back in the Select Block dialog box, AutoCAD gives you two options:

- **Select the drawing as a block** — this is your choice if the drawing *is* the block.
- **Choose blocks in the drawing file** — this is your choice when the drawing *contains* blocks.

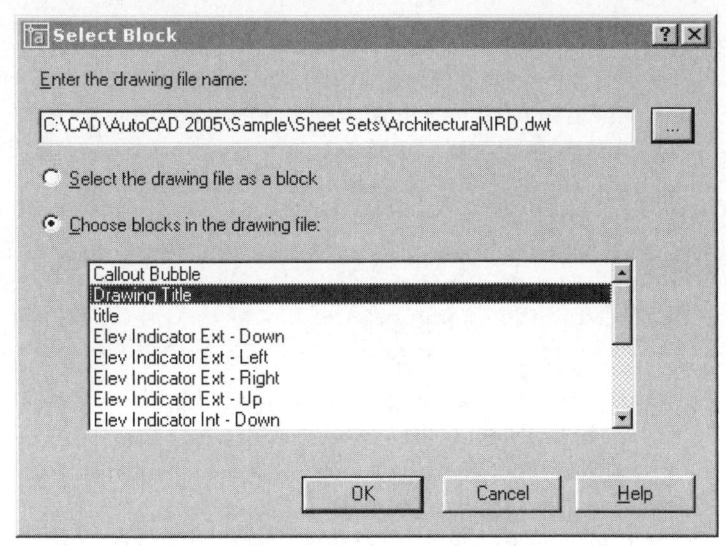

Selecting blocks from a drawing.

6. Choose the **Choose blocks in the drawing file** option.
   Select "Drawing Title," and then click **OK**.
7. Repeat to copy the "Callout Bubble" block for the **Callout Blocks** field.

The sheet set now has access to the view blocks, which AutoCAD requires in a later tutorial.

TUTORIAL 4: ADDING VIEWS

You can place views in sheets, just like placing viewports in layouts.

Views-in-sheets have the advantage that AutoCAD keeps better track of them, allowing for automated labels. The drawback is that this process takes more steps: (1) create named views in a drawing's *model space* (not in layouts!); (2) if necessary, add the drawing to the sheet list; (3) create a new sheet; and (4) drag the model-space views from the Sheet Set Manager onto the new sheet.

Create Named Views in Model Space

Open the drawing, and create named views, as follows:

1. In the Sheet Set Manager's **Sheet List** tab double-click "MKMPlan-S2.2". Notice that the drawing is opened.
2. Click **Model** tab. (Don't create views in a layout tab.)
   AutoCAD notes, "Switching to: Model."
3. From the **View** menu, select **Named Views**.
   In the Views dialog box, click **New**.
4. Create a named view using these settings:

View name	**NW Quadrant**
View category	**Quarter Views**

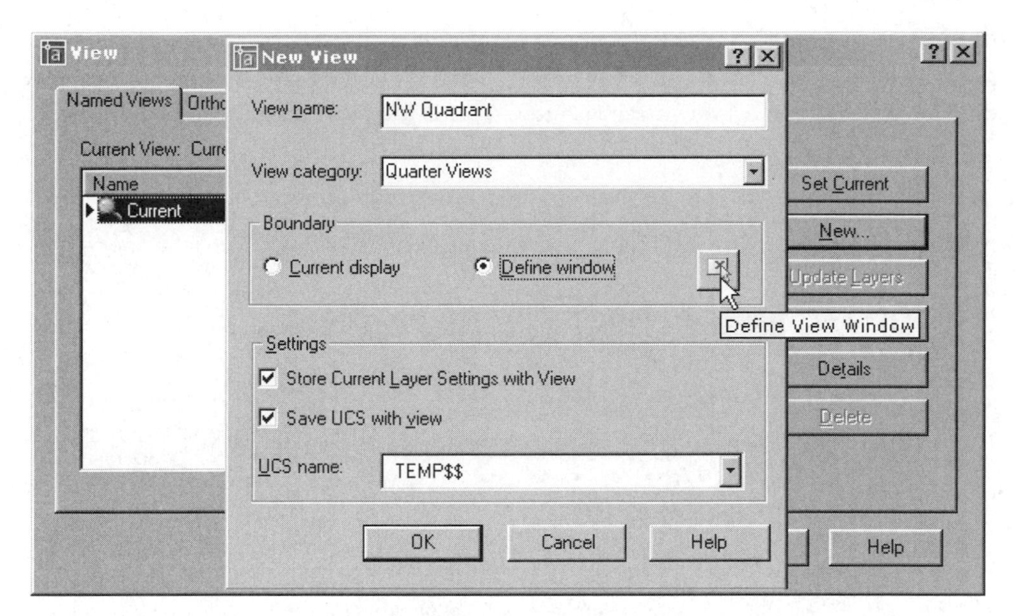

Creating named views.

Define window *(Click the **Define Window** radio button, and then pick two points encompassing the area shown in white, below.)*

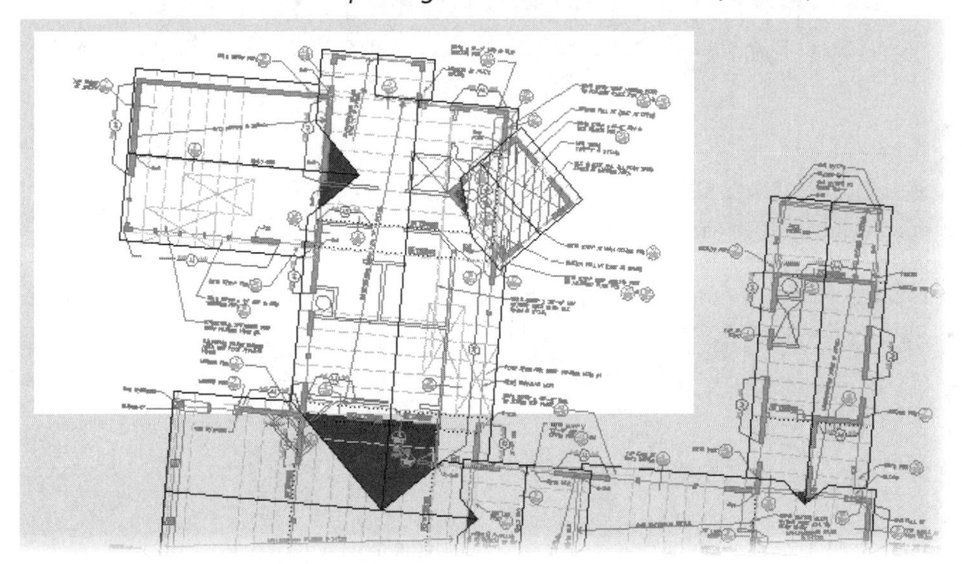

The white rectangle shows the extent of the windowed view.

Note: The DEFAULTVIEWCATEGORY system variable, undocumented by Autodesk, stores the default text for the **Category** droplist in the New View dialog box.

5. Click **OK**.

   Repeat, creating three more views under the "Quarter Views" category:

View name 2	**NE Quadrant**	*(upper right of drawing)*
View name 3	**SW Quadrant**	*(lower left of drawing)*
View name 4	**SE Quadrant**	*(lower right of drawing)*

   Exit the View dialog box.

6. Save and exit the drawing with the **CLOSE** command.

   (If no drawing is open in AutoCAD, create a new one with the **QNEW** command. There must be at least one drawing open in AutoCAD for the next stage to work.)

Create New Sheets

In the Sheet Set Manager's **Sheet List** tab, create a new blank sheet, as follows:

1. Right-click "MKM Expansion Project," and then select **New Sheet** from the shortcut menu.
2. In the New Sheet dialog box, enter the following information:

   Number **Q-01**

   Sheet Title **Quadrangle Views**

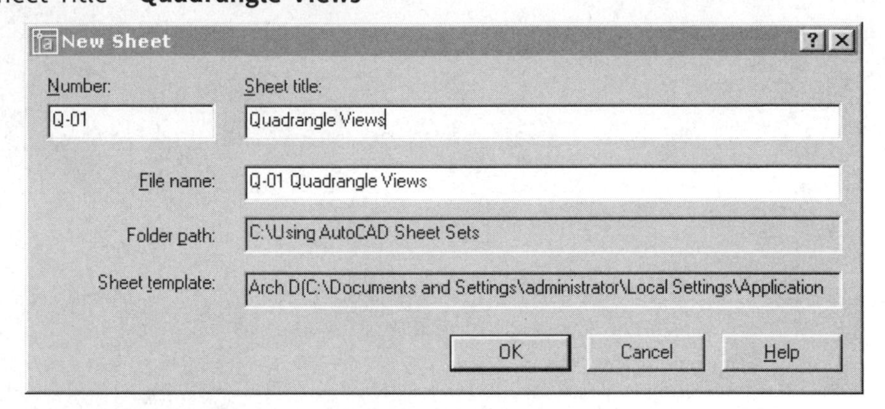

Naming a new sheet.

   Click **OK**.

   Notice that AutoCAD adds the sheet name to the sheet list.

3. Double click "Q-01-Quadrangle Views."

   Notice that AutoCAD opens a new drawing, blank except for the drawing border. (Ensure the sheet is in layout mode; if not, click the layout tab named "Q-01-Quadrangle Views.") The title block has the drawing number filled in, "Q-01."

The title block is numbered automatically by AutoCAD.

4. In the Sheet Set Manager, choose the **Resource Drawings** tab.

   (If the list is blank, double-click **Add New Location**, and then choose the *\Using AutoCAD Sheet Sets* folder.)

5. Click the **+** next to "MKMPlan.dwg" to expand the list. Notice the named views.
6. Place the view on the sheet using these steps:

   a. Select a named view.

   b. Drag the view on the sheet.

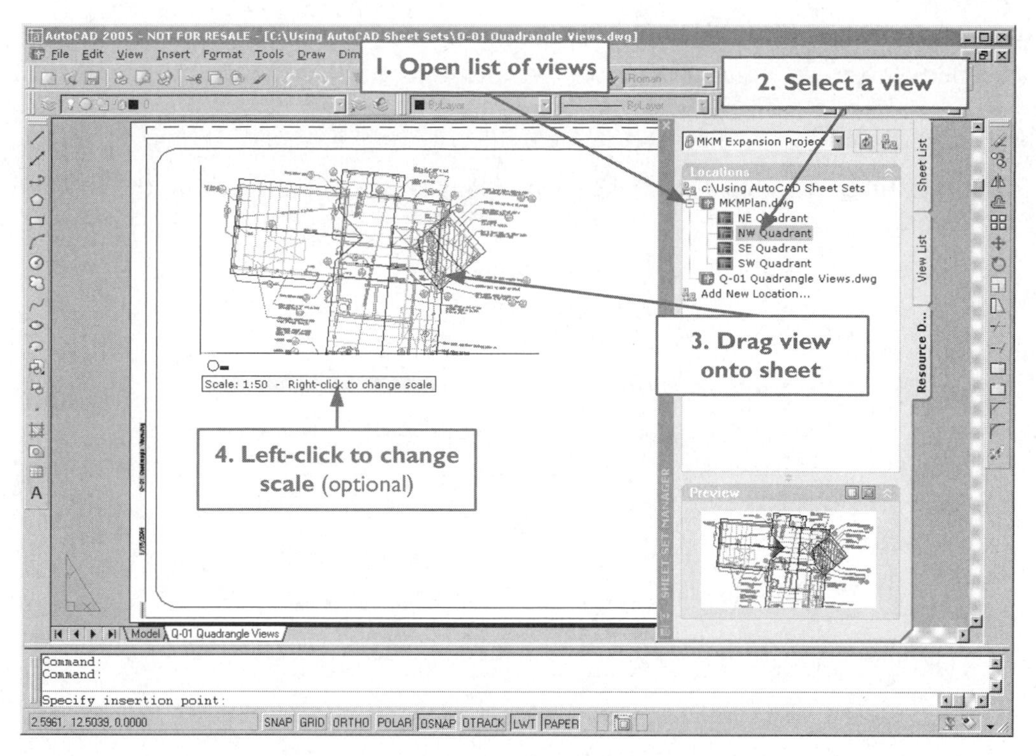

Placing views in sheets.

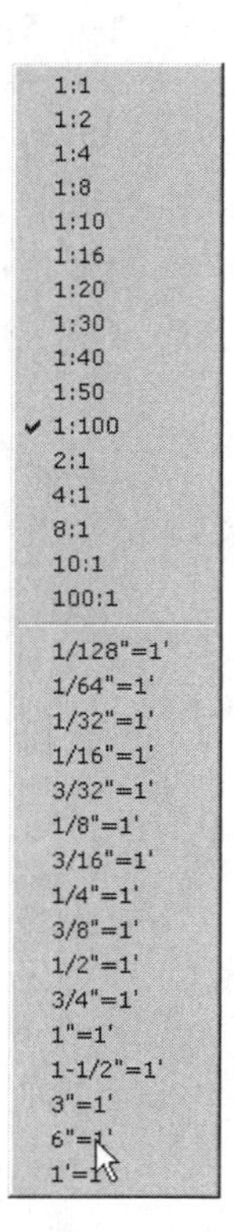

Select a scale factor for the view.

c. (*Optional.*) Notice the tooltip advising you of the scale. If you wish to change the scale, click the right mouse button. AutoCAD displays a shortcut menu listing scale factors. Select a suitable scale factor.

d. Locate the view in the sheet:

Specify insertion point: *(Pick a point in the sheet.)*

Notice that AutoCAD automatically labels the view with the view label block.

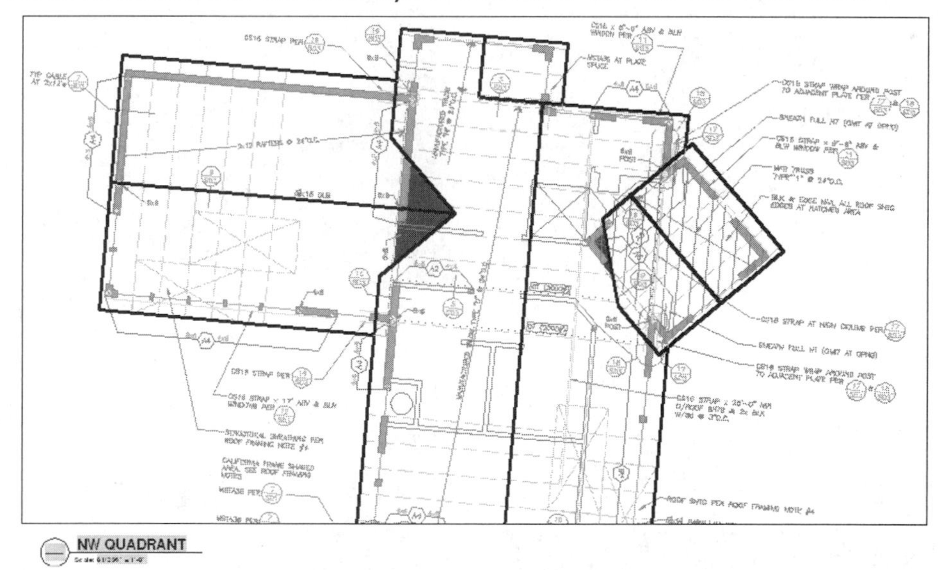

Views are automatically labeled.

7. Place the three other views on the sheet.
8. Save the sheet with the **QSAVE** command.

TUTORIAL 5: CREATING TITLE SHEETS

The Sheet Set Manager creates "sheet list tables" semiautomatically. (This is Autodesk's name for *title pages* or *cover sheets*.) The process is semiautomatic in this sense: after you select the information you want, AutoCAD generates the table filled in with the data.

Creating title sheets takes two steps: (1) make a new, blank sheet; and (2) execute the **Insert Sheet List Table** command.

1. Right-click the sheet set name, and then select **New Sheet**.
2. In the New Sheet Dialog box, enter a number and name for the sheet:

   Number **01**

   Sheet Title **Title Sheet**

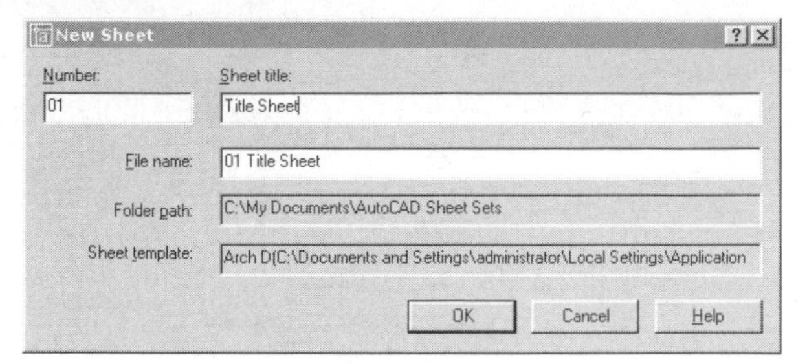

Naming a new sheet.

3. Click **OK**.
4. Notice that AutoCAD places the new sheet at the bottom of the list. Drag "01-Title Sheet" to the top, just under the name of the sheet set.
5. Double-click the "01 - Title Sheet" sheet to open it in AutoCAD.
6. In the Sheet Set Manager, right-click the *sheet set name* "MKM Expansion Project" (*not* "01-Title Sheet"!), and then select the **Insert Sheet List Table** command.

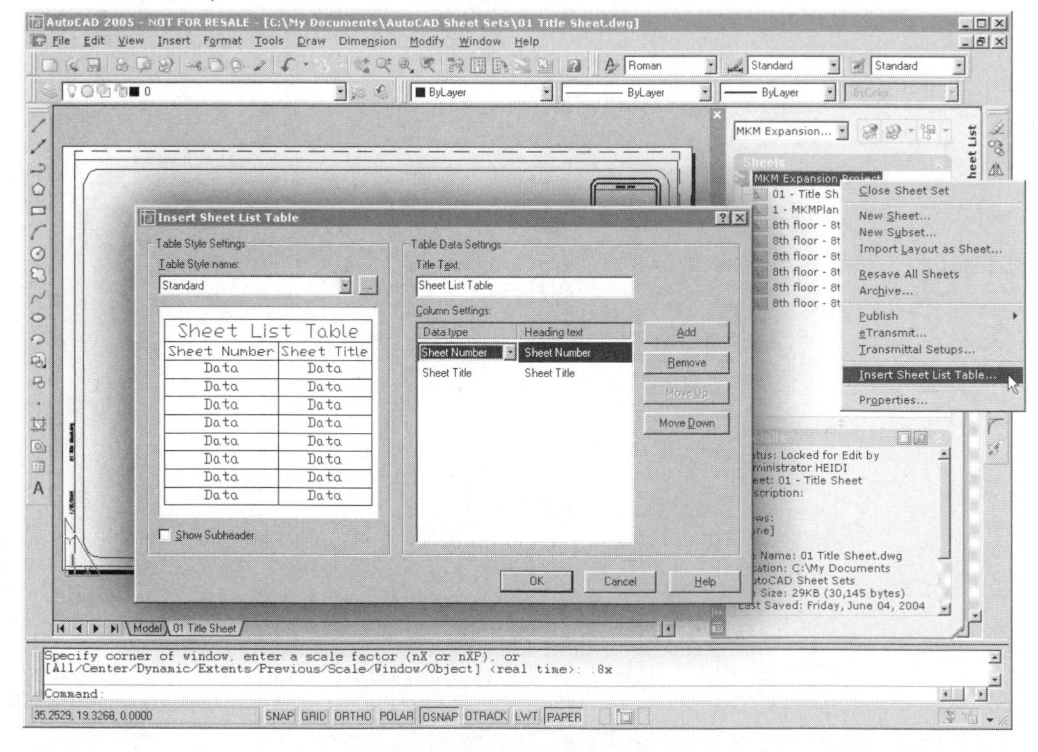

Selecting the command to create sheet list tables.

7. The Insert Sheet List Table dialog box is similar to the dialog box displayed by the **TABLE** command, but is customized for sheet sets. For this tutorial, specify the following settings:

   Table Style Name *(Select a predefined table style. To create a style, click ... .)*

   Title Text **Sheet List Table**

8. There is much information that can be included with the sheet list table. To see the options, click **Add**, and then double-click on "Sheet Number." Notice the list of options.

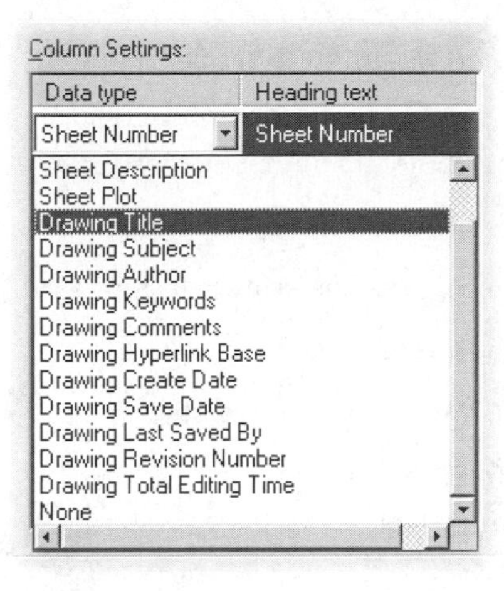

Selecting the headers for the sheet list table.

   Select an option, such as "Drawing Save Date." Notice that the preview image updates to include the heading.

9. Heading titles can be edited. Click "Drawing Save Date," and then change the wording to:

   Drawing Last Saved

10. The order of heading titles can be changed. Select the heading name to move, and then click the **Move Up** or **Move Down** buttons to move the headings left or right. Notice that the preview image updates automatically.

11. Click **OK**. AutoCAD prompts you to locate the table in the drawing:

    Specify insertion point: *(Pick a point.)*

   The number of sheets listed in the table varies according to that in the sheet set. As sheets are added and removed, the table updates itself.

   Double-click a sheet name to jump to that sheet. (The field text consists of hyperlinks.)

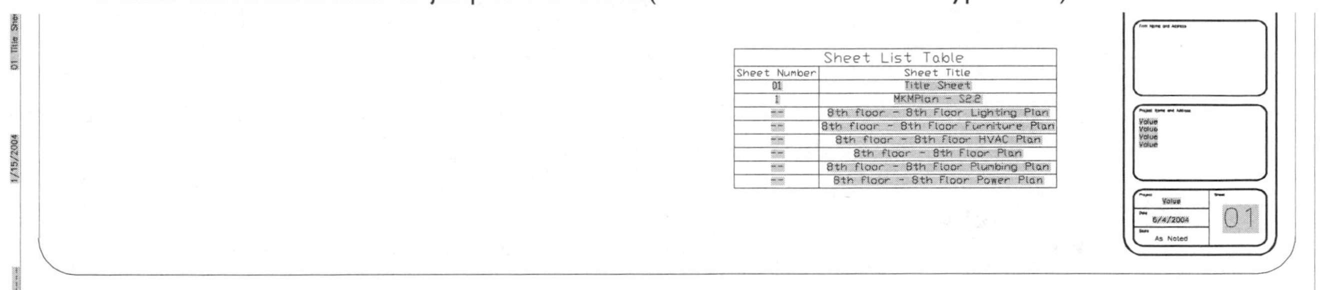

Sheet list table inserted in blank sheet.

SHEETSET

The user interface of the Sheet Set Manager looks clean and unencumbered, because it hides more than one hundred commands, options, and operations in shortcut menus and toolbars. Here is a brief guide to some of them.

The **SHEETSET** command opens the Sheet Set Manager window.

1. To view the Sheet Set Manager, start the **SHEETSET** command:
   - From the **Tools** menu bar, choose **Sheet Set Manager**.
   - From the **Standard** toolbar, choose the **Sheet Set Manager** button.
   - At the 'Command:' prompt, enter the **sheetset** command.
   - As an alternative, enter the **ssm** alias.
   - Alternatively, press the **CTRL+4** shortcut keystroke.

   Command: **sheetset** *(Press* ENTER.*)*

   AutoCAD displays the Sheet Set Manager window.

TOOLBARS

The toolbar of the Sheet Set Manager changes, depending on the tab selected.

Sheet List tab

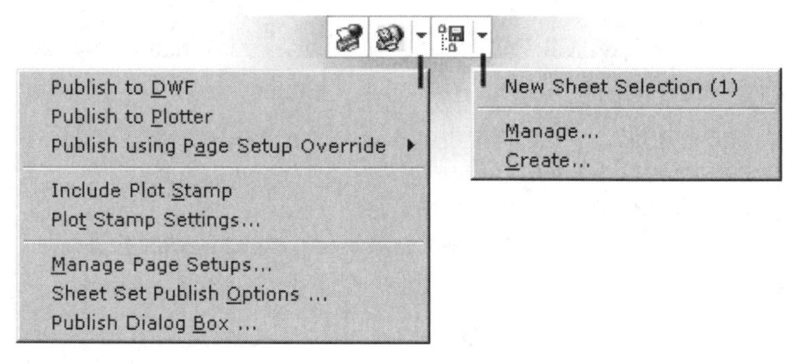

The toolbar of the Sheet List tab.

Publish to DWF prompts you for a file name, and then publishes all sheets as a single *.dwf* file.

Publish displays a menu of choices for publishing sheets. See the **PAGESETUP**, **PLOT**, **PUBLISH**, and **PLOTSTAMP** commands in *Using AutoCAD: Basics*.

Sheet Selections displays a menu for managing and creating sheet selections. These allow you to plot subsets of sheets, such as all architectural sheets for the architect.

View List tab

New View Category displays the View Category dialog box. Specify a name for the category, and then select the callout blocks to be used by the category.

Resource Drawing tab

Refresh checks resource locations for new and removed drawing files, checks drawing files for new and removed model space views, and updates information in the *.dst* sheet set data file. (You can also press **F5** to force refreshes.)

Add New Location displays the Browse for Folder dialog box. Select a folder, and then click **Open**. The pathed folder is added to the list of locations.

SHORTCUT MENUS

Many commands and options are found in shortcut menus, whose content varies depending on where you right-click. Most commands are self-explanatory, such as **Open** and **Rename**. Others repeat commands found elsewhere in AutoCAD, such as **PUBLISH** and **ETRANSMIT**. The following sections describe commands more difficult to understand and that are unique to the Sheet Set Manager.

Sheet List tab

Shortcut menus on the Sheet List tab:

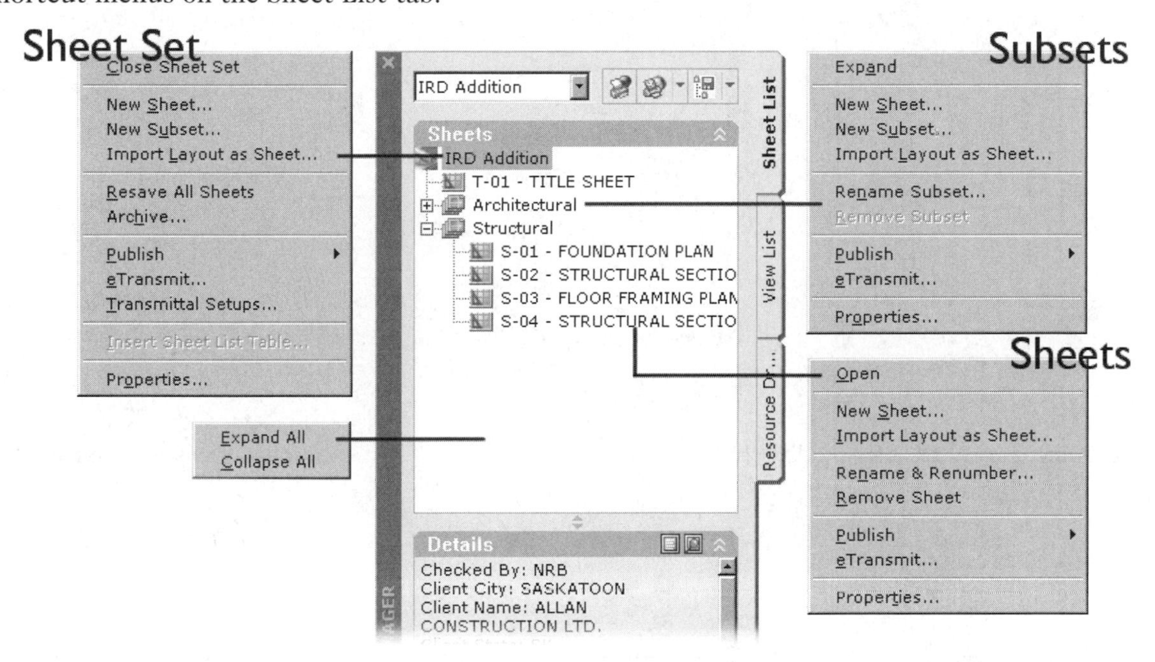

The shortcut menus of the Sheet List tab.

SheetSet:

New Sheet — displays New Sheet dialog box. Enter a number and name. AutoCAD adds it as a blank new sheet (well, blank except for any objects added by the master template file).

Import a Layout as Sheet — displays the Import Layouts as Sheets dialog box. Select a drawing, and then select one or more layouts to add as sheets.

Insert Sheet List Table — creates a sheet list table on the currently open sheet. This option is available only when a sheet is open.

Sheets:

Rename and Renumber — displays the same dialog box as New Sheet, but allows you to change the sheet's number and name.

Remove Sheet — deletes the sheet from the set, but does not erase the drawing file.

View List tab

Shortcut menus on the View List tab:

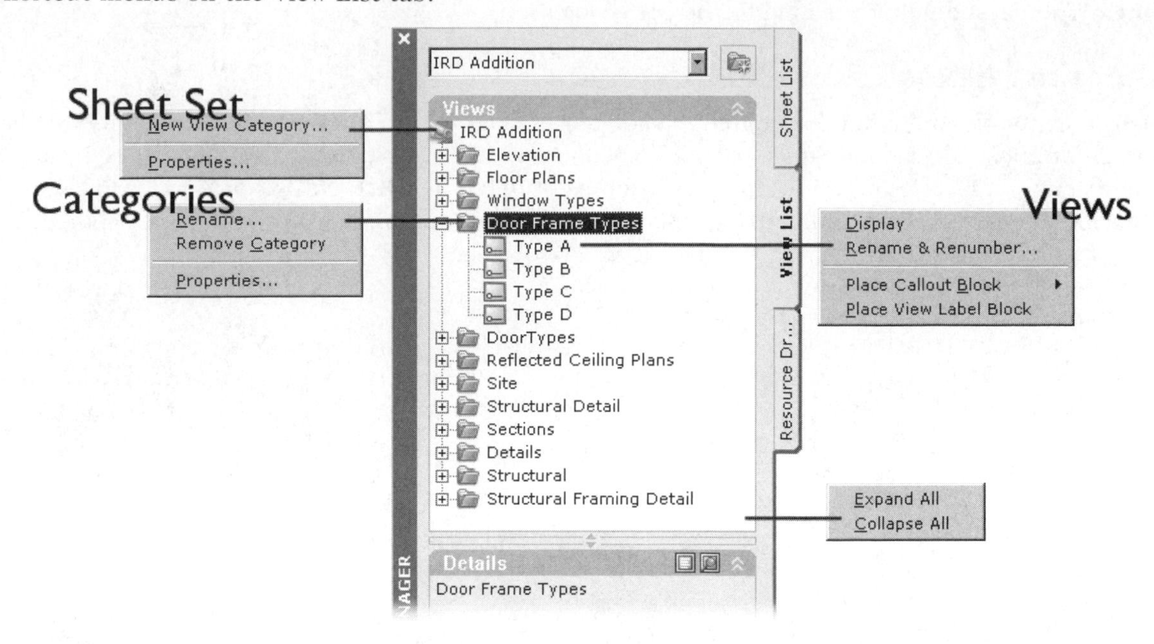

The shortcut menus of the View List tab.

Views:

Place View Label Block — adds the label block to views. AutoCAD numbers views automatically.

Place Callout Block — adds the callout block to reference other sheets.

Resource Drawings tab

Shortcut menus on the Resource Drawing tab:

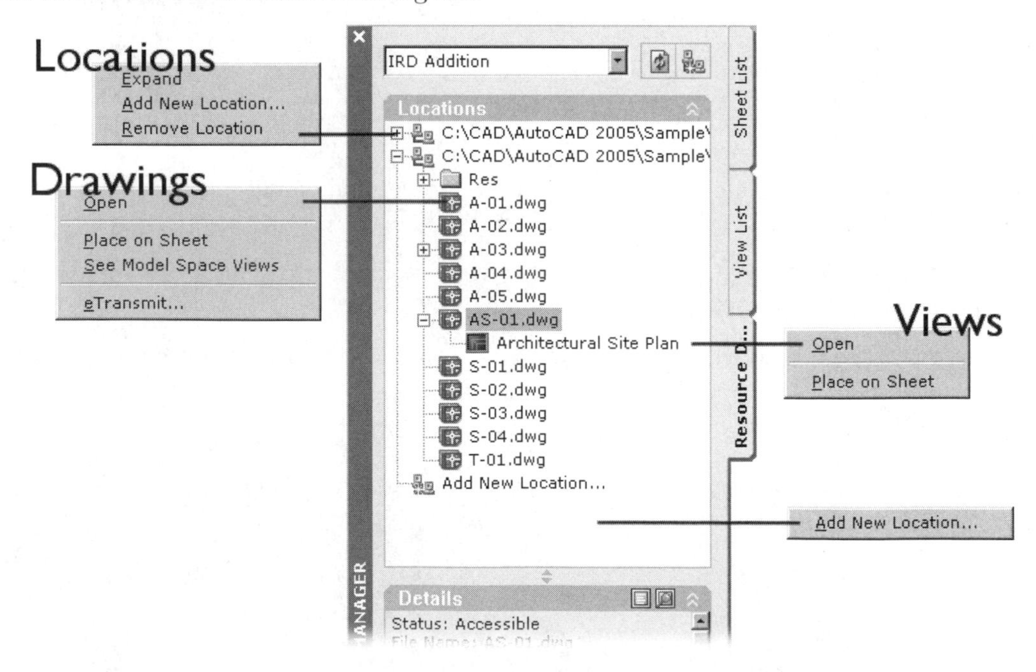

The shortcut menus of the Resource Drawings tab.

SHEETSETS: ADDITIONAL METHODS

The **SHEETSET** command is supported by several other commands, system variables, and command-line startup switches:

- **OPENSHEETSET** command opens *.dst* sheet set data files.
- **SHEETSETHIDE** command closes the Sheet Set Manager window.
- **SSFOUND, SSLOCATE, SSMAUTOOPEN,** and **SSMSTATE** system variables report on sheet sets.
- **/NOSSM** and **/SET** switches control the Sheet Set Manager window when AutoCAD starts up.
- **UPDATETHUMBSNOW** command regenerates the preview images displayed by the Sheet Set Manager.
- **UPDATETHUMBNAILS** system variable determines how preview images are updated.

Let's look at them all.

OpenSheetSet and SheetSetHide

The **OPENSHEETSET** command opens *.dst* sheet set data files. This command is meant for use with macros and programs. From the **File** menu, select **Open Sheet Set**. AutoCAD displays the Open Sheet Set dialog box. Select a *.dst* file, and then click **Open**.

The **SHEETSETHIDE** command closes the Sheet Set Manager window. You can later reopen the window with the **SHEETSET** command.

SsLocate, SsmAutoOpen, SsFound, and SsmState

These system variables control when the Sheet Set Manager is opened, and report on its status.

The **SSLOCATE** system variable determines whether AutoCAD opens the associated sheet set when a drawing is opened.

The **SSMAUTOOPEN** system variable determines whether AutoCAD displays the Sheet Set Manager when a drawing associated with a sheet is opened.

The **SSFOUND** system variable reports the path and file name of the current sheet set. For example:

```
SSFOUND = "C:\AutoCAD Sheet Sets\MKM Expansion Project.dst" (read only)
```

The **SSMSTATE** system variable reports whether the Sheet Set Manager window is open.

/NoSsm and /Set

These command-line startup switches control sheet sets.

The **/nossm** switch (short for "NO Sheet Set Manager") prevents the Sheet Set Manager from opening when AutoCAD starts.

The **/set** switch loads the specified *.dst* sheet set file when AutoCAD starts.

To use command-line switches, you edit AutoCAD's command line. On the Windows desktop, right-click the AutoCAD icon, and then select **Properties** from the shortcut menu. In the Properties dialog box, select the **Shortcut** tab. In the Target text box, add the text shown in boldface below:

Target: "c:\autocad 2005\acad.exe **/set c:\sheet sets\mkm expansion project.dst**"

Desktop shortcut icons launch AutoCAD with different Sheet Set Manager collections.

Click **OK** to close the dialog box. When you double-click the AutoCAD icon, AutoCAD now automatically loads the associated sheet set data file. You can make copies of the icon, customizing each for a different project's sheet set.

UpdateThumbsNow and UpdateThumbnail

The Sheet Set Manager displays thumb-size preview images of every drawing and view. When these change, the thumbnails need updating. The **UPDATETHUMBSNOW** command forces the update of the previews of every sheet, sheet view, and model space view.

The related **UPDATETHUMBNAIL** system variable determines which thumbnails are updated. The default setting is **15**, which means that settings 1, 2, 4, and 8 are turned on.

UpdateThumbnail	*Meaning*
0	Thumbnail previews are not updated.
1	Model view thumbnails are updated.
2	Sheet view thumbnails are updated.
4	Sheet thumbnails are updated.
8	Thumbnails are updated when sheets and views are created, modified, or restored.
16	Thumbnails are updated when the drawing is saved.

ARCHIVE

The **ARCHIVE** command archives sheet sets as *.dwf* or *.zip* files. This is commonly done at the end of each project phase, or to create a backup set for off-site storage.

1. To archive sheet sets, start the **ARCHIVE** command:
   - In the Sheet Set Manager, right-click the sheet set name, and then select **Archive**.
   - At the 'Command:' prompt, enter the **archive** command.

   Command: **archive** *(Press* ENTER.*)*

   AutoCAD displays the Archive a Sheet Set dialog box. Notice that it lists the names of the sheets used by the sheet set.

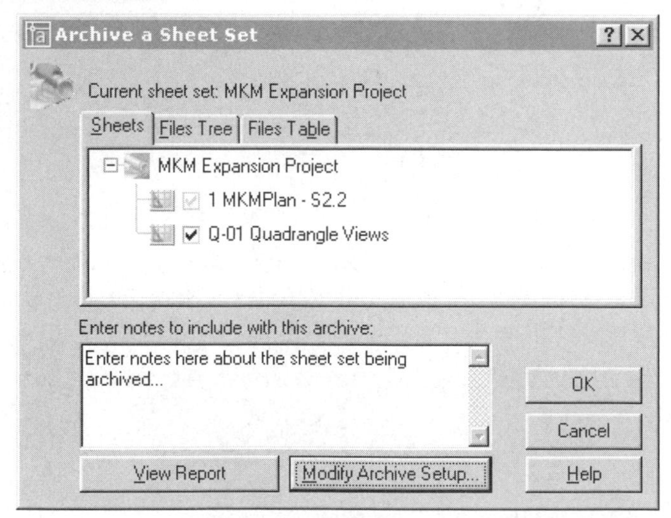

Selecting the files to include with the sheet set archive.

2. The Files Tree and the Files Table tabs provide different views of the files associated with the sheet set.

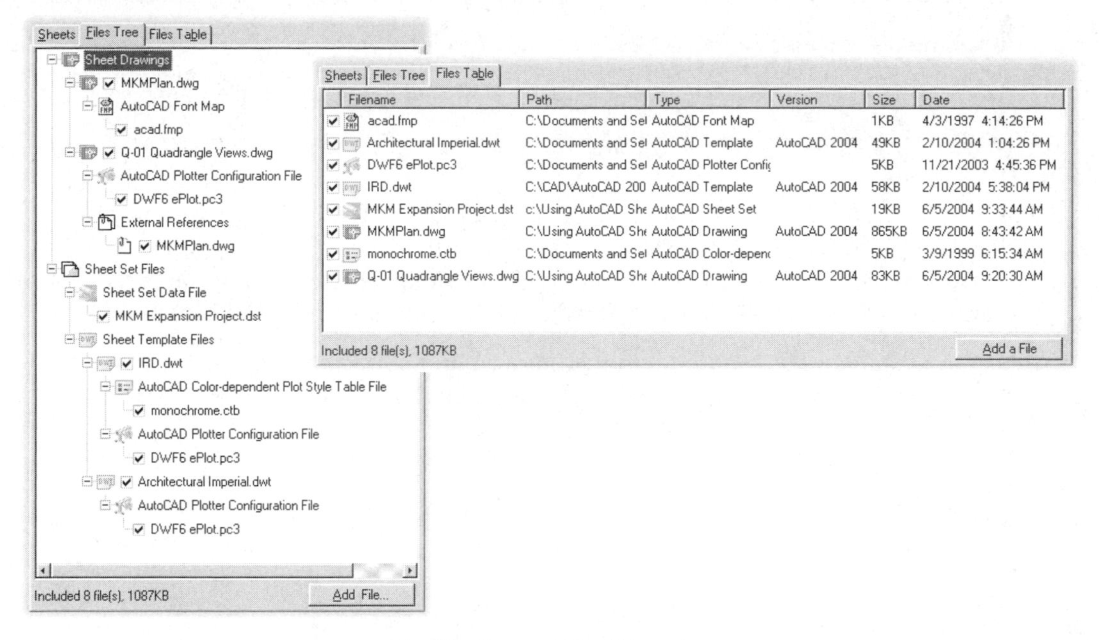

The Files Tree and Files Table tabs.

To remove files from the archive, click the checkbox beside the file name so that the checkmark disappears.

To add files, click the **Add File** button, and then select files from the Add File to Archive dialog box.

3. To specify archiving parameters, click the **Modify Archive Setup** button.

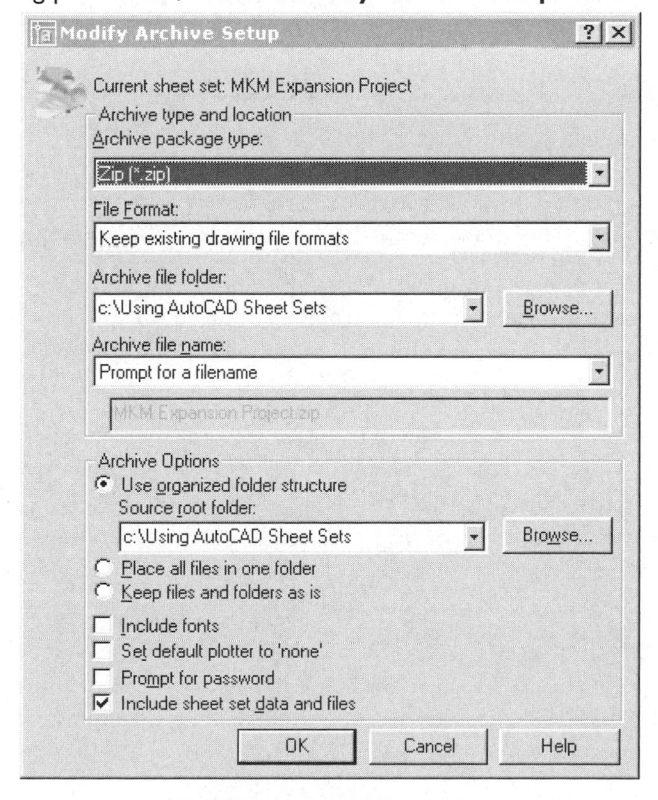

Specifying options for the archive.

Some of the more important options include:

Archive package type — saves as compressed *.zip* files, executable PkZip files, or in folders.

File format — saves in AutoCAD 2004 or 2002 format.

Archive options — keeps folder structure, or moves all files into a single folder.

4. Click **OK** to exit the dialog boxes. If prompted, provide a file name for the archive.

After the archive is created, store the files in a secure location. For instance, you may want to make copies, and then store them in several locations — in your office, at an off-site location, and on an FTP site on the Internet. By making multiple copies (at least three), you are assured that at least one copy survives disasters, such as fire or flood, and accidental or even deliberate erasure.

Ensure that you are able to read the archive. If stored as a PkZip file, archive a copy of WinZip or other software capable of reading *.zip* files. If stored in DWF format, include a copy of DWF Viewer.

THE FUTURE OF SHEET SETS

The Sheet Set Manager is one stage of a master plan that Autodesk calls "lifecycle management." (Other CAD vendors call it PLM — product lifecyle management.)

The master plan is to manage the products over their lifetime, from the initial concept, on to design and construction, through to maintenance, and finally to their destruction. You won't see it all at once, because PLM is a huge software programming project that will take years to complete.

Some elements are already in place. DesignCenter provides visual access to drawings located anywhere, as well as parts of drawings, such as blocks, layers, and linetypes. The **SECURITYOPTIONS** command provides simple password and digital-rights protection. The **PUBLISH** command plots groups of drawings. (See *Using AutoCAD: Basics* for information about these commands.) The Sheet Set Manager now provides a way to gather and control these groups of drawings.

In the future, Autodesk is expected to add its Vault technology to AutoCAD, for tracking reusable blocks and drawings. Other enhancements may include greater control over who can view, print, and edit drawings. I imagine that the Sheet Set Manager of the future will work with non-AutoCAD documents, images, and drawings.

EXERCISES

1. Create a new folder called *\my sheet set*. From AutoCAD's *\sample* folder, copy the drawing files beginning with "8th floor."
   Use these drawings to create a sheet set named "Floor 8 Renovations."

2. Use the **VIEW** command to create several named views in one of the drawings.
   Add the views to a new, blank sheet.

3. Publish the sheet set in DWF format.

4. Archive the sheet set in ZIP format.

CHAPTER REVIEW

1. Describe the purpose of the Sheet Set Manager.
2. What is the difference between the **OPENSHEETSET** and **SHEETSET** commands?
3. How would you keep the Sheet Set Manager from opening automatically?
4. How do you open a drawing listed in the Sheet Set Manager?
5. List at least three things that must be done before creating a sheet set:
   a.
   b.
   c.
6. What are *.dst* files?
7. Why would you store *.dst* files on a disk drive accessible over the network?
8. What is the purpose of a *sheet selection*?
9. When would you use the **ARCHIVE** command?
10. Views are created with the **VIEW** command. In which space must the drawing be before named views can be defined?
11. If the sheet set seems out of date, what are three things you can do to update it?
    a.
    b.
    c.
12. When are categories used?
    When are subsets used?
13. Place the following category names in AutoCAD's sort order:
    T-1
    T-32
    T-12
    T-2

Accessing the Internet

You are probably already familiar with many uses for the Internet, such as sending email (electronic mail) and browsing the Web (short for "World Wide Web"). Email lets users exchange messages and documents at very low cost. The Web brings together text, graphics, audio, and movies in an easy-to-use format.

AutoCAD allows you to interact with the Internet in several ways. It can open and save files located on the Internet. It includes a simple Web browser, and can also use the default browser on your computer. It saves drawings in DWF format for viewing on Web pages.

Sometimes you need to share drawings with other offices and clients. AutoCAD provides several commands for publishing drawing sets, packaging drawings for transmittal by email, and creating Web pages for displaying drawings.

In this chapter, you learn to work with these Internet-related tools:

SEND opens email client software, and attaches the current drawing.

ETRANSMIT packages drawings and support files for transmittal by courier or Internet.

PUBLISH eplots collections of drawings (changed in AutoCAD 2005).

PUBLISHTOWEB creates Web pages for posting drawings on Web sites.

BROWSER opens the default Web browser with the specified Web page.

HYPERLINK places and edits links to documents in drawings.

HYPERLINKSOPTIONS sets options for using hyperlinks in AutoCAD.

HYPERLINKOPEN, BACK, FWD, and **STOP** navigates hyperlinks in drawings.

FINDING THE COMMANDS

On the **STANDARD** and **WEB** toolbars:

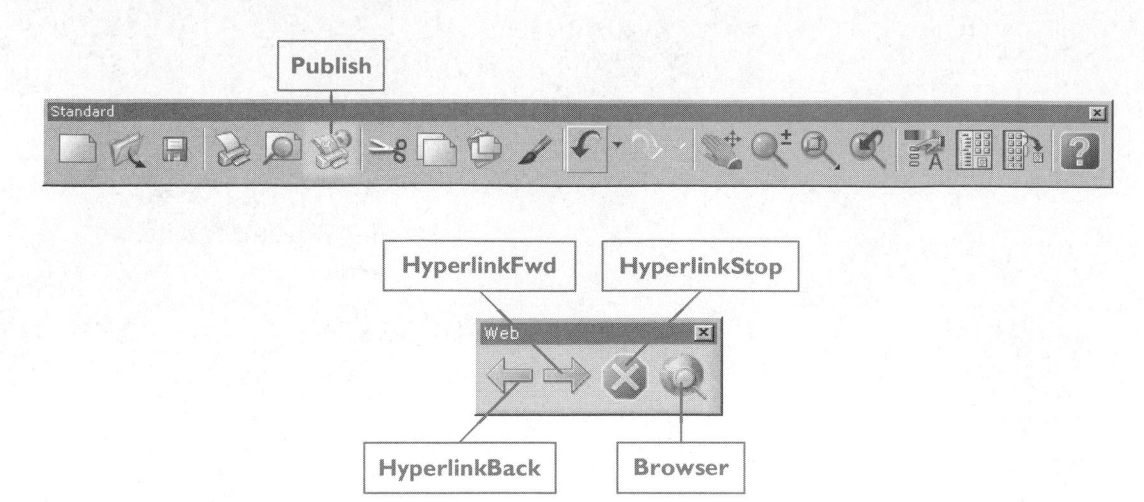

On the **FILE** and **INSERT** menus:

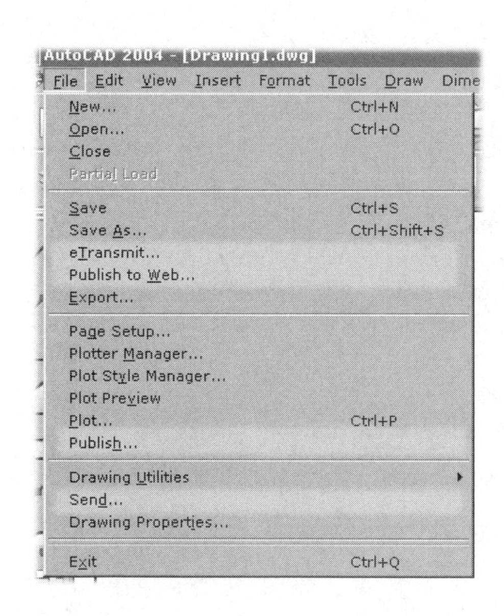

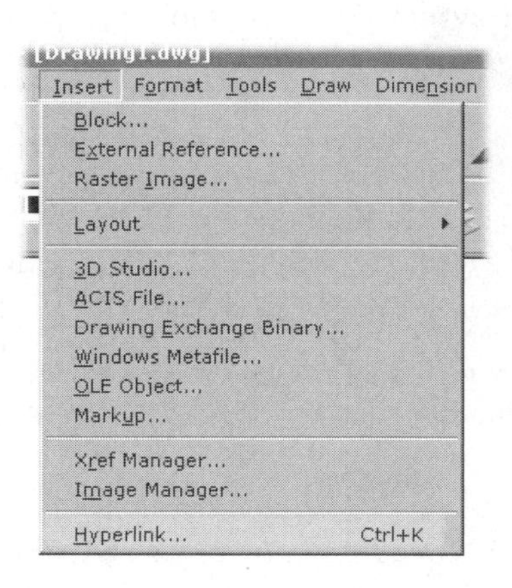

SEND

The **FILE | SEND** menu selection opens your computer's email client, and then attaches the current drawing as an attachment. (I have tested this command to work with Eudora and Outlook Express; it may not work with other email clients, such as Web-based mail services.)

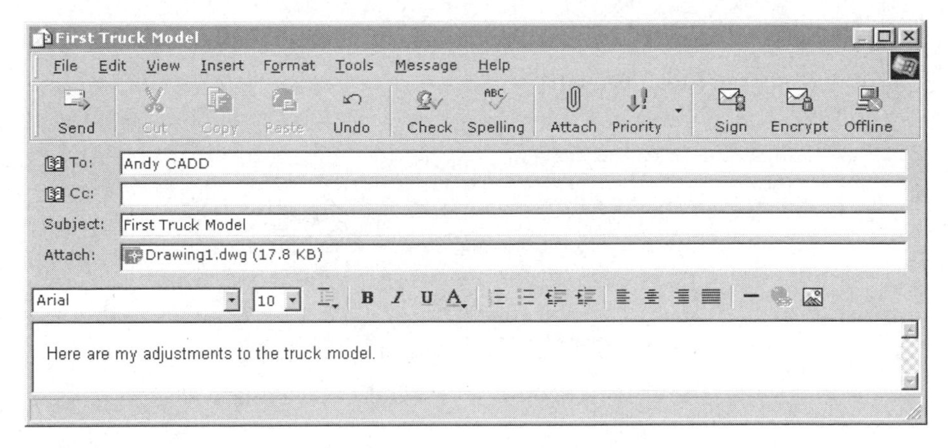

Send attaches the current drawing to email messages.

When the New Message window appears, fill out the TO: and SUBJECT: fields, include a note, and then click **Send**.

The advantage of this command is that it is quick. The drawback is that it does not include support files, such as xrefs and font files. When you need to include support files, use the **ETRANSMIT** command.

ETRANSMIT

The **ETRANSMIT** command packages drawings and support files for transmittal by courier or Internet.

Emailing a drawing is not quite as easy as emailing, say, a Word document. That's because an AutoCAD drawing is not an island; it usually connects with a number of other files. These files include fonts, font mapping, raster images, and externally-referenced drawings.

Thus, Autodesk added the **ETRANSMIT** command to check for additional files that may need to accompany drawing files. This command performs the following tasks:

- Finds all the files associated with the drawing.
- Collects the files into a single compressed file, a self-extracting file, or a folder of files; it also allows you to lock the collection with a password.
- Strips paths from xrefs and image files (optional).
- Provides an area for you to enter notes for the recipient.
- Saves the drawing in AutoCAD 2000 or 2004 format.
- Generates a Web page with a link to the files, instead of sending the package by email.
- Produces a report, which includes instructions to the recipient.

(Historically, the **ETRANSMIT** command was introduced with AutoCAD 2000i as an expanded version of the **Pack'n Go** command, a "bonus" command provided with AutoCAD Release 14. Many Internet-related commands were added to AutoCAD 2000i — hence the "i" tacked on to the end of "2000." Earlier, Autodesk began adding Internet features halfway through Release 14, with the addition of DWF export and hyperlinking.)

BASIC TUTORIAL: PACKAGING DRAWINGS FOR EMAIL

For this tutorial, start AutoCAD with the *truckmodel.dwg* file found on the CD-ROM.

1. To package a drawing and its support files for distribution by email, start the **ETRANSMIT** command:
   - From the **File** menu, choose **eTransmit**.
   - At the 'Command:' prompt, enter the **etransmit** command.

     Command: **etransmit** *(Press* ENTER.*)*

   If the drawing is not saved, AutoCAD asks you do so now. (This ensures that AutoCAD is aware of all changes made to the drawing.)

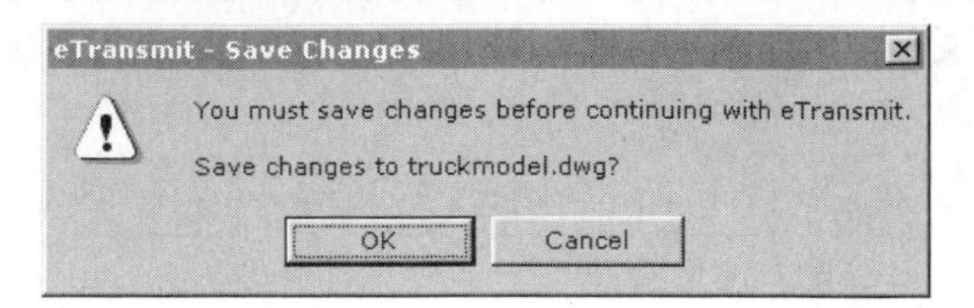

Changed drawings must be saved before etransmitting.

AutoCAD displays the Create Transmittal dialog box.

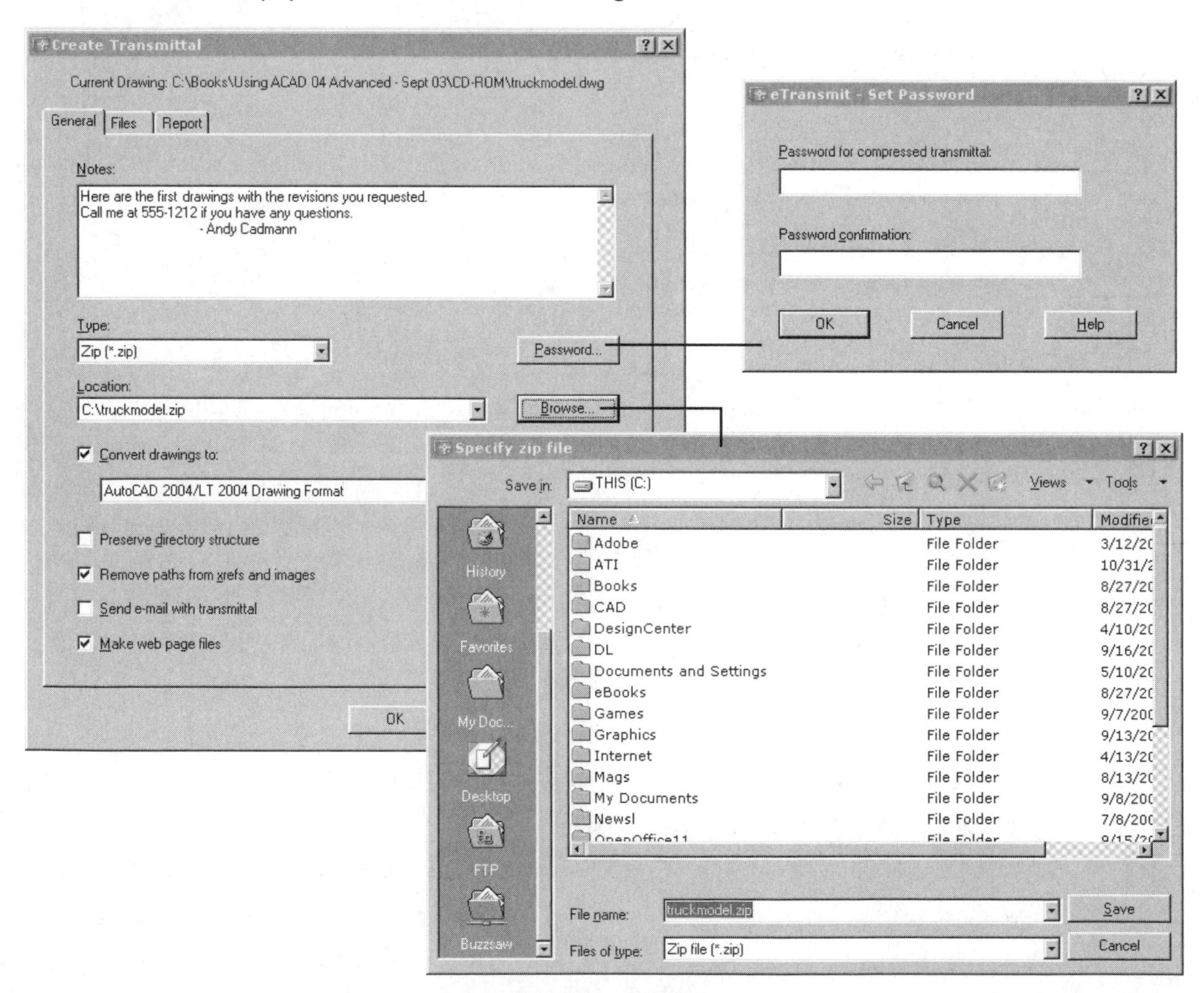

Create Transmittal dialog boxes.

2. In the **Notes** section, type a message to your recipient, such as:

    Here are the first drawings with the revisions you requested.
    Call me at 555-1212 if you have any questions.

    - Andy Cadmann

    You can copy text (**CTRL+C**) from other documents, and then paste (**CTRL+V**) them into the Notes field.
3. In the lower half of the dialog box, select these options:

Type:	**Zip (*.zip)**
Password:	*(Leave blank)*
Location:	**C:**
Convert drawing to:	**AutoCAD 2004**
Preserve directory structure:	☐ *(No)*
Remove paths from xrefs and images:	☑ *(Yes)*
Send email with transmittal:	☐ *(No)*
Make Web page files:	☑ *(Yes)*

Type specifies how files are compressed:

Zip (*.zip) means the files are compressed into a single *.zip* file. (You do not need the PkZip or WinZip programs on your computer; AutoCAD includes the compression function.)

Self-extracting executable means that the files will be compressed into a single file with the extension *.exe*. The email recipient double-clicks the file to extract (uncompress) the files.

The benefit: recipients don't need the PkUnzip or WinZip software on their computer.

The drawback: viruses could hide in the *.exe* file; even if there are no viruses, some email and anti-virus systems may reject your message, because it contains the *.exe* attachment.

Folder (set of files) means the files are not compressed, but are stored in a single folder. This is the preferred option when sending the files on a CD-ROM or other storage media.

Password locks the transmittal so that only people who know the password can extract the files. It provides a simple level of security. Be careful how you transmit the password.

Location specifies where the transmittal file(s) should be stored on your computer. The C:\ location is handy for finding it easily. You may, however, want a separate folder to store all transmittals.

Convert drawing to saves the drawing in the current (2004) or previous (2000, 2000i, and 2002) versions of AutoCAD. The benefit to saving in 2000 format is that clients with older versions of AutoCAD and AutoCAD LT can read the file; the drawback is that objects specific to 2004 might be erased or modified to a simpler format.

Preserve directory structure means that files are extracted to the same folders as they were collected from. This option should be turned on only when recipients have an identical folder structure on their computers to your own.

Remove paths from xrefs and images means recipients do not have to worry about xrefs and images being located in a specific folder.

Send email with transmittal launches Windows' built-in email software automatically. If you use an alternative email program, turn off this option.

Make Web page files generates a Web page. This lets recipients download the transmittal files with their Web browser. This is useful in several cases: (1) when recipients don't have email access, such as when out in the field; (2) when the recipients email system blocks them from downloading the file (due to limits on file size, or virus/spam blocking); or (3) when you want transmittals available on a more public basis. Recall that the files can be locked via the password to prevent unauthorized access.

4. Click the **Files** tab.

   Notice the tree list of associated files. AutoCAD determines that the drawing depends on other files. (Imagine figuring out that list on your own.)

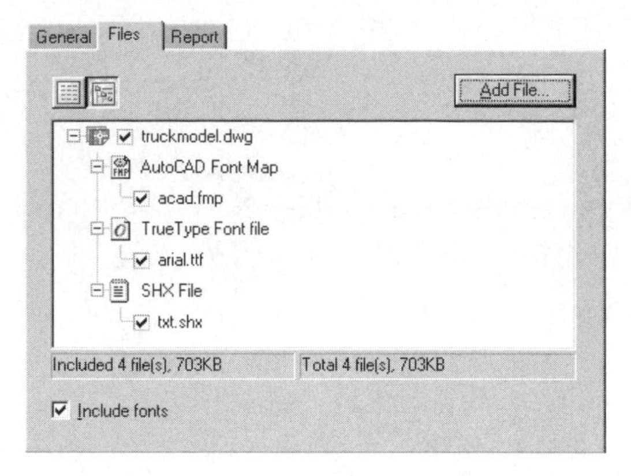

Files tab lists the support files required by the drawing.

   Near the bottom of the dialog box you can view the summary:

   Included 4 file(s), 703KB

   The size given, 703KB, is the uncompressed total; once the files are compressed, the file size should become smaller.

Add File allows you to include additional files with the transmittal. These files could be other documents, spreadsheets, and digital photographs.

Include Fonts toggles whether font files (*.shx* and *.ttf*) are included in the transmittal. Normally, you leave on this option. You would turn off the option for these reasons: (1) some fonts you use are be copyrighted, and thus cannot be copied legally; (2) the recipient already has the fonts; and (3) not including the fonts saves some file space, which takes less time to transmit.

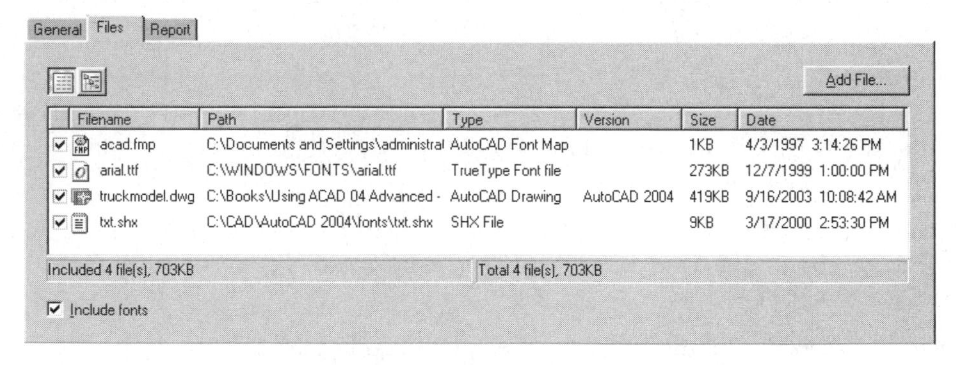

Filename	Path	Type	Version	Size	Date
acad.fmp	C:\Documents and Settings\administrat	AutoCAD Font Map		1KB	4/3/1997 3:14:26 PM
arial.ttf	C:\WINDOWS\FONTS\arial.ttf	TrueType Font file		273KB	12/7/1999 1:00:00 PM
truckmodel.dwg	C:\Books\Using ACAD 04 Advanced -	AutoCAD Drawing	AutoCAD 2004	419KB	9/16/2003 10:08:42 AM
txt.shx	C:\CAD\AutoCAD 2004\fonts\txt.shx	SHX File		9KB	3/17/2000 2:53:30 PM

List view shows details of each file in the transmittal set.

List View shows details of the files included with the transmittal. The check box next to each filename allows you to include (☑) and exclude (☐) files.

5. Click the **Report** tab.

Notice that AutoCAD automatically generates a report describing the files being sent, how to deal with external files (xrefs and so on), and other notes. Scroll through the report to familiarize yourself with its contents.

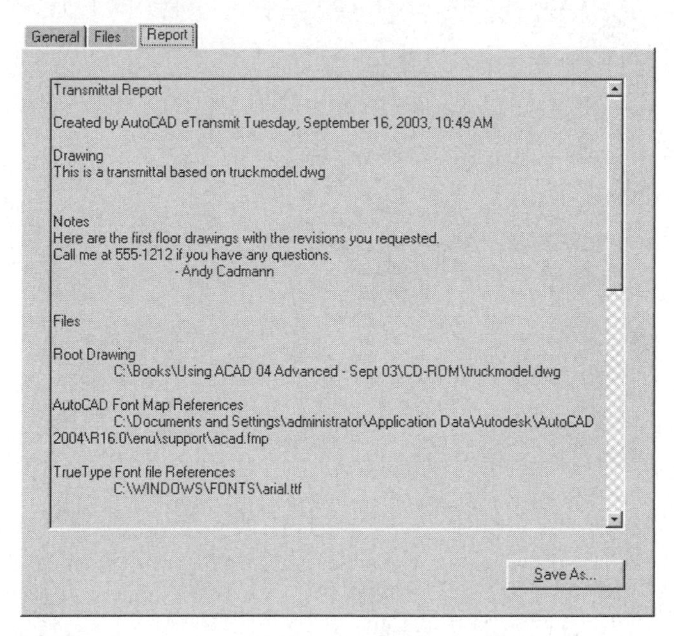

The Report tab provides instructions to the recipient regarding the content of the transmittal.

The report is included in the transmittal package as *truckmodel.txt.* To save the report as a text file for yourself, click **Save As**.

6. Choose **OK**.

   AutoCAD spends a few seconds putting together the transmittal package, as well as the Web page.

7. When done, you have three choices for sending the transmittal:

   - Email the files to your recipient as attachments to the message.

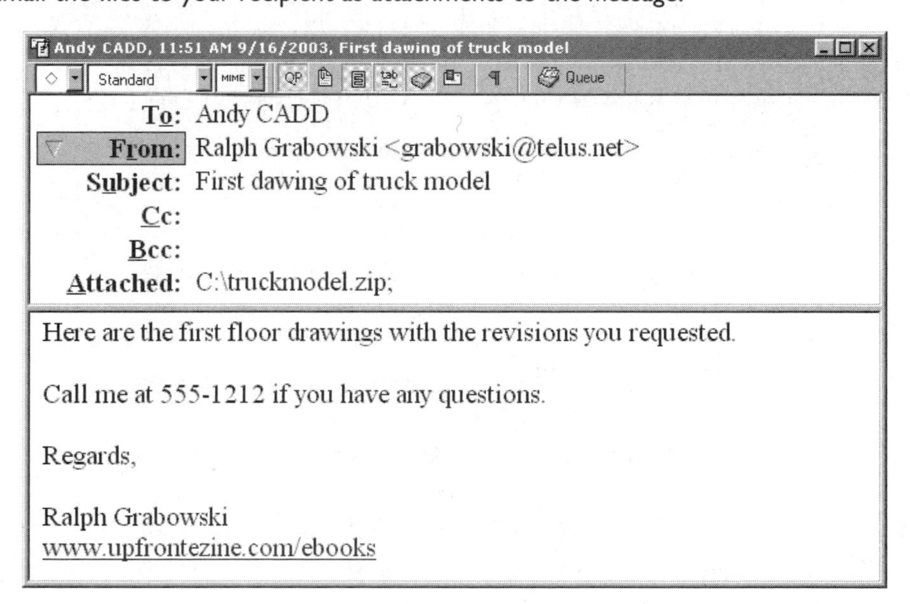

Attach .zip or .exe file to email message.

   - Upload the files to a Web site via FTP (short for "file transfer protocol"). Your recipients then use their Web browsers to access the files.

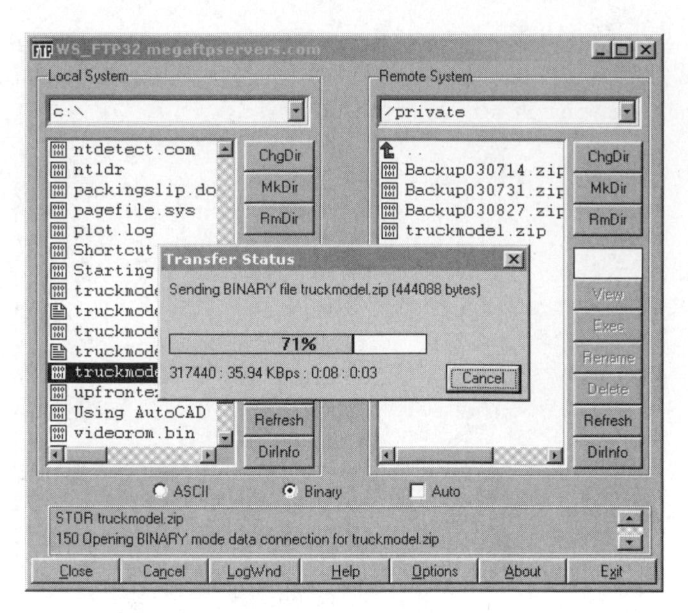

FTP is the primary method for uploading files to Web sites.

- Or copy the files to diskette, CD-R disc, or Jaz cartridge, and then send it by courier to the recipient.

Notes: Before using this command, AutoCAD requires that you save the drawing.

Very large drawings can take a long time to send or receive over slow Internet connections, such as modems.

Including TrueType fonts (TTF) is a touchy issue, because making copies might infringe on the owner's copyright. Many TrueType can be freely copied, such as those included with the AutoCAD package; others, however, cannot be copied without first paying a license fee.

Recipients sometimes have difficulty opening files sent by email. Files are occasionally corrupted travelling through email systems. First, try to solve the problem by resending the files to the recipients. If they continue to have problems, you may need to change a setting in your email software. For example, try changing the attachment encoding method from BinHex or Uuencode to MIME.

If recipients cannot open the transmittals, confirm whether they uncompressed ("un-zipped") the files.

FTP

FTP is a method of sending and receiving files over the Internet (short for "file transfer protocol"). You may be familiar with sending files through email at *attachments*. The drawback is that email was never designed to handle very large files, and some email providers limit attachments to 1.5MB or smaller.

FTP allows you to send files of any size, even entire CDs and DVDs of files — 650MB, 4.7GB, and more. FTP is the method used to upload Web pages to Web sites. The drawback is that FTP needs to be initially set up (with a user name, password, and other data).

Clicking on **FTP** in the Places list of file-related dialog boxes results in an empty list. You need to first add a site, as described by the following tutorial.

TUTORIAL: SENDING DRAWINGS BY FTP

1. In the **SAVEAS** command's Save Drawing As dialog box, choose **Tools**, and then choose **Add/Modify FTP Locations**.

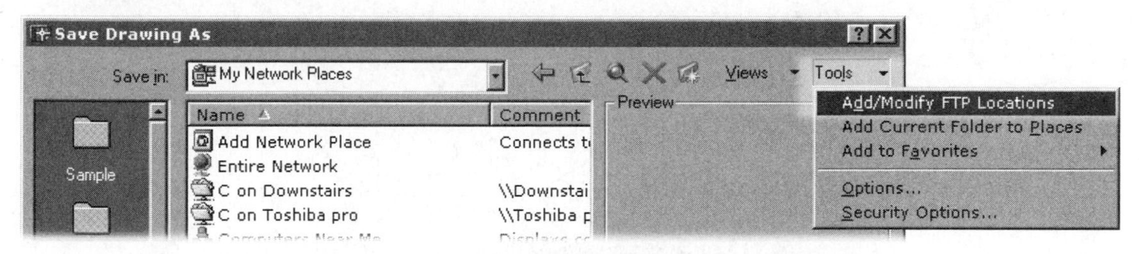

Creating a new FTP location.

AutoCAD displays the Add/Modify FTP Locations dialog box.

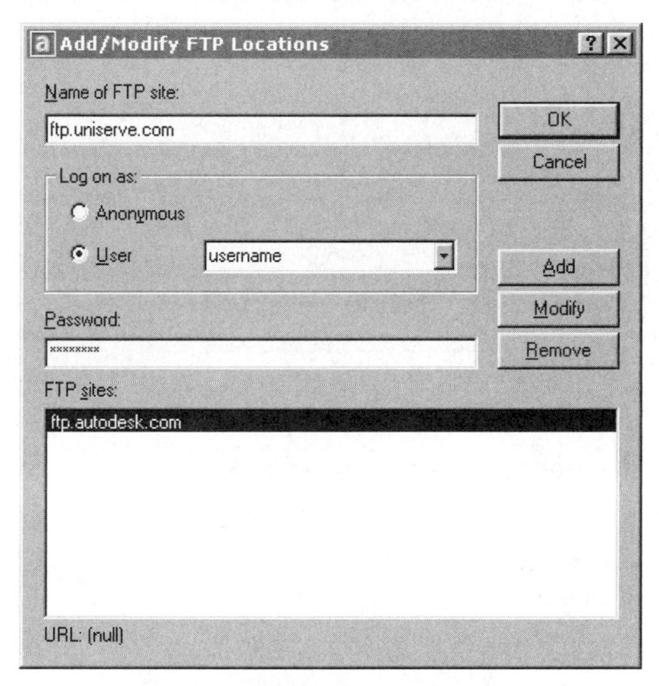

Add FTP Locations dialog box.

There are two types of FTP sites: anonymous and password.

- *Anonymous* sites allow anyone to access them (also known as "public FTP sites"). All you need is to provide your email address as the password; a username is not needed.
- *Password* sites require that you have a username and a password, usually provided by the person running the FTP site.

2. Enter the details for the FTP site. For example, enter the following parameters to access Autodesk's public FTP site:

Name of FTP site:	**ftp.autodesk.com**
Logon as:	**Anonymous**
Password:	**email@email.com**

3. Click **Add**, and then click **OK**.

   The FTP item has been created. Notice that AutoCAD adds the FTP site to the list in the dialog box.

4. To save the drawing, double-click the FTP site.

   Notice that the dialog box displays a list of folders.

5. Select a folder, and then click **Save**.
   A dialog box shows the progress as AutoCAD transfers the file to the site. Expect it to be slower than saving to disk.

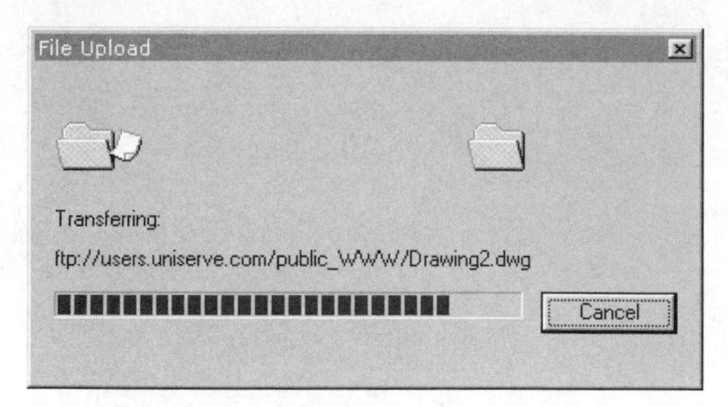

6. Some sites require that you enter a *username* and *password* before accepting files. If so, a dialog box appears requesting the information.
   Fill in the missing information, and click **OK**.

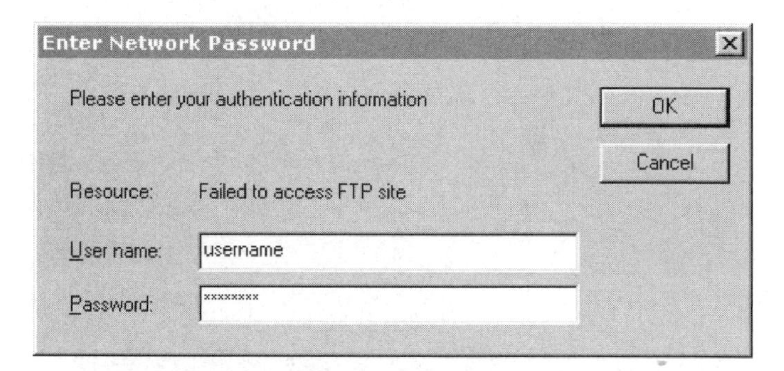

ADDITIONAL OPTIONS: RECEIVING ETRANSMITTALS

In this tutorial, you learn how to "read" the transmittal.

1. To open the transmittal, double-click the file:
   - If the file is named *truckmodel.**exe***, it opens itself.

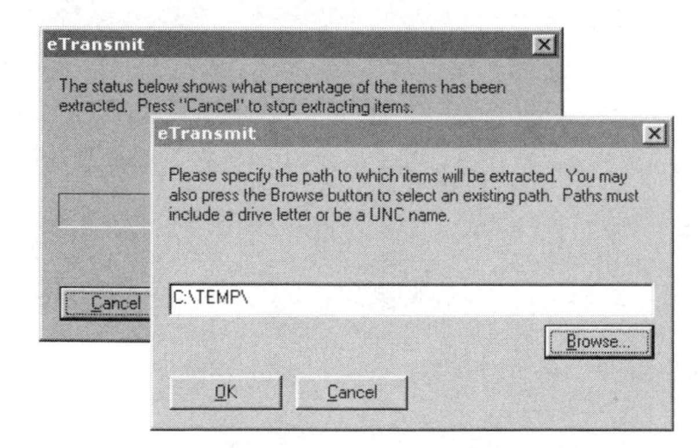

Select a folder in which to unpack the files.

   A dialog box asks where to extract the files. I suggest selecting a temporary folder, such as C:\temp, or setting up a folder for all incoming transmittals specific to projects.
   Choose **OK**.
   - If the file name has an extension of ".zip", such as *truckmodel.**zip***, your computer needs the PkUnzip or WinZip programs to open the file. (WinZip is US$29 from www.winzip.com.)

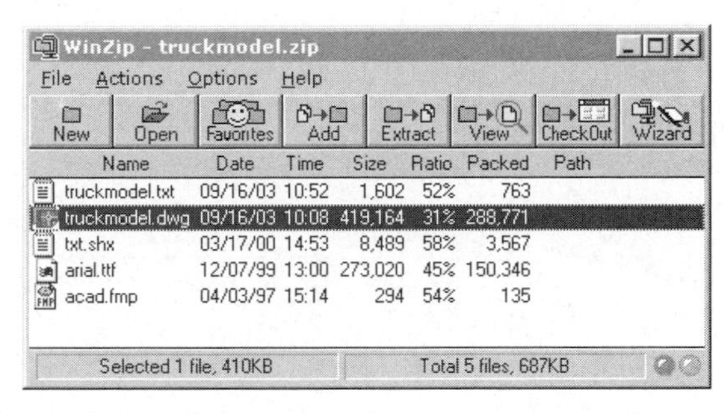

WinZip displays the files stored in the compressed transmittal file.

Click **Extract**, and then specify **All Files**. Specify a folder in which to place the files. Alternatively, drag the files from WinZip to a folder in Window Explorer.

2. Use Windows Explorer to view the contents of the folder to which you extracted the transmittal files.
   - To read the report file, double-click *truckmodel.txt.*
   - To open the drawing in AutoCAD, double-click *truckmodel.dwg*.
   - To use a Web browser to view the Web page generated by the **ETRANSMIT** command, double-click the *truckmodel.htm* file. The page contains a couple of links:

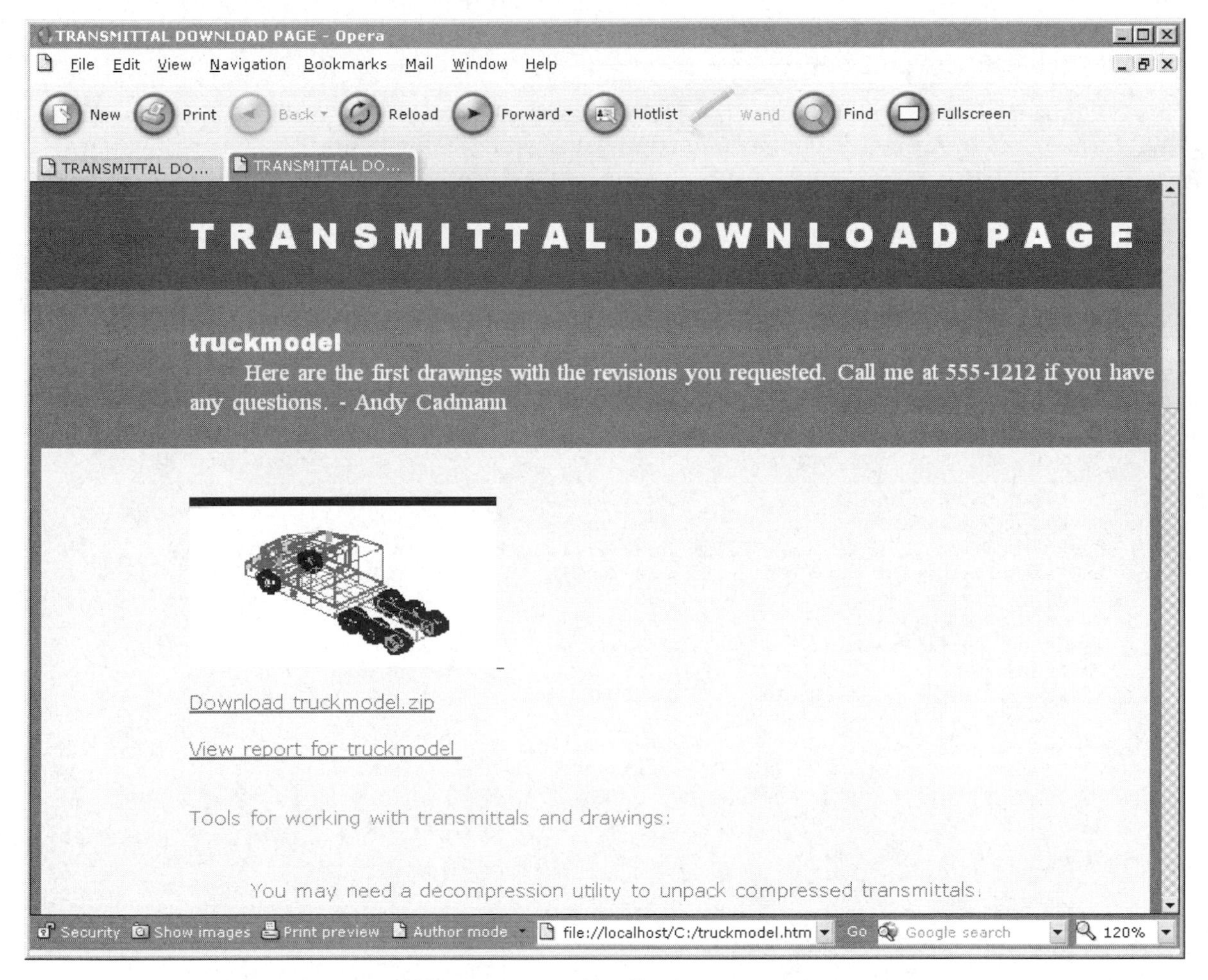

Web page generated by eTransmit command.

Download *truckmodel.exe* (or *.zip*) downloads the compressed transmittal file to the folder you specify. (Alternatively, click the image of the drawing.) Follow steps 1 and 2 above to uncompress the files.

View report for *truckmodel* scrolls down the Web page to show the rest of the report.

Technically, AutoCAD generates the *.htm* file using these template tags in the form of HTML comments:

Template Tag	***Meaning***
##ETRANSMIT##NAME##	Transmittal name.
##ETRANSMIT##NOTES##	Transmittal notes.
##ETRANSMIT##IMAGE##	BMP image of drawing.
##ETRANSMIT##HREF##	Hyperlink to *.zip* or *.exe* package file.
##ETRANSMIT##REPORTLINK##	Text and link to *.txt* report file.
##ETRANSMIT##UNZIP##	Text and link to unzip utility program.
##ETRANSMIT##REPORT##	Transmittal report.

PUBLISH

The **PUBLISH** command plots and "e-plots" collections of drawings.

A drawing *set* consists of two or more drawings and layouts. They are plotted in order during the same plot session. Engineering and architectural offices often use drawing sets to combine all drawings belonging to projects.

This command allows you to create drawing sets, and then export them in DWF format. Once in DWF format, the drawing set can be emailed to other offices and clients. They view the drawing set with Autodesk's free DWF Viewer software. If they have DWF Composer, they can also mark up (redline) the drawing set with their comments, and then return them to you. (This command also plots drawings by using printers and plotters.)

AutoCAD 2005 changed the user interface of the Publish dialog box, with icons replacing buttons and a general rearrangement.

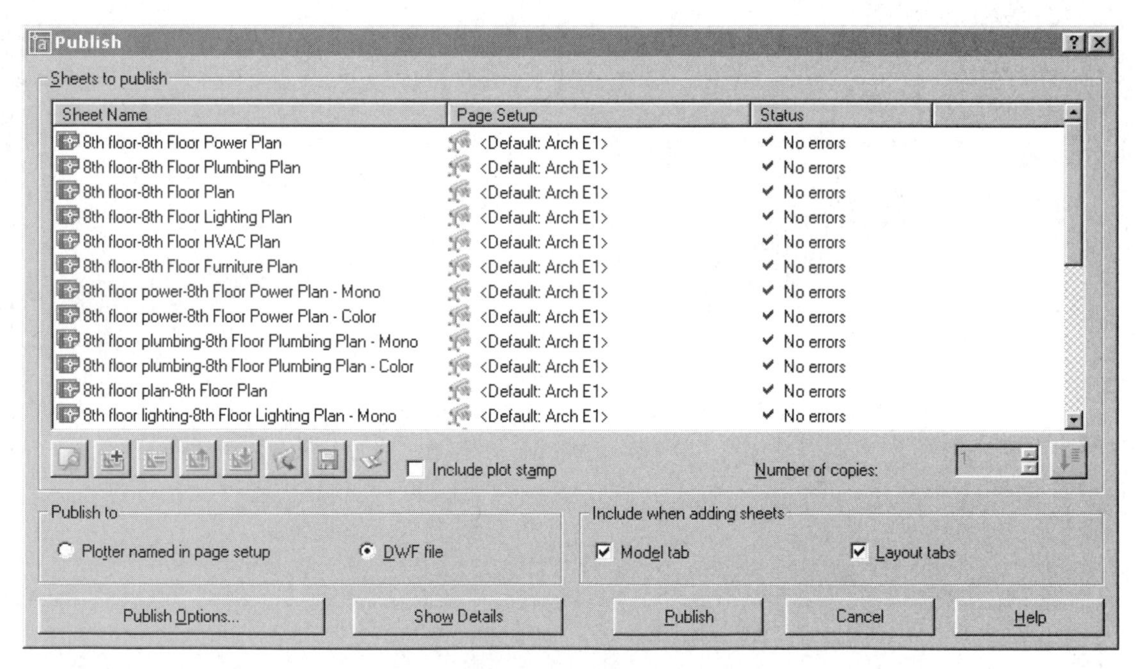

Names of drawings and layouts that will be plotted as sets.

TUTORIAL: PUBLISHING DRAWING SETS

1. To set create drawing sets, start the **PUBLISH** command:
   - From the menu bar, choose **File**, and then **Publish**.
   - From the **Standard** toolbar, choose the **Publish** button.
   - At the 'Command:' prompt, enter the **publish** command.

   Command: **publish** *(Press* **ENTER***.)*

   In all cases, AutoCAD displays the Publish Drawing Sheets dialog box. Notice that it lists the names of the drawings and layouts currently open in AutoCAD.
2. To add more drawings, click **Add Sheets**. Select one or more *.dwg* files — in the manner of the **OPEN** command — and then click **Select**.
3. Drawing sets are usually plotted in a specific order, so use the **Move Up** and **Move Down** buttons to change the order of the layouts. (Click **Remove** to remove them from the list.)
4. Drawing sets are published in DWF format for transmittal by email or posting to Web site. Each drawing or layout is a page in the *.dwf* file. In the Publish To section, select **Multi-sheet DWF file** to save all the layouts in a single *.dwf* file. You have the option of protecting the file with a password to stymie unauthorized viewing.
5. If you want to save the list of drawing and layout names for reuse, click **Save List**.
6. Click **Publish**.

   AutoCAD loads each drawing, and then generates the page.

Note: When layouts are not initialized (have no plotters assigned), AutoCAD displays an error message and does not generate that sheet. If a drawing or layout cannot be found, AutoCAD skips it and publishes the next layout.

When done, AutoCAD asks if you want to view the *.dwf* file in the DWF Viewer software provided with AutoCAD 2005.

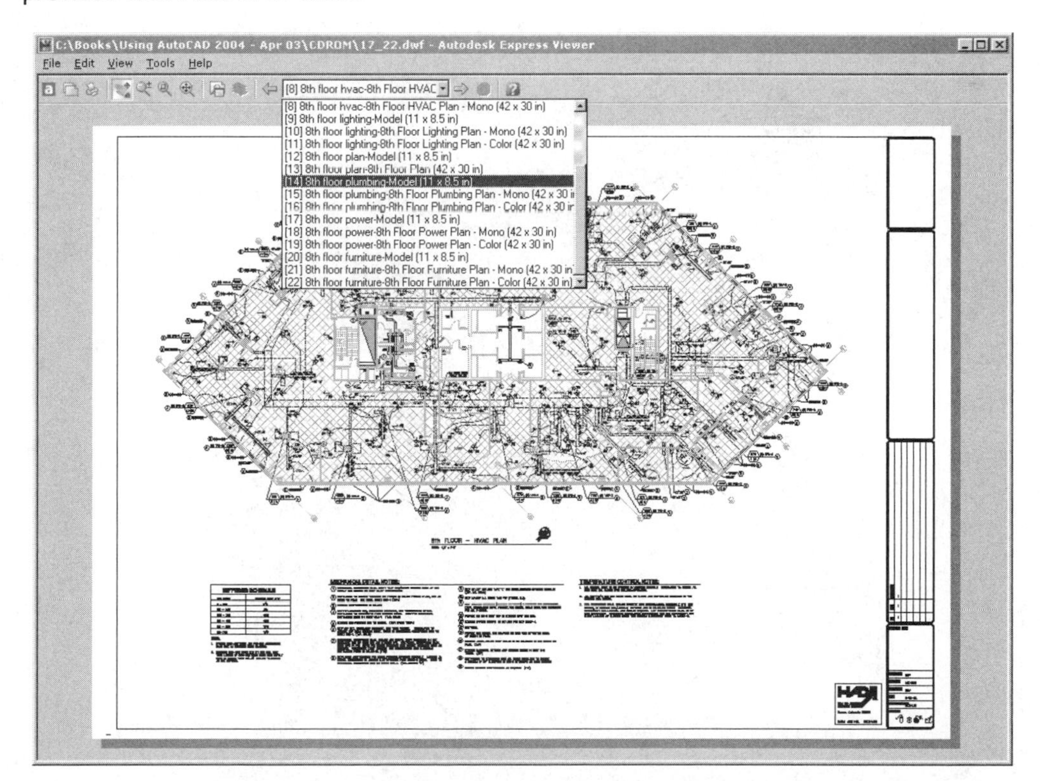

DWF Viewer displaying a drawing set generated in DWF format.

Setting DWF Properties

Files saved in DWF format can have a limited number of properties, such as differing levels of compression. These properties are set with the Plotter Manager.

1. From the **File** menu, select **Plotter Manager**.
2. Double-click "DWF6 ePlot.pc3." This is the file that configures the *.dwf* files output by the **PLOT**, **PUBLISH**, and **PUBLISHTOWEB** commands.
3. Click the Device and Document Settings tab.

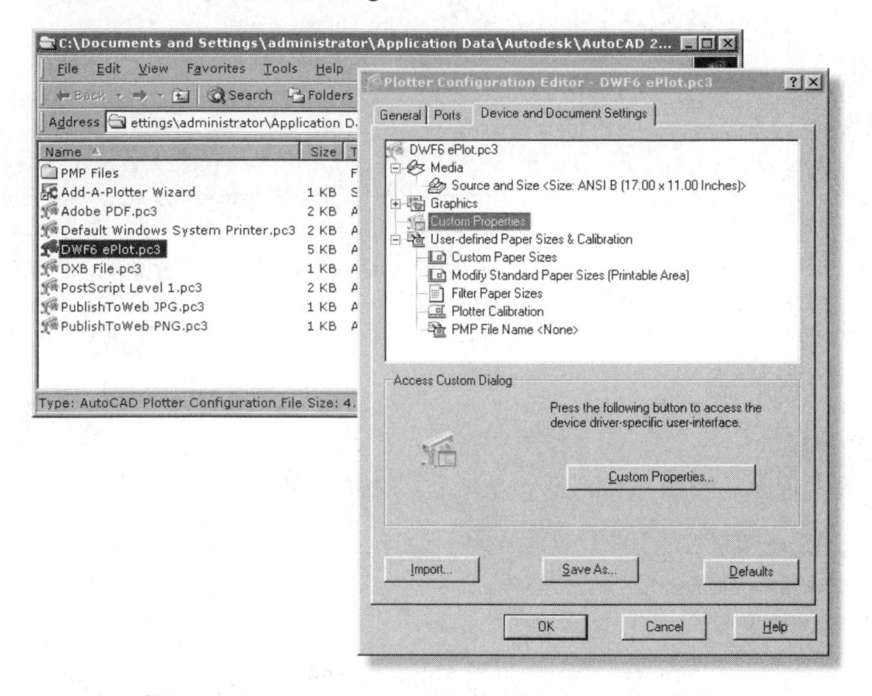

AutoCAD sets plotter settings in .pc3 files, including those for DWF output.

4. In the tree list, select **Custom Properties**.
5. In the lower half of the dialog box, choose the **Custom Properties** button to see the options available for DWF creation.

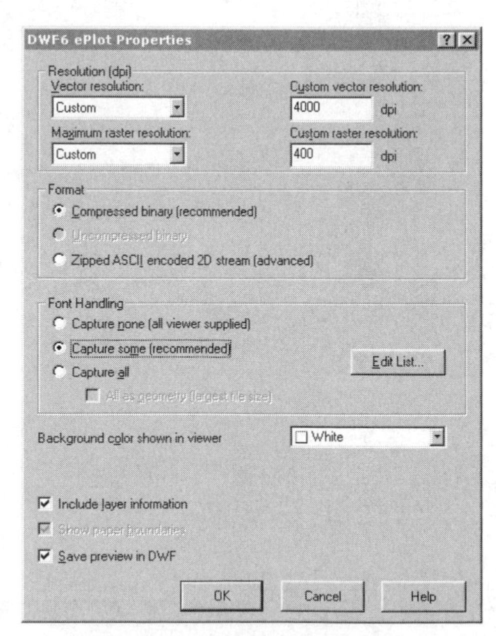

Properties of DWF that can be changed.

ABOUT THE DESIGN WEB FORMAT

To display AutoCAD drawings on the Internet, Autodesk invented the "design Web format" (DWF). DWF has benefits and drawbacks over using drawing files (DWG).

DWF is compressed to make it as much as eight times smaller than the original *.dwg* drawing file, so that it takes less time to transmit over the Internet, particularly vital with relatively slow telephone modem connections. The DWF format is more secure, because the original drawing is not being sent and viewed. Unlike raster images (JPEG and GIF), DWF contains vector data, and can be zoomed and panned. Layers can be toggled, and the drawing printed.

DWF has, however, some drawbacks. You must go through the extra step of translating from DWG to DWF. Be aware that earlier versions of DWF do not handle paper space objects (version 2.x and earlier), or linewidths and non-rectangular viewports (version 3.x and earlier). The current release as of this book is DWF v6.

Future releases of DWF will support 3D drawings, as well as rendered and shaded drawings.

To view DWF images on the Internet, your Web browser needs a *plug-in* (a software extension). The plug-in should have been added to your browser when AutoCAD was installed on your computer. If not, Autodesk makes the plug-in freely available as part of the DWF Viewer software at www.autodesk.com/dwfviewer. It's a good idea regularly to check for updates to the DWF plug-in, which is updated about twice a year.

DWF Viewer, a free stand-alone viewer, displays and prints DWF files. DWF Composer is needed to mark up DWF files. DWF Writer is a printer driver that allows any software to produce *.dwf* files.

Resolution. You can specify a range of resolution separately for vector and raster images. (Technically, vector images are independent of resolution, but DWF files are based on integers, unlike AutoCAD, which uses real numbers. The resolution defines the number of integers used to provide the level of desired accuracy.)

The raster resolution is used for raster images in drawings, which includes images placed with the **IMAGEATTACH** command, OLE objects pasted from Clipboard, and AutoCAD-generated renderings.

Resolution is stated as *dpi* (short for "dots per inch"). DWF allows the following range of resolution:

Resolution	Vector	Raster
Minimum	150	150
Default	400	400
Maximum	100,000,000	400

Autodesk recommends that the resolution match that of your printer. Check your printer's Setup dialog box for the available resolution. Keep in mind that the higher the resolution, the better the print quality, but the longer the printing time.

Printer	Typical Resolution Range
Inkjet	300 - 2400 dpi
Laser	300 - 1200 dpi

For viewing DWF files with DWF Viewer, Autodesk recommends a vector resolution of 2400 dpi, and discourages resolution settings higher than 40,000 dpi. Setting the vector resolution changes the raster resolution automatically to its limit of 400 dpi.

The figure shows close-ups of a drawing exported in DWF. The portion at left was saved at the minimum resolution of 150 dpi and shows some bumpiness. The portion at right was saved at 2400 dpi. (DWF Viewer was unable to open a *.dwf* file generated at 100 million dpi.)

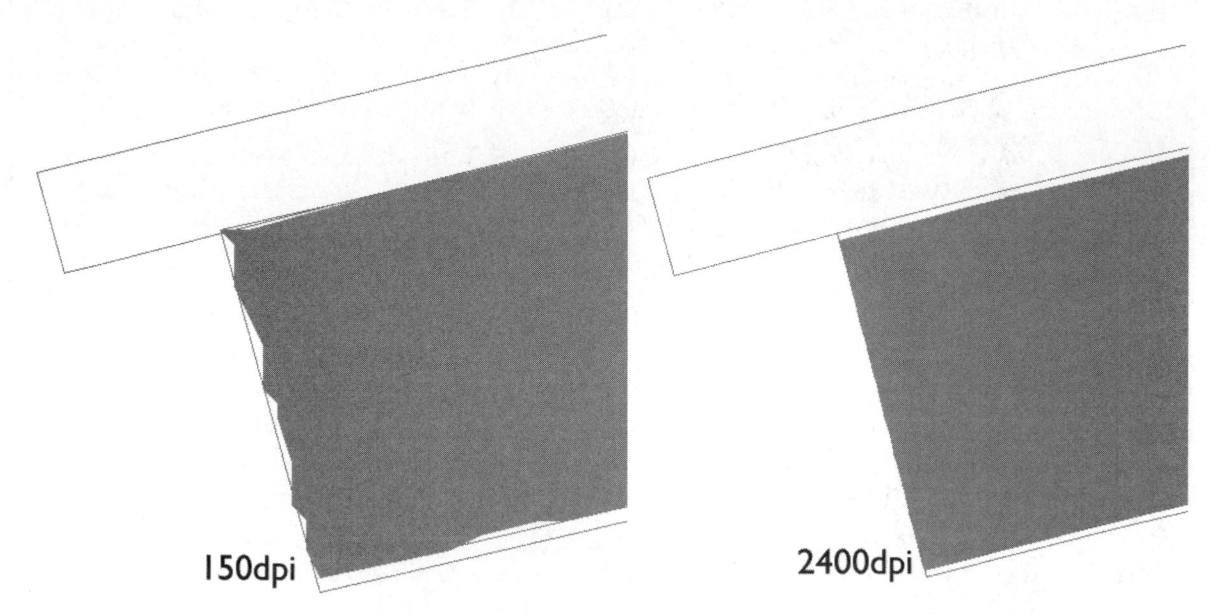

***Left**: Detail of drawing output as DWF at 150 dpi.*
***Right**: And at 2400 dpi.*

Also significant is the difference in file size. When creating a DWF from the *Hummer Elevation.dwg* file (462KB), the size varies depending on the resolution:

Resolution	***DWF File Size***
150 dpi	55KB
2400 dpi	57KB
100,000,000 dpi	4,138KB

Format selects the type of compression to reduce the size of the DWF file:

- **Compressed Binary** is recommended by Autodesk, saving the drawing using lossless compression in a binary format
- **ASCII coded 2D stream** saves the drawing in ASCII format, and then compresses it in Zip format. Use PkUnzip or Winzip to uncompress the file. This option is recommended if you need to read the file output.
- **Uncompressed Binary** is not available in this release of AutoCAD.

Font Handling determines how fonts are included:

- **Capture None** means that no fonts are included in the *.dwf* file. This option allows the file to be smaller, and assumes that the necessary fonts are available on all computers where the file will be viewed. This is a valid assumption when drawings use fonts included with AutoCAD, or the three standard TrueType fonts (Arial, Courier New, and Times New Roman). When needed fonts are missing, DWF Viewer substitutes an available font, so that text can still be read.

- **Capture Some** is recommended by Autodesk, and includes only those fonts that are not usually found on computers running the Windows operating system. Fonts installed with Windows, AutoCAD, DWF Composer, and DWF Viewer are not included by default. Click the **Edit List** button to determine which TrueType fonts will (and won't) be included; changes made here affect all *.dwf* files.
- **Capture All** means all fonts found in the drawing are included in the DWF. This ensures that all text is displayed with the correct font.
- **All as Geometry** converts the fonts to geometry. This ensures the *.dwf* file can be displayed on any system, including those that do not support TrueType fonts.

Background Color Shown in Viewer selects the color to use as the background, although white is probably the best background color.

Include Layer Information includes layers, which allows you to toggle layers off and on. Only layers turned on and thawed are included. Turn off this option if you do not want layer settings tampered with. Autodesk suggests turning off this option for better performance when working with xrefs.

Show Paper Boundaries cannot be turned off in AutoCAD 2005.

Save preview in DWF stores a thumbnail view, much like those for drawings. Autodesk recommends turning on this option when DWFs are stored at its www.buzzsaw.com Web site.

PUBLISHTOWEB

The **PUBLISHTOWEB** command creates Web pages for posting drawings on Web sites.

You may want to create a central Web site that provides drawings for all contractors working on a project. Or, your firm may want to advertise its drafting and design abilities. Posting examples of your drawings to a Web site allows others to see your work. Or, you may want to share drawings and symbol libraries with others in your office (via a local intranet) and your clients (via an extranet). The **PUBLISHTOWEB** command allows you to do this easily.

This command is a "wizard" that takes you through the stages to output drawings as Web pages. (If you wish to place hyperlinks in drawing, read about them later in this chapter.)

BASIC TUTORIAL: CREATING WEB PAGES OF DRAWINGS

1. To convert drawings into Web pages, start the **PUBLISHTOWEB** command:
   - From the **File** menu, choose **Publish to Web**.
   - At the 'Command:' prompt, enter the **publishtoweb** command.
   - Alternatively, type the alias **ptw** at the keyboard.

   Command: **publishtoweb** *(Press* ENTER.*)*
2. In all cases, AutoCAD displays the Publish to Web wizard:
   - **Create new Web page** guides you through creating a new Web page.
   - **Edit existing Web page** guides you through editing a Web page previously created by this wizard.

   Select **Create new Web page**, and then click **Next**.

Note: If you change the settings for the Template file location in the Options dialog box, this wizard might not work. The Template folder *must* contain the *\ptw template* folder, its five subfolders, and all their files.

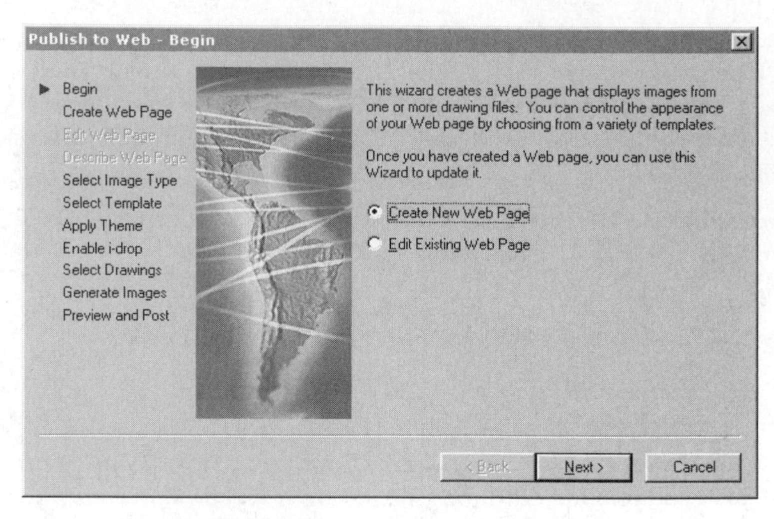

The PublishToWeb command creates and edits Web pages.

3. For this exercise, enter the following options:

Specify the name of your Web page:	**Using AutoCAD**
Specify the parent directory:	**C:\temp** *(To find it more easily.)*
Provide a description:	**Example Web page created by AutoCAD**

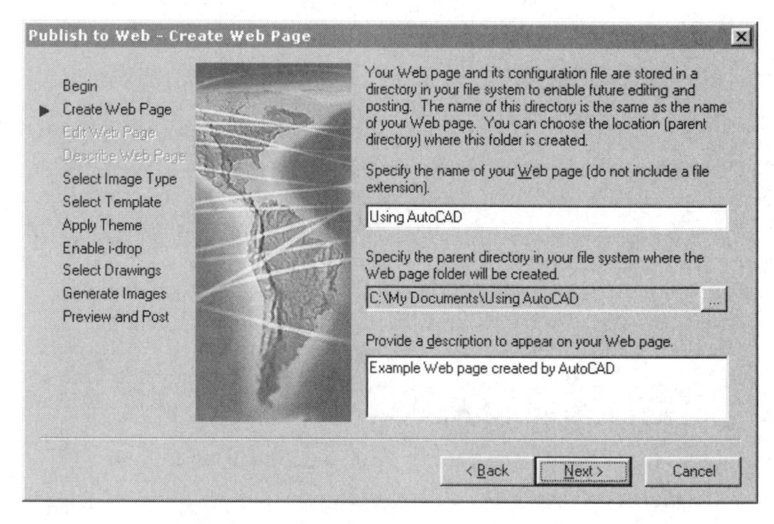

Specify the name and description of the Web page.

To later edit the Web page, AutoCAD stores parameters in a file found in a folder, whose name varies depending on the version of Windows your computer employs. In this step, AutoCAD asks for the name it will give the file.

On the Web page, the **name** appears at the top, while the **description** appears just below.

Click **Next**.

4. Select an image type from the drop list:
   - **DWF** (drawing Web format) is a vector format that displays cleanly, and can be zoomed and panned; Web browsers can only display DWF if the appropriate plug-in is first installed.
   - **JPEG** (joint photographic experts group) is the raster format that all Web browsers display; it may create *artifacts* (details that don't exist).
   - **PNG** (portable network graphics) is a raster format that does not suffer the artifact problem; some older Web browsers do not display PNG.

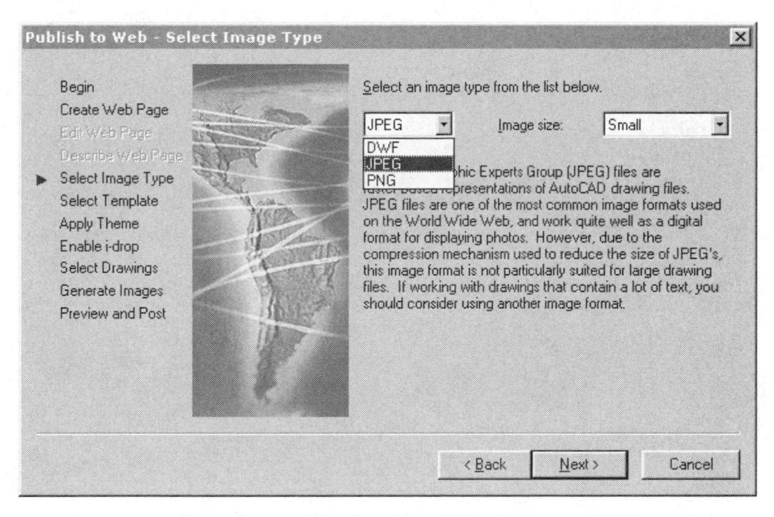

Drawings are represented by images in JPEG, PNG, or DWF formats.

Larger images provide more detail, but take longer to transmit. The size of raster image available for JPEG and PNG are:

Image Size	***Resolution***	***Equivalent Megapixels***
Small	789 x 610	0.5
Medium	1009 x 780	0.8
Large	1302 x 1006	1.3
Extra Large	1576 x 1218	2.0

For this exercise, select **JPEG**.
From the Image size drop list, select **Small**.
Click **Next**.

5. The Select Template page provides pre-designed formats for the Web page.
   Select **List plus Summary**, and then choose **Next**.

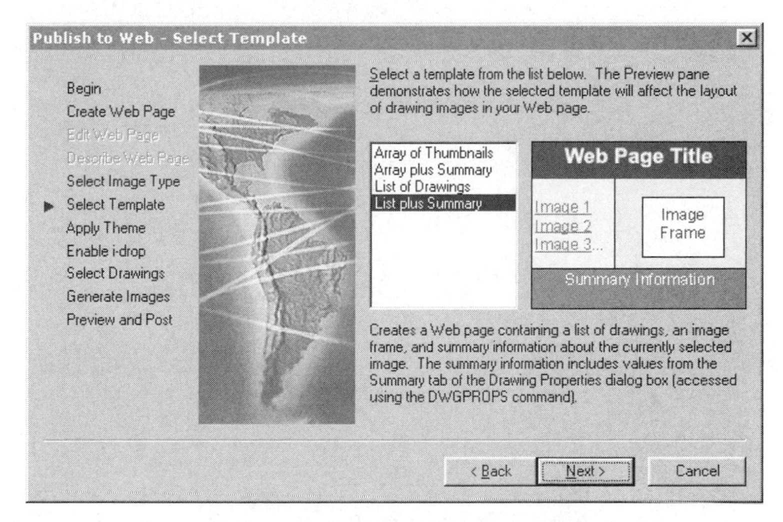

Select one of the four Web page styles.

6. Select a theme to apply to the Web page, such as "Autumn Fields."
   Click **Next**.

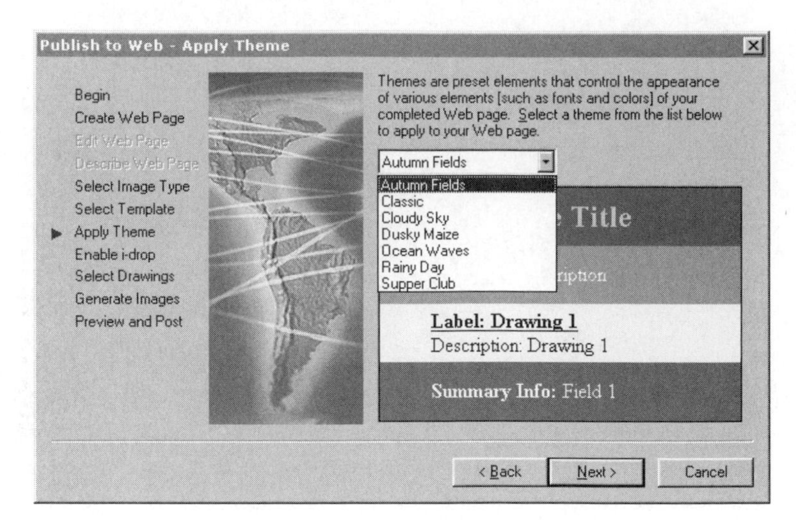

Select a Web page theme.

7. As an option, you can enable Autodesk's "i-drop" feature in the Web page. For this exercise, leave the **Enable i-drop** box unchecked, and then click **Next**.

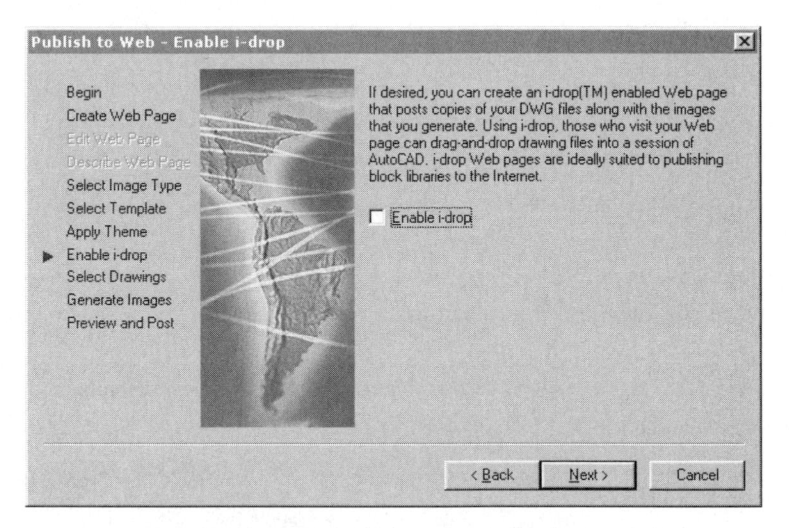

Enable — or disable — i-drop technology.

8. Select the drawing(s) and specify related parameters.

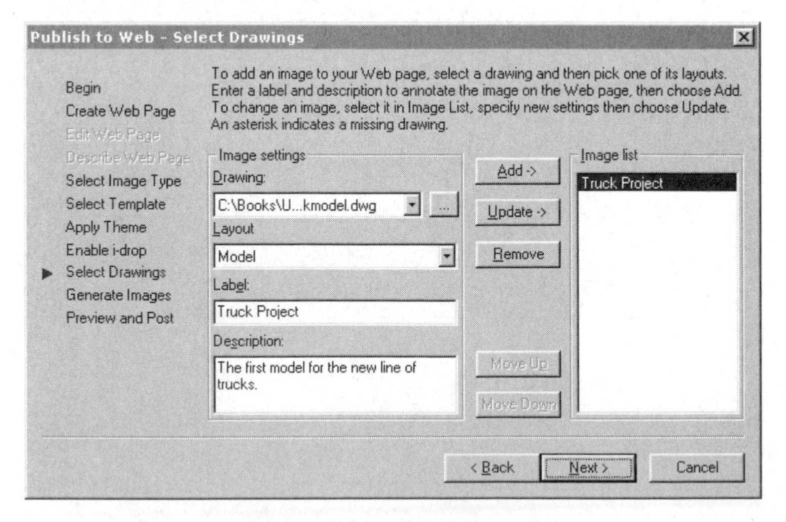

Select the drawings and layouts.

For this tutorial, enter the following:

Drawing: *truckmodel.dwg*

Layout: **Model**

Label: **Truck Project**

Description: **The first model for the new line of trucks.**

Drawing selects the drawing. If one or more drawings are open, their names are listed. You can also select drawings by clicking the **...** button, which displays a file dialog box.

Layout selects a model or a layout mode. If the drawing contains one or more layouts, they are listed here.

Label provides a name by which the drawing is known on the Web page. You could use the drawing's file name, or another more descriptive name.

Description provides a description that appears with the drawing on the Web page.

When done, click the **Add** button. Notice that drawings are referred to by their Label.

If you change your mind, you can remove drawings from the list, or move drawings up and down the list. When done, click **Next**.

9. To ensure the drawings are up-to-date, AutoCAD regenerates them. Select the **Regenerate all images** option, unless you have an exceptionally slow computer or a large number of drawings to process.
   Wait while AutoCAD regenerates the images.
   Click **Next**.

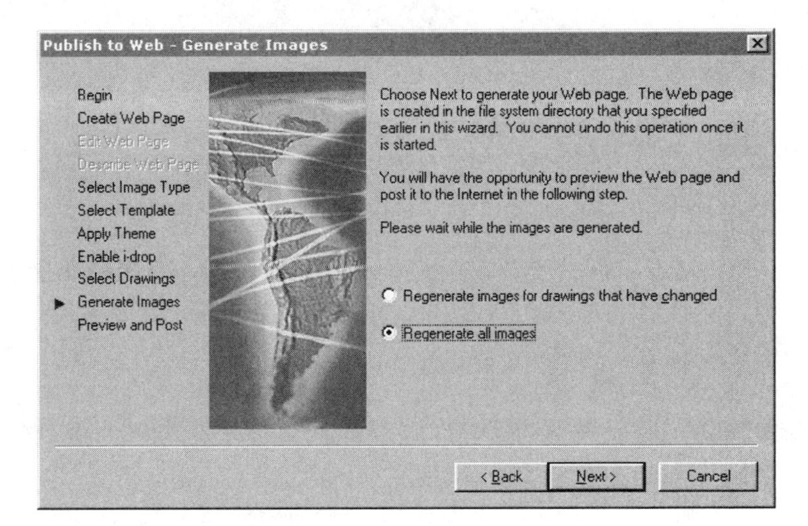

Regenerate images from drawings.

10. Click **Preview** to see what the Web page will look like. AutoCAD launches your computer's default Web browser.

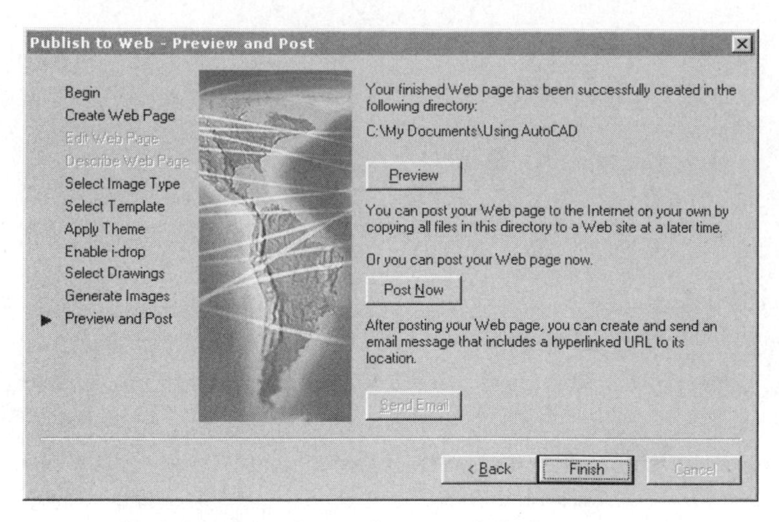

Check the results of your selections with the Preview button.

11. If the page is not to your liking, click the **Back** button, and make changes.

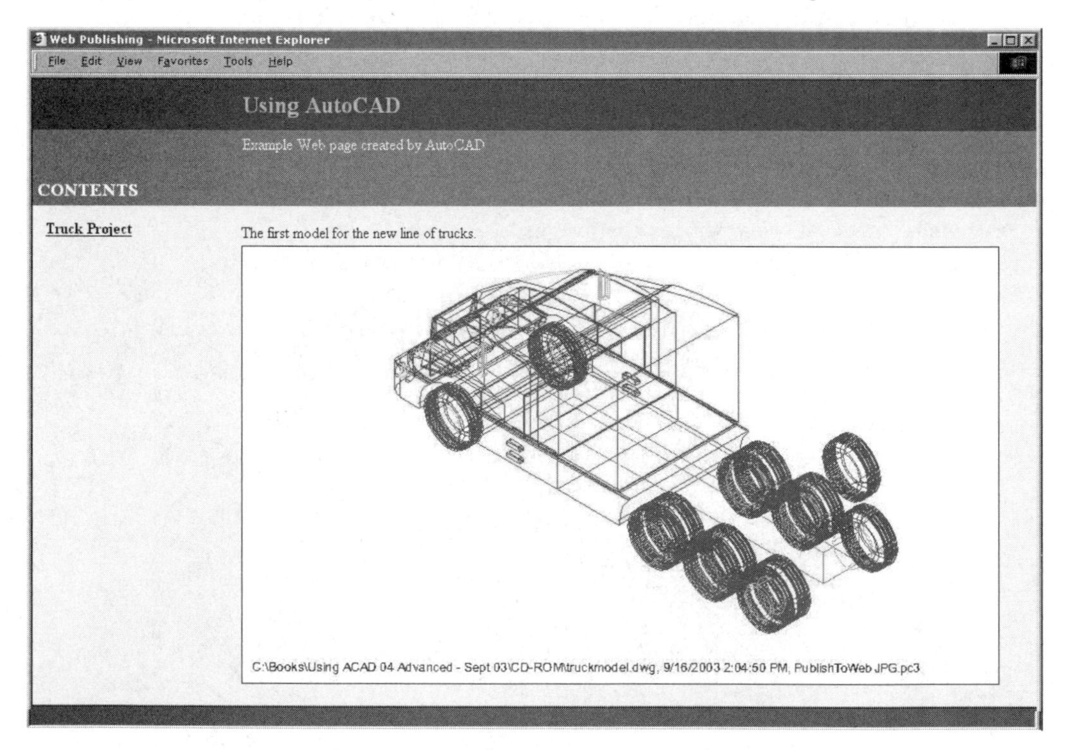

Previewing AutoCAD-generated Web page in browser.

Notes: When you click the **Preview** button to launch the Web browser, you might not see the drawing, because the HTML code created by AutoCAD requires that JavaScript be turned on. Make sure your Web browser's JavaScript option is turned on.

AutoCAD creates 17 files to support the Web page. Wondering which one to drag into a Web browser? Try *acwebpublish.htm*.

If the image is in DWF format, the browser displays it only when Autodesk's DWF Viewer software has been added to the Web browse. You can download the software free from www.autodesk.com/DWFviewer.

To change the view of a DWF image, right-click the image to see a shortcut menu that allows you to zoom and pan, toggle layers and named views, print, and so on.

The **Post Now** option works only if you have correctly set up the FTP (file transfer protocol) parameters. If so, AutoCAD directly uploads the HTML files to Web sites. If not, use a separate FTP program to upload the files from the folder defined in step 3, above.

If the Web page fails to display drawings and links, the problem is due to stricter error checking in Microsoft's Internet Explorer 6. (This problem occurs when you select the "List of Drawings" or "List Plus Summary" options.) To solve the problem, download updated templates from usa.autodesk.com/getdoc/id=DL404058.

12. Exit the We browser.
    In AutoCAD, click **Finish** to complete the wizard.

ADDITIONAL OPTIONS: EDITING WEB PAGES

In the previous tutorial, you used the Publish to Web wizard to create a Web page with a drawing. In this tutorial, you edit the Web page using the same wizard.

1. Start the **PUBLISHTOWEB** command.
2. Select the **Edit Existing Page** option, then click **Next**.
3. Notice that the wizard displays a list of previously-published Web pages.
   Select the Web project you created in the previous exercise, and then click **Next**.

If the Web page created by AutoCAD is not on the list, click **Browse**, and then select the *.ptw* project file. (If you cannot remember what the Web page looks like, click **Preview**. AutoCAD launches your computer's default browser, and displays the selected Web page.)

4. Change the title and the description of the Web page.
   If necessary, edit the wording, and then click **Next**.
5. Change the format of the Web page.
   Click **Next**.
6. Change the template, and then click **Next**.
7. Change the Web theme, and then click **Next**.
8. Toggle whether i-drop is enabled, and then click **Next**.
9. Here you can change the descriptions of drawings, as well as select different drawings.
   If you wish, make changes, and then click **Next**.
10. Wait for AutoCAD to regenerate the drawings, then click **Preview** to check the changes.
11. Exit the Web browser.
    In AutoCAD, click **Finish** to exit the wizard.

Notes: If you cannot find the Web page generated by AutoCAD, look for it in the folder named *\Documents and Settings\username\Application Data\Autodesk,* unless you specified a different folder.

The folder holds numerous HTML files generated by AutoCAD. The primary HTML file is called *acwebpublish.htm*. Drag this file into the Web browser to view the page and its drawing(s).

The Publish to Web wizard supports browsers that support JavaScript. I have successfully tested it with Opera 5-7, Netscape 4-6, and Internet Explorer 5.5-6. It may not work with browsers found on handheld devices, which tend not to support JavaScript.

BROWSER

The **BROWSER** command opens the default Web browser with the specified Web page.

This command starts a Web browser, and goes to the Web address you specify.

BASIC TUTORIAL: OPENING THE WEB BROWSER

1. To open the Web browser, start the **BROWSER** command:
   - At the 'Command:' prompt, enter the **browser** command.

     Command: **browser** *(Press* ENTER.*)*
2. AutoCAD prompts you to enter a Web address:

   Enter Web location (URL) <http://www.autodesk.com>: *(Type a Web address, or press* ENTER *to go to Autodesk's Web site.)*

   AutoCAD opens the Web address in your computer's default browser.

OPENING THE WEB BROWSER: ADDITIONAL METHODS

A number of AutoCAD commands access AutoCAD's build-in Web browser. In addition, a system variable stores values associated with this command.

- **OPEN** command opens drawings from the Internet or intranet.
- **INSERT** command accesses blocks (symbols) stored on the Internet.
- **INETLOCATION** system variable holds the default Internet location.

UNDERSTANDING URLS

URL (short for "uniform resource locator") is the file naming system of the Internet. URLs allow you to find any resource on the Internet. Examples of resources include text files, Web pages, program files, audio and movie clips — in short, anything you might also find on your own computer. The primary difference is that these resources are located on somebody else's computer. Some typical URLs are listed below:

Example URL	*Meaning*
http://www.autodesk.com	Autodesk Primary Web Site
news://adesknews.autodesk.com	Autodesk News Server
ftp://ftp.autodesk.com	Autodesk FTP Server
http://www.autodeskpress.com	Autodesk Press Web Site
http://www.upfrontezine.com	Editor Ralph Grabowski's Web site

The http:// prefix is not required, because most of today's Web browsers automatically add the routing prefix, which saves you a few keystrokes. (The // characters indicate a network address.) The following table gives templates for typing the URL to open a drawing file:

Drawing Location	*Example URL*
Web or HTTP Site	http://servername/pathname/filename.dwg
FTP Site	ftp://servername/pathname/filename.dwg
Local File	drive:\pathname\filename.dwg
Network File	\\localhost\drive:\pathname\filename.dwg

Let's look at them all.

INetLocation

The **INETLOCATION** system variable holds the default Internet location employed by the **BROWSER** command.

To change the default Web page that your browser starts with, change the **INETLOCATION** system variable.

Command: **inetlocation**

Enter new value for INETLOCATION <"http://www.autodesk.com">: *(Enter a URL, such as* **www.autodeskpress.com**.*)*

Open

The **OPEN** command can open drawings from the Internet (or your firm's intranet).

When a drawing is stored on the Internet, you access it from within AutoCAD using the standard **OPEN**, **INSERT**, **SAVE**, and other file-related commands. Instead of specifying the file's location with the usual drive-folder-file name format, such as *c:\autocad 2005\filename.dwg*, use the URL format. (Recall that the URL is the Internet's universal file naming system to access any file located on any computer hooked up to the Internet.)

In the upper right of the dialog box, notice the **Search the Web** button.

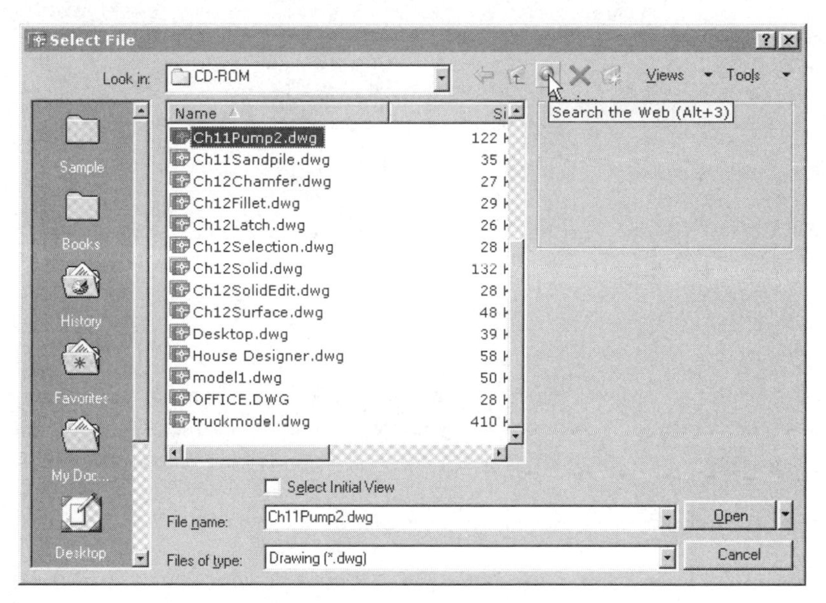

Open command accesses drawings from the Internet.

Click the **Search the Web** button, and AutoCAD opens the **Browse the Web** window, a simple Web browser that lets you to browse files at a Web site. (By default, this window displays the contents of the Web address stored by the **INETLOCATION** system variable.)

Along the top, the window has six buttons:

- **Back** returns to the previous Web address.
- **Forward** goes forward to the next Web address.
- **Stop** halts displaying the Web page, useful if the connection is slow or the page is very large.
- **Refresh** redisplays the current Web page.
- **Home** returns to the location specified by **INETLOCATION**.
- **Favorites** lists stored Web pages or bookmarks.

If you have previously used Internet Explorer, you find all your favorites listed here. Favorites are stored in the *\Favorites* folder on your computer.

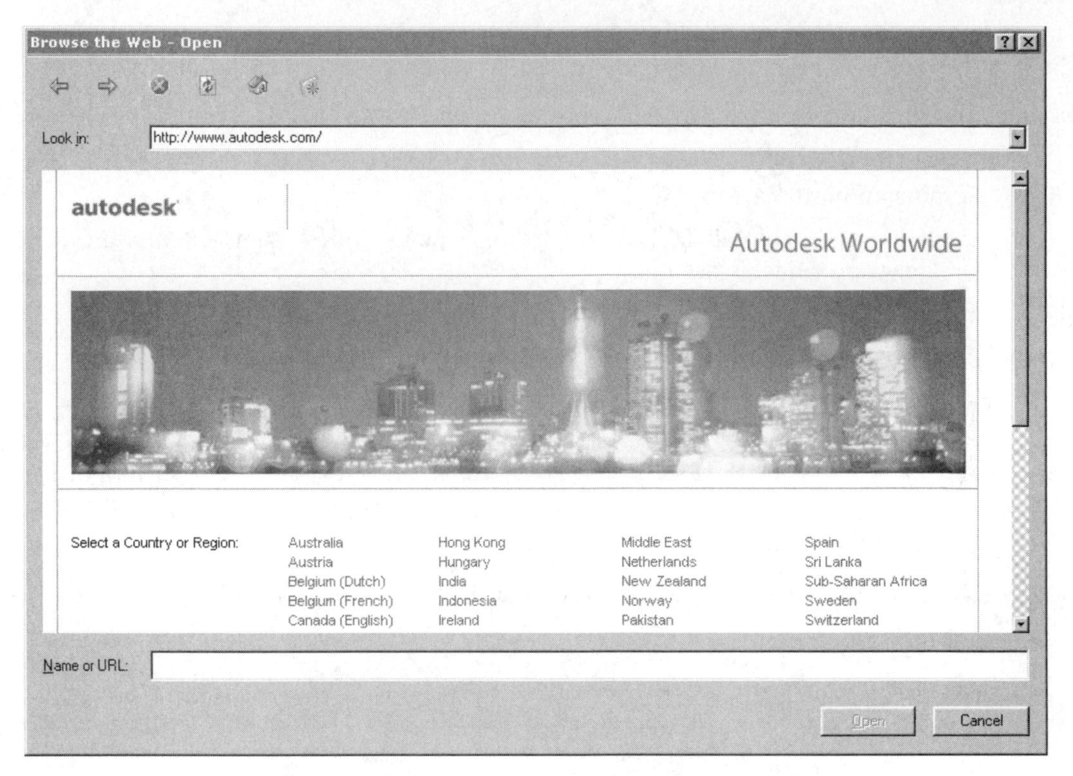

AutoCAD's built-in Web browser.

The **Look in** field allows you to type the Web address. Alternatively, click the down arrow to select a previous destination. If you have stored Web site addresses in the Favorites folder, then select a Web address from that list.

Notes: Opening a drawing from the Internet takes longer than opening it from your computer. During the file transfer, AutoCAD displays a dialog box to report the progress. If your computer uses a 56Kbps modem, you should allow about two to five minutes per megabyte of drawing file size. When your computer has access to a faster DSL or cable connection to the Internet, you should expect a transfer speed of about a half-minute to a minute per megabyte.

It may be helpful to understand that **OPEN** does not open drawings directly from the Internet. Instead, it copies the files to your computer's designated temporary folder, and then loads the drawing from the hard drive into AutoCAD. (This is known as "caching," which helps speed up the processing of the drawing, because the drawing file is now located on your computer's fast hard drive, instead of the relatively slow Internet.)

Insert

The **INSERT** command can access blocks (symbols) stored on the Internet.

When the Insert dialog box appears, choose the **Browse** button to display the Select Drawing File dialog box identical to the dialog box discussed above. After selecting the *.dwg* file, AutoCAD downloads it, and then continues with the **INSERT** command's familiar prompts.

The process is identical for accessing external reference (with the **XATTACH** command) and raster image files (with the **IMAGEATTACH** command). Other files that AutoCAD can access over the Internet include: 3D Studio, SAT (ACIS solid modeling), DXB (drawing exchange binary), and WMF (Windows metafile). All of these commands are found in the **Insert** menu on the menu bar.

Most other file-related dialog boxes allow you to access files from the Internet or intranet. This allows your firm or agency to have a central location that stores drawing standards. When you need to use a linetype or hatch pattern, for example, you access the *.lin* the*.pat* files over the Internet. More than likely, you would keep the location of those files stored in the Favorites list.

Support File	*Access Through*
Linetypes	From the menu bar, choose **Format | Linetype**.
	In Linetype Manager dialog box, choose **Load**, **File**, and **Look in Favorites**.
Hatch Patterns	Use the Web browser to copy *.pat* files from the remote location to your computer.
Multiline Styles	From the menu bar, choose **Format | Multiline Style**.
	In Multiline Styles dialog box, choose **Load**, **File**, and **Look in Favorites**.
Layer Names	From the menu bar, choose **Express | Layers | Layer Manager**.
	In the Layer Manager dialog box, choose **Import** and **Look in Favorites**.
LISP and ARX Apps	From the menu bar, choose **Tools | Load Applications**.
Scripts	From the menu bar, choose **Tools | Run Scripts**.
Menus	From the menu bar, choose **Tools | Customize Menus**.
	In Menu Customization dialog box, choose **Browse** and **Look in Favorites**.
Images	From the menu bar, choose **Tools | Displays Image | View**.

You cannot access text files, text fonts (SHX and TTF), color settings, lineweights, dimension styles, plot styles, OLE objects, or named UCSs over the Internet.

Save

The **SAVE** command can save the drawing to the Internet.

When you are finished editing a drawing in AutoCAD, you can save it to a file server on the Internet with the **SAVE** command. If you had inserted the drawing from the Internet (with the **INSERT** command) into a default drawing, such as *drawing1.dwg*, AutoCAD insists you first save the drawing to your computer's hard drive.

When a drawing of the same name already exists at that Web site, AutoCAD warns you — just as it does when you use the **SAVSEAS** command. Recall from the description of the **OPEN** command that AutoCAD uses your computer's temporary folder, hence the reference to it in the dialog box.

HYPERLINK

The **HYPERLINK** command attaches and edits links to objects a dialog box; the **-HYPERLINK** command also allows links attached to areas.

These commands create, edit, and remove hyperlinks (a.k.a *URLs*, *Web addresses*, or just plain *links*). AutoCAD allows you to add hyperlinks to any object in drawings; you can attach one or more hyperlinks to one or more objects.

BASIC TUTORIAL: ATTACHING HYPERLINKS

1. To attach hyperlinks to objects, start the **HYPERLINK** command:
   - From the **Insert** menu, choose **Hyperlink**.
   - At the 'Command:' prompt, enter the **hyperlink** command.
   - Alternatively, press the **CTRL+K** keyboard shortcut.

   Command: **hyperlink** *(Press* ENTER.*)*

2. In all cases, AutoCAD prompts to select objects to which hyperlinks (Web addresses) should be attached.

   Select objects: *(Select one or more objects.)*

   Select objects: *(Press* ENTER *to end object selection.)*

3. AutoCAD displays the Insert Hyperlink dialog box.

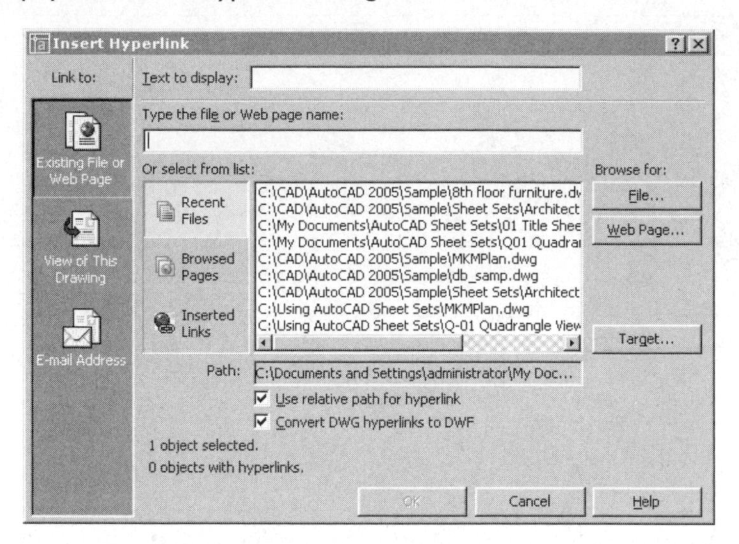

This dialog box allows all kinds of URLs to be constructed.

4. Enter a URL in the **Type the file or Web page name** text field. (If you don't know the URL, click **File** for file names or **Web page** for locations on the Internet.)

   If you like, add a line of description in the **Text to display** field.

5. Click **OK**.

Note: Hyperlinks are normally invisible in AutoCAD drawings. Pass the cursor over objects. When objects have hyperlinks, the cursor changes to the "linked Earth" icon, and displays a tooltip describing the link.

ATTACHING HYPERLINKS: ADDITIONAL METHODS

The HYPERLINK command is supported by a number of options that attach and remove Web addresses from objects. In addition, a system variable stores values associated with URLs.

- **-HYPERLINK** command attaches hyperlinks to objects and areas.
- **GOTOURL** command opens the hyperlink associated with selected objects.
- **PASTEASHYPERLINK** command pastes hyperlinks from Clipboard to objects.
- **DETACHURL** command removes hyperlinks through the command line.
- **HYPERLINKSOPTIONS** command sets options for using hyperlinks in AutoCAD.
- **HYPERLINKOPEN**, **BACK**, **FWD**, and **STOP** commands navigate hyperlinks in drawings.
- **HYPERLINKBASE** system variable holds the path for relative hyperlinks in drawings.

Let's look at them all.

-Hyperlink

The **-HYPERLINK** command attaches and removes hyperlinks from objects and areas.

This command displays prompts at the command line, and is useful for scripts, macros, and AutoLISP routines. In addition, this command allows you to create *hyperlink areas,* which are rectangles that can be thought of as "2D hyperlinks." (Note that the dialog box-based **HYPERLINK** command does not create these areas). When you choose the **Area** option, the rectangle is placed automatically on layer URLLAYER and colored red.

Command: **-hyperlink**

Enter an option [Remove/Insert] <Insert>: **i**

Enter hyperlink insert option [Area/Object] <Object>: *(Type* **o**.*)*

The prompts for attaching hyperlinks to objects are:

Select objects: *(Pick one or more objects.)*

1 found Select objects: *(Press* **ENTER** *to end object selection.)*

Enter hyperlink <current drawing>: *(Enter URL.)*

Enter named location <none>: *(Enter the name of a bookmark, or an AutoCAD view.)*

Enter description <none>: *(Optionally, enter a description of the hyperlink.)*

EDITING HYPERLINKS

To edit hyperlinks and related data, use the **HYPERLINK** command. If necessary, first use the **SELECTURL** or **QSELECT** command to find the object with the specific hyperlink.

Command: **hyperlink**

Select objects: *(Select objects with the same URL; do not include objects with different URLs.)*

Select objects: *(Press* **ENTER** *to end object selection.)*

The Edit Hyperlink dialog box that appears looks identical to the Insert Hyperlink dialog box. Make the changes, and then choose **OK**.

REMOVING HYPERLINKS

To remove URLs from objects, use the **HYPERLINK** command. When the Edit Hyperlink dialog box appears, notice the **Remove Hyperlink** button. Click it.

You can remove rectangular area hyperlinks with the **ERASE** command.

If an object has two or more hyperlinks, but you wish to remove just one of them, then use the **-HYPERLINK** command:

Command: **-hyperlink**

Enter an option [Remove/Insert] <Insert>: **r**

Select objects: *(Select object with URLs.)*

Select objects: *(Press* **ENTER**.*)*

1. www.autodesk.com

2. www.autodeskpress.com

Enter number, hyperlink, or * for all: **1**

1 hyperlink deleted.

The prompts for attaching hyperlinks to areas are the same, except that AutoCAD asks you to pick the two corners of the rectangle.

First corner: *(Pick a point.)*

Other corner: *(Pick another point.)*

This command, however, does not allow you to edit a hyperlink. To do this, use its **Insert** option to respecify the hyperlink, or use the dialog box of the **HYPERLINK** command.

GoToUrl

The **GOTOURL** command opens the hyperlink associated with the selected object.

Command: **gotourl**

Select objects: *(Select one or more objects.)*

Select objects: *(Press* **ENTER** *to end object selection.)*

_.browser Enter Web location (URL) <http://www.autodesk.com>: www.opendwg.org

AutoCAD sends the URL to your computer's default Web browser, which opens the page.

As an alternative, you can select an object, and then right-click. From the shortcut menu, select **Hyperlink** to see the submenu:

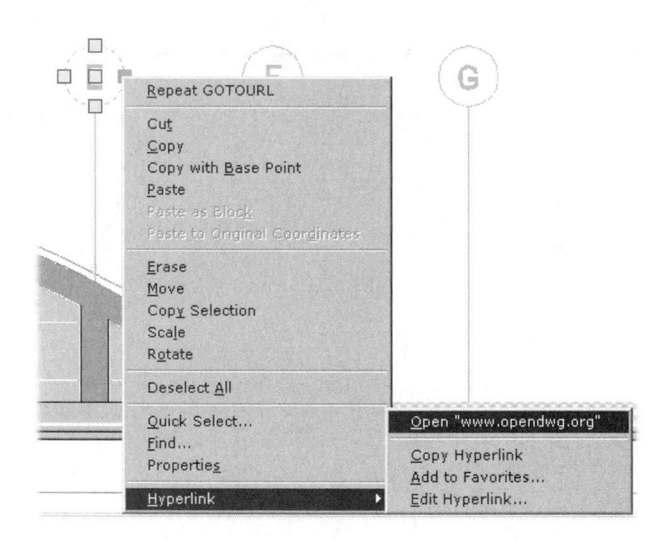

Right-click a hyperlinked object for this shortcut menu.

Open "*url*" opens the Web page in the browser.

Copy Hyperlink copies the URL to Clipboard; it can then be pasted to another object with the **PASTEASHYPERLINK** command.

Add to Favorites adds the URL to the Explorer's Favorites list.

Edit Hyperlink displays the Edit Hyperlink dialog box, where you edit or remove the link.

PasteAsHyperlink

The **PASTEASHYPERLINK** command pastes hyperlinks from Clipboard to objects.

This undocumented command seems to work only when a URL has been copied to Clipboard with the **Copy Hyperlink** shortcut menu item, noted above.

Command: **pasteashyperlink**

Select objects: *(Select one or more objects.)*

Select objects: *(Press* **ENTER** *to end object selection.)*

HIGHLIGHTING HYPERLINKS

Hyperlinks attached to objects are invisible; the exception is the red rectangle of area hyperlinks. AutoCAD provides several methods for locating objects with hyperlinks, and for locating specific hyperlinks.

SelectURL

To select all objects with hyperlinks, use the undocumented **SELECTURL** command.

Command: **selecturl**

AutoCAD highlights all objects that contain hyperlinks (URLs), and places them in the **Previous** selection set.

QSelect

To select objects with a specific hyperlink, use the **QSELECT** command. From the menu bar, choose **Tools | Quick Select**. In the Quick Select dialog box, enter these specifications:

Apply to: **Entire drawing**

Object type: **Multiple**

Properties: **Hyperlink**

Operator: **= Equals**

Value: *(Enter a specific URL, or a URL with wildcards.)*

Choose **OK**, and AutoCAD highlights all objects wtih the specific hyperlink.

DetachURL

The **DETACHURL** command removes hyperlinks through the command line.

Command: **detachurl**

Select objects: *(Select one or more objects.)*

Select objects: *(Press **ENTER** to end object selection.)*

1 hyperlink deleted.

To remove all hyperlinks, use the **All** option at the "Select objects" prompt.

HyperlinkOptions

The **HYPERLINKSOPTIONS** command sets options for using hyperlinks in AutoCAD.

Command: **hyperlinkoptions**

Display hyperlink cursor and shortcut menu? [Yes/No] <Yes>: *(Type **Y** or **N**.)*

Display hyperlink tooltip? [Yes/No] <Yes>: *(Type **Y** or **N**.)*

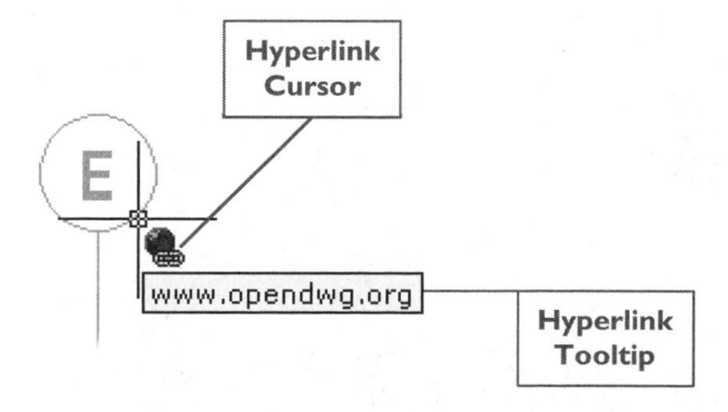

HyperlinkBack, **HyperlinkFwd**, and **HyperlinkStop**

The HYPERLINKBACK, FWD, and STOP commands navigate hyperlinks in drawings.

These commands are associated with the Web toolbar.

Web toolbar.

HYPERLINKBACK returns to the previous hyperlinked page.

HYPERLINKFWD goes to the next hyperlinked page; it can be used only after HYPERLINKBACK.

HYPERLINKSTOP stops displaying the Web page.

HyperlinkBase

The HYPERLINKBASE system variable holds the path for relative hyperlinks in drawings.

Command: **hyperlinkbase**

Enter new value for HYPERLINKBASE, or . for none <"">: *(Enter a new URL.)*

When no path is specified (as shown by ""), AutoCAD uses the drawing path for relative hyperlinks.

EXERCISES

1. Start AutoCAD with the *8th floor plan.dwg* file found in AutoCAD's *\sample* folder. (If necessary, click the Model tab to display the drawing in model space.)
   Place the following text in the drawing:

   Site Plan

   Lighting Specs

   Electrical Bylaw

   Attach the following hyperlinks to each line of text:

Text	Link drawing	Text to display	Target
Site Plan	*SPCA Site Plan.dwg*	Click to view the site plan	ANSI D Plot
Lighting Specs	*timesrvr.txt* (found in *AutoCAD 2005* folder).	Contract text	*none*
Electrical Bylaw	*augi.htm* (found in *AutoCAD 2005\help* folder).	Related Web site	*none*

2. From the CD-ROM, open the *House Designer.dwg* drawing file.
   Use the **ETRANSMIT** command to package the drawings and related support files in ZIP format.

3. From the CD-ROM, open the following files:
   Ch11Pump2.dwg
   Ch11Pump.dwg
   Use the **PUBLISH** command to create a drawing set in DWF format.

4. Use the **PUBLISHTOWEB** command to create a Web page with the following drawings:
   Ch02WoodLake.dwg
   Ch02SalisburyHill.dwg

CHAPTER REVIEW

1. Can you launch a Web browser from within AutoCAD?
2. What is "DWF" short for?
3. Briefly describe the purpose of *.dwf* files?
4. What is "URL" is short for?
5. Which of the following URLs are valid?
   a. www.autodesk.com
   b. http://www.autodesk.com
6. What is "FTP" short for?
7. What is a "local host"?
8. Are hyperlinks active in AutoCAD drawings?
9. Explain the purpose of URLs.
10. What happens to hyperlinks in blocks that are scaled unevenly, stretched, or exploded?
11. Can you can attach a URL to any object in AutoCAD drawings?
12. The **-HYPERLINK** command allows you to attach a hyperlink to"
    a.
    b.
13. Which command shows the locations of hyperlinks in drawings?
14. Can the **ETRANSMIT** command create Web pages.
15. Do compressed *.dwf* files take longer to transmit over the Internet than uncompressed ones?
16. List some commands that create *.dwf* files from within AutoCAD.
    Can AutoCAD open *.dwf* files?
17. What must be done before a Web browser can view *.dwf* files?
18. Can Web browsers view DWG drawing files over the Internet?
19. Briefly describe the purpose of the **ETRANSMIT** command.
20. What does a "self-extracting" executable mean?
21. Might it be illegal under your country's copyright act to transmit certain font files.
22. Explain the purpose of the **PUBLISHTOWEB** command.
23. Can Web pages created by the Publish To Web wizard be edited?

Connecting with Other Programs

Importing and Exporting Files

You save most of your AutoCAD drawings in DWG format. Sometimes, however, you need to provide drawings in other formats. This involves *exporting* the drawing in a vector or raster format.

Other times, you may need to view images, or read drawings from other CAD packages. This involves *importing* them into AutoCAD.

In this chapter, you learn these commands for exporting drawings and importing other files:

SAVEAS and **OPEN** export and import drawings as DXF files.

EXPORT exports drawings in WMF, ACIS, STL, EPS, BMP, and 3DS raster and vector formats.

INSERT menu imports files in 3DS, ACIS, DXF, WMF, and RML formats.

SAVEIMG and **REPLAY** save and view drawings in BMP, TGA, or TIFF raster formats.

RENDER saves renderings to file in BMP, PCX, PostScript, Targa, and TIFF raster formats.

PLOT exports drawings to file in DWF, DXB, and HPGL vector formats, as well as many raster formats.

IMAGEATTACH command places images in BMP, RLE, DIB, CALS, FLIC, Geospot, GIF, IGS, IG4, JFIF (JPEG), PCX, Targa, and TIFF raster formats.

MSLIDE and **VSLIDE** make and view slides of the current viewport.

COPYCLIP copies drawings to Clipboard in native AutoCAD, picture, and bitmap formats.

PASTESPEC pastes data from other applications as AutoCAD entities, image entities, text, pictures, and bitmap formats.

JPGOUT, **PNGOUT**, and **TIFOUT** export the viewport in JPEG, PNG, and TIFF raster formats.

PSIN imports PostScript files.

FINDING THE COMMANDS

On the **INSERT** toolbar:

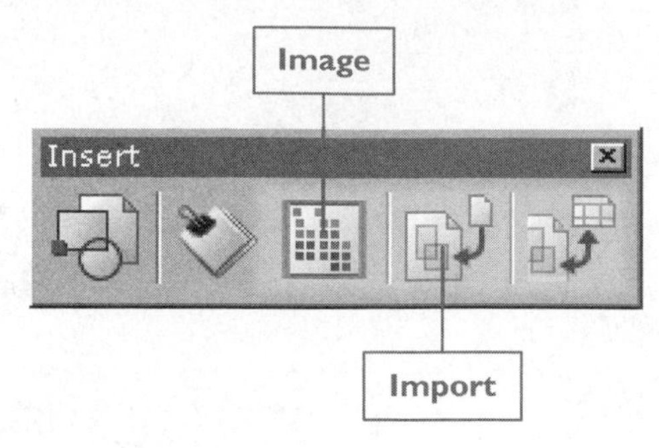

On the **FILE**, **EDIT**, **INSERT**, and **TOOLS** menus:

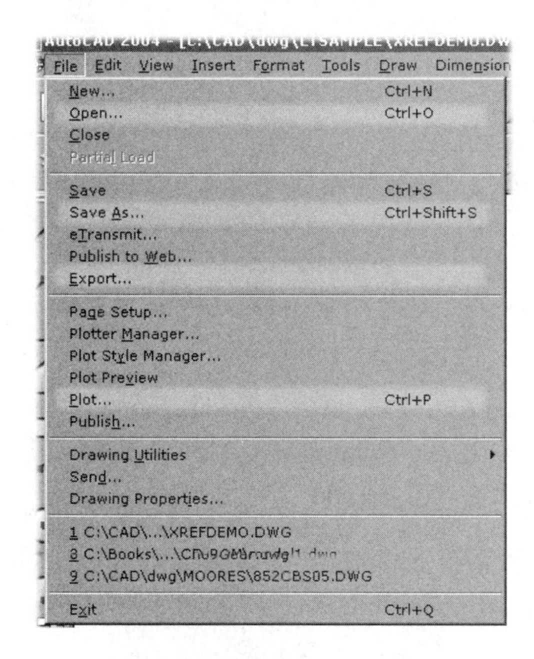

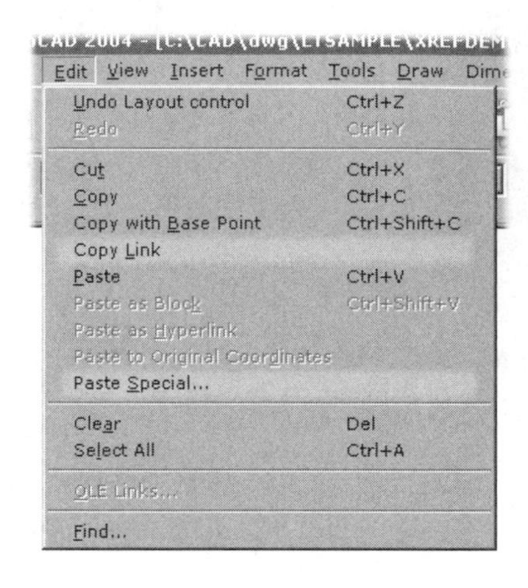

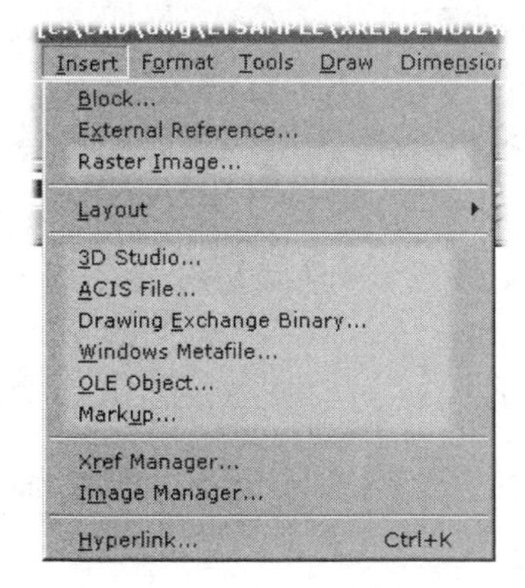

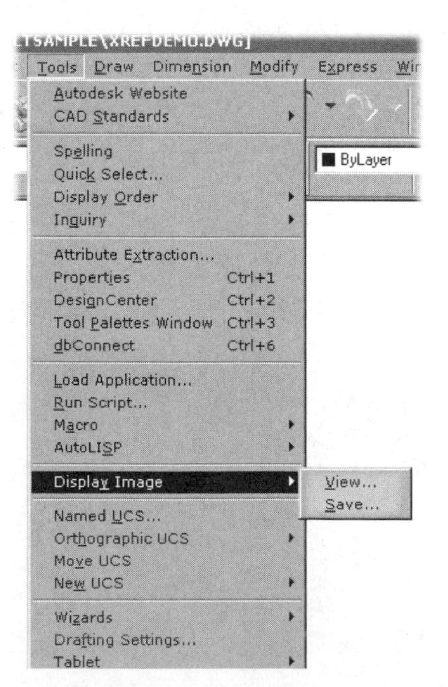

VECTOR AND RASTER IMPORT/DISPLAY/EXPORT FORMATS

AutoCAD supports the importing, displaying, and exporting of many vector and raster formats. Below is a brief overview. Some formats have the notation that AutoCAD cannot import or export them. In all cases, products available from third-party developers (at a cost $$$) allow AutoCAD access to these formats.

VECTOR FORMATS

AutoCAD exports and imports the following vector formats, with some exceptions, as noted below. "Export" means the drawing is converted from AutoCAD format to the destination's format. "Import" means the data is converted to AutoCAD format.

3D Studio

***.3ds:** exchanges drawings with Autodesk's 3D Studio software.

- Export: **3DSOUT** (exports only 3D surface and solid models).
- Import: **3DSIN** (imports as surface models).

ACIS

***.sat:** exchanges drawings with other CAD programs that use the ACIS solid modeler. ACIS is thought to be short for "Andy Charles Ian System." SAT is short for "Save As Text." Autodesk replaced ACIS with ShapeManager in AutoCAD 2004.

- Export: **ACISOUT** (exports only 2D regions and 3D solid models).
- Import: **ACISIN** (imports as solid models).

DWF

***.dwf:** saves drawings in a compressed format suitable for emailing and viewing on the Web; developed by Autodesk. DWF was recently re-acryonymed as "Design Web Format"; it originally stood for "*Drawing* Web Format." It cannot be imported (may change with a future release of AutoCAD).

- Export: **DWFOUT** or **PLOT** (does not retain 3D information; may change with a future release).

DXB

***.dxb:** saves drawings in an extremely simple vector format; developed by Autodesk. DXB is short for "Drawing eXchange Binary." This format was designed for importing data generated by their CAD\camera raster-to-vector conversion software (no longer available).

- Export: **PLOT** (does not retain 3D information; explodes all objects into line segments).
- Import: **DXBIN**.

DXF

***.dxf:** saves the entire *.dwg* drawing database in a fully-documented format; developed by Autodesk. DXF is short for "Drawing *interchange* Format," not drawing *Exchange* format, as is commonly believed. Designed by Autodesk to make the entire content of drawings available to third-party developers; it is also used by other programs to import AutoCAD drawings.

- Export: **DXFOUT** (saves in ASCII or binary format, as well as in AutoCAD Release 12 format).
- Import: **DXFIN**.

Enhanced Metafile

***.emf:** this updated version of WMF; solves problems with fonts and rotated text. This format is commonly represented as "Picture (EMF)" in the Paste Special dialog box. It cannot be exported.

- Import: **PASTESPEC**.

HPGL

.hpg or ***.hgl*** or ***.plt:*** plotter control language developed by Hewlett-Packard that is sometimes used to exchange vector drawing between dissimilar software applications. It converts all objects into short lines, and is sometimes used as a "recovery" file of last resort, because it is written in ASCII format and easily decoded. HPGL is short for "Hewlett-Packard Graphics Language," which has been superceded by the more-compact HPGL/2. It cannot be imported by AutoCAD.

- Export: **PLOT** (after configuring with HPGL plotter driver and plot-to-file).

PostScript

.ps or ***.eps:*** format designed by Adobe to create very high quality printed output that mixes vector and raster formats. Because most applications cannot display PostScript, files often include a preview image in TIFF format. It can only be printed on printers designed to handled PostScript. EPS is short for "encapsulated PostScript."

- Export: **PSOUT** or **PLOT**
- Import: **PSIN** (undocumented by Autodesk).

RML

.rml: redline markup file saved by Volo View; developed by Autodesk. RML is short for "Redline Markup Language," and is based on XML (extended markup language). It cannot be exported by AutoCAD; this may change in a future release of AutoCAD.

- Import: **RMLIN** (all objects places on layer _MARKUP_).

Slide

.sld: an early method of saving images of AutoCAD drawings; developed by Autodesk.

- Export: **MSLIDE** (exports a 2D image of the current viewport).
- Import: **VSLIDE** (displays in the current viewport until the next redraw or regen-related command).

STL

.stl: meant for use with "3D plotters," devices that create plastic 3D prototypes. STL is short for "Stereolithography ." Cannot be imported by AutoCAD.

- Export: **STLOUT** (exports only 3D solid models).

Windows Metaformat

.wmf: a format that mixes raster and vector elements; developed by Microsoft as a standard for Windows, and is based on CGM (computer graphics metafile). This format is commonly represented as "Picture (WMF)" in the Paste Special dialog box. The **WMFOPTS** command provides some import control.

- Export: **WMFOUT**.
- Import: **WMFIN**.

RASTER FORMATS

AutoCAD exports, displays, and attaches the following raster formats, with some exceptions, as noted below. "Export" means the drawing is converted from AutoCAD format to the destination's raster format. "Display" means the image is displayed in the drawing or in the current viewport; "Attach" means the image is placed in the drawing, like an xref, with the **IMAGEATTACH** command. AutoCAD no longer imports or converts raster files into vector objects.

Bitmap

.bmp: raster format created by Microsoft as a standard for Windows, but is generally avoided because it is typically uncompressed, and hence creates huge disk files. BMP is short for "bitmap"; DIB is short for "Device Independent Bitmap"; and RLE is short for "run length encoding."

.dib: same as BMP, but colors are independent of monitors and printers.

.rle: same as BMP, but with RLE compression.

- Export: **BMPOUT**.
- Attach: **IMAGEATTACH**.
- Display: **REPLAY**.

CALS

.cal or ***.rst*** or ***.gp4*** or ***.cg4*** or ***.mil:*** raster format designed by the US government. It is often used with scanned images. CALS is short for "Computer Aided Acquisition and Logistics Support."

- Export: **PLOT** (after configuration with Raster plotter driver).
- Attach: **IMAGEATTACH**.

Flic

.flc or ***.fli:*** animation format saved by Autodesk's (now defunct) Animator software. Just the first frame of the animation is imported. "Flic" is the common pronunciation of the format's file extension. It cannot be exported by AutoCAD.

- Attach: **IMAGEATTACH**.

Geospot and IGS

.bil or ***.igs:*** format commonly used to provide aerial photographs for GIS (geographic information systems) and mapping software. BIL is short for "band interleaved by line." IGS is short for "image systems grayscale." Neither can be exported by AutoCAD.

- Attach: **IMAGEATTACH**

Aerial photograph of ↑ author's house.

GIF

.gif: compact file format developed by CompuServe for use on the Internet. It is limited to 256 colors. Until 2003, many software products avoided GIF due to licensing terms from compression patent owner Unisys. GIF is short for "Graphic Interchange Format," and I've thought it to be the reverse of "fig" or figure. It cannot be exported by AutoCAD.

- Attach: **IMAGEATTACH**.

JPEG or JFIF

***.jpg*:** popular format for showing images on Web pages and storing pictures in digital cameras. This format's strength is its huge compression capability, meaning images take little storage space; its downside is that the strong compression results in *artifacts* — an incorrect representation of the image and loss of detail. It is not recommended for use with AutoCAD drawings. JPEG is short for "Joint Photographic Experts Group," named after the organization that defined the format. JFIF is short for "JPEG File Interchange Format."

- Export: **JPGOUT** (exports selected objects in a single viewport or all viewports) or **PLOT**.
- Attach: **IMAGEATTACH**.

PC Paint

***.pcx*:** format used by PC Paint, one of the original paint programs for the IBM Personal Computer.

- Export: **PLOT** (when configured with the Raster plotter driver).
- Attach: **IMAGEATTACH**.

PICT

***.pct*:** standard raster format used on Macintosh systems; Pict is short for "Picture." It cannot be exported by AutoCAD.

- Attach: **IMAGEATTACH**.

PNG

***.png*:** developed as a royalty-free alternative to JPEG. It features lossless compression, but has failed to replace JPEG. PNG is short for "Portable Network Graphics."

- Export: **PNGOUT**.
- Attach: **IMAGEATTACH**.

RLC

***.rlc*:** often used for faxes, scans, and other black/white and grayscale images. RLC is short for "Run Length Coded," a description of how the file is compressed. It cannot be exported by AutoCAD.

- Attach: **IMAGEATTACH**.

Targa

***.tga*:** file format popular for high-quality images that use 24 bits (or more) to describe color, transparency, and other effects.

- Export: **SAVEIMG**
- Attach: **IMAGEATTACH**
- Display: **REPLAY**

TIFF

***.tif*:** format popular with desktop publishing and high-quality digital camera photographs; comes in many varieties, which can make it difficult for some software applications to decode correctly. It was developed jointly by Adobe and Microsoft. TIFF is short for "Tagged Image File Format."

- Export: **TIFOUT** or **SAVEIMG**.
- Attach: **IMAGEATTACH**.
- Display: **REPLAY**.

SAVEAS AND OPEN

The **SAVEAS** and **OPEN** commands export and import drawings as DXF files.

BASIC TUTORIAL: EXPORTING DXF FILES

1. To export drawings in DXF format, start the **DXFOUT** command:
   - From the **File** menu, choose **Save As**.
   - At the 'Command:' prompt, enter the **dxfout** command.

     Command: **dxfout** *(Press* ENTER*.)*

   AutoCAD displays the Save Drawing As dialog box.
2. Enter a file name, select a folder, and click **Save**.
   The drawing is saved in DXF format.

EXPORTING DXF FORMAT: ADDITIONAL METHODS

The **DXFOUT** command has a number of options that control the DXF file being created. In the Save Drawing As dialog box, click **Tools | Options**, and then select the **DXF Options** tab.

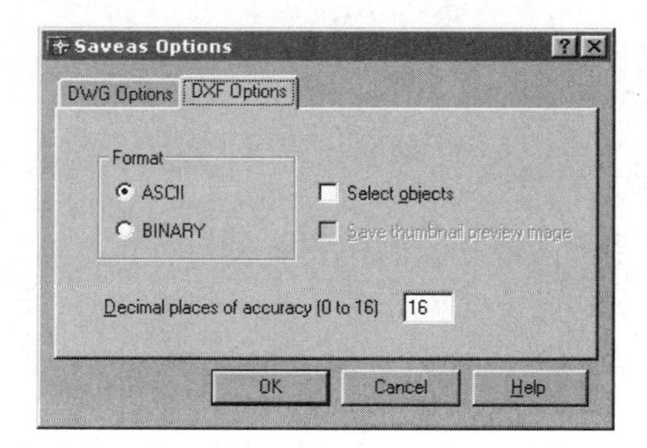

Options for exporting drawings in DXF format.

The **Format** option selects ASCII or binary format:

- **ASCII** can be read by humans, and a larger number of software applications, but creates larger files that take up more disk space.
- **Binary** is more compact and is created faster than ASCII.

The **Select Objects** option determines how much of the drawing is exported:

☑ You are prompted to select objects.

Command: **dxfout**

Select objects: *(Select one or more objects.)*

Select objects: *(Press* ENTER *to end object selection.)*

In addition, the *.dxf* file contains only block references, and no block definition tables.

☐ The entire drawing is output in the *.dxf* file.

The **Save Thumbnail Preview Image** option determines whether a preview image is included with the *.dxf* file. If this option is unavailable (grayed out), the **RASTERPREVIEW** system variable has been turned off (set to 0).

The **Decimal Places of Accuracy** option determines the number of decimal places saved for real numbers in ASCII versions only of the *.dxf* file. Lower precision results in a smaller file; some applications, such as CAM systems, require no more than 4 decimal places.

TUTORIAL: IMPORTING DXF FORMAT

1. To import drawings in DXF format, start the DXFIN command:
   - From the **File** menu, choose **Open**.
   - At the 'Command:' prompt, enter the **dxfin** command.

     Command: **dxfin** *(Press* ENTER.*)*

   AutoCAD displays the Select Files dialog box.
2. Select a folder and a file, and then click **Open**. (There are no options.)
   The drawing is opened in a separate window.

EXPORT

The EXPORT command exports drawings in WMF, ACIS, STL, EPS, BMP, and 3DS raster and vector formats.

This command is a "shell" for AutoCAD commands that actually do the exporting, such as WMFOUT for exporting drawings in WMF format.

BASIC TUTORIAL: EXPORTING DRAWINGS

1. To export drawings, start the EXPORT command:
   - From the **File** menu, choose **Export**.
   - At the 'Command:' prompt, enter the **export** command.
   - Alternatively, enter the alias **exp** at the keyboard.

     Command: **export** *(Press* ENTER.*)*

   In all cases, AutoCAD displays the Export Data dialog box.
2. From the **Files of type** drop list, select a format:

File of Type	*Command*	*Options*	*Selection Prompt*	*Objects*
3D Studio (*.3ds)	3DSOUT	Yes	"Select objects"	Surface and solid models.
ACIS (*.sat)	ACISOUT	Yes	"Select objects"	2D regions and 3D solids.
Bitmap (*.bmp)	BMPOUT	*none*	"Select objects or <all objects and viewports>"	All types.
Encapsulated PS	PSOUT	Yes	*none*	All types.
Lithography (*.stl)	STLOUT	*none*	"Select objects"	3D solids.
Metafile (*.wmf)	WMFOUT	*none*	"Select objects"	All types.

Notes:
"DXX Extract (*.dxx)" extracts attribute data; see Chapter 6 for EATTEXT.
"Block (*.dwg)" is equivalent to using the WBLOCK command.

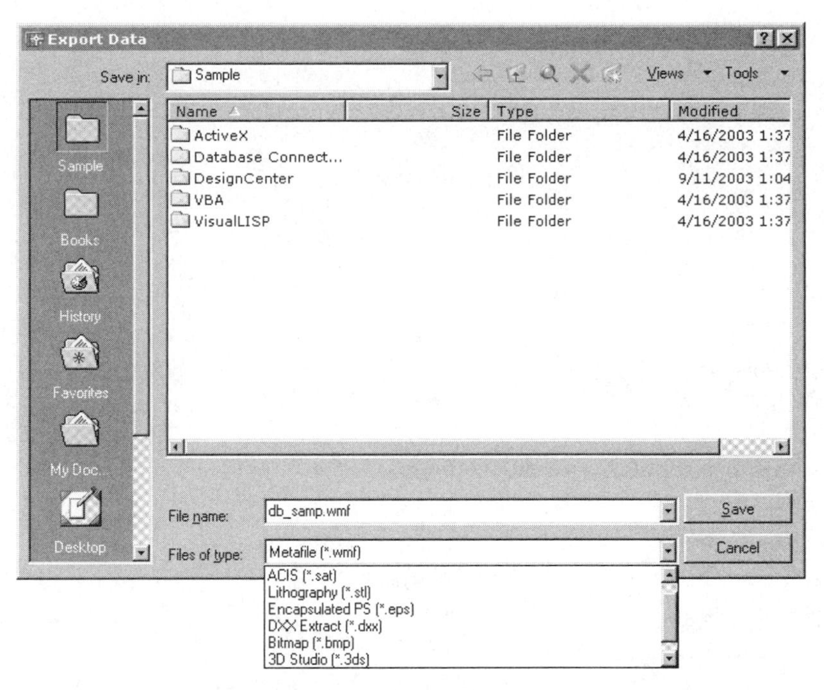

The Export Data dialog box.

3. If you select "Bitmap (*.bmp)" you have the choice of selecting objects or viewports:

   Select objects or <all objects and viewports>: *(Pick objects, or press* **ENTER**.*)*

   - **Select objects** selects all or some objects visible in the current viewport.
   - **All objects and viewports** saves the entire drawing area, including all viewports.

4. Give the file a name, chose a folder, and then click **Save**.

EXPORTING DRAWINGS: ADDITIONAL METHODS

Two of the file formats handled by the **EXPORT** command have options associated with them:

- **ACISOUTVER** system variable sets the ACIS version (15-18, 20, 21, 30, 40, 50, 60, and 70).
- "Encapsulated PS (*.eps)" options are accessed through **Tools | Options.**
- "3D Studio (*.3ds)" displays an options dialog box automatically after you select objects.

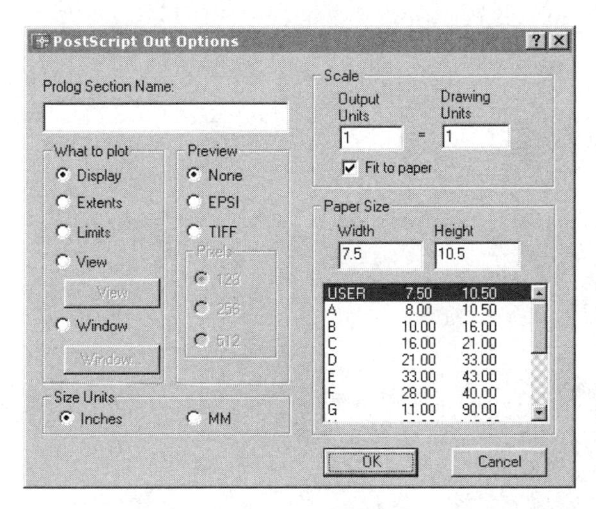

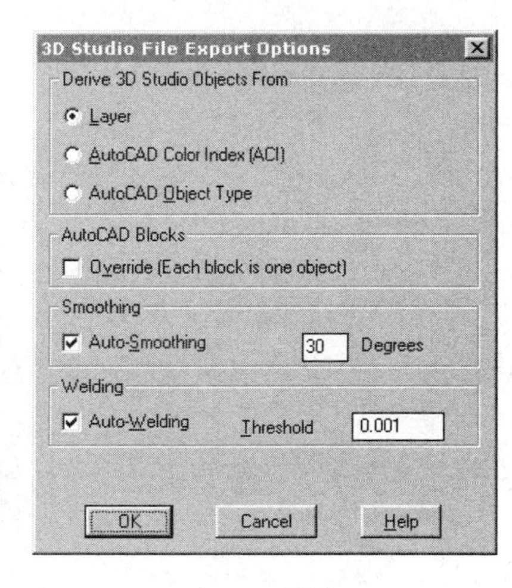

Left: *Options for creating PostScript files.*
Right: *Options for creating 3D Studio files.*

Menu	Command	Vector/Raster	Ex/Import/Display
3D Studio (**.3ds*)			
File | Export	3DSOUT	Vector	Export
Insert | 3D Studio	3DSIN		Import
ACIS (**.sat*)			
File | Export	ACISOUT	Vector	Export
Insert | ACIS File	ACISIN		Import
Bitmap (**.bmp, *.dib, or *.rle*)			
File | Export	BMPOUT	Raster	Export
File | Plot[2]	PLOT		Export
Tools | Display Image | Save	SAVEIMG		Export
View | Render | Render[4]	RENDER		Export
Insert | Raster Image	IMAGEATTACH		Attach
Edit | Paste Special	PASTESPEC		Insert
Tools | Display Image | View	REPLAY		Display
CALS (**.cal, *.rst, *.gp4, *.cg4, or *.mil*)			
File | Plot[2]	PLOT	Raster	Export
Insert | Raster Image	IMAGEATTACH		Attach
Design Web Format (**.dwf*)			
File | Plot	DWFOUT	Vector	Export
Drawing eXchange Binary (**.dxb*)			
File | Plot	PLOT	Vector	Export
Insert | Drawing Exchange Binary	DXBIN		Import
Drawing Interchange Format (**.dxf*)			
File | Save As	DXFOUT	Vector	Export
File | Open	DXFIN		Import
Enhanced Metafile or Picture (**.emf*)			
Edit | Paste Special	PASTESPEC	Vector	Insert
Flic (**.flc or *.fli*)			
Insert | Raster Image	IMAGEATTACH	Raster	Attach
Geospot (**.bil*)			
Insert | Raster Image	IMAGEATTACH	Raster	Attach
Graphic Interchange Format or **CompuServe** (**.gif*)			
Insert | Raster Image	IMAGEATTACH	Raster	Attach
Hewlett-Packard Graphics Language (**.hpg or *.hgl or *.plt*)			
File | Plot[3]	PLOT	Vector	Export
Image Systems Grayscale (**.igs or *.ig4*)			
Insert | Raster Image	IMAGEATTACH	Raster	Attach
Joint Photographic Experts Group or **JFIF** (**.jpg*)			
...	JPGOUT	Raster	Export
File | Plot[2]	PLOT		Export
Insert | Raster Image	IMAGEATTACH		Attach
Object Linking and Embedding			
Edit | Copy	COPYCLIP	Vector	Export
Insert | OLE Object	INSERTOBJ	Vector/Raster	Insert

Menu	*Command*	*Vector/Raster*	*Ex/Import/Display*
PC Paint (**.pcx*)			
File | Plot[2]	PLOT		Export
View | Render | Render[4]	RENDER		Export
Insert | Raster Image	IMAGEATTACH	Raster	Attach
Macintoish Picture (**.pct*)			
Insert | Raster Image	IMAGEATTACH	Raster	Attach
Portable Network Graphics (**.png*)			
...	PNGOUT	Raster	Export
File | Plot[2]	PLOT		Export
Insert | Raster Image	IMAGEATTACH		Display
PostScript (**.ps* or **.eps*)			
File | Export	PSOUT	Vector	Export
File | Plot	PLOT		Export
View | Render | Render[4]	RENDER		Export
...	PSIN[1]		Import
Run Length Coded (**.rlc*)			
Insert | Raster Image	IMAGEATTACH	Raster	Attach
Redline Markup Language (**.rml*)			
Insert | Markup	RMLIN	Vector	Import
Slide (**.sld*)			
...	MSLIDE	Vector	Export
...	VSLIDE		Display
Stereolithography (**.stl*)			
File | Export	STLOUT	Vector	Export
Targa (**.tga*)			
Tools | Display Image | Save	SAVEIMG	Raster	Export
File | Plot[2]	PLOT		Export
View | Render | Render[4]	RENDER		Export
Insert | Raster Image	IMAGEATTACH		Attach
Tools | Display Image | View	REPLAY		Display
Tagged Image File Format (**.tif*)			
...	TIFOUT	Raster	Export
Tools | Display Image | Save	SAVEIMG		Export
File | Plot[2]	PLOT		Export
View | Render | Render[4]	RENDER		Export
Tools | Display Image | View	REPLAY		Display
Insert | Raster Image	IMAGEATTACH		Attach
Windows Metaformat (**.wmf*)			
File | Export	WMFOUT	Vector	Export
Insert | Windows Metafile	WMFIN		Import
Edit | Paste Special	PASTESPEC		Insert

[1] Command undocumented by Autodesk.
[2] Available after configuring AutoCAD with the Raster plotter driver.
[3] Available through the PLOT command's plot-to-file option.

IMAGEATTACH

The **IMAGEATTACH** command attaches images in BMP, RLE, DIB, CALS, FLIC, Geospot, GIF, IGS, IG4, JFIF (JPEG), PCX, Targa, and TIFF raster formats. Details are provided in Chapter 15.

INSERT Menu

The **INSERT** menu item imports files in 3DS, ACIS, DXF, WMF, and RML formats. In addition, the undocumented **PSIN** command imports PostScript files.

BASIC TUTORIAL: INSERTING FILES

1. To import graphical files into drawings, click the **Insert** menu.

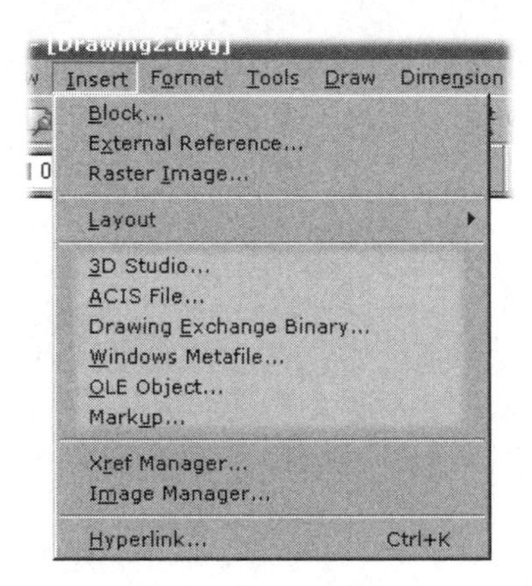

Insert menu.

2. From the **Insert** menu, select a format:

Menu Option	Command	Options	Objects Created
3D Studio	3DSIN	Yes	Surface models.
ACIS File	ACISIN	*none*	2D regions and 3D solids.
Drawing Exchange Binary	DXBIN	*none*	Lines.
Window Metafile	WMFIN	Yes	Polylines.
OLE Object	INSERTOBJ	Yes	OLE objects.
Markup	RMLOUT	Yes	All objects on layer _Markup_.

3. Select a folder, a file, and then click **Open**.
   Some file types display options, as noted above, and described below.
   The contents of the file are converted to AutoCAD objects, and then placed in the drawing.

INSERTING FILES: ADDITIONAL METHODS

Most of the file formats handled by the **Insert** menu have options associated with them:

3D Studio

Inserting a *.3ds* file displays a dialog box of options:

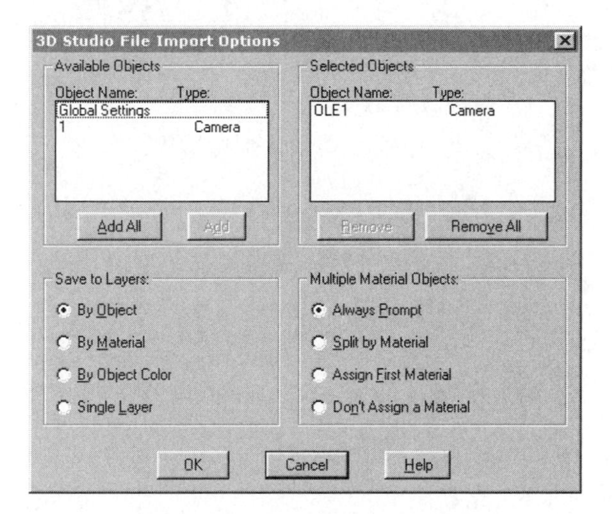

Options for importing 3D Studio files.

Available Objects

The Available Objects list displays the names of objects in the *.3ds* file. Click **Add All**, or select up to 70 objects.

Save to Layers

The **Save to Layers** options control how 3D Studio objects are assigned to layers:

By Object creates a layer for each object. Each layer is given the name of the associated object.

By Material creates a layer for each material; objects with that material are placed on the associated layer. Each layer is given the name of the material.

By Object Color creates a layer for each color; objects with that color are placed on the associated layer. Each layer is named "COLOR*nn*", where *nn* is the 3D Studio color index number. (If the objects have no colors, the layer is named "COLORNONE".)

Single Layer creates a layer called "AVLAYER"; all objects on that layer.

Multiple Material Objects

3D Studio can assign materials to faces, elements, and objects, whereas AutoCAD assigns materials to objects. Thus, you have to tell the translator how to handled 3D Studio objects assigned multiple materials:

Always Prompt displays the Material Assignment Alert dialog box for each object with multiple materials.

Split by Material splits objects with multiple materials into multiple objects to preserve material assignments, but creates larger files.

Assign First Material assigns the first material assigned to the entire object.

Don't Assign a Material assigns no material to each multiple-material object, but takes on AutoCAD's default material.

Window Metafile

Inserting a *.wmf* file displays the Import WMF dialog box. In the dialog box, select **Tools | Options** to view the options for importing the file:

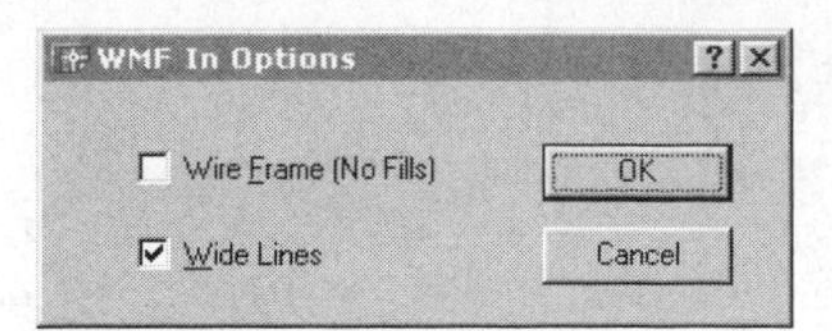

WMF In Options dialog box.

Wire Frame (No Fills)

☑ Imports objects as wireframes.

☐ Imports objects as filled objects (default).

Wide Lines

☑ Lineweights are preserved (default).

☐ Lines have zero width; lineweights are lost.

Click **OK** to exit the options dialog box.

Select a *.wmf* file, and then click **Open**. AutoCAD displays prompts on the command line similar to that of the **INSERT** command:

Command: _wmfin Specify insertion point or

[Scale/X/Y/Z/Rotate/PScale/PX/PY/PZ/PRotate]: *(Pick an insertion point.)*

Enter X scale factor, specify opposite corner, or [Corner/XYZ] <1>: *(Specify the scale.)*

Enter Y scale factor <use X scale factor>: *(Press* ENTER*.)*

Specify rotation angle <0>: *(Specify the angle.)*

AutoCAD places the WMF object on the current layer.

OLE Object

Inserting an OLE object displays a dialog box of options:

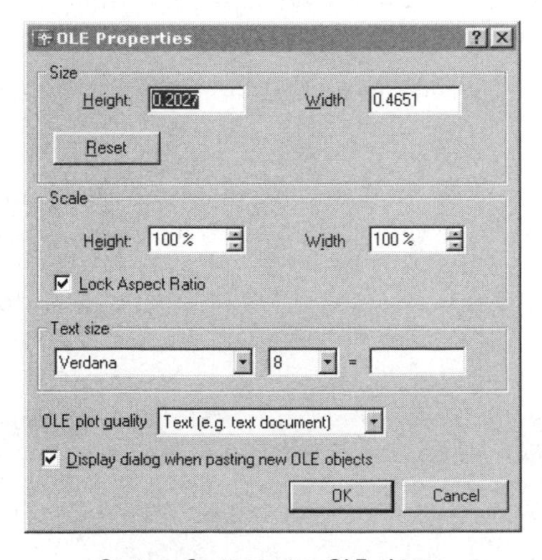

Options for importing OLE objects.

Size

The **Size** option specifies the size of the imported OLE object in units. When **Lock Aspect Ratio** is turned on, the **Width** and **Height** values change to maintain the aspect ratio between them.

Reset returns the OLE object to its original size.

Scale

The **Scale** option changes the scale of the inserted OLE object.

Text Size

The **Font** option changes the font of text in the OLE object. You are limited to selecting font names from the droplist.

The **Point Size** option lists the points sizes of fonts in the OLE object. Use this option in conjunction with the **Text Height** option to set the size of the text in drawing units. (There are 72 points to the inch.) If the OLE object contains text of other size, they are adjusted by a similar amount.

OLE Plot Quality

The **OLE Plot Quality** option determines the plot quality of the OLE object:

Line art quality is meant for use with embedded spreadsheets.

Text quality is meant for use with embedded documents.

Graphics quality is meant for simple graphics, such as embedded charts.

Photograph quality is meant for embedded images.

High quality photograph quality is meant when images print poorly. In fact, Hewlett Packard recommends this setting as the best choice in most cases, saying it has "no significant impact in printing speed."

Markup

Inserting an *.rml* file displays a dialog box of options:

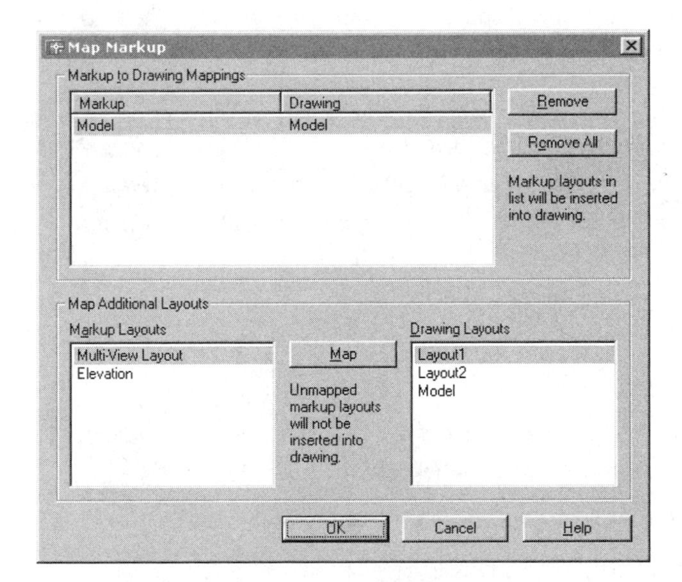

Options for importing markup files from Volo View.

The **Markup to Drawing Mappings** section lists how Volo View's markup layouts are mapped to AutoCAD's drawing layouts.

Select a layout name in **Markup Layouts**, and then a matching layout under **Drawing Layouts**. Click **Map**, and repeat for all other layouts.

SAVEIMG AND REPLAY

The SAVEIMG command saves whatever is displayed in the current viewport — whether a 2D drawing, hidden-line removed, or rendered — in BMP, TGA, or TIFF raster formats. (To save all viewports or selected objects, use the TIFOUT command.) Lineweights are not recorded.

The REPLAY command displays the opened image in the current viewport, temporarily. The image disappears, however, with the next redraw- or regen-related command.

TUTORIAL: SAVING IMAGES

1. To save the current viewport as a raster image, start the SAVEIMG command:
   - From the **Tools** menu, choose **Display Image**, and then **Save**.
   - At the 'Command:' prompt, enter the **saveimg** command.

     Command: **saveimg** *(Press* ENTER.*)*

   In all cases, AutoCAD displays the Save Image dialog box.

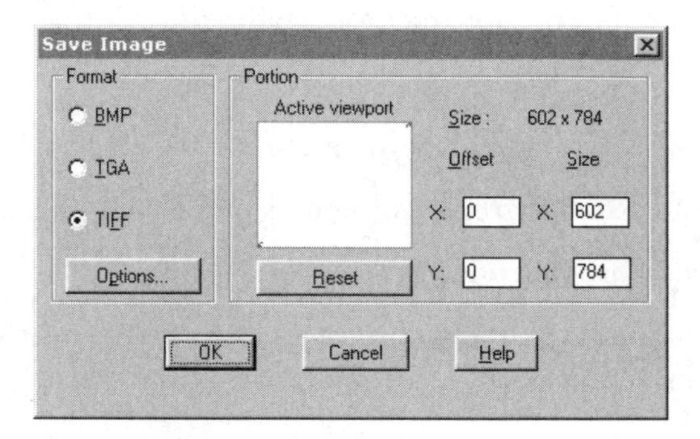

Save Image dialog box.

2. Select a file format:
   - **BMP** saves in uncompressed 32-bit bitmap format.
   - **TGA** saves in 32-bit Targa format.
   - **TIFF** saves in 32-bit RGBA TIFF format.
3. The **Options** button specifies whether the TGA and TIFF files will be compressed. In general, compression should be turned on.
4. The **Portion** option saves an area smaller than the current viewport.
   Enter values for the offset (from lower left corner) and overall size.
5. Click **OK**.
   AutoCAD displays the Image File dialog box.
6. Enter a file name, select a folder, and then click **Save**.

TUTORIAL: REPLAYING IMAGES

1. To display temporarily a raster image in the current viewport, start the REPLAY command:
   - From the **Tools** menu, choose **Display Image**, and then **View**.
   - At the 'Command:' prompt, enter the **replay** command.

     Command: **replay** *(Press* ENTER.*)*

   In all cases, AutoCAD displays the Replay Image dialog box.
2. From the **Files of type** list box, select a file format.
3. Click **Open**.

4. The Image Specifications dialog box allows you to adjust the image to fit the viewport.

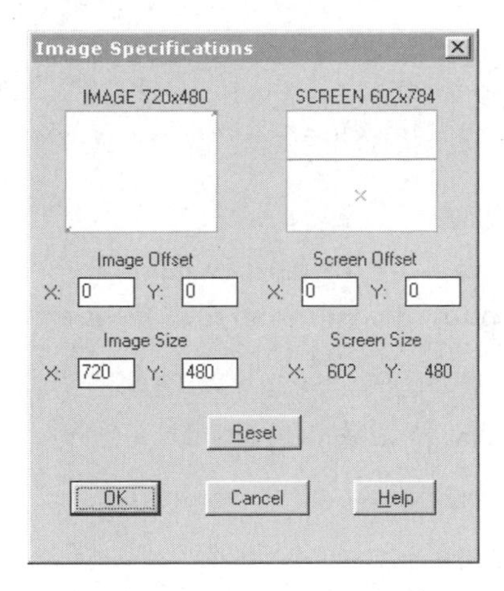

The Image Specifications dialog box.

5. Click **OK**.
   AutoCAD displays the image in the current viewport, covering up whatever else is there. If the image does not fit, a portion of the viewport is shown in black.

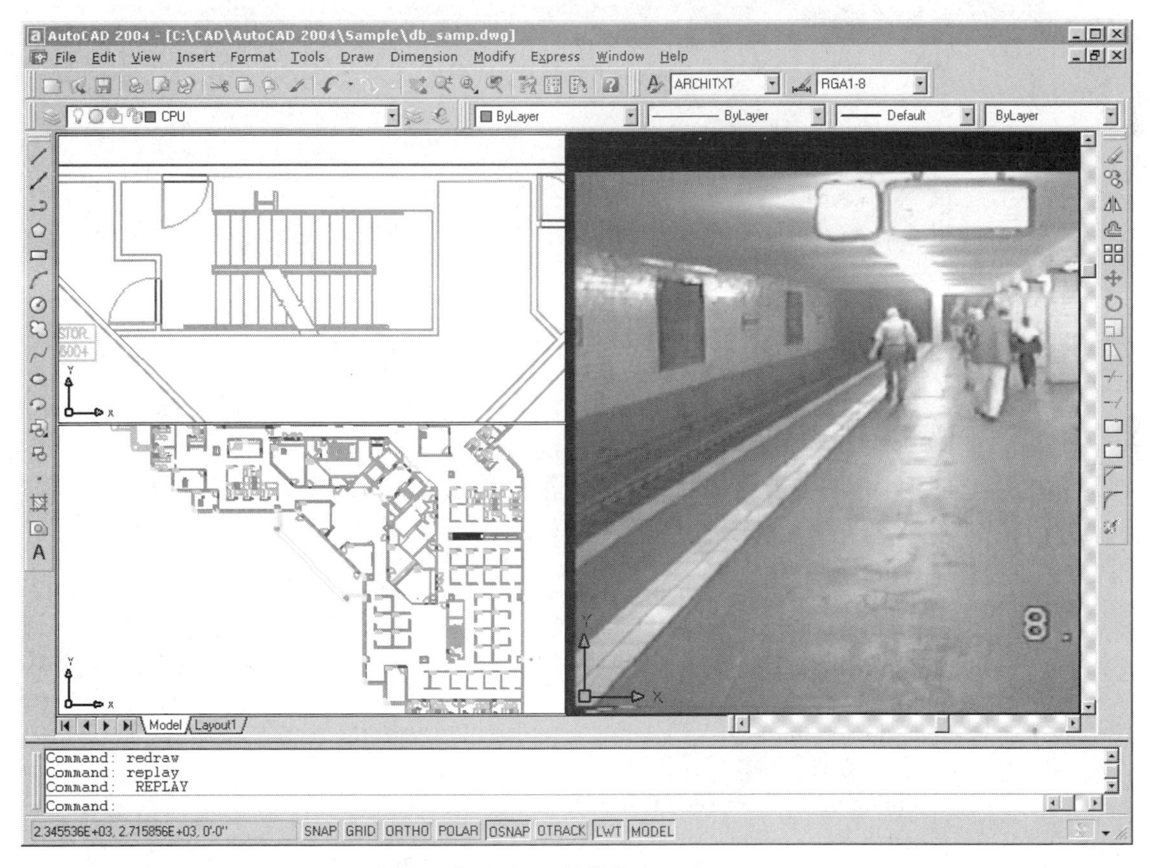

Replay displaying a BMP file in a viewport.

6. To remove the image, enter **REDRAW**.

RENDER

The **RENDER** command saves renderings to file in BMP, PCX, PostScript, Targa, and TIFF raster formats. More details on rendering follow in Chapter 14.

TUTORIAL: RENDERING TO FILE

1. To render 3D drawings to file, start the **RENDER** command:
   - From the **View** menu, choose **Render**, and then **Render**.
   - At the 'Command:' prompt, enter the **render** command.
   - Alternatively, enter the alias **rr** at the keyboard.

   Command: **render** *(Press* ENTER.*)*

   In all cases, AutoCAD displays the Render dialog box.
2. In the **Destination** area, select "File" from the drop list.
3. To select a format, click **More Options**.

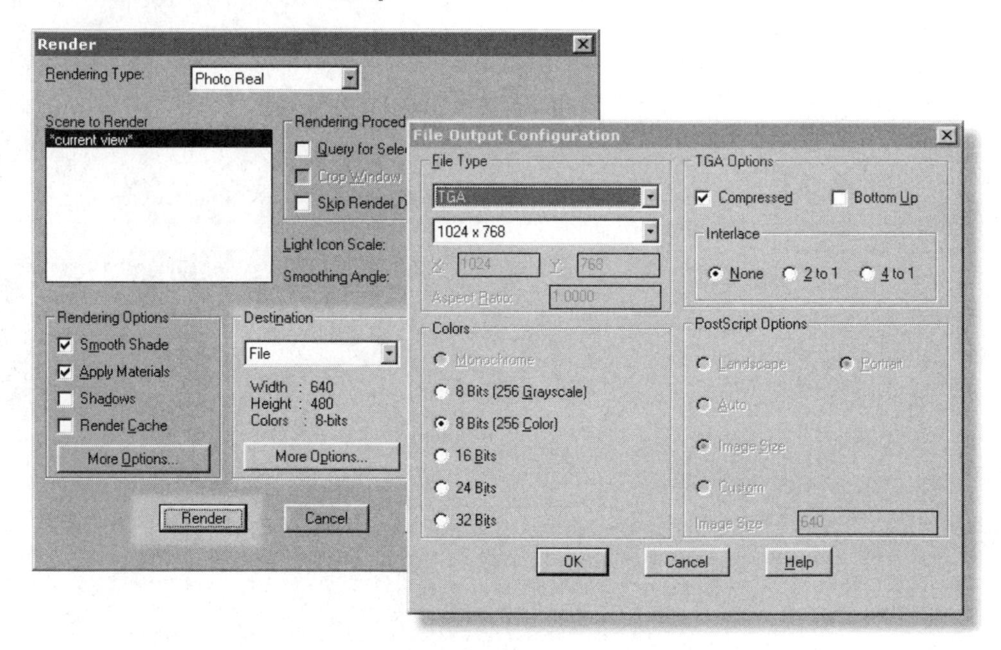

Selecting "render to file" options.

4. In the File Type area, select a format and a preset size, such as 1024x768.

   To set a custom size, select "User Defined," and then enter values for **X** and **Y** (measured in pixels).
5. Select the number of colors, as well as other options (if available).
6. To save the options, click **OK**.
7. Back in the Render dialog box, click **Render**.

   AutoCAD displays the Rendering File dialog box.
8. Enter a file name, select a folder, and then click **Save**.

   AutoCAD generates the file, muttering to itself:

   Using current view.

   Default scene selected.

   100% complete, 480 of 480 scan lines

PLOT

The **PLOT** command exports drawings to file in DWF, DXB, and HPGL vector formats, as well as numerous raster formats.

The format is controlled by the plotter configuration. In some cases, you may need to use the **PLOTTERMANAGER** command to create new configurations.

BASIC TUTORIAL: PLOTTING TO FILE

1. To plot the drawing to file, start the **PLOT** command:
   - From the **File** menu, choose **Plot**.
   - At the 'Command:' prompt, enter the **plot** command.
   - Alternatively, press the **CTRL+P** keyboard shortcut.

   Command: **plot** *(Press* **ENTER**.*)*

   In all cases, AutoCAD displays the Plot dialog box.
2. Click the **Plot Device** tab.
3. In the Plotter Configuration section, select a plotter configuration from the **Name** drop list. For specific formats, select these configurations:

Format	*Name*
DWF	"DWF6 ePlot.pc3."
DXB	"DXF File.pc3."
HPGL	*Select an HP plotter configuration.*
Raster	*Select a raster configuration.*

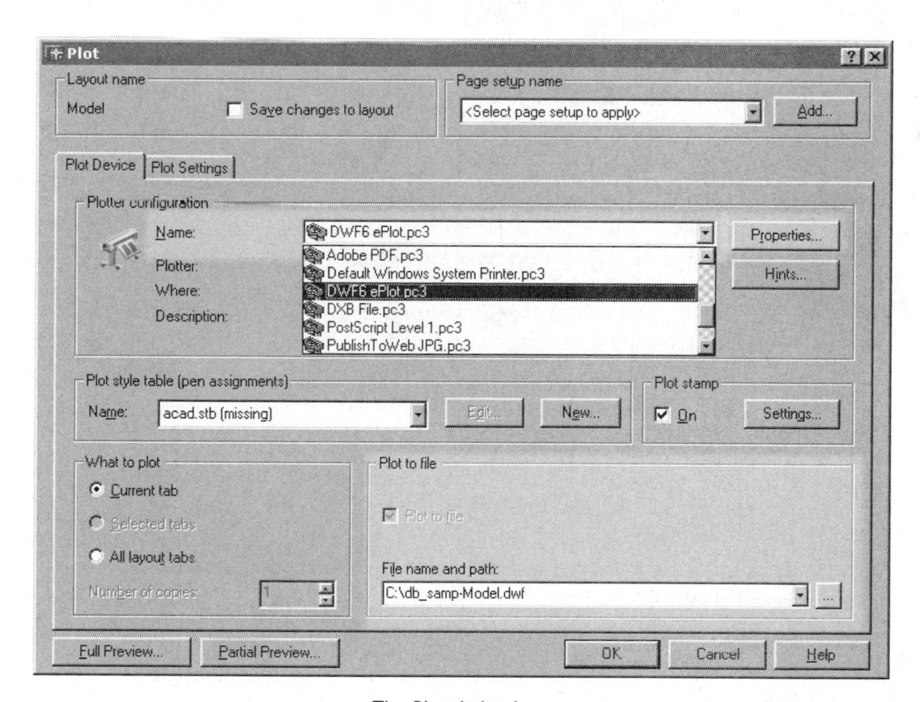

The Plot dialog box.

4. If necessary, click the **Properties** button to change the properties of the plotter configuration. (See Chapter 19 of *Using AutoCAD: Basics* for details on configuring plotters and creating plots.)

5. In the **Plot to File** area, specify the name for the exported file. Click **...** to select a drive and folder.

   If you select an HP plotter, you must also turn on the **Plot to File** option. (The other formats automatically turn on the option.)
6. Change other settings, as needed, such as **Plot Scale** and the **Plot Area**.
7. To check that the plot will turn out all right, click the **Full Preview** button.

   When done previewing, press **ENTER** to return to this dialog box.
8. Click **OK**.

   Wait as AutoCAD generates the plot file.

PLOTTING TO FILE: ADDITIONAL METHODS

Some formats need to be prepared with the PLOTTERMANAGER command.

1. To prepare a plotter configuration, start the **PLOTTERMANAGER** command:
   - From the **File** menu, choose **Plotter Manager**.
   - At the 'Command:' prompt, enter the **plottermanager** command.

     Command: **plottermanager** *(Press* **ENTER**.*)*

   In all cases, AutoCAD displays a window listing the file names of current plot configurations (**.pc3* files).
2. Double-click the **Add a Plotter Wizard**, and then click **Next**.
3. Select **My Computer**, and then click **Next**.

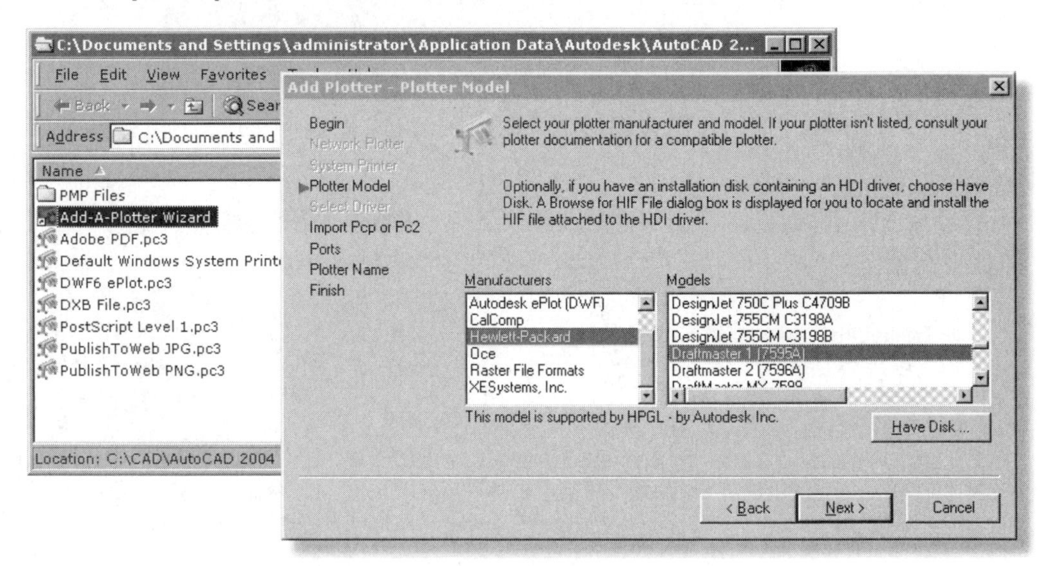

Configuring new plotters.

4. In the Plotter Model dialog box, select the format:
   - **HPGL**: Under **Manufacturers**, select "Hewlett-Packard."

     Under **Models**, select any plotter model number that supports HPGL.

     For this tutorial, choose "DraftMaster I," because it supports plots as large as E-size. (Do *not* select HPGL/2, because it is a compressed binary format that most software cannot read.)
   - **Raster**: Under **Manufacturers**, select "Raster File Formats."

     Under **Model**, select any specific format you are interested in.

     For this tutorial, choose "CALSMIL-R..."

   Click **Next**, twice.
5. In the Ports dialog box, select **Plot to File**, and then click **Next**.

   (Raster configurations automatically select the **Plot to File** option.)

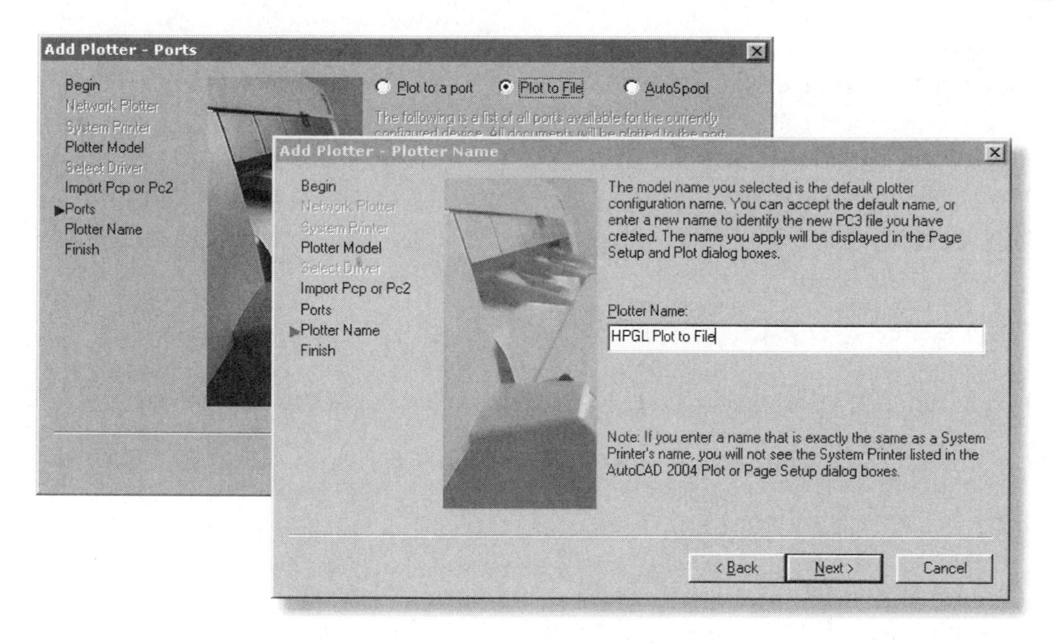

Specifying "plot to file."

6. In the Plotter Name dialog box, enter a descriptive name, such as:

   Plotter Name: **HPGL Plot to File**

   Plotter Name: **CALS MIL-R Plot to File**

   Click **Next**.
7. Click **Edit Plotter Configuration**.

   Most plot options do not apply when plotting to file, but you may want to consider:
   - **Custom Paper Sizes** allows you to size the plot to fit a specific destination, such as a report or other document.
   - **Graphics** allows you specify colors, shades of gray, or monochrome output.
   - **Custom Properties** includes features specific to the plotter configuration, such as background color.
8. When done, click **OK**, and then **Finish**.

The new plotter configuration shows up in the **Name** list the next time you use the **PLOT** command.

MSLIDE AND VSLIDE

The **MSLIDE** and **VSLIDE** command make and view slides of drawings.

Slides are stored in *.sld* files. In model space, slides are made of the current viewport; objects on frozen and off layers are not captured. In paper space, slides are made of the entire layout — all viewports and their content. Lineweights are not preserved. (**MSLIDE** is short for "make slide," and **VSLIDE** is short for "view slide.")

TUTORIAL: MAKING AND VIEWING SLIDES

1. To make slides of drawings, start the **MSLIDE** command:
   - At the 'Command:' prompt, enter the **mslide** command.

     Command: **mslide** *(Press* ENTER.*)*

   AutoCAD displays the Make Slide File dialog box.
2. Enter a file name, select a folder, and then click **Save**.

3. To view the slide, enter the **VSLIDE** command.
   AutoCAD displays the Select Slide File dialog box.
4. Select a folder and a file.
5. Click **Open**.
   AutoCAD displays the slide in the current viewport until the next command that causes a redraw or regen.

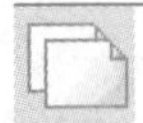 COPYCLIP AND PASTESPEC

The COPYCLIP command copies drawings to Clipboard in native AutoCAD, picture, and bitmap formats. See Chapter 2 for more details.

The PASTESPEC command pastes data from other applications as AutoCAD entities, image entities, text, pictures, and bitmap formats.

TUTORIAL: PASTING FROM CLIPBOARD

1. To paste from the Clipboard in specific formats, start the **PASTESPEC** command:
   - From the **Edit** menu, choose **Paste Special**.
   - At the 'Command:' prompt, enter the **pastespec** command.
   - Alternatively, enter the alias **pa** at the keyboard.

   Command: **pastespec** *(Press* ENTER.*)*

   AutoCAD displays the Paste Special dialog box. The contents vary, depending on the data copied to Clipboard. The figure below illustrates the formats available after copying a Visio drawing to Clipboard.

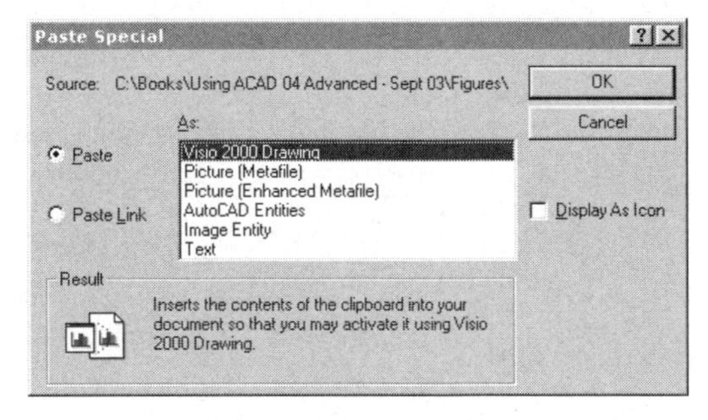

Contents of the Paste Special dialog box vary, according to the copied data.

2. From the list under **As**, select a format.
3. Click **OK**.
   The location and type of the object depend on its format:

Format	**Location**	**Type**
Native Drawing	Upper left corner of current viewport	OLE object
Picture (Metafile)	Upper left corner of current viewport	OLE object
Bitmap	Upper left corner of current viewport	OLE object
AutoCAD Entities	At insertion point	Polylines and text
Image Entities	At insertion point	Raster image
Text	Upper left corner of current viewport	Mtext

Note: Limited editing of OLE objects is available by right-clicking the object:

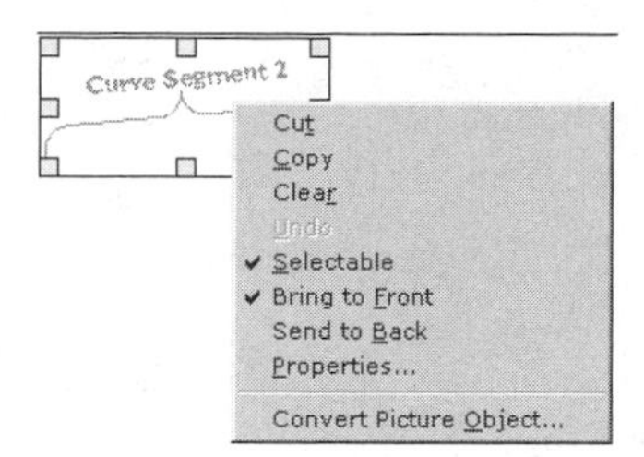

Right-click OLE objects to access editing commands.

JPGOUT, PNGOUT, AND TIFOUT

The **JPGOUT**, **PNGOUT**, and **TIFOUT** commands export views in JPEG, PNG, and TIFF raster formats.

These three commands operate identically. There are no options, however, for controlling the quality or compression style.

TUTORIAL: EXPORTING IN RASTER FORMATS

1. To export the drawing as a raster image, start one of the commands:
   - **JPGOUT** for JPEG files.
   - **PNGOUT** for PNG files.
   - **TIFOUT** for TIFF files.
2. In all cases, AutoCAD displays the Create Raster File dialog box.
   Enter a file name, select a folder, and then click **Save**.
3. AutoCAD prompts you to select objects:

   Select objects or <all objects and viewports>: *(Select objects to export.)*

   Select objects or <all objects and viewports>: *(Press* **ENTER** *to end object selection.)*

   If the drawing screen consists of two or more viewports, you can capture them all by pressing **ENTER** at the "all objects and viewports" prompt.

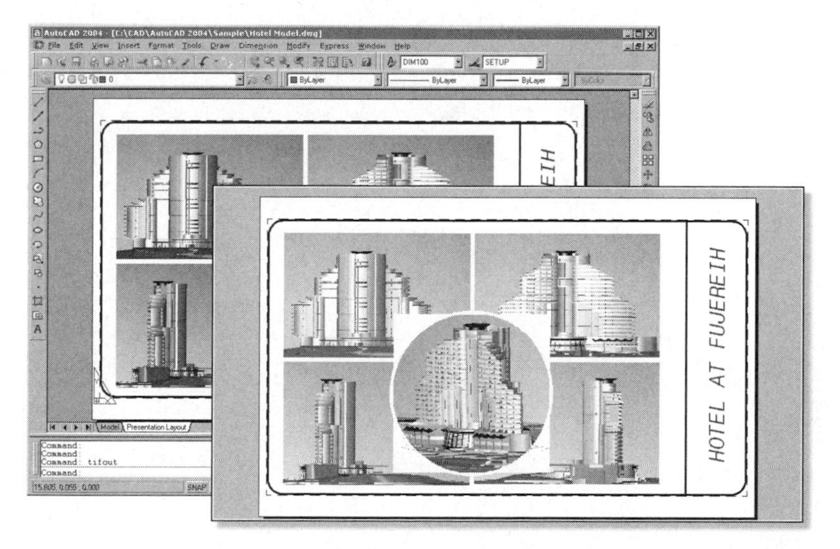

The "all objects and viewports" option captures the entire drawing area.

EXERCISES

1. From the CD-ROM, open the *Ch12Surface.dwg* file, a 3D drawing of a plastic housing.
   a. Export the drawing in WMF format.
   b. Start a new drawing, and import the WMF file.
   Does the drawing look different?

2. Use the **SAVEIMG** command, and save the screen image in TGA format.
   a. Change the view of the drawing.
   b. Use the **REPLAY** command to display TGA file.
   How do you remove the image?

3. From the CD-ROM, open the *Model1.dwg* file.
   a. Create a plotter configuration with the "TrueVision Targa" raster driver.
   b. Plot the drawing to file.
   c. View the *.tga* file in AutoCAD.
   Which command did you use to view the file?

CHAPTER REVIEW

1. Why would you want to save a drawing in DXF format?
2. Which of the following are raster, and which are vector formats?
   ACIS
   STL
   BMP
   TIFF
3. When might you want to import *.rml* files?
4. What command removes slide images from the screen?

Attaching Attribute Data

Attributes are text data — any kind of text — attached to blocks. The data could include description, part number, and price. Each time you insert the block, you are prompted to enter data, which is stored in the attributes. Then you can extract the attribute data and use another program — such as a spreadsheet — to process the data.

Attributes are one way to get text data out of AutoCAD. Some other ways include the **LIST** and **DXFOUT** commands, as well as **DBCONNECT**, which uses SQL (structured query language). Attributes, however, give you the best "price-performance" of all these methods.

In this chapter, you learn about the following commands:

ATTDEF defines attribute tags, prompts, and default values (enhanced in AutoCAD 2005).

BLOCK attaches attributes to symbols, and **INSERT** inserts attributed blocks.

ATTREQ toggles whether attributes are filled in automatically.

ATTDISP controls the display of attributes in the drawing.

EATTEDIT edits attribute values and properties of individual block insertions.

BATTMAN edits attributes and properties in global block definitions.

EATTEXT extracts attributes from drawings to data files.

FINDING THE COMMANDS

On the **MODIFY II** toolbar:

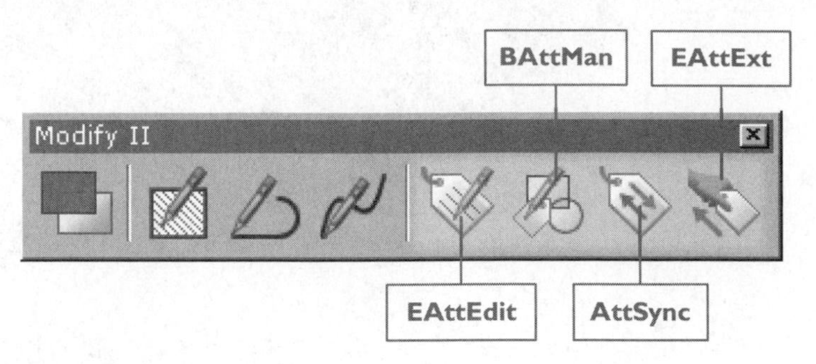

On the **DRAW** and **MODIFY** menus:

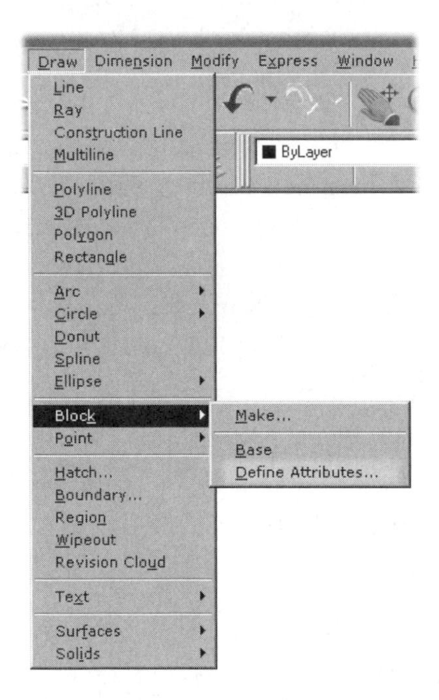

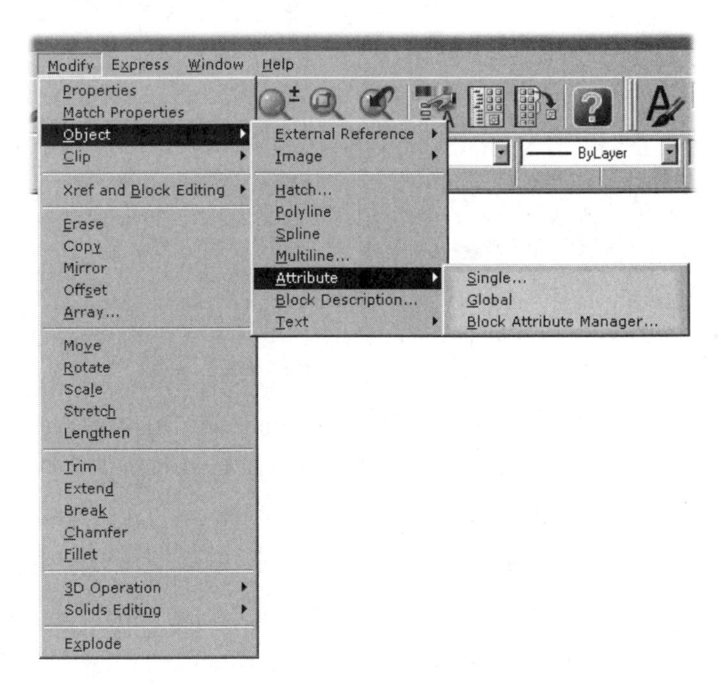

ABOUT ATTRIBUTES

AutoCAD allows you to add attributes to blocks. Attributes can be considered "labels" attached to blocks. The labels contains any information you desire. It's up to you to determine the information requested by AutoCAD, the wording of the prompts, and the default values, if any. For example, you could create a chair block with attributes that define its owner, the manufacturer, the model number, its fabric type, the date purchased, and purchase price.

Attributes are placed in drawings when you insert the blocks to which they are attached. When the blocks are inserted, AutoCAD requests from you values for the attributes.

The information can be extracted from each block in the drawing, and used in other software, such as database and spreadsheet programs.

The steps for creating and defining attributes are:

Step 1: Draw symbols.

Step 2: Define attribute tags, prompts, and values with the **ATTDEF** command.

Step 3: Attach attributes to symbols with the **BLOCK** command.

Step 4: Insert blocks in drawings with the **INSERT** command, answering the prompts for attributes data.

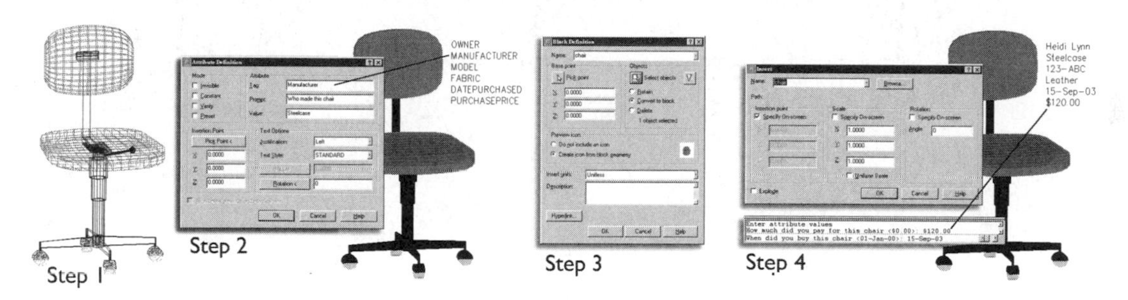

Steps in creating attributes.

Step 5: Edit attributes: use the **BATTMAN** command to edit attribute definitions, and the **EATTEDIT** command to edit attribute data.

Step 6: Extract attribute data from drawings to data files with the **EATTEXT** command.

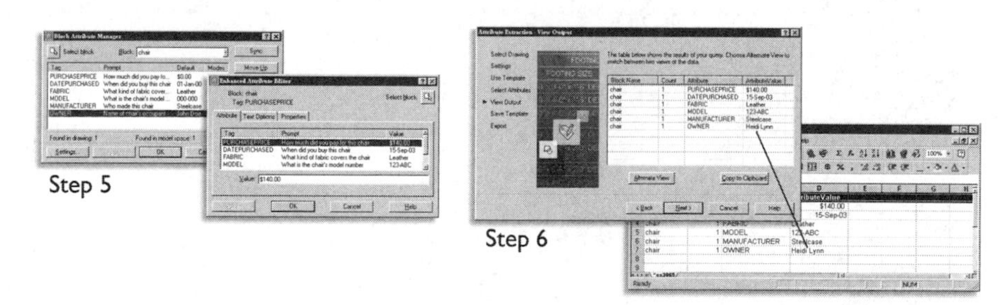

Steps in editing and extracting attributes.

Step 7: Import data files into spreadsheets or database programs for further processing.

Note: Attributes can only be part of blocks and no other objects. It is possible to have blocks that consist solely of attributes. To attach data to any object, use "xdata" (not discussed in this text), or create links to external database tables (discussed in the next chapter).

EXAMPLES OF ATTRIBUTE USAGE

Attributes are commonly used to create BOMs (bills of material) and to fill in title blocks.

For example, piping designers use valve blocks that contain attributes. As the designers insert valves, they enter data relating to the valve number, its size, and so on. When the attribute data is extracted, designers use spreadsheets to add up the occurrence of each valve type, producing a BOM.

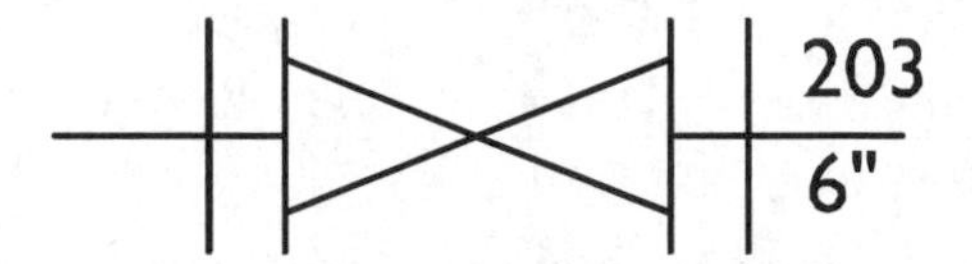

Valve block with attribute data.

Facilities managers apply attributes to drawings of floor plans to keep track of square footage, capital assets (furniture), and resources (telephone and network access). One system keeps track of buildings, floors, type of use within floors, occupants, and description.

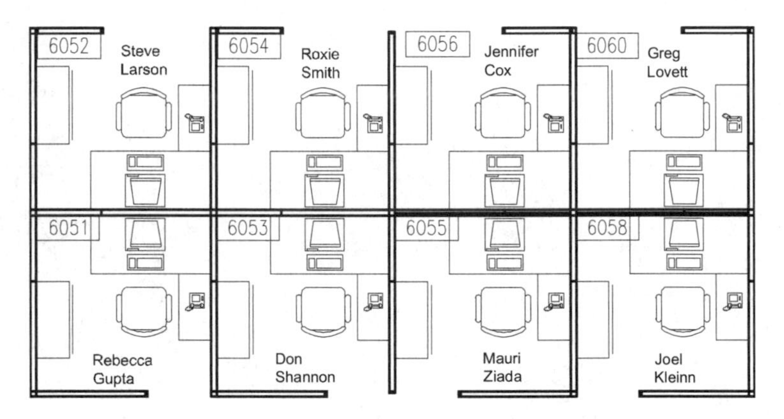

Attributes keeping track of building facilities.

The template drawings provided with AutoCAD contain attributes in their title blocks. Open a template file, such as *ISO A3 -Named Plot Styles.dwt*. Double-click the title block, and AutoCAD displays the attribute editor. Select an attribute, such as USER, and enter a value, such as your name. Click **OK**, and your name appears in the title block.

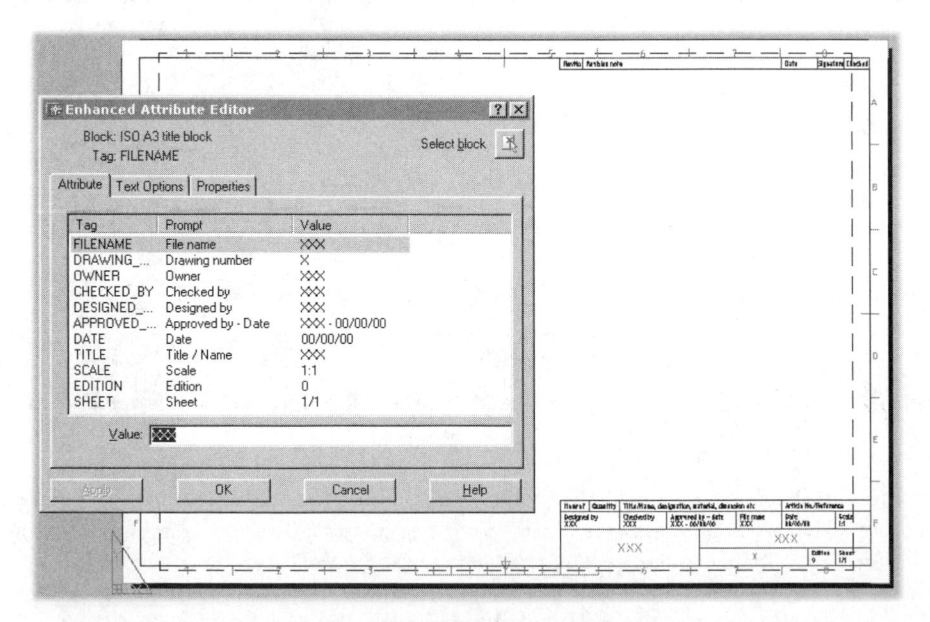

Editing attributes by double-clicking title blocks.

ATTDEF

The **ATTDEF** command defines attribute tags, prompts, and default values (short for "attribute definition").

Let's take the example of adding attributes to office chair symbols. The *tag* identifies the attribute. It consists of a single word, although you can use dashes and underscores. AutoCAD automatically converts all the text to uppercase. The tags could be:

Owner

Manufacturer

Model

Fabric

DatePurchased

PurchasePrice

The *prompt* reminds you of the attribute's purpose. It is a sentence, and is allowed spaces between words. The prompts could be:

Employee chair assigned to

Manufacturer

Model number

Fabric

Date of purchase

Price

The *value* is the default data. When attributes are inserted in the drawing, pressing **ENTER** accepts the default value, or else enter another value. And the values could be:

Heather MacKenzie

Steelcase

98S0197

Leather

2003 September 15

$495.00

Taken together, we get the following:

Tag	***Prompt***	***Data***
Owner	Employee chair assigned to	Heather MacKenzie
Manufacturer	Manufacturer	Steelcase
Model	Model number	98S0197
Fabric	Fabric	Leather
DatePurchased	Date of purchase	2003 09 15
PurchasePrice	Price	$495.00

Notice the format of the date: year-month-day. This format makes it easy later to sort the data by date.

BASIC TUTORIAL: CREATING ATTRIBUTES

1. To create attribute definitions, start the **ATTDEF** command:
   - From the **Draw** menu, choose **Block**, and then **Define Attributes**.
   - At the 'Command:' prompt, enter the **attdef** command.
   - Alternatively, enter the alias **att** at the keyboard.

   Command: **attdef** *(Press* **ENTER***.)*

   In all cases, AutoCAD displays the Attribute Definition dialog box.

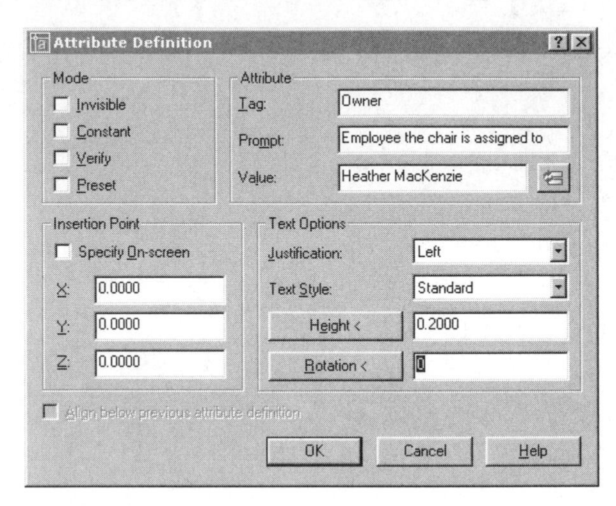

Attribute Definition dialog box.

2. In the **Tag** field, enter a tag that identifies the attribute. (Spaces are not allowed in tags.)

   Tag: **Owner**

Note: Tags, prompts, and values can each be up to 256 characters long. AutoCAD converts all text to uppercase. Tags cannot have spaces. To start a prompt or default value with a space, use a backslash (\ ); to start with a backslash, start with two backslashes (\\ ).

3. In the **Prompt** field, enter a prompt that reminds the user of the purpose of the attribute. (This field is not available when **Constant** mode is turned on.) You can leave this field blank; AutoCAD uses the tag text as the prompt.

   Prompt: **Employee the chair assigned to**

   AutoCAD automatically adds the colon (:) to the end of the prompt.

4. In the **Value** field, enter a default value.

   Value: **Heather MacKenzie**

   When **Constant** mode is turned on, AutoCAD uses this value as the attribute.

   AutoCAD 2005 adds the **Insert Field** button, which allows you to use field text in attributes.

5. To locate the attribute in the drawing, turn on the **Specify On-screen** option, or else enter coordinates for **X**, **Y**, and **Z**. (Attributes are usually located near or in the block.)

6. Click **OK**. Notice that AutoCAD displays the attribute *tag* at the pick point. (The attribute is not part of the block until you use the **BLOCK** command.)

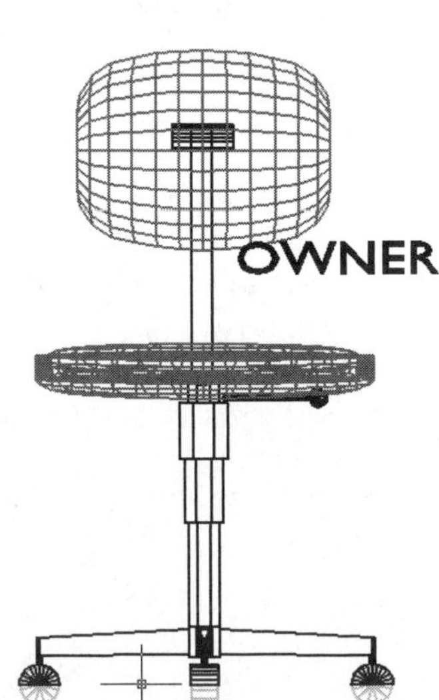

Attributes placed near or in blocks.

CREATING ATTRIBUTES: ADDITIONAL METHODS

A number of options and a system variable affect attribute modes.

- **Align Below Previous Attribute Definition** option lines up multiple attributes nicely.
- **Mode** option specifies how attributes are inserted and displayed.
- **AFLAGS** system variable presets mode options.
- **Text Options** option specifies the text style, height, and rotation of attributes.

Let's look at them.

Align Below Previous Attribute Definition

The **Align Below Previous Attribute Definition** option lines up multiple attributes nicely. When this option is turned on, AutoCAD places subsequent attribute tags below the previous one. Turning on this options turns off the **Insertion Point** and **Text Options** options.

Mode

The **Mode** option specifies how attributes are inserted and displayed:

- **Invisible** determines the attribute's visibility:
  - ☑ Attribute is invisible in drawings. The **ATTDISP** command can force the display of invisible attributes.
  - ☐ Attribute is visible (default).
- **Constant** fixes the attribute's value:
  - ☑ Attribute value is constant. (The **Prompt** field becomes unavailable.) During block insertion, AutoCAD does not prompt you to enter values. *Beware!* When you designate an attribute as constant, you cannot change its value.
  - ☐ Attribute value can be edited (default).
- **Verify** asks for verification during attribute insertion:
  - ☑ Attribute values verified. After entering attribute values, AutoCAD prompts to verify them:

    Verify attribute values

    Size of Valve <6">: *(Press* ENTER *if correct, or enter a different value.)*
  - ☐ Attribute value is not verified (default).
- **Preset** presets attribute values:
  - ☑ Attribute value is preset to the default. AutoCAD uses the default as the attribute, and during block insertion, does not prompt you to enter a value.
  - ☐ Attribute value can be edited (default).

AFlags

The **AFLAGS** system variable presets mode options.

Changing the setting of this system variable changes the default settings of the **Mode** section in the Attribute Definition dialog box. Mode settings of existing attributes are unaffected.

AFlags	*Meaning*
0	No attribute mode selected.
1	Invisible.
2	Constant.
4	Verify.
8	Preset.

Command: **aflags**

Enter new value for AFLAGS <0>: *(Enter the sum of values, such as* **3***.)*

The value stored in aflags is the sum of the numbers. For example, enter **3** if you want all attributes to be invisible (= 1), and have constant values (= 2) : 1 + 2 = 3.

Text Options

The **Text Options** option specifies the text style, height, and rotation, in a manner similar to the TEXT command. If the attributes are invisible, you may not want to make the text look nice.

Justification specifies the justification of the attribute text — left justified, aligned, fitted between two points, centered, and so on.

Text Style specifies a text style, as previously defined by the **STYLE** command.

Height specifies the height of the text. You can enter a value, or click the **Height** button to pick two points that indicate the height. (This option is not available when the style has a fixed text height, or when the justification is set to "Align.")

Rotation specifies the angle of the text. You can enter a value, or click the **Rotation** button to pick two points that indicate the angle. (Option not available when justification set to "Align" or "Fit.")

BLOCK

The BLOCK command attaches attribute definitions to blocks, after ATTDEF defines the attributes.

TUTORIAL: ATTACHING ATTRIBUTES TO BLOCKS

1. To attach attributes to blocks, start the **BLOCK** command:
   - From the **Draw** menu, choose **Block**, and then **Make**.
   - At the 'Command:' prompt, enter the **block** command.
   - Alternatively, enter the alias **b** at the keyboard.

   Command: **block** *(Press* **ENTER***.)*

   In all cases, AutoCAD displays the Block Definition dialog box.

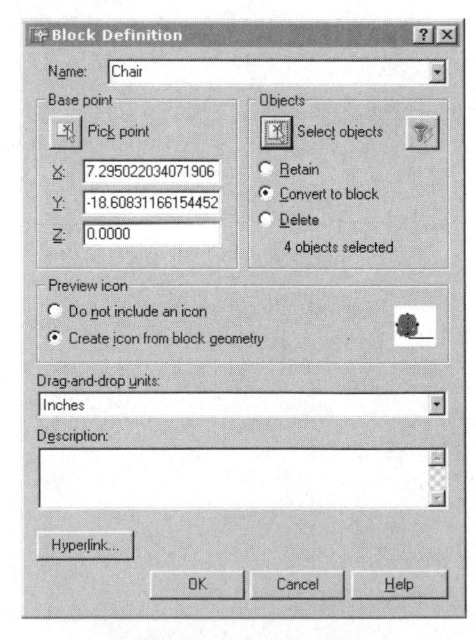

Block Definition dialog box.

2. Enter a name, and select a base point for the block.
3. When selecting objects, it is important to include the attribute definitions in the selection set. You may include blocks in the selection set.

   Select objects: *(Select objects, including attributes. Select each attribute in the order in which you wish them to be prompted during insertion.)*

   Select objects: *(Press* ENTER *to return to dialog box.)*
4. Click **OK**.

See Chapter 6 of *Using AutoCAD: Basics* for detailed information on creating blocks.

INSERT

The INSERT command inserts attributed blocks.

After the BLOCK command binds attribute definitions to objects (creating blocks), this command places the block in the drawing, and prompts you to fill in the attribute data.

BASIC TUTORIAL: INSERTING ATTRIBUTES IN DRAWINGS

1. To insert attributes in drawings, start the INSERT command:
   - From the **Insert** menu, choose **Block**.
   - At the 'Command:' prompt, enter the **insert** command.
   - Alternatively, enter the alias **i** at the keyboard.

   Command: **insert** *(Press* ENTER*.)*

   In all cases, AutoCAD displays the Insert dialog box.

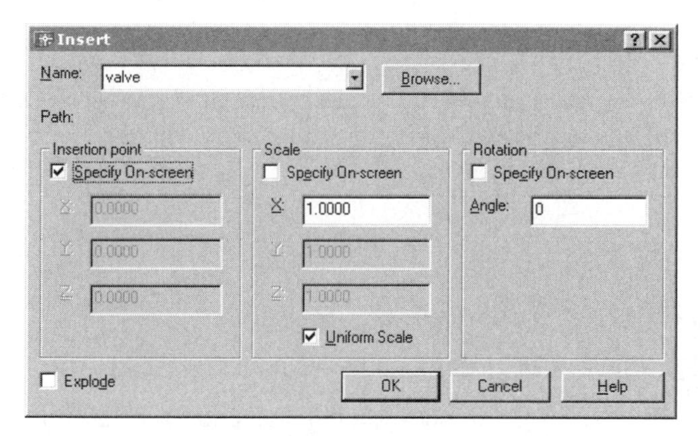

Insert dialog box.

2. From the **Name** drop list, select the name of a block.
3. I generally set the other options as follows:

Insertion Point	☑ **Specify on-screen**
Scale	**X** *(specify in dialog box)*
	☑ **Uniform scale**
Rotation	*(Varies, depending on type of block.)*
Explode	☐ *(Off)*

4. Click **OK**.

   What happens next depends on the options you selected in the Insert dialog box:

   - When the **Specify On-screen** option is off for all three (Insertion Point, Scale, and Rotation), AutoCAD displays the Edit Attributes dialog box. It displays those attributes not in **Constant** mode.

     Enter values for the attributes, and then click **OK**. (Clicking **Cancel** cancels the entire **INSERT** command.)

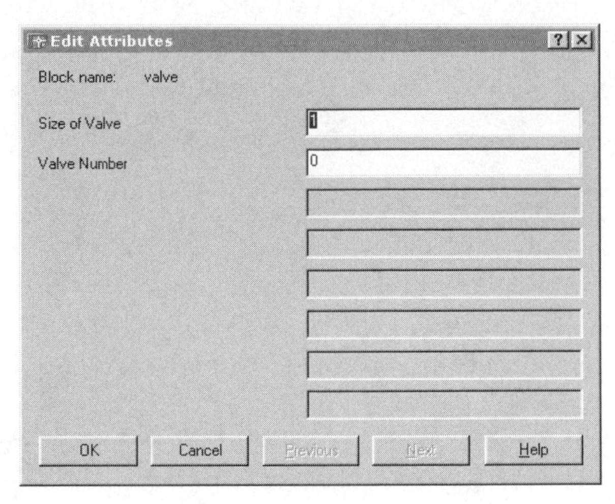

Edit Attributes dialog box.

   - When the **ATTDIA** system variable is set to 0, *and* at least one **Specify On-screen** option is turned on AutoCAD displays prompts at the command line:

     Specify insertion point or [Scale/X/Y/Z/Rotate/PScale/PX/PY/PZ/PRotate]: *(Pick a point.)*

     Enter attribute values

     Size of Valve <1>: **6"**

     Valve Number <0>: **123**

   Notice that AutoCAD inserts the block and associated attributes. If not set to invisible, the attributes appear near the block. In some cases, the attributes may be so tiny you that you need to zoom in to view them.

See Chapter 6 of *Using AutoCAD: Basics* for detailed information on inserting blocks.

INSERTING ATTRIBUTES: ADDITIONAL METHOD

The **INSERT** command is affected by these system variables:

- **ATTREQ** system variable determines whether attributes are filled in with default values.
- **ATTDIA** system variable determines whether the Edit Attributes dialog box is displayed.

Let's look at them.

AttReq

The **ATTREQ** system variable toggles whether attributes are filled in automatically (short for "attribute request").

This system variable suppresses attribute requests. When set to 0, the **INSERT** command does not ask for attribute values; all attributes are set to their default values. This is similar to using **Preset** mode in the **ATTDEF** command's Attribute Definition dialog box.

Command: **attreq**

Enter new value for ATTREQ <1>: **0**

The default setting is 1, which causes the **INSERT** and **-INSERT** commands to prompt for attribute values.

AttDia

The **ATTDIA** system variable determines whether the Edit Attributes dialog box is displayed (short for "attributes dialog").

AttDia	*Meaning*
0	Attribute prompts displayed on command line.
1	Attribute prompts displayed in dialog box.

It may be a bug that AutoCAD ignores the setting of this system variable when the **Specify On Screen** option is turned off for all three items — insertion point, scale, and rotation.

ATTDISP

The **ATTDISP** command controls the display of attributes in drawings (short for "attribute display").

You may not always want attributes to show, because the text can clutter drawings. This command determines the visibility of the attributes:

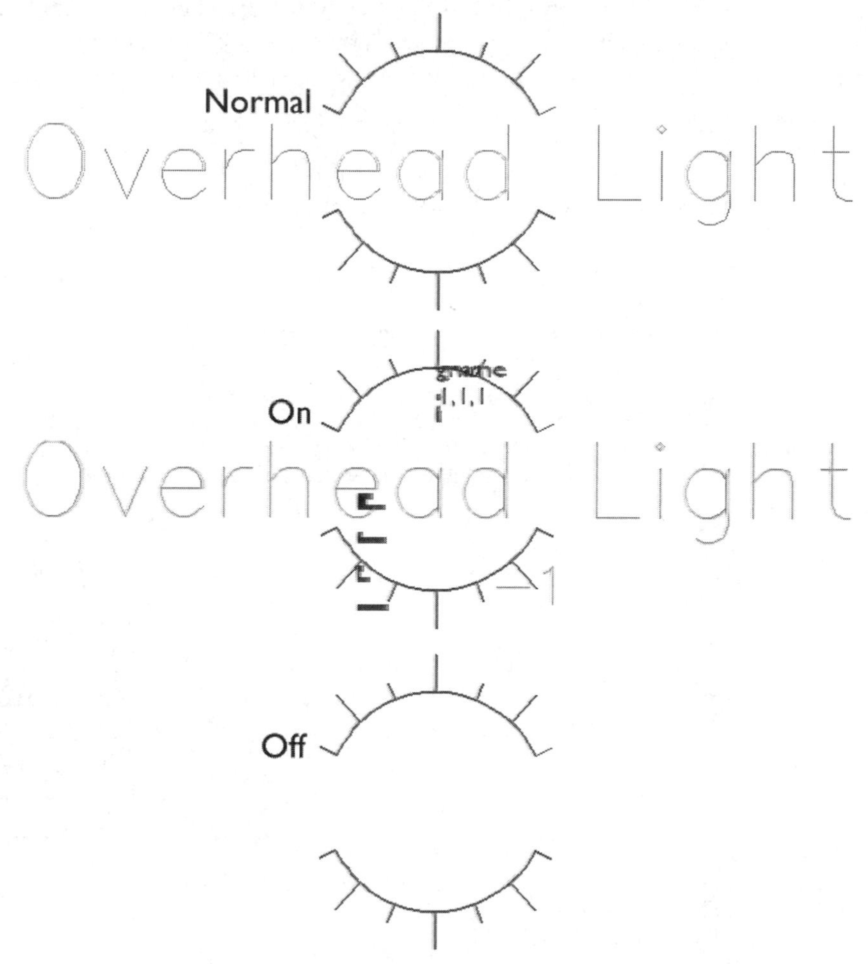

Top: *Normal does not display invisible attribute text.*
Center: *On displays all attribute text.*
Bottom: *Off hides all attribute text.*

Command: **attdisp**

Enter attribute visibility setting [Normal/ON/OFF] <Normal>: *(Enter an option.)*

The meaning of the options is:

- **Normal** displays attributes, unless set to **Invisible** mode by the **ATTDEF** command .
- **On** displays all attributes, even those set to **Invisible** mode.
- **Off** turns off the display of all attributes to reduce screen clutter.

After exiting the **ATTDISP** command, AutoCAD regenerates the display automatically to show the new state of the attributes (unless the **REGENAUTO** is turned off).

EATTEDIT

The **EATTEDIT** and **PROPERTIES** commands edit the values and properties of attribute insertions.

AutoCAD has two primary commands for editing attributes:

- To edit attributes in a block *insertion*, use the **EATTEDIT** command (short for "enhanced attribute editor"). It affects only the values and properties in the block you edit, and so is sometimes called "local attribute editing."
- To edit attributes in a block *definition*, use the **BATTMAN** command (short for "batch attribute manager"). It affects the attributes in all blocks inserted with that definition, and so is sometimes called "global attribute editing." It's like using the **ATTDEF** command all over again.

In summary, **EATTEDIT** affects only the blocks you select to edit, while **BATTMAN** affects all blocks of the same name.

EATTEDIT edits attribute values, text options, and general properties, such as layer and color. The **PROPERTIES** command edits only attribute values; the other properties belong to the block, not to the attribute. The figure below illustrates the different capabilities of the two commands.

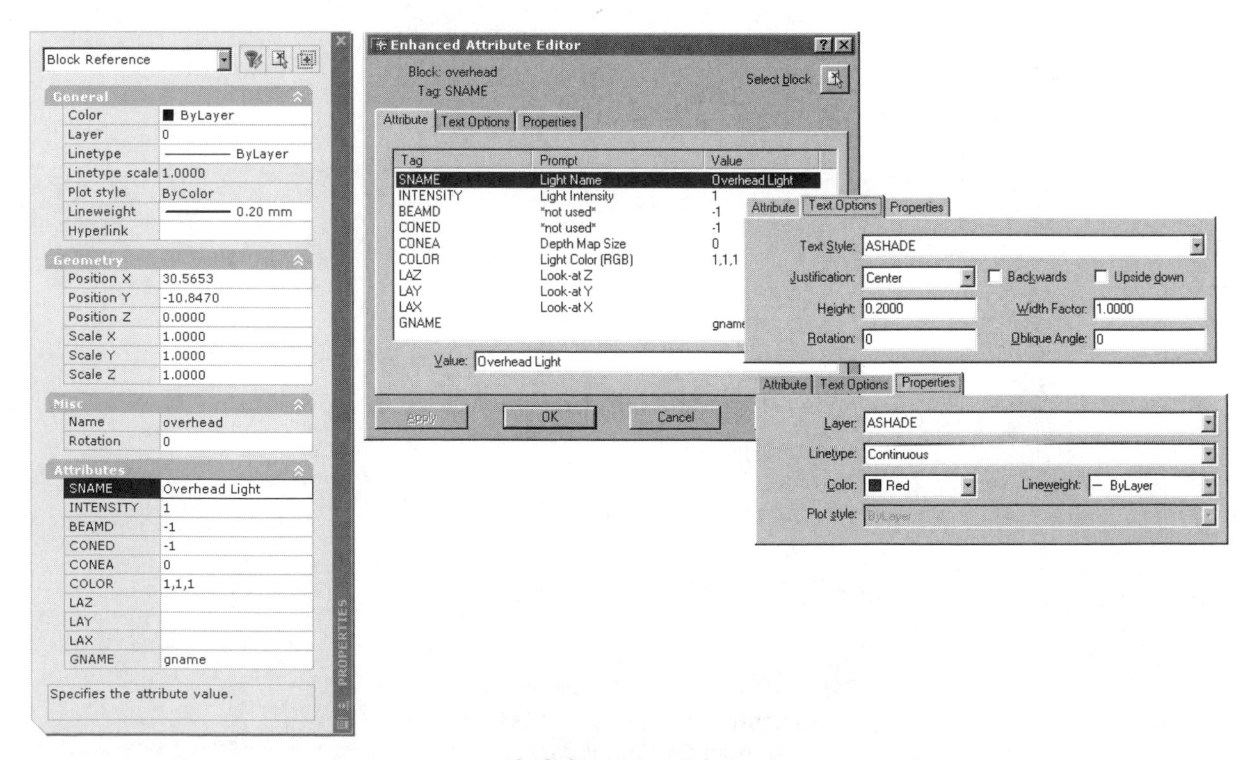

***Left**: Properties window.*
***Right**: Enhanced Attribute Editor dialog box.*

TUTORIAL: EDITING ATTRIBUTE VALUES

1. To edit the values of attributes in a single block, start the **EATTEDIT** command:
   - From the **Modify** menu, choose **Object**, **Attribute**, and then **Single**.
   - At the 'Command:' prompt, enter the **eattedit** command.

   Command: **eattedit** *(Press* ENTER.*)*
2. AutoCAD prompts you to select a block:

   Select a block: *(Pick a block.)*

   This command works with a single block at a time.
3. Notice that AutoCAD displays the Enhanced Attribute Editor dialog box, which has three tabs:
   - **Attributes** tab changes the value of attributes. The tags and prompts can only be changed with the **BATTMAN** command.
   - **Text Options** tab changes the text style, justification, height, and other properties of the attribute text. The font can only be changed with **STYLE** command.
   - **Properties** tab changes the layer, linetype, color, lineweight, and plot style of the attribute text. The properties of the block can be changed with the **PROPERTIES** command.
4. Once you finish changing the attributes of one block, click the **Select block** button to edit another one.
5. Choose **OK**.

BATTMAN

The BATTMAN command edits attributes and properties in block definitions (short for "block attribute manager").

This command makes global changes to attributes in all insertions of a single block. In addition, it can remove attributes, and can change the order in which attributes are prompted. (If you selected the attributes by windowing during the BLOCK command, AutoCAD curiously displays attribute prompts in the reverse order from which you created them.)

Note: The **BATTMAN** command changes block and attribute *definitions.* This means that any changes made by this command affect all insertions of the block.

BASIC TUTORIAL: CHANGING ATTRIBUTES GLOBALLY

1. To insert attributes in drawings, start the **BATTMAN** command:
   - From the **Modify** menu, choose **Objects**, **Attributes**, and then **Global**.
   - At the 'Command:' prompt, enter the **battman** command.

   Command: **battman** *(Press* ENTER.*)*

   AutoCAD displays the Block Attribute Manager dialog box, along with the attributes of the first block (in alphabetical order).
2. To view the attributes of another block, select its name from the list.

   To see more data at one time, stretch the dialog box by grabbing an edge with the cursor, and dragging.

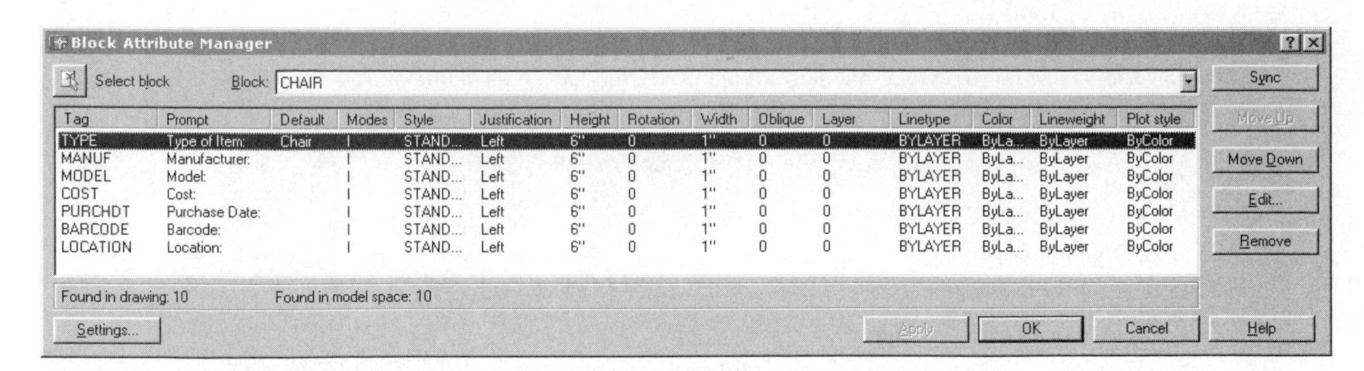

Block Attribute Manager dialog box.

3. The buttons along the right provide important control over attributes:
   - **Sync** applies the changes you make to all insertions (a.k.a. instances) of the block. This button provides the same function at the **ATTSYNC** command. Upon exiting this dialog box with the **OK** button, AutoCAD automatically synchronizes block instances to reflect the changes you made (although this behavior can be modified by the Settings button, described below).
   - **Move Up** and **Move Down** allow you to change the order in which the attribute values appear. This is a manual fix to AutoCAD's flaw in prompting for attribute data in reverse order, which forces the **INSERT** command to display attributes in the "correct" order.
   - **Edit** displays a dialog box for changing the attribute data and mode, text, and properties.

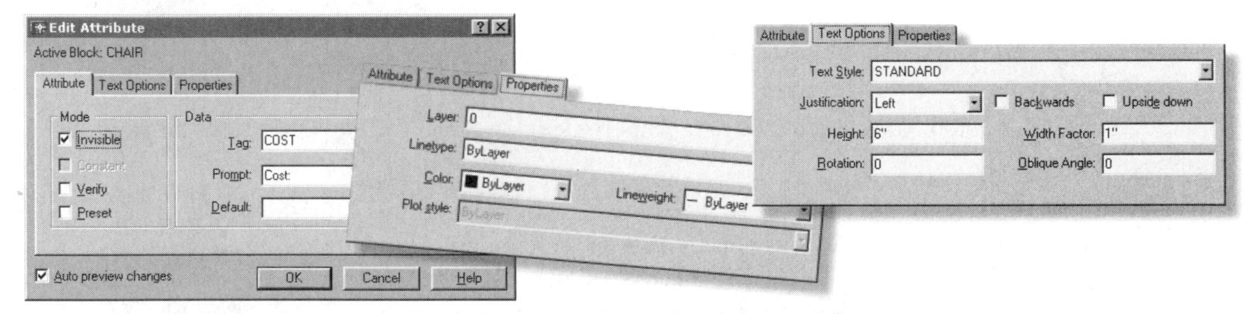

Editing block and attribute properties.

   - **Remove** deletes the selected attribute from the block. If you change your mind and want to restore the attribute, click **Cancel**, or use the **U** command after exiting this dialog box.
4. The buttons along the bottom provide control over the **BATTMAN** command:
   - **Apply** applies the changes to the block definition without exiting the dialog box.
   - **Settings** displays a dialog box that lets you decide which properties to display. For example, if you clear the check mark next to **Rotation**, the rotation angle is no longer displayed in the list of attributes.

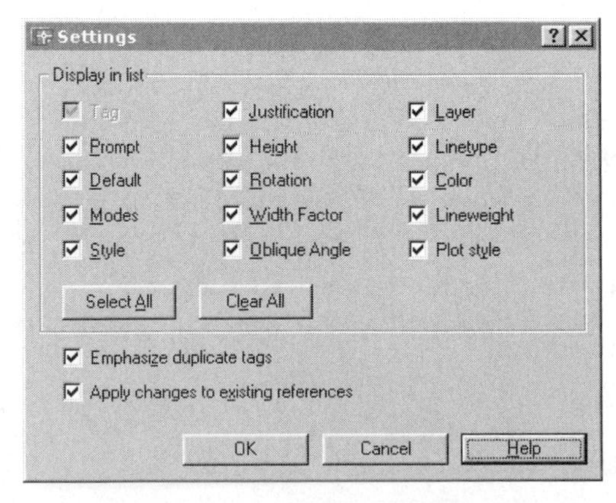

Settings options for the Battman command.

- **Emphasize duplicate tags** causes AutoCAD to display duplicates in red. This lets you find duplicate tag names more easily. (I don't understand why AutoCAD would allow duplicates in the first place.)
- **Apply changes to existing references** determines what happens to changes when you exit this dialog box:
  ☑ All instances of the block are updated with the changes.
  ☐ Only new instances of the block are affected by the changes; existing blocks are unaffected. This setting however, is overridden by the **Sync** button.

5. Choose **OK**.

Note: **BATTMAN** cannot add attributes to block definitions. To add attributes, explode the block, and then use the **ATTDEF** and **BLOCK** commands or the **ATTREDEF** command.

EATTEXT

The EATTEXT command extracts attributes from drawings to data files (short for "enhanced attribute extraction").

This command extracts attribute data in several formats: Excel spreadsheet, Access database, ASCII comma-delimited, and tab-delimited formats. The Excel and Access formats are available only if the software is installed on your computer; if you prefer use alternatives to Microsoft-branded software, no worries: the "Copy to Clipboard" option makes this command compatible with just about any software. (The "old" ATTEXT command also extracts attribute data in partial-DXF format.)

BASIC TUTORIAL: EXTRACTING ATTRIBUTES

1. To extract attribute data from drawings, start the **EATTEXT** command:
   - From the **Tools** menu, choose **Attribute Extraction**.
   - At the 'Command:' prompt, enter the **eattext** command.

     Command: **eattext** *(Press* ENTER.*)*

   AutoCAD displays the Attribute Extraction wizard, which guides you through the steps needed to select drawings, blocks, attributes, and export formats.
2. AutoCAD allows you to select the blocks from which to extract attributes:
   - **Select objects** selects specific blocks in the current drawing.
   - **Current drawing** selects *all* blocks in the current drawing.
   - **Select drawings** selects all blocks in other drawings, such as those on a network drive.

   Recall that attributes are found in blocks only. That's why you select blocks to extract attributes.

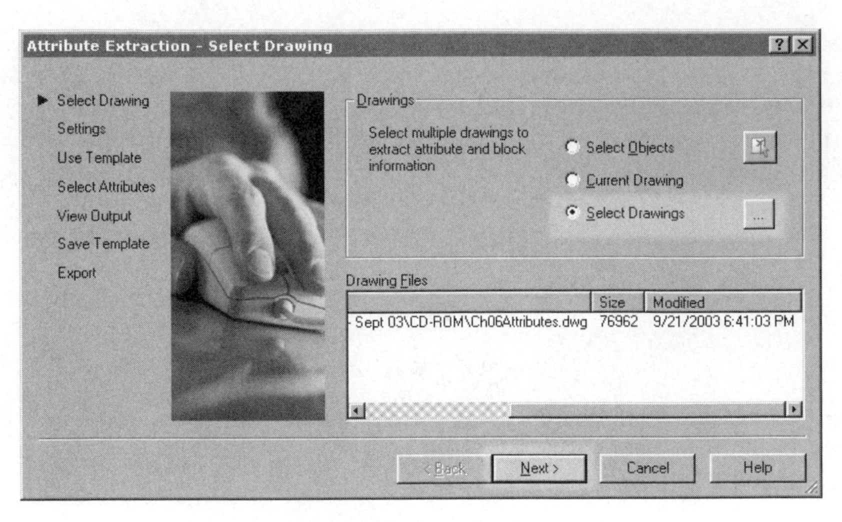

Selecting blocks or drawings.

For this exercise, click **Select Drawing**, and then open the *Ch06Attributes.dwg* file from the CD-ROM.

Click **Next**.

3. AutoCAD can optionally extract attributes from nested blocks (blocks in blocks) and blocks stored in xrefs.

   In this exercise, the drawing has neither xrefs nor nested blocks. Uncheck both options, and then click **Next**.

4. Like a drawing template, the attribute extraction *template* predefines which blocks and attributes are extracted. The **Use Template** step does not, however, apply the first time you use this command, because templates are not created until later.

   For this exercise, select **No template**, and click **Next**.

5. The next step is to select the blocks and attributes to be extracted; the list of blocks and attributes can seem overwhelming.

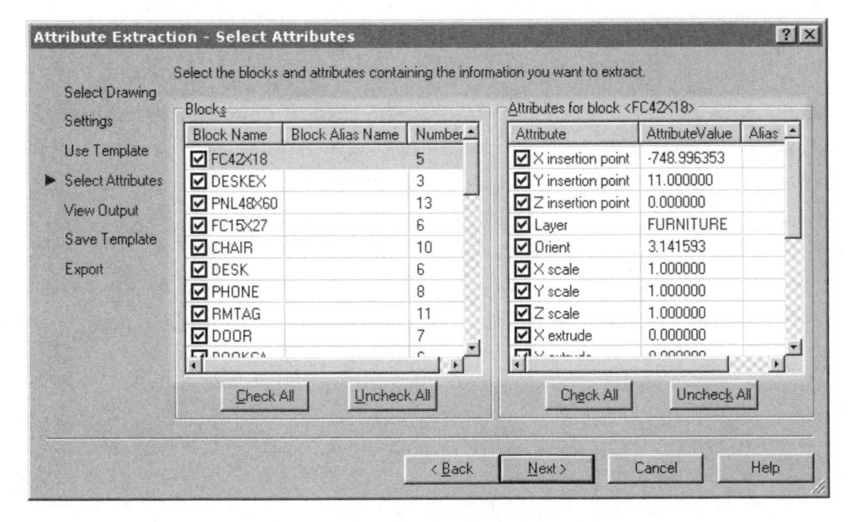

All the blocks, and all the attributes.

- **Blocks** lists the names of blocks found in the drawing. The Block Alias column allows you to assign alias names to block, while the Number column indicates the number of occurrences of each block.
- **Attributes** column lists the attributes for the highlighted block name. Notice that there are additional attributes that you did not create earlier in this exercise, such as "X insertion point" and "Layer."

Note: Block aliases give prettier names to blocks in extracted data, such as in spreadsheets. In the list above, the first block is a 42"x18" filing cabinet named "FC42X18." This appears as the block name in the spreadsheet. After assigning an alias, such as "Cabinet, Filing, 42x18," the alias appears in place of the name.

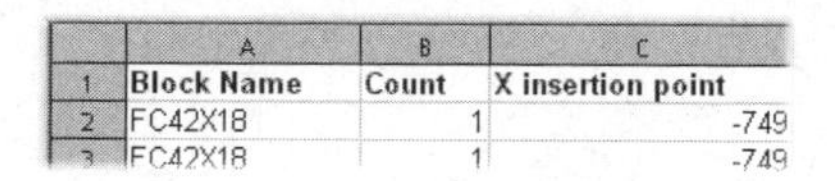

	A	B	C
1	Block Name	Count	X insertion point
2	FC42X18	1	-749
3	FC42X18	1	-749

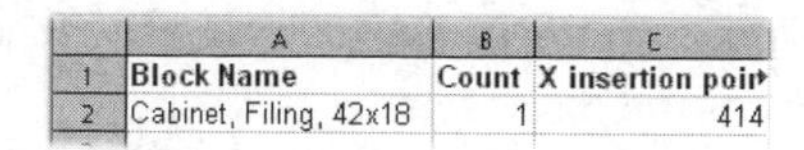

	A	B	C
1	Block Name	Count	X insertion poir
2	Cabinet, Filing, 42x18	1	414

***Left**: Block name appears in spreadsheet.*
***Right**: Block alias appears in place of block name.*

To assign an alias, click the blank area under Block Alias Name, and then type in the name. I recommend a logical name structure to sort blocks by function more easily: "Cabinet, Filing, 42x18" represents type of furniture, type of cabinet, size.

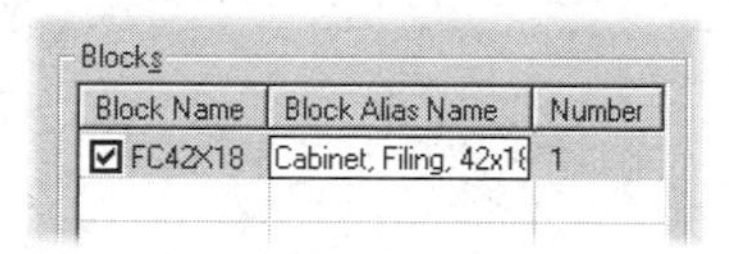

Blocks

Block Name	Block Alias Name	Number
☑ FC42X18	Cabinet, Filing, 42x1{	1

Assigning aliases to block names.

Similarly, you can assign aliases to the attribute tags.

The attributes assigned to block FC42X18 are Location, Barcode, PurchDt, Cost, Model, Manuf, and Type.

Attributes for block <FC42X18>

Attribute	AttributeValue	Alias
☑ X insertion point	-748.996353	X-coordinate
☑ Y insertion point	11.000000	Y-coordinate
☐ Z insertion point	0.000000	
☐ Layer	FURNITURE	
☐ Orient	3.141593	
☐ X scale	1.000000	
☐ Y scale	1.000000	
☐ Z scale	1.000000	
☐ X extrude	0.000000	
☐ Y extrude	0.000000	
☐ Z extrude	1.000000	
☑ TYPE	File cab. 42x18	
☑ MANUF	Sierra Furnitu...	Manufacturer
☑ MODEL	F3357	Model No.
☑ COST	226.00	
☑ PURCHDT	07-03-92	Purchase Date
☑ BARCODE	03010	
☑ LOCATION	2201	

Check All Uncheck All

User-defined and AutoCAD-generated attributes.

When you examine the list in the Select Attributes dialog box, there seem to be many more attributes. That's because AutoCAD automatically generates additional attributes that provide additional information helping to describe the block. These are:

X, Y, Z insertion point are the x,y,z coordinates of the block's insertion point, which pinpoint the location of the block in the drawing. In 2D drawings, the z-coordinate is almost always 0.

Layer is the name of the layer on which the block resides.

Orient is the rotation angle of the block.

X, Y, Z scale are the scale factors of the block in the x, y, and z directions.

X, Y, Z extrude are the extrusion distances in the x, y, and z directions. The curious thing is that there is no such thing in AutoCAD as an x or y extrusion distance, so they always equal 0.

All attributes are initially selected.

For this exercise, click **Uncheck All**, and then select the following blocks.

Give them the aliases listed below:

Block	***Alias***
FC42X18	Cabinet, Filing, 42x18
DESKEX	Desk, Executive
PNL48x60	Partition, Panel, 48x60
FC15X27	Cabinet, Filing, 15x27

For some of the blocks listed above, select the following additional attributes, and then assign aliases:

Attribute	***Alias***
X insertion point	X-coordinate
Y insertion point	X-coordinate
Location	
Barcode	
PurchDt	Purchase Date
Cost	
Model	Model No.
Manuf	Manufacturer
Type	

It's tedious work, but someone's gotta do it. And then choose **Next**.

6. AutoCAD lists the attribute data in a spreadsheet-like format: each row lists the attribute data to be extracted for each block. In the figure below, notice the use of alias names. (To see all the data, stretch the dialog box by grabbing the edge with the cursor, and then dragging.)

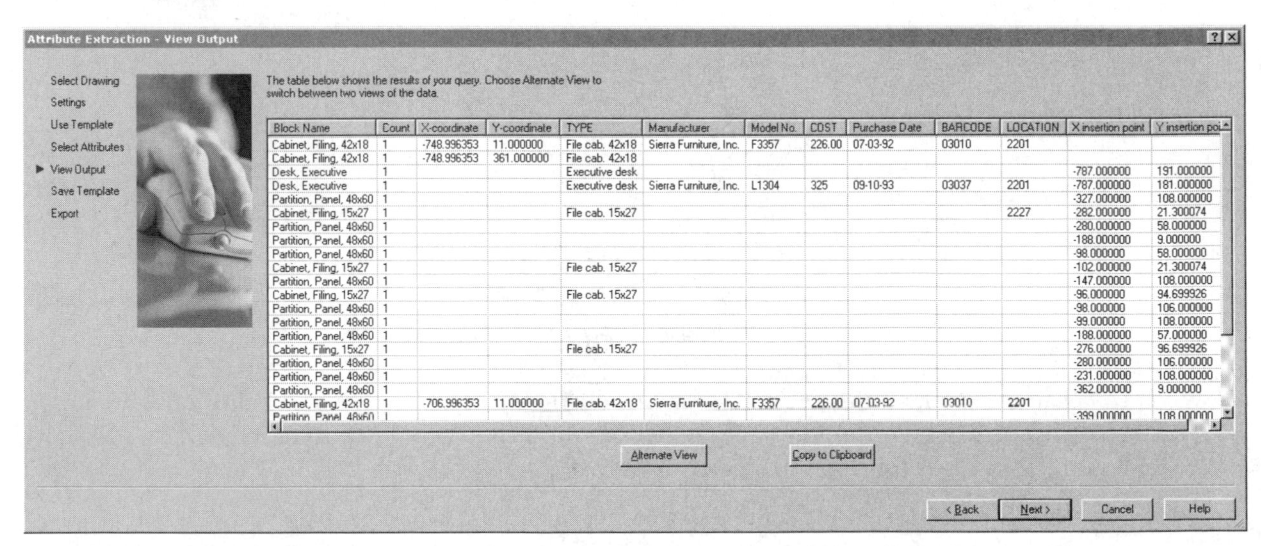

Data to be extracted.

7. I find the easiest way to get the data into a spreadsheet is as follows:
   a. Click **Copy to Clipboard**. (You cannot select a subset to copy; you need to do that in the previous step, "Select Attributes.")

b. Switch to the spreadsheet program. (A free one is available from www.openoffice.org.)
c. Press **CTRL+V** to paste the data into the spreadsheet.
d. If you wish, apply formatting to the presentation.

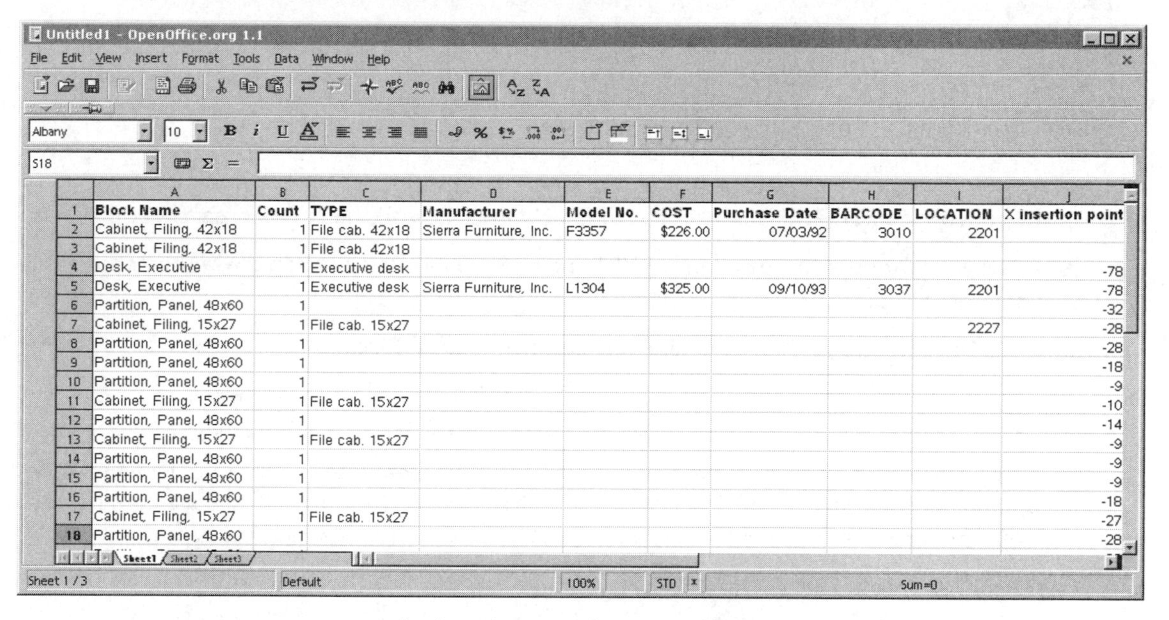

	Block Name	Count	TYPE	Manufacturer	Model No.	COST	Purchase Date	BARCODE	LOCATION	X insertion point
2	Cabinet, Filing, 42x18	1	File cab. 42x18	Sierra Furniture, Inc.	F3357	$226.00	07/03/92	3010	2201	
3	Cabinet, Filing, 42x18	1	File cab. 42x18							
4	Desk, Executive	1	Executive desk							-78
5	Desk, Executive	1	Executive desk	Sierra Furniture, Inc.	L1304	$325.00	09/10/93	3037	2201	-78
6	Partition, Panel, 48x60	1								-32
7	Cabinet, Filing, 15x27	1	File cab. 15x27						2227	-28
8	Partition, Panel, 48x60	1								-28
9	Partition, Panel, 48x60	1								-18
10	Partition, Panel, 48x60	1								-9
11	Cabinet, Filing, 15x27	1	File cab. 15x27							-10
12	Partition, Panel, 48x60	1								-14
13	Cabinet, Filing, 15x27	1	File cab. 15x27							-9
14	Partition, Panel, 48x60	1								-9
15	Partition, Panel, 48x60	1								-9
16	Partition, Panel, 48x60	1								-18
17	Cabinet, Filing, 15x27	1	File cab. 15x27							-27
18	Partition, Panel, 48x60	1								-28

Attribute data pasted into spreadsheet.

Note: AutoCAD copies the data to Clipboard in tab-delimited ASCII format. You can paste the data in (almost) any other Windows application — including AutoCAD drawings — by using the **Edit | Paste** command.

8. To see the data in another view, click **Alternate View**. This view shows every attribute in a single column. If you click **Copy to Clipboard**, the data is copied in this format.
   Choose **Next**.
9. If you wish to save the settings, such as the aliases and selections, click **Save Template**. AutoCAD stores the settings to a **.blk* file (block template), which you can use the next time you invoke this command with this drawing.
   Choose **Next**.
10. If you used the Copy to Clipboard option earlier, you can skip this step.
    If not, then choose one of the available formats: :
    - **CSV (Command delimited) (*.csv)** creates an ASCII file where the attribute values are separated by commas:

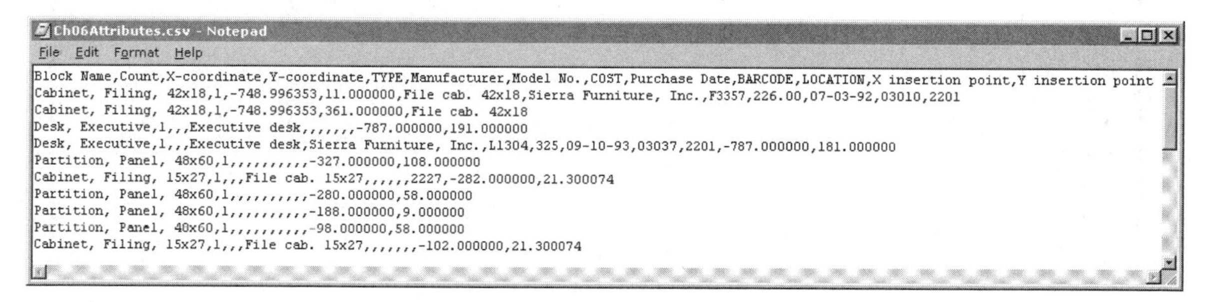

```
Block Name,Count,X-coordinate,Y-coordinate,TYPE,Manufacturer,Model No.,COST,Purchase Date,BARCODE,LOCATION,X insertion point,Y insertion point
Cabinet, Filing, 42x18,1,-748.996353,11.000000,File cab. 42x18,Sierra Furniture, Inc.,F3357,226.00,07-03-92,03010,2201
Cabinet, Filing, 42x18,1,-748.996353,361.000000,File cab. 42x18
Desk, Executive,1,,,Executive desk,,,,,,,-787.000000,191.000000
Desk, Executive,1,,,Executive desk,Sierra Furniture, Inc.,L1304,325,09-10-93,03037,2201,-787.000000,181.000000
Partition, Panel, 48x60,1,,,,,,,,,,,-327.000000,108.000000
Cabinet, Filing, 15x27,1,,,File cab. 15x27,,,,,,2227,-282.000000,21.300074
Partition, Panel, 48x60,1,,,,,,,,,,,-280.000000,58.000000
Partition, Panel, 48x60,1,,,,,,,,,,,-188.000000,9.000000
Partition, Panel, 40x60,1,,,,,,,,,,,-98.000000,58.000000
Cabinet, Filing, 15x27,1,,,File cab. 15x27,,,,,,,-102.000000,21.300074
```

Comma-separated values.

- **Tab Delimited File (*.txt)** creates an ASCII file where the attribute values are separated by tabs:

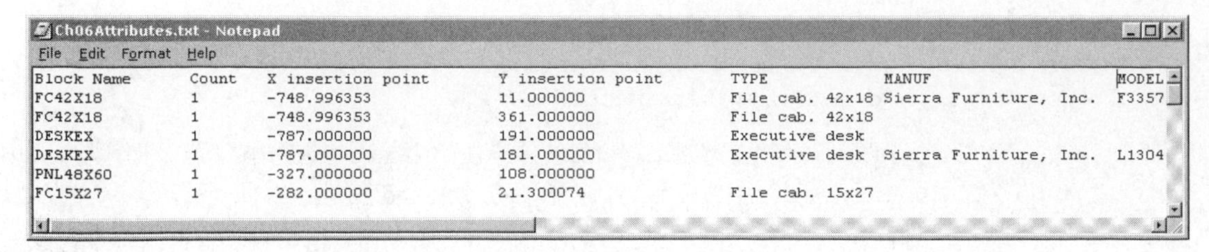

Ch06Attributes.txt - Notepad

Block Name	Count	X insertion point	Y insertion point	TYPE	MANUF	MODEL
FC42X18	1	-748.996353	11.000000	File cab. 42x18	Sierra Furniture, Inc.	F3357
FC42X18	1	-748.996353	361.000000	File cab. 42x18		
DESKEX	1	-787.000000	191.000000	Executive desk		
DESKEX	1	-787.000000	181.000000	Executive desk	Sierra Furniture, Inc.	L1304
PNL48X60	1	-327.000000	108.000000			
FC15X27	1	-282.000000	21.300074	File cab. 15x27		

Tab-delimited format.

- **Microsoft Excel (*.xls)** creates an *.xls* file compatible with the Excel spreadsheet program.
- **Microsoft Access Database (*.mdb)** creates an *.mdb* file compatible with Microsoft's Access database program.

Select a file format, and then enter a file name. Then choose **Finish**.

AutoCAD extracts the attribute data, and then deposits it in the file.

Note: This command allows you to extract attributes from more than one drawing at a time. When the drawings have the same name for a block, AutoCAD adds a tilde (~), the path, and the drawing file name. When drawing *Office.dwg* and *Shop.dwg* both have block "FC42X18," it is renamed as follows:

FC42X18~C:\drawings\office.dwg

FC42X18~C:\drawings\shop.dwg

ATTRIBUTE TUTORIAL

Suppose you are the manager of an engineering department that uses AutoCAD. (See how your knowledge of CAD has already helped your career?) Your department has several CAD workstations, each with a desk. You want to use attributes to keep track of the desks, also known as "capital assets."

To prepare for this tutorial, start AutoCAD with a new drawing. Set the limits of 0,0 and 252,132, and then **ZOOM All**. Set the grid and snap to 12". With the **MLINE** command, draw a 240" x 120" (20'x10') room outline. Save the drawing as "attributetutorial.dwg."

20'x10' office plan.

STEP 1: CREATING THE SYMBOL

For this tutorial, the symbol is a standard office desk.

1. Draw a 24" x 36" desk with the **RECTANG** command; set the polyline width to 1".
   It may help to zoom in a bit with the **ZOOM Window** command.

Standard-size office desk.

STEP 2: DEFINING ATTRIBUTE TAGS, PROMPTS, AND VALUES

The ATTDEF command defines attributes by tag, prompt, and default value. For this exercise, the attributes are: the name of employee, the type of computer, and the telephone extension number.

Planning the Tags, Prompts, and Values

Before creating attributes, it is important to preplan attributes:

How many would be useful, how many overkill?

What are the tags?

Do the prompts provide sufficient information?

What format should the default value be in?

And, in which order should the attributes be presented?

When answering these questions, keep in mind that data has to be input (therefore, perhaps a nuisance to CAD operators) and output (therefore must appear logical in the spreadsheet or database to users).

The *tag* identifies the attribute. Use these tags for the attributes in this tutorial:

Employee_name
Computer
Tel_ext

Notice that underscores are used in place of spaces, because blank spaces are not allowed in tag names. AutoCAD automatically converts all text to uppercase.

The *prompt* reminds you of the attribute's purpose. Use these prompts for this tutorial:

Name of the employee
Computer brand name
Telephone extension number

You do not need to include a colon (:) at the end of the prompts, because AutoCAD adds them automatically.

The *value* is the default data. These values are used:

John Doe
IBM PC
100

With the attribute tags, prompts, and values planned out, go ahead and create them.

Defining the Attributes

1. To create the attributes, start the **ATTDEF** command (from the menu, select **Draw | Block | Define Attributes**):

    Command: **attdef**

2. **Tag** identifies the attribute. For this exercise, enter the tag:

    Tag: **Employee_name**

3. **Prompt** specifies the text that appears on the command line when the block is inserted. Enter the prompt:

    Prompt: **Name of the employee**

    (To make the prompt the same as the tag, press **TAB;** do not press **ENTER**, because that dismisses the dialog box).

4. **Value** specifies the default value of the attribute. Enter the value:

    Value: **John Doe**

    This value is displayed as the default later when you insert the block.

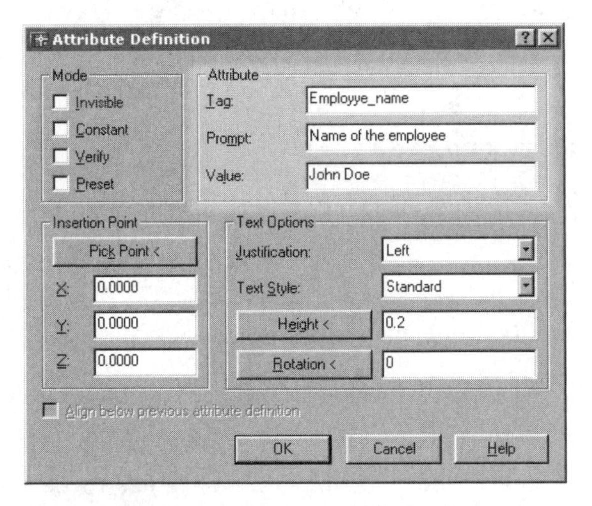

Filling in tag, prompt, and default value.

5. Dealing with the remainder of the dialog box consists of filling in a series of fields much like placing text. The location and text size you specify become the location of the attribute text with the inserted block.

    Enter a text **Height** of 1.0.

6. To locate the attribute text in the drawing, choose the **Pick Point** button.

    AutoCAD clears the dialog box.

    Turn off snap (click the **SNAP** button on the status bar), and then select a point inside the desk.

    Start point: *(Pick a point.)*

    AutoCAD returns the dialog box.

7. You have finished defining the first attribute. Choose **OK** to exit the dialog box.

Notice that the EMPLOYEE_NAME attribute tag appears with the desk.

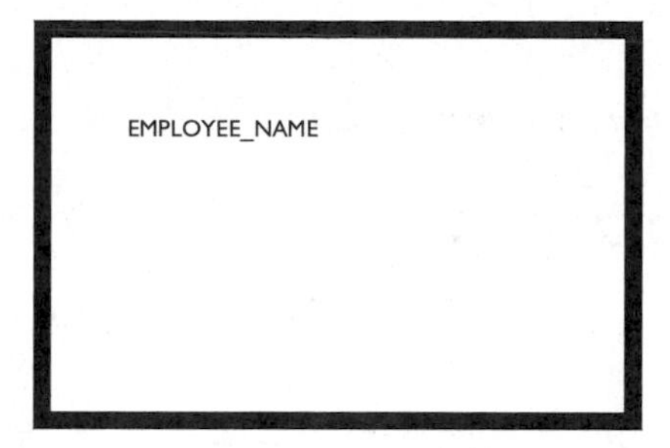

First attribute definition.

8. To create the two other attributes, repeat the **ATTDEF** command:

   Command: *(Press spacebar.)*

   This time, enter the attribute data for the computer:

   Tag: **Computer**

   Prompt: **Computer brand name**

   Value: **IBM PC**

9. Choose the **Align below previous attribute definition** option to turn it on.

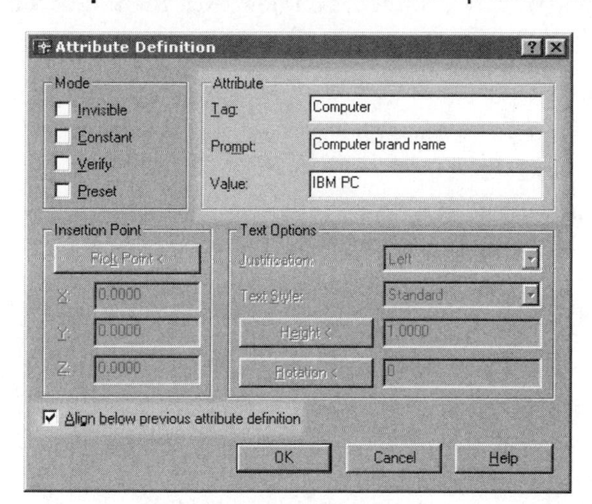

Lining up attributes.

10. Click **OK**, and notice that the COMPUTER tag appears below the previous tag.

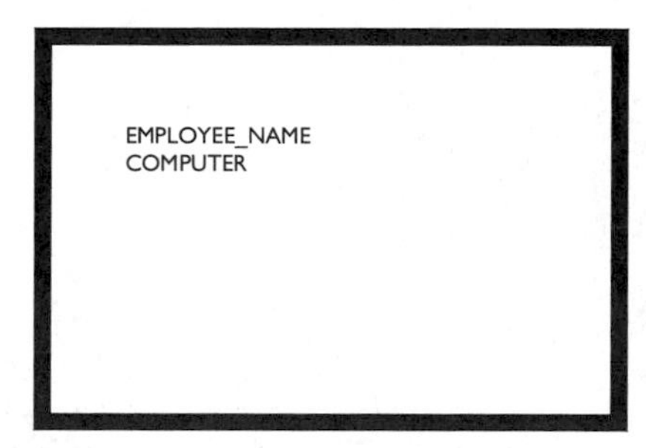

Second attribute definition.

11. Now, for the telephone extension number. Press the spacebar to repeat the **ATTDEF** command, and enter the following:

    Tag: **Tel_ext**

    Prompt: **Telephone extension number**

    Value: **100**

12. After choosing **OK**, you see the three tags inside the desk.

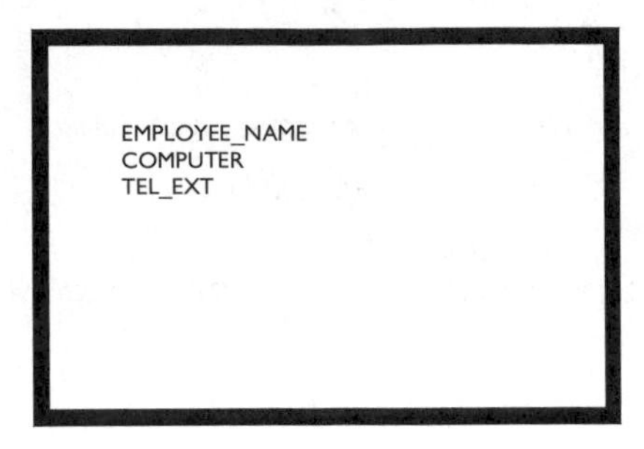

Third attribute definition.

STEP 3: CREATING THE BLOCK

With the desk symbol and attributes defined, you now turn them into a block.

1. From the menu, select **Draw | Block | Make**, or enter the command:

   Command: **block**

   AutoCAD displays the Block Definition dialog box.

2. Enter a name for the block:

   Name: **workstation**

3. To specify the insertion point of the block, click **Pick Point**.

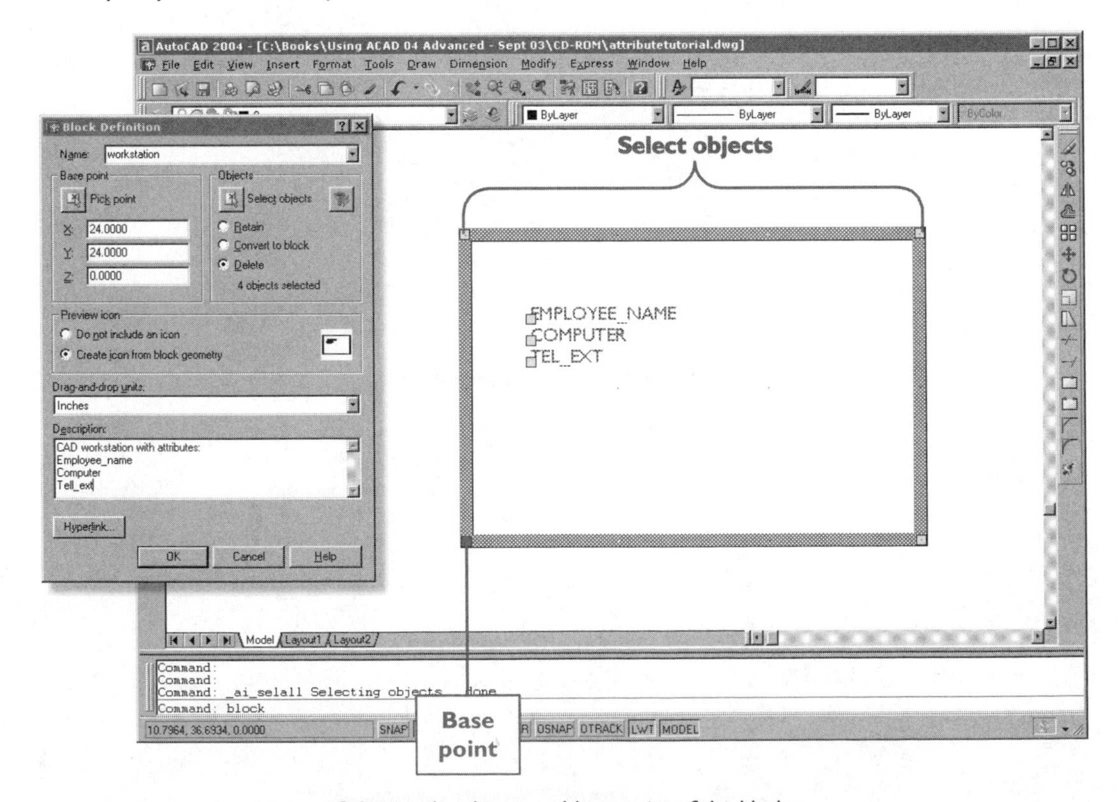

Selecting the objects and base point of the block.

The dialog box disappears, and AutoCAD prompts:

Specify insertion base point: **int**

of *(Select the lower left corner of the desk.)*

The dialog box returns. Notice that AutoCAD has filled in the x, y, z coordinates of the pick point.

4. To tell AutoCAD which objects are part of the block, click the **Select objects** button. Again, the dialog box disappears, and AutoCAD prompts you to select the objects that form the block. Use Crossing mode to select the desk and three attribute tags:

   Select objects: **c**

   Specify first corner: *(Pick a point.)*

   Specify opposite corner: *(Pick another point.)*

   Select objects: *(Press* ENTER *to end object selection, and return to the dialog box.)*

Note: If you want the attributes to appear in a specific order (when you later insert the block), do not use a Window or Crossing selection mode at the "Select objects" prompt. Instead, select symbol, and then select the attribute definitions — one by one — in the same order you want them presented.

5. When finished selecting the objects, press **ENTER** to return to the dialog box. Notice that the dialog box reports "4 objects selected." Specify the following options in the dialog box:

Objects:	**Delete**
Create icon from block geometry:	☑ *(Yes.)*
Insert units:	**Inches**
Description:	**CAD workstation with attributes:**
	Employee_name
	Computer
	Tel_ext

6. Choose the **OK** button to dismiss the dialog box, and define the block.

   Do not be alarmed when the symbol and attribute definitions disappear; they are stored safely in the drawing.

   You are now ready to insert the block with its attributes in the drawing.

STEP 4: INSERTING THE BLOCK

1. Before inserting the Workstation block, ZOOM **All** to see the entire office.
2. From the menu bar, select **Insert | Block**, or enter the command:

   Command: **insert**

   The Insert dialog box appears.
3. Ensure the name of the block is Workstation.

   Respond to the other options presented by the dialog box as follows:

Insertion point:	**Specify on-screen**
X Scale:	**1.0**
Uniform Scale:	☑ *(On)*
Rotation Angle:	**0**
Explode:	☐ *(Off)*

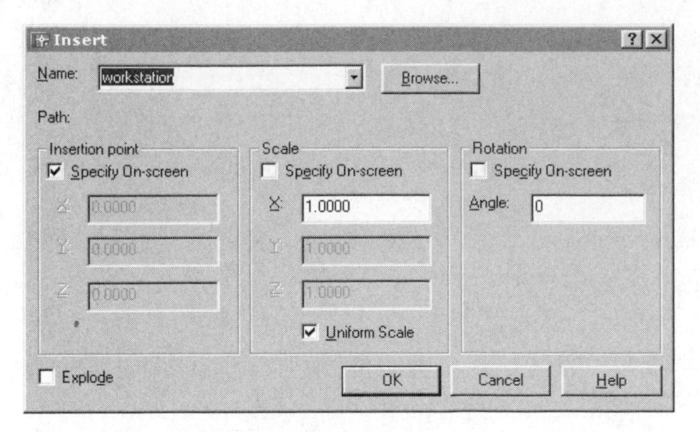

Insert dialog box properties.

4. Choose **OK**.
5. AutoCAD switches to the command line, prompting you for the block's insertion point.

   Turn on snap mode, and then pick a point inside the office walls:

   Specify insertion point or [Scale/X/Y/Z/Rotate/PScale/PX/PY/PZ/PRotate]: *(Pick a point anywhere inside the office walls.)*
6. The final step is to provide values for the attributes.

   If **ATTDIA** is set to 1, AutoCAD asks for the attribute values in a dialog box, otherwise at the command line. (Curiously, the **WBLOCK** command reverses the attribute order.)

   Enter attribute values, as follows. (They may appear in a different order than listed here):

   Telephone extension number <100>: **124**

   Computer brand name <IBM PC>: *(Press* ENTER.*)*

   Enter employee name <John Doe>: **Heidi Lynn**

   Notice that the default values are provided in angle brackets; to accept them, press **ENTER**. AutoCAD places the desk, and shows the attribute values.

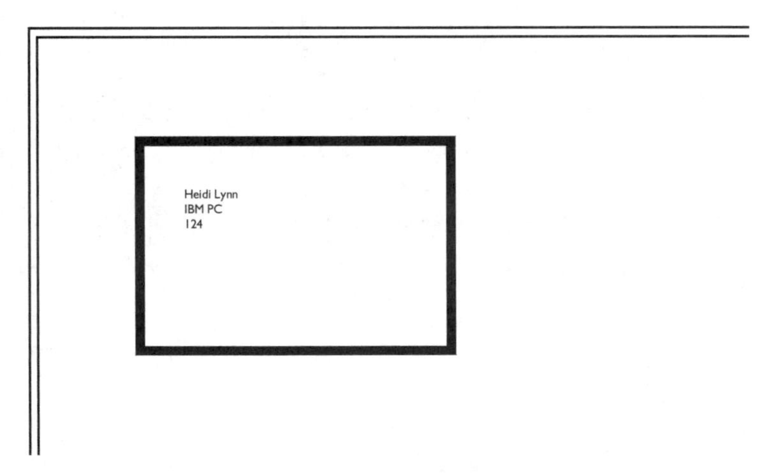

Inserted block with attribute data.

7. Place several more desk symbols.

   You can use any values you wish, or follow along with these values:

Employee	*Computer Name*	*Tel. Ex.*
Stefan Scott	Compaq	115
Bill Fane	Dell	103
Stephen Dunning	Gateway	107
Kyrie Eleison	Palm	111
Katrina Nicole	Macintosh	112

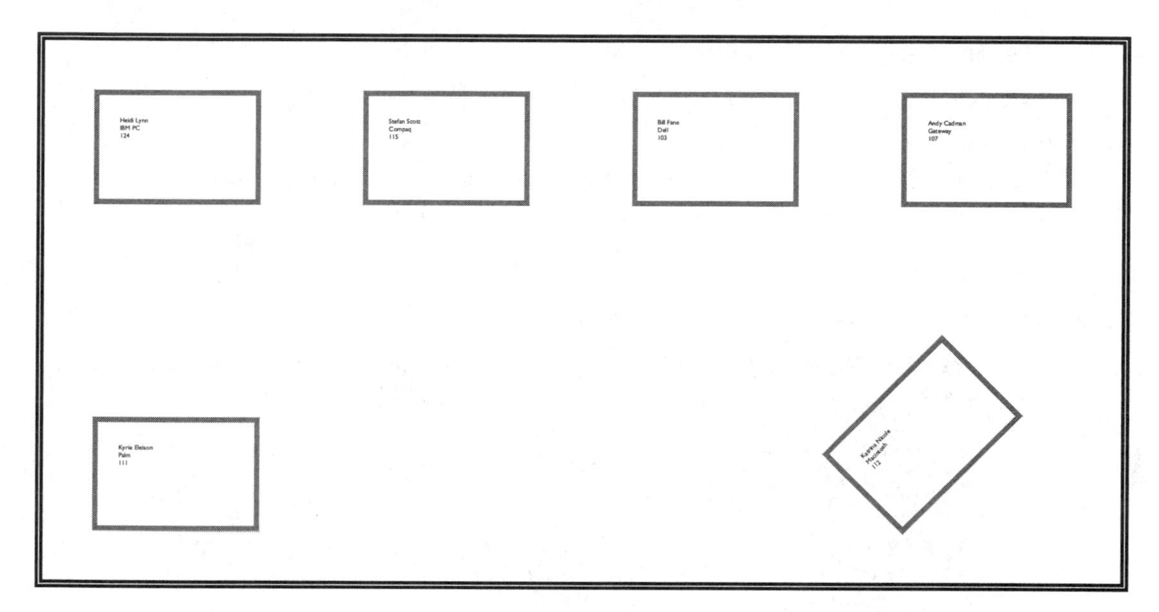

Completed drawing.

8. Save the drawing.

STEP 5: EDITING ATTRIBUTES

Now that everything is all set up properly, suppose Stephen Dunning leaves, and Andy Cadmann is hired to take his place. You have to change the attribute database.

1. Start the **EATTEDIT** command — or, double-click the attribute.

   In either case, AutoCAD prompts you to select a block:

   Select a block: *(Select Stephen Dunning's desk.)*

   If you need a closer look to figure out which desk belonged to the late Dr Dunning, use the transparent **'ZOOM** command.

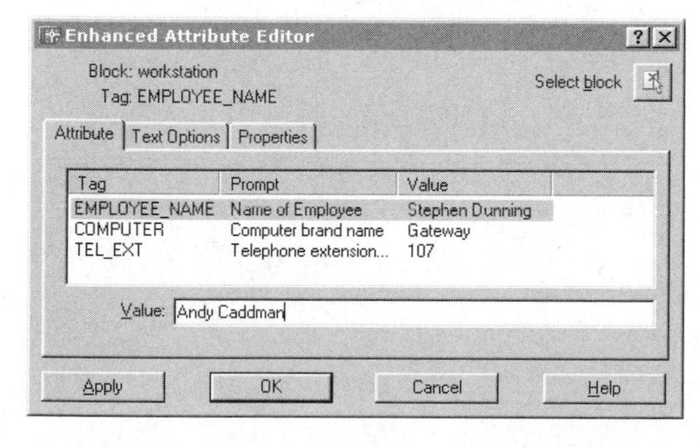

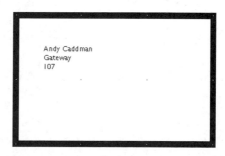

Editing attributes.

2. In the **Attribute** tab, select the EMPLOYEE_NAME tag. Notice the name in the **Value** text box.
   Replace "Stephen Dunning" with "Andy Cadmann." Notice that the attribute changes in the desk block.
3. Click **OK** to exit the dialog box.
4. Your boss decides the new color inkjet printers should be used to their fullest potential, and asks you to make the attributes colorful.
   Start the **BATTMAN** command, and then click the **Properties** tab.
5. Change the color of the attributes, as follows:

EMPLOYEE_NAME	**Red**
COMPUTER	**Green**
TEL_EX	**Blue**

6. When done, click OK to exit the dialog box.
   Notice that all attributes in the drawing change their color.
7. Save the drawing.

STEP 6: EXTRACTING ATTRIBUTES

To export the attributes, use the EATTEXT command.

1. Start the **EATTEXT** command, and follow the wizard to extract the Workstation block and the attributes you defined; exclude the AutoCAD-generated attribute data.

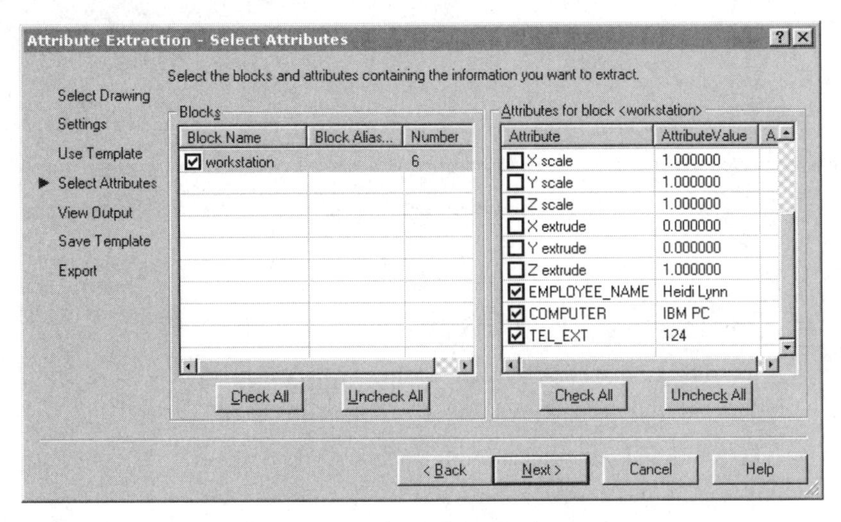

Extracting attributes from the drawing.

2. Click **Next**, and then click **Copy to Clipboard**.
3. Start a spreadsheet program, text editor, or other software application.
   Press **CTRL+V** to paste the attribute data.

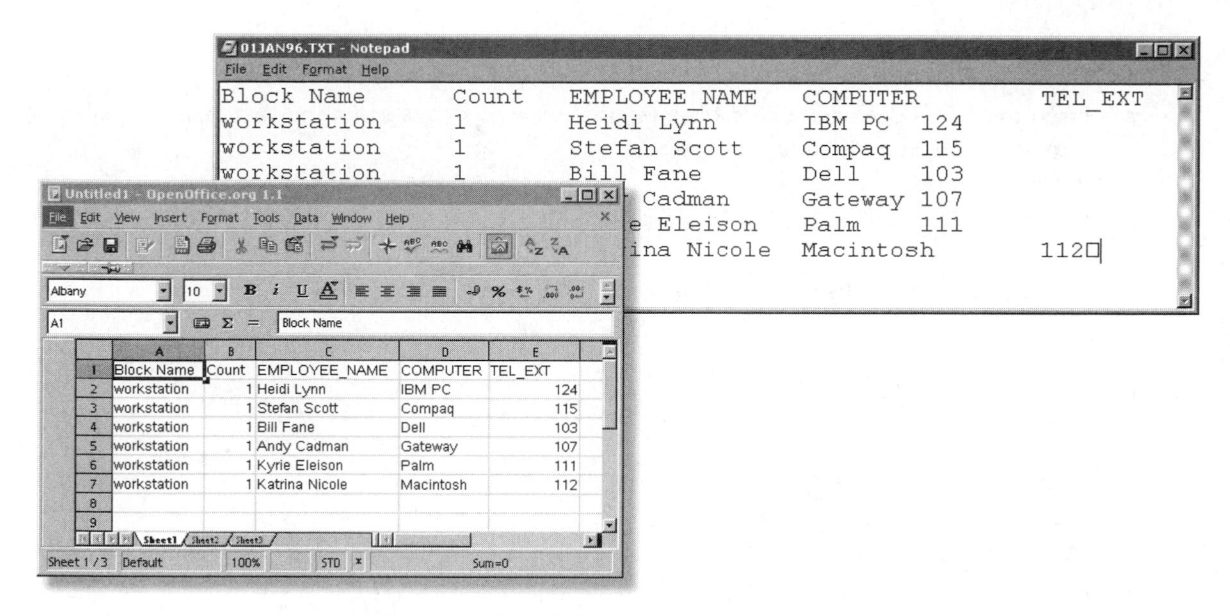

Pasting attribute data into other applications.

4. In AutoCAD, you press **Cancel** to exit the command.

EXERCISES

1. From the CD-ROM, open *Ch06facility.dwg*, and use the **ATTDEF** command to attach attributes to some of the furniture blocks.

2. From the CD-ROM, open *Ch06Extract.dwg*, and use the **EATTEXT** command to extract the attribute data.

   Save this in a spreadsheet or word processing document.

CHAPTER REVIEW

1. What is an attribute?
2. What is an attribute tag?
3. What command defines attributes?
4. What is the attribute prompt?
5. How would you suppress attribute prompts?
6. List two ways to change the display of attributes in drawings:

   a.

   b.
7. Describe several ways to edit attributes.
8. What parts of attributes can be changed?
9. What is the difference between the **BATTMAN** and **EATTEDIT** commands?
10. How would you obtain a file of all the attribute values in your drawing?
11. Can attribute data be extracted for use by Excel spreadsheets and Access databases?

Introducing Database Links

by Kevin Standiford

The previous chapter introduced you to attributes, which store data in blocks. Attributes have their drawbacks: data can be attached to blocks only, and not to any other objects. It can be difficult to identify specific objects with attributes, especially when the data is invisible. The data can be exported to external programs, but not in real-time; you must extract the data manually with the **EATTEXT** command. Extracted data must be viewed by external programs, such as spreadsheets and word processors; it is not easily viewed within AutoCAD. And the data can only be viewed in two forms, listed by blocks or by attributes.

Autodesk's solution to these drawbacks is the **DBCONNECT** command, which connects data to any object in drawings, permits real-time updating of changes to data, and, when you select a data link, zooms in to show the linked object close-up. The database allows many views of data.

The subject of databases is huge. There are entire books written on the subject, including the 300-page *AutoCAD Database Connectivity* by Scott McFarlane (Autodesk Press). For that reason, this chapter is a mere introduction to linking AutoCAD drawings with databases.

In this chapter, you learn about the following command:

DBCONNECT connects objects in drawings with external database files.

FINDING THE COMMANDS

On the **DBCONNECT MANAGER** toolbar (appears in the dbConnect Manager window):

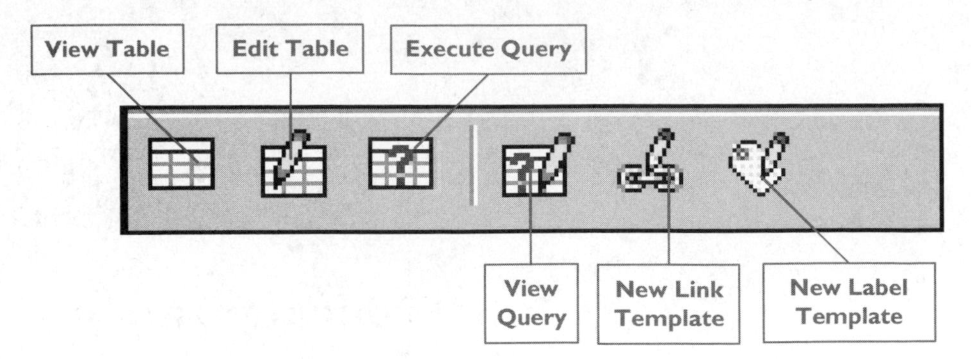

On the **TOOLS** and **DBCONNECT** (appears after dbConnect is started) menus:

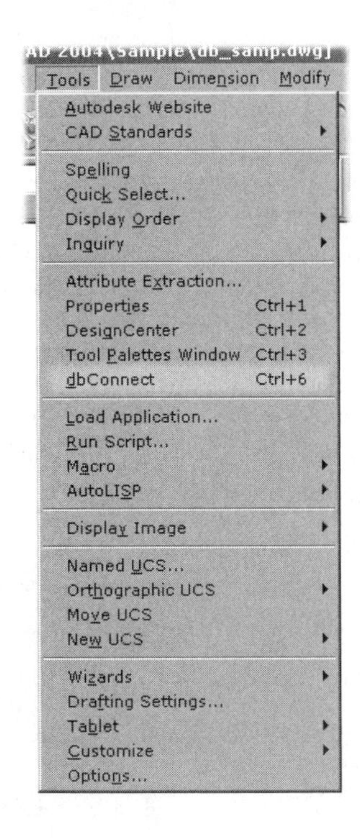

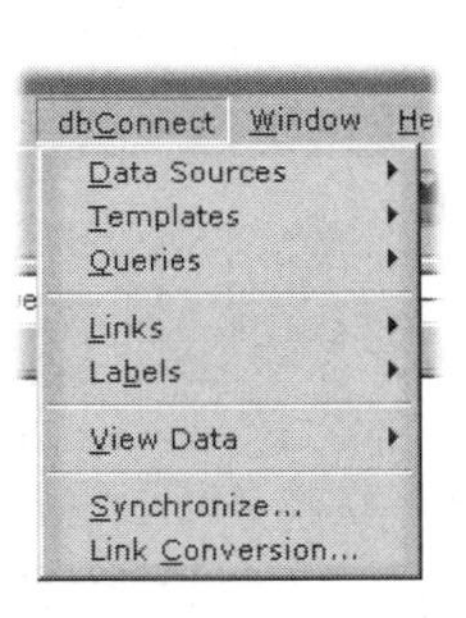

INTRODUCTION TO DATABASE LINKS

The ability to track information is essential to the engineering community. Often, this information includes part numbers, dimensions, costs, material strengths, or any other data that may be associated with objects. The most commonly used tool for managing critical data is the *database*. A database is a collection of text and numerical data stored in a list created and managed by an application external to AutoCAD, called a "database management system" (a.k.a. DBMS).

Technically, AutoCAD is itself a database manager that stores information about objects: entity types, coordinates, layer names, styles, and so on. The data is stored in drawing files (*.dwg*). In addition to managing its own data, AutoCAD also manages database files created by other DBMSs, and links these databases to objects in drawing files. This is accomplished in AutoCAD with the dbConnect Manager.

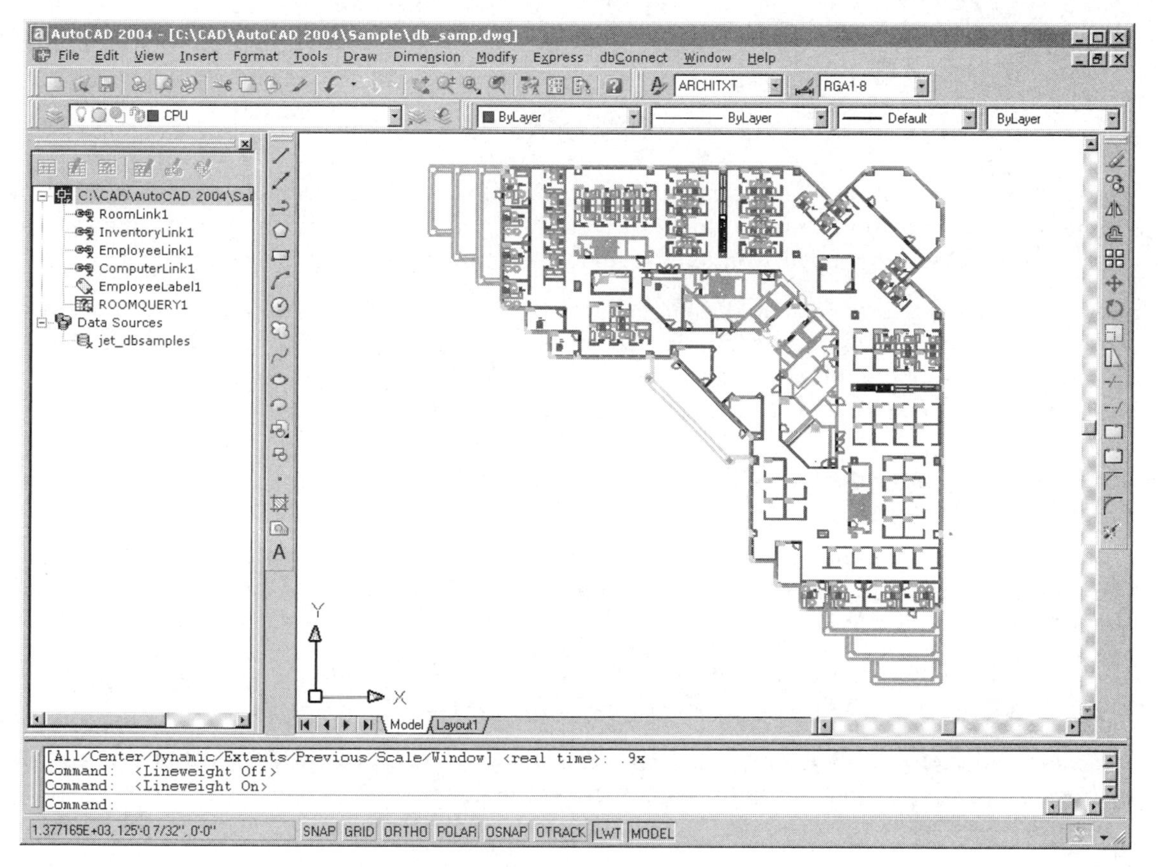

AutoCAD's dbConnect Manager (docked at left).

For example, suppose a manufacturer of spur gears creates a database containing the following information: number of teeth, pitch diameter, pressure angle, hole depth, chordal addendum, chordal thickness and working depth. The data can be linked to AutoCAD drawings of groups of gears, allowing the drawings to display the specifications as text.

ANATOMY OF DATABASES

A database is a collection of textual and numerical data arranged in spreadsheet-like *tables*. These tables are made of rows and columns. Rows run horizontally; columns, vertically.

The rows are called "records." Records contain one or more pieces of data called "fields." (Think of *cells* in spreadsheet parlance.)

When one or more columns are used to identify specific rows, then the columns are called "keys."

Keys (Unique identifiers)
Field (Cell)
Column
Record (Row)
Database Table

Tag_Number	Manufacturer	Equipment_Description	Item_Type	Room
24352	MITSUBISHI	HL6605ATK, COLOR, 16"	MONITOR COLOR 16	6090
24417	MITSUBISHI	HL6605ATK, COLOR, 16"	MONITOR COLOR 16	6079
24463	SUN	GDM-1662B V, COLOR, 19"	MONITOR COLOR 19	6190
24857	SUN	GDM-1604, COLOR, 15"	MONITOR COLOR 15	6190
25422	NEC	MULTISYNC 3FGX, COLOR, 15"	MONITOR COLOR 15	6023
25452	SUN	GDM-1962B, COLOR, 19"	MONITOR COLOR 19	6190
26072	NEC	[illegible]4FG, COLOR, 15"	MONITOR COLOR 15	6028
26174	AST	[illegible], COLOR, 14"	MONITOR COLOR 14	6022
26240	NEC	[illegible]3FGX, COLOR, 15"	MONITOR COLOR 15	6067
26578	TELEBIT	T3SA-WB1, MODEM, EXT.	MODEM	6190
26591	NEC	MULTISYNC 4FG, COLOR, 15"	MONITOR COLOR 15	6076
26712	NEC	MULTISYNC 4FG, COLOR, 15"	MONITOR COLOR 15	6078
27042	SKUTCH	AS-2500-2	NETWORK REMOTE RESET	6190
27394	APC	SMART UPS 900XLPE	UPS CONTROLLER	6190

Database tables consist of rows (records) and columns.

Suppose you had a drawing of the facilities in your office. The database can display lists of all equipment in specific rooms, or of rooms with ranges of equipment. It can show which employees are assigned to which rooms, or what equipment is assigned to a group of employees.

There are drawbacks, however: the dbConnect Manager can be difficult to understand, and you need to know SQL (short for "structured query language") to make full use of the database. The good news is that SQL is a standard; once you learn it for AutoCAD, you can use it with other software: IBM's DB2, Informix, Sybase, Oracle, Microsoft's Access and SQL Server, and others.

INTRODUCTION TO SQL

The purpose of SQL is to ask questions of the database in a structured manner. Here is an example of a query that asks the database to select the data ("Room," "Room_Type," and "Entity_Handle") for all rooms that are type **OFF-STD** (standard offices) or **CONF-F** (conference rooms) from database table "Room."

```
SELECT Room, Room_Type, Area, Entity_Handle FROM Room
WHERE (Room_Type = 'OFF-STD') OR  (Room_Type = 'CONF-M')
```

Words shown in **color** are SQL *operators*. Power users tend to type in their queries, but **DBCONNECT** also provides a user interface to help you construct the query.

Although it is not the intention of this book to provide you with in-depth coverage of the SQL language, it is essential that you have a basic understanding of four commonly used keywords: **SELECT**, **WHERE**, **ORDER BY**, and **DISTINCT**.

(Because DBMS software is developed by different vendors, the dialects of SQL used to communicate with these systems may vary. Before they are considered to be in compliance with ANSI standards, however, their dialect of SQL must support all major SQL keywords in a similar fashion.)

SELECT Keyword

The SELECT keyword selects column data from tables, and then displays the result in lists called "result-sets."

The syntax for this keyword is:

SELECT *column-name(s)* **FROM** *table-name*

For example, to produce the result-set illustrated by the figure below, the following SQL statement is entered:

SELECT FirstName, LastName, Title **FROM** Employees

The result-set consists of a table with three columns: First Name, Last Name, and Title. The records in the result-list have data in those three columns.

First Name	Last Name	Title
Nancy	Davolio	Sales Representative
Andrew	Fuller	Vice President, Sales
Janet	Leverling	Sales Representative
Margaret	Peacock	Sales Representative
Steven	Buchanan	Sales Manager
Michael	Suyama	Sales Representative
Robert	King	Sales Representative
Laura	Callahan	Inside Sales Coordinator
Anne	Dodsworth	Sales Representative
Nancy	Davolio	Sales Representative

Result-set with all employees.

WHERE Keyword

The WHERE keyword specifies a search condition, thereby restricting the number of rows returned.

The syntax for this key word is:

WHERE < *search_condition* >

For example, to produce a result-set showing only employees whose title is Sales Representative, the following SQL statement is entered:

SELECT FirstName, LastName, Title **FROM** Employees WHERE Title = 'Sale Representative'

The result-set is a shorter table that contains only the records of employees who are sales representatives.

First Name	Last Name	Title
Nancy	Davolio	Sales Representative
Janet	Leverling	Sales Representative
Margaret	Peacock	Sales Representative
Michael	Suyama	Sales Representative
Robert	King	Sales Representative
Anne	Dodsworth	Sales Representative

Result-set with sales representatives only.

ORDER BY Keyword

The **ORDER BY** keyword specifies the order in which records are presented in the result-set.

The syntax for this key word is

ORDER BY { *order_by_expression* [**ASC** | **DESC**] } [,...*n*]

The two arguments **ASC** and **DESC** specify whether the results are shown in *ascending* or *descending* order. If ascending, then the **ASC** argument is used; if descending, then the **DESC** argument is used.

For example, to produce a result-set in which the information is sorted in ascending order by the employees' last names, enter the following SQL statement:

```
SELECT EMPLOYEES.FirstName, EMPLOYEES.LastName, EMPLOYEES.Title FROM
    EMPLOYEES WHERE title = 'Sales Representative' ORDER BY LastName ASC.
```

The result-set is a table that is sorts the sales representatives alphabetically by last name.

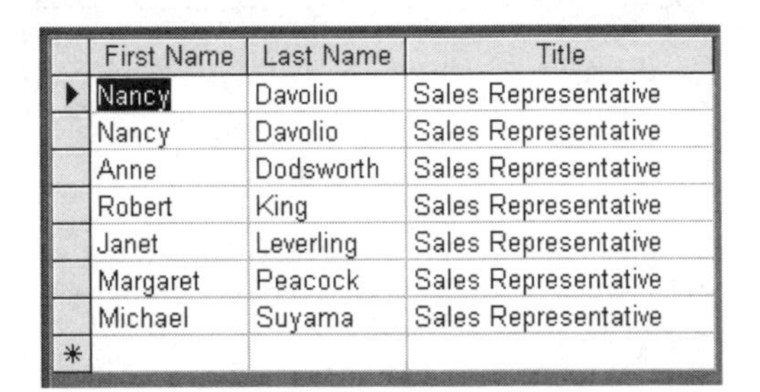

First Name	Last Name	Title
Nancy	Davolio	Sales Representative
Nancy	Davolio	Sales Representative
Anne	Dodsworth	Sales Representative
Robert	King	Sales Representative
Janet	Leverling	Sales Representative
Margaret	Peacock	Sales Representative
Michael	Suyama	Sales Representative
*		

Result-set with surnames sorted alphabetically.

DISTINCT Keyword

The **DISTINCT** key word removes duplicate rows from the result-set.

Notice that the result-set illustrated above has a duplicate record ("Nancy Davlio"). To remove duplicate records, enter the following SQL statement:

```
SELECT DISTINCT EMPLOYEES.lastName, EMPLOYEES.firstName, EMPLOYEES.Title FROM
    EMPLOYEES WHERE title = 'Sales Representative' ORDER BY LASTName ASC.
```

The duplicate record is removed from the result-set.

First Name	Last Name	Title
Nancy	Davolio	Sales Representative
Anne	Dodsworth	Sales Representative
Robert	King	Sales Representative
Janet	Leverling	Sales Representative
Margaret	Peacock	Sales Representative
Michael	Suyama	Sales Representative

Result-set with duplicate records removed using the DISTINCT key word

DBCONNECT

The **DBCONNECT** command connects objects in drawings with external database files.

This command opens the dbConnect Manager window, which displays two basic types of information: drawings and data sources.

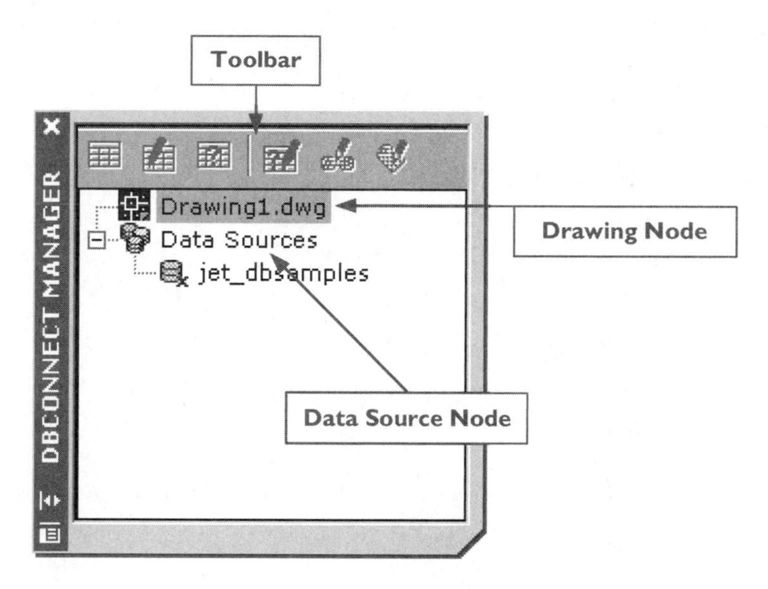

dbConnect Manager window.

The *drawing node* displays all open drawings and the database objects associated with them. New drawings have no associated database objects.

The *data sources node* displays all configured data sources available on your computer. By default, AutoCAD displays a sample Microsoft Access database file, called *jet_dbsamples*. This is a special file that allows you to learn database connectivity without needing a database program installed on your computer.

As additional drawings are opened, they are listed by the dbConnect Manager, whether they have database connections or not. (In fact, you can use the dbConnect Manager simply to switch between open drawings!)

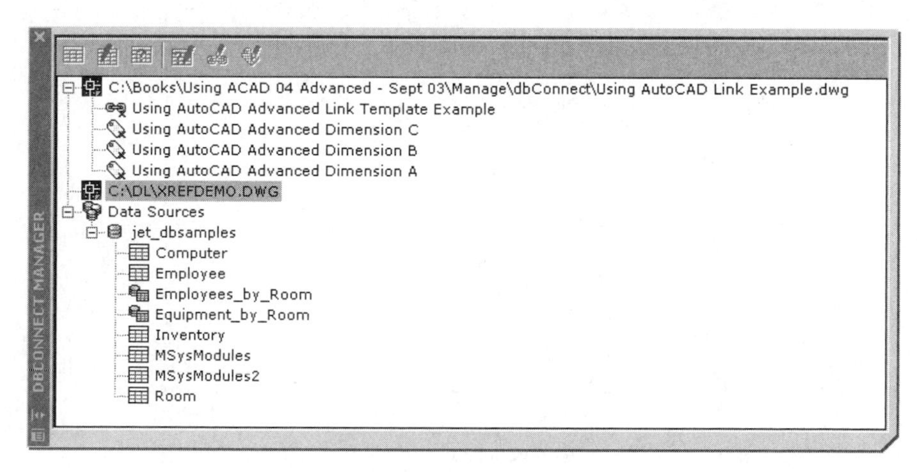

dbConnect Manager lists drawings open in AutoCAD.

BASIC TUTORIAL: CONNECTING WITH DATABASES

For this tutorial, open AutoCAD with the *db_sample.dwg* drawing file, available in AutoCAD 2005's *sample* folder and on the companion CD-ROM.

1. To connect drawings with databases, start the **DBCONNECT** command:
   - From the **Tools** menu, choose **dbConnect**.
   - At the 'Command:' prompt, enter the **dbconnect** command.
   - Press the **CTRL+6** keyboard shortcut.
   - Alternatively, enter the alias **dbc** at the keyboard.

   Command: **dbconnect** *(Press* ENTER.*)*

   In all cases, AutoCAD displays the dbConnect Manager window.

   Notice that a number of predefined links appear below the drawing node. If a small red x is shown in each icon, this indicates the drawing is not yet connected to a database.

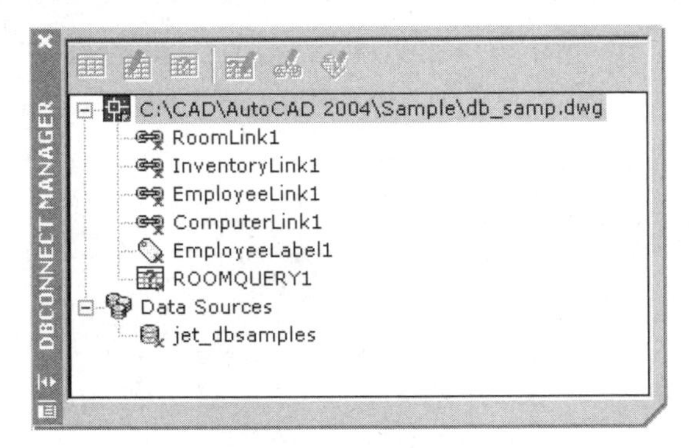

dbConnect Manager window.

   In addition, notice that AutoCAD adds the **dbConnect** menu to the menu bar.

2. To connect the drawing to the database, right-click "jet_dbsamples," and then select **Connect** from the shortcut menu.

   Notice that a number of predefined tables appear below the Data Sources node.

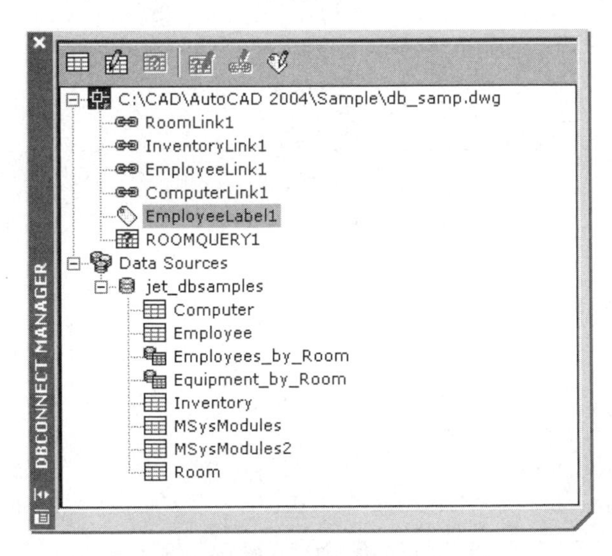

Connected database with predefined tables.

The icons in the window have the following meaning:

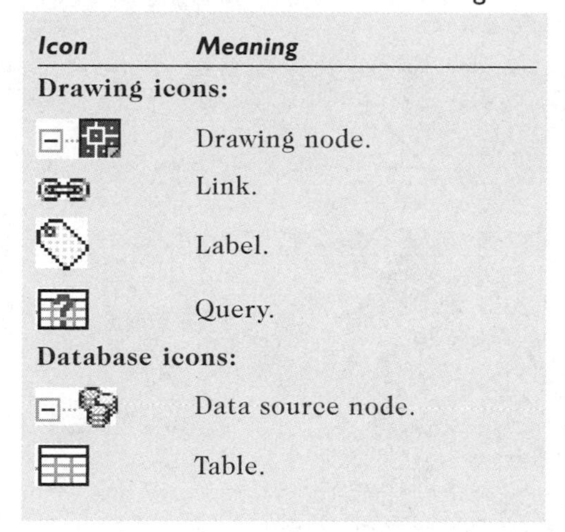

Icon	*Meaning*
Drawing icons:	
	Drawing node.
	Link.
	Label.
	Query.
Database icons:	
	Data source node.
	Table.

3. To get a feel for AutoCAD's built-in database capabilities, right-click "RoomLink1," and then select **View Table** from the shortcut menu.

   Notice that AutoCAD displays a spreadsheet-like window called "Data View." (In addition, AutoCAD adds the **Data View** menu to the menu bar.)

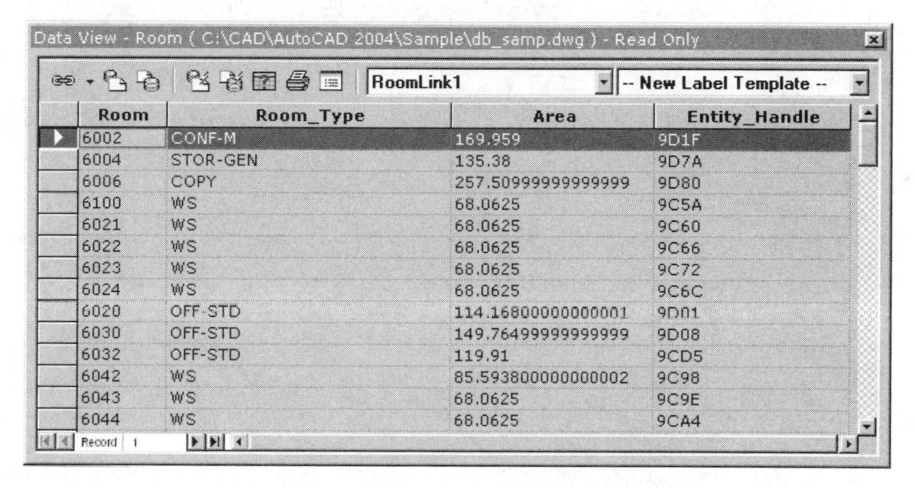

Data View - Room (C:\CAD\AutoCAD 2004\Sample\db_samp.dwg) - Read Only

RoomLink1 | -- New Label Template --

Room	Room_Type	Area	Entity_Handle
6002	CONF-M	169.959	9D1F
6004	STOR-GEN	135.38	9D7A
6006	COPY	257.50999999999999	9D80
6100	WS	68.0625	9C5A
6021	WS	68.0625	9C60
6022	WS	68.0625	9C66
6023	WS	68.0625	9C72
6024	WS	68.0625	9C6C
6020	OFF-STD	114.16800000000001	9D01
6030	OFF-STD	149.76499999999999	9D08
6032	OFF-STD	119.91	9CD5
6042	WS	85.593800000000002	9C98
6043	WS	68.0625	9C9E
6044	WS	68.0625	9CA4

Record 1

Data View window displays records and fields.

This data view shows you all linked room numbers in the *db_sample.dwg* drawing, along with their type, area, and entity handle. (*Entity handles* are unique identifiers AutoCAD automatically assigns to every object at the moment it is created in the drawing.)

Notice that all the records are gray, indicating you cannot change them. They are in view, or read-only mode.

4. To show that database records are indeed linked to objects in the drawing:
   a. Select any record in the Data View window.
   b. Right-click the record.
   c. From the shortcut menu, select **View Linked Records**.

Notice that AutoCAD highlights the connected object in the drawing. (When set up correctly, AutoCAD also zooms in to show the room number.)

Viewing objects linked to records.

5. The records in the Data View window can be sorted, hidden, and moved about.
   - To sort the records by column, double-click a header such as "Room" or "Area." The first time you double-click, the column is sorted in alphabetical order (A - Z); the second double-click sorts in reverse alphabetical order (Z - A).
   - To sort by multiple columns, right-click any column header. From the shortcut menu, select **Sort**. Select the order in which you wish the data to be sorted.

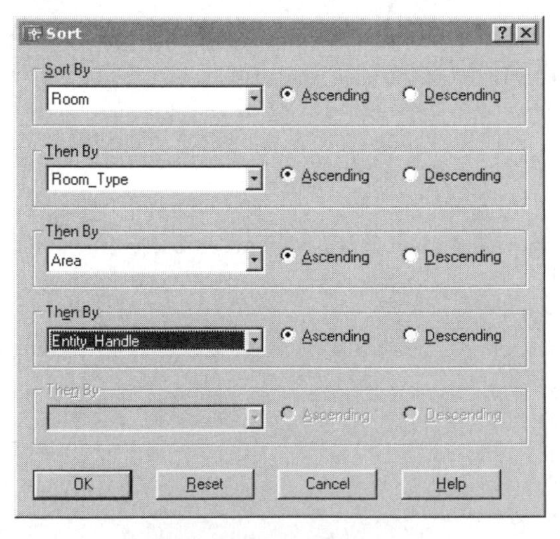

Sorting database columns.

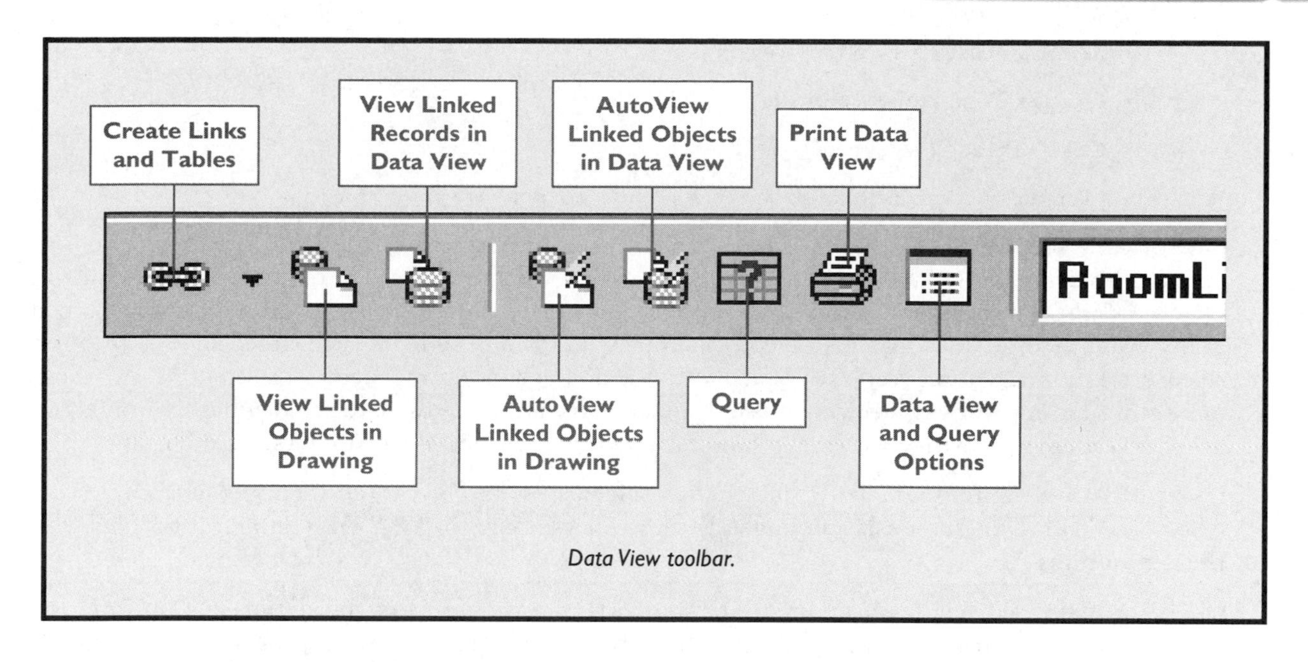

Data View toolbar.

- To hide a column, right-click and select **Hide**. To return the column to visibility, right-click one of the other columns, and select **Unhide All.**
- To move columns about, drag the column by its header. A vertical red line shows where the column ends up. In addition, you can change the width of columns by dragging the vertical line between headers. Hold down the CTRL key to select more than one column at a time.

6. To exit the data table, click the small **x** in the upper right corner of the window.
7. To edit the data, right-click "RoomLink1," and then select **Edit Table** from the shortcut menu.

   Notice that the Data View window reappears, but this time all rows are white. This indicates that you can change the data.

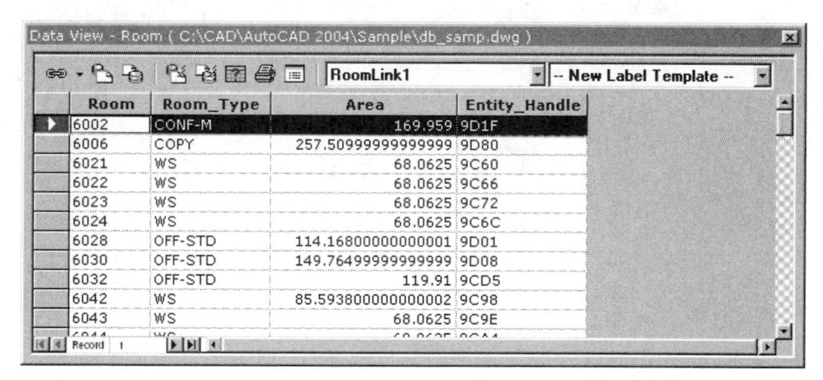

Data View - Room (C:\CAD\AutoCAD 2004\Sample\db_samp.dwg)

RoomLink1 | -- New Label Template --

Room	Room_Type	Area	Entity_Handle
6002	CONF-M	169.959	9D1F
6006	COPY	257.50999999999999	9D80
6021	WS	68.0625	9C60
6022	WS	68.0625	9C66
6023	WS	68.0625	9C72
6024	WS	68.0625	9C6C
6028	OFF-STD	114.16800000000001	9D01
6030	OFF-STD	149.76499999999999	9D08
6032	OFF-STD	119.91	9CD5
6042	WS	85.593800000000002	9C98
6043	WS	68.0625	9C9E

Editable database table.

CONFIGURING DATABASE CONNECTIONS

Before AutoCAD can access a database other than the "jet_dbsamples" sample, AutoCAD must first be configured for the database program. This procedure includes creating a *data source* and providing the necessary driver information using ODBC or OLE DB programs. (ODBC is short for Open Database Connectivity, while OLE DB is short for Object linking and embedding for databases.)

ODBC is the Windows standard that allows just about any software to communicate with databases, even Excel and Notepad. AutoCAD 2005 supports the following database systems via ODBC:

- dBASE's dBase Plus www.dbase.com
- Oracle's Database www.oracle.com
- Corel's Paradox www.corel.com
- Micrsoft's Access, Excel, SQL Server, and Visual FoxPro www.microsoft.com

(Historically, dBASE II was the first database program for personal computers powerful enough to be used by companies. The software was first named "Vulcan," after the race of logical beings in the *Star Trek* television series. The company formed to market the software, Ashton-Tate, was named after owner George Tate and his parrot, Ashton. When the product was renamed dBase, the first release was Roman-numbered "II" to make the brand-new program seem more mature. All these marketing tactics worked, because dBase II quickly became the standard for databases running on PCs; its leadership position did not, however, survive the transition to Windows.)

If your DBMS software is Oracle Database, Microsoft Access, or SQL Server, then it comes with direct drivers for OLE DB. Autodesk recommends that you use only the OLE DB configuration files with these programs.

Once the database has been configured, AutoCAD can then access the database's information regardless of its format or the platform that created it. Each database program is configured differently;, therefore a single procedure for configuring databases cannot be provided by this book. However, information regarding the procedures for configuring the different databases can be obtained in the help files provided with ODBC and OLE DB. (From the **Help** menu, select **Help**. In the **Contents** tab, select **Drivers & Peripherals**, and then **Configure Ext. Database**.)

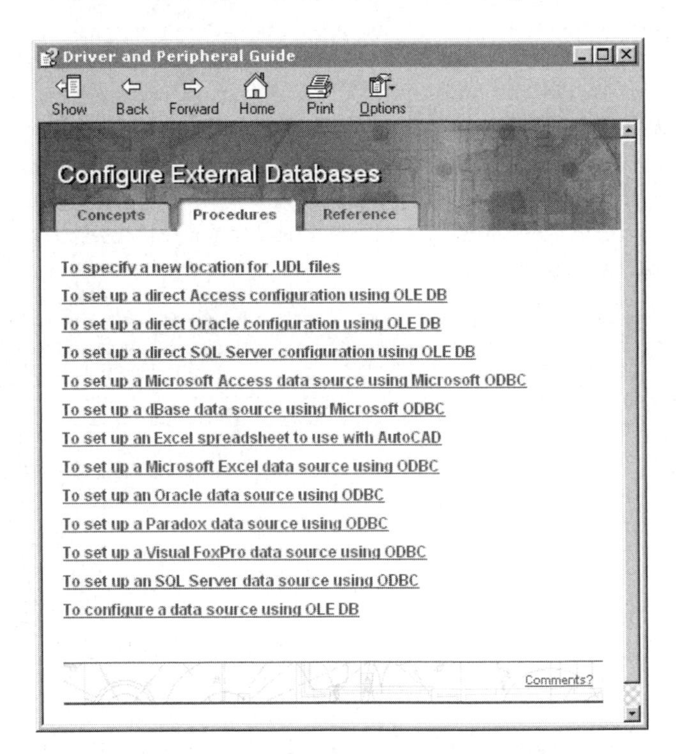

Autodesk's online help for setting up database connections.

Once a database has been successfully configured, a configuration file is produced containing the extension *.udl* (short for "Universal Data Link"). You can read more about the purpose of UDL files at kandkconsulting.tripod.com/VB/Tutorials/udl_file_tutorial.htm. (This Web address is case-sensitive.)

By default, older releases of AutoCAD search for configuration files in its *\data links* folder. In AutoCAD 2005, the folder has been hidden, and relocated to *\documents and settings\administrator\application data\autodesk\autocad\r16.1\enu\data links*.

A different location may be specified through the **OPTIONS** command. In the Options dialog box, click the **Files** tab, and then look for **Data Sources Location**. To change the location, click **Browse**, and then select a different drive and folder. Changes to the path do not take effect until you exit and restart AutoCAD.

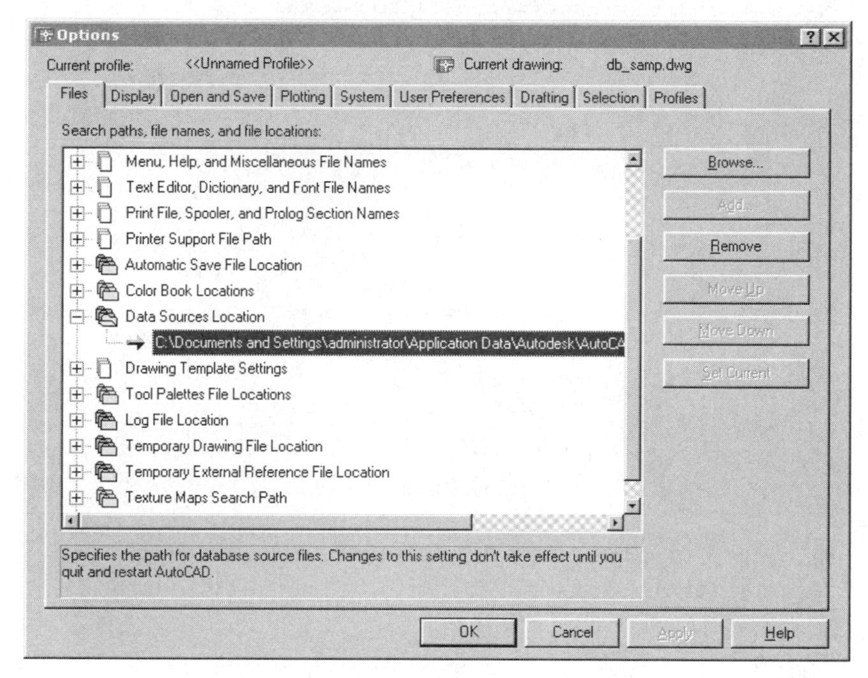

Options dialog box

TUTORIAL: RECONFIGURING DATABASES

The Data Link Properties dialog box allows you to configure data sources. The **Provider** tab lists the different database types supported by AutoCAD. Selecting the provider then determines the options available in the **Connection** tab.

1. Using the *db_sample.dwg* drawing file from the CD-ROM, open the Configure a Data Source window:
   - From the **dbConnect** menu, choose **Data Sources**, and then **Configure**.
   - Right-click **Data Sources** node in the dbConnect Manager window.

   AutoCAD displays the Configure a Data Source window.

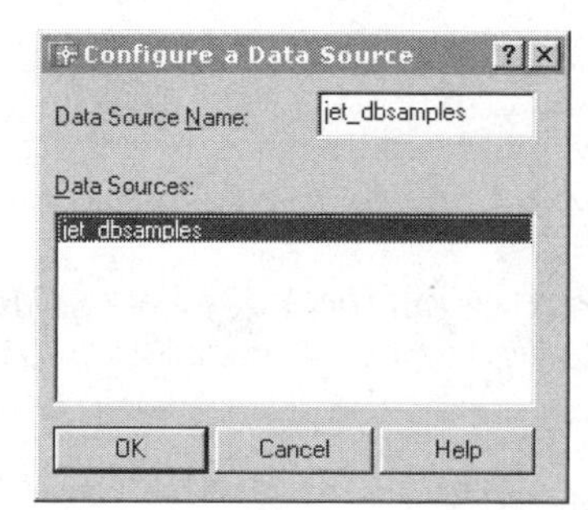

Configure a Data Source dialog box.

2. To reconfigure the data source, double-click its name in the **Data Sources** list.

   AutoCAD displays the Data Link Properties dialog box, which displays the data stored in the related *.udl* file.

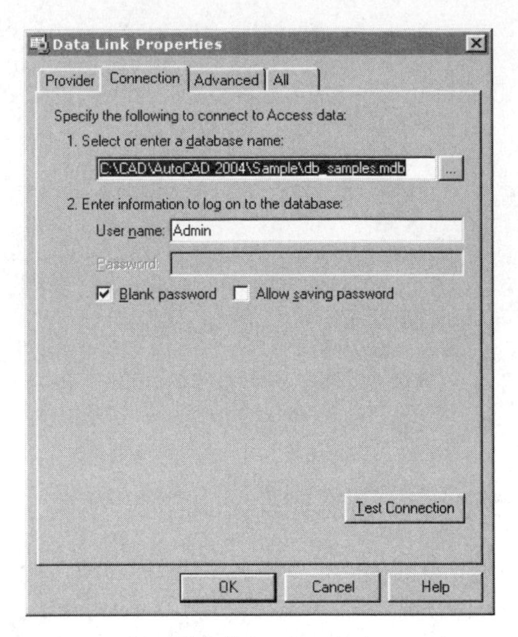

Data Link Properties window.

3. To change the database name and location, select the **...** button under the **Select or Enter Database Name** option.

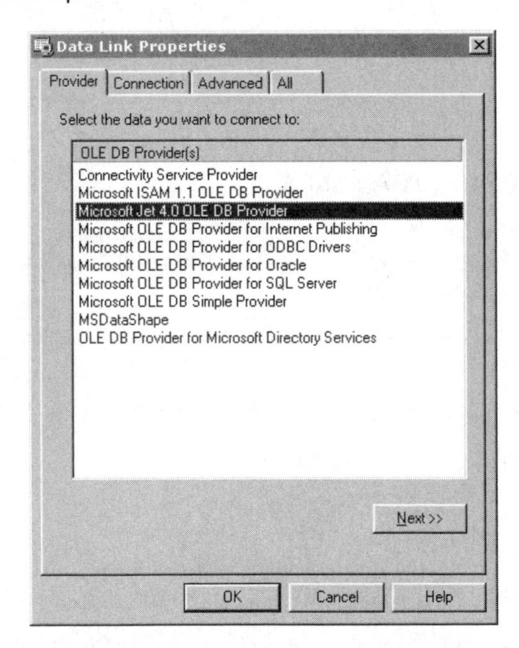

Provider tab lists database programs supported by AutoCAD.

4. To change the database driver, click the **Provider** tab, and then select a database driver name.
5. Click **Next** to configure the driver.

6. Click **Test Connection** to test the connection between AutoCAD and the database. The **Advanced** and **All** tabs are meant for advanced database users.

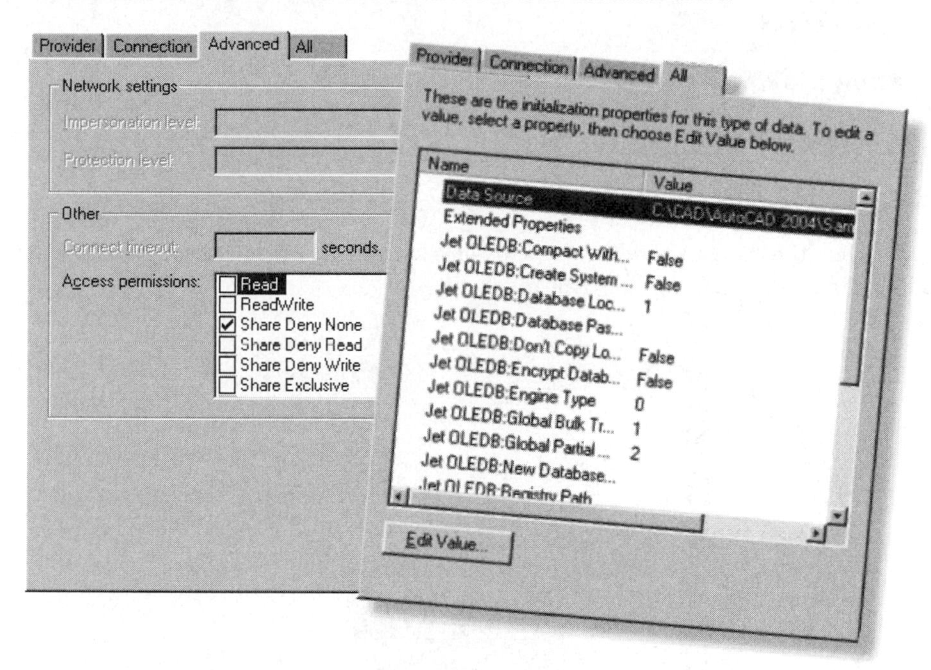

Advanced and All tabs.

7. Click **OK** to commit the changes.

CREATING LINKS TO GRAPHICAL OBJECTS

The primary objective of AutoCAD's database connectivity feature is to associate database information with AutoCAD objects. The association is created through *links*.

Links can be created with graphical objects only (lines, polylines, and so on), and not with named objects, such as styles, layers, or views. Once links are created, they are tightly associated with the objects to which they are connected.

The links are *dynamic*. When information changes in a database table linked to an AutoCAD object, then the information stored in the drawing is updated to match the database table.

But before links can be established, a link *template* must be created. Link templates identify the fields from the database table associated with the links connected to the AutoCAD graphical objects.

In summary, three steps are needed to link database tables to objects:

Step 1: Create the link templates.

Step 2: Create the label templates.

Step 3: Link to objects.

TUTORIAL: LINKING TO GRAPHICAL OBJECTS

Open AutoCAD with the *Ch07Links.dwg* drawing from the CD-ROM. Ensure that the *Test Data Source* data source is connected.

Step 1: Creating Link Templates

1. To create link templates:
   - From the **dbConnect** menu, choose **Templates**, and then **New Link Template**.
   - Select the **Link** button in the Data View dialog box's toolbar.
   - Right-click the database table to which a link template is to be associated, and then select **New Link Template** from the shortcut menu.

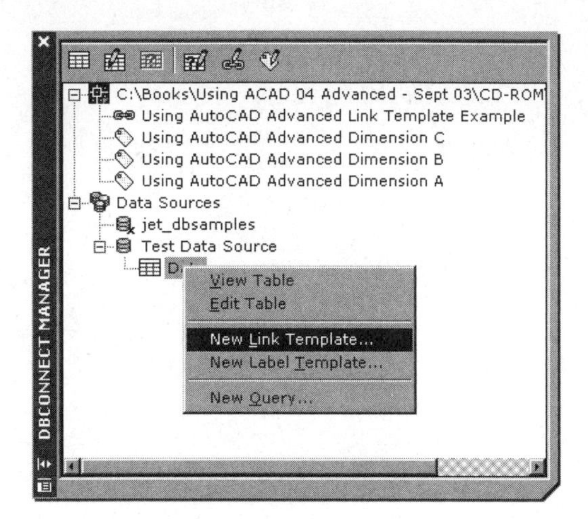

Create a new link template by right-clicking the table.

AutoCAD displays the New Link Template dialog box.

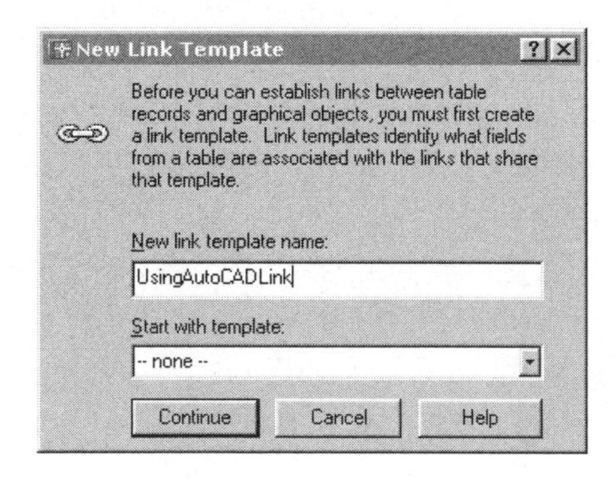

New Link Template dialog box.

2. Name the new link template. For this tutorial, use the following name:

   UsingAutoCADLink

   (The name must not have spaces in it.)
3. Click **Continue**.
4. In the Link Template dialog box, select a field as a *key* field.

   When selecting a key field, always try to select a field that contains unique values.

For this tutorial, use the **ID** field as the key field, because it contains a unique value for each row in the table.

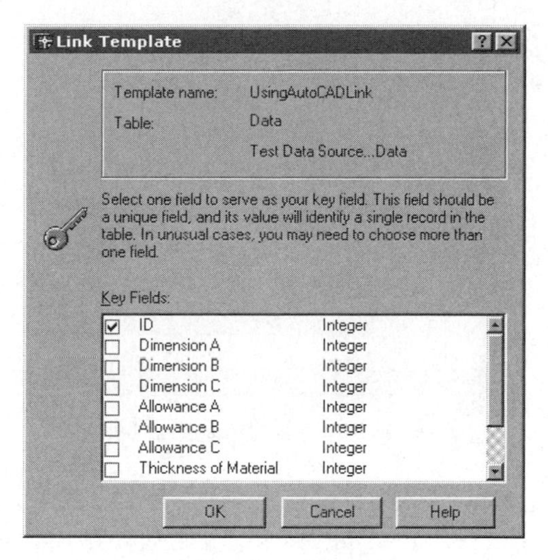

Link Template dialog box.

5. Click **OK**.
   The "UsingAutoCADLink" link template is added to the list under the drawings node in the dbConnect Manager window.

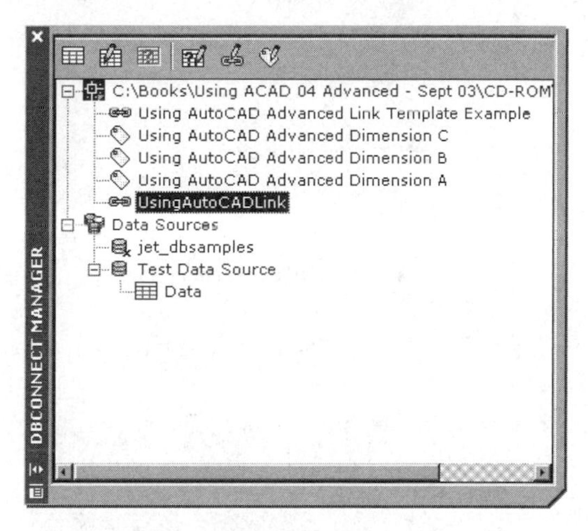

Link template added to dbConnect Manager.

Step 2: Creating Label Templates

Before selected fields can be displayed as labels on an AutoCAD drawing, a *label template* must first be created. The label template defines the field(s) that will be displayed in the label, as well as the format in which they are to be shown.

1. To create a label template:
   - From the **dbConnect** menu, choose **Templates**, and then **New Label Template**.
   - Right-click the database table that a link template is to be associated with ("Data"), and then select **New Label Template** from the shortcut menu.

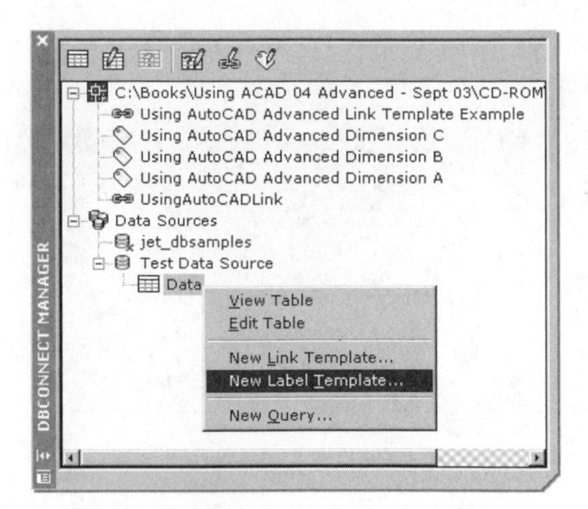

Selecting New Label Template command.

AutoCAD displays the New Label dialog box.

2. Create the label template by naming it. For this tutorial, use the following name:

   UsingAutoCAD Advanced Dimension A

New Label Template dialog box.

3. AutoCAD displays the Label Template dialog box. It establishes the format of the label, as well as which fields to display.

   To specify the field(s) to display in the label, select "Dimension A" from the Field droplist, and then click **Add**.

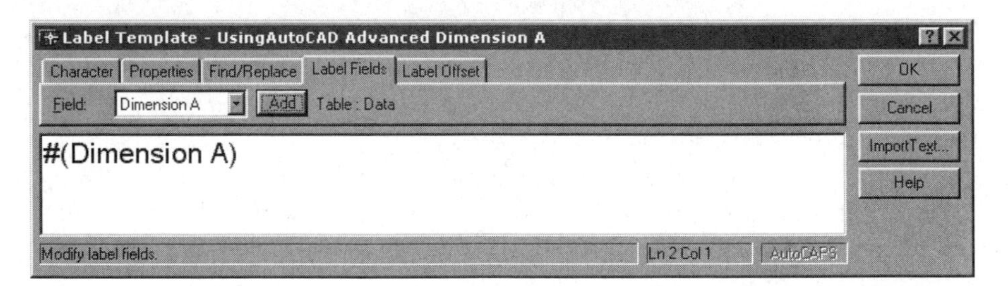

label Template dialog box.

4. Click **OK**.

Step 3: Linking to Objects

With the link and label template created, the link can be established to a graphical object in AutoCAD.

1. Open the table in which the template was created: in the dbConnect Manager window, right-click Data, and then select **Edit Table** from the shortcut menu.

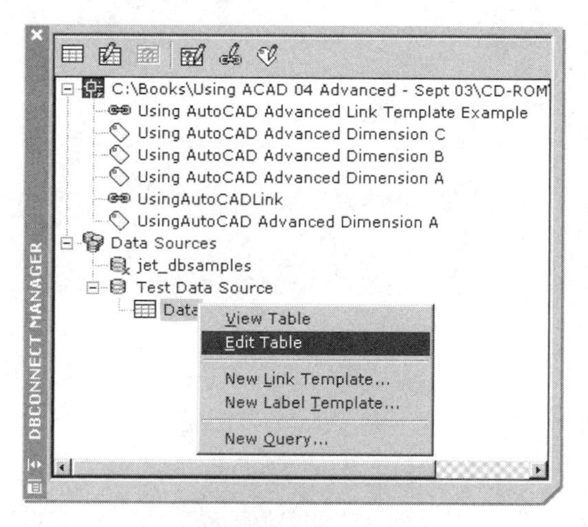

Opening the table for editing.

2. Select the link icon (in the upper left corner of the Data View dialog box).

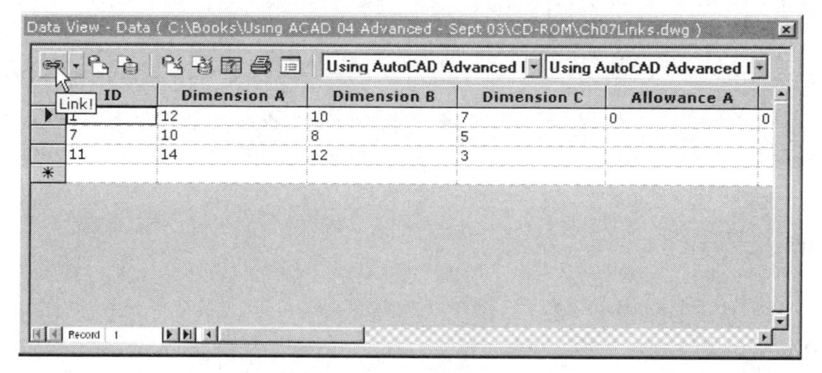

Choosing the Link icon.

3. In the drawing, select the object to link.

Notice that AutoCAD generates label 12 in the drawing.

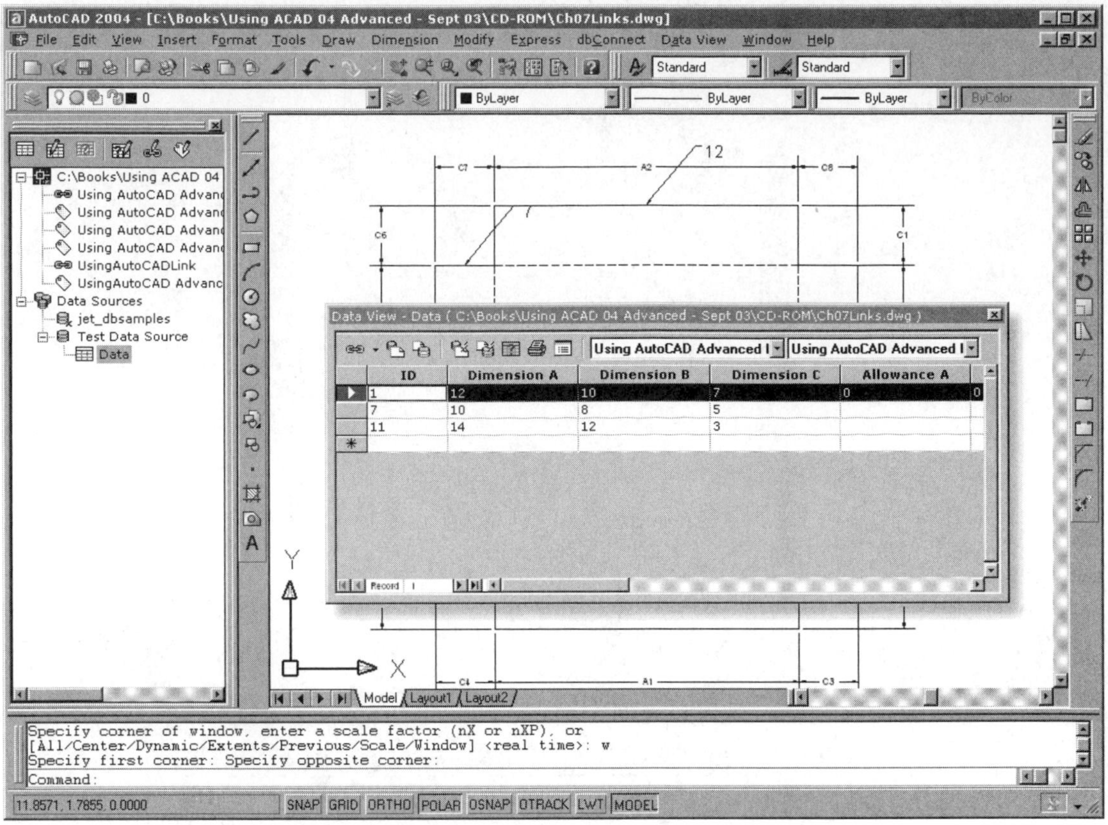

Label 12 added to an AutoCAD drawing

CHAPTER REVIEW

1. On which Microsoft technology is dbConnect based?
2. How do you display the Data View window for a specific table?
3. What must be created before you can start linking objects to a table?
4. When are changes to data in the Data View window actually written to the database?
5. How can you quickly sort a single column in the Data View window?
6. What constitutes a good primary key?

Isometric Drafting

Isometric drawing is one of the standards in the drafting profession. These drawings look 3D, but use special 2D construction techniques to create the illusion. This chapter describes how to construct isometric drawings, and how to set isometric text and dimension styles.

AutoCAD provides the following commands to assist in isometric drafting:

DSETTINGS switches AutoCAD into isometric drafting mode.

ISOPLANE switches the cursor between the three isoplanes.

ELLIPSE draws isometric circles through its **Isocircle** option.

STYLE creates isometric text styles through its **Oblique** option.

TEXT makes text look isometric through its **Rotate** option.

DIMSTYLE creates isometric dimension styles through its **Align** option.

DIMEDIT converts dimensions to isometric angles through its **Rotate** and **Oblique** options.

ABOUT ISOMETRIC DRAFTING

There are three primary styles of engineering pictorial drawings: oblique, axonometric, and perspective. (Isometric drawings are one type of axonometric drawing.) The figure illustrates an example of each.

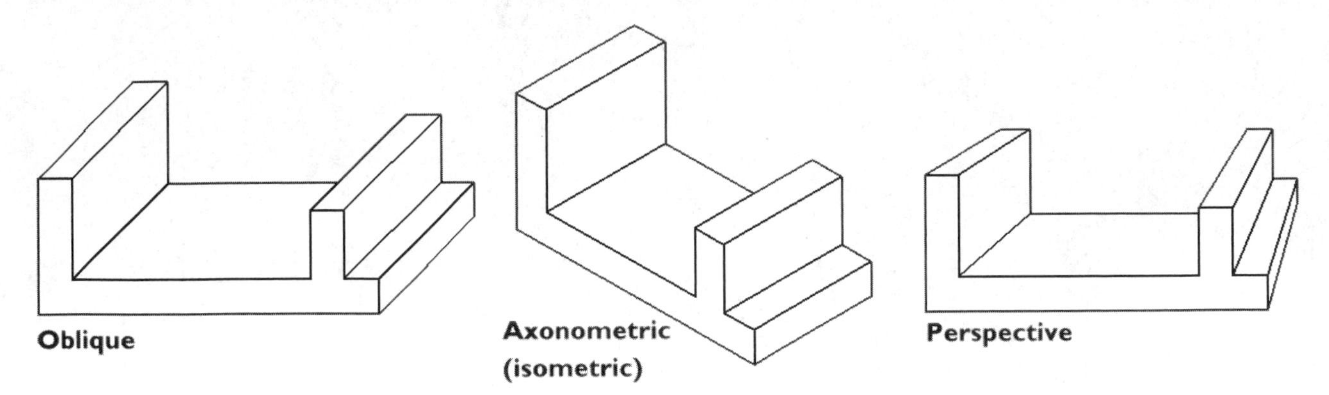

The same part drawn in oblique, axonometric (isometric), and perspective views.

Oblique drawings are primarily used for quick design. The front face is shown in-plane, and the sides are drawn back at an angle. This makes oblique drawings easy to construct from an elevation view of the front face.

Axonometric (isometric) drawings are most-common for engineering drawings. Each face of the object is shown in true length, allowing you to measure each length. AutoCAD provides a mode for drawing isometric drawings. This mode displays three drawing planes, which you use to construct the drawings.

Perspective drawings are the most realistic. Through the use of one or more vanishing points, the drawings show objects as they would appear to the eye from a specified location and distance. Although perspectives look more realistic, most lengths in drawings are not true lengths, and cannot be measured accurately. (AutoCAD provides perspective viewing of 3D drawings through the **3DORBIT** command.)

PRINCIPLES OF ISOMETRIC DRAWING

Isometric drawings show three sides of objects at the same angle. To accomplish this, AutoCAD draws lines at 30 degrees from the horizontal. The figure on the following page illustrates a box drawn in isometric view.

All elements of isometric drawings are drafted in one of three isometric planes: top, left, and right, as illustrated in the figure. (Isometric planes are called "isoplanes" for short.) You draw isometrically in one of these three planes.

Because everything is "slanted" in isometric drawings, circles must be drawn as ellipses to "look right." Similarly, text and dimension text must be slanted by 30 degrees (known as the obliquing angle in AutoCAD), as must the extension lines of dimensions.

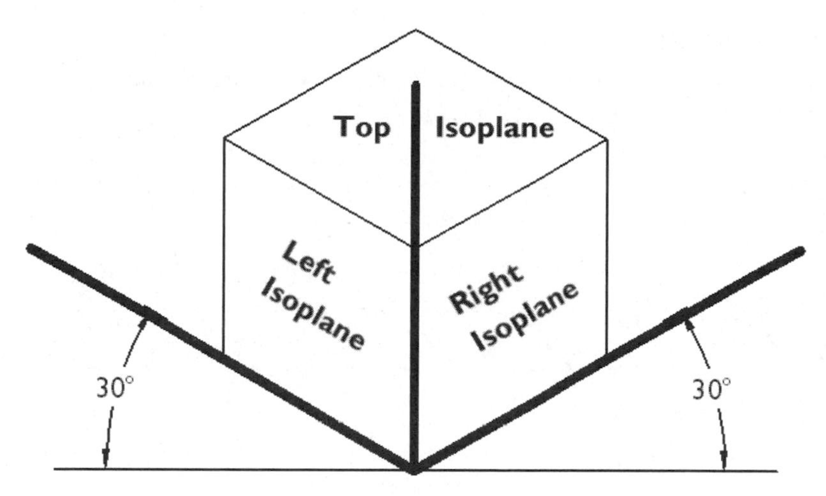

Box in isometric view.

AutoCAD supports isometric drafting by providing:

- Isometric cursor and drawing planes.
- Isometric grid display, snap mode, and ortho mode.
- Isometric circles.

AutoCAD does *not* provide isometric text and dimension styles; these you have to provide yourself, as described later in this chapter.

To switch to isometric drafting mode, turn on a setting "hidden" in the SNAP command.

TUTORIAL: ENTERING ISOMETRIC MODE

1. To enter isometric mode via a dialog box:
   - On the status bar, right-click SNAP and select **Settings**.
   - Alternatively, from the **Tools** menu, choose **Drafting Settings**, and then **Snap and Grid** tab.
2. In the Snap Type & Style section, click the radio button next to **Isometric snap**.

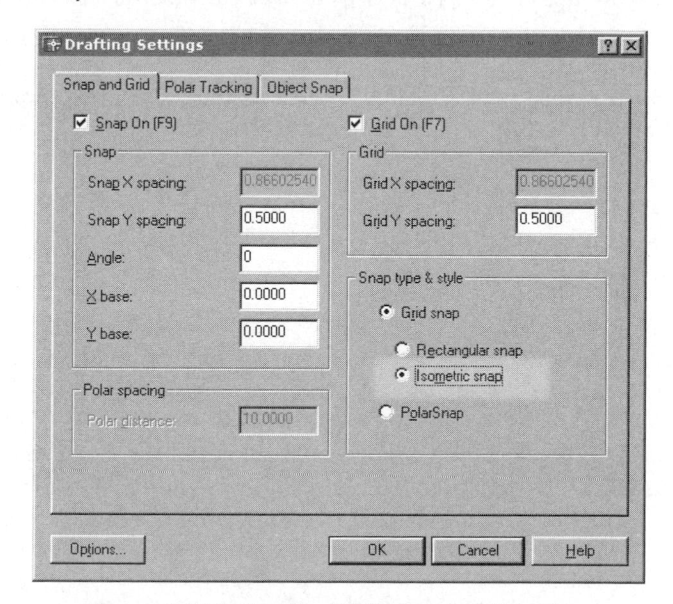

Snap and Grid tab of the Drafting Settings dialog box.

3. Ensure **Snap** and **Grid** are turned on.

   Notice that the **Snap X** and **Grid X spacing** text fields are grayed out. You cannot have an aspect ratio during isometric drafting.

4. Click **OK**.

   Notice that the crosshairs and grid are displayed at an angle. (The angle changes, depending on the current isometric plane.)

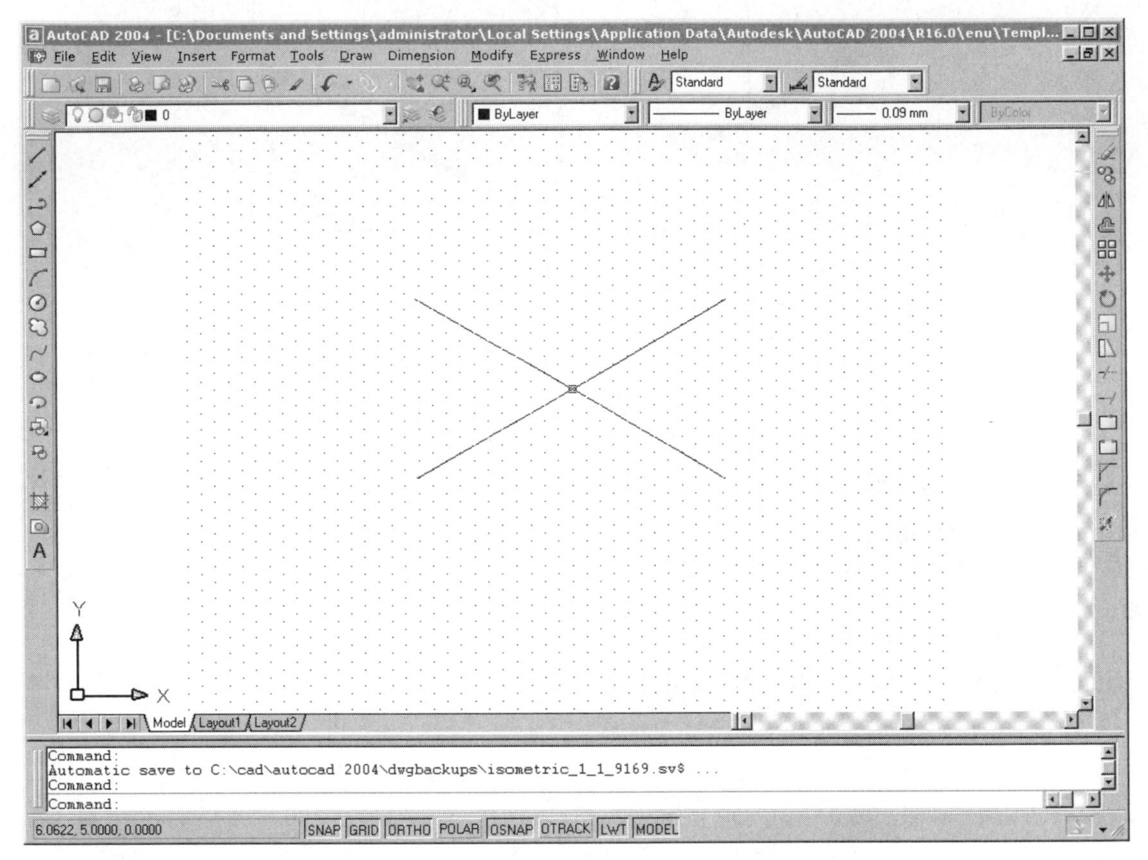

In isometric mode, cursor crosshair and grid are displayed at an angle.

Note: To enter isometric mode via the command line, at the 'Command:' prompt, enter the **SNAP** command. (Alternatively, enter the alias **sn** at the keyboard.)

Command: **snap** *(Press* ENTER.*)*

At the command prompt, enter **s** to select the **Style** option:

Specify snap spacing or [ON/OFF/Aspect/Rotate/Style/Type] <0.5000>: **s**

Enter **i** to select the **Isometric** option:

Enter snap grid style [Standard/Isometric] <S>: **i**

To turn off isometric mode, select the **Standard** option for snap grid style.

I recommend that you keep ortho mode turned on while creating isometric drawings. Choose the **ORTHO** button on the status line to turn on ortho mode. When AutoCAD is in isometric mode, it adjusts ortho mode by 30 degrees automatically to reflect the current isoplane.

ISOPLANE

The **ISOPLANE** command changes the display between three isometric planes.

Isometric drafting draws in any of the three isoplanes. To draw in a particular isoplane, you must change between the planes.

TUTORIAL: SWITCHING ISOPLANES

1. To switch between isometric drafting planes, start the **ISOPLANE** command:
   - At the 'Command:' prompt, enter the **isoplane** command.
   - Alternatively, press **CTRL+4** or function key **F5** on the keyboard.

   Command: **isoplane** *(Press* ENTER.*)*
2. AutoCAD reports the current isoplane, and then prompts you to select another plane:

   Current isoplane: Left

   Enter isometric plane setting [Left/Top/Right] <Top>: *(Type an option.)*

   Alternatively, press **ENTER** to go to the next plane, indicated by the name shown in angle brackets, such as <Top>. The isoplanes are displayed in a rotating fashion: left, top, right, and back to left again.

You may find it faster to switch between the isoplanes transparently by pressing **CTRL+E** or **F5**. Each time you press **CTRL+E**, the isoplane changes to the next one — top, left, or right — in a cycle.

When the grid is on, the grid display reflects the isoplane angles.

ISOMETRIC DRAFTING TECHNIQUES

After setting up isometric mode, you can use any of AutoCAD's commands to draft and edit the isometric drawing.

The general method for constructing isometric drawings is first to "box in" the general shape of the object. Then add and trim the lines and isocircles to create the final shape. The figure illustrates how isometric drafting proceeds from a plain box to greater detail:

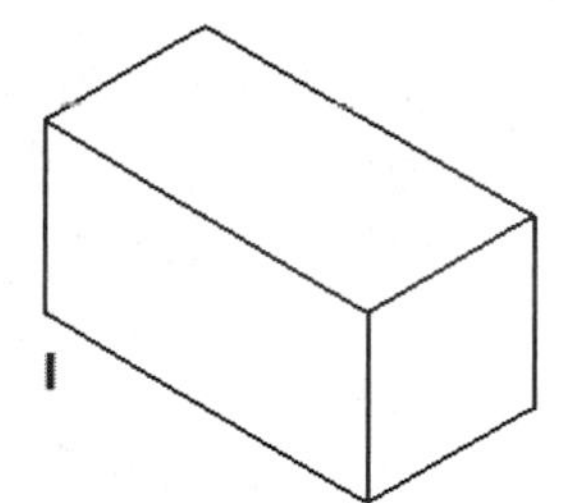

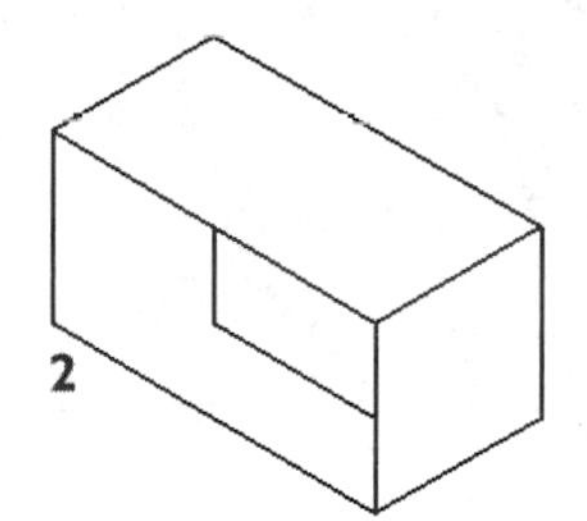

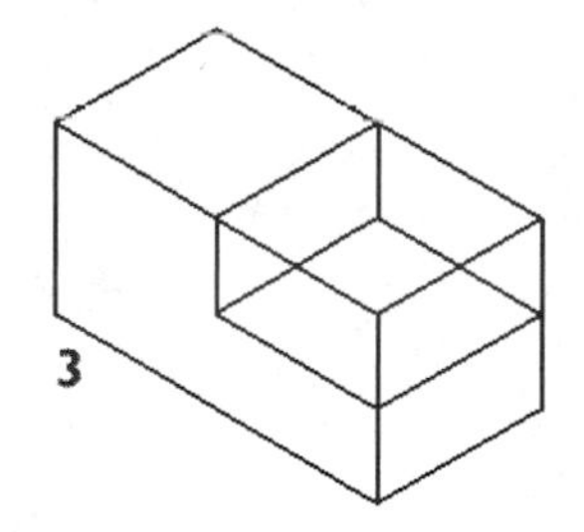

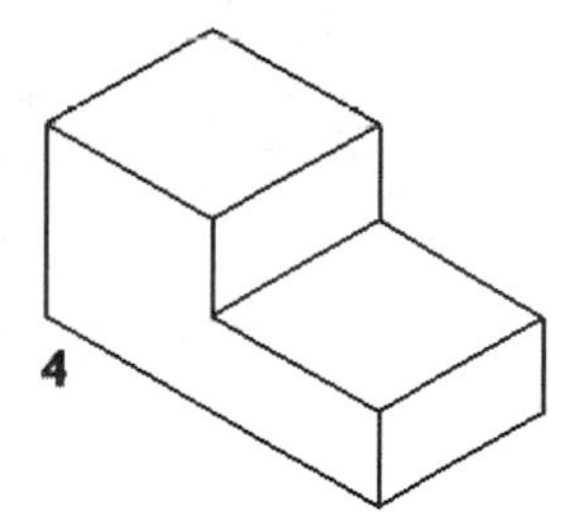

***1:** Box defines general shape.*
***2:** Define cutouts on front face.*
***3:** Project isometric lines into box.*
***4:** Trim lines to create final drawing.*

Direct distance entry makes it easier to draw lines in isometric mode. Tracking efficiently moves the starting point of a line.

Isometric Circles

In isometric drafting, circles are drawn as 30-degree ellipses. In AutoCAD, isometric circles are placed with the **ELLIPSE** command's **Isocircle** option. (Don't be surprised if you have never seen the option: it appears only when isometric mode is turned on.) Why not just draw ellipses? Because AutoCAD automatically adjusts the angle of the isocircle to look correct in the current isoplane.

TUTORIAL: DRAWING ISOCIRCLES

1. To draw with isometric circles, start the **ELLIPSE** command:
   - From the **Draw** menu, choose **Ellipse**.
   - On the **Draw** toolbar, click the **Ellipse** button.
   - At the 'Command:' prompt, enter the **ellipse** command.
   - Alternatively, enter the alias **el** at the keyboard.

   Command: **ellipse** *(Press* ENTER.*)*
2. In all cases, enter **i** to select the **Isocircle** option:

   Specify axis endpoint of ellipse or [Arc/Center/Isocircle]: **i**
3. Pick the center of the isocircle. Ensure you are drawing on the correct face for the current isoplane. If necessary, first press **F5** to switch to the correct isoplane.

   Specify center of isocircle: *(Pick point 1.)*
4. Specify a point on the radius or diameter of the isocircle:

   Specify radius of isocircle or [Diameter]: *(Pick point 2.)*

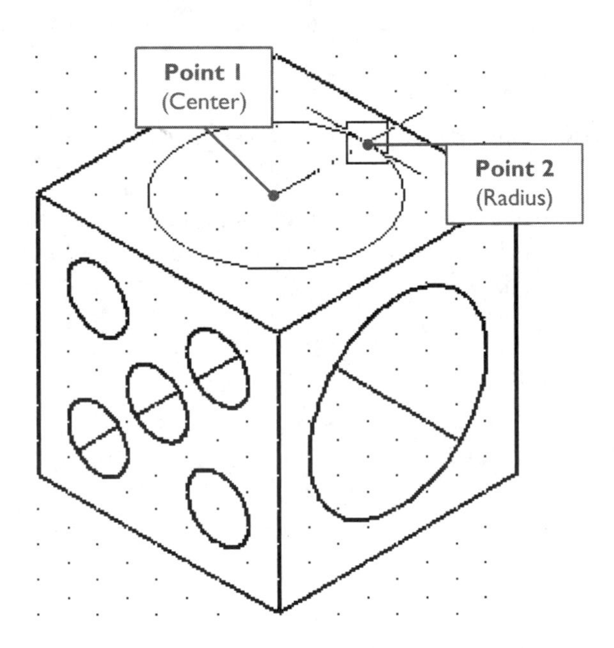

Isometric circles are ellipses drawn at 30 degrees.

Isometric Text

Text must be slanted to make it look "right" in each of the three isoplanes. The approach is to: (1) create three text styles, each slanting the text correctly for each of the isoplanes; and (2) place text rotated at the correct angle. AutoCAD does not provide text styles appropriate for isometric drawing. (The support files that include the term "ISO" refer to the International Organization for Standardization, and not isometric drawing standards.)

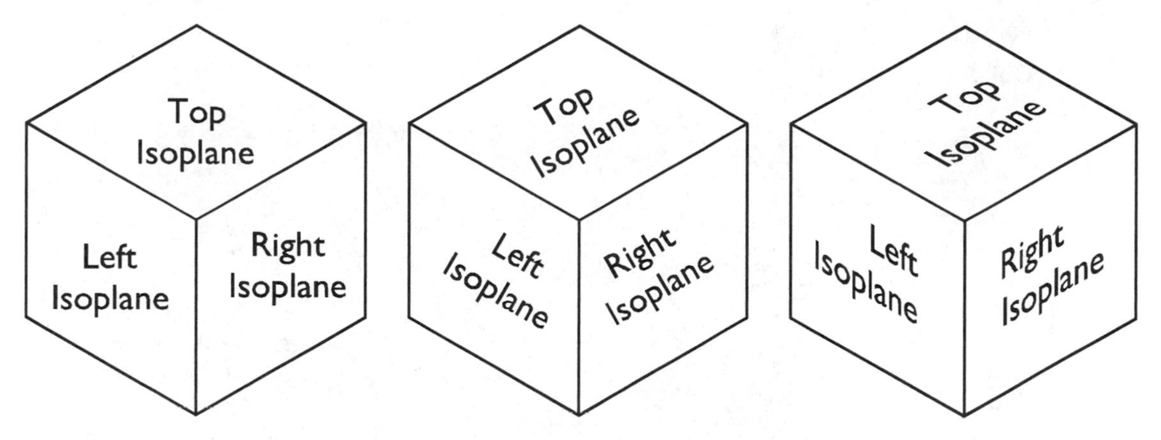

Left: *Normal text looks wrong in isometric drawings.*
Center: *Text looks slanted when rotated by 30 degrees to align with isoplanes.*
Right: *Text looks correct after applying an obliquing angle of 30 degrees.*

The following table summarizes the combinations of positive and negative 30-degree angles required to place the text correctly:

Isoplane	***Oblique Angle***	***Text Rotation***
Left	-30°	-30°
Top	-30°	30°
Right	30°	30°

The oblique angle is set with the STYLE command, while the text rotation is handled by the TEXT command.

TUTORIAL: CREATING ISOMETRIC TEXT STYLES

1. To create text styles for isometric drafting, start the STYLE command:
   - From the **Format** menu, choose **Text Style**.
   - At the 'Command:' prompt, enter the **style** command.
   - Alternatively, enter the alias **st** at the keyboard.

   Command: **style** *(Press* ENTER.*)*

   In all cases, AutoCAD displays the Text Style dialog box.
2. In the dialog box, click **New**.
3. In the New Text Style dialog box, enter the name of the text style:

   Style Name: **isotop**

   Click **OK**, and the Text Style dialog box returns.
4. Specify the following options:

   Font Name: **simplex.shx**

   Oblique Angle: **-30**

   And then click **Apply**.
5. You have created the first of three text styles. Two more to go.
   Click **New** to start the second text style.

   Style Name: **isoright**

   Click **OK**.

Font Name: **simplex.shx**

Oblique Angle: **30**

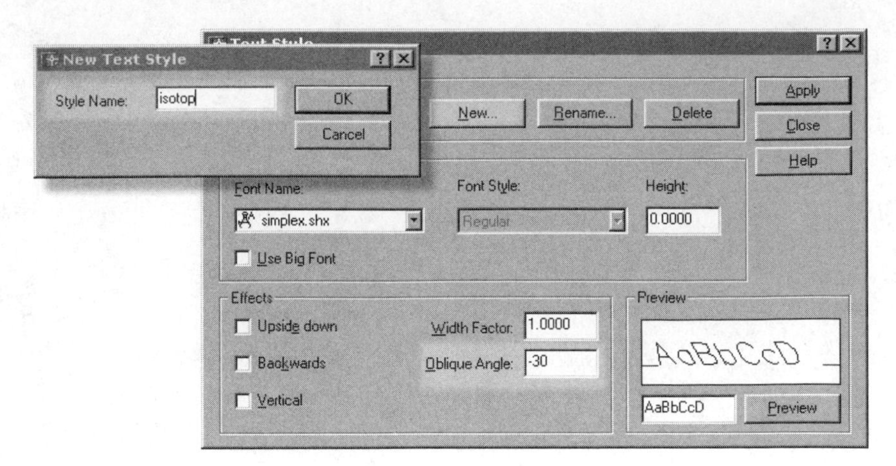

Creating new text styles.

6. Click **Apply**, and then click **New** to start the third text style.

   Style Name: **isoleft**

   Click **OK**.

   Font Name: **simplex.shx**

   Oblique Angle: **-30**

   Click **Apply**, and then click **Close**.

7. Save the styles in a template file for reuse later.
   a. From the **File** menu, select **Save As**.
   b. In the **Files of type** droplist, select "AutoCAD Template File."
   c. Name the file:

      File name: **isometric**

   d. Click **Save**.

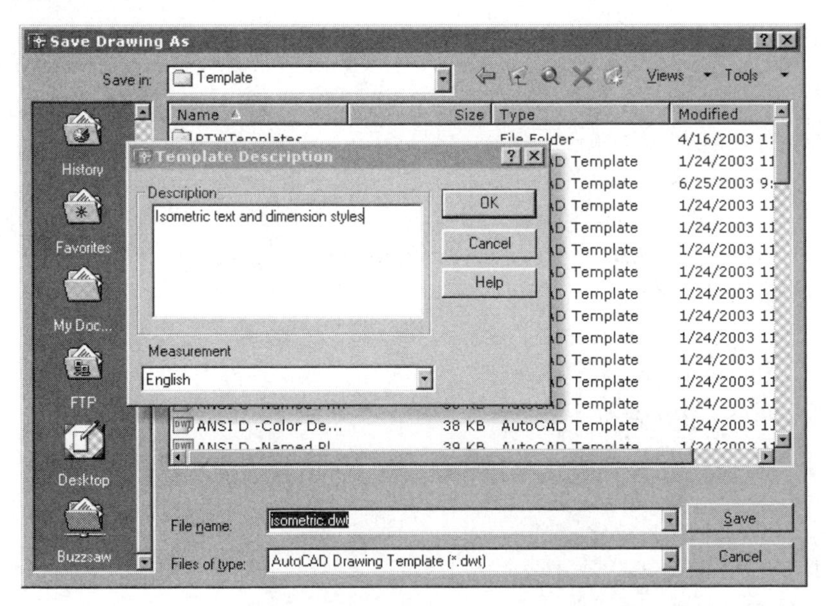

Saving the text styles in a template file.

e. In the Template Description dialog box, enter:
   Description: **Isometric text and dimension styles.**

f. Click **OK**.

You have the three standard isometric text styles saved in a template file. But before starting isometric drafting, you also need to create three isometric dimension styles.

Isometric Dimensions

Dimensioning an isometric object requires that the extension lines and dimension text align properly in the appropriate isometric plane. To dimension an isometric drawing takes three steps: (1) create three dimension styles, one for each isoplane, with the **DIMSTYLE** command; (2) place aligned dimensions with the **DIMALIGNED** command; and (3) adjust the angle of extension lines with the **DIMEDIT** command's **Oblique** option.

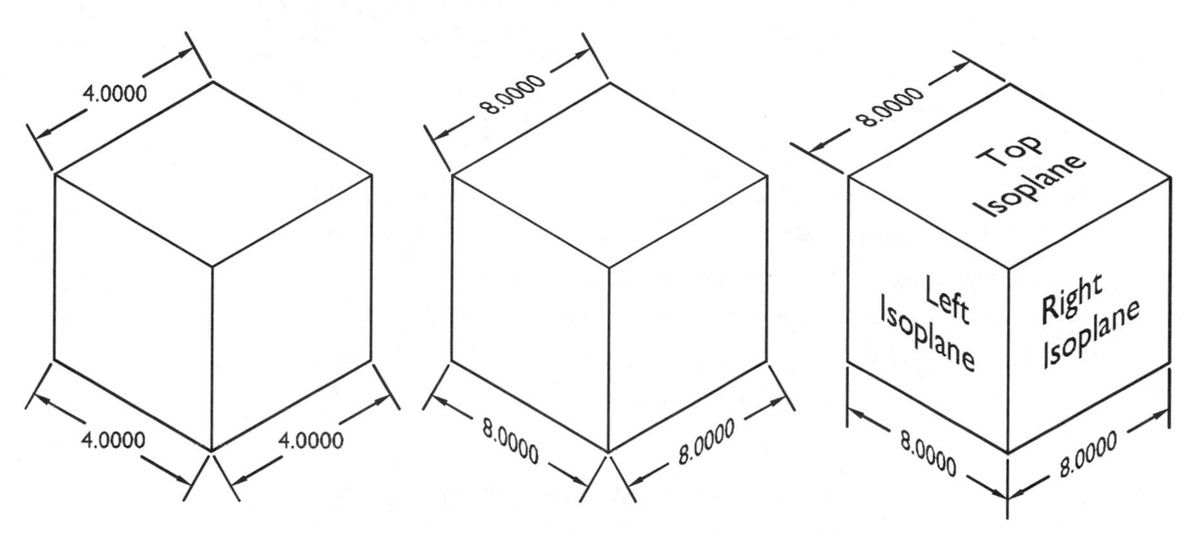

Left: *Dimensions look slanted, but are aligned with isoplanes.*
Center: *Text looks correct after applying isometric text styles, but not the extension lines.*
Right: *Extension lines look correct after being obliqued by 30 degrees.*

The following table summarizes the combinations of positive and negative 30-degree angles required to place dimensions correctly:

Isoplane	*Text Oblique*	*Extension Lines Oblique*
Left	-30°	90°
Top	-30°	-30°
Right	30°	90°

TUTORIAL: CREATING ISOMETRIC DIMENSION STYLES

This tutorial makes use of the three isometric text styles created in the previous tutorial. Continue with the *isometric.dwt* template drawing:

1. To create dimension styles for isometric drafting, start the **DIMSTYLE** command:
   - From the **Format** menu, choose **Dimension Style**.
   - At the 'Command:' prompt, enter the **dimstyle** command.
   - Alternatively, enter the alias **d** at the keyboard.

   Command: **dimstyle** *(Press* ENTER.*)*

   In all cases, AutoCAD displays the Dimension Style Manager dialog box.

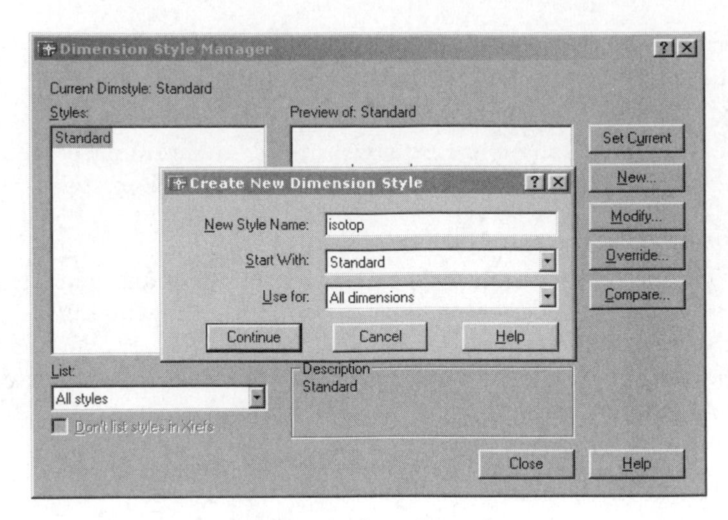

Creating new dimensions styles.

2. To create a new dimension style, click the **New** button.
3. In the Create New Dimension Style dialog box, give the new dimstyle its name.

   New Style Name: **isoleft**

   Click **Continue**, and the Lines and Arrows tab appears.
4. Choose the **Text** tab, and set the following options:

   Text Style: **isoleft**

   Text alignment: **Aligned with dimension line**

   Click **OK**.

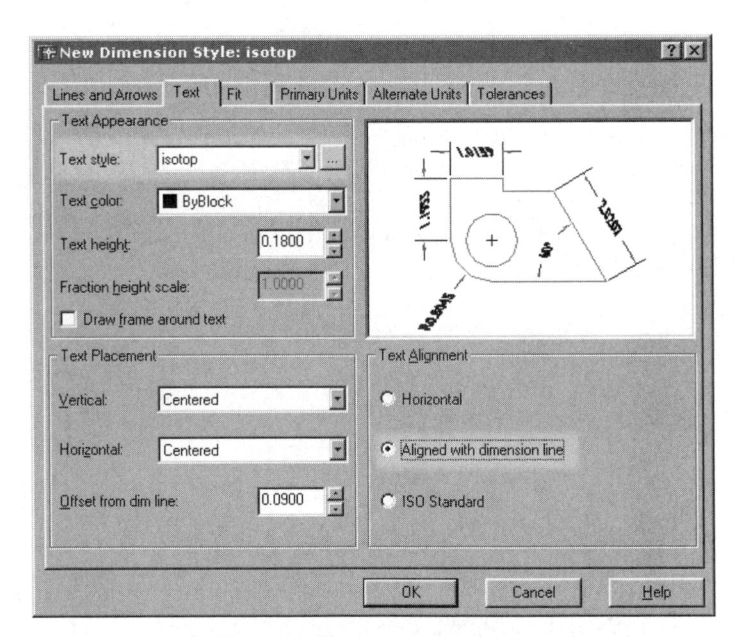

Creating new dimensions styles.

AutoCAD unfortunately does not allow the dimstyle to set the angle of the extension lines; these are adjusted later while drafting the isometric drawing.

5. You have now created one of the three dimension styles. Repeat twice more for the other isoplanes:

   Click **New**.

   New Style Name: **isotop**

Click **Continue**, and in the Text tab:

Text Style: **isotop**

Text alignment: **Aligned with dimension line**

6. Click **OK**, and then click **New**.

   New Style Name: **isoright**

   Click **Continue**, and in the Text tab:

   Text Style: **isoright**

   Text alignment: **Aligned with dimension line**

   Click **OK**, and then click **Close**.
7. Save the template drawing with **CTRL+S**.

ISOMETRIC DRAFTING TUTORIAL

In this tutorial, we create an isometric drawing of an L-bracket, and then dimension it. The figure shows the completed project.

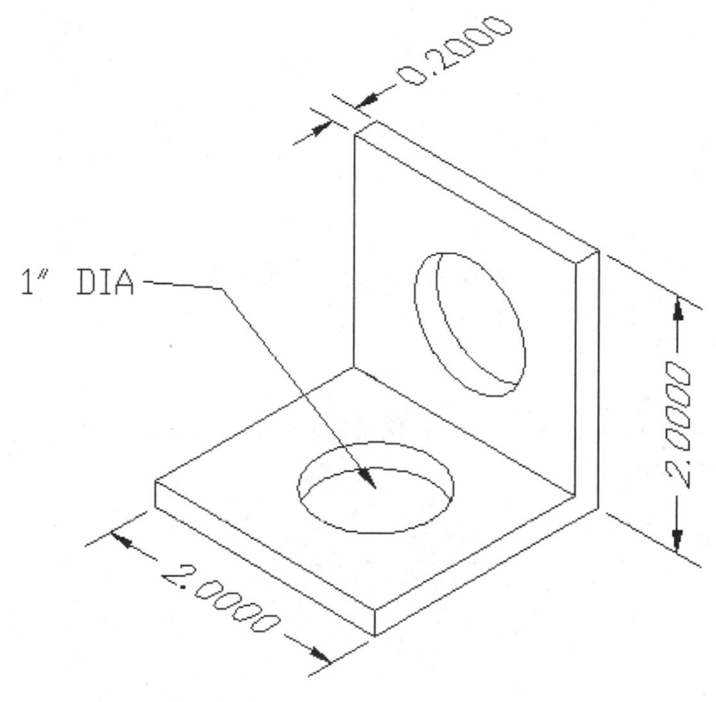

Isometric drawing created by this tutorial.

1. Start AutoCAD with the *isometric.dwt* template drawing.
2. Prepare the drawing with the following settings:
   - Turn on snap, grid, and ortho modes.
   - Set the snap y spacing to 0.1, and the grid to 0.5.
   - Press **F5** until the left isoplane is displayed.
   - Set the limits to 0,0 and 6,6.
   - Perform the **ZOOM All** command.

Now you are ready to create the isometric drawing. The easiest way to draw in isometric mode is first to draw a cube; the cube represents the bulk of the model. It is made of four lines, each 2 units long. Use direct distance entry to assist you.

3. Draw the cube with the **LINE** command, starting with the left face:

   Command: **line**

   Specify first point: *(Pick point 1 on top of a grid mark.)*

   Specify next point or [Undo]: *(Move cursor northwest to 2.)* **2**

   Specify next point or [Undo]: *(Move the cursor north to 3.)* **2**

   Specify next point or [Close/Undo]: *(Move the cursor southeast to 4.)* **2**

   Specify next point or [Close/Undo]: **c**

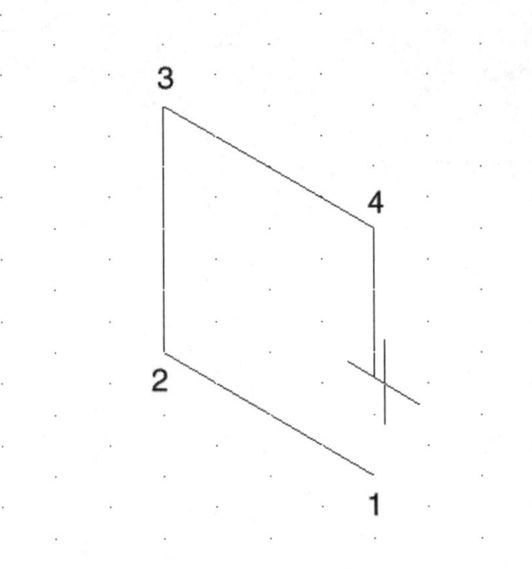

Drawing the left face of the cube.

4. To draw the top face, first press **F5** to switch to the top isoplane.

The top face is made of three lines; each is also 2 units long. Use INTersection object snap to start and end the lines accurately.

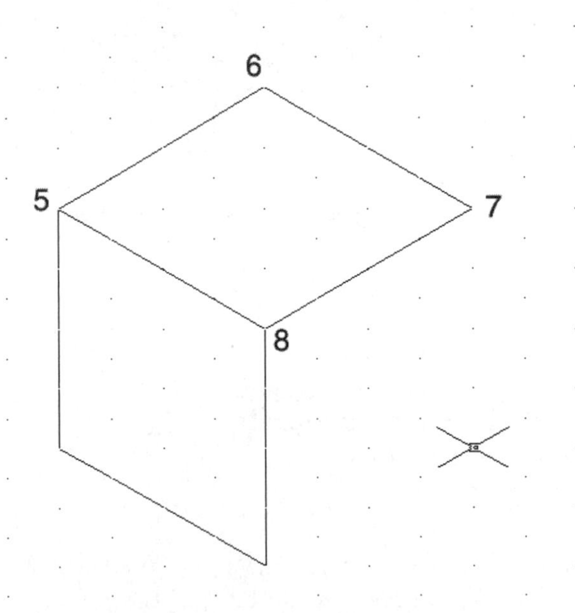

Drawing the top face of the cube.

5. Draw the top face of the cube.

   Command: *(Press* **ENTER** *to repeat the* LINE *command.)*

   LINE Specify first point: **int**

   of *(Move cursor over the upper-left corner of the left face at 5 and click).*

   Specify next point or [Undo]: *(Move cursor to the northeast at 6.)* **2**

   Specify next point or [Undo]: *(Move cursor southeast at 7.)* **2**

   Specify next point or [Close/Undo]: *(Move cursor to the southwest at 8.)* **2**

   Specify next point or [Close/Undo]: *(Press* **ENTER.***)*

   The drawing now has the left and top face of the cube.
6. Press **CTRL+E** to switch the to the right isoplane.
7. Draw the two lines that form the right face of the cube:

   Command: *(Press* **ENTER** *to repeat the* **LINE** *command.)*

   LINE Specify first point: **int**

   of *(Move cursor over the rightmost corner of the top face at 9 and click.)*

   Specify next point or [Undo]: *(Move cursor south at 10.)* **2**

   Specify next point or [Undo]: *(Move cursor to the southwest at 11.)* **2**

   Specify next point or [Close/Undo]: *(Press* **ENTER.***)*

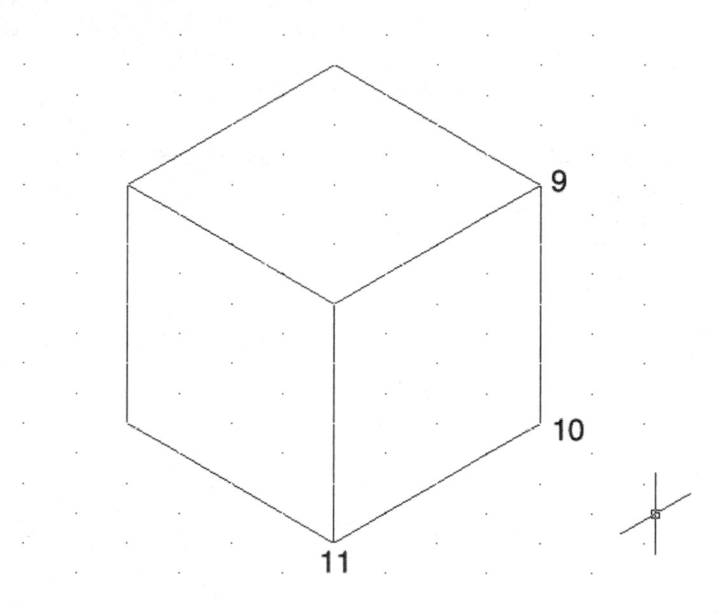

Draw the right face of the cube.

8. The isometric cube is complete. Save your work with **CTRL+S**, naming the drawing "isobracket."

The next stage is to draw the "inner" parts that define the legs of the L-bracket, which are 0.2 units thick. This time you employ tracking to help you start drawing the lines.

9. Return to the left isoplane, and continue drawing.

   Command: *(Press* **ENTER***, and then press* **F5** *to return to the left isoplane.)*

   <Isoplane left> LINE Specify first point: **tk**

   First tracking point: **int**

   of *(Pick the bottommost corner at 12.)*

Next point (Press enter to end tracking): *(Move cursor up to 13.)* **0.2**

Next point (Press enter to end tracking): *(Press* ENTER.*)*

Specify next point or [Undo]: *(Move cursor northwest to 14.)* **2**

Specify next point or [Undo]: *(Press* ENTER.*)*

You have drawn the left face of the bracket's lower leg.

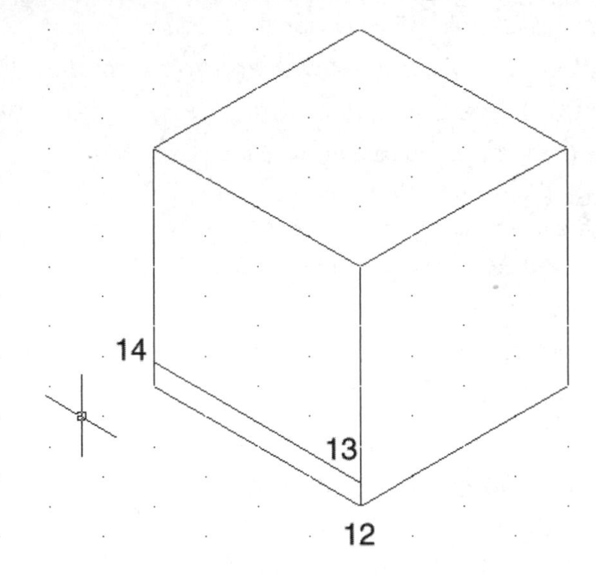

Drawing the left face of the bracket.

10. Draw the remainder of the bracket's lower leg. Press **F5** to switch to the upper isoplane.

    Command: **line**

    Specify first point: **int**

    of *(Pick the just completed line endpoint at 15.)*

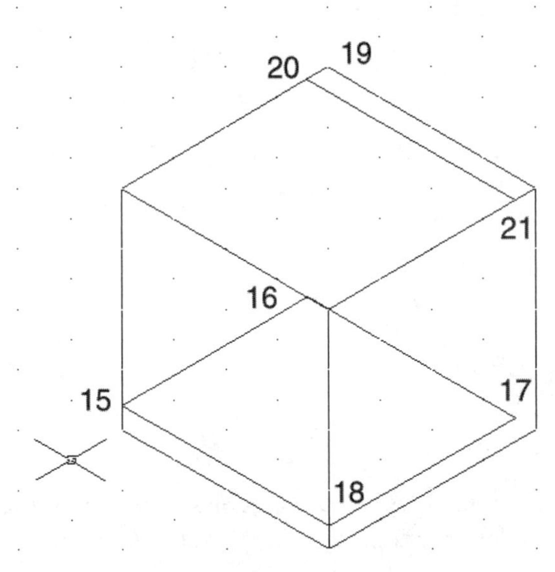

Continuing the bracket.

Specify next point or [Undo]: *(Move cursor northeast at 16.)* **1.8**

Specify next point or [Undo]: *(Move cursor southeast to 17.)* **2**

Specify next point or [Close/Undo]: *(Move cursor southwest to 18.)* **1.8**

Specify next point or [Close/Undo]: *(Press* **ENTER***.)*

That completes the lower leg.

11. While still in the top isoplane, draw the top face of the upright leg:

    Command: *(Press* **ENTER***.)*

    LINE Specify first point: **tk**

    First tracking point: **int**

    of *(Pick uppermost corner of the cube at 19.)*

    Next point (Press enter to end tracking): *(Move cursor southwest to "20".)* **0.2**

    Next point (Press enter to end tracking): *(Press* **ENTER***.)*

    Specify next point or [Undo]: *(Move cursor southeast to 21.)* **2**

    Specify next point or [Undo]: *(Press* **ENTER***.)*

12. Finish the leg by switching to the right isoplane: press **F5**. There are two more lines to draw:

    Command: **line**

    Specify first point: **int**

    of *(Pick end of the line you just drew at 22.)*

    Specify next point or [Undo]: *(Move cursor south to 23.)* **1.8**

    Specify next point or [Undo]: *(Press* **ENTER** *to end the command.)*

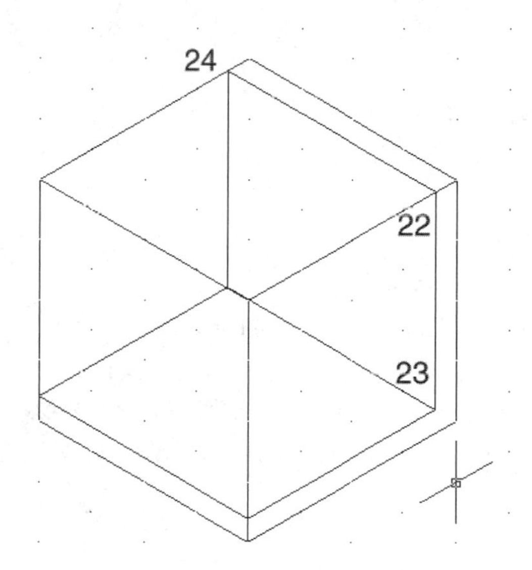

Completing the bracket.

13. Use the **COPY** command to draw the final line, using the **Last** option to select the line just drawn:

    Command: **copy**

    Select objects: **l**

    1 found Select objects: *(Press* **ENTER** *to end object selection.)*

    <Base point or displacement>/Multiple: **int**

    of *(Pick end of line just drawn at 22.)*

    Specify second point of displacement or <use first point as displacement>: *(Move cursor.)*

Oops! If you find you cannot move the line to where you want it to go, it's because you are in the wrong isoplane. Press **CTRL+E** to switch isoplanes, and then move the cursor.

Specify second point of displacement or <use first point as displacement>: *(Press* **F5**.*)*

<Isoplane Left> **nea**

of *(Pick at 24.)*

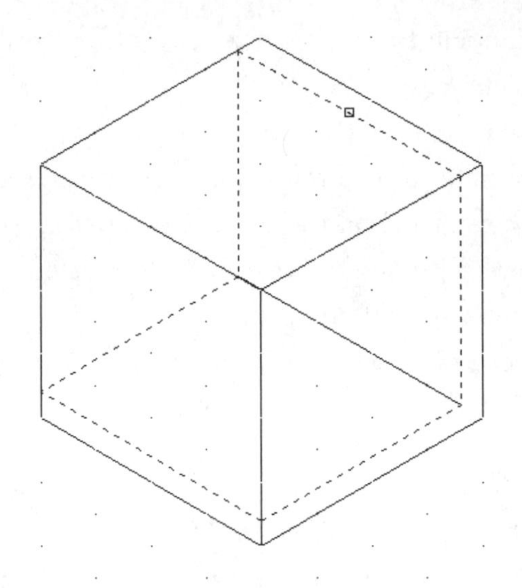

Bracket inside the cube.

The drafting of the two legs is complete. The next stage is to erase the unneeded parts of the cube with the **TRIM** command.

14. Trim unneeded parts of the drawing:

    Command: **trim**

    Current settings: Projection=UCS Edge=None

    Select cutting edges...

    Select objects: *(Pick a line.)*

    1 found Select objects: *(Pick another line.)*

    Select the six lines shown dotted in the figure. These form the cutting edges.

    Cutting Edges for Trim Command Select objects: *(Press* **ENTER**.*)*

    Select object to trim or [Project/Edge/Undo]: *(Pick an unneeded line of the cube.)*

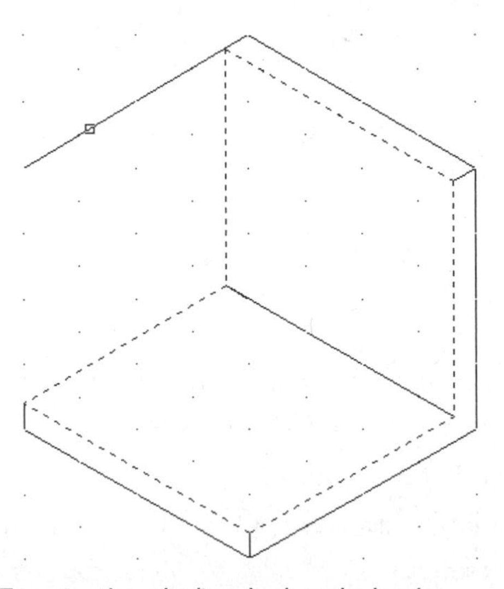

Trimming the cube lines back to the bracket.

15. Select the five unneeded lines, as illustrated by the figure.

    Select object to trim or [Project/Edge/Undo]: *(Press* **ENTER**.*)*

The "L" of the bracket is now complete. The next step is to draw the holes with the **ELLIPSE** command. The center of the hole is 1,1 from the front corner of the lower leg. It can be tricky finding the exact middle of isometric objects, so employ tracking again.

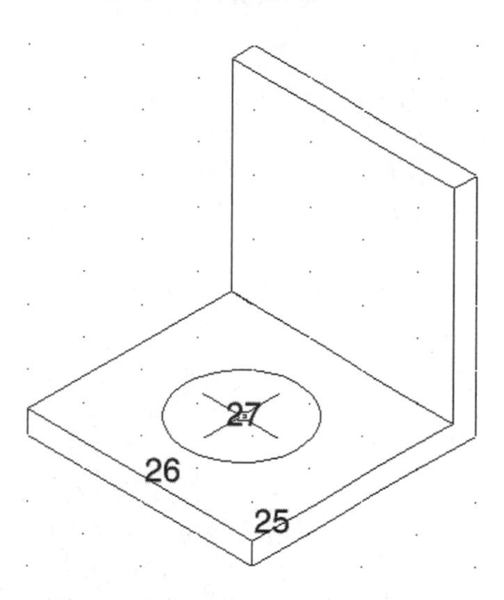

Drawing the top part of the hole.

16. Draw the circles with the **ELLIPSE** command, together with tracking and INTersection object snap.

    First, though, press **F5** to reach the top isoplane.

    Command: *(Press* **F5**.*)*

    <Isoplane top> **ellipse**

    Arc/Center/Isocircle/<Axis endpoint 1>: **i**

    Specify center of isocircle: **tk**

    First tracking point: **int**

    of *(Pick corner at 25 on the lower leg.)*

    Next point (Press enter to end tracking): *(Move northwest to 26.)* **1**

    Next point (Press enter to end tracking): *(Move northeast to 27.)* **1**

    Next point (Press enter to end tracking): *(Press* **ENTER**.*)*

    Specify radius of isocircle or [Diameter]: **0.5**

17. To draw the "bottom" of the hole, use the **COPY** command. Because the copying must take place in the vertical direction, press **F5** to get the right isoplane, which allows movement in the up-down direction.

    Command: *(Press* **F5**.*)*

    <Isoplane right> **copy**

    Select objects: **l**

    1 found Select objects: *(Press* **ENTER**.*)*

<Base point or displacement>/Multiple: **nea**

to *(Pick the isocircle.)*

Second point of displacement: *(Move cursor south.)* **0.2**

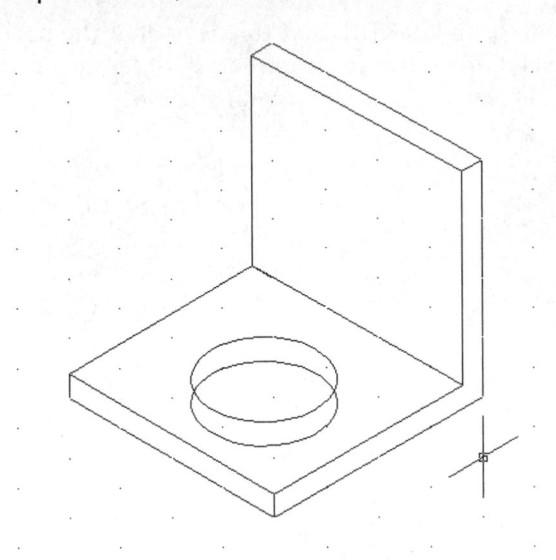

Drawing the bottom part of the hole.

18. Use the **TRIM** command to erase the parts of the lower isocircle that are hidden from view.

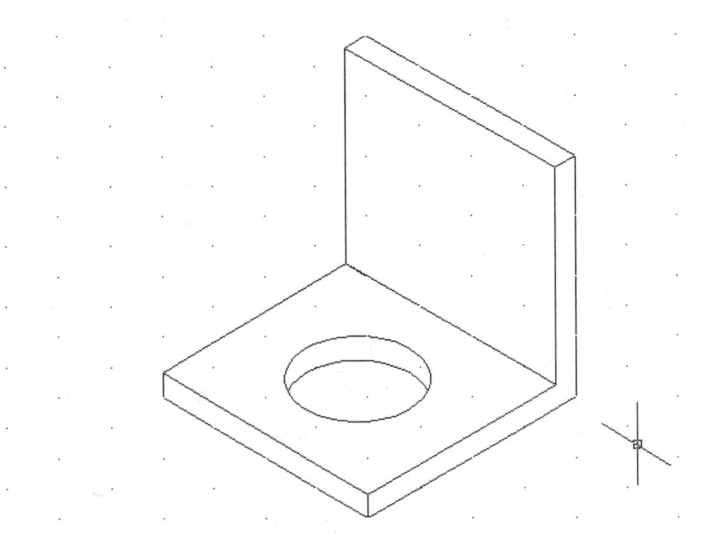

Trimming away unnecessary lines.

Command: **trim**

Current settings: Projection=UCS Edge=None

Select cutting edges ...

Select objects: *(Pick the upper circle.)*

1 found Select objects: *(Press* **ENTER.***)*

Select object to trim or [Project/Edge/Undo]: *(Pick unwanted portion of lower circle.)*

Select object to trim or [Project/Edge/Undo]: *(Press* **ENTER.***)*

You've now given the lower bracket a hole that looks quite realistic!

19. The easiest way to add the hole to the upright leg is with the **MIRROR** command. It will be easier if you first switch to the left isoplane:

    Command: *(Press* **F5***.)*

    <Isoplane left> **mirror**

    Select objects: *(Pick the isocircle.)*

    1 found Select objects: *(Pick the isoarc.)*

    1 found Select objects: *(Press* **ENTER***.)*

    Specify first point of mirror line: **int**

    of *(Pick one end of "fold" line.)*

    Specify second point of mirror line: **int**

    of *(Pick other end of "fold" line, as shown in Figure 21.25.)*

    Delete source objects? [Yes/No] <N>: *(Press* **ENTER***.)*

    Instant second hole!
20. Before going on to dimension the bracket, save your work with **CTRL+S**.

The final stage is to place dimensions. The best approach is always to use the **DIMALIGNED** command, because it aligns the dimension along the iso-axes. Do all dimensioning in one isoplane, and then dimension in the next isoplane.

21. Ensure AutoCAD is showing the left isoplane by pressing **F5** until <Isoplane left> shows up at the 'Command:' prompt. Start placing dimensions with the **DIMALIGNED** command.

    Command: **dimaligned**

    First extension line origin or press enter to select: **int**

    of *(Pick corner at 28.)*

    Second extension line origin: **int**

    of *(Pick other corner at 29.)*

    Dimension line location (Mtext/Text/Angle): *(Pick a point.)*

    Dimension text=2.0000
22. Does the dimension look odd to you? To correct it, you need to skew the extension lines. Apply the **DIMEDIT** command:

    Command: **dimedit**

    Dimension Edit (Home/New/Rotate/Oblique) <Home>: **o**

    Select objects: *(Pick the dimension.)*

    1 found Select objects: *(Press* **ENTER***.)*

    Enter obliquing angle (press enter for none): **30**

    Now the isometric dimension looks proper.

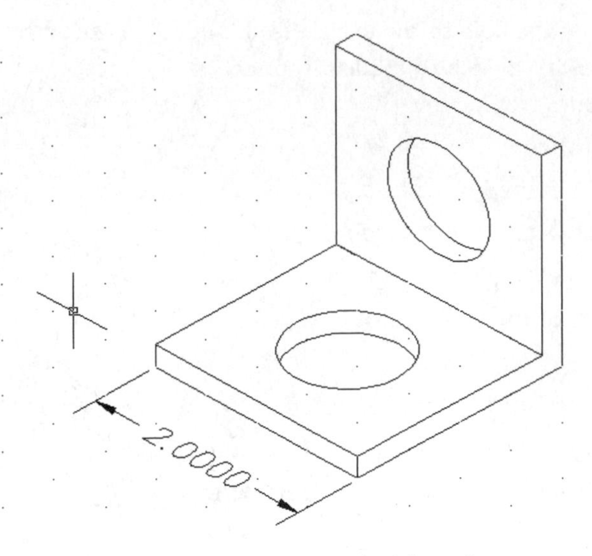

Adding dimensions to the left isoplane.

23. Continue dimensioning on the top isoplane:
    - Press **F5** to switch to the top isoplane.
    - Use the **DIMSTYLE** command to select the "isotop" dimension style.
    - Place the dimension with the **DIMALIGNED** command and INTersection object snap.
    - Apply the **DIMEDIT** command's **Oblique** option to skew the extension lines by 30 degrees.
24. Finally, place dimensions in the right isoplane with the "isoright" dimension style and skew the extension lines by -30 degrees.

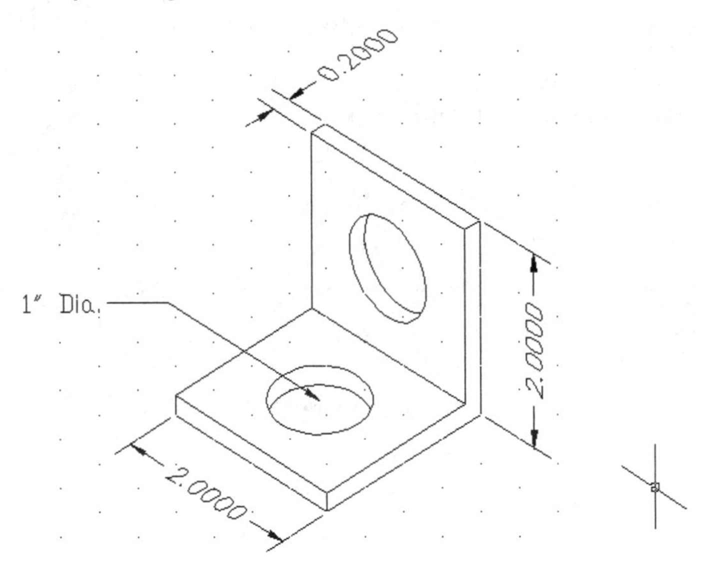

Dimensioning completed

25. To place leaders, use the Standard dimstyle and Standard text style.
26. Save your work!

EXERCISES

1. Start AutoCAD with a new drawing.
   Set up the drawing for isometric drafting.
   Draw a cube with a circle on each face.
   Tip: Use ortho mode to assist drawing isometric lines perfectly along the isometric axes.
2. Draw the isometric objects shown in the figures below.
   a.

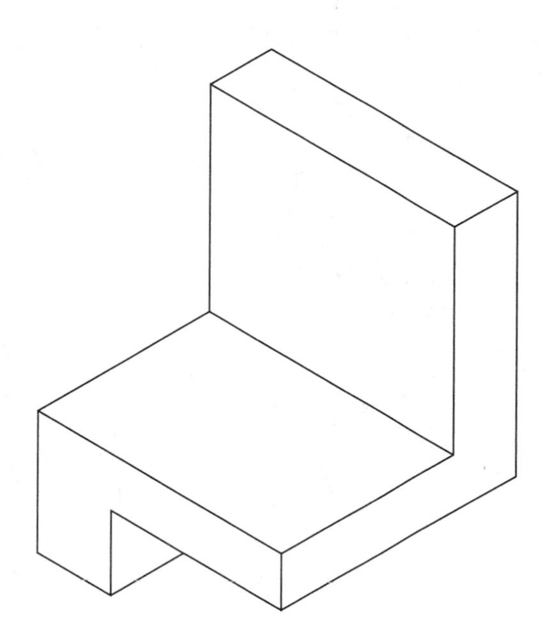

b.

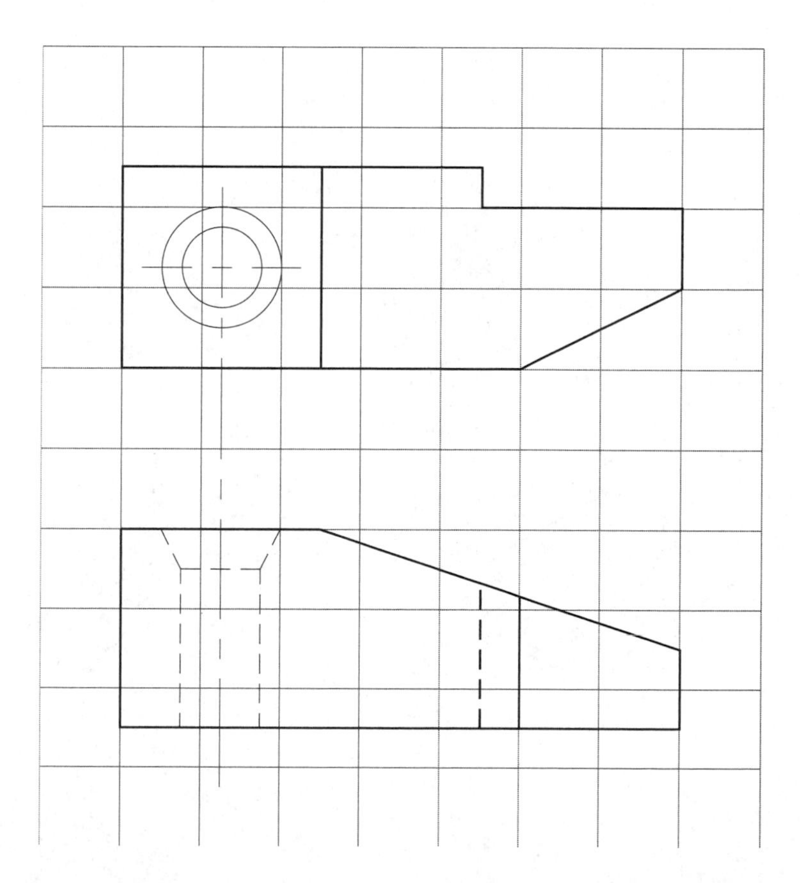

c.

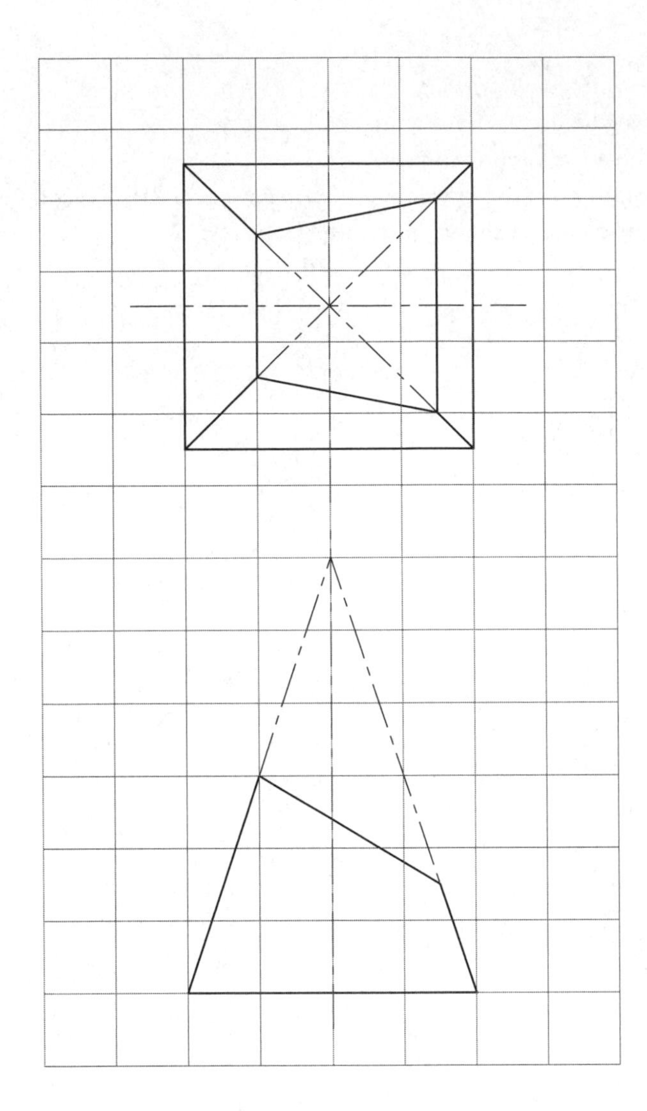

d.

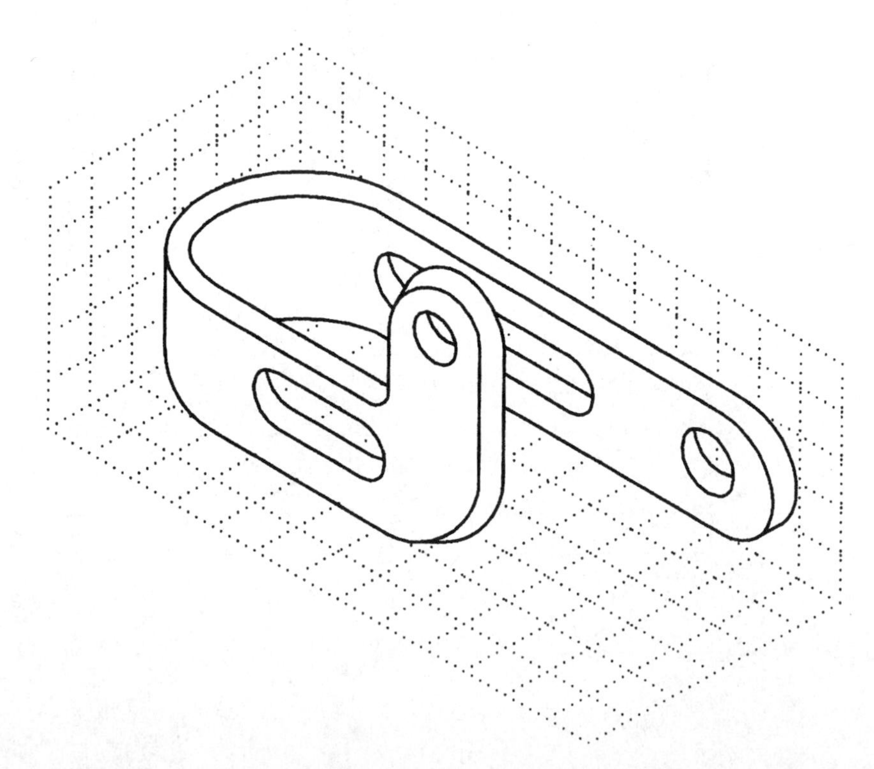

e.

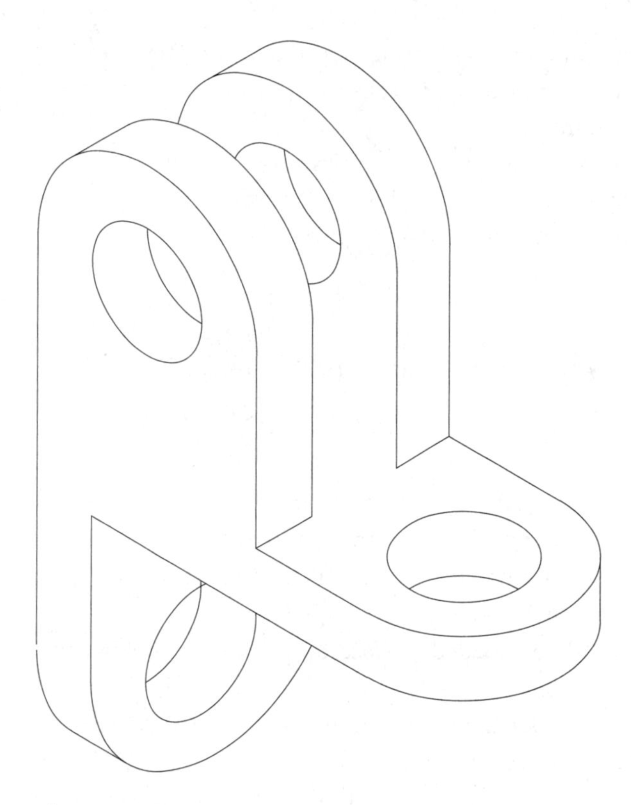

3. Dimension the objects drawn in the previous exercise.

   Tip: Dimensioning is made easier by setting the snap spacing to an increment of the object's dimensions, and the object snap to ENDpoint.

CHAPTER REVIEW

1. What are three types of engineering pictorial drawings?
   Which is the most realistic in appearance?
   What type is isometric?
3. How many axes are used in isometric drawing?
   How many degrees above horizontal are the isometric axes?
4. List the names of the three isometric planes:
   a.
   b.
   c.
5. Name three ways to change isometric planes in AutoCAD:
   a.
   b.
   c.
6. What commands place AutoCAD into isometric mode?
7. Describe how circles are drawn in isometric mode.
8. What command(s) are needed to set text for isometric drafting?
   To set up dimensions?
9. What command(s) are needed to *place* text in isometric drawings?
   To place dimensions?
10. Which parts of AutoCAD are affected by isometric mode?
11. How do you exit isometric mode?

UNIT III

Three Dimensional Design

9

Viewing in 3D

Computers are inherently two-dimensional. The screen is flat, the mouse moves on a pad, and the output is on paper — all are 2D. This makes it difficult to work with three-dimensional drawings in AutoCAD. Either the third dimension must be simulated (for example, with shading and perspective views), or the 2D interface must be adapted to interact with 3D objects.

AutoCAD has numerous commands for manipulating the 3D viewpoint. Some let you see 3D models from a static viewpoint; some allow you interactively to rotate the view; others create multiple viewports so that you can see the 3D model from several directions at once; and some change the orientation of the x,y-plane — the working plane that most 2D commands require.

In this chapter, you learn about the following commands:

VPOINT changes the 3D viewpoint.

3DORBIT changes the 3D viewpoint interactively.

PLAN returns the viewpoint to the plan view.

UCS New creates new coordinate systems.

UCSICON controls the look and position of the UCS icon.

UCS manages coordinate systems.

VPORTS creates multiple viewports.

FINDING THE COMMANDS

On the **UCS** and **UCS II** toolbars:

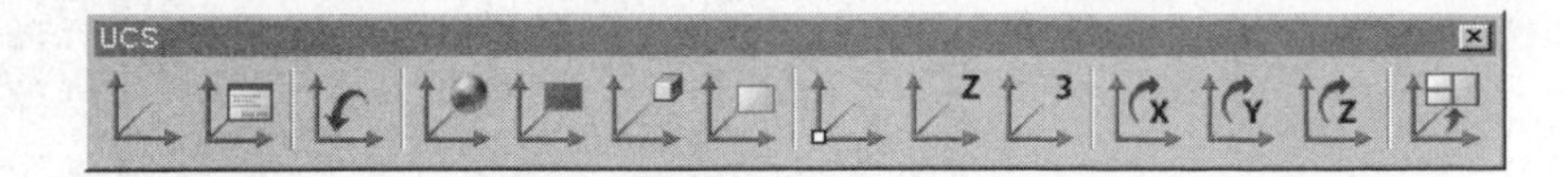

On the **VIEW** menu:

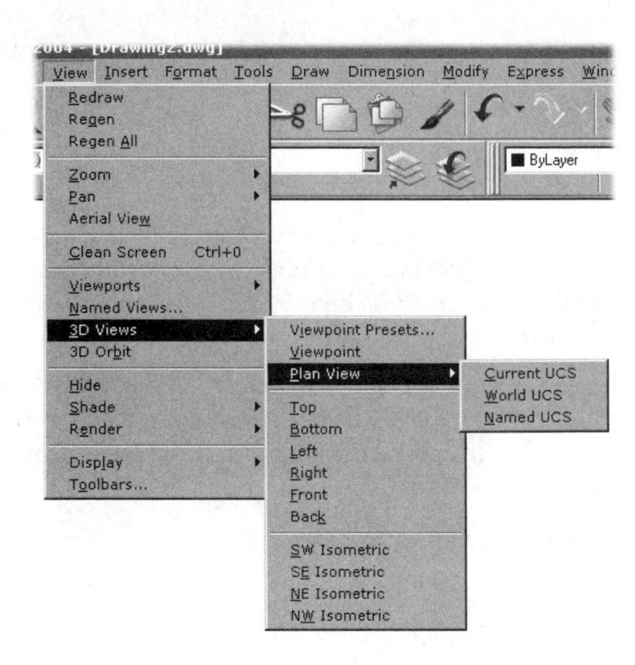

INTRODUCTION TO 3D

Until now you have been working with 2D drawings, drawing and editing in the x,y-plane. In AutoCAD, all drawings are 3D; all along, the z-coordinate has been 0. One way to think of the x, y, z-coordinate system is to picture the *plan view* of your drawing lying in the x, y-plane.

Think of the room you are in. Its floor represents the x,y-plane. Now look into the corner. The corner represents the z-axis, as it "goes up." One wall is the x,z-plane — the wall over the x axis. The other wall is the y,z-plane, the one over the y-axis.

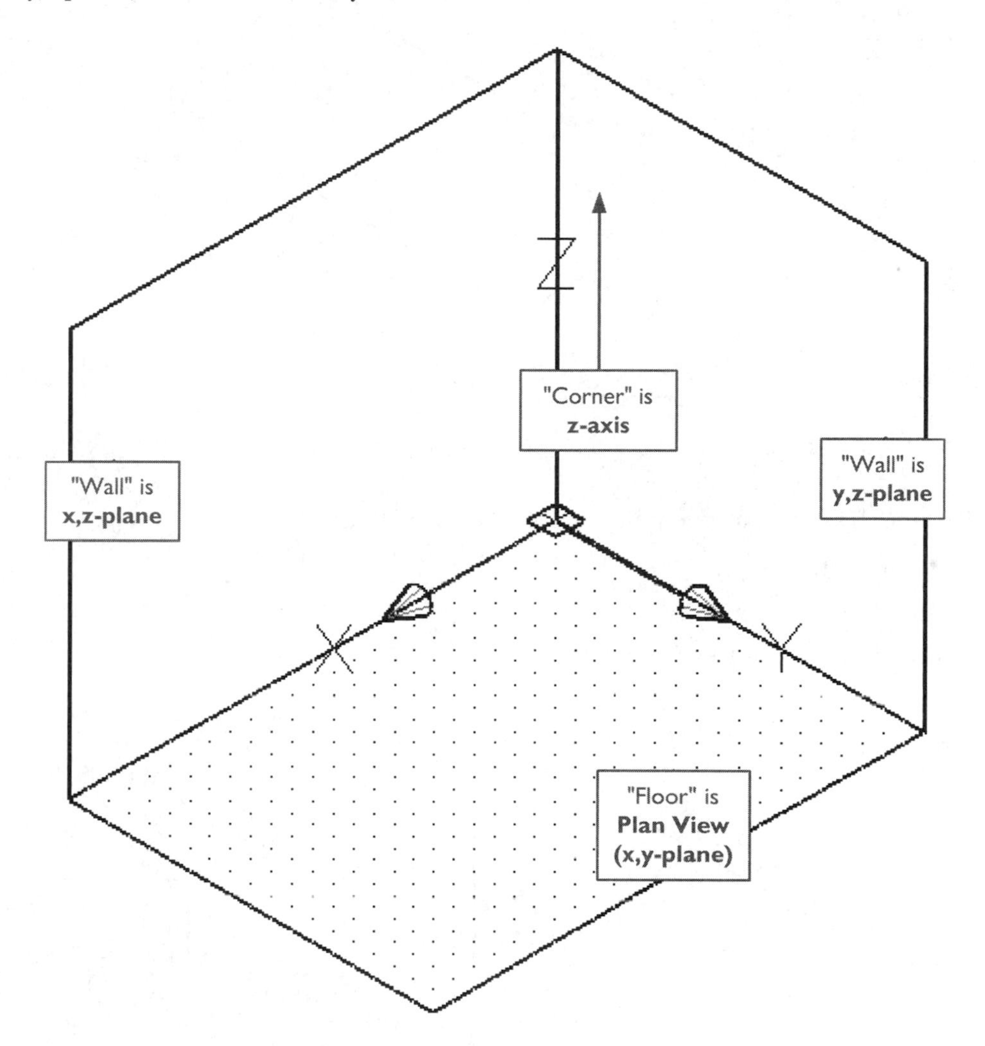

Z-axis turns 2D drawings into 3D.

Take the example of a box with dimensions of 4x3x2. The figure shows the box sitting in the positive quadrants of the x,y,z-coordinate system: four units along the x axis, three units along the y axis, and two units along the z axis.

As in 2D drawings, each intersection of the box has coordinates. In the case of 3D drawings, the coordinates are represented in the x,y,z format, such as 4,0,2.

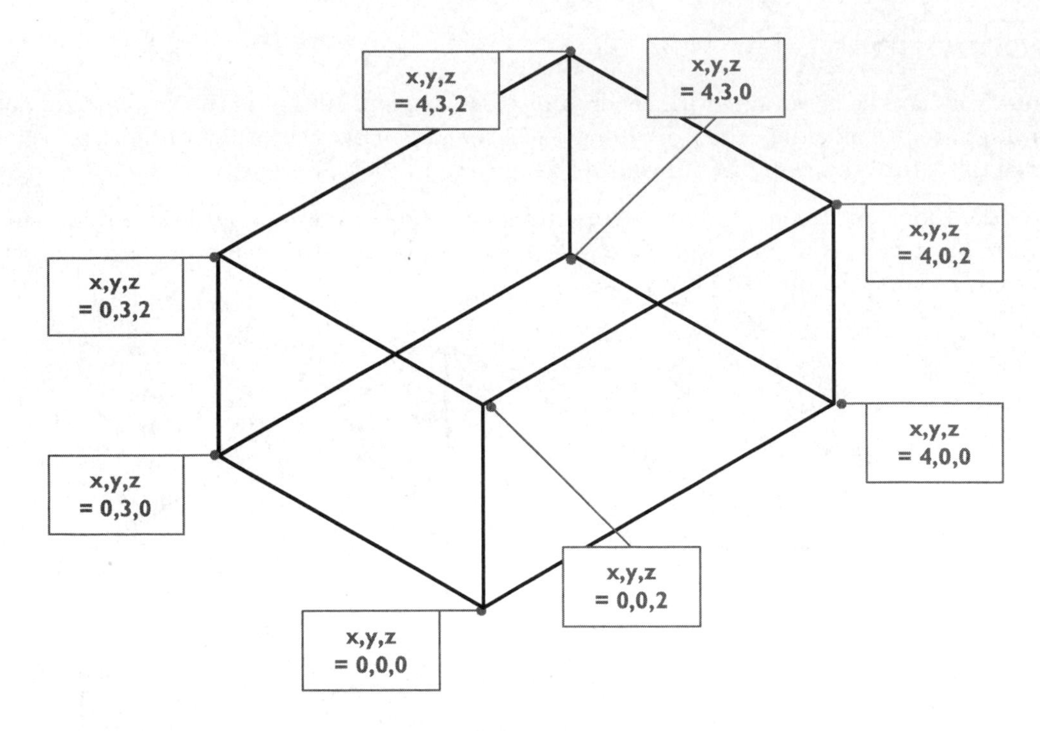

3D coordinates

Drawing lines in 3D involves specifying the third coordinate, z — it's that simple.

Command: **line**

Specify first point: **1,2**

Specify next point or [Undo]: **3,4,5**

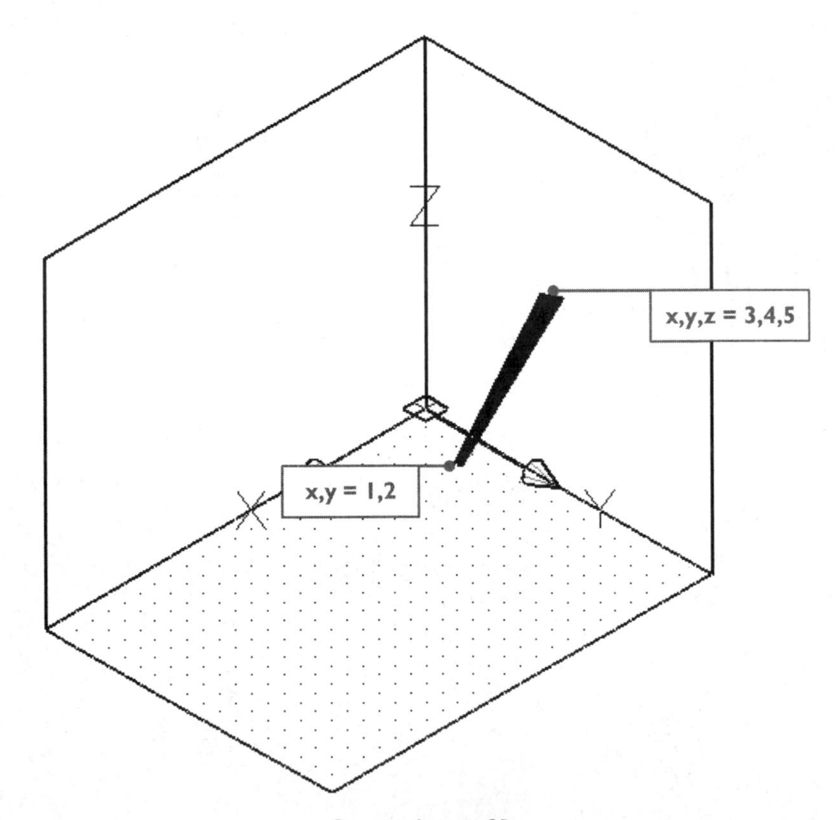

Drawing lines in 3D.

When you leave out the value of the z coordinate, AutoCAD assumes it is 0. (The exception is when the value of the **ELEVATION** system variable is set to a value other than 0.)

The x,y,z coordinates shown above called "Cartesian." AutoCAD also works with cylindrical and spherical 3D coordinates.

Cylindrical coordinates define 3D points by measuring two distances and an angle in 3D space. The first distance lies in the x,y-plane; the second distance measures the distance along the z-axis; the angle is measured from 0 degrees. Use this notation: **1<23,4** — where 1 is the polar distance, <23 is angle from 0 degrees, and 4 is the height in the z direction. (The **<** indicates an angle.) To draw a line using cylindrical coordinates:

Command: **line**

Specify first point: **1<23,4**

Specify next point or [Undo]: **5<67,8**

Spherical coordinates define 3D points by measuring one distance and two angles from the origin. The first angle lies in the x,y-plane, while the second angle is up (or down) from the x,y-plane. The distance lies in the x,y,z-plane. To draw a line using spherical coordinates:

Command: **line**

Specify first point: **1<23<45**

Specify next point or [Undo]: **6<78<90**

Note: It can get confusing to figure out the orientation of the x, y, and z axes. The key is to know the right-hand rule. Follow these steps:

1. Hold out your right hand.
2. Point your thumb up, the forefinger straight ahead, and bend your middle finger to the left.
3. The three fingers point to the three axes:

Thumb	**x**
Forefinger	**y**
Middle finger	**z**

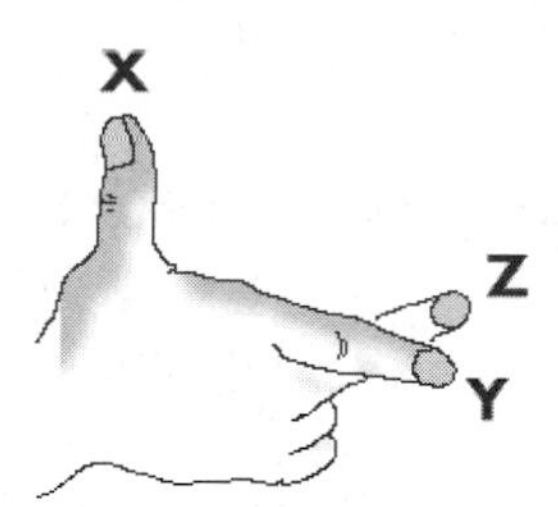

Right-hand rule determines the positive direction of three Cartesian axes.

Try it: rotate your hand so that the thumb (x) points up. Notice that your forefinger points out (y) and your middle finger points sideways (z). Your fingers indicate the direction of the three axes. Knowing the direction of any one axis, the right-hand rule determines the direction of the other two. Remember: this works only with your right hand; no slight intended to left handed people!

VPOINT AND **PLAN**

The **VPOINT** command changes the 3D viewpoint, while the **PLAN** command returns the viewpoint to the plan (2D) view.

BASIC TUTORIAL: CHANGING THE 3D VIEWPOINT

1. To change the 3D viewpoint, start the **VPOINT** command:
   - From the **View** menu, choose **3D Views**, and then **Viewpoint**.
   - At the 'Command:' prompt, enter the **vpoint** command.
   - Alternatively, enter the alias **-vp** at the keyboard.

   Command: **vpoint** *(Press* **ENTER**.*)*
2. In all cases, AutoCAD reports the current view direction, and then prompts you for the new viewpoint.

   Current view direction: VIEWDIR=0.0,0.0,1.0

   Specify a view point or [Rotate] <display compass and tripod>: *(Enter x,y,z coordinates, such as* **-1,-1,1**.*)*

   Regenerating model.

You enter x,y,z coordinates that specify a point of view looking toward the origin at 0,0,0. A viewpoint of 0,0,1 looks down at the drawing in plan view — directly down along the z axis. A negative coordinate value places the viewpoint at the negative end of the axis. Thus, a viewpoint of 0,0,-1 would look up at the drawing directly from below. To view a 3D drawing in isometric mode, enter coordinates -1,–1,1.

Other useful viewpoint coordinates are:

X	Y	Z	*Viewpoint*
0	0	1	Plan or top view
-1	0	0	Front view
0	-1	0	Side view
-1	-1	1	Isometric view

3. To return to the "2D" view, enter the **PLAN** command, and press **ENTER** twice:

   Command: **plan** *(Press* **ENTER**.*)*

   Enter an option [Current ucs/Ucs/World] <Current>: *(Press* **ENTER**.*)*

CHANGING THE 3D VIEWPOINT: ADDITIONAL METHODS

The **VPOINT** command has a pair of options that change the 3D viewpoint, in addition to menu selections and other commands.

- **Rotate** option rotates the viewpoints.
- **Display compass and tripod** option displays simple visual aids.
- **DDVPOINT** command displays a dialog box for selecting view angles.
- **View | 3D Views** menu provides preset viewpoints.
- **-VIEW** command has undocumented options for isometric views.

Let's look at them all.

Rotate

The **Rotate** option specifies the 3D view in angles, like spherical coordinates.

Specify a view point or [Rotate] <display compass and tripod>: **r**

Enter angle in XY plane from X axis <0>: *(Enter an angle, such as **225**.)*

Enter angle from XY plane <90>: *(Enter an angle, such as **35**.)*

The first prompt rotates the view about the z axis, while the second prompt rotates the view above or below the x,y-plane. Some useful viewpoint rotation angles are:

X Axis	*XY Plane*	*Viewpoint*
270°	90°	Plan or top view
180°	0°	Front view
270°	0°	Side view
225°	35°	Isometric view

Display Compass and Tripod

The **Display compass and tripod** option displays simple visual aids.

Specify a view point or [Rotate] <display compass and tripod>: *(Press* **ENTER**.*)*

The drawing screen temporarily displays a *tripod* and the *compass*, a flattened globe.

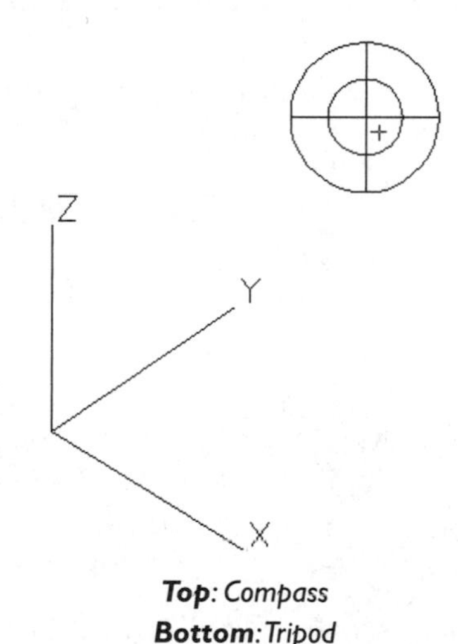

Top: *Compass*
Bottom: *Tripod*

To set the view, move the mouse. This action rotates the tripod, which represents the rotation of each axis. Think of your drawing as lying in the x,y-plane of the tripod. As you move the mouse, the drawing tilts and rotates.

The compass is a 2D representation of the globe, with the center point as the North Pole. The middle circle is the equator, and the outer circle is the South Pole. Moving the cursor above the equator produces an "above-ground" view; below the equator, a "below-ground" view.

The four quadrants of the globe are the direction of the view. For example, selecting the lower right quadrant produces a view that is represented from the lower right of the plan view.

DdVpoint

The **DDVPOINT** command displays a dialog box for selecting view angles.

Enter the **DDVPOINT** command, or from the menu select **View | 3D Views | Viewpoint Presets**. Select the degree settings shown in the dialog box to set the viewing angle by the preset degrees. Click inside the circle/arc for a finer selection.

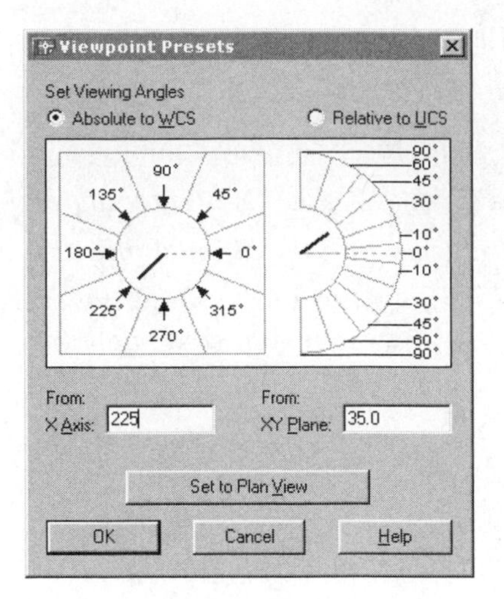

DdVPoint dialog box.

As an alternative, enter the angle of view in the **From X Axis** and **From XY Plane** text boxes.

To return the 3D view to plan view, click the **Set to Plan View** button.

View | 3D Views

The **View** | **3D Views** menu provides preset viewpoints. From the **View** menu, select **3D Views**.

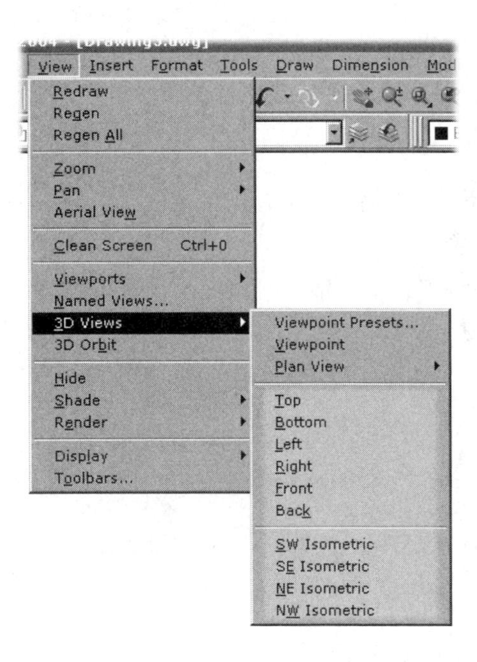

View | 3D Views menu contains preset viewpoints.

All the items on the submenu, from **Top** to **SW Isometric**, are preset views. They are often used in 3D engineering drawings.

-View

The **-VIEW** command has four undocumented options that change the 3D viewpoint to the four standard isometric views:

Command: **-view**

Enter an option [?/Orthographic/Delete/Restore/Save/Ucs/Window]: *(Enter an option, such as* **swiso***.)*

The undocumented options are:

-View Option	Meaning
swiso	South west isometric view.
seiso	South east isometric view.
nwiso	North west isometric view.
neiso	North east isometric view.

Note: Yet another alternative method of specifying 3D viewpoints is through the Viewports dialog box. From the menu bar, select **View | Viewports | New Viewports**. In the dialog box, select the following items:

Standard Viewports:	**Single**
Apply to:	**Display**
Setup:	**3D**
Change view to:	*(Select a predefined views, such as "Right" or "SW Isometric.")*

Click **OK**, and AutoCAD displays the 3D viewpoint.

3DORBIT

The **3DORBIT** command interactively sets 3D viewpoints.

Unlike **VPOINT** and other viewpoint-related commands, this command tilts and rotates the drawing in real time as you move the cursor. It also sets clipping planes, perspective mode, and a variety of rendering modes.

The drawback is that you cannot edit objects while this command is active. You can, however, activate **'3DORBIT** as a transparent command during editing commands.

BASIC TUTORIAL: INTERACTIVE 3D VIEWPOINTS

1. To change the 3D viewpoint interactively, start the **3DORBIT** command:
   - From the **View** menu, choose **3D Orbit**.
   - At the 'Command:' prompt, enter the **3dorbit** command.
   - Alternatively, enter **3do** or **orbit** aliases at the keyboard.

   Command: **3dorbit** *(Press* ENTER.*)*

   In all cases, AutoCAD displays the arc ball, and the following prompt:

   Press esc or enter to exit, or right-click to display shortcut menu.

The *arc ball* is like a "3D controller" for rotating the viewpoint of 3D objects; it helps orient you in 3D space. (I'm not sure why it is called "arc" ball, because it consists of five circles, not arcs.)

The center of the arcball is the *target point*, which is the point *at* which you are looking. The point *from* which you are looking (that is, your eyeball) is called the *camera point*. The target stays stationary, while the camera location moves around the target.

You control the arc ball by dragging anywhere on the screen; different locations rotate the view in different manners:

- **Top and bottom small circles** tilt the view about the x axis.

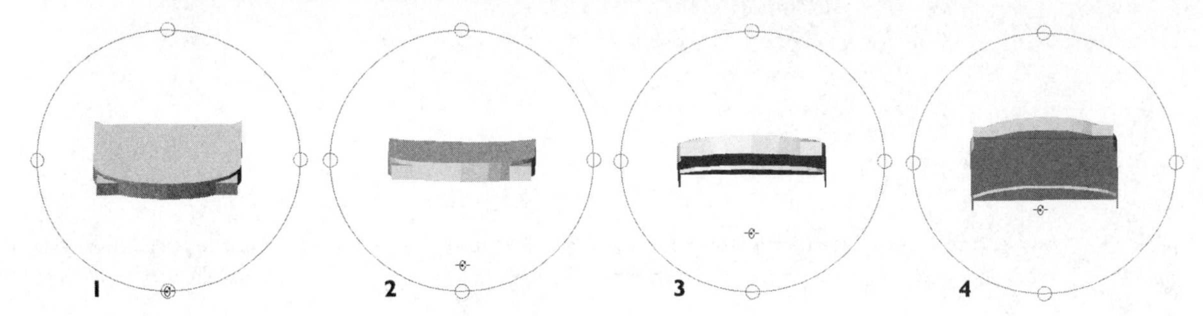

Tilting the viewpoint about the x-axis.

- **Left and right small circles** twist the view about the y axis.
- **Outside the large circle** rolls the view about the viewpoint axis (from the eye to the center of the object).
- **Inside the large circle** rotates the view about all axes; this is the most difficult view to control, but the most versatile.

As you move the cursor about the arc ball, the cursor changes to indicate the style of viewpoint rotation.

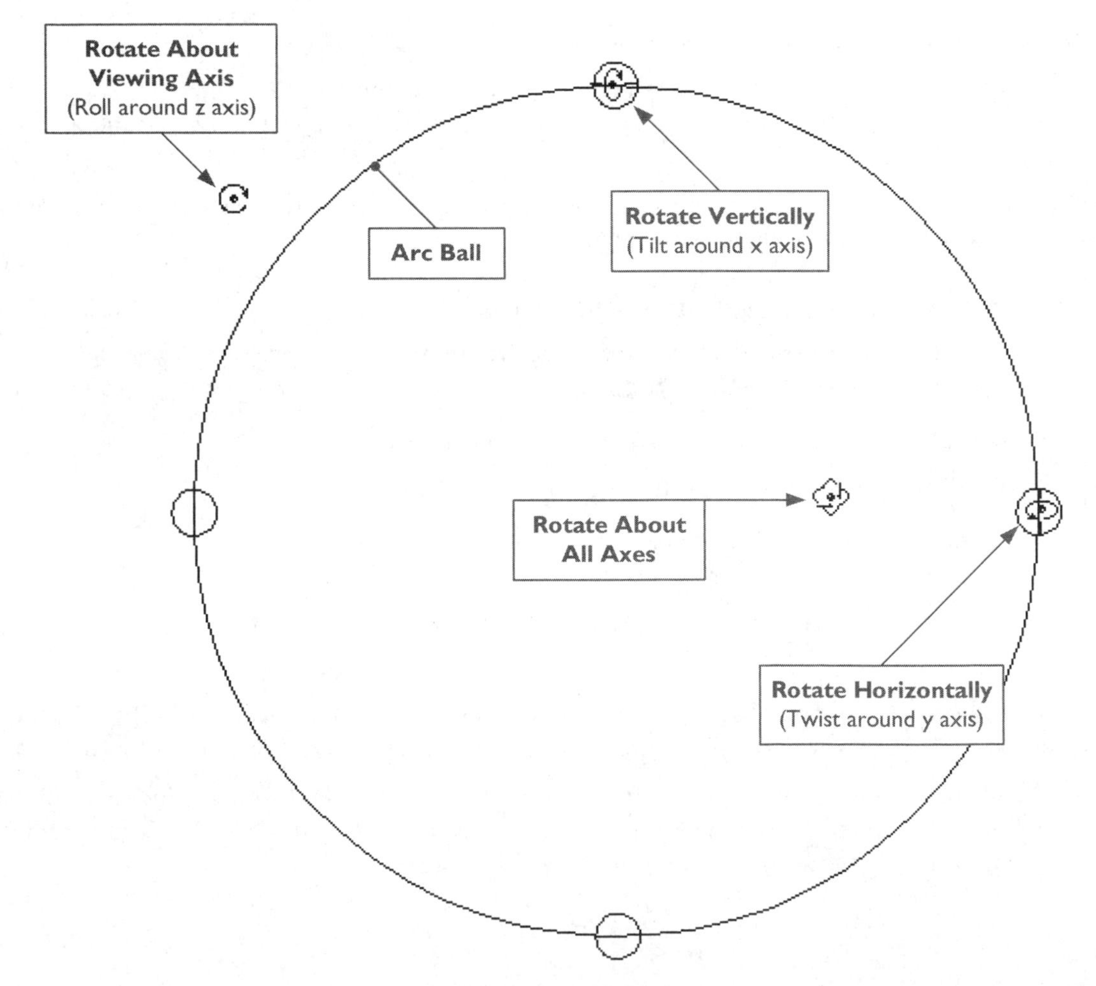

3dOrbit command's arc ball.

In addition to displaying the arc ball, this command makes other changes to the drawing area. The UCS icon changes to a colorful, "3D" 3-axis icon: the x axis becomes red; y axis, green; and the z axis, blue. The drawing loses its lineweight, because AutoCAD switches automatically to 3D wireframe view.

THE CURSOR MENU

The **3DORBIT** user interface has many commands "hidden" in its shortcut menu. Right-click anywhere in the drawing area to see the menu:

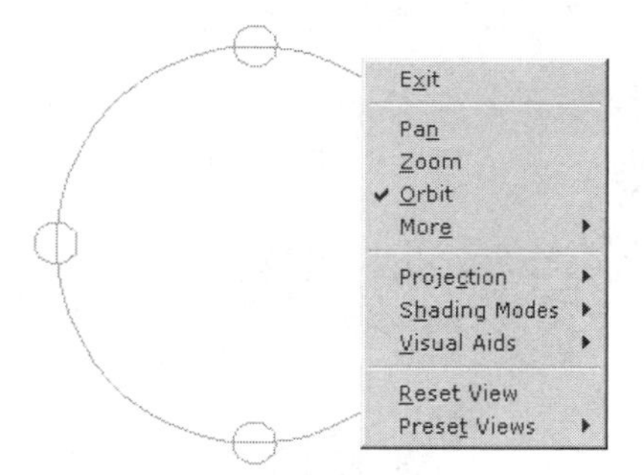

3dOrbit command's shortcut menu.

Exit exits 3D orbit viewing mode. As an alternative, you can press **ESC**. You must exit 3D orbit viewing mode to continue editing the drawing. (No other command operates while AutoCAD is in 3D orbit mode.)

Pan invokes real-time panning like the **PAN** command. The cursor changes to a hand; move the cursor to move the view.

Zoom switches to real-time zooming like the **ZOOM** command. The cursor changes to a magnifying glass; move the cursor vertically to zoom in and out.

Orbit returns to the default mode, which rotates the view with the aid of the arc ball.

More

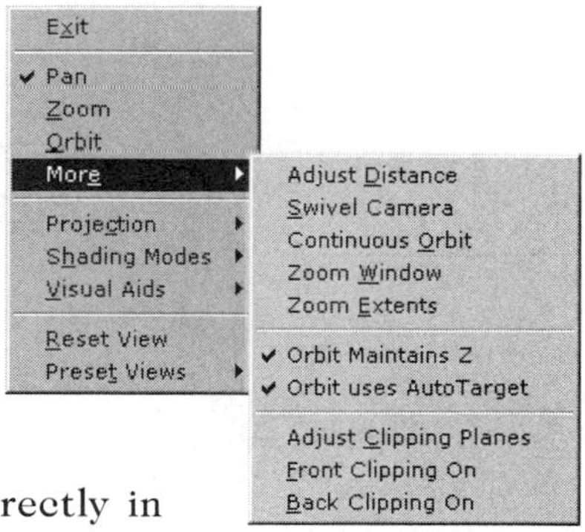

Adjust Distance moves the camera (your viewpoint) closer to or farther from the target point.

Swivel Camera turns the camera.

Continuous Orbit (pretty cool) rotates the view continuously when you drag the cursor across the screen, then let go. The faster you drag the cursor, the faster the model rotates.

Zoom Window drags a window around the area to zoom in to, like the **ZOOM** command's **Window** option.

Zoom Extents sizes the view to display all objects; it may not work correctly in perspective mode.

 Adjust Clipping Planes

Clipping planes cut off parts of 3D models. For example, you may want to cut away part of a building to see the interior. You can think of clipping planes as transparent walls that slice through parts of 3D drawings and remove everything in front or behind them. The cutting plane is parallel to the viewing plane.

(Clipping planes also control the fog effect of the **RENDER** command; see Chapter 14.)

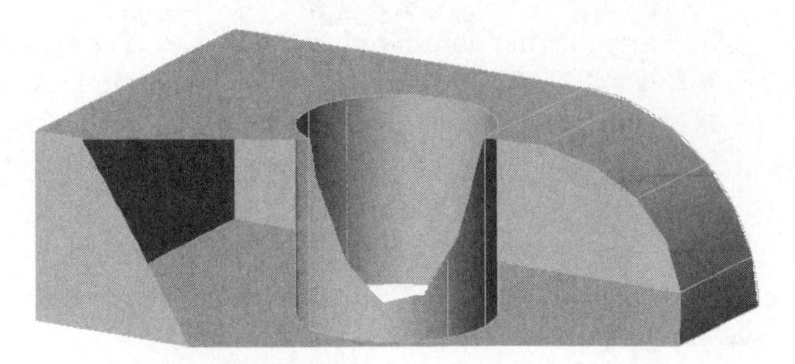

Front cutting plane shows inside of 3D model.

There is a *front* and a *back* clipping plane. Either or both planes can be positioned in the drawing with slider bars, in either perspective or non-perspective mode.

Adjust Clipping Planes window.

Adjust Clipping Planes option opens the Adjust Clipping Planes window to let you move the front and back clipping planes. Alternatively, this window is accessed through the **3DCLIP** command.

The window has a toolbar with buttons that assist you in changing the clipping plane:

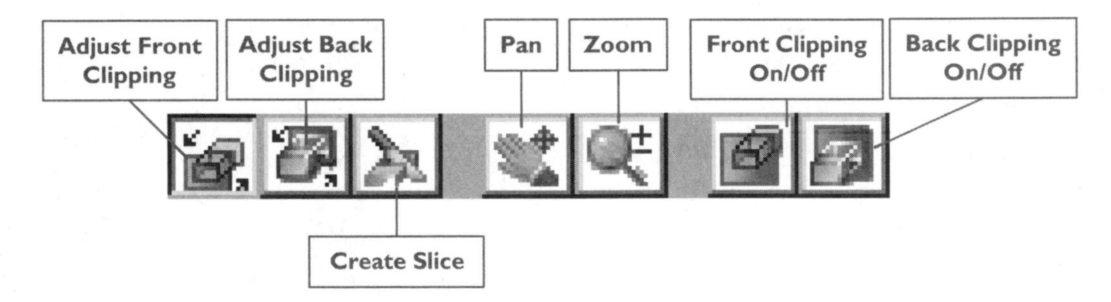

Adjust Clipping Planes toolbar

Adjust Front Clipping button turns on front clipping. With the cursor, move the black horizontal line to adjust the front clipping plane. As you move the line, you see the front of the object clipped off (disappearing from view).

Adjust Back Clipping button turns on back clipping. With the cursor, move the green horizontal line to adjust the back clipping plane. (The front and back clipping planes cannot be adjusted independently at the same time.)

Create Slice button locks the front and back clipping planes so that they move together. This allows you to "slice" through the objects. I suggest you adjust the front and back clipping planes independently before slicing.

Pan button pans the image in realtime inside the window. Hold down the left button, and then move the mouse about to pan.

Zoom button zooms the image in realtime larger and smaller. Hold down the left button, and then move the mouse up to zoom in, and down to zoom out.

Front Clipping On button toggles the front clipping plane. When off, the object is not clipped; when on, the object is clipped.

Back Clipping On button toggles the back clipping plane in the same manner.

To close the Adjust Clipping Planes window, select the small x button in the upper right corner of the window. The effect of the clipping planes remains after the window is closed. You may want to rotate the object to see inside of it; the clipping planes remain static. As you rotate the object, it may disappear entirely if it comes wholly in front of (or behind) a clipping plane.

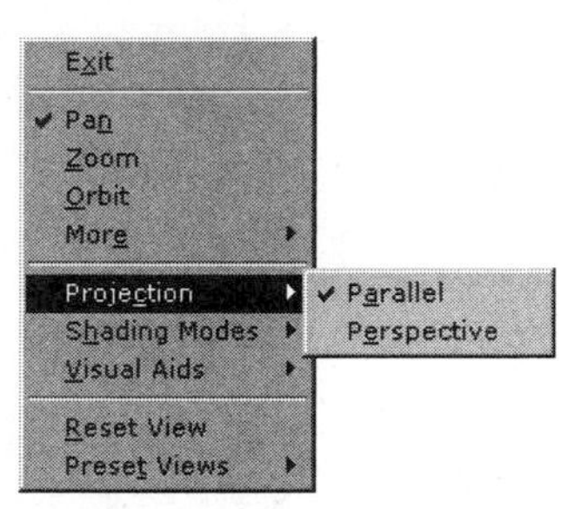

Projection

Parallel projection is the mode most-used for viewing large objects, such as buildings, because perspective projection makes them look distorted. Two parallel lines never converge at a single point.

Perspective projection causes parallel lines to converge at one point, called the "vanishing point." Objects recede into the distance; small objects can become very distorted. You cannot edit a drawing if perspective mode is on. AutoCAD complains, "You cannot point within a Perspective view."

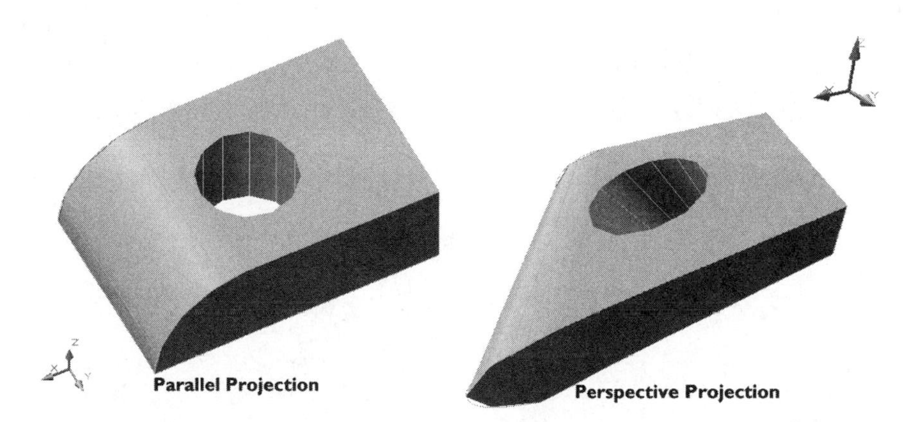

Perspective projection can look either more realistic or more exaggerated

Shading Modes

During and after the **3DORBIT** command is in effect, the drawing can be displayed in one of several shading modes — ranging from traditional wireframe to Gouraud shading. You can edit the drawing while it is fully shaded. When you draw, the objects are displayed in shaded mode — when appropriate. A 2D line is unshaded, because it is just a segment. A 2D line with thickness, however, is shaded.

The more advanced the shading, the more slowly AutoCAD updates the screen of large 3D drawings on slow computers. (I don't notice any slowdown on my 2.4GHz desktop computer.) Thus, while Gouraud shading is more realistic, it may slow down operation of AutoCAD; if this is a problem, switch to an "easier" form of shading, such as flat. (Shading modes are also controlled by the **View | Shade** menu selection, and the Shade toolbar. See Chapter 14.)

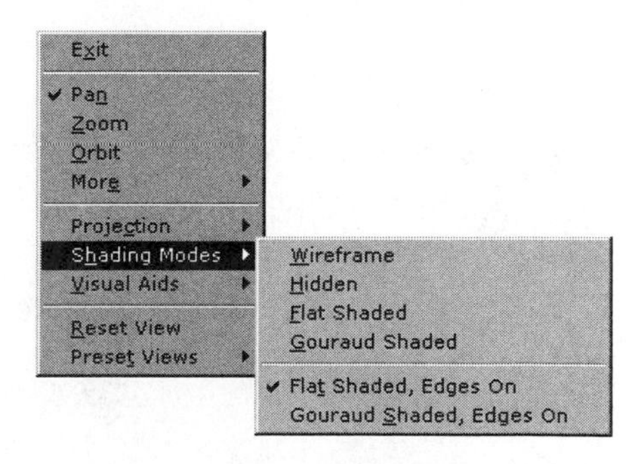

Visual Aids

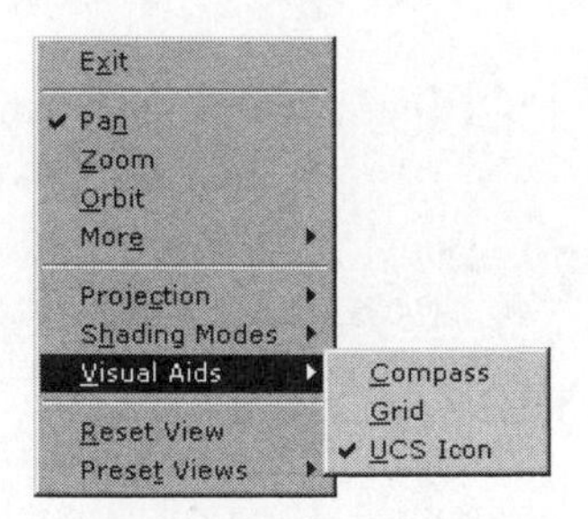

In addition to the arc ball described earlier, the **3DORBIT** command displays several visual aids to help you find your place in three-dimensional space — because it is easy to get lost in space.

Compass is like a gimbal: three intersecting circles represent the x, y, and z axes.

Grid is similar to the grid in 2D drafting, but instead of dots, this grid consists of lines (like graph paper). Like the 2D grid, this grid lies in the x,y-plane, is constrained by the drawing limits, and is affected by the settings of the **DSETTINGS** dialog box. The spacing is between major grid lines; there are ten minor grid lines drawn horizontally and vertically between major grid lines. (Technical editor Bill Fane notes that conditions must be just right to see the minor grid lines: in a hidden or shaded mode, you must be zoomed just right.)

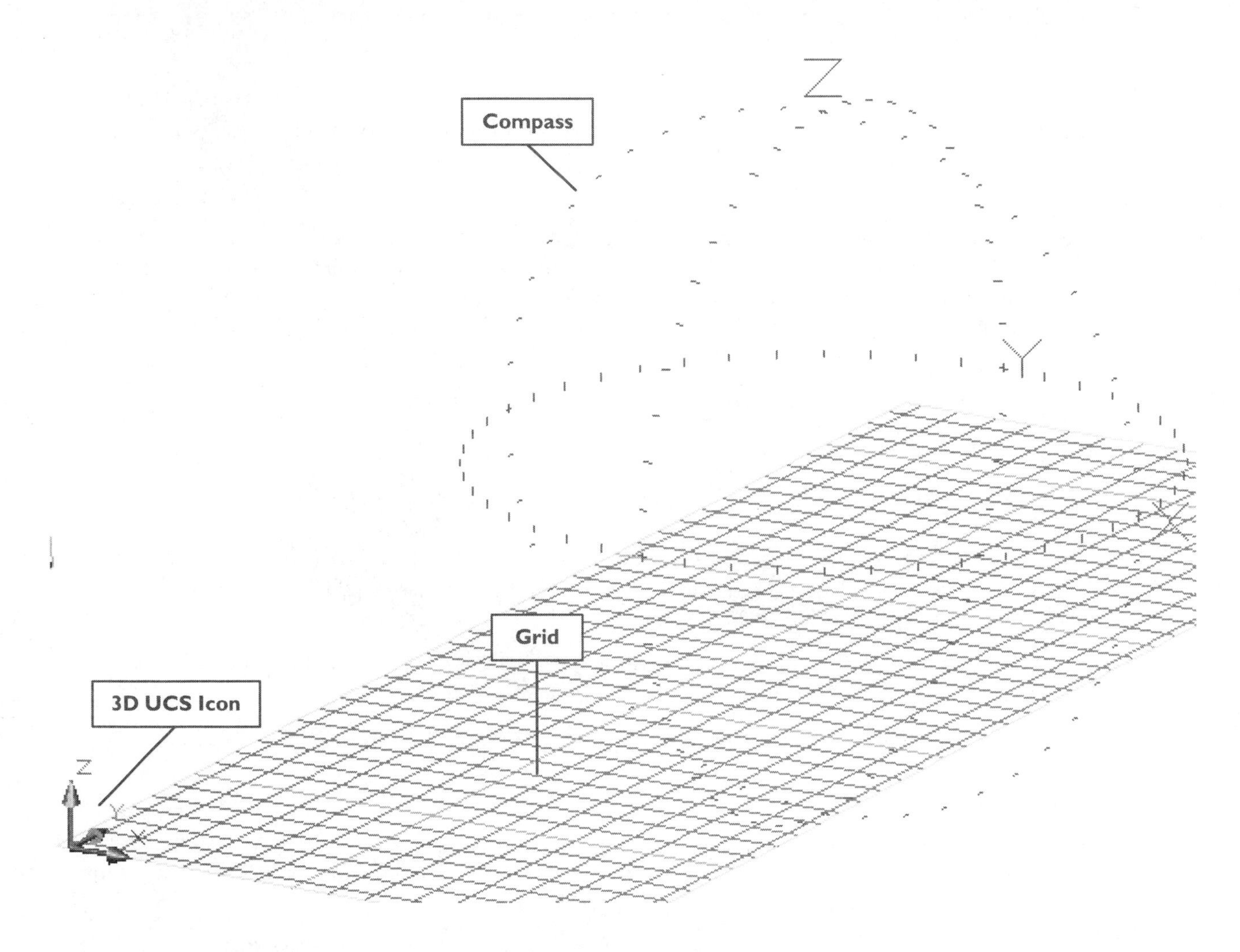

3D UCS icon, compass, and lined grid.

UCS Icon is the shaded 3D icon, with axes colored red (x), green (y), and blue (z). In perspective mode, the icon takes on a perspective look. The **UCSICON** command controls whether the icon displayed at the drawing's origin (0,0,0).

Reset View is perhaps the most important, because it is easy to get "lost" when viewing in 3D. It resets the view to what it was when you began the **3DORBIT** command.

Preset Views

Preset Views displays a list of standard views, such as Top, Bottom, and SE Isometric.

Saved Views displays a list of named views. If no views are saved in the drawing, this option is not shown. (You must have previously used the **VIEW** command to create one or more named views.)

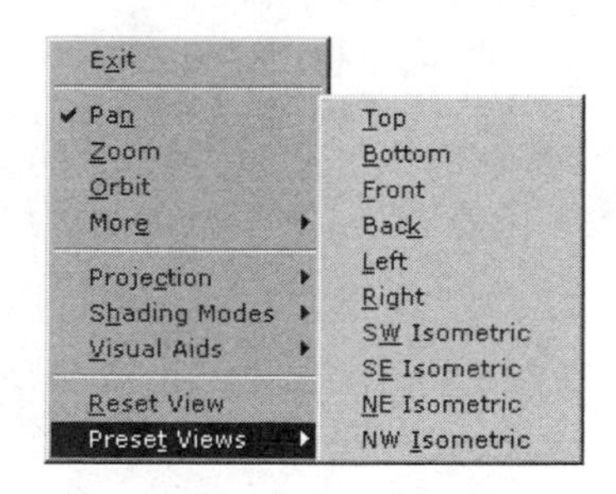

INTRODUCTION TO UCS

To create and edit 3D drawings, you must get comfortable with changing the point of view, and working at any angle in space.

AutoCAD provides a number of tools that create and save 3D viewpoints. The commands include **UCS** for creating 3D viewpoints, and **VIEW** for saving and restoring them. Even so, AutoCAD does not make it particularly easy moving around in 3D space. (*UCS* is short for *user-defined coordinate system*.)

If there are coordinate systems that can be defined by you, then there must be another, absolute coordinate system to which they refer. There is: it is named the "world coordinate system" (WCS, for short). When you start AutoCAD, it is in WCS mode: you look down on the x,y-plane, with the z-axis coming out of the monitor toward your face. You can change the viewpoint, and it is still the WCS. But when you change the coordinate system, you define a new one — a UCS.

Why would you care about defining a new coordinate system? The answer: to orient yourself in 3D space. Think about drawing on the surface of a sloped roof: the task is hard if you could not reorient the drawing plane. And that's the problem with most CAD software: it is still so 2D oriented that drawing in 3D is hard. The workaround is to create UCSs.

UCSs are created with the **UCS** command, and managed with the **UCSMAN** command. Before working with UCSs, though, you should understand the UCS icon, because it orients you in 3D space. When you first start AutoCAD, it shows the direction of the x and y axes in the WCS:

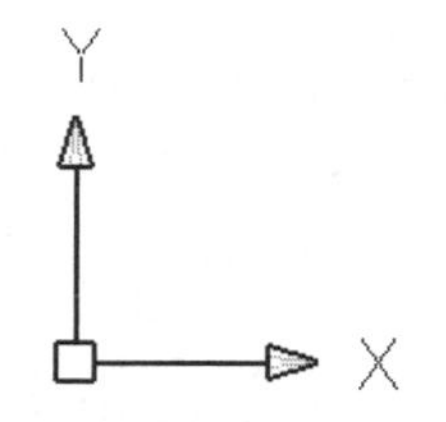

UCS icon in WCS mode.

The small square means AutoCAD is in WCS (world coordinate system), the default coordinates in every new drawing. You are looking straight down on the x,y-plane, so you do not see the z axis. When you change to a 3D viewpoint, the z axis makes its appearance. Enter **VPOINT 1,2,3** to see the following:

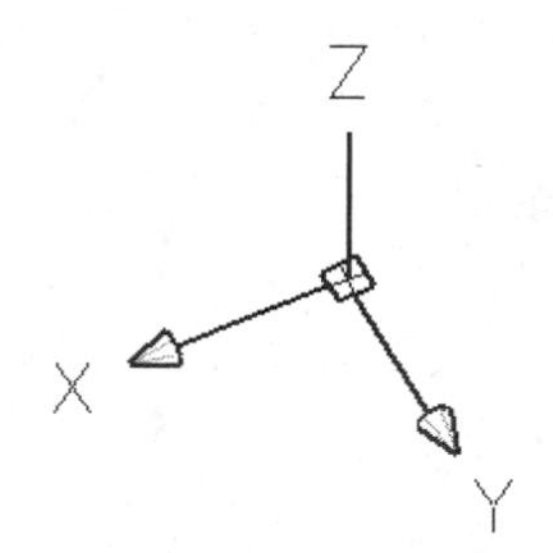

UCS icon viewed at an angle.

Look closely at the square and you may be able to see a plus sign. That indicates the UCS icon is located at the origin (0,0,0); no plus sign indicates that the UCS icon is located at the lower left corner of the viewport.

When the viewpoint goes underneath the x,y-plane, that z axis is negative. The z-portion of the UCS icon becomes dotted. Enter **VPOINT 1,1,-1** to see this:

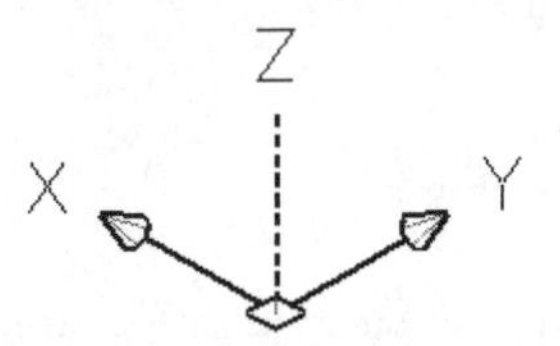

Z-axis is dashed when viewed from underneath.

The icon changes shape, sometimes due to your request, and sometimes because of AutoCAD. Below are several forms of UCS icon. The original UCS icon shows only the direction of the x and y axes, and no z axis. (It was introduced in 1990, along with full 3D in AutoCAD Release 10.) It can be turned on with the **UCSICON** command's **Properties** option.

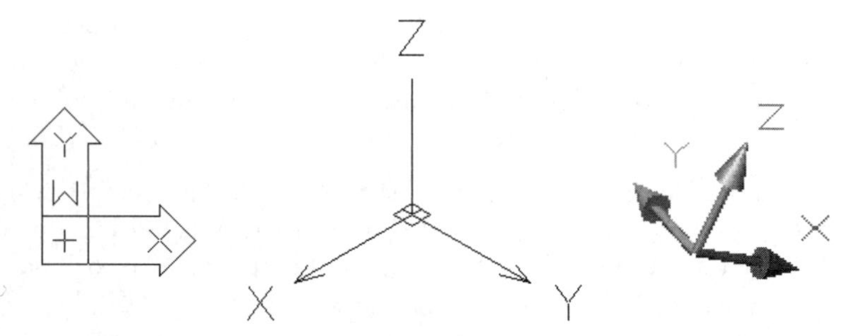

***Left**: Original 2D UCS icon.* ***Center**: Standard 3D UCS icon.* ***Right**: 3D Orbit UCS icon.*

The icon illustrated in the center (above) the default 3D icon in wireframe model space. The icon at the right is displayed by AutoCAD when the viewport is in a shade mode, such as flat shading.

When you switch to layouts (paper space), the icon is strictly 2D and strictly ornamental, because there is no 3D in paper space. (Historically, the paper space UCS icon signalled to the user that the drawing area was in paper space. With layouts, this is no longer necessary.)

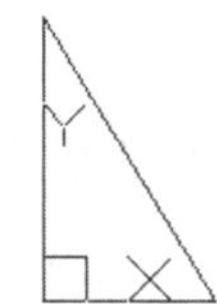

UCS icon in paper space.

When you switch to Model mode in layouts, the 3D UCS icon appears in the current viewport (or in all viewports, if you use the **All** option of the **UCSICON** command.)

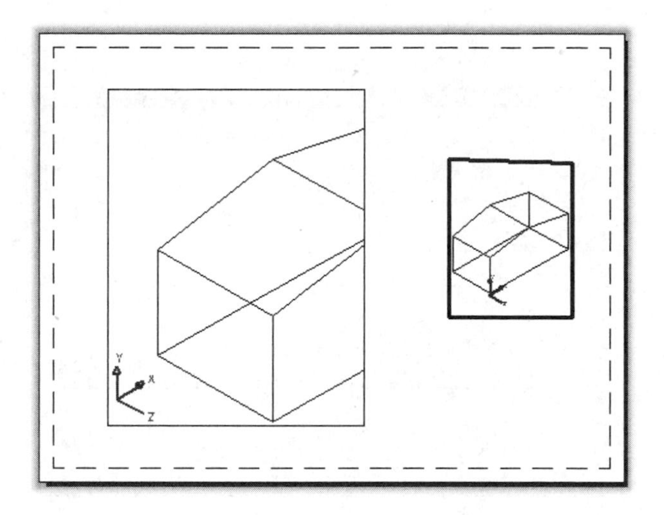

UCS icons in model space viewports in layouts (paper space).

UCSICON

The UCSICON command controls the look and visibility of the UCS icon.

TUTORIAL: CONTROLLING THE UCS ICON

1. To control the UCS icon, start the **UCSICON** command:
   - From the **View** menu, choose **Display**, and then **UCS Icon**.
   - At the 'Command:' prompt, enter the **ucsicon** command.

     Command: **ucsicon** *(Press* ENTER.*)*
2. Enter an option:

   Enter an option [ON/OFF/All/Noorigin/ORigin/Properties] <ON>: *(Enter an option.)*

   ON and **OFF** turn on and off the display of the UCS icon. It makes sense to turn off the icon during 2D drafting; otherwise, always leave it on.

   All specifies that changes made with this command apply to all viewports; ignore this option, and the changes apply to the UCS icon of the current viewport only.

   ORigin forces the UCS to the origin (0,0,0); when AutoCAD cannot display the icon at the origin (because it is off the edge of the display), the icon is displayed at the lower-left corner. A plus sign in the icon indicates the icon is at the origin.

   Noorigin always displays the UCS icon at the lower left corner of the viewport.

Properties displays the UCS Icon dialog box:

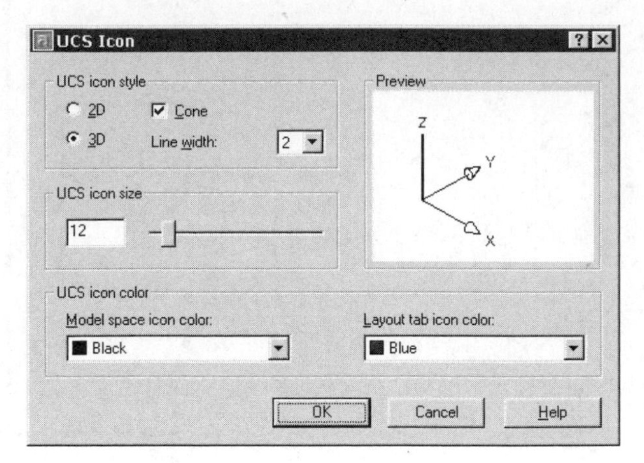

UCS Icon dialog box.

UCS Icon Style

The **UCS Icon Style** section specifies the appearance of the UCS icon.

The **2D** option displays the traditional 2D icon, which lacks the z axis. The sole reason for selecting this option is nostalgia.

The **3D** option displays all three axes (default).

The **Cone** option displays 3D cones on the x and y axes; when turned off, 2D arrowheads are displayed instead. I find the cones look nicer than the arrowheads.

The **Line Width** option thickens the axes lines; I like the setting of 2 the best.

UCS Icon Size

The **UCS Icon size** option adjusts the size of the icon. The value is a percentage of the viewport size, and ranges from 5% to 95%. (This means the icon size changes as the viewport becomes larger and smaller.) I find the default of 12% is just fine.

UCS Icon Color

The **UCS Icon Color** section specifies the color of the icon independently in model space and paper space.

The **Model Space Icon Color** option specifies the color of the UCS icon in model space viewports (whether in Model or Layout mode). You can choose any color; black is fine with me.

The **Layout Tab Icon Color** option specifies the color of the icon in paper space (layouts). I find the default color, blue, annoying, and tend to change it to black.

3. Make changes to the dialog box, and then click **OK**.

UCS

The UCS command creates and modifies user coordinate systems.

New coordinate systems are created with the **New** option of the UCS command. To create a new coordinate system, you need to tell AutoCAD two things: (1) the x,y,z coordinates of the origin, and (2) the direction of the x,y,z axes.

BASIC TUTORIAL: CREATING A UCS

Before starting this tutorial, turn on snap and grid. Draw a square, and then rotate the viewpoint:

Command: **viewpoint**

Specify a view point or [Rotate] <display compass and tripod>: **1,1,1**

The square and the UCS icon look rotated:

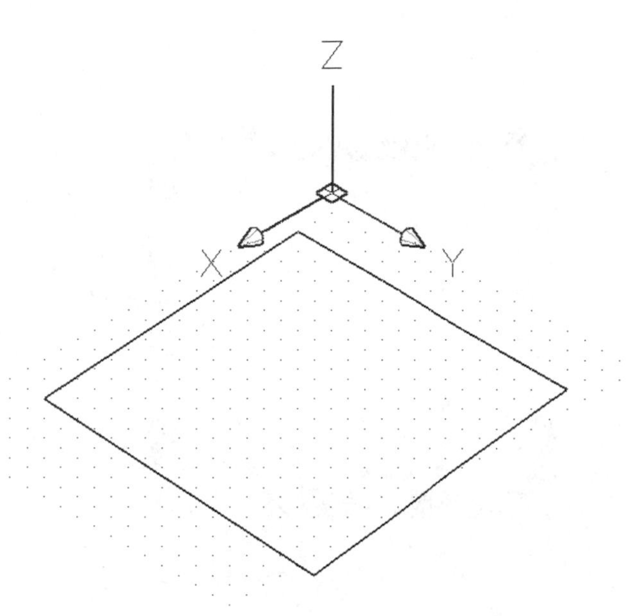

WCS with viewpoint 1,1,1.

1. To create new UCSs, start the **UCS** command:
   - From the **Tools** menu, choose **New UCS**.
   - At the 'Command:' prompt, enter the **UCS** command.

   Command: **ucs** *(Press* **ENTER***.)*
2. AutoCAD reports the name of the current UCS, and then displays a long prompt.

   Current ucs name: *WORLD*

   Enter an option

   [New/Move/orthoGraphic/Prev/Restore/Save/Del/Apply/?/World] <World>: **n**

   Enter "n" to create a new UCS. (Shortcut: you don't need to enter "n"; you can directly enter one of the options listed next, such as ZA or OB.)
3. There are no fewer than nine ways to specify a new UCS.

   Specify origin of new UCS or [ZAxis/3point/OBject/Face/View/X/Y/Z] <0,0,0>: **v**

   To align the coordinate system with the current viewpoint, enter "v". Think of this as aligning the UCS to the current view. (This is probably the easiest option, but perhaps the least useful, because it does not necessarily align with features on 3D objects.)

Notice that the UCS icon rotates to show just the x and y axes, just as if you were looking at the drawing in plan view. In addition, the grid "faces you."

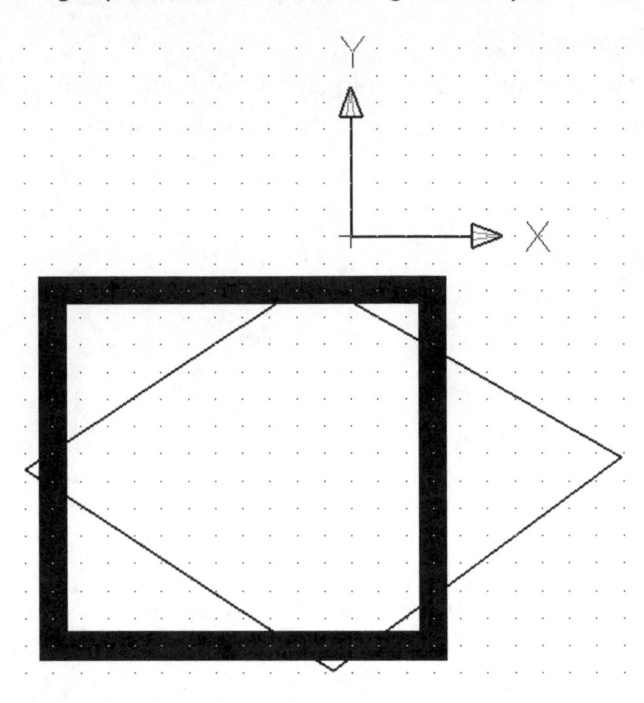

UCS matches viewpoint.

4. Draw another square, this time with the **RECTANG** command. Give the polyline a width of 0.5 to help distinguish it from the first square. Notice that you are drawing in the new x,y-working plane.
5. To return to the WCS, enter the **World** option:

   Command: **ucs**

   Enter an option [New/Move/orthoGraphic/Prev/Restore/Save/Del/Apply/?/World] <World>: **w**

By changing the coordinate system to match the viewpoint, you now draw just as if you were in plan view.

CREATING A UCS: ADDITIONAL METHODS

The UCS command has many ways to create user coordinate systems. The options are:

- **Specify origin of new UCS** asks for a new origin; the direction of the x,y,z axes are unchanged. Think of this as the "move" option.

Specify origin of new UCS <0,0,0>: *(Pick a point, or specify x,y,z coordinates.)*

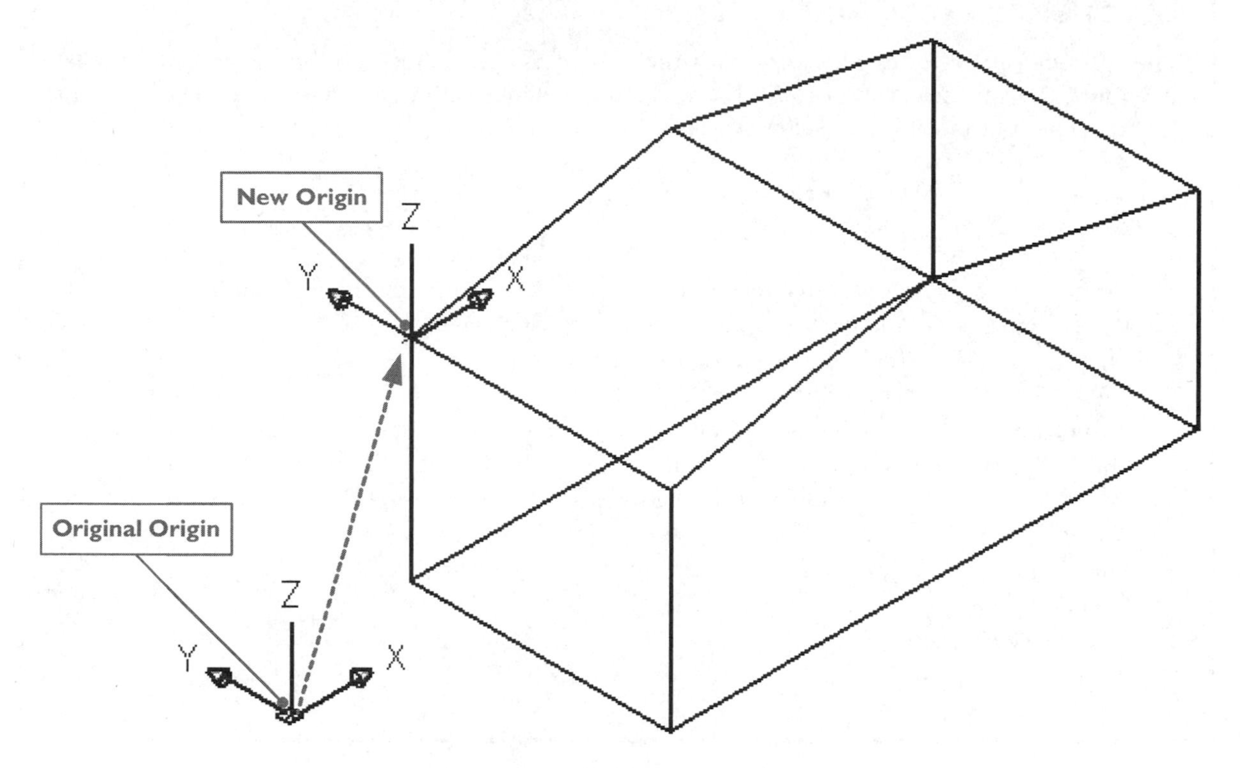

Creating a new UCS by moving the origin.

- **ZAxis** asks for a new origin, and then asks for a new direction of the z axis, rotating the x,y-plane about the z axis. Think of this as the "tilt" option, or as a newly defined extrusion direction. When you specify a new direction for the z axis, AutoCAD determines the direction of the x,y-plane by the right-hand rule.

  Specify new origin point <0,0,0>: *(Pick a point, specify x,y,z coordinates, or press* ENTER.*)*

  Specify point on positive portion of Z-axis <0,0,1>: *(Pick a point on the z-axis.)*

- **3point** asks for a new origin, and then asks for new directions for the x and y axes. There is no need to specify the z axis, because the right-hand rule fixes it in place. Think of this as the "universal" option, because it creates completely arbitrary UCSs. (See the tutorial following this section.)

- **OBject** bases the UCS on the selected object. See boxed text for details.

  Select object to align UCS: *(Pick a single object.)*

- **Face** bases the UCS on the selected planar face of a 3D solid object (curved faces need not apply); further options locate the UCS on adjacent faces and optionally rotate the UCS by 180 degrees about the x or y axes.

  Select face of solid object: *(Pick the flat face of a 3D solid model.)*

  Enter an option [Next/Xflip/Yflip] <accept>: *(Press* ENTER*, or enter an option.)*

OBJECT OPTION

The **OBject** option does not work with many objects: 3D solids (use the **Face** option), 3D polylines, 3D meshes, viewports, mlines, regions, splines, ellipses, rays, xlines, leaders, and mtext. Objects that can be used with this option include:

Object	*UCS Origin*	*X Axis Direction*
Arc	Center of arc.	Passes through arc's endpoint closest to pick point.
Circle	Center of circle.	Passes through the pick point.
Dimension	Midpoint of the dimension text.	Unchanged.
Line	Endpoint nearest pick point.	Along the line.
Point	The point.	
2D polyline	Start point of the polyline.	First segment.
Solid	First point of the solid.	Along the line of the first two points.
Trace	Start point of the trace.	Centerline from the start point.
3D face	The first point.	The first two points.
Shape, text, block reference, attribute definition	Insertion point.	Extrusion direction.

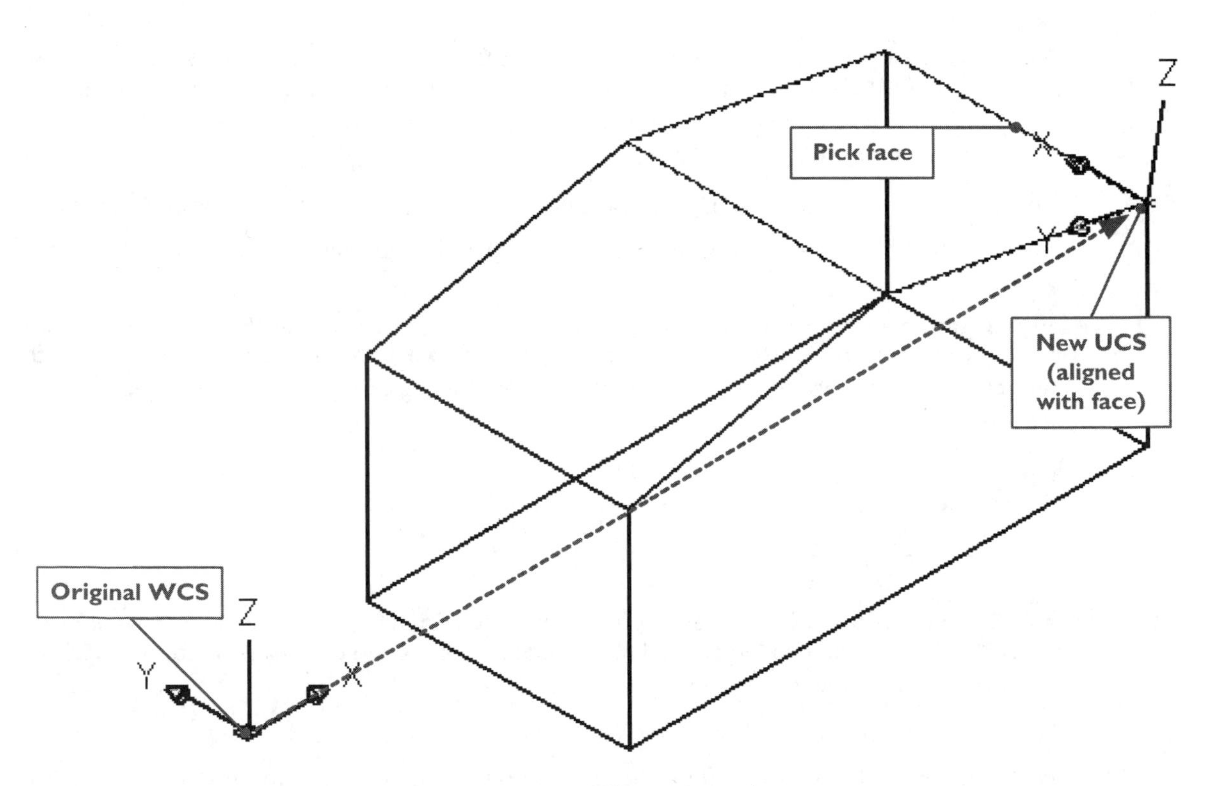

Creating a new UCS by picking a face.

- **X**, **Y**, and **Z** ask for the angle to rotate the UCS about one of the three axes. Think of this as the "rotate" option. To rotate the x,y working plane about the z axis, enter Z:

  Specify rotation angle about Z axis <90>: *(Specify an angle.)*

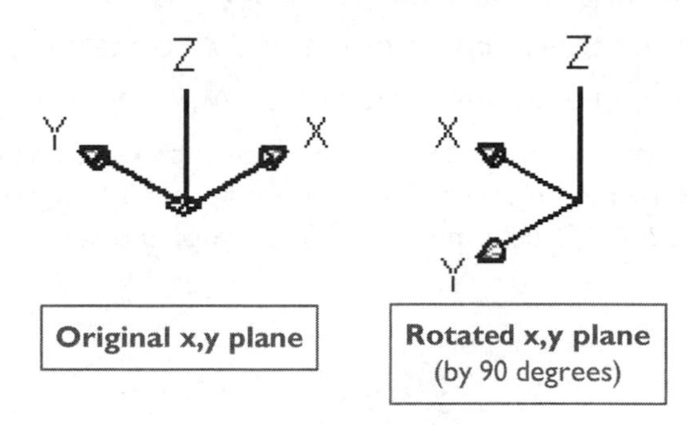

Creating a new UCS by rotating the x,y-plane.

TUTORIAL: CREATING A WORKING PLANE

In this tutorial, you learn that UCSs make it easier to draw at strange angles.

1. Open AutoCAD with the *Ch08UCS.dwg* drawing file from the CD-ROM. The drawing is of a simple 3D house with pitched roof.
2. With the **TEXT** command, attempt to place text on one of the roof slopes. You will find it is impossible, because AutoCAD only places text in the x,y-plane of the current coordinate system.
3. The solution is to reorient the coordinate system so that the roof becomes the new x,y-plane. Create a UCS oriented to one of the roof slopes with the **3point** option, as follows:

   Command: **ucs**

   Current ucs name: *WORLD*

   Enter an option [New/Move/.../Del/Apply/?/World] <World>: **3**

   Notice that I avoided the **New** option and entered **3** (short for "3point") directly.
4. Use temporary object snap modes to make your picks accurate:

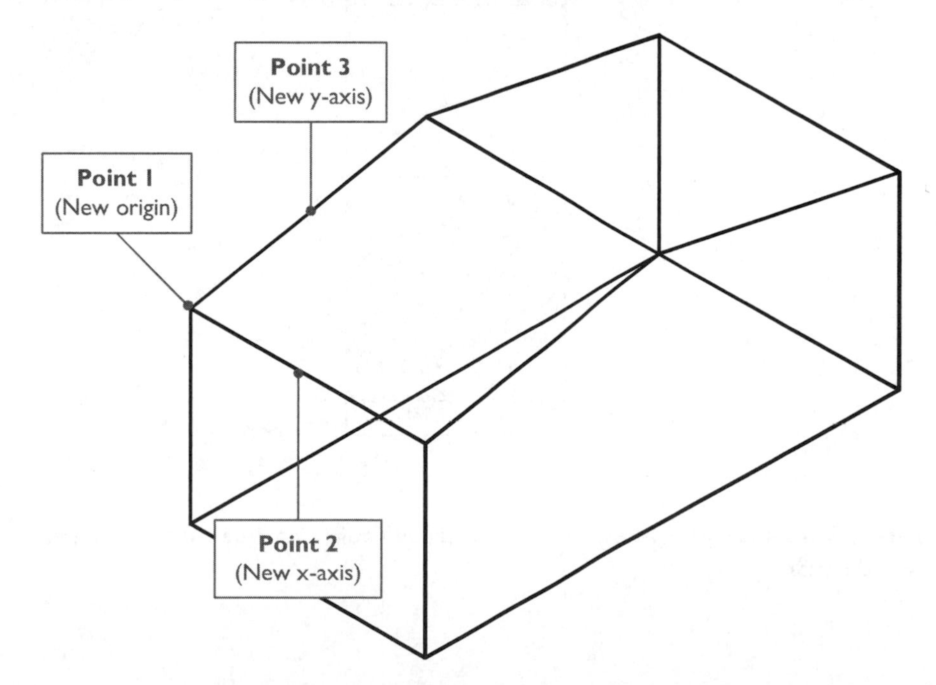

Creating a new coordinate system by picking three points.

Specify new origin point <0,0,0>: **end**

of *(Pick point 1, the origin for the new UCS.)*

Specify point on positive portion of X-axis <1,25,20>: **nea**

to *(Pick point 2 to show the direction of the new x-axis.)*

Specify point on positive-Y portion of the UCS XY plane <1,25,20>: **nea**

to *(Pick point 3 to pin down the angle of the new x,y-plane.)*

Notice that the UCS icon relocates to the new origin. Its axes show the direction of the x,y-plane (highlighted by me in gray) and z. There is no need to specify the direction of the z-axis, because AutoCAD determines it by the right-hand rule.

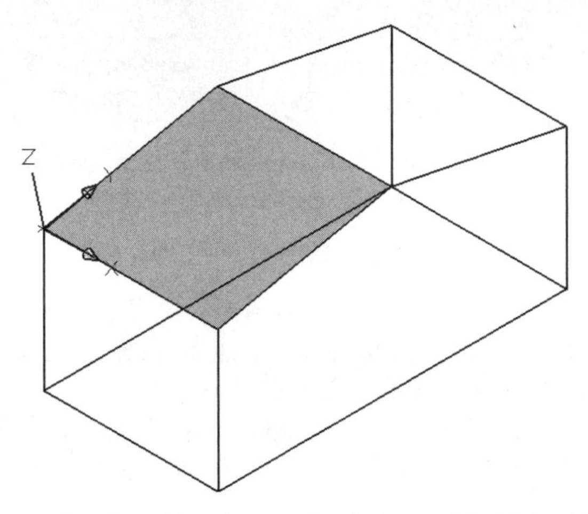

UCS orients the x,y working plane to the sloping roof (highlighted in gray).

5. You created a new coordinate system, a *user*-defined coordinate system (UCS). That's the first step; the second step is to view the new x,y-plane with the **PLAN** command.

   Command: **plan**

   Enter an option [Current ucs/Ucs/World] <Current>: *(Press* **ENTER**.*)*

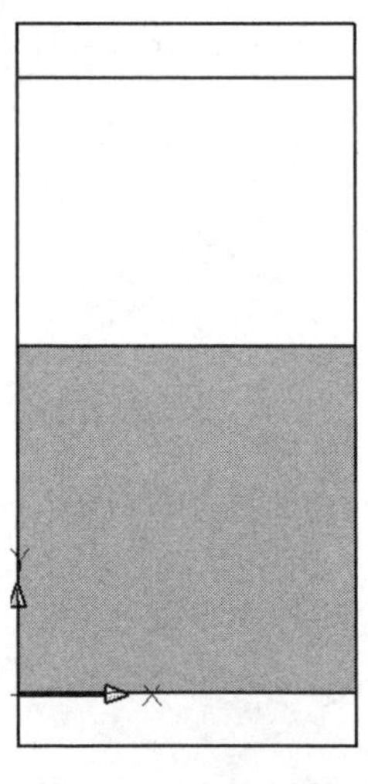

Plan view of the UCS.

You are now looking straight down on the inclined roof (a.k.a. *true plane*). In the figure, I've shaded the plane gray.

6. With the roof oriented in the drawing plane, you can now draw and edit to your heart's content. Try placing text in the new working plane.
7. After the text is in place, use the **ZOOM Previous** command to see the text-on-the-roof from another viewpoint.

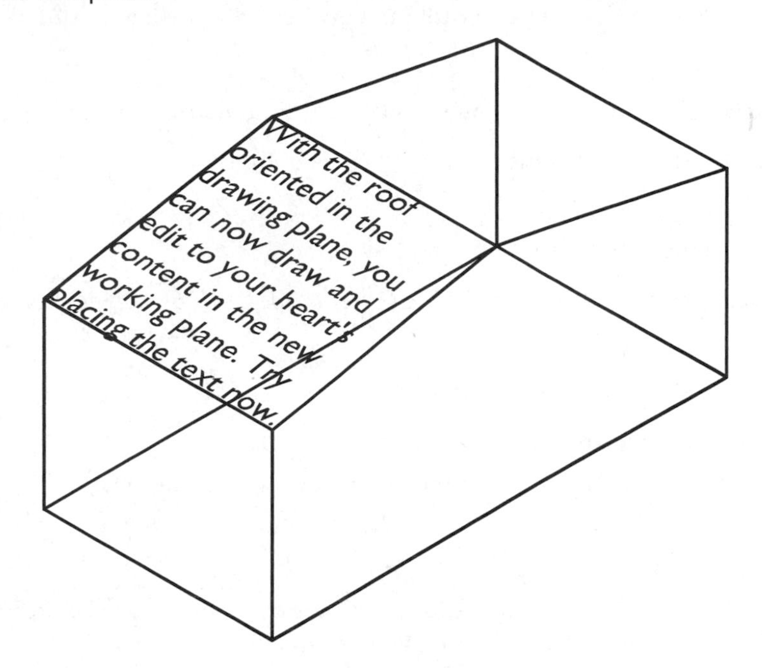

Text placed on the pitched roof.

SETTING THE UCS: ADDITIONAL METHODS

In addition to the **New** option, the **UCS** command allows you to name and edit UCSs.

Command: **ucs**

Current ucs name: *NO NAME*

Enter an option [New/Move/orthoGraphic/Prev/Restore/Save/Del/Apply/?/World] <World>: *(Enter an option.)*

Move

The **Move** option shifts the UCS's origin and/or changes its z depth. Moving the origin moves the UCS in the x,y-plane. (Autodesk notes that this option adds the UCS to the Previous list.)

Specify new origin point or [Zdepth] <0,0,0>: *(Pick a point, or type* **z***.)*

Specify Zdepth <0>: *(Enter a positive or negative distance.)*

Changing the z depth moves the UCS up and down the z axis. The change is applied to the current viewport only.

Orthographic

The **orthoGraphic** option changes the UCS to one of the six standard views:

Enter an option [Top/Bottom/Front/BAck/Left/Right] <Top>: *(Enter an option.)*

The "Top" view is initially the default, because you are looking down in the plan view. AutoCAD changes the UCS relative to the WCS, unless the **UCSBASE** system variable contains the name of a UCS.

Prev

The **Prev** option restores the previous UCS. AutoCAD remembers the last ten coordinate systems for each of model and paper space. There are a number of cases where AutoCAD does not add a change in UCS to its list, such as when switching between viewports with different UCSs.

Save

The **Save** option saves the current UCS once you give it a name of up to 255 characters.

Enter name to save current UCS or [?]: *(Type a name.)*

Restore

The **Restore** option makes current a UCS saved by name.

Enter name of UCS to restore or [?]: *(Enter name of UCS, or type* **?** *to for list of UCSs.)*

Del

The **Del** option removes UCSs from the drawing.

Enter UCS name(s) to delete <none>: *(Enter a name.)*

If the UCS is current, AutoCAD deletes the name, and then renames the UCS "*UNNAMED*."

Apply

The **Apply** option applies the current UCS to selected viewports. (The **UCSVP** system variable determines whether the UCS is saved with viewports.)

Pick viewport to apply current UCS or [All]: *(Pick a viewport, or type* **A** *for all viewports.)*

World

The **World** option sets the user coordinate system to the world coordinate system, the default system.

MANAGING THE UCSs

There are many more aspects to user coordinate systems. Here is some additional information on managing them.

Naming UCSs

It's not really clear how to save UCSs by name. When using the command-line oriented **UCS** command, follow these steps:

1. Create the UCS (or switch to it).
2. Use the **UCS** command's **Save** option.
   AutoCAD prompts you for a name, and then saves the current UCS by that name.

Unlimited UCSs

Drawings can have an unlimited number of UCSs. In other words, you can create as many UCSs as you need. Only one UCS, however, can be active at a time in the current viewport. (Each viewport can have a different UCS.)

To keep track of UCSs, give them names; otherwise AutoCAD eventually "forgets" earlier UCSs. AutoCAD does remember up to ten previous UCSs, which you access with the **UCS Previous** command — just like **ZOOM Previous**.

Switching Between Named UCSs

There is no keystroke shortcut for switching between UCSs, as there is for viewports (use **CTRL+R**). Although Autodesk's documentation suggests using the **UCSMAN** command's dialog box, there is a faster method: the UCS II toolbar.

The advantage is that the toolbar is always available, whereas you need to keep opening and closing the UCS Manager dialog box — a nuisance— as you edit.

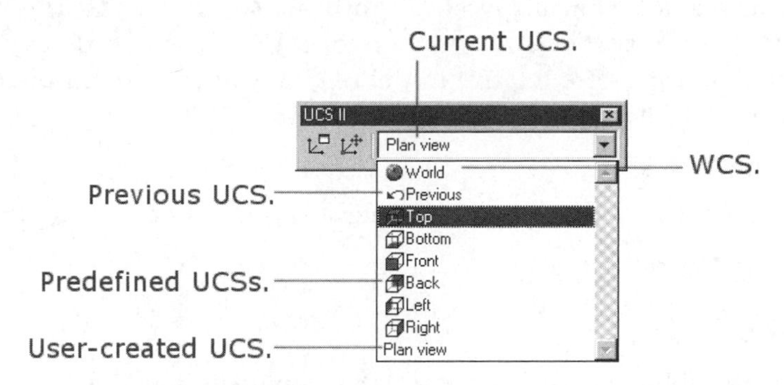

UCS II toolbar lists names of all UCSs.

1. To open the UCS II toolbar, right-click any toolbar.
2. From the shortcut menu, select **UCS II.** Notice the toolbar opens, and has a drop list containing the names of all UCSs in the drawing, plus a number of predefined UCSs (such as Bottom and Left), and the WCS (called "World").
3. To change to another UCS, select the UCS name from the list.

Saving UCSs with Views

AutoCAD can save views by name, which is handy for quickly seeing portions of a drawing. For example, the plan of a house might have views defined for the kitchen area, the master bedroom, and so on. This saves you from having to zoom in and out, and pan about.

You can also save named UCSs with named views:

1. From the menu bar, select **View | Named Views**.
2. In the View dialog box, click **New**.
3. In the New View dialog box, ensure **Save UCS with View** is turned on.
   Select the named UCS to save.

Grid, Snap, Plan, etc. Behavior in UCSs

Once you set a UCS, many commands operate relative to the UCS, and not to the WCS. For example, the grid, snap, and ortho modes orient themselves to the current UCS.

The PLAN command gives you the choice of showing the plan view of the current UCS, the WCS, or any named UCS.

Applying UCSs to Viewports

Each viewport can have its own UCS, whether in model space or paper space. Just click the viewport to make it current, and then use the UCS command to set up the UCS.

To copy a UCS from one viewport to another, follow these steps:

1. Pick the viewport with the UCS to be copied.
2. From the menu bar, select **Tools | New Ucs | Apply.**
3. Pick the viewport to which the UCS should be applied.
   Use the **All** option to apply the UCS to all viewports.

Sharing UCSs — Not

You cannot, unfortunately, share UCSs between drawings as you can with, for example layers or paper space layouts using the AutoCAD DesignCenter.

WCS Coordinates in a UCS

The WCS (world coordinate system) is the default UCS. It is the coordinate system in effect when you start new drawings, and it remains in effect until you set up a UCS. The WCS does not change with changes in 3D viewpoints. Once AutoCAD is in a UCS, the coordinates you enter — at the keyboard, or with the mouse — are relative to the UCS. If you want to enter coordinates relative to the WCS, prefix them with the asterisk, like this:

From point: ***1,2,3**

The asterisk also works for polar, relative, and other forms of coordinate input:

From point: ***20<45**

RELATED SYSTEM VARIABLES

The following system variables are used by AutoCAD in conjunction with UCSs, as listed by Autodesk's documentation:

UCS Controls

The **UCSICON** system variable determines how the UCS icon is displayed.

The **UCSFOLLOW** system variable generates the plan view automatically each time you change the UCS.

The **PUCSBASE** system variable stores the named UCS that defines orthographic UCS settings in paper space.

The **UCSORTHO** system variable determines whether an orthographic UCS setting is restored automatically when an orthographic view is restored.

The **UCSVIEW** system variable determines whether the current UCS is saved with a named view.

The **UCSVP** system variable determines whether the UCS in viewports remains fixed or changes to reflect the UCS of the current viewport.

UCS Properties

The **UCSBASE** system variable stores the name of the UCS defining the origin and orientation for orthographic UCSs.

The **UCSNAME** system variable stores the name of the current UCS for the current viewport.

The **UCSORG** system variable stores the origin of the current UCS.

The **UCSAXISANG** system variable stores the default angle when rotating the UCS around one of its axes.

The **UCSXDIR** system variable stores the x direction of the current UCS.

The **UCSYDIR** system variable stores the y direction of the current UCS.

VIEWPORTS

The **VIEWPORTS** command divides the drawing area into several windows called "viewports." Each viewport can display a different view of the current drawing.

ABOUT VIEWPORTS

Only one viewport is active at a time. It is called the "current viewport." When you draw in the current viewport, the crosshair cursor is displayed in that viewport, and all command activities are performed normally.

You can create dozens of viewports per drawing. There is a limit of 64 viewports in which the drawing is displayed — which out to be plenty.

Viewports act differently, depending on whether they are in model or paper space (layouts).

Model Viewports	***Layout (Paper space) Viewports***
Rectangular only	Rectangular, and any other shape.
Tiled, no gaps	Tiled, overlapped, or separated.
Edited with the VPORTS command only	Edited with most AutoCAD commands.
Visible border	Visible or invisible border.

BASIC TUTORIAL: CREATING VIEWPORTS

For this tutorial, start AutoCAD with the *Ch12SolidEdit.dwg* drawing file from the CD-ROM.

1. To create viewports, start the **VPORTS** command:
   - From the **View** menu, choose **Viewports**, and then **New Viewports**.
   - At the 'Command:' prompt, enter the **vports** command.
   - Alternatively, enter the alias **viewports** at the keyboard.

     Command: **vports** *(Press* ENTER.*)*

   In all cases, AutoCAD displays the Viewports dialog box with the New Viewports tab.
2. From the Setup drop list, select **3D**.
3. In the Standard Viewports list, select "Four: Equal."

   Notice that AutoCAD previews the view direction in each viewport, such as Top and SE Isometric.

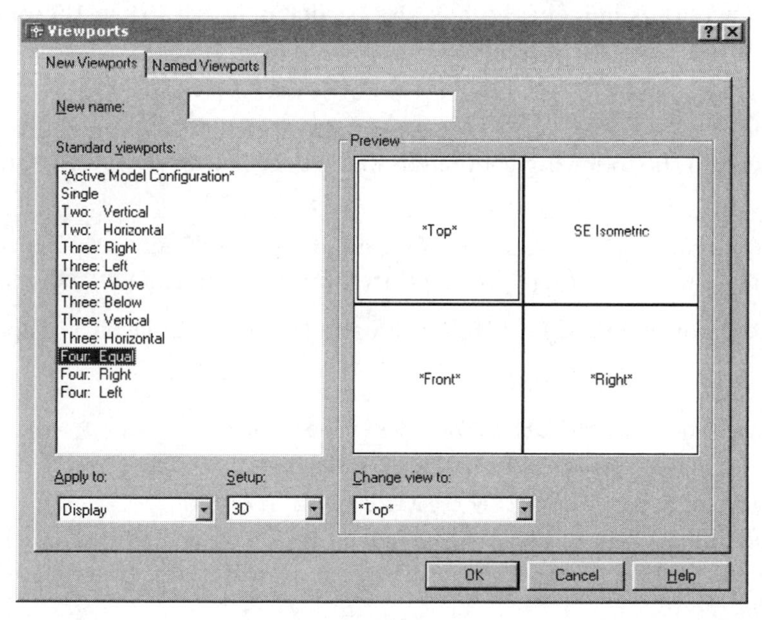

Viewports dialog box.

4. Click **OK** to exit the dialog box.

   Notice that AutoCAD splits the drawing area into four viewports, each showing a different "engineering" viewpoint.

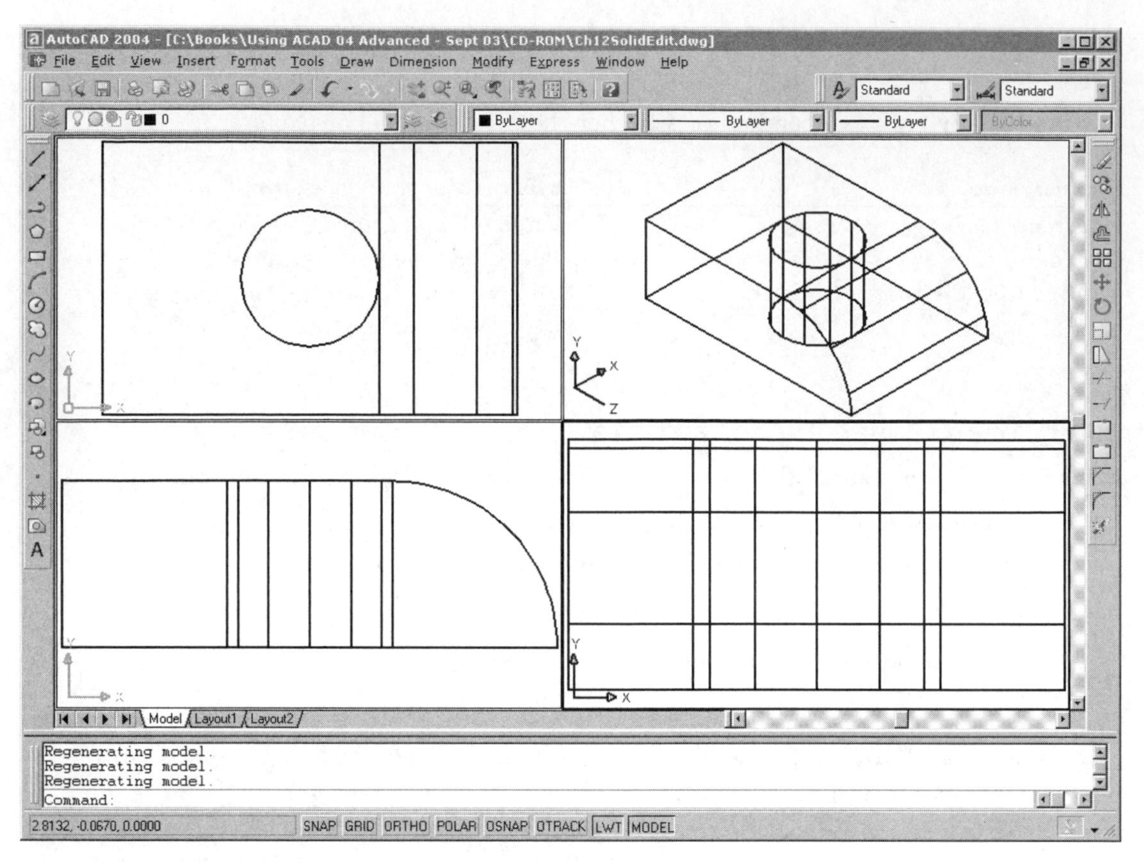

Model space split into four viewports.

The lower right viewport is current, because its border is heavier than the other three. To make another viewport current, move the cursor into it and click. (When you move the crosshair cursor into another viewport, it becomes an arrow.)

As an alternative to making another viewport current, press **CTRL+R** on the keyboard. You cannot change viewports while the following commands are active: **DVIEW**, **VPOINT**, **VPORTS**, **PAN**, **ZOOM**, and **SNAP.**

Viewports created in model space are static. You cannot change them, except through the **VIEWPORTS** command. In contrast, viewports in paper space are dynamic: you can edit them like any other object.

Let's continue the tutorial, and take a look at creating viewport in layouts (paper space).

5. On the status line, click MODEL to switch to paper space.
   When the Page Setup dialog box appears, click **Cancel**.
   Notice that AutoCAD automatically creates a single viewport.
   (Historically, older releases of AutoCAD didn't create a viewport upon switching to paper space for the first time, leaving users scratching their heads over where the drawing had disappeared to.)
6. The automatically created viewport is in the way, so erase it: select its border, and then press **DELETE** — just like any other object in the drawing.

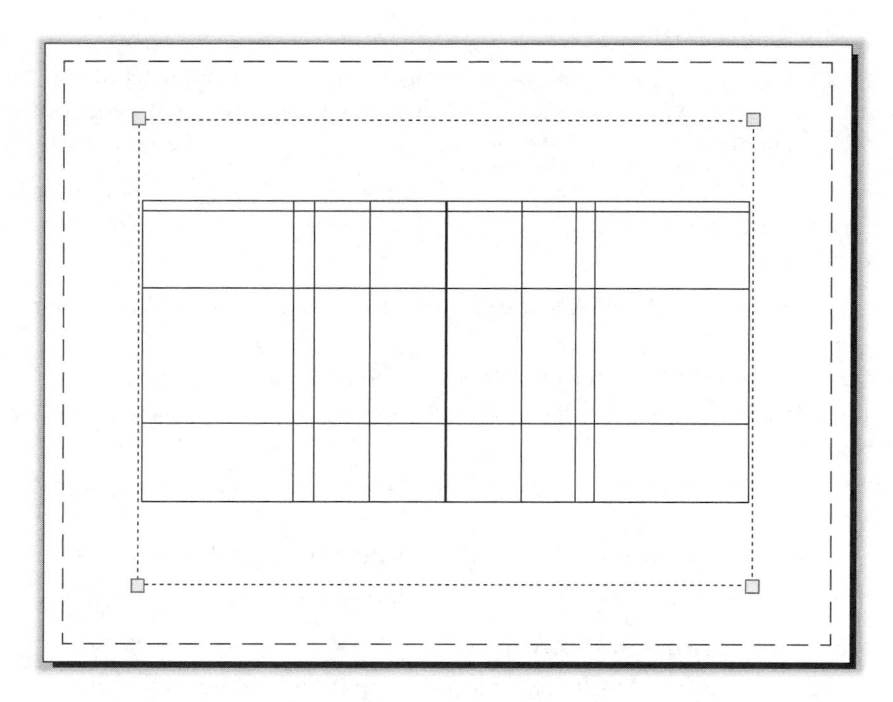

Paper space viewports are objects.

7. Repeat the **VIEWPORTS** command, and again create "Four:Equal" viewports using the "3D" Setup.
8. In addition, change the viewport spacing to 0.5. This creates a gap between them. (You can enter a gap value between 0 and 1 units.)
9. After you click **OK**, AutoCAD wants to know where to place the viewports, because they can be placed anywhere. The easiest response to let AutoCAD fit the viewports within the page margins:

   Specify first corner or [Fit] <Fit>: *(Press* **ENTER** *to fit the viewports.)*
10. You can use grips editing to move, copy, resize, and delete the viewports.

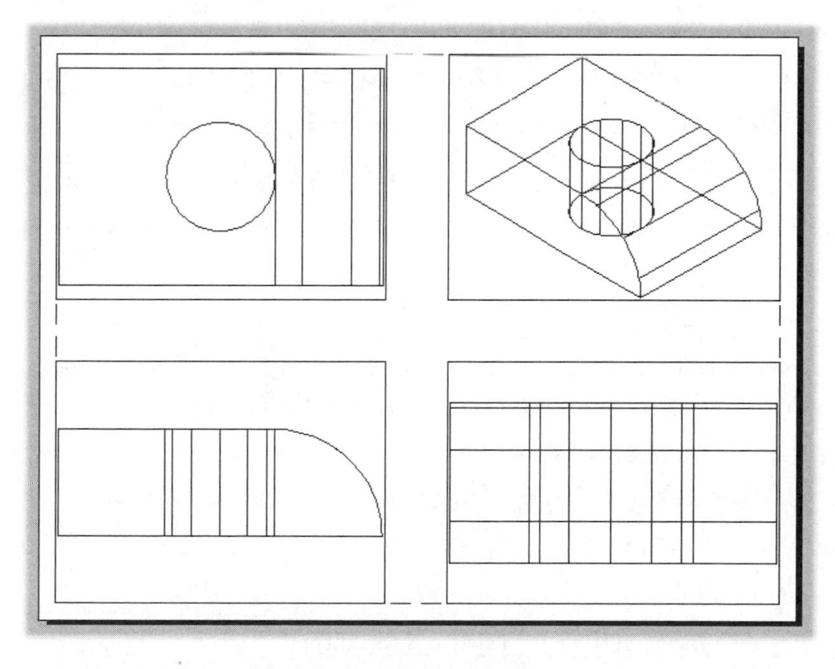

Engineering viewports in paper space.

Drawing In and Between Viewports

When drawing and editing, you work in one viewport. After you complete a drawing command, the result appears in the other viewports as well. When you select and edit objects in one viewport, they show as being selected and edited in all other viewports (where the object is visible).

AutoCAD allows you to draw from one viewport to another while in model space in layouts. For example, you may want to connect a line from the front view of one viewport to the southwest isometric view in another.

1. On the status line, click the PAPER button to return to model space (don't click the Model tab!).
2. Start the **LINE** command, and then pick a point in the current viewport.
3. Move the cursor into the other viewport, and then click to make it current (the border thickens).
4. The cursor changes to crosshairs, and you can complete the line by picking a point in the new current viewport.

You can set the grid, snap, ortho, and UCS icon independently in each viewport. Other modes cannot be set independently: object snaps, polar and object tracking, lineweights, and model/paper space.

Redraws and Regenerations in Viewports

The **REDRAW** and **REGEN** (short for "regeneration") commands affect only the current viewport. If you wish either to redraw or regenerate all viewports at the same time, use **REDRAWALL** or **REGENALL**, respectively.

CREATING VIEWPORTS: ADDITIONAL METHODS

The dialog box displayed by the **VPORTS** command has a number of options that control viewports being created. Some options are available in model or paper space only. Viewports are also affected by a pair of system variables.

- **New Name** option saves viewport configurations by name (in model space only).
- **Apply to** option applies settings to entire display or current viewport (in model space only).
- **Viewport Spacing** option places gaps between viewports (in paper space only).
- **Setup** option specifies 2D or 3D viewpoints.
- **Change View To** option changes 3D viewpoint of viewports.
- **New Viewports** tab selects a saved viewport configuration.
- **MAXACTVP** system variable limits the number of viewports.
- **CVPORT** system variable reports AutoCAD's ID number of the current viewport.

Let's look at them all.

New Name (MODEL SPACE ONLY)

The **New Name** option saves viewport configurations by name. If you leave this blank, the viewport configuration is applied, but not saved for later reuse. Enter a name, such as:

New Name: **Engineering Views**

In addition, AutoCAD saves the following data with the viewport definition:

- ID number of viewport.
- Viewport position.
- Grid and snap mode settings, and spacings for each.

- **VIEWRES** mode.
- **UCSICON** setting and UCS name.
- Views set by the **DVIEW** and **VPOINT** commands.
- **DVIEW** perspective mode and clipping planes.

Apply To (MODEL SPACE ONLY)

The **Apply to** option determines where the viewports are placed:

- **Display** applies viewports to the entire drawing area (default). Existing viewport configurations are replaced.
- **Current Viewport** applies viewports to the current viewport only. This allows you to subdivide viewports.

Display replaces the current configuration with the new viewport configuration. For example, suppose the AutoCAD window has two horizontal viewports. If you select "Three Vertical," the two are replaced by the three.

Current Viewport subdivides viewports further. For example, if AutoCAD has two horizontal viewports, and you select "Three Vertical," then the current viewport is subdivided into three vertical viewports — for a total of four viewports.

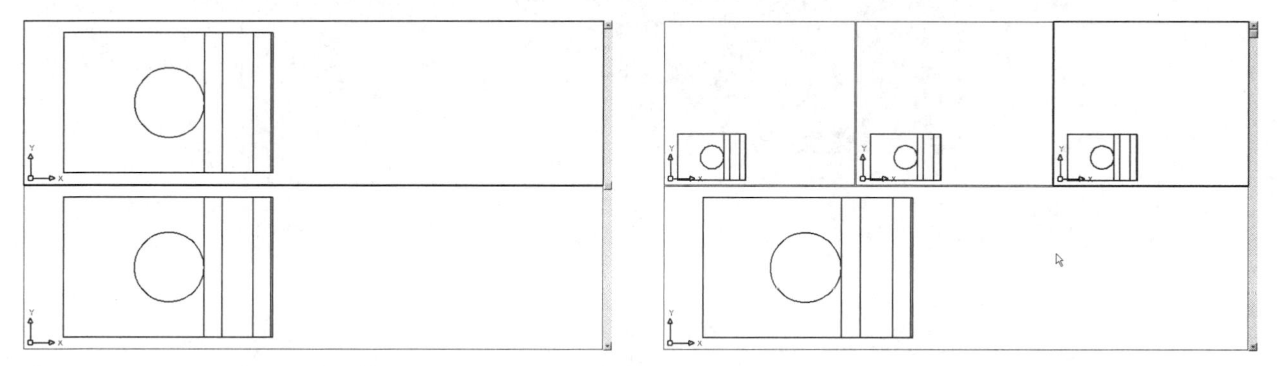

Apply to Current Viewport option subdivides viewports in model space.

Viewport Spacing (PAPER SPACE ONLY)

The **Viewport Spacing** option specifies the distance between viewports. The default is 0, meaning no gap. The units are in inches or mm, depending on the setting of the **MEASUREINIT** system variable.

Setup

The **Setup** option selects either 2D or 3D:

- **2D** creates new viewports with the same viewpoint as the current viewport.
- **3D** creates a different orthogonal viewpoint in each viewport, as far as possible.

When creating viewports in 3D mode, AutoCAD assigns these viewpoints to the viewports:

Number of Viewports	*Views Created*
Single	Current view.
Two	Top, SE Isometric.
Three	Top, Front, SE Isometric.
Four	Top, Front, Right, SE Isometric.

Change View To

The **Change View to** option allows you to change viewpoints in viewports. In the Preview window, click on a viewport, and then select an option from the list. In addition to named viewports, you can choose from these preconfigured views:

Current	
Top	Bottom
Front	Back
Left	Right
SW Isometric	SE Isometric
NW Isometric	NW Isometric

Only user-defined and the *Current* viewpoints are available in 2D mode. Asterisks surrounding the names indicate AutoCAD-generated viewpoints.

New Viewports

The **New Viewports** tab lets you select a saved viewport configuration. (If you haven't saved any, then the only name available is *Active Model Configuration*, which is the current viewport configuration.)

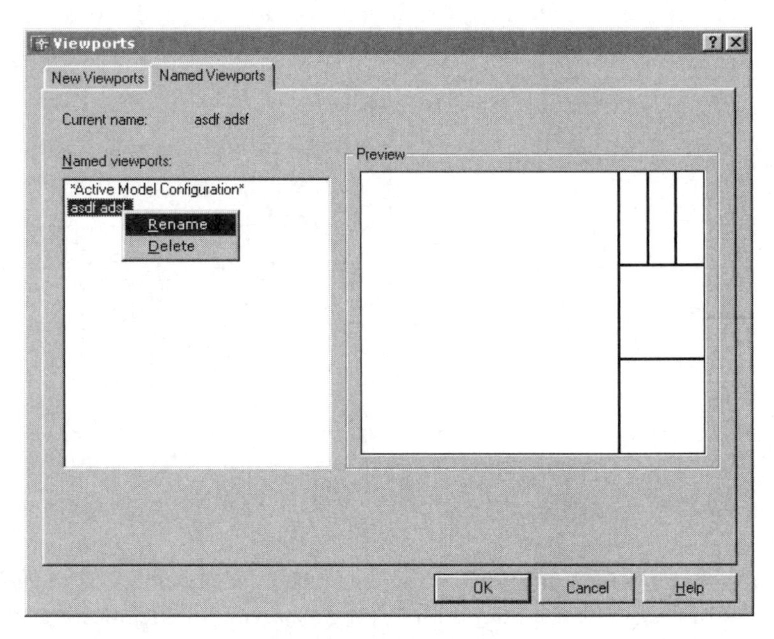

Named Viewports tab.

If you have saved configurations (using the **New name** option), their names are listed under Named Viewports. Choose a name to see its configuration in the Preview window.

To rename or delete viewport configurations, right-click the name. From the shortcut menu, select **Rename** or **Delete**.

Choose **OK** to close the dialog box, and AutoCAD reconfigures the entire display with the named viewport configuration.

MaxActVp

The MAXACTVP system variable specifies the maximum number of viewports that AutoCAD displays. This variable was important in the days of slower computers, but it now appears to the technical editor and me that AutoCAD ignores MAXACTVP.

Note: In layouts, viewports are objects. This means you can change their color, layer, linetype, linetype scale, and lineweight. You can attach hyperlinks and assign plot styles.

The quickest method to changing drawings back to a single viewport is to click the **Single Viewport** button on the Viewports toolbar.

An alternative is to use the **-VPORTS** command with the **Single** option, as follows:

Command: **-vports**

Enter an option [Save/Restore/Delete/Join/SIngle/?/2/3/4] <3>: **si**

The **SIngle** viewport option displays the view of the viewport that was current before you began this command.

CONTROLLING VIEWPORTS: ADDITIONAL METHODS

The **-VPORTS** command provides several added functions not found in the dialog box version of the command.

- **ON*** and **OFF*** options toggle the display of the viewport contents.
- **Lock*** option locks the viewport (doesn't work for me!)
- **Shadeplot*** option specifies that the viewport contents are shaded when plotted.
- **Object** option converts closed objects into viewports.
- **Polygonal** option creates irregularly shaped viewports.
- **LAYER Freeze** command hides viewport boundaries.

* Options also available in the Properties window. Other options — **Restore**, **2**, **3**, **4**, and **Fit** — operate identically to the Viewports dialog box.

Let's look at the new options.

On and Off

The **ON and OFF** options toggle the display of the *contents* of viewports.

Select objects: *(Pick one or more viewports.)*

The border remains visible; to turn off the border, use **LAYER Freeze**.

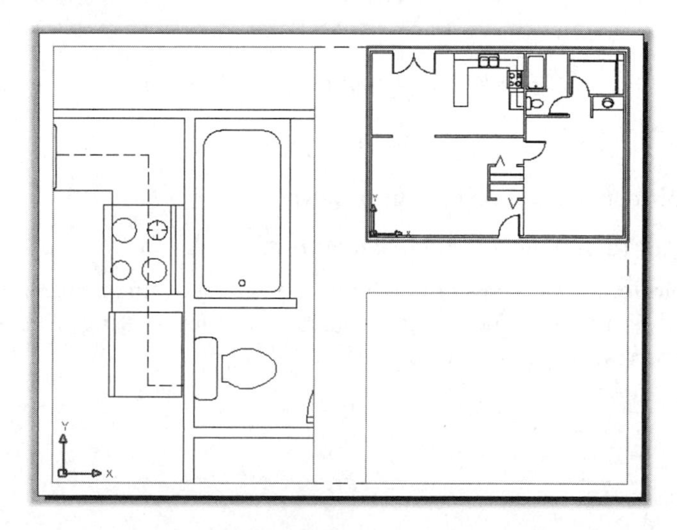

Viewport display in lower right is turned off.

Lock

The **Lock** option locks the zoom ratio between model and paper space. When your drawing is in model mode of paper space, then the entire layout pans and zooms, instead of the model window.

Shadeplot

The **Shadeplot** option specifies that the viewport contents should be shaded when plotted.

Shade plot? [As displayed/Wireframe/Hidden/Rendered] <As displayed>: *(Enter option.)*

The options are:

- **As Displayed** plots the viewport as it is displayed. If the viewport is rendered, it is plotted rendered; if wireframe, then wireframe.
- **Wireframe** plots viewport contents as wireframe, and ignores other shade or rendering settings.
- **Hidden** removes hidden lines during plotting, regardless of the display.
- **Rendered** plots rendered.

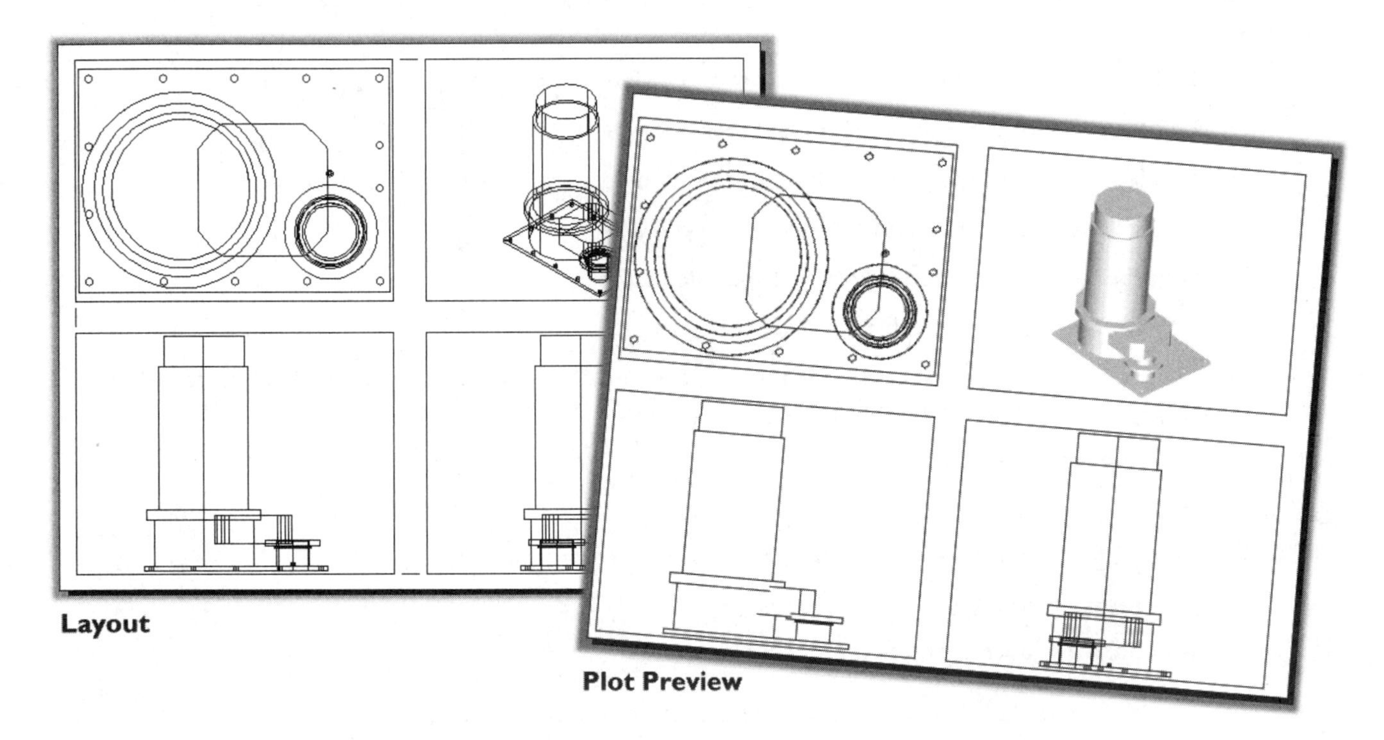

Viewplots can plot differently from their display.

Object

The **Object** option converts closed objects into viewports.

Select object to clip viewport: *(Pick an object.)*

The object can be a circle, ellipse, closed polyline (consisting of lines and/or arcs), spline, or region. Unlike with the **XCLIP** command, the polyline can be self-intersecting. Naturally, you need to draw the object *before* converting it into a viewport.

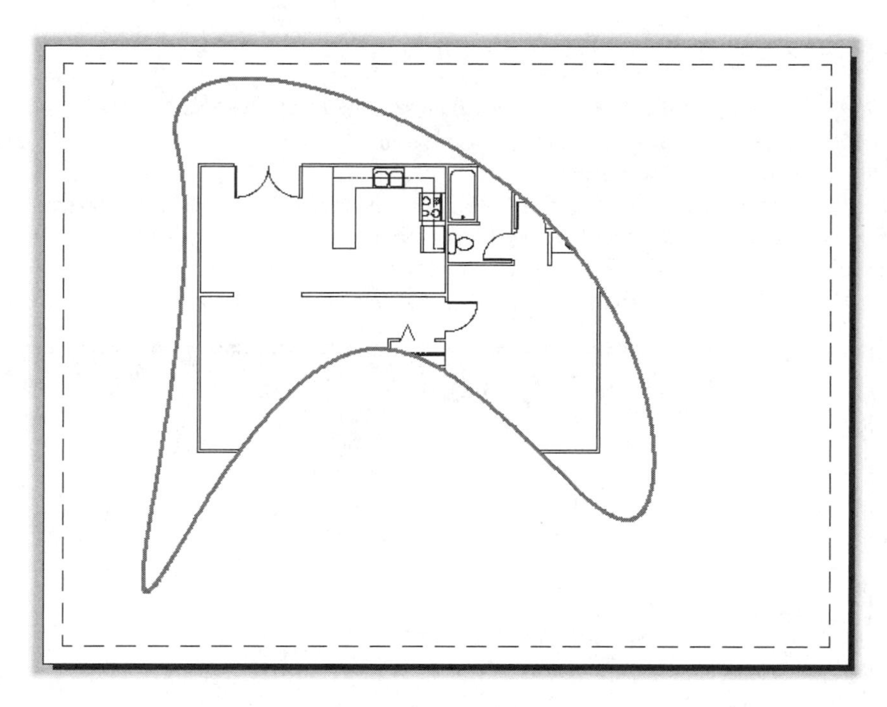

Viewport converted from spline.

Polygonal

The **Polygonal** option creates irregularly shaped viewports in a manner similar to drawing polylines with the PLINE command. (This command is similar to the XCLIP command described in Chapter 2.) The **Arc** option allows you to create round viewports.

Specify start point: *(Pick a point.)*

Specify next point or [Arc/Length/Undo]: *(Pick another point, or enter an option.)*

Layer Freeze

The **LAYER** command hides viewport boundaries. This works correctly when you place viewports on their own layer. Use the command's **Freeze** option to hide the boundary; the content of the viewport continues to display normally.

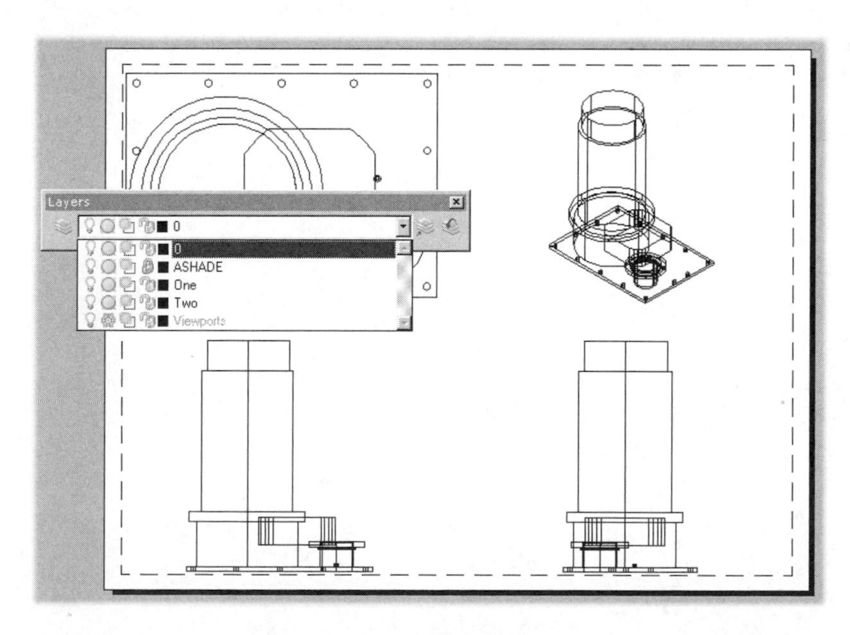

Viewport border frozen.

EXERCISES

1. Start a new drawing, and set up three equal viewports in model space.
   a. Draw several circles with the **CIRCLE** command.
      Where do the circles appear?
   b. Use the **ZOOM** command in each viewport to make the circles different sizes.
   c. Start the **LINE** command and, with CENter object snap, select the center of a circle.
      Move to another viewport, and place the line's endpoint at the center of another circle.
2. From the CD-ROM, open the *psplan.dwg* file, a drawing of a floor plan.

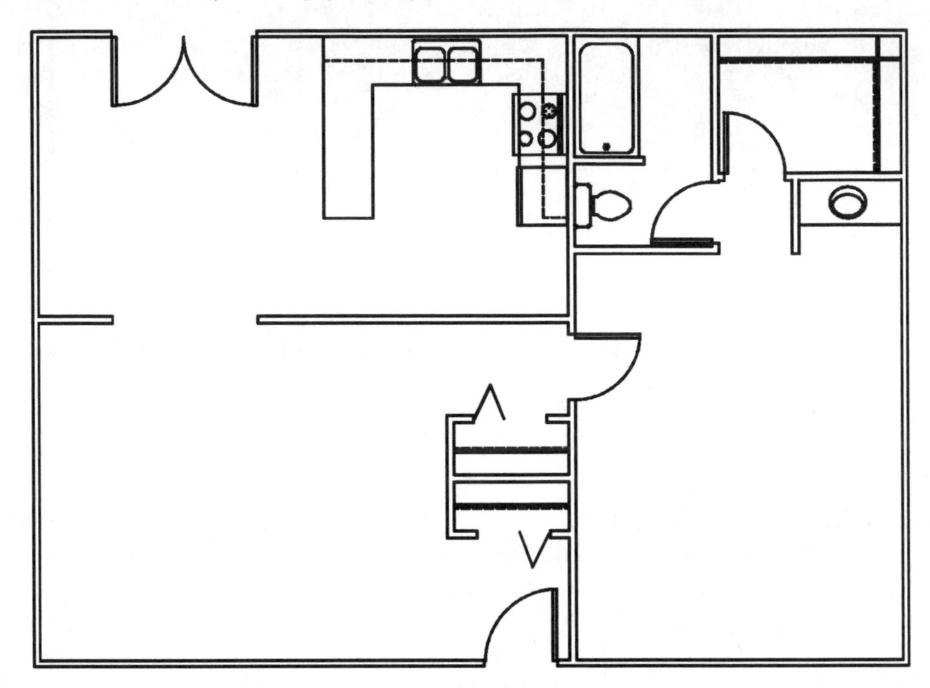

Floor plan drawing.

The drawing is in architectural units and is set up to be plotted at 1/4" =1'0" scale on a C-size (24"x18") sheet of paper.

a. In layout mode, create a viewport that shows the entire drawing.
b. Create a second viewport that shows just the bathroom.
c. Edit the second viewport to make it non-rectangular.
d. Use the **TEXT** command to label each viewport.

REVIEW QUESTIONS

1. What are the three axes in the Cartesian coordinate system?
2. In which direction from a page does the z axis normally project?
3. What is a clipping plane?
   How many clipping planes are there?
   Name each.
4. What form of 3D viewing most closely approximates the view of the human eye?
5. In AutoCAD, how would you enter a coordinate that has values of X=3, Y=5, Z=9?
6. In which plane are two-dimensional drawings constructed?
7. Which axis normally represents the height of objects?
8. What commands generate 3-dimensional views of drawings?
9. What is the **VPOINT** command meant for?
10. What does the **3DORBIT** command do?
11. What is the *target point*?
12. Where is the *camera* located?
13. How do you turn on perspective mode?
14. Can you edit in perspective mode?
15. Can you edit in shaded mode?
16. How do you return 3D views to plan view?
17. What does Gouraud shading mode do?
18. Briefly explain the purpose of the **UCS** command.
19. Can the UCS icon be turned off?
    If so, how?
20. Can the names of user-defined coordinate systems be changed?
21. Match the finger with its axes based on the right-hand rule:

a. Thumb	i. Points to nothing of importance.
b. Forefinger	ii. Points to the x-axis.
c. Middle finger	iii. Points to the y-axis.
d. Little finger	iv. Points to the z-axis.

22. How many UCSs can a drawing contain?
23. To enter a WCS coordinate while in a UCS, prefix the coordinate with _______.
24. Which command aligns the view with the current UCS?
25. Can you can associate a UCS with a named view?
26. Can you switch between viewports during drawing commands?
27. How do you save viewport configurations that you want to reuse?
28. Describe how to place a new viewport that fills the entire page in layouts.
29. List some of the differences between viewports created in model and paper space.

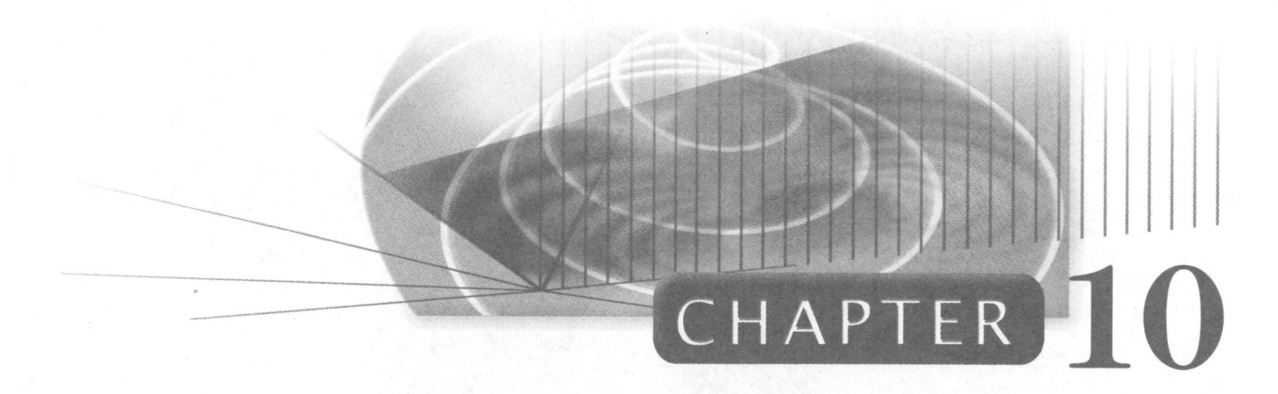

Basic 3D Drafting and Editing

AutoCAD provides several methods of drawing in 3D: adding thickness and elevation to 2D objects, drawing with 3D surfaces, and modeling with 3D solid objects.

In this chapter, you learn about the following commands and techniques:

ELEV locates 2D and 3D objects in the z axis, and gives 2D objects depth.

Drawing 2D objects with 3D coordinates.

Editing objects with 3D coordinates.

3DPOLY draws polylines in 3D space.

ALIGN, **3DARRAY**, **MIRROR3D**, and **ROTATE3D** edit objects in 3D space.

WMFIN converts 3D drawings to "2D."

FINDING THE COMMANDS

On the **DRAW** and **MODIFY** menu:

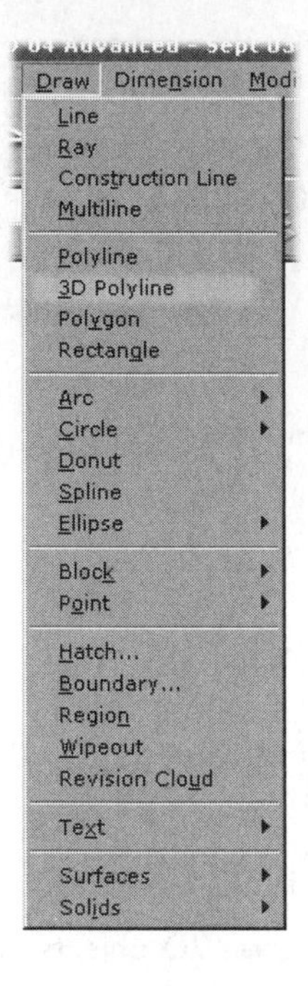

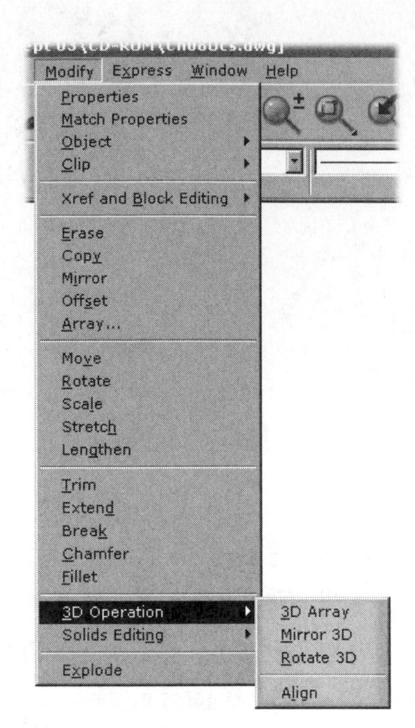

ABOUT 3D DESIGN WITH AUTOCAD

AutoCAD provides several methods of creating 3D objects, each of which has its advantages and drawbacks. (Historically, each method was added as each release of AutoCAD improved its ability to work in three dimensions.)

AutoCAD v2.1 gave certain 2D objects thickness and elevation using the **ELEV** command, as described in this chapter. This is sometimes called "2-1/2D," because the extrusion always results in a flat top. Adding thickness is easy to understand and edit, but is limited in capability and sometimes renders incorrectly.

AutoCAD Release 9 created 3D wireframe objects with the **3DLINE** command; later the **3DPOLY** command was added, and many editing commands were extended to allow 3D editing. Wireframe objects are difficult to construct, and do not render at all.

AutoCAD Release 10 created 3D objects from surface meshes with commands like **3DMESH**, **RULESURF**, and **3D**. (See Chapter 11.) Meshes are triangular and rectangular planes joined to create 3D surfaces. They are popular for styling automobiles and consumer products. Surface meshes are difficult to edit, and AutoCAD provides little assistance in manipulating surfaces.

AutoCAD Release 11 created 3D objects from solid models with commands such as **BOX**, **EXTRUDE**, and **SOLIDEDIT**. (See Chapters 12 and 13.) Over the years, Autodesk switched the solid modeling engine from PADL to ACIS and now to ShapeManager. Solids are popular for designing mechanical devices. AutoCAD provides numerous commands for creating and editing solids.

To supplement AutoCAD's limited 3D capabilities, Autodesk markets additional software for 3D design in specialized areas. Architectural Desktop and Revit provide 3D design of buildings; Land Desktop provides 3D design of earthworks; 3D Studio provides 3D design for games; and Mechanical Desktop and Inventor provide 3D design for machines. (Products with the word "Desktop" in the name are add-ins to AutoCAD; the others are stand-alone programs.)

ELEV

The **ELEV** command specifies the thickness and elevation of 2D objects, turning them into 3D objects.

The *thickness* is an extrusion above or below the current elevation. The *elevation* is the distance above or below the x,y-plane. (Elevation is equivalent to distance along the z-direction: elevation = z.) The figure below illustrates how different combinations of thickness and elevation conspire to extrude and locate a rectangle in 3D space.

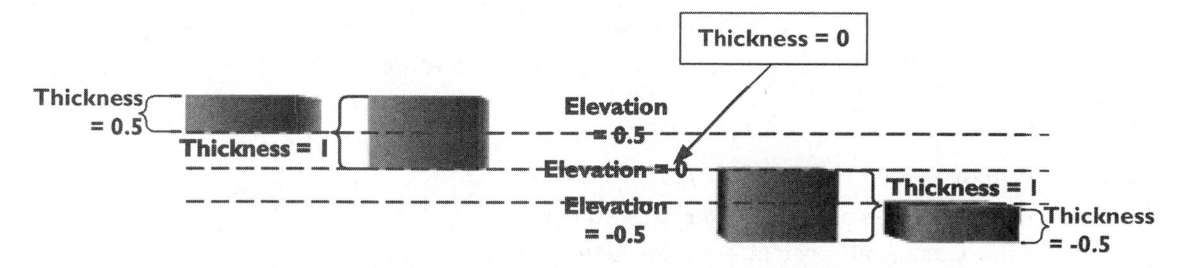

Thickness and elevation effects on a 2D rectangle.

The **ELEV** command sets the thickness and elevation in the z direction; more accurately, the settings are made along the z axis of the current UCS. This means you can draw extruded objects at any angle by changing the UCS.

This command is not retroactive. Once the thickness and elevation are set with this command, objects drawn thereafter take on those settings. The thickness and elevation can be changed with the **CHANGE** and **PROPERTIES** commands.

Thickness can be applied to points, lines, circles, arcs, 2D solids, and most forms of polyline — polygons, boundaries, donuts, and splined polylines. Extruding 2D objects results in new 3D objects:

2D Object	*3D Extrusion*
Point	Line
Line	Plane
Rectangle	Box
Arc	Curved plane
Circle	Cylinder
Donut	Tube
Polygon	Prism

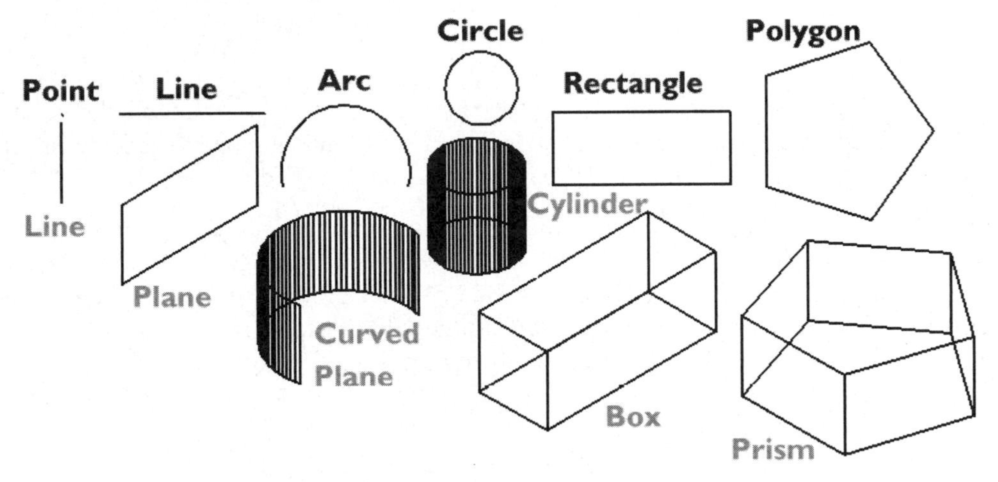

Extruding 2D objects into 3D.

Note: The **THICKNESS** system variable has no effect on the **RECTANG** command, curiously enough. Instead, use the **Thickness** option provided with the **RECTANG** command:

Command: **rectang**

Specify first corner point or [Chamfer/Elevation/Fillet/Thickness/Width]: **t**

Specify thickness for rectangles <0.0000>: *(Enter a value.)*

BASIC TUTORIAL: DRAWING 2D OBJECTS AS 3D

1. To draw 2D objects as 3D, start the **ELEV** command:
   - At the 'Command:' prompt, enter the **elev** command.

   Command: **elev** *(Press* ENTER.*)*

2. AutoCAD prompts you to specify the elevation:

   Specify new default elevation <0.0>: *(Enter a value such as* **2**, *or press* ENTER *to keep the current setting.)*

3. And for the thickness:

Specify new default thickness <0.0>: *(Enter a value such as* **1***, or press* **ENTER** *to keep the current setting.)*

4. No change occurs until you draw the next objects.
   Use the **POLYGON** command, for example, or any command other than **RECTANGLE**.
5. To see the object in 3D, change the viewpoint to isometric.
   From the **View** menu, select **3D Views**, and then **SW Isometric**.

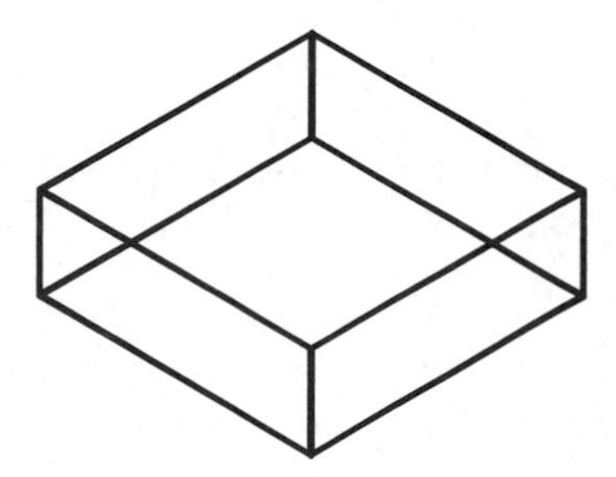

Extruded square polygon seen from southwest isometric viewpoint.

DRAWING 2D AS 3D: ADDITIONAL METHODS

The elevation and thickness are controlled by namesake system variables, and can be retroactively changed by a pair of commands.

For new objects:

- **ELEVATION** system variable sets the elevation.
- **THICKNESS** system variable sets the thickness.

For existing objects:

- **CHANGE** command changes thickness and elevation at the command line.
- **PROPERTIES** command changes thickness and the elevation through a dialog box.

Let's look at them all.

Elevation

The **ELEVATION** system variable sets the elevation. You can use this in place of the **ELEV** command, when you want to change only the elevation.

Command: **elevation**

Enter new value for ELEVATION <0.0>: *(Enter a value.)*

Thickness

The **THICKNESS** system variable sets the thickness. You can use this in place of the **ELEV** command, when you want to change only the thickness.

Command: **thickness**

Enter new value for THICKNESS <0.0>: *(Enter a value.)*

Change

The **CHANGE** command changes thickness and elevation at the command line. Use this command retroactively to change objects. (The **CHPROP** command changes the thickness only, and not elevation.)

Command: **change**

Select objects: *(Select one or more objects.)*

Select objects: *(Press* **ENTER** *to end object selection.)*

Enter the **e** option to change the elevation:

> Specify change point or [Properties]: **e**
>
> Specify new elevation <0.0>: *(Enter a value.)*

Enter the **t** option to change the elevation:

> Enter property to change [Color/Elev/LAyer/LType/ltScale/LWeight/Thickness]: **t**
>
> Specify new thickness <0.0>: *(Enter a value.)*

Press **ENTER** or **ESC** to exit the command:

> Enter property to change [Color/Elev/LAyer/LType/ltScale/LWeight/Thickness]: *(Press* **ENTER** *to exit the command.)*

Properties

The **PROPERTIES** command changes thickness and elevation through a dialog box. To change the thickness, click the **Thickness** item, and then enter a new value.

Changing the elevation is a bit tricky. For most objects, including lines and circles, the elevation is not listed; instead, use the **Start Z** and **End Z** items for lines, and **Center Z** items for arcs and circles. For text, it is **Position Z**, and for xlines it is **Basepoint Z**.

For other objects, such as polylines, click the **Elevation** item, and then enter a new value. In summary, look for Elevation; if it is missing, look for a Geometry item ending in "Z."

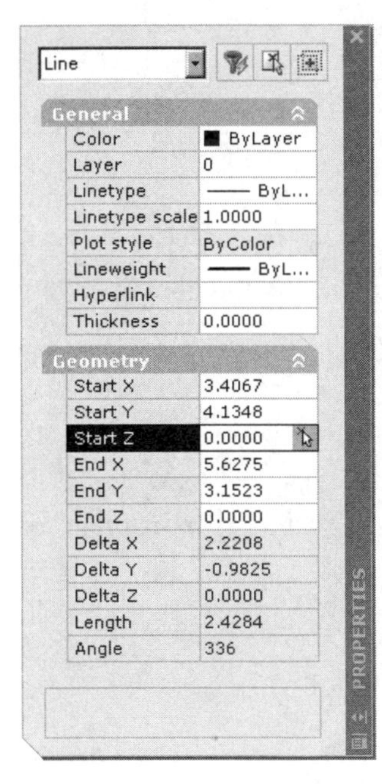

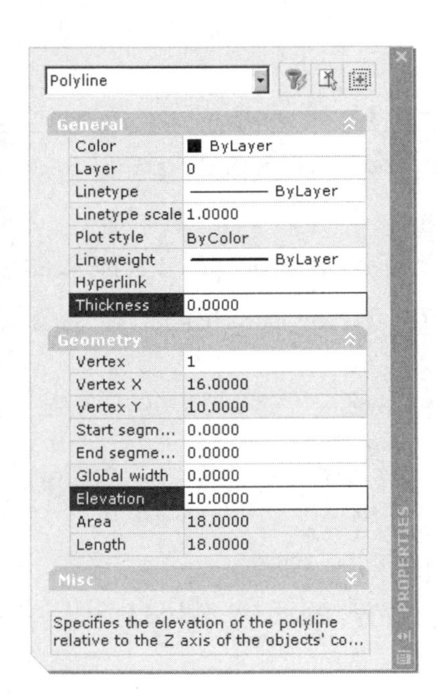

Left: *Elevation for lines is called Start Z and End Z*
Right: *Elevation for polylines is called Elevation.*

Notes: Grips cannot be used to change the thickness or elevation.

The **MATCHPROP** command matches thickness, but not elevation.

ELEVATION and **THICKNESS** have no effect on objects drawn with the **RECTANGLE** command. Use the **PROPERTIES** or **CHANGE** commands instead. (The **RECTANGLE** command has its own settings for thickness and elevation.)

The **LIST** command reports the elevation as the z coordinate, as in:
X= 0.0 Y= 0.0 Z= 1.0

TUTORIAL: DRAWING A TABLE IN 3D

In this tutorial, you complete the 3D drawing of a wood coffee table using the **ELEV** command, among others. The table is 34" x 34", and has legs 17-1/4" high.

1. Start AutoCAD, and then from the CD-ROM open the *Ch09Thickness.dwg* file, a partially completed drawing of a coffee table. The drawing consists only of a profile of the four wooden legs.

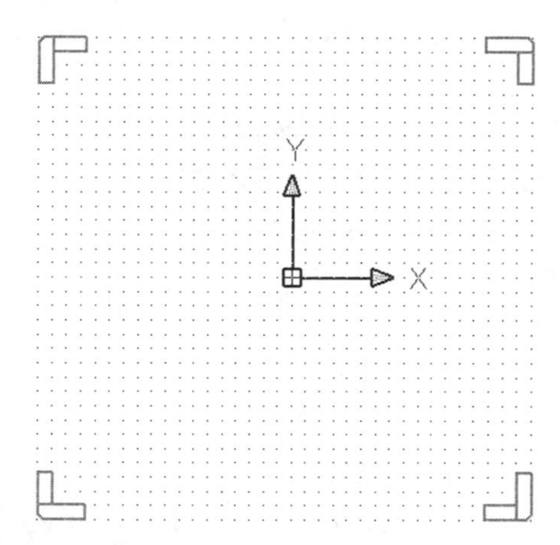

Ch09Thickness.dwg shows profile of legs.

2. To give the legs their height of 17-1/4", use the **CHPROP** command (short for "change properties") to change the thickness:

   Command: **chprop**

   Select objects: **all**

   Select objects: *(Press* **ENTER** *to end object selection.)*

   Enter property to change [Color/LAyer/LType/ltScale/LWeight/Thickness]: **t**

   Specify new thickness <0.0000>: **17.25**

   Enter property to change [Color/LAyer/LType/ltScale/LWeight/Thickness]: *(Press* **ENTER** *to exit the command.)*

3. To see the 3D view of the legs, from the **View** menu select **3D Views | SW Isometric**. Alternatively, use the **VPOINT** command, and then enter the coordinates for the southwest isometric viewpoint:

   Command: **vpoint**

   Specify a view point or [Rotate] <display compass and tripod>: **-1,-1,1**

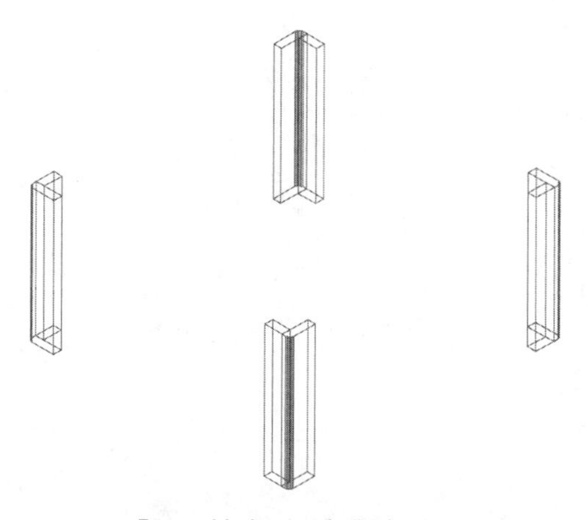

Four table legs with thickness.

4. Before drawing the 3/4" table top, work out its thickness and elevation settings:

   Elevation: ________"

   Thickness: ________"

5. Use the **ELEV** command to preset the elevation and thickness, as follows:

   Command: **elev**

   Specify new default elevation <0.0000>: **17.25**

   Specify new default thickness <0.0000>: **0.75**

6. To draw the table top, use the **POLYGON** command. (It can be tricky locating the table top precisely, so I've provided the coordinates.)

   Command: **polygon**

   Enter number of sides <5>: **4**

   Specify center of polygon or [Edge]: **e**

   Specify first endpoint of edge: **-17,-17**

   At this point, it may appear that AutoCAD will draw the polygon at elevation = 0, but never fear: the correct height appears in a moment.

   Specify second endpoint of edge: **17,-17**

7. Use the **HIDE** command to reveal the problem with applying elevation and thickness to objects: they have no "top." (Perhaps you can convince your client this is a glass-top coffee table.)

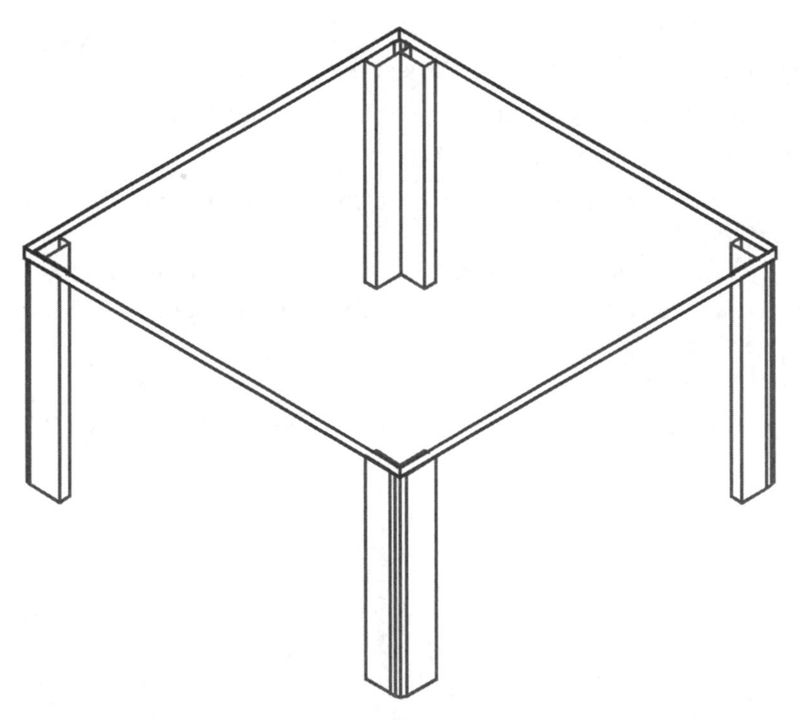

Completed table, with top.

DRAWING WITH Z COORDINATES

Only some of AutoCAD's 2D drawing commands can be used to draw in 3D space: **POINT**, **LINE**, **SPLINE**, **RAY**, and **XLINE**. Drawing in 3D with these commands means you enter coordinate *triplets* (x,y,z) to place the lines anywhere in 3D. For example:

Command: **line**

Specify first point: **1,2,3**

Specify next point or [Undo]: **9,8,7**

AutoCAD draws the line in 3D space from 1,2,3 to 9,8,7. (The old **3DLINE** command no longer exists; **LINE** now handles both 2D and 3D line drawing.)

PLANAR DRAWING COMMANDS

Many of AutoCAD's other 2D drawing commands are *planar*. This means they are drawn "flat" in the current x,y-plane, at a specified elevation. These commands are **MLINE**, **PLINE**, **POLYGON**, **CIRCLE**, **SPLINE**, **RECTANG**, **ARC**, **DONUT**, **ELLIPSE**, **TEXT**, **TRACE**, **SOLID**, and **MTEXT**.

You can set their elevation with the **ELEV** command, or else specify a z coordinate as the first coordinate triplet. (When you leave out the z coordinate, AutoCAD assumes that z equals the value of the **ELEVATION** system variable.)

Here is an example with the **PLINE** command:

Command: **pline**

Specify start point: **1,2,3**

Specifying 1,2,3 as the start point means the *entire* polyline is drawn at elevation = 3. Below, notice how AutoCAD objects to the next coordinate triplet:

Specify next point or [Arc/Halfwidth/Length/Undo/Width]: **9,8,7**

2D point or option keyword required.

The solution is to specify just x, y coordinates for the remainder of the command:

Specify next point or [Arc/Halfwidth/Length/Undo/Width]: **9,8**

You are not limited to planar drafting of polylines, circles, and so on. There are several work-arounds: rotate the drafting plane, rotate the object, or use 3D-specific drafting commands:

- Before drawing the object, rotate the x,y-drawing plane with the **UCS** command, as described in the previous chapter.
- After drawing the object, rotate it in 3D space with the **ROTATE3D** and **ALIGN** commands, as described later in this chapter.
- The 3D-specific polylines are drawn with the **3DPOLY** command, which draws polyline segments in 3D space. It lacks most of the **PLINE** command's options, such as width and arcs. The 3D polyline can be splined with the **PEDIT** command, but not curve-fitted.

Blocks can be 3D, placed in 3D space, and scaled differently in the x, y, and z directions. Other objects are strictly 2D, such as xrefs, images, shapes, and attributes. Object snaps snap to 3D points, when appropriate.

3D Text

During a trade show several years ago, Autodesk showed a preview version of AutoCAD manipulating text in 3D. (The effects were similar to those created by Microsoft's WordArt.) The text effects were removed, however, and never shipped to customers.

Text created with *.shx* files, such as Simplex and RomanT, can be given a thickness. TrueType fonts cannot, however. There are, however, a couple of work-arounds.

The Express Tool **ARCTEXT** command places text along an arc. From the **Express** menu, select **Text** and then **Arc-Aligned Text**.

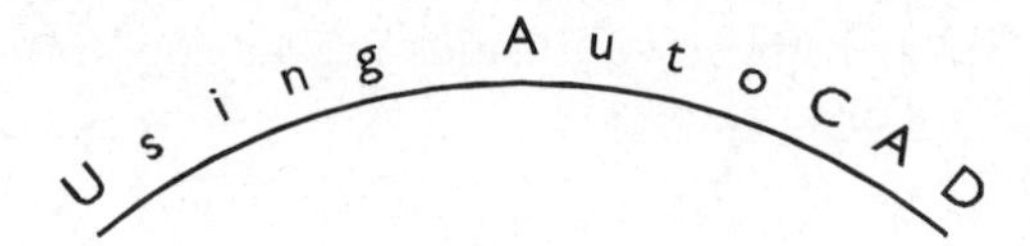

Text placed along an arc.

Another Express Tool command, **TXTEXP**, explodes TrueType text into polylines (**Express | Text | Explode Text**). You can then add thickness to create 3D "text." The text can no longer be edited with a text editor.

3D effect created by exploding TrueType text, then adding thickness.

BASIC TUTORIAL: DRAWING AND EDITING 3D POLYLINES

1. To draw polylines in 3D, start the **3DPOLY** command:
   - From the **Draw** menu, choose **3D Polyline**.
   - At the 'Command:' prompt, enter the **3dpoly** command.
   - Alternatively, enter the alias **3p** at the keyboard.

   Command: **3dpoly** *(Press* ENTER.*)*
2. In all cases, AutoCAD prompts you to specify a start point.

   Specify start point of polyline: *(Pick a point.)*

   Specify endpoint of line or [Undo]: *(Pick another point.)*

   You can pick points, or enter coordinate triplets, such as 1,2,3.

3. Once the 3D polyline is drawn, it can be edited with the **PEDIT** command, although it has limited functions.

   Command: **pedit**

   Select polyline or [Multiple]: *(Select the 3D polyline.)*

   Enter an option [Open/Edit vertex/Spline curve/Decurve/Undo]: *(Enter an option.)*

   You can also edit the 3D polyline with grips editing and AutoCAD's other editing commands.

EDITING WITH Z COORDINATES

Only two of AutoCAD's 2D editing commands can edit in 3D space: **COPY** and **MOVE**. This means you can enter coordinate triplets (x,y,z) when specifying the displacement. For example:

Command: **move**

Select objects: *(Select objects.)*

Select objects: *(Press ENTER.)*

Specify base point or displacement: **1,2,3**

Specify second point of displacement or <use first point as displacement>: **9,8,7**

There are four commands specific to editing in three dimensions: **3DARRAY**, **MIRROR3D**, **ROTATE3D**, and **ALIGN**, as described below. In addition, the **FILLET** and **CHAMFER** commands create 3D fillets and chamfers on solid models only, as described in Chapter 13.

Grips editing does not work in the z direction.

Planar Editing Commands

Most of AutoCAD's other 2D editing commands are *planar* meaning they edit only in the current x,y-plane. These commands include **MIRROR**, **OFFSET**, **ARRAY**, **ROTATE**, **SCALE**, **FILLET**, **CHAMFER**, **STRETCH**, **LENGTHEN**, **TRIM**, **EXTEND**, and **BREAK**.

Apparent Intersections

AutoCAD has one object snap mode designed for 3D: APParent intersection snaps to the visual intersection of two objects. They do not have to physically cross in 3D space, but appear to cross from the current viewpoint.

The related object snap, EXTended apparent intersection, snaps to where the objects would intersect visually if they were extended.

These object snaps work with two lines, mlines, arcs, circles, ellipses, elliptical arcs, polylines, rays, splines, and xlines.

ALIGN

The **ALIGN** command moves, transforms, scales, and rotates objects in 3D.

This command is tricky to use, because its editing function depends on the number of points you enter.

- To **move** objects in 3D, enter one set of source and destination points. (The **MOVE** command performs the same function.)
- To **rotate** and **scale** objects in 3D, enter two sets of source and destination points.
- To **transform** (move and rotate) objects in 3D, enter three sets of source and destination points.

BASIC TUTORIAL: ALIGNING 3D OBJECTS

1. To move, rotate, scale, and align objects in 3D space, start the **ALIGN** command:
   - From the **Modify** menu, choose **3D Operation**, and then **Align**.
   - At the 'Command:' prompt, enter the **align** command.
   - Alternatively, enter the alias **al** at the keyboard.

   Command: **align** *(Press ENTER.)*

2. In all cases, AutoCAD prompts you to select objects.

   Select objects: *(Select one or more objects.)*

   Select objects: *(Press* **ENTER** *to end object selection.)*

Moving Objects

3. Pick a point that identifies the source:

   Specify first source point: *(Pick point 1.)*

4. Pick a destination point, the point where the source point should end up:

   Specify first destination point: *(Pick point 2.)*

5. When you press **ENTER** at the following prompt, AutoCAD moves the selected objects.

   Specify second source point: *(Press* **ENTER***.)*

   ... and the command ends.

Rotating Objects

6. If, however, you want to rotate or scale the object, pick more source and destination points:

   Specify second source point: *(Pick point 3.)*

   Specify second destination point: *(Pick point 4.)*

   Notice that AutoCAD draws a guideline so that you can "see" where the object will end up.

7. Press **ENTER** at the following prompt:

   Specify third source point or <continue>: *(Press* **ENTER***.)*

8. If you wish only to rotate objects, enter "n":

   Scale objects based on alignment points? [Yes/No] <N>: **n**

   AutoCAD does not ask you for a rotation angle; instead, it determines the angle between the second source and destination points (points 3 and 4). AutoCAD rotates the object, and the command ends.

Scaling Objects

9. If, however, you want also to scale the object, answer "y":

   Specify third source point or <continue>: **y**

   Specify third destination point: *(Pick point 5.)*

   AutoCAD does not ask you for a scale factor; instead, it determines the scale from the distance between the first and second destination points (points 2 and 4). AutoCAD rotates and scales the object, and the command ends.

Aligning Objects

10. If, however, you want to transform (align) the object, pick one more destination point:

    Specify third source point or <continue>: *(Pick point 6.)*

    Specify third destination point: *(Pick point 7.)*

    AutoCAD uses your pick points to determine how to align the object:

    - Move from source point 1 to destination point 2.
    - Rotate one angle between source points 1, 3 and destination points 2, 4.
    - Rotate a second angle between source points 3, 5 and destination points 4, 6.

3DARRAY

The **3DARRAY** command copies objects in all three directions for rectangular arrays, and at an angle for polar arrays.

This command is similar to the **-ARRAY** command, but has an additional prompt for the z direction, which AutoCAD calls "levels" (for rectangular arrays) and the rotation angle (for polar arrays).

BASIC TUTORIAL: ARRAYING IN 3D

1. To array objects in 3D, start the **3DARRAY** command:
   - From the **Modify** menu, choose **3D Operation**, and then **3D Array**.
   - At the 'Command:' prompt, enter the **3darray** command.
   - Alternatively, enter the alias **3a** at the keyboard.

   Command: **3darray** *(Press* **ENTER***.)*
2. In all cases, AutoCAD prompts you to select objects.

   Select objects: *(Select one or more objects.)*

   Select objects: *(Press* **ENTER** *to end object selection.)*
3. Decide on a polar or rectangular array:

   Enter the type of array [Rectangular/Polar] <R>: **r**

Rectangular Array

4. Specify the number of rows (x direction) and columns (y direction):

   Enter the number of rows (---) <1>: *(Enter a number, such as* **3***.)*

   Enter the number of columns (| | |) <1>: *(Enter a number, such as* **4***.)*
5. And specify the number of levels (z direction):

   Enter the number of levels (...) <1>: *(Enter a number, such as* **5***.)*
6. Specify the distance between rows, columns, and levels. Enter a negative distance to copy the objects in the negative direction.

   Specify the distance between rows (---): *(Enter a distance, such as* **2.5***.)*

   Specify the distance between columns (| | |): *(Enter a distance, such as* **3.6***.)*

   Specify the distance between levels (...): *(Enter a distance, such as* **4.7***.)*

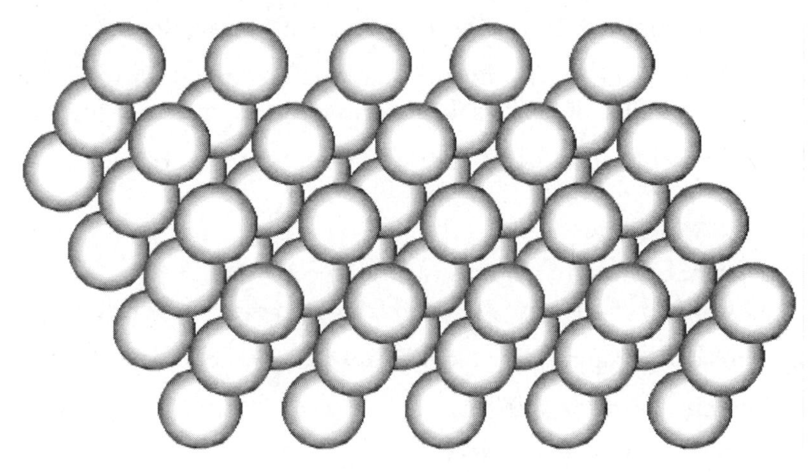

Spheres arrayed in 3D space.

Polar Array

Repeat steps 1 and 2 from above.

3. To draw a polar array, enter "p":

   Enter the type of array [Rectangular/Polar] <R>: **p**

4. Specify the parameters of the array:

   Enter the number of items in the array: *(Enter a number, such as* **9***.)*

   Specify the angle to fill (+=ccw, -=cw) <360>: *(Enter an angle, such as* **360***.)*

5. And, do you want the entire polar array rotated in 3D space?

   Rotate arrayed objects? [Yes/No] <Y>: **y**

   Specify center point of array: *(Pick a point.)*

   Specify second point on axis of rotation: *(Pick a second point.)*

The second point can have a different z value from the center point, tilting the array.

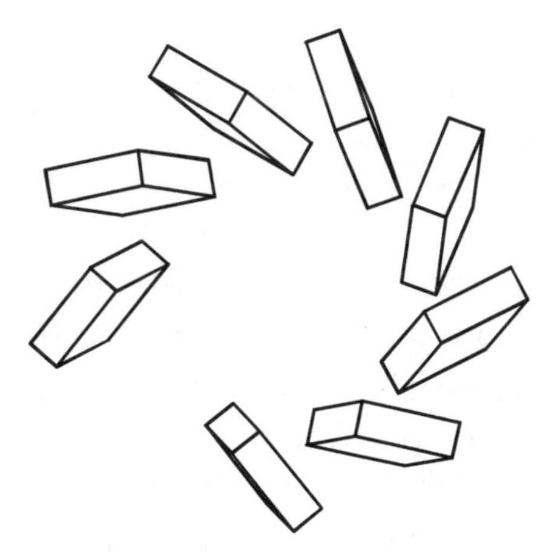

Polar array of boxes in 3D space.

MIRROR3D

The **MIRROR3D** command makes mirror copies of objects at any angle in 3D space.

Instead of mirroring 2D objects about a mirror line, this command mirrors objects about a mirror plane. There are a large number of options for determining the mirror plane.

The **MIRRTEXT** system variable determines whether or not text is mirrored.

BASIC TUTORIAL: MIRRORING IN 3D

1. To mirror objects at any angle, start the **MIRROR3D** command:
   - From the **Modify** menu, choose **3D Operation**, and then **Mirror 3D**.
   - At the 'Command:' prompt, enter the **mirror3d** command.

   Command: **mirror3d** *(Press* ENTER*.)*

2. AutoCAD prompts you to select objects.

   Select objects: *(Select one or more objects.)*

   Select objects: *(Press* ENTER *to end object selection.)*

3. Select an option for specifying the mirror plane:

   Specify first point of mirror plane (3 points) or

   [Object/Last/Zaxis/View/XY/YZ/ZX/3points] <3points>: *(Enter an option.)*

4. Decided whether to keep or erase the selected objects:

   Delete source objects? [Yes/No] <N>: *(Type* **Y** *or* **N***.)*

MIRRORING IN 3D: ADDITIONAL METHODS

The MIRROR3D command has a number of options that determine the angle of the mirror plane.

Object

The **Object** option takes the plane of a selected object as the mirror plane.

Select a circle, arc or 2D-polyline segment: *(Select an object.)*

Last

The **Last** option mirrors the selected objects about the previous mirror plane.

Z Axis

The **ZAxis** option determines the mirror plane from two points: one defines a plane, and the other defines the normal to the plane. Because the normal is at right angles (90 degrees) to any plane, just two points are needed to define the mirror plane.

Specify point on mirror plane: *(Pick a point.)*

Specify point on Z-axis (normal) of mirror plane: *(Pick another point.)*

View

The **View** option aligns the mirror plane with the current view, and asks for a point to define the depth of the mirror plane.

Specify point on view plane <0,0,0>: *(Pick a point.)*

XY/YZ/ZX

The **XY**, **YZ**, and **ZX** options align the mirror plane with one of the three planes, and ask for a point to define the depth of the mirror plane.

Specify point on (XY, YZ, ZX) plane <0,0,0>: *(Pick a point.)*

3 Points

The **3points** option defines the mirror plane by three points.

Specify first point on mirror plane: *(Pick a point.)*

Specify second point on mirror plane: *(Pick a point.)*

Specify third point on mirror plane: *(Pick a point.)*

ROTATE3D

The ROTATE3D command rotates objects at any angle in 3D space.

Like the 2D version of this command, ROTATE3D command rotates objects about an axis; the difference is that the axis can be anywhere in 3D space. There are a large number of options for determining the axis.

BASIC TUTORIAL: ROTATING IN 3D

1. To rotate objects in 3D space, start the **ROTATE3D** command:
   - From the **Modify** menu, choose **3D Operation**, and then **Rotate 3D**.
   - At the 'Command:' prompt, enter the **rotate3d** command.

   Command: **rotate3d** *(Press ENTER.)*
2. AutoCAD prompts you to select objects.

   Select objects: *(Select one or more objects.)*

   Select objects: *(Press ENTER to end object selection.)*
3. Select an option for specifying the axis:

   Specify first point on axis or define axis by

   [Object/Last/View/Xaxis/Yaxis/Zaxis/2points]: *(Enter an option.)*
4. Specify the rotation angle:

   Specify rotation angle or [Reference]: *(Enter an angle.)*

ROTATING IN 3D: ADDITIONAL METHODS

The **ROTATE3D** command has a number of options that determine the axis of rotation.

Object

The **Object** option aligns the axis with a selected object:

Select a line, circle, arc or 2D-polyline segment: *(Pick an object.)*

The angle of the axis depends on the type of object.

Object	*Axis of Rotation Alignment*
Line	The line.
Circle	3D axis at center of circle and perpendicular to its plane.
Arc	3D axis at center of arc and perpendicular to its plane.
Polyline	Segment of 2D polyline.

Last

The **Last** option reuses the previously-defined axis.

View

The **View** option aligns the axis with the current view, and asks for a point on the axis.

Specify a point on the view direction axis <0,0,0>: *(Pick a point.)*

X Axis, Y Axis, Z Axis

These options aligns the axis with the x, y, or z axis passing through a selected point.

Specify a point on the (X, Y, or Z) axis <0,0,0>: *(Pick a point.)*

2 Points

The **2points** option defines the axis by two points.

Specify first point on axis: *(Pick a point.)*

Specify second point on axis: *(Pick another point.)*

CONVERTING 3D TO 2D

Sometimes you may want to flatten 3D drawings, to make them 2D. There are several ways to convert them to 2D. The method you employ depends on the type of 3D objects in the drawings.

For objects drawn with elevations, a free AutoLISP routine sets the z-coordinate to 0. *Flatten.lsp* works with objects created as 3D faces, arcs, attribute definitions, circles, dimensions, ellipses, hatches, inserted blocks, lines, lwpolylines, mtext, points, polylines, 2D solids, and text. (The routine was written by Mark Middlebrook, and is available on the CD-ROM; check www.markcad.com for updates.)

For objects drawn with solid models, AutoCAD includes the **SOLPROF** command for generating 2D profiles and sections of 3D drawings. See Chapter 13 for details.

WMF FLATTENING

AutoCAD has a universal flattener that works with all objects: (1) export 3D drawings in WMF format; and then (2) import the *.wmf* file back into AutoCAD. This method has two advantages: everything in the drawing is flattened to 2D, and you can preserve a specific 3D viewpoint in 2D. The drawback is that all objects in the drawing are converted to polylines; all other drawing information is lost, such as layers and attributes.

1. Establish the drawing's 3D viewpoint.
2. Start the **WMFOUT** command.
3. In the Create WMF File dialog box, click **OK**. (AutoCAD gives the *.wmf* file the same name as the drawing.)
4. AutoCAD prompts you to select objects:

   Select objects: **all**

   Select objects: *(Press* **ENTER** *to end object selection.)*

   AutoCAD creates the *.wmf* file.
5. To import the *.wmf* file, select the **Insert** menu, and then **Windows Metafile**.
   Or just use the **WMFIN** command.
6. From the Import WMF dialog box, choose the file, and then click **Open**.
7. AutoCAD asks you for the insertion point, and then asks questions identical to the **INSERT** command:

   Specify insertion point or

   [Scale/X/Y/Z/Rotate/PScale/PX/PY/PZ/PRotate]: *(Pick a point.)*

   Enter X scale factor, specify opposite corner, or [Corner/XYZ] <1>: *(Press* **ENTER***.)*

   Enter Y scale factor <use X scale factor>: *(Press* **ENTER***.)*

   Specify rotation angle <0>: *(Press* **ENTER***.)*

   AutoCAD places the *.wmf* file as a block called "WMF0" on the current layer. (The block is made up of polylines.)
8. Change the viewpoint to prove to yourself the 3D drawing is now 2D.
   You can edit the block like any other AutoCAD block.

Note: If you want the 2D image to always face the 3D viewpoint, use the following technique:

1. Press **CTRL+A** and then press **CTRL+C** to copy all objects to Clipboard.
2. From the **Edit** menu, select **Paste Special**.
3. In the Paste Special dialog box, select "Picture (Metafile)".

AutoCAD pastes the image as a metafile. To edit, right-click the image.

WIREFRAME TUTORIAL

In this tutorial, you draw a small shop building in 3D using lines and UCSs.

1. Start AutoCAD with a new drawing.
2. Set up the drawing with these parameters.

   Units: **Architectural**

   Limits: **0,0** and **70',50'** *(remember the* **ZOOM All** *command)*

   Snap: **1'**

   Grid: **10'**
3. Start by drawing the floor as a polyline. The dimensions are 50 feet in length and 30 feet in width. (Do not dimension the plan.) When entering coordinates, remember to type 50' and 30' (feet) and not just 50 and 30, which represents inches.

Drawing the floor as a polyline.

4. To view the drawing in 3D, change the viewpoint with **VPOINT -1,1,0.3**.

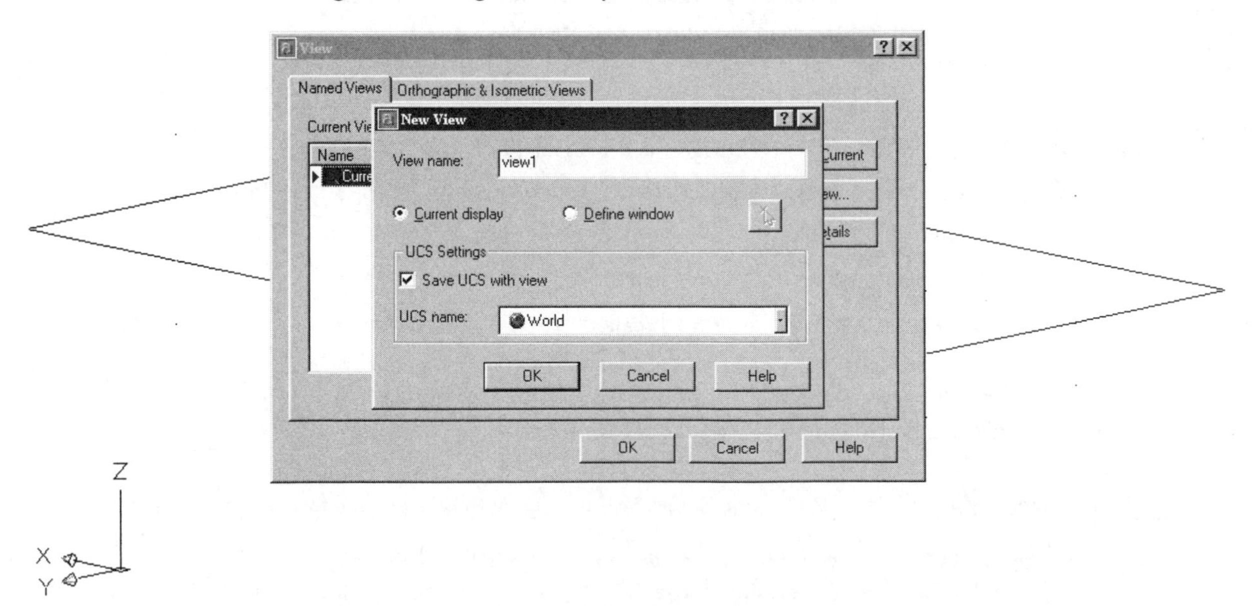

Saving the viewpoint.

5. Save this view for recall later: start the **VIEW** command, and then click **New**.

   View name: **view1**

   Click **OK** twice to save the view, and to exit the dialog boxes.
6. Check whether the UCS icon moves to each new origin as you set it. From **View** menu, select **Display | UCS Icon | Origin**. If **Origin** has a check mark, the option is turned on; leave it turned on.
7. Before drawing the walls, the coordinate system must be changed so that you can draw "vertically."

   From the **Tools** menu, select **New UCS | Object**. AutoCAD prompts:

   Select object to align UCS: *(Select the line at the nearest corner, going right.)*

   Notice that the UCS icon moves to the corner of the building; observe the orientation of the axes.

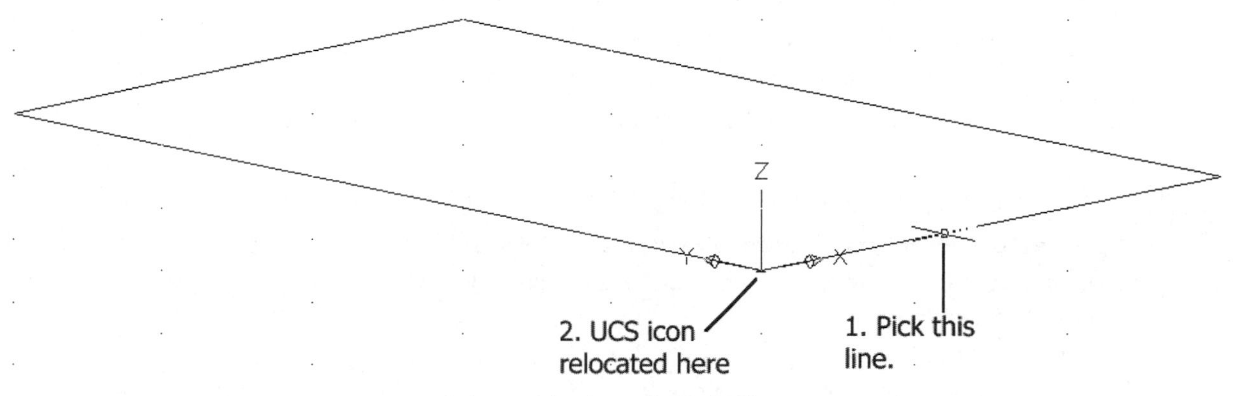

Relocating the drawing plan (UCS).

8. To draw the 12'-high walls, start with the first vertical corner line. Enter the **LINE** command, and then use point filters and object snaps, as follows:

   Command: **line**

   Specify first point: **.xy**

   of **int**

   of (need Z): **12'**

   Specify next point: **int**

   of *(Pick the same corner.)*

   Specify next point: *(Press ENTER.)*

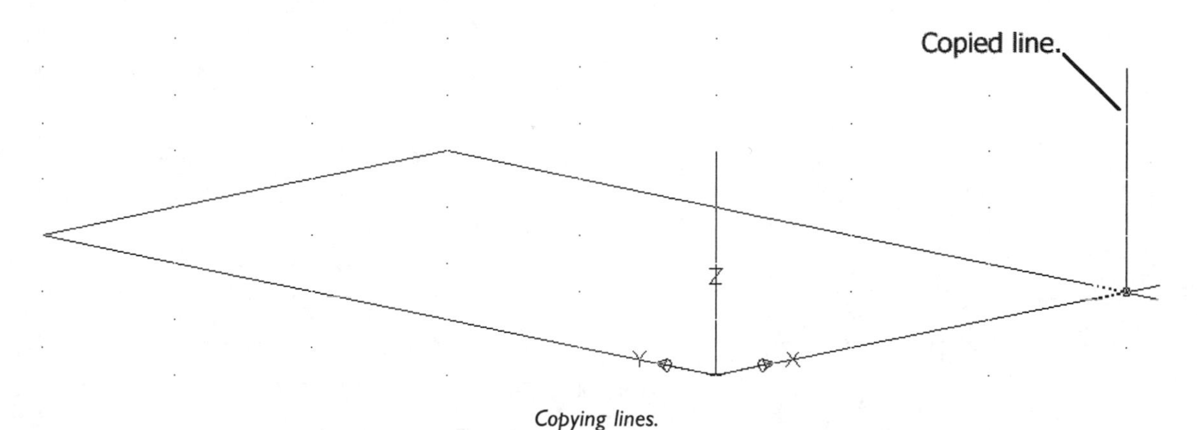

Copying lines.

9. To draw the other vertical line, use the **COPY** command with INTersection object snap. Copy the first vertical line to the other end wall corner.
10. Rotate the UCS so that you can draw directly on the end wall.

    Command: **ucs**

    Enter an option [New/.../World] <World>: **x**

    The **X** option rotates the UCS about the x axis. The right-hand rule determines that the UCS should be rotated by 90 degrees.

    Specify rotation angle about X axis <90>: **90**

    Notice that the UCS icon rotates.
11. To save this UCS for later recall, enter the **UCS** command with the **Save** option:

    Command: *(Press spacebar to repeat the command.)*

    Enter an option [New/.../Save/Del/Apply/?/World] <World>: **s**

    And then enter a descriptive name, like "end-wall":

    Enter name to save current UCS or [?]: **end-wall**
12. To construct the roof, switch the viewpoint to the plan view of the current UCS.

    Command: **plan**

    Enter an option [Current ucs/Ucs/World] <Current>: *(Press* **ENTER**.*)*

    Zoom to a comfortable working size, such as **ZOOM 0.9x**.
13. Construct the roof angles as illustrated by the figure below. Create the first line with the **LINE** command:

    Command: **line**

    From point: **end**

    of *(Pick the top end of the left vertical line.)*

    Specify next point: **@16'<15**

    Specify next point: *(Press* **ENTER**.*)*

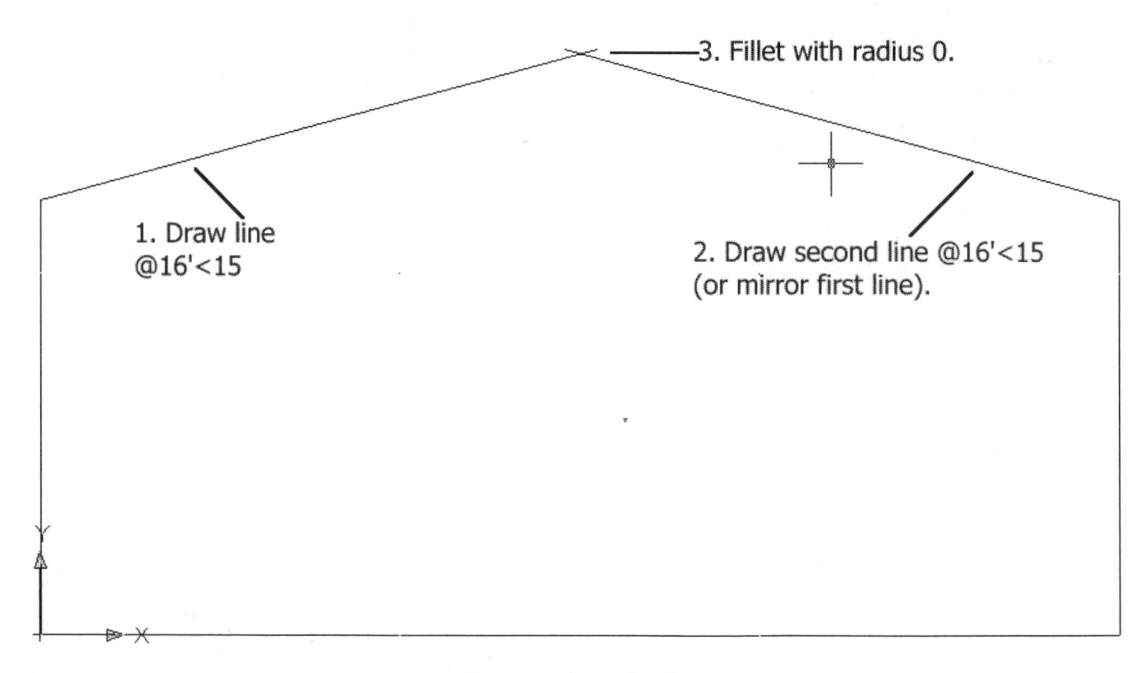

Trimming the roof peak.

14. To place the other line, use the **MIRROR** command. For the mirror point, use the MIDdle object snap.
15. Finish off the peak by filleting the two roof lines with a radius of 0. The **FILLET** command trims the ridge peak to a perfect intersection.
16. To copy the end-wall "panel" to the other end of the barn, restore the previously-stored view through the View dialog box's **Set Current** button: select "view1."

    Copy the four lines making up the end wall panel to the opposite end. Use the INTersection object snap for exact placement.

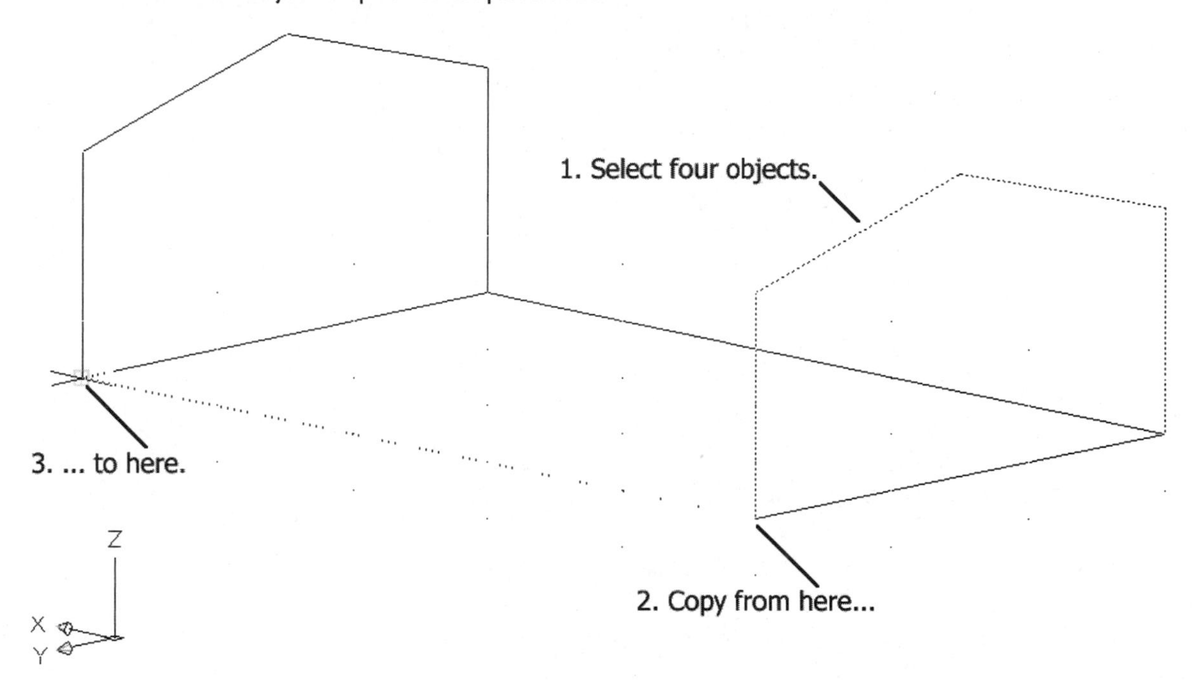

Copying the end wall.

Save the drawing with the **SAVEAS** command, naming the drawing "shop.dwg."

17. To draw the roof soffits, change the viewpoint to -2,-2,0.75.

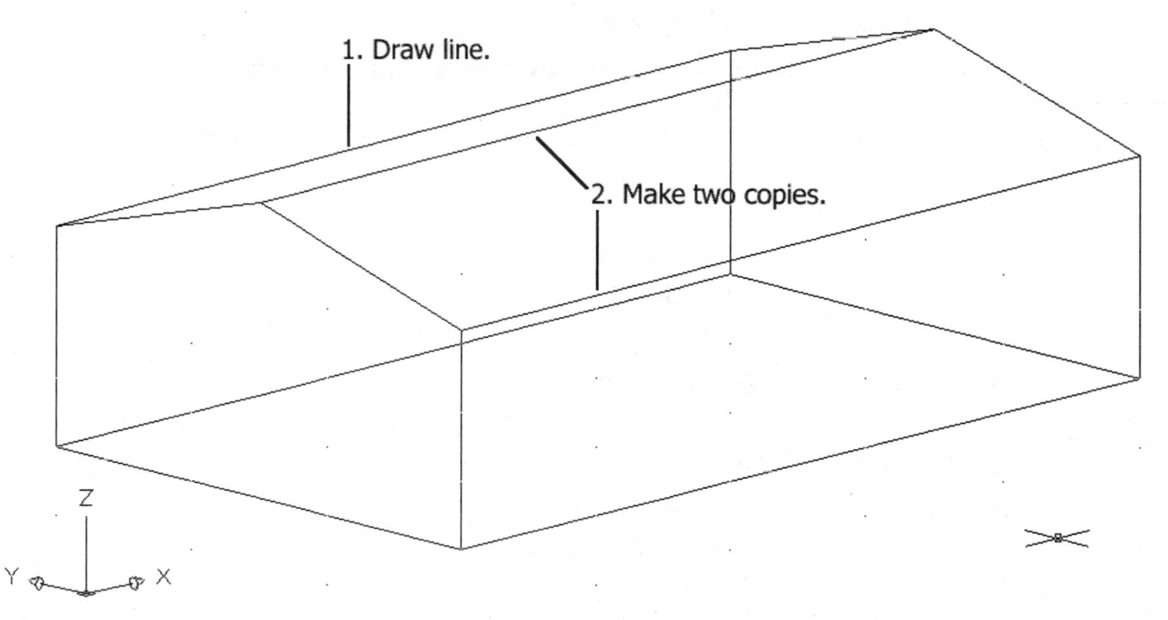

Drawing the soffits.

Use the **LINE** command with the INTersection object snap to draw one line. (See the figure above.) Place the other two lines with the **COPY** command's **Multiple** option.

18. To draw lines on the roof representing rafters, change the UCS:

    Command: **ucs**

    Enter an option [New/.../World] <World>: **3**

    Entering "3" is a shortcut option.

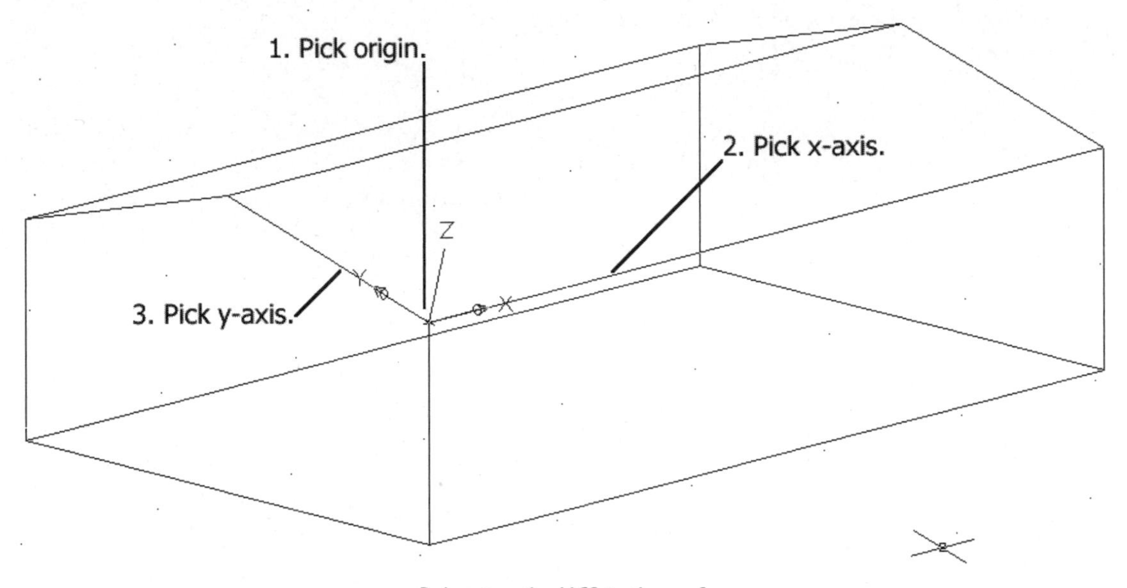

Relocating the UCS to the roof.

Specify new origin point <0,0,0>: **int**

of *(Corner of roof.)*

Specify point on positive portion of X-axis <5.3,2.8,0.0>: **nea**

of *(Pick a point along the soffit.)*

Specify point on positive-Y portion of the UCS XY plane <4.7,3.7,0.0>: **nea**

of *(Pick a point along the roof peak.)*

Notice that the UCS icon relocates to the edge of the roof and rotates to match the slope.

19. To complete the roof rafters, use the **ARRAY** command to array the edge lines across the roof.

    Command: **array**

    In the dialog box, enter the following options:

Array:	**Rectangular**
Rows:	**1**
Columns:	**25**
Row offset:	**0**
Column offset:	**24"**
Select objects:	*(Pick the two roof lines.)*

Click **Preview** to ensure the array turns out okay.

Choose **Accept**.

20. Before drawing "doors and windows" on the walls, use the **UCS** command to change the drafting plane to the front wall:

    Command: **ucs**

    Enter an option [New/.../World] <World>: **3**

    Specify new origin point <0,0,0>: **int**

    of *(Lower corner of wall.)*

    Specify point on positive portion of X-axis <5.3,2.8,0.0>: **nea**

    of *(Pick a point along the floor.)*

    Specify point on positive-Y portion of the UCS XY plane <4.7,3.7,0.0>: **nea**

    of *(Pick a point along the wall edge.)*

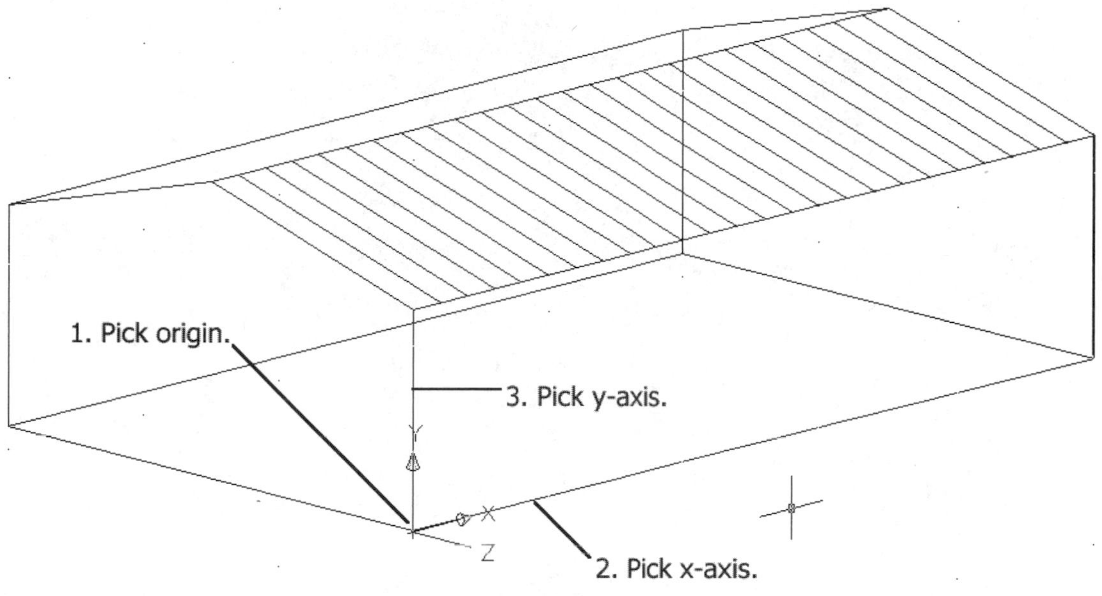
1. Pick origin.
3. Pick y-axis.
2. Pick x-axis.
Y
X
Z

Relocating the UCS to the front wall.

To see the view in plan, use the **PLAN** command.

21. To draw the doors and windows, use the **RECTANG** command. (It may be easier with osnap turned off.) Draw any shape of door and window you wish on the side of the barn.

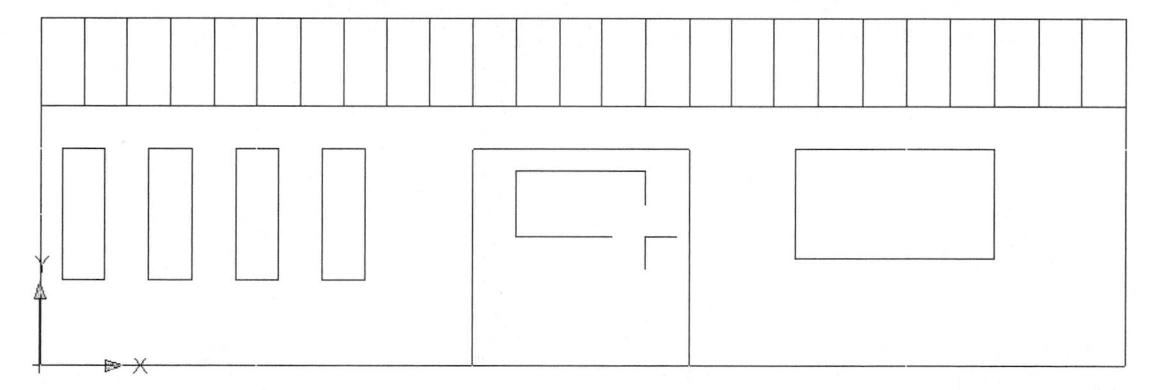
Y
X

Adding doors and windows.

22. The drawing is complete. Save your work with the **QSAVE** command.
    If you wish, use the **3DORBIT** command to view the drawing. Explore its options, such as **Perspective** and **Zoom**.

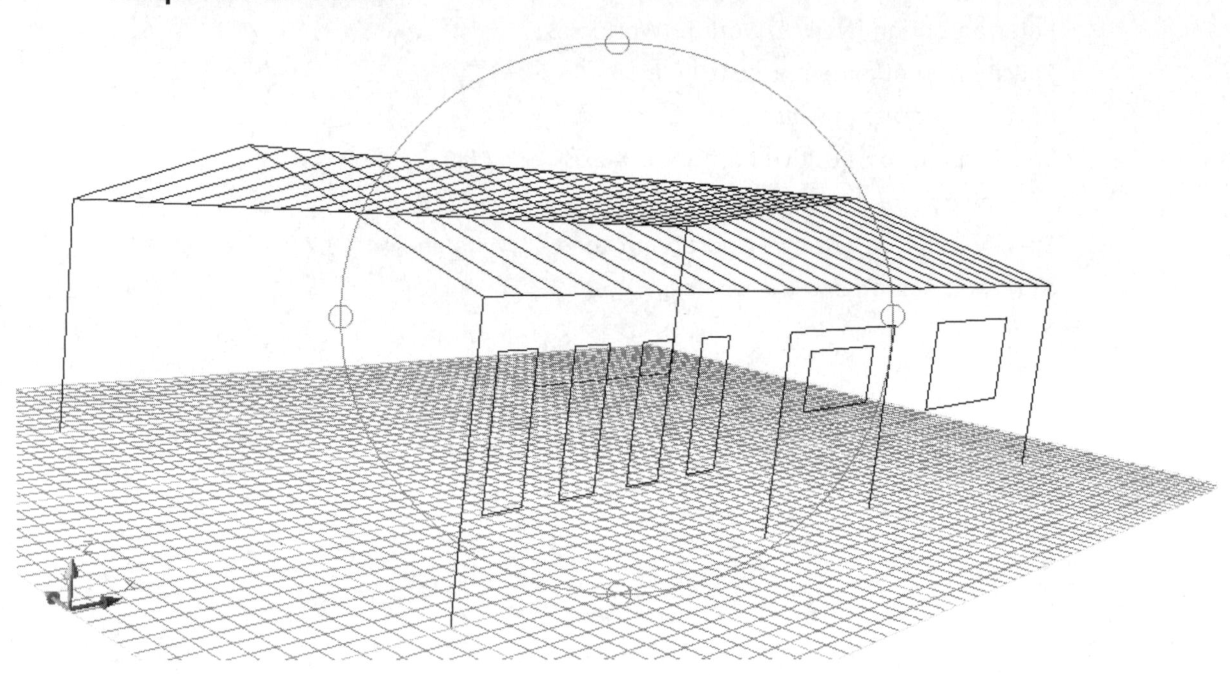

Viewing the barn in 3D.

EXERCISES

1. Draw the following objects (to any size) with the elevation and thickness indicated.

Object	*Elevation*	*Thickness*
Circle	0	10
Line	5	5
Point	-1	3
Donut	-5	8

2. Change the view to SW isometric, and then change the UCS to view.
   Repeat drawing exercise #1.
3. From the CD-ROM, open the *Ch09Gear.dwg* file, a drawing of a gear.
   Use the **CHANGE** or **PROPERTIES** commands to give the gear a thickness of 0.5 units.

Ch09Gear.dwg of a gear.

4. Given a chair with these dimensions:

4 Legs	1" wide x 1" deep x 15" high
1 Seat	15" wide x 15" deep x 3" thick
1 Back	15" wide x 1" deep x 15" high

   a. Calculate the elevation and thickness settings:

	Elevation	Thickness
Leg	0 "	________"
Seat	________"	________"
Back	________"	________"

   b. Draw the chair in 3D using simple rectangular forms, and the commands learned in this and previous chapters.
5. In a new drawing, use the **TEXT** command to place some text with an AutoCAD SHX font.
   Type your name or the name of your city. Give the text a thickness of 1 unit.
6. If Express Tools are installed with AutoCAD, place some text using a TrueType font. Use the **TXTEXP** command to explode the text, and then give the text a thickness of 1 unit.
   Change the viewpoint to confirm the text has an extrusion.
7. In a new drawing, draw an object, and then use the **3DARRAY** command to create:
   a. A 3 x 4 x 2 rectangular array.
   b. An 11-element polar array at 45 degrees.

REVIEW QUESTIONS

1. The **LIST** command shows the following coordinates for a circle:

   X = 12.45 Y = -23.65 Z = 10.00

   Which one is equivalent to the circle's elevation?
2. Which axis direction is an extrusion normally projected into?
3. Briefly explain:
   a. Elevation
   b. Thickness
4. How could you change the thickness of an existing object?
5. After rotating the viewpoint, how do you return to the plan view?
6. Describe the change made by thickness to the following objects. (For example, a point becomes a line.)
   a. Line
   b. Circle
   c. Donut
   d. Polygon
7. How do you apply thickness to objects drawn by the **RECTANGLE** command?
8. How do you change the elevation when the **Elevation** option does not appear in the Properties window?
9. Can you use the following commands to change the elevation of objects already in the drawing?
   a. **ELEV**
   b. **CHANGE**
   c. **CHPROP**
   d. **PROPERTIES**
   e. **MATCHPROP**
10. Can grips editing change the elevation and thickness of objects?
11. Can an arc be drawn with different z coordinates at each end?
12. Which of the following commands allow editing in 3D space:
    a. **MOVE**
    b. **OFFSET**
    c. **COPY**
    d. **ROTATE**
13. Which object snap allows you to snap to intersections that look as if they intersect from a 3D viewpoint?
14. Describe two methods of converting 3D drawings to 2D.

3D Surface Modeling

Surface modeling defines the surfaces of objects — car bodies, consumer products, and terrain maps. AutoCAD's commands for creating and editing 3D surfaces are not particularly powerful for defining the surfaces of products, but are sufficient for digital terrain mapping.

In this chapter, you learn about the following commands and techniques:

3D constructs basic 3D surface objects, such as spheres, cones, and dishes.

EDGESURF fits surfaces between four connected edges.

RULESURF stretches surfaces between two disconnected edges.

TABSURF raises surfaces from path curves and direction vectors.

REVSURF revolves surfaces from path curves and axes of rotation.

3DFACE draws 3- and 4-sided faces.

3DMESH and **PFACE** draw 3D meshes and polyface meshes.

PEDIT edits 3D meshes.

FINDING THE COMMANDS

On the SURFACES toolbar:

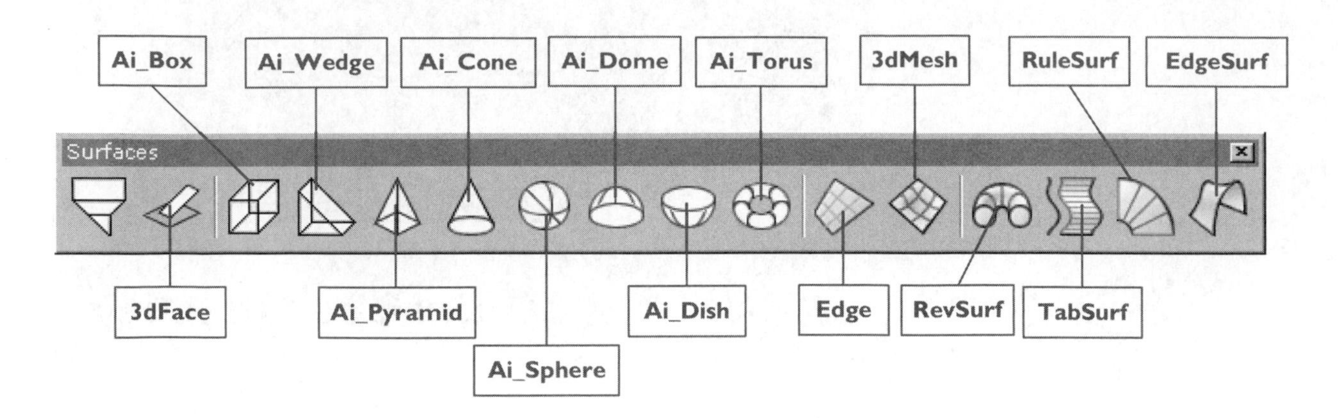

On the DRAW menu:

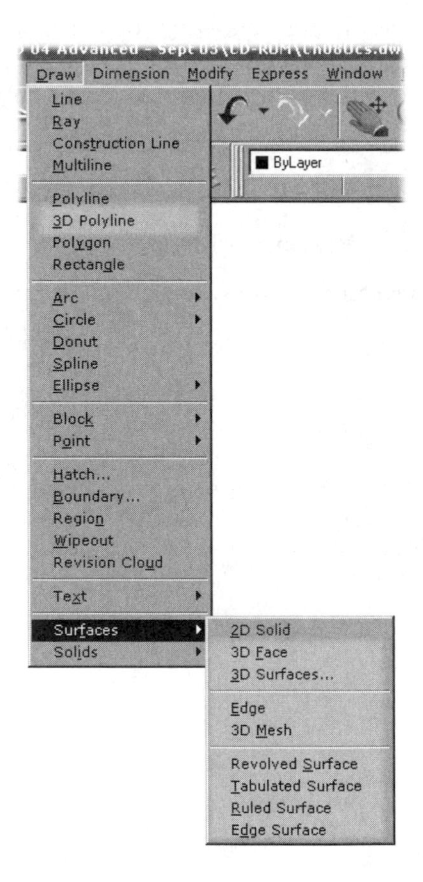

3D SURFACE OBJECTS

Surface models define objects by *edges* and the *surfaces* between edges. They are better than the 3D wireframe models described in the previous chapter, because they render correctly. Unlike solid models, surfaces have no "insides," because they have no thickness; surfaces simulate curves with many flat planes, which are either triangles or quadrilaterals. (AutoCAD can make edges invisible.)

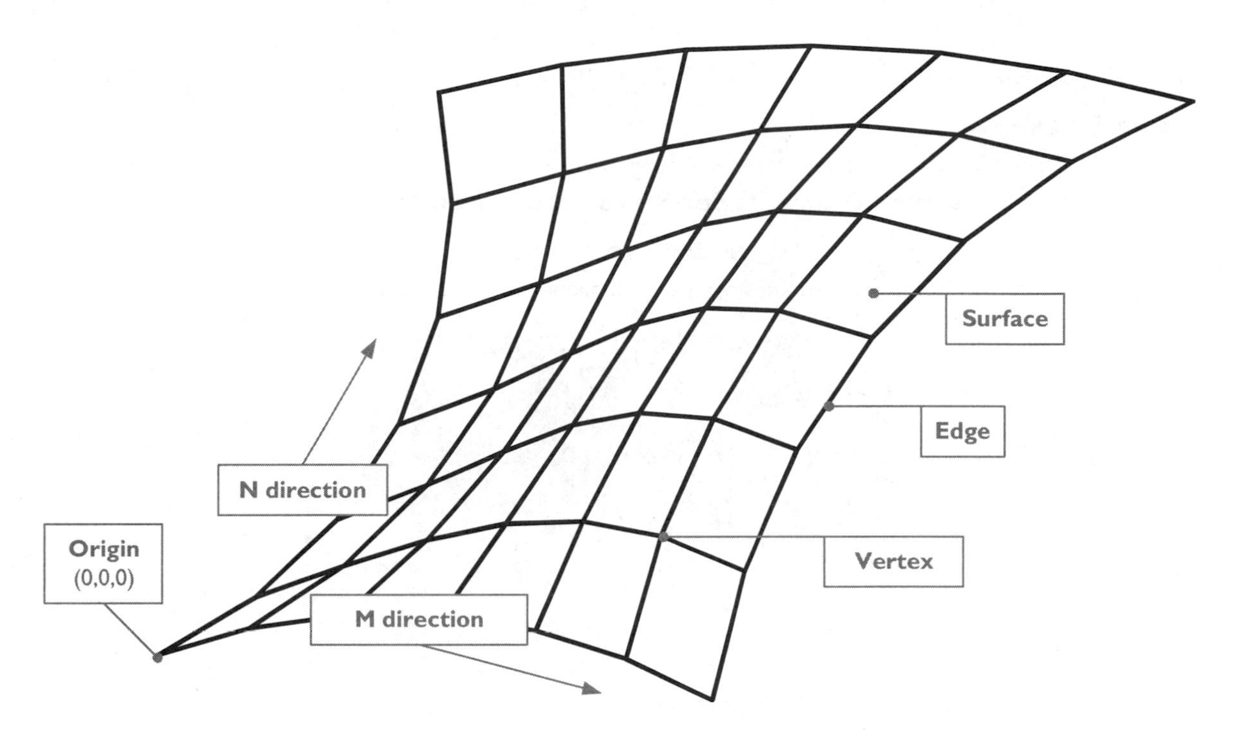

Curved surfaces are approximated by many flat quadrilaterals.

Surface models are edited with grips, or with the PEDIT command.

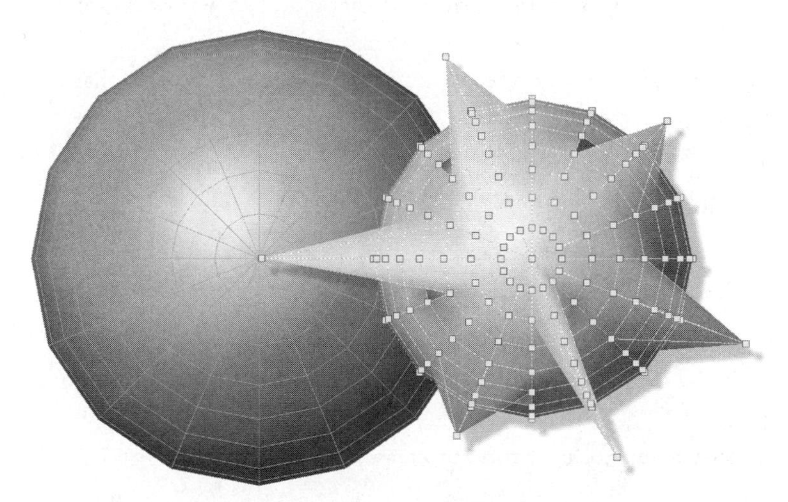

Editing a sphere surface model with its grips.

Use surface modeling when you are concerned about the *look* of your design; use solids when you need to *analyze* the design.

3D

The **3D** command draws basic surface objects, such as spheres, cones, and dishes.

This command is a "shell" that displays a dialog box illustrating the objects: you select a surface object, and then it launches the associated 3D modeling command There are two versions of this command: (1) the dialog box is displayed when you start the command from the menu bar, (2) prompts are displayed when you start the command at the 'Command:' prompt.

BASIC TUTORIAL: DRAWING 3D SURFACE OBJECTS

Dialog Box Version

1. To draw 3D surface objects, start the **3D** command:
   - From the **Draw** menu, choose **Surfaces**, and then **3D Surfaces**.
2. AutoCAD displays the 3D Objects dialog box.
   (Historically, this was called an "image tile," a predecessor to today's dialog boxes.)

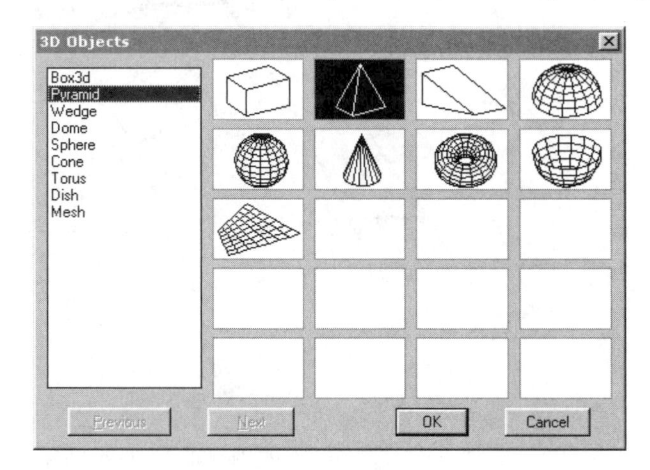

3D Objects dialog box.

3. Select a surface object by its name (Box3D, Pyramid, and so on) or by its icon.
4. Click **OK**.
   AutoCAD prompts you to draw the object, as described later in this chapter.

Command-line Version

1. As an alternative:
   - At the 'Command:' prompt, enter the **3d** command.

     Command: **3d** *(Press* ENTER.*)*
2. AutoCAD prompts you on the command-line:

   Enter an option

   [Box/Cone/DIsh/DOme/Mesh/Pyramid/Sphere/Torus/Wedge]: *(Enter an option.)*

   Enter an option.
   AutoCAD prompts you to draw the object, as described next.

DRAWING 3D SURFACE MODELS: ADDITIONAL METHODS

The **3D** command executes commands that construct basic 3D surface shapes. The commands are undocumented by Autodesk:

- **Ai_Box** command constructs rectangular and square boxes.
- **Ai_Cone** command constructs regular and truncated cones, and cylinders.

- **Ai_Dish** command constructs the top half of spheres.
- **Ai_Dome** command constructs the bottom half of spheres.
- **Ai_Pyramid** command constructs tetrahedrons, and regular and truncated pyramids with three- or four-sided bases.
- **Ai_Sphere** command constructs spheres.
- **Ai_Torus** command constructs donuts.
- **Ai_Wedge** command constructs wedges.

Ensure you use the "Ai_" prefix; leaving out the prefix creates a solid model, rather than a surface model. Let's look at each of these.

Ai_Box

The **AI_BOX** command constructs rectangular and cubic boxes:

Command: **ai_box**

Specify corner point of box: *(Pick a point, or enter x,y,z coordinates, such as* ***0,0,0****.)*

Specify length of box: *(Enter a length, such as* ***4****.)*

Specify width of box or [Cube]: *(Enter a width, such as* ***3****.)*

When you select the **Cube** option, a cube is constructed with all edges equal in length.

Specify height of box: *(Enter a height, such as* ***2****.)*

Specify rotation angle of box about the Z axis or [Reference]: *(Enter a rotation angle, such as* ***0****, or type* ***R****.)*

The **Reference** option allows you to align the box with other objects in the drawing. It operates identically to the **ROTATE** command's **Reference** option.

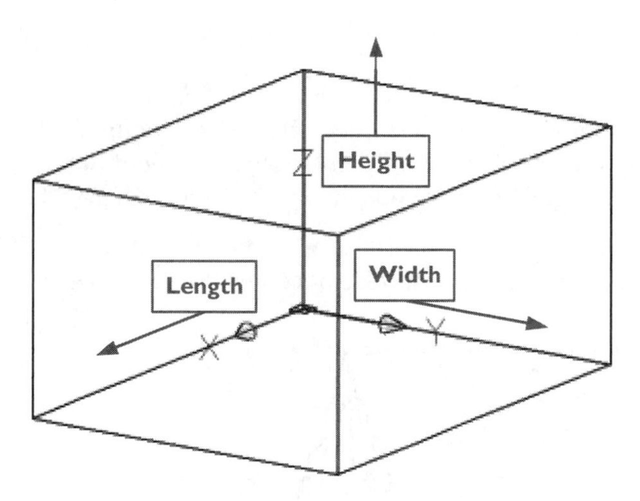

Surface model box.

Note: The *length* is in the **x** direction, *width* is in the **y** direction, and *height* is in the **z** direction.

Ai_Cone

The **AI_CONE** command constructs cones, truncated cones, and cylinders.

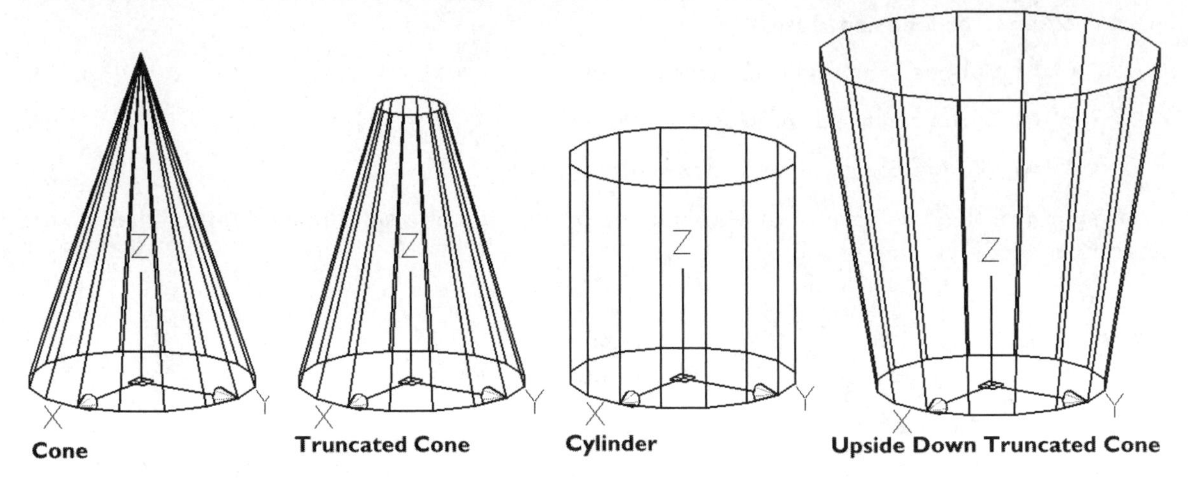

Cones and cylinders constructed by Ai_Cone command.

Command: **ai_cone**

Specify center point for base of cone: *(Pick a point, or enter x,y,z coordinates, such as* **2,2,2***.)*

Specify radius for base of cone or [Diameter]: *(Enter a radius, such as* **1***.)*

The radius and diameter options specify the top and bottom radius or diameter of the cone. (The diameter is double the radius.) For a cone with a pointy top, enter 0 for the "top of cone" radius; for a cylinder, make the top radius the same as the bottom radius:

Specify radius for top of cone or [Diameter] <0>: *(Enter another radius, such as* **0***.)*

The cone is drawn with its base parallel to the x,y plane.

Specify height of cone: *(Enter a height, such as* **4***.)*

The curved surfaces of cones (and other surface objects) are defined by *segments*, which act much like **VIEWRES** for 2D objects and **ISOLINES** for solid objects. The difference here is that the number of segments cannot be changed after the objects are constructed. The default value of 16 segments is usually good enough.

Enter number of segments for surface of cone <16>: *(Press* **ENTER***.)*

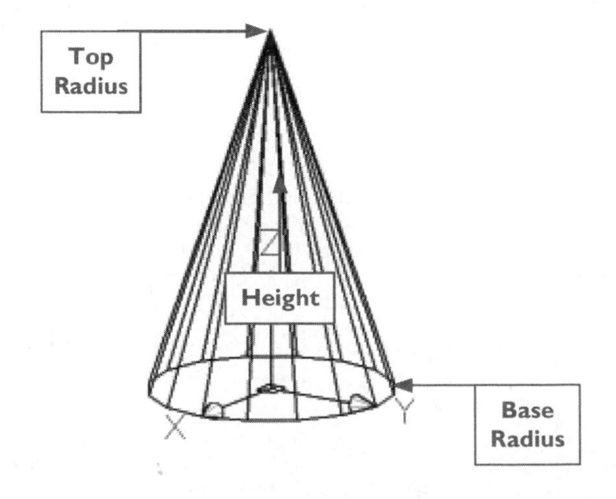

Surface model cone.

Ai_Dish

The AI_DISH command creates a *dish*, the bottom half of a sphere:

Command: **ai_dish**

Specify center point of dome: *(Pick a point, or enter x,y,z coordinates, such as* **1,3,-1**.*)*

Specify radius of dome or [Diameter]: *(Enter a radius, such as* **3**.*)*

Enter number of longitudinal segments for surface of dome <16>: *(Press* ENTER.*)*

Enter number of latitudinal segments for surface of dome <8>: *(Press* ENTER.*)*

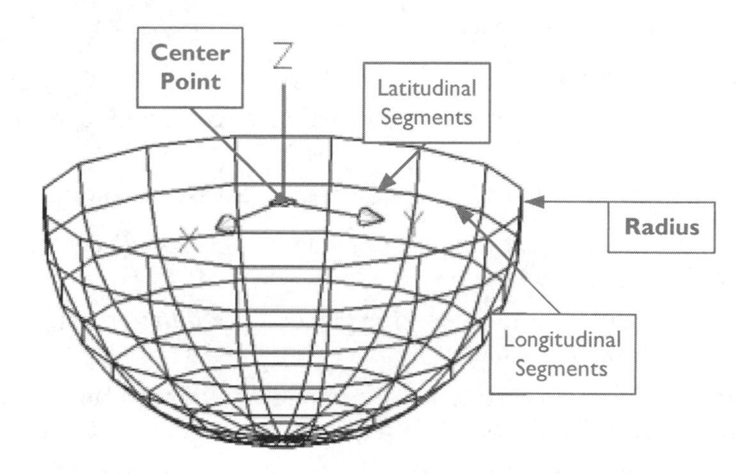

Surface model dish.

Ai_Dome

The AI_DOME command creates a *dome*, the top half of a sphere, and the other half of the dish.

Command: **ai_dome**

Specify center point of dome: *(Pick a point, or enter x,y,z coordinates, such as* **3,2,1**.*)*

Specify radius of dome or [Diameter]: *(Enter a radius, such as* **3**.*)*

Enter number of longitudinal segments for surface of dome <16>: *(Press* ENTER.*)*

Enter number of latitudinal segments for surface of dome <8>: *(Press* ENTER.*)*

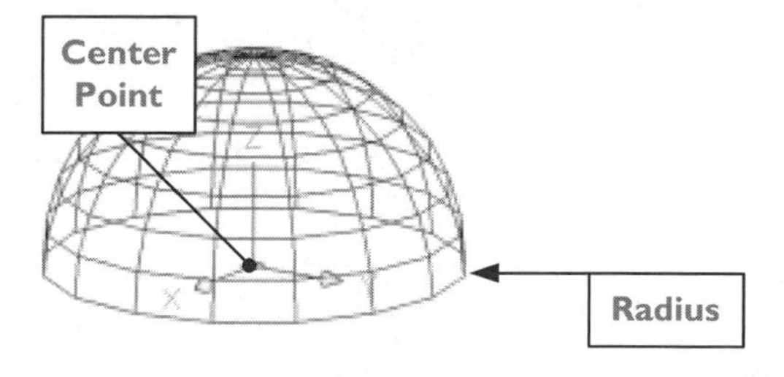

Surface model dome.

Ai_Pyramid

The **AI_PYRAMID** command constructs a variety of pyramidal shapes:

- Tetrahedrons with three-cornered bases.
- Pyramids with four-cornered bases.
- Tops with points, ridges, or truncations.

This command does not include a rotate option, as with **AI_BOX**, but you can specify differing z coordinates for the base, resulting in a tipped pyramid.

The command sequence for drawing "true" pyramids is:

Command: **ai_pyramid**

Specify first corner point for base of pyramid: *(Pick a point, or enter x,y,z coordinates, such as* **0,0,0***.)*

Enter the coordinates for the base in the correct order (clockwise or counterclockwise).

Specify second corner point for base of pyramid: *(Pick another point, or enter x,y,z coordinates, such as* **4,0,0***.)*

Specify third corner point for base of pyramid: *(Pick a third point, or enter x,y,z coordinates, such as* **4,4,0***.)*

Specify fourth corner point for base of pyramid or [Tetrahedron]: *(Pick a fourth point, or enter x,y,z coordinates, such as* **0,4,0***.)*

As you pick points for the base, AutoCAD highlights its "lines" in yellow. These are only indicators, not actual lines, so you cannot osnap to them.

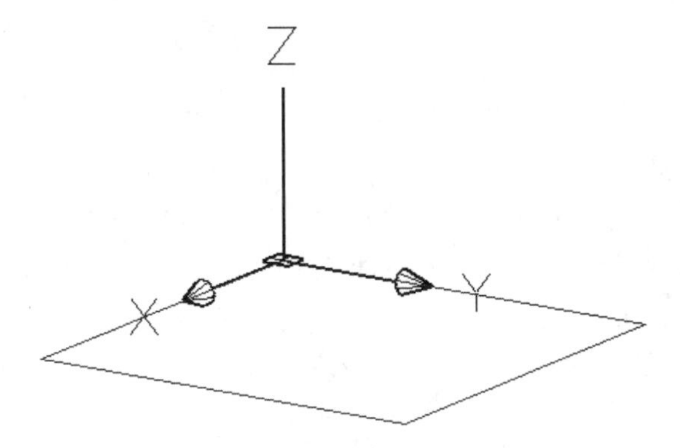

AutoCAD highlighting base.

If you "crossover" the corner points, AutoCAD draws a pyramid with a bow-tie base.

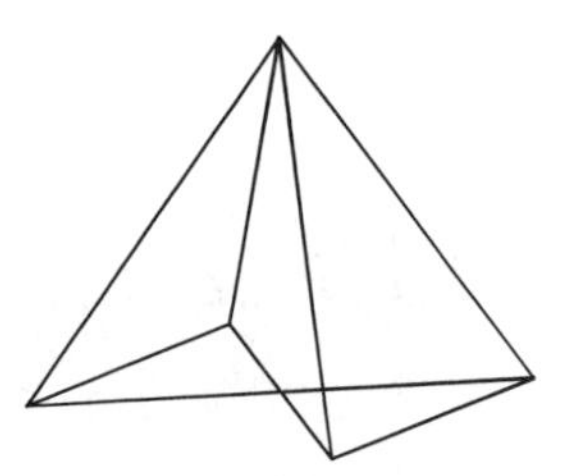

Bow-tie base resulting from picking corner points in incorrect order.

Leave out the z coordinate to create a flat pyramid.

Specify apex point of pyramid or [Ridge/Top]: *(Pick a point, or enter x,y,z coordinates, such as* **2,2,3***.)*

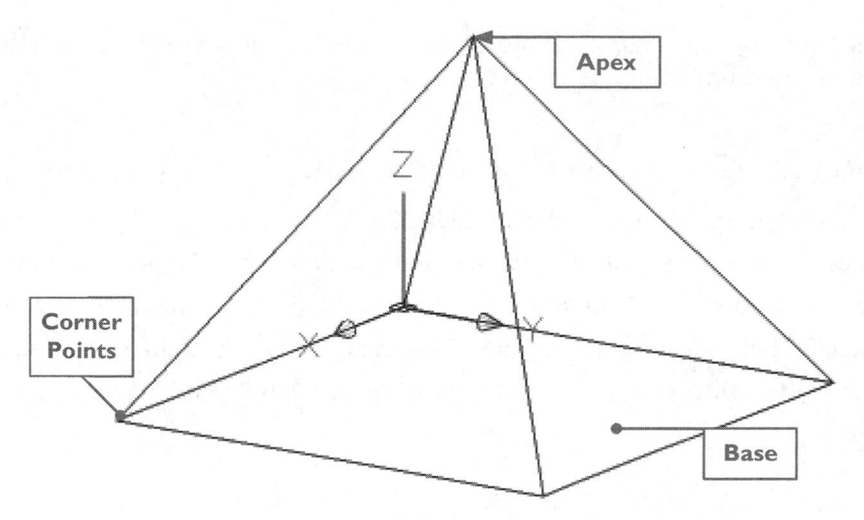

Surface model pyramid with four-corner base.

Tetrahedron

To draw the tetrahedron, a pyramid with a triangular base and four faces, enter "t" at the prompt:

Specify fourth corner point for base of pyramid or [Tetrahedron]: **t**

Specify apex point of tetrahedron or [Top]: *(Pick a point, or enter x,y,z coordinates.)*

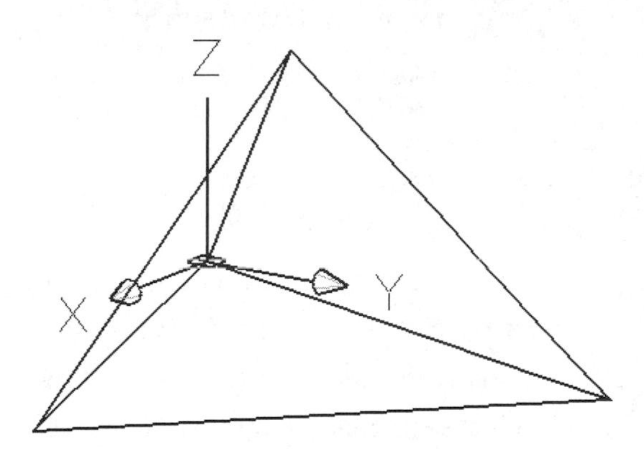

Surface model tetrahedron.

Ridge

To draw a pyramid with a ridge, like the roof of a house, enter "r" at the prompt:

Specify apex point of pyramid or [Ridge/Top]: **r**

Specify first ridge end point of pyramid: *(Enter coordinates, such as* **1.5,0.5,2**.*)*

Specify second ridge end point of pyramid: *(Enter coordinates, such as* **1.5,2.5,2**.*)*

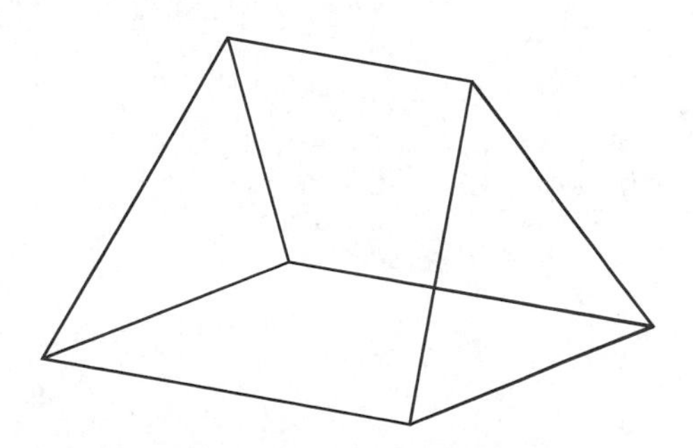

Surface model ridged pyramid.

For the **Ridge** and **Top** options, it is important to provide the coordinates in the correct order; otherwise the pyramid folds upon itself.

Top

To draw a truncated pyramid (one with a slice off the top), enter "t" at the prompt:

Specify apex point of pyramid or [Ridge/Top]: **t**

Specify first corner point for top of pyramid: *(Enter coordinates, such as* **0.5,0.5,2**.*)*

Specify second corner point for top of pyramid: *(Enter coordinates, such as* **0.5,2.5,2**.*)*

Specify third corner point for top of pyramid: *(Enter coordinates, such as* **2.5,2.5,2**.*)*

Specify fourth corner point for top of pyramid: *(Enter coordinates, such as* **2.5,0.5,2**.*)*

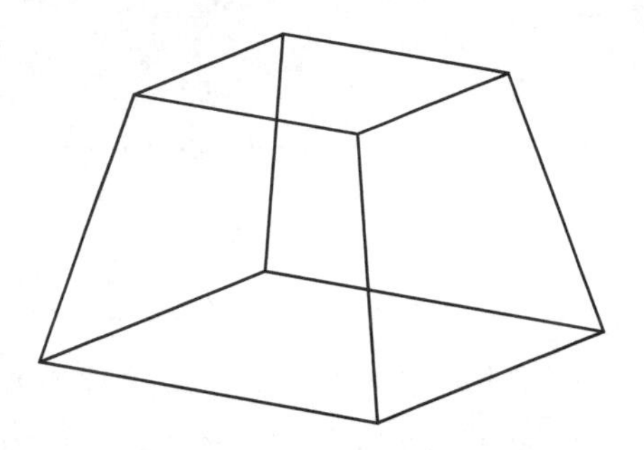

Surface model truncated pyramid.

Ai_Sphere

The **AI_SPHERE** command creates spheres:

Command: **ai_sphere**

Specify center point of sphere: *(Pick a point, or enter x,y,z coordinates, such as* **7,8,9**.*)*

Specify radius of sphere or [Diameter]: *(Enter a radius, such as* **4**.*)*

Enter number of longitudinal segments for surface of sphere <16>: *(Press* **ENTER**.*)*

Enter number of latitudinal segments for surface of sphere <16>: *(Press* **ENTER**.*)*

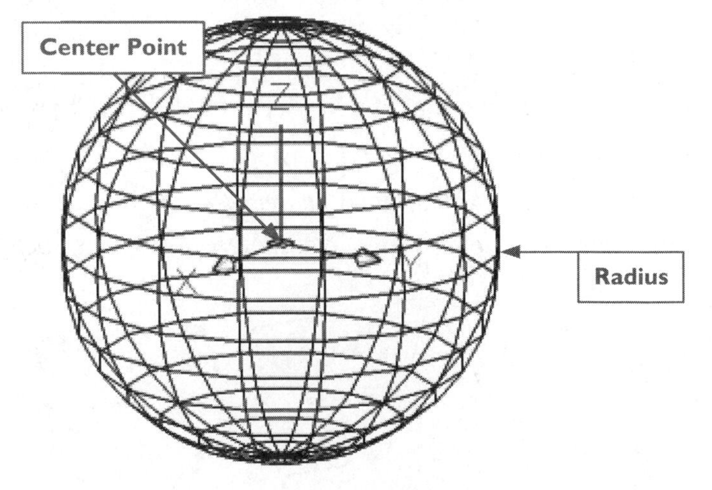

Surface model sphere.

Ai_Torus

The **AI_TORUS** command creates a torus (donut), which consists of a tube rotated about an axis:

Command: **ai_torus**

Specify center point of torus: *(Pick a point, or enter x,y,z coordinates, such as* **1,2,3**.*)*

Specify radius of torus or [Diameter]: *(Enter a radius, such as* **5**.*)*

When you enter a tube radius greater than the torus radius, AutoCAD complains, "Tube radius cannot exceed torus radius." Try again.

Specify radius of tube or [Diameter]: *(Enter another radius, such as* **1**.*)*

Enter number of segments around tube circumference <16>: *(Press* **ENTER**.*)*

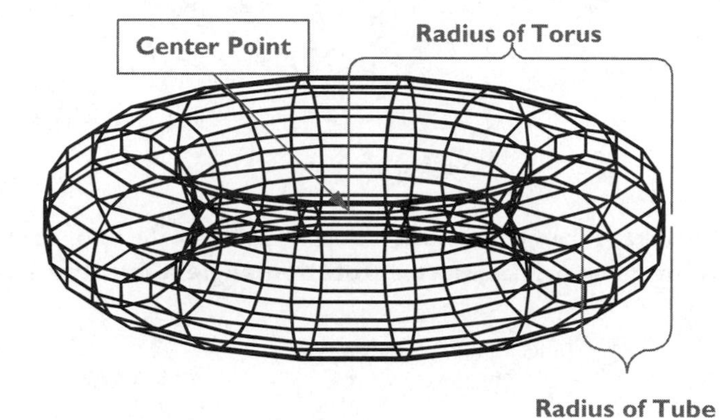

Surface model torus.

Ai_Wedge

The **AI_WEDGE** command creates wedges:

Command: **ai_wedge**

Specify corner point of wedge: *(Pick a point, or enter x,y,z coordinates, such as* **4,5,6**.*)*

Specify length of wedge: *(Enter a length, such as* **4**.*)*

Specify width of wedge: *(Enter a width, such as* **3**.*)*

Specify height of wedge: *(Enter a height, such as* **2**.*)*

Specify rotation angle of wedge about the Z axis: *(Enter rotation angle, such as* **0**, *or type* **R**.*)*

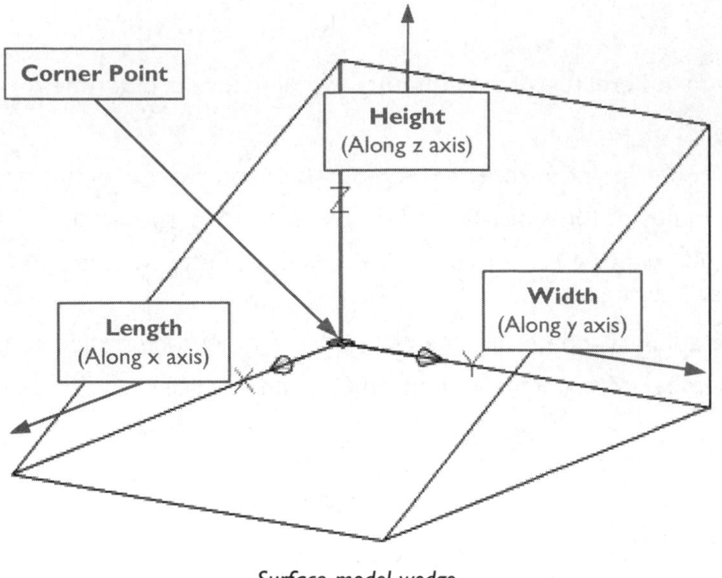

Surface model wedge

Note: AutoCAD draws the wedge so that it slopes down from the initial point along the X-axis. If you wish, you can rotate it into position with the **Rotation angle about the Z axis** option.

EDGESURF

The **EDGESURF** command (short for "edge surface") constructs surfaces between four *edges*. The edges are defined by open 2D and 3D objects: lines, arcs, splines, and open 2D or 3D polylines. The figure illustrates the Coons patch between two lines and two arcs.

Important: the four objects *must* connect at their endpoints, and there must be precisely four objects — no more, no less.

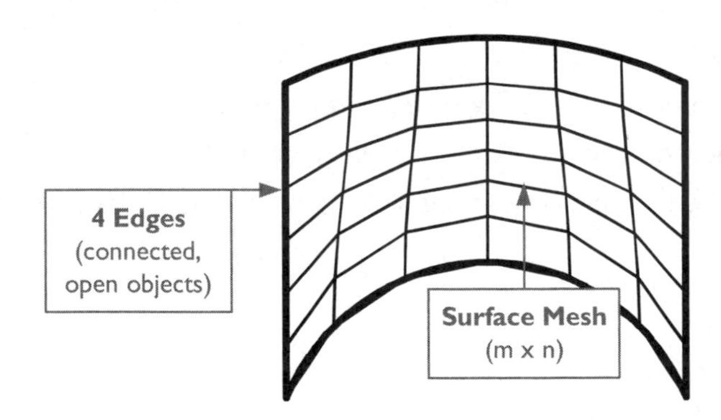

Coons patch constructed by Edgesurf command.

The surface follows the contour of each edge, forming meshes that approximate the surfaces attached to all points of each edge. (AutoCAD uses the Coons patch to construct the sphere and torus with the **AI_SPHERE** and **AI_TORUS** commands.)

(Technically, AutoCAD interpolates a bicubic surface between four general space curves. Historically, the patch was developed by Steven Coons, co-developer of the first CAD system. In 1963, he envisioned "the designer seated at the console [early term for keyboard] drawing a sketch of his proposed device on the screen of an oscilloscope tube [early form of monitor] with a light pen modifying his sketch into a perfect drawing." www.huc.ca/Pages/Portfolio/CAD_Paradox.pdf.)

BASIC TUTORIAL: SURFACING BETWEEN FOUR EDGES

Before using this command, you must first draw four connected edges.

1. To construct surfaces between four objects, start the **EDGESURF** command:
   - From the **Draw** menu, choose **Surfaces**, and then **Edge Surfaces**.
   - At the 'Command:' prompt, enter the **edgesurf** command.

   Command: **edgesurf** *(Press* ENTER.*)*
2. AutoCAD reports the current setting for the **SURFAB1** and **SURFAB2** system variables, and then prompts you to select the first surface edge:

   Current wire frame density: SURFTAB1=6 SURFTAB2=6

   Select object 1 for surface edge: *(Pick object.)*

   Your selection must consist of a single pick; as an alternative, you can also enter "g" and select a group, or "cl" and select a class. (*Classes* are named groups of similar objects found in add-ons to AutoCAD, such as Map and Land Desktop; classes are not found in AutoCAD.)
3. Pick the other surface curves:

   Select object 2 for surface edge: *(Pick object.)*

   Select object 3 for surface edge: *(Pick object.)*

   Select object 4 for surface edge: *(Pick object.)*

You may select the objects in any order. If one edge is not connected to the adjacent edge, AutoCAD complains "Edge *x* does not touch another edge." You can force edges to touch with the **FILLET** or **TRIM** and **EXTEND** commands; then restart the **EDGESURF** command.

DRAWING SURFACES: ADDITIONAL METHODS

The smoothness of surfaces depends on two system variables.

- **SURFTAB1** system variable sets tabulations, and mesh density in the m direction.
- **SURFTAB2** system variable sets mesh density in the n direction.

Coons patches (**EDGESURF**) and revolved surfaces (**REVSURF**) construct surfaces as *meshes*. Ruled surfaces (**RULESURF**) and tabulated surfaces (**TABSURF**) construct surfaces as *tabulations*, as illustrated below.

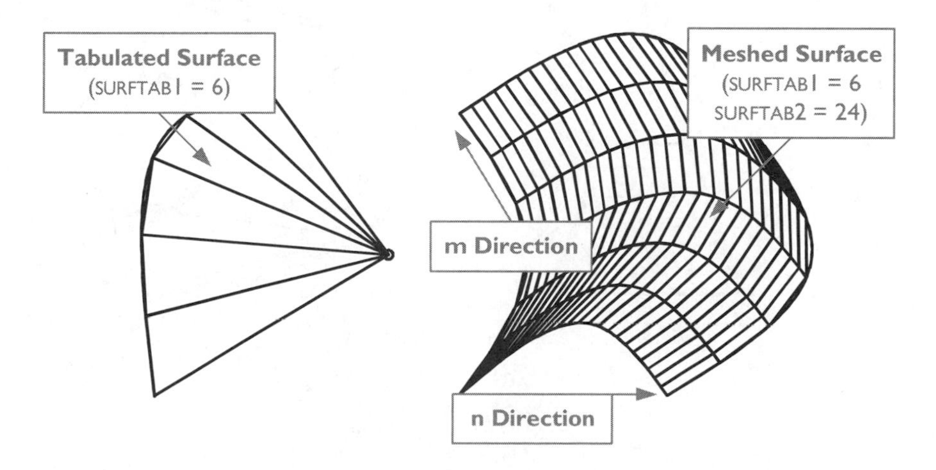

System variables control tabulations and mesh density.

The object selected first defines the *m* direction; the object connected to the first defines the *n* direction (not necessarily at right angles). The letter "m" and "n" are used instead of x and y, because the direction of the meshes doesn't necessarily correspond with the x and y axes.

The default values of **SURFTAB1** and **SURFTAB2** are 6, which is far too low. Change the value to at least 16, or higher; the range is between 2 and 32766. In general, the larger the surface, the bigger the values of **SURFTAB1** and **SURFTAB2** should be.

SurfTab1

The **SURFTAB1** system variable sets the number of tabulations constructed by the **RULESURF** and **TABSURF** commands, and the mesh density in the *m* direction constructed by the **EDGESURF** and **REVSURF** commands.

Command: **surftab1**

Enter new value for SURFTAB1 <6>: *(Enter a higher value, such as **24**.)*

SurfTab2

The **SURFTAB2** system variable sets the mesh density in the *n* direction by the **EDGESURF** and **REVSURF** commands.

Command: **surftab2**

Enter new value for SURFTAB2 <6>: *(Enter a higher value, such as **24**.)*

Warning! When you change the value of these system variables, the change is not retroactive. To increase the tabulations and mesh densities of surfaces already in the drawing, you must first erase them, and then construct the surfaces again, which is highly inefficient.

RULESURF

The **RULESURF** command (short for "ruled surfaces") creates meshes that span two objects — lines, points, arcs, circles, any objects drawn as a 2D polyline, and 3D polylines.

These objects are called "defining curves" — even though some eligible objects are lines and points. "Curve" is a general term in mathematics that includes straight lines or segments, and curves. AutoCAD uses ruled surfaces to construct cones with the **AI_CONE** command.

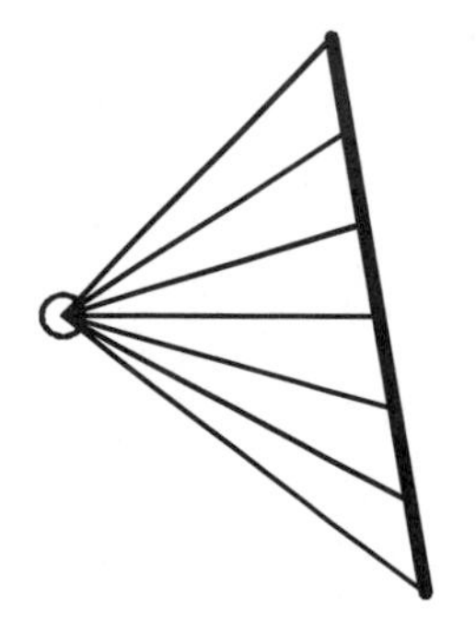

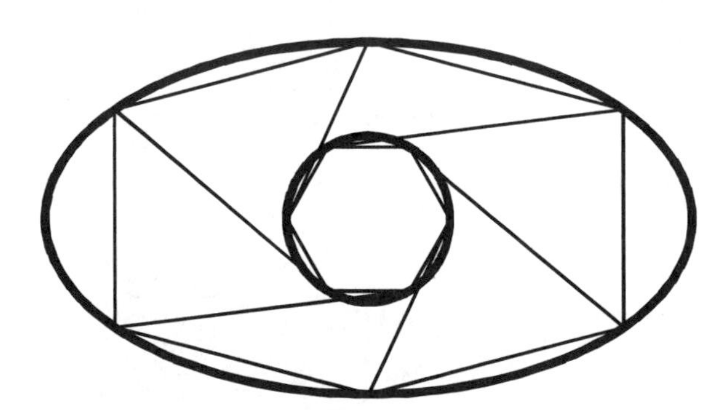

Ruled surfaces between:
***Left**: Point and line.*
***Middle**: Line and arc.*
***Right**: Circle and ellipse.*

TUTORIAL: SURFACING BETWEEN TWO CURVES

Before using this command, you must first draw the two curves.

1. To construct ruled surfaces between two objects, start the **RULESURF** command:
   - From the **Draw** menu, choose **Surfaces**, and then **Ruled Surfaces**.
   - At the 'Command:' prompt, enter the **rulesurf** command.

     Command: **rulesurf** *(Press* ENTER.*)*
2. AutoCAD reports the current setting for the **SURFAB1** system variable, and then prompts you to select the first defining curve:

   Current wire frame density: SURFTAB1=6

   Select first defining curve: *(Pick object.)*

   Your selection must consist of a single pick; as an alternative, you can enter "g" (and then select a group) or "cl" (and then select a class).
3. Pick the second defining curve:

   Select second defining curve: *(Pick object.)*

   AutoCAD constructs the ruled surface. The objects defining the curve remain in place.

Defining Starting Points

AutoCAD uses these rules to determine the starting point of the ruled surface:

For Closed Objects

Circles begin at 0 degrees. This angle is affected by the direction of the x axis and the value of SNAPANG.

Closed polylines begin at the last vertex, and then go backward to the first vertex.

In effect, ruled surfaces are drawn on circles and closed polylines in opposite directions, creating the iris effect illustrated on the previous page. It's better to use round polylines instead of circles.

For Open Objects

AutoCAD draws the ruled surface from the endpoint nearest to the pick point on the object. Selecting opposite ends creates ruled surfaces that cross.

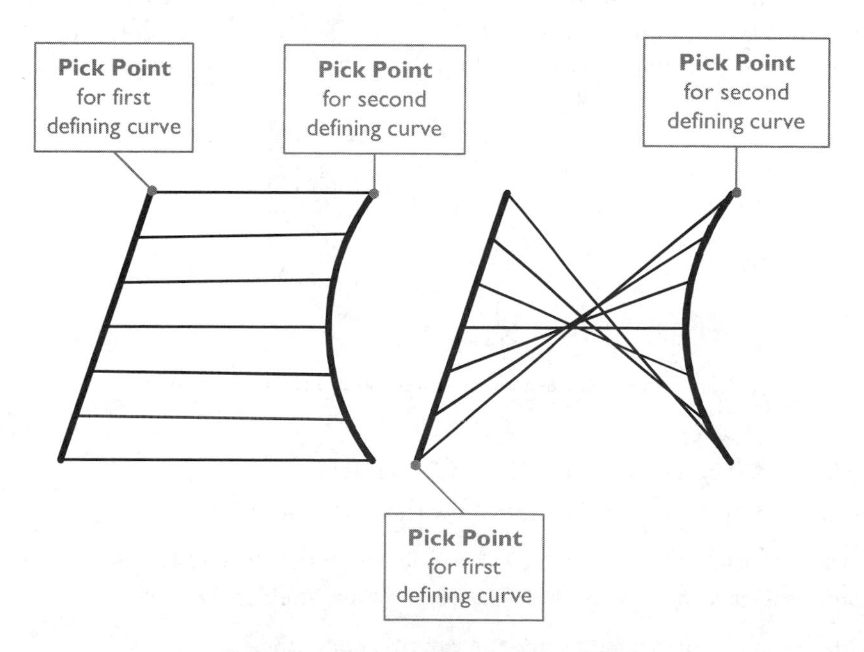

Pick points affect outcome of ruled surfaces.

TABSURF

The **TABSURF** command constructs mesh surfaces defined by paths and direction vectors. The effect is similar to extrusions described in the previous chapter, except that the surface is meshed.

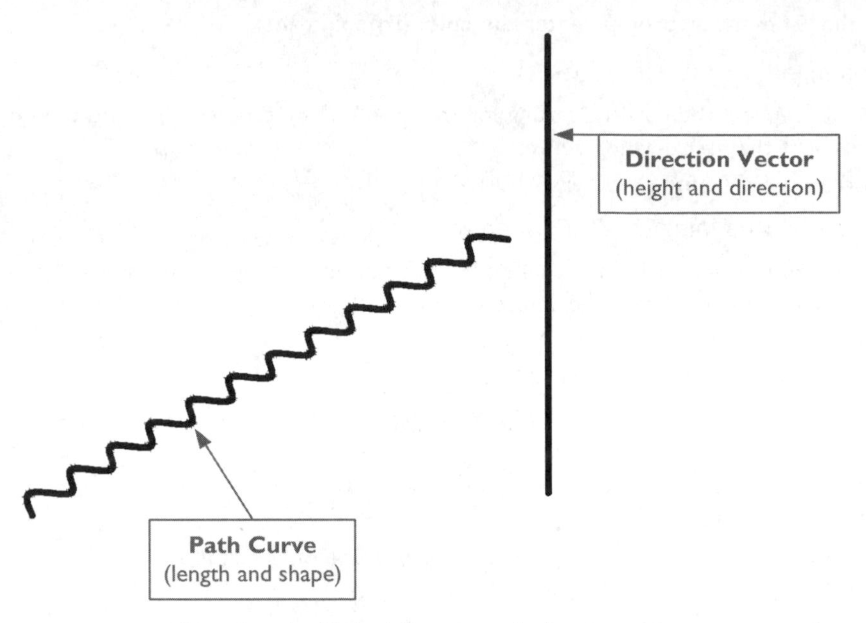

Tabsurf needs objects defining its path, direction, and size.

Constructing a tabulated surface takes these steps: (1) draw the *path curve* from which the surface is calculated; (2) draw the *direction vector* that defines the direction and length of the mesh; (3) start the **TABSURF** command, and then pick the path curve and direction vector.

Path curves can be lines, arcs, circles, ellipses, splines, and 2D or 3D polylines — open or closed. Direction vectors are lines and open 2D or 3D polylines. On polylines, AutoCAD uses only the first and last points to determine the direction vector; intermediate vertices, curves, and splines are ignored.

The number of tabulations on the surface is controlled by the **SURFTAB1** system variable. AutoCAD ignores this system variable along polyline segments: tabulation lines are drawn at the ends of straight segments. (Polyline arc segments are divided by the number of tabulations specified by **SURFTAB1**.)

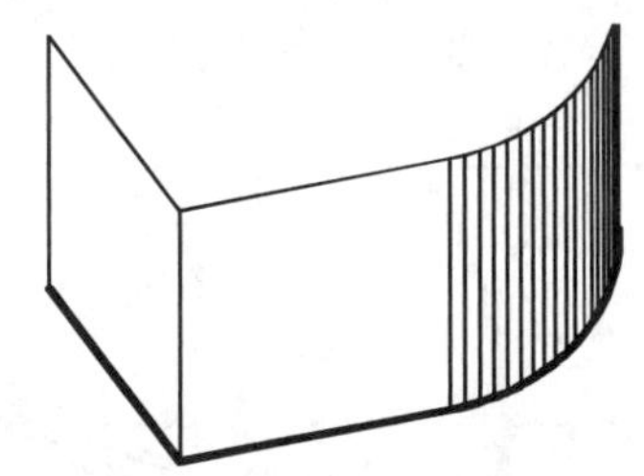

Single tabulation placed at vertices of polyline segments.

TUTORIAL: SURFACING BY PATH AND DIRECTION

Before using this command, you must first draw the path and direction objects.

1. To construct surfaces from paths and direction vectors, start the **TABSURF** command:
   - From the **Draw** menu, choose **Surfaces**, and then **Tabbed Surfaces**.
   - At the 'Command:' prompt, enter the **tabsurf** command.

Command: **tabsurf** *(Press* ENTER.*)*

2. AutoCAD reports the current setting for the **SURFAB1** system variable, and then prompts you to select the path curve:

   Current wire frame density: SURFTAB1=6

   Select object for path curve: *(Pick object.)*

   Your selection must be either a single pick or the name of a group.
3. Pick the second defining curve:

   Select object for direction vector: *(Pick either end of object.)*

AutoCAD constructs the tabulated surface. The objects defining the surface remain in place.

The point you pick on the direction vector determines the direction that the mesh is projected. Select a point closer to one end of the direction vector, and the tabulation is projected in that direction.

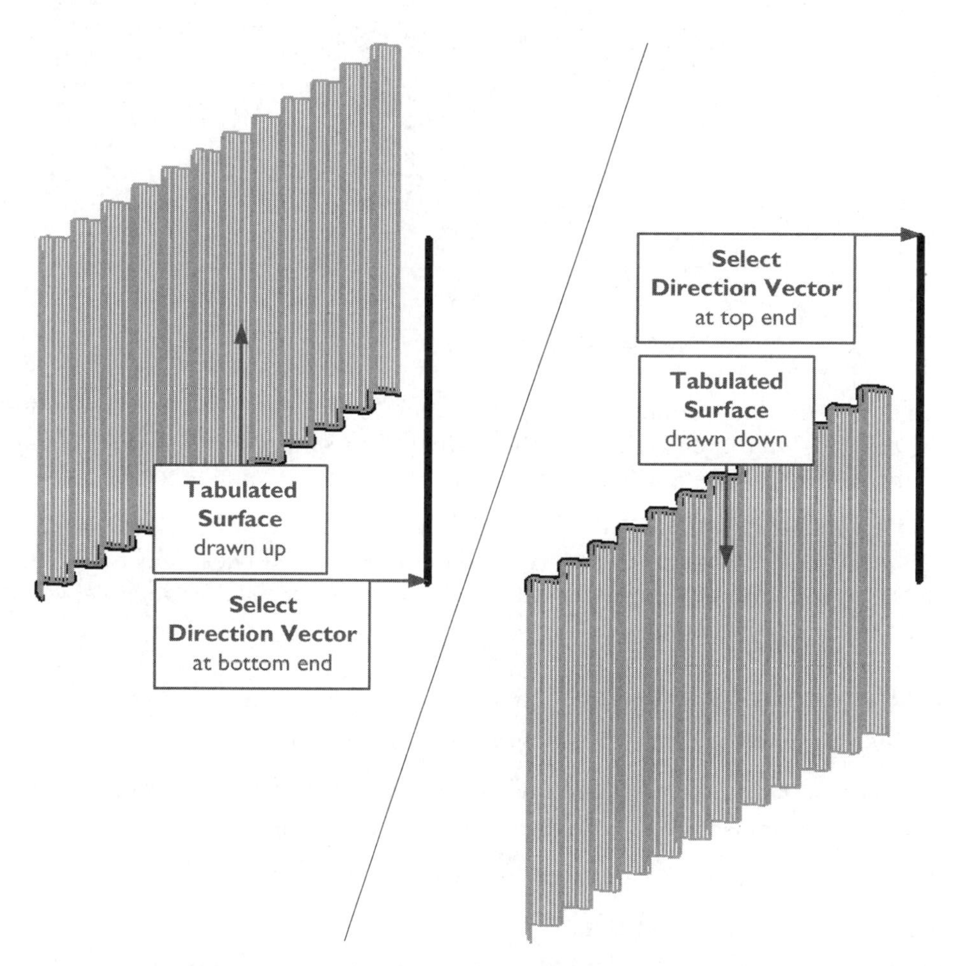

Pick point on the direction vector determines direction of tabulated surface.

Note: To draw a vertical line in plan view, use point filters, as follows:

Command: **line**

Specify first point: *(Pick a point.)*

Specify next point or [Undo]: **.xy**

of *(Pick another point, or even the same point.)*

(need Z): *(Enter a height, such as* **10***.)*

REVSURF

The **REVSURF** command (short for "revolved surface") constructs meshed surfaces by rotating objects about an axis.

To create the revolved surfaces, you must: (1) draw a *path curve*, the object to be revolved; (2) draw an *axis of revolution*, an object about which the surface revolves; and (3) use the **REVSURF** command to construct the meshed surface.

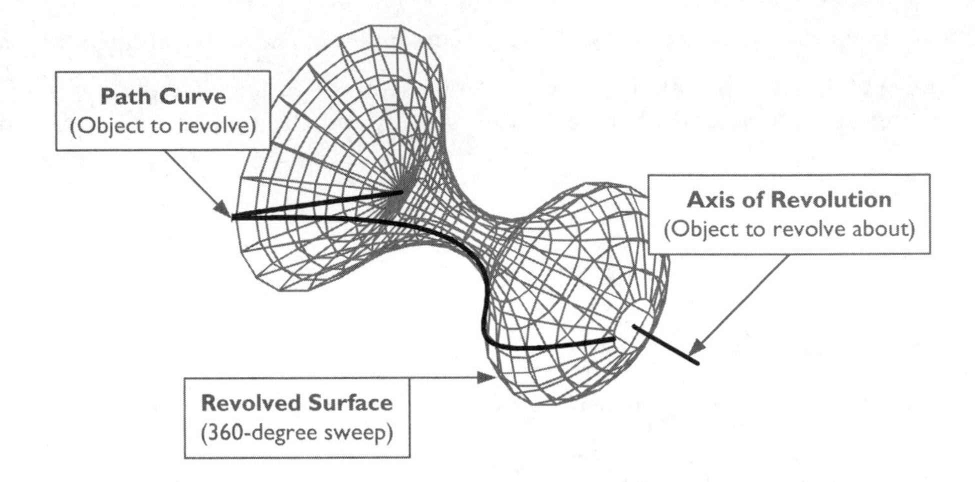

Revolved surface defined by path curve and axis.

The path curves can be lines, circles, arcs, splines, and 2D or 3D polylines — they must be open. The axes can be lines and open polylines (2D or 3D). The same considerations apply as for the **TABSURF** command. The surface mesh is controlled by the **SURFTAB1** and **SURFTAB2** system variables, with the direction of the axis defining the *m* direction.

AutoCAD uses this kind of surface to construct spheres, dishes, domes, and tori with the related surface modeling commands. A sphere, for example, is an arc rotated 360 degrees, while a donut is a circle rotated through 360 degrees.

TUTORIAL: SURFACING BY PATH AND AXIS

Before using this command, you must first draw the path and axis objects.

1. To construct surfaces from paths and axes of revolution, start the **REVSURF** command:
   - From the **Draw** menu, choose **Surfaces**, and then **Revolved Surfaces**.
   - At the 'Command:' prompt, enter the **revsurf** command.

     Command: **revsurf** *(Press* ENTER*.)*
2. AutoCAD reports the current setting for the **SURFAB1** and **SURFTAB2** system variables, and then prompts you to select the object to be revolved (path curve):

     Current wire frame density: SURFTAB1=6 SURFTAB2=6

     Select object to revolve: *(Pick object.)*

   Your selection must consist of a single pick, or enter "g" (for groups) or "cl" (for classes).
3. Pick the objects that define the axis:

     Select object that defines the axis of revolution: *(Pick object.)*

4. The start and included angles allow you to construct a revolution of less than 360 degrees (full circle). A start and included angle of 0 and 285, for example, allow you to see inside the 3D object.

   Specify start angle <0>: *(Press* **ENTER**, *or specify an angle, such as* **15**.*)*

   Specify included angle (+=ccw, -=cw) <360>: *(Press* **ENTER**, *or specify an angle, such as* **285**.*)*

The start angle is defined by the location of the path curve, not by the x axis. The right-hand rule determines the direction; specify a negative angle for the surface to revolve in the opposite direction.

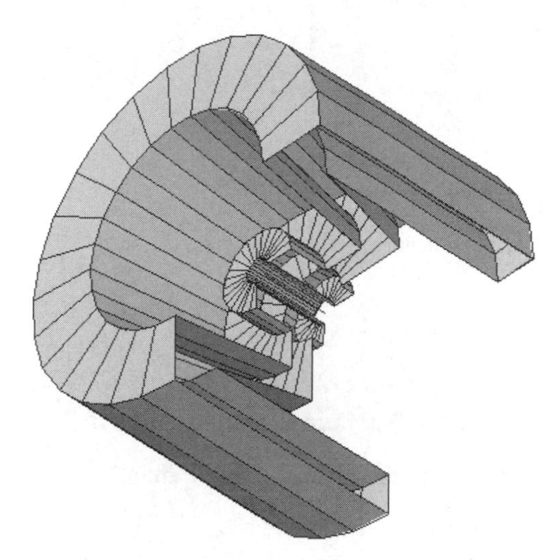

Three-quarters of a revolved surface.

 3DFACE

The **3DFACE** command creates triangular and quadrilateral 3D faces, including nonplanar or warped planes when each corner has a different z coordinate. The AutoCAD sample drawing of the Sydney Opera House was constructed from 3D faces.

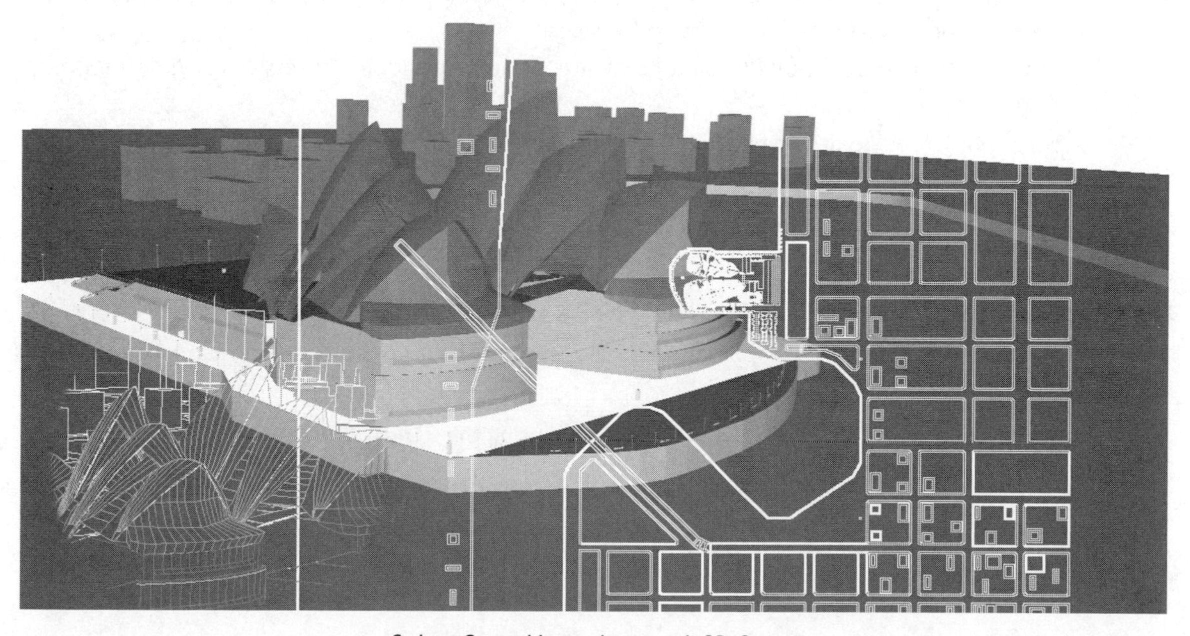

Sydney Opera House drawn with 3D faces.

AutoCAD uses this command to construct boxes, wedges, and pyramids with their namesake commands. Drawing a cube, for example, requires six faces, while a wedge needs three rectangular faces and two triangular faces.

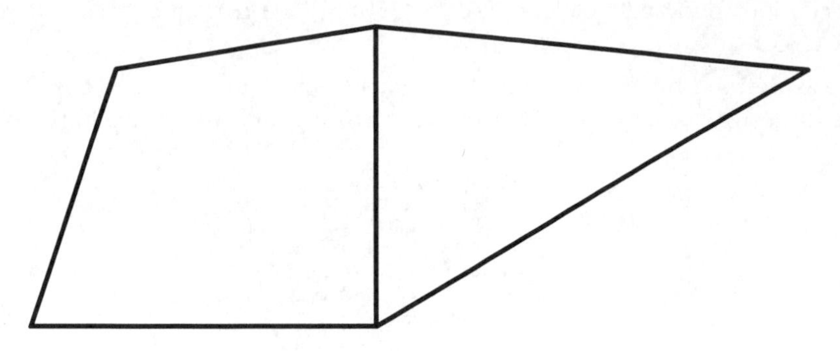

Quadrilateral and triangular 3D faces.

BASIC TUTORIAL: SURFACING BY FACES

1. To construct surfaces from 3- and 4-corner faces, start the **3DFACE** command:
   - From the **Draw** menu, choose **Surfaces**, and then **3D Faces**.
   - At the 'Command:' prompt, enter the **3dface** command.
   - Alternatively, enter the alias **3f** at the keyboard.

   Command: **3dface** *(Press* ENTER*.)*
2. AutoCAD prompts you to pick points:

   Specify first point or [Invisible]: *(Pick a point, or specify x,y,z coordinates.)*

   Specify second point or [Invisible]: *(Pick another point.)*

   Specify third point or [Invisible] <exit>: *(Keep picking points, or press* ENTER *to exit the command.)*

   Exiting after three points draws a triangular face; AutoCAD closes the triangle.
3. Pick another point to keep on drawing faces. Press **ENTER** at the 'create three-sided face" prompt for — you guessed it — triangular faces.

   Specify fourth point or [Invisible] <create three-sided face>: *(Press* ENTER *for triangular faces.)*
4. After picking four points, AutoCAD draws a quadrilateral face, and then starts drawing adjacent faces.

   Specify third point or [Invisible] <exit>: *(Keep picking points, or press* ENTER *to exit the command.)*

DRAWING 3D FACES: ADDITIONAL METHODS

The edges of 3D faces can be visible or invisible, as determined by the following option, system variable, and command:

- **Invisible** option hides individual face edges during the **3DFACE** command.
- **EDGE** command toggles display of edges following the **3DFACE** command.
- **SPLFRAME** system variable toggles display of visible and invisible edges.

Let's look at them.

Invisible

The **Invisible** option hides individual face edges. Enter "i", press **ENTER**, and AutoCAD repeats the prompt so that you can place the point:

> Specify first point or [Invisible]: **i**
>
> Specify first point or [Invisible]: *(Pick a point.)*

You reverse the display of edges with the **EDGE** command, and force the display of all invisible edges with the **SPLFRAME** system variable.

Edge

The **EDGE** command toggles the display of edges following the **3DFACE** command. If the selected edge is visible, this command makes it invisible, and vice versa.

> Command: **edge**
>
> Specify edge of 3dface to toggle visibility or [Display]: *(Pick an edge.)*
>
> Specify edge of 3dface to toggle visibility or [Display]: *(Press **ENTER** to end the command.)*

The **Display** option, oddly named, determines whether you select individual edges (**Select** option), or AutoCAD selects all edges (**All** option).

> Enter selection method for display of hidden edges [Select/All] <All>: *(Type **S** or **A**.)*

Autodesk suggests using AutoSnap to find invisible faces.

SplFrame

The **SPLFRAME** system variable toggles the display of visible and invisible edges.

> Command: **splframe**
>
> Enter new value for SPLFRAME <0>: **1**
>
> Command: **regen**

When 3D faces have invisible edges, you can make all of them visible by setting this system variable to 1, followed by the **REGEN** command.

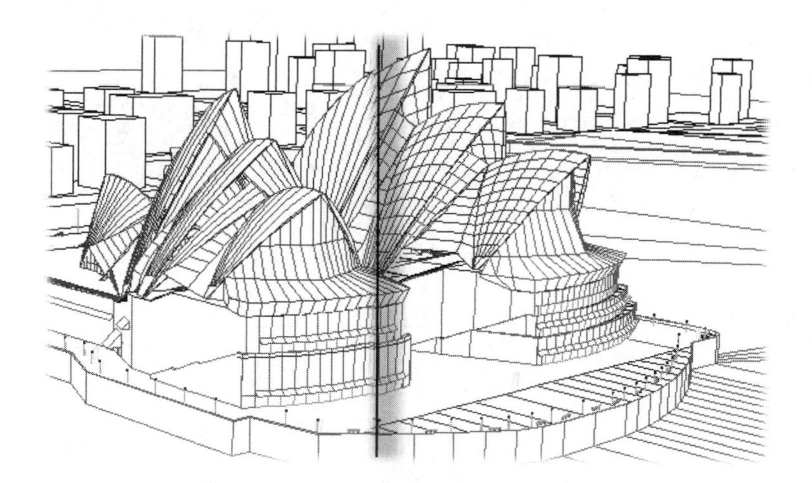

Left: *Splframe = 0 displays edges normally.*
Right: *Splframe = 1 displays invisible edges.*

3DMESH AND PFACE

Th **3DMESH** command constructs a 3D mesh by stipulating the number of vertices in each (M and N) direction, then specifying the *X,Y, Z* coordinate of each of the vertices. The **PFACE** command (short for "polygon mesh") draws a polygonal mesh.

It should be noted that constructing 3D meshes is very tedious. In most situations, it is more efficient to construct a mesh of the same type with another command. The **3DMESH** command is best utilized in programming routines.

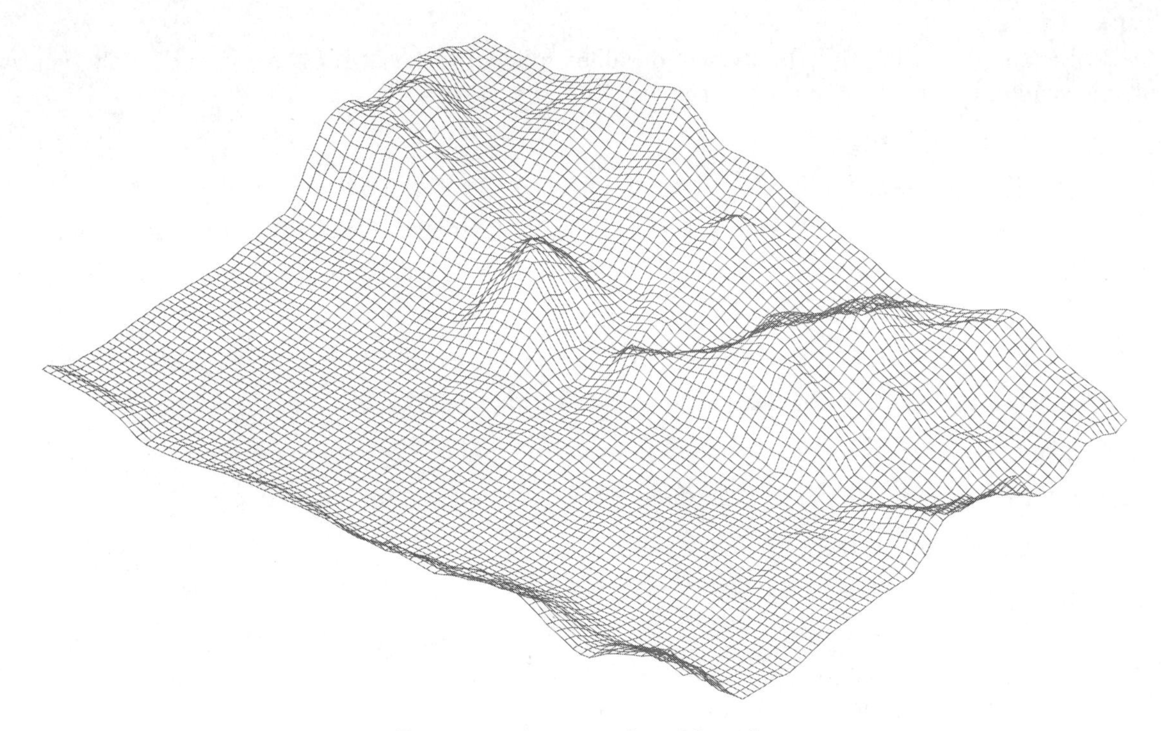

Terrain map drawn as a large 3D mesh.

TUTORIAL: LARGE SURFACE MESHES

This command is meant for use by programmers.

1. To construct large-area surface meshes, start the **3DMESH** command:
   - From the **Draw** menu, choose **Surfaces**, and then **3D Meshes**.
   - At the 'Command:' prompt, enter the **3dface** command.

   Command: **3dmesh**

   Enter size of mesh in M direction: **3**

   Enter size of mesh in N direction: **3**

   Specify location for vertex (0,0): **10,10,-1**

   Specify location for vertex (0,1): **10,20,1**

   Specify location for vertex (0,2): **10,30,3**

   Specify location for vertex (1,0): **20,10,1**

   Specify location for vertex (1,2): **20,20,0**

   Specify location for vertex (1,3): **20,30,-1**

   Specify location for vertex (2,0): **30,10,0**

Specify location for vertex (2,1): **30,20,1**

Specify location for vertex (2,2): **30,30,2**

The order of entry is: one column at a time, and then on to the next column.

PFACE

The **PFACE** command draws polygonal meshes. Unlike the **3DFACE** command, this command is not limited to 3- or 4-sided faces; the faces can have as many edges as needed. This command is meant for use with programs that automatically enter many data points, such as topographic data. The meshes are defined by designating each vertex in x,y,z coordinates.

This command is meant for use by programmers.

Command: **pface**

Specify location for vertex 1: *(Pick point.)*

Specify location for vertex 2 or <define faces>:

... and so forth, until you press **ENTER** to close the mesh. Press **ENTER** a second time to terminate the command.

PEDIT

The **PEDIT** command edits 3D meshes created by the **3DMESH**, **RULESURF**, **TABSURF**, **REVSURF**, and **EDGESURF** commands. (It does not edit 3D faces created by the **3DFACE** or 3D polyfaces created by **PFACE**.)

As an alternative to this command, you can use grips to edit 3D meshes, 3D faces, and 3D polyfaces.

BASIC TUTORIAL: EDITING SURFACE MESHES

1. To edit 3D surface meshes, start the **PEDIT** command:
   - From the **Modify** menu, choose **Object**, and then select **Polyline**.
   - At the 'Command:' prompt, enter the **pedit** command.
   - Alternatively, enter the alias **pe** at the keyboard.

   Command: **pedit** *(Press ENTER.)*
2. AutoCAD prompts you to specify select a polymesh.

   Select polyline or [Multiple]: *(Select polygon mesh objects.)*

   Enter an option [Edit vertex/Smooth surface/Desmooth/Mclose/Nclose/Undo]: *(Enter an option.)*

PEdit Options

Edit Vertex edits each vertex of the mesh individually. It display a submenu of options:

Enter an option [Next/Previous/Left/Right/Up/Down/Move/REgen/eXit <current>: *(Enter an option, or type* **X** *to return to previous level of options.)*

Next moves the X marker to the next vertex.

Previous moves the X marker to the previous vertex.

Left moves the X marker to the previous vertex in the n direction.

Right moves the X marker to the next vertex in the n direction.

Up moves the X marker to the next vertex in the m direction.

Down moves the X marker to the previous vertex in the *m* direction.

Move repositions the vertex, and moves the editing mark, prompting:

Specify new location for marked vertex: *(Specify a point.)*

Regen regenerates the mesh to see the effect of editing changes.

Exit exits **Edit Vertex** options, and returns to the previous prompt.

Smooth Surface smoothes the mesh, depending on the setting of the **SURFTYPE** system variable:

SurfType	*Meaning*
5	Quadratic B-spline surface.
6	Cubic B-spline surface.
8	Bezier surface.

Desmooth restores the mesh to its original, unsmooth state.

Mclose closes the mesh in the *m* direction. (When the mesh is closed, **Mclose** or **Nclose** are replaced by **Mopen** or **Nopen**.)

Nclose closes the mesh in the *n* direction.

Undo undoes operations back to the start of this command.

EXERCISES

1. Use the **3D** command to draw 3D surface models of the following objects:

Object	*Dimensions*		
Box	Length = 5	Width = 4	Height = 3
Wedge	Length = 1	Width = 1.5	Height = 2
Dome	Radius = 3.5		
Sphere	Radius = 2.25		
Cone	Radius = 1.5	Height = 3	
Dish	Diameter = 4.25		

2. Draw a cylinder as a 3D surface model:

   Diameter = 3", height = 4"

   Which command did you use?

3. Set **SURFTAB1** and **SURFTAB2** to 24.

   Draw a torus with these dimensions:

   Diameter = 4", tube diameter = 0.5"

4. Use the **REVSURF** command to draw a sphere:

   Diameter = 2.5"

   What object did you use as the defining curve?

5. Draw four connected surface edges using the **SPLINE** command.

   Apply a Coons patch between them.

   Use the **REVSURF** command to generate a 360-degree 3D surface model.

6. Use the **TABSURF** command to draw a surface with a hole in it.
   Hint: The defining curves are both polylines.
   How must you draw the polylines to keep the tabulations from twisting?

Hole in a surface.

7. Use the **REVSURF** command to draw a wineglass. Use these dimensions:
   Base diameter = 3"
   Stem diameter = 0.5"
   Top diameter = 4"
   Height = 5.75"
8. From the CD-ROM, open the *Ch10RevSurf.dwg file,* a partial drawing of a plastic wheel. Use the **REVSURF** command to generate a 360-degree 3D surface model.

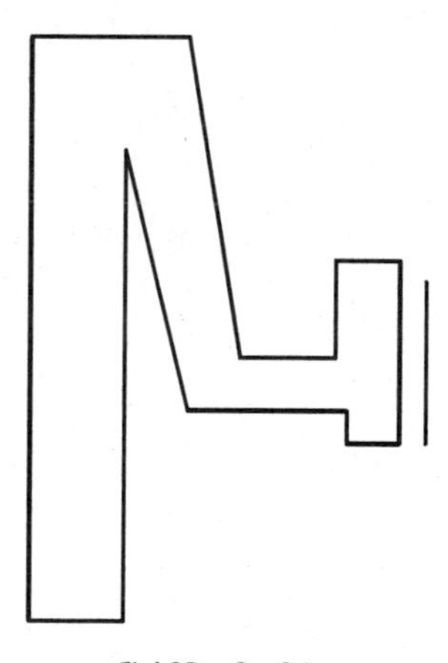

Ch10RevSurf.dwg

9. Using the commands learned so far in this book, draw a 3D surface model of a building block similar to a Lego® or Duplo® brick.

REVIEW QUESTIONS

1. Which commands could you use to draw a 3D surface model of a cylinder?
2. How does a *dish* differ from a *dome*?
3. Explain the difference between:
   a. Wireframe and surface models:
   b. Surface and solid models:
4. Describe a use for:
   a. Surface models
   b. Solid models
5. How can surface models be edited?
6. What happens when you execute the **BOX** command, instead of **AI_BOX**?
7. How do you place a 3D face without showing the edges?
8. Describe a use for the following commands in constructing surfaces:

   RULESURF

   EDGESURF

   TABSURF

   REVSURF
9. The density of two-direction 3D surfaces is described by the values of *m* and *n*. Which system variables control the values of each?
10. Why must you be careful about the order in which you select the corners of a pyramid's base?
11. Define the following shapes:
    a. *Tetrahedron*
    b. *Sphere*
    c. *Torus*
12. Which command creates a "Coons patch"?
13. How do **TABSURF** and **REVSURF** differ?

    How are they the same?

Creating Solid Models

In 2D drafting, you draw and edit 2D objects — such as lines, arcs, and other basic objects — into complete drawings. In 3D design, the process is similar: you draw and edit basic boxes, spheres, and other 3D objects into completed drawings.

The difference between 2D and 3D lies in the nature of the basic object, the kinds of editing tools available, and the extra dimension (z or depth). You can add or subtract solids from each other. For example, holes are created by subtracting cylinders from plates.

Solid modeling is used primarily in mechanical engineering, because models of different materials can be easily analyzed. Solid modelling is also sometimes used by architects to represent complex intersections, such as building roofs. After the model is constructed, the design is usually "converted" to a standard set of 2D drawings for construction.

You can use a limited form of grips editing on solid models. The move, copy, mirror, and rotate options work, but AutoCAD cannot stretch or manipulate parts of solid models with grips. The workaround is to use the **SOLIDEDIT** command, as described in the chapter following.

All object snaps work on solid models, with the exception of INTersection. In its place, use ENDpoint to snap to the intersections of edges and faces.

In this chapter, you learn the following commands for constructing solid models:

BOX, CONE, CYLINDER, SPHERE, TORUS, and **WEDGE** create basic 3D solid bodies.

EXTRUDE and **REVOLVE** create 3D solid bodies from 2D designs.

UNION, SUBTRACT, and **INTERSECT** create new 3D bodies through Boolean operations.

SECTION creates 2D regions from 3D solids.

SLICE cuts 3D bodies in two.

INTERFERE creates 3D bodies from intersecting solids.

MASSPROP reports the properties of 3D bodies.

FINDING THE COMMANDS

On the **SOLIDS** and **SOLIDS EDITING** toolbars:

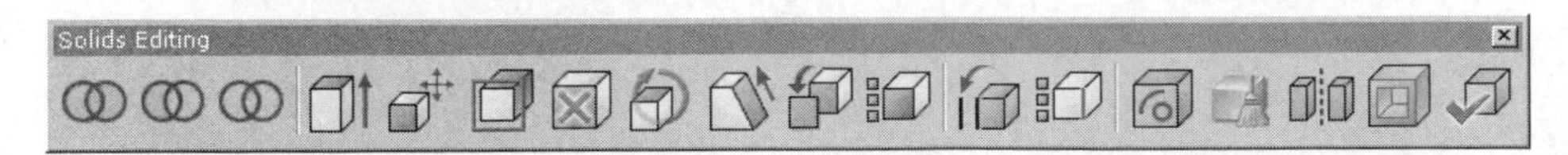

On the **DRAW** and **MODIFY** menus:

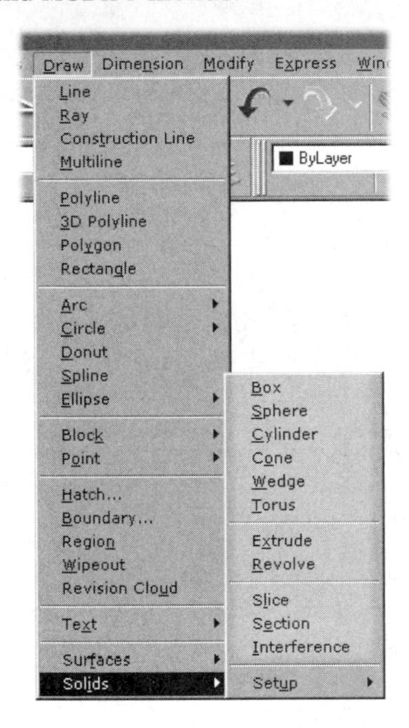

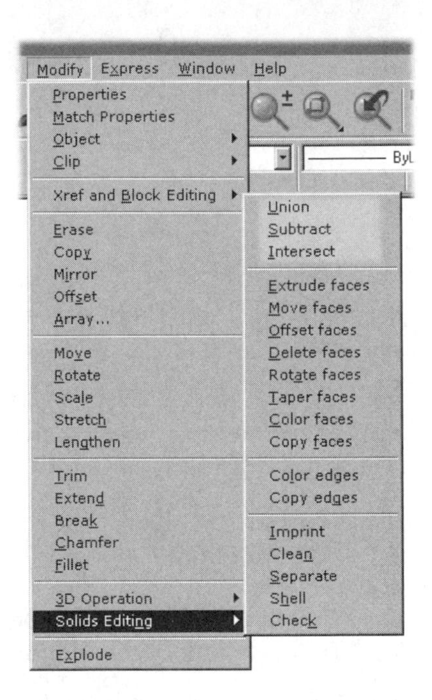

SPHERE

The **SPHERE** command creates solid model balls.

This command draws the easiest of solid objects. All AutoCAD needs to know is a location for the center of the sphere, and its radius (or diameter).

BASIC TUTORIAL: DRAWING SOLID SPHERES

1. To draw 3D solid spheres, start the **SPHERE** command:
   - From the **Draw** menu, choose **Solids**, and then **Sphere**.
   - At the 'Command:' prompt, enter the **sphere** command.

   Command: **sphere** *(Press* ENTER.*)*
2. In all cases, AutoCAD reports the current isoline setting, and then prompts you to specify a point for the sphere's center. You can pick the point with the cursor, or enter x,y,z coordinates.

   Current wire frame density: ISOLINES=4

   Specify center of sphere <0,0,0>: *(Pick point 1, or enter x,y,z coordinates.)*
3. Indicate the size of the sphere by specifying its radius. Enter a value, or show the radius by picking a point with the cursor.

   Specify radius of sphere or [Diameter]: *(Type a value, or pick point 2.)*

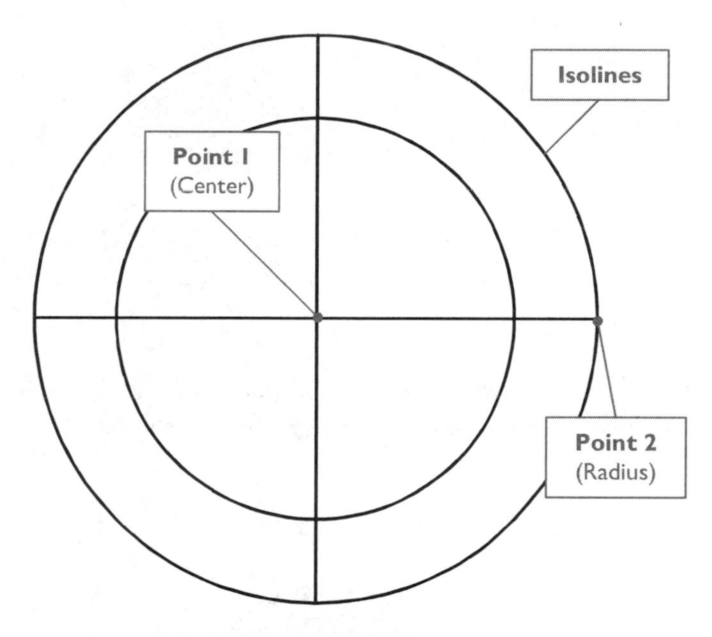

Spheres specified by a center point and radius (or diameter).

AutoCAD positions spheres so that their central axis is parallel to the z-axis of the current UCS. The latitudinal isolines are parallel to the x,y-plane.

DRAWING SPHERES: ADDITIONAL METHODS

The **SPHERE** command has one option that controls the size of sphere. In addition, a system variable controls the look of all curved solids.

- **Diameter** option specifies the diameter of the sphere.
- **ISOLINES** system variable specifies the number of isolines on curved surfaces of solid models.

Let's look at them.

Diameter

The **Diameter** option specifies the diameter of the sphere.

Specify radius of sphere or [Diameter]: **d**

Specify diameter: *(Type a value, or pick a point.)*

The diameter is twice the radius.

Isolines

The **ISOLINES** system variable specifies the number of isolines on curved surfaces of solid models.

Curved surfaces have *isolines*, the vertical and curved lines that show the curvature of solid models and faces. Technically, an isoline indicates a constant value. An isoline on a solid model connects points of equal value. In the case of spheres, isolines shows a constant radius from the center.

Isolines are a display aid, like grid dots. You can vary the number of isolines displayed on curved solids with the **ISOLINES** system variable, followed by the **REGEN** command. You can enter a value ranging from 0 to 2047, but a value of 12 or 16 looks best.

Command: **isolines**

Enter new value for ISOLINES <4>: **12**

Command: **regen**

Regenerating model.

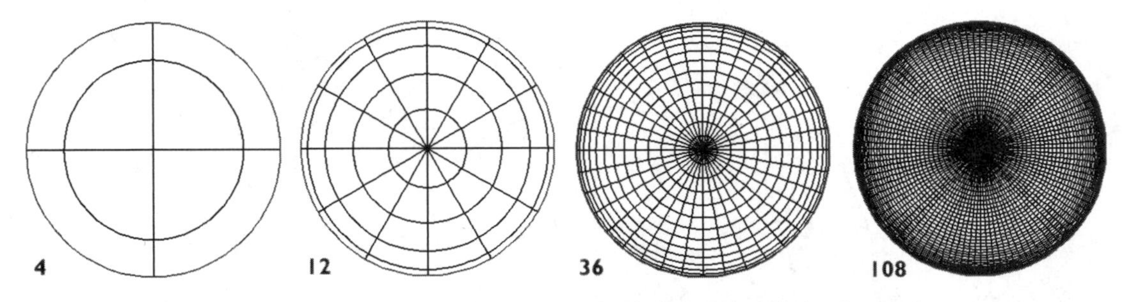

Tripling the number of isolines from 4 (default) through to 108 (in plan view).

Note: You look at the plan view (straight down) of new drawings. To get a better idea of what 3D objects look like, change the view point away from plan to something like 1,1,1.

Command: **vpoint**

Specify a view point or [Rotate] <display compass and tripod>: **1,1,1**

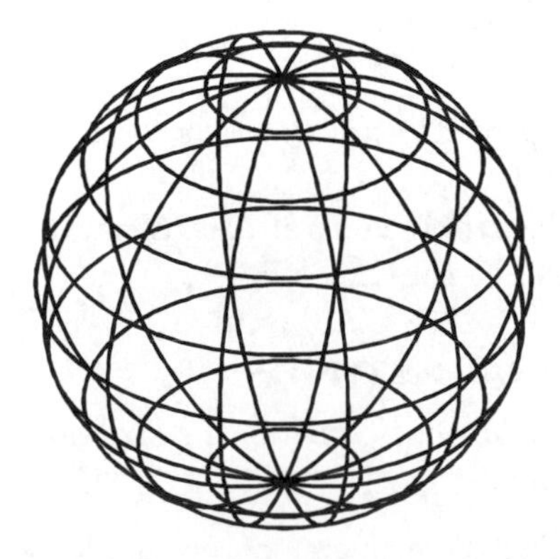

View point rotated to 1,1,1.

BOX

The BOX command creates solid model boxes that are cubistic or rectangular in shape.

BASIC TUTORIAL: DRAWING SOLID BOXES

1. To draw 3D solid boxes, start the **BOX** command:
   - From the **Draw** menu, choose **Solids**, and then **Box**.
   - At the 'Command:' prompt, enter the **box** command.

   Command: **box** *(Press* **ENTER**.*)*
2. In all cases, AutoCAD prompts you for the location of a corner:
   Specify corner of box or [CEnter] <0,0,0>: *(Pick point 1, or enter x,y coordinates.)*
3. Locate the other corner to form the base of the box:
   Specify corner or [Cube/Length]: *(Pick point 2, or enter x,y coordinates.)*
4. Indicate the height:
   Specify height: *(Show the height, or enter a value.)*

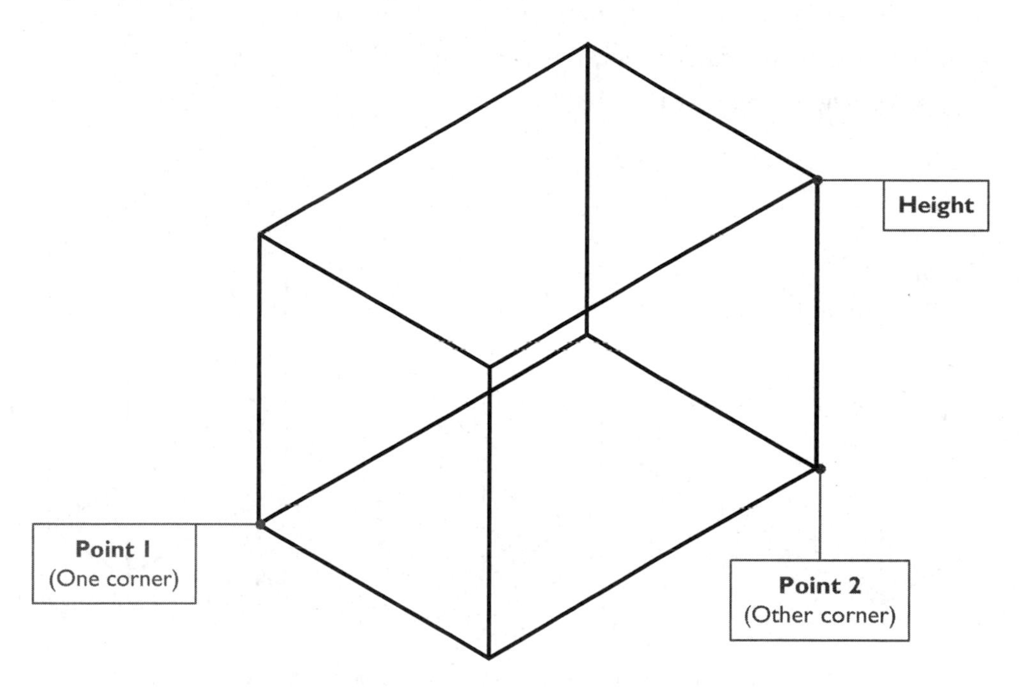

Boxes defined by the area of the base and their height.

DRAWING BOXES: ADDITIONAL METHODS

The BOX command has a number of options that control the size of the box.

- **Center** option specifies the center of the box.
- **Cube** option draws a cube.
- **Length** option specifies the lengths of all three sides.

Let's look at them.

Center

The **CEnter** option specifies the center of the box. The "center" is the 3D center, including the z axis.

> Specify corner of box or [CEnter] <0,0,0>: **ce**
>
> Specify center of box <0,0,0>: *(Pick a point, or enter x,y,z coordinates.)*
>
> Specify corner or [Cube/Length]: *(Pick another point, or enter coordinates.)*
>
> Specify height: *(Indicate the height.)*

Cube

The **Cube** option draws a cube, where all three sides have the same length.

> Specify corner or [Cube/Length]: **c**
>
> Specify length: *(Indicate a length.)*

Length

The **Length** option specifies the lengths of all three sides.

> Specify corner or [Cube/Length]: **l**
>
> Specify length: *(Indicate the length of the box's base.)*
>
> Specify width: *(Indicate the width of the base.)*
>
> Specify height: *(Indicate the height.)*

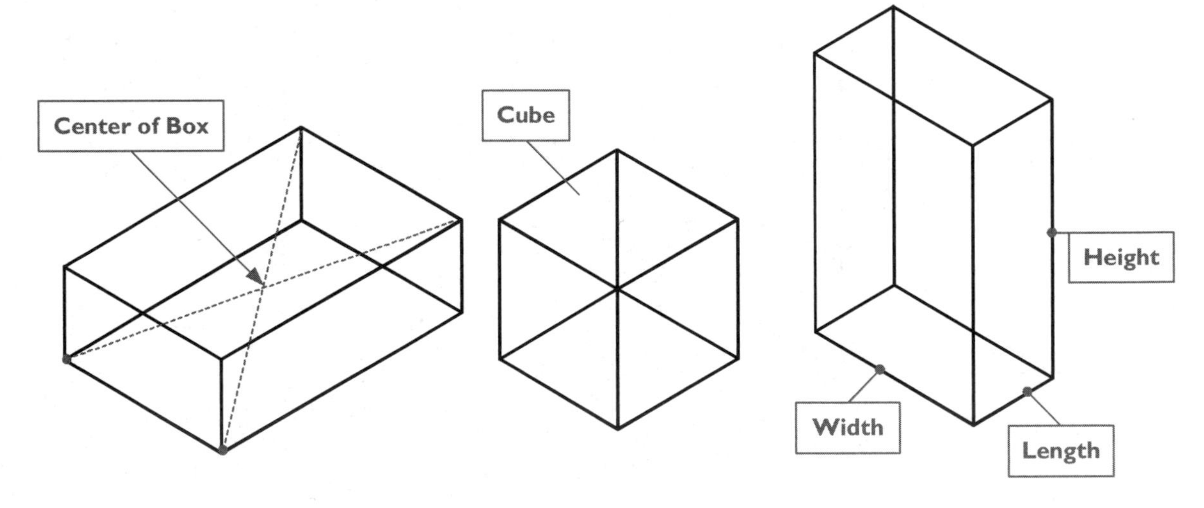

Boxes are defined by their center, as cubes, or by their sides.

Notes: AutoCAD draws boxes so that their sides are parallel to the x, y, and z axes of the current UCS. Enter negative values for coordinates and for the height to draw the box downward and/or to the left.

You can only change the size of boxes by extruding their faces using the **SOLIDEDIT** command, described in the chapter following.

Isolines do not apply to boxes, because they have no curved surfaces; boxes are displayed by their edges.

CYLINDER

The **CYLINDER** command creates solid model cylinders, both round and elliptical, as well as straight and slanted.

BASIC TUTORIAL: DRAWING SOLID CYLINDERS

1. To draw 3D solid cylinders, start the **CYLINDER** command:
   - From the **Draw** menu, choose **Solids**, and then **Cylinder**.
   - At the 'Command:' prompt, enter the **cylinder** command.

   Command: **cylinder** *(Press* ENTER.*)*
2. In all cases, AutoCAD reports the current isoline setting, and then prompts you to specify a point for the cylinder's center. You can pick the point with the cursor, or enter x,y,z coordinates.

   Current wire frame density: ISOLINES=12

   Specify center point for base of cylinder or [Elliptical] <0,0,0>: *(Pick point 1, or enter x,y,z coordinates.)*
3. Indicate the radius of the cylinder by entering a value, or show the radius by picking a point with the cursor.

   Specify radius for base of cylinder or [Diameter]: *(Type a value, or pick point 2.)*
4. Indicate the height of the cylinder. Enter a negative value to draw the cylinder downwards.

   Specify height of cylinder or [Center of other end]: *(Type a value, or show the height.)*

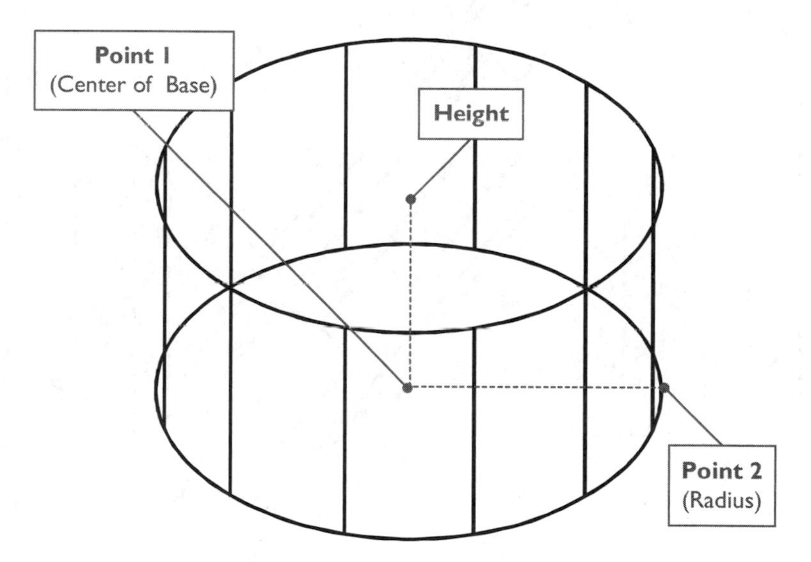

Cylinders are define by the radius of the base and the height.

DRAWING CYLINDERS: ADDITIONAL METHODS

The **CYLINDER** command has a number of options that control the look and size of the cylinders.

- **Diameter** option specifies the diameter of the cylinder.
- **Center of other end** option creates a slanted cylinder.
- **Elliptical** option creates a cylinder with an elliptical base.

Let's look at them all.

Diameter

The **Diameter** option specifies the diameter of the cylinder.

Specify radius for base of cylinder or [Diameter]: **d**

Specify diameter for base of cylinder: *(Type a value, or pick a point.)*

Center of other end

The **Center of other end** option creates slanted cylinders.

Specify height of cylinder or [Center of other end]: **c**

Specify center of other end of cylinder: *(Pick a point, or enter x,y,z-coordinates.)*

Elliptical

The **Elliptical** option creates cylinders with elliptical bases. Some of the prompts are similar to those of the ELLIPSE command.

Specify center point for base of cylinder or [Elliptical] <0,0,0>: **e**

Specify axis endpoint of ellipse for base of cylinder or [Center]: *(Pick a point.)*

Specify second axis endpoint of ellipse for base of cylinder: *(Pick another point.)*

Specify length of other axis for base of cylinder: *(Pick a point.)*

Specify height of cylinder or [Center of other end]: *(Indicate the height.)*

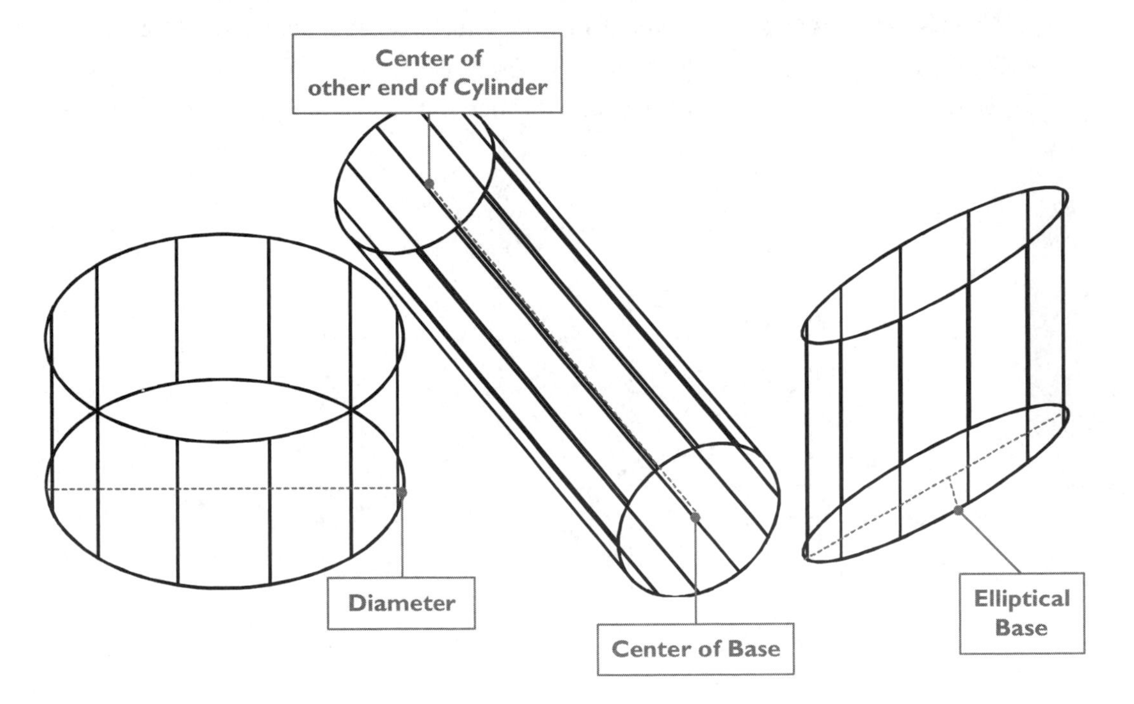

Cylinders can be drawn by diameter, offset center, and elliptical base.

WEDGE

The **WEDGE** command creates solid model wedges, with square or rectangular bases.

AutoCAD doesn't allow you to pick the location of the sloping face, so you need to memorize its location: "The sloped face faces the second corner."

TUTORIAL: DRAWING SOLID WEDGES

1. To draw 3D solid wedges, start the **WEDGE** command:
   - From the **Draw** menu, choose **Solids**, and then **Wedge**.
   - At the 'Command:' prompt, enter the **wedge** command.

   Command: **wedge** *(Press* ENTER.*)*
2. The prompts are similar to those of the **BOX** command:

   Specify first corner of wedge or [CEnter] <0,0,0>: *(Pick point 1, or enter x,y coordinates.)*
3. Locate the other corner to form the base of the box:

   Specify corner or [Cube/Length]: *(Pick point 2, or enter x,y coordinates.)*
4. Indicate the height:

   Specify height: *(Show the height, or enter a value.)*

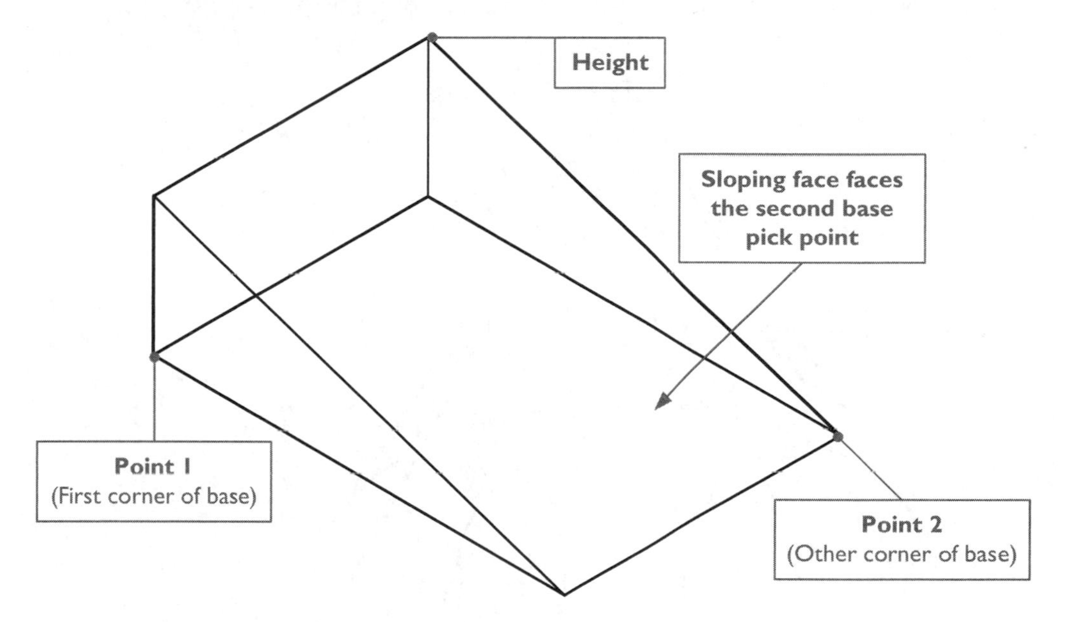

Wedges are defined by the size of the base and their height.

The **CEnter**, **Cube**, and **Length** options are identical to those of the **BOX** command.

Note: When you specify a 3D point for the wedge's second corner, AutoCAD extracts the x,y-values and uses them as the second point. The z value is applied to the height.

Snapping to a 3D point has the same result. The wedge does not anchor to the snapped point.

CONE

The **CONE** command creates straight and slanted solid model cones, with round and elliptical bases.

BASIC TUTORIAL: DRAWING SOLID CONES

1. To draw 3D solid cones, start the **CONE** command:
   - From the **Draw** menu, choose **Solids**, and then **Cone**.
   - At the 'Command:' prompt, enter the **cone** command.

   Command: **cone** *(Press* **ENTER***.)*
2. In all cases, AutoCAD prompts you:

   Current wire frame density: ISOLINES=12

   Specify center point for base of cone or [Elliptical] <0,0,0>: *(Pick point 1, or enter x,y,z coordinates.)*
3. Indicate the radius of the cone's base by entering a value, or show the radius by picking a point with the cursor.

   Specify radius for base of cone or [Diameter]: *(Type a value, or pick point 2.)*
4. Indicate the height of the cone. Enter a negative value to drawn the cone downwards.

   Specify height of cone or [Apex]: *(Type a value, or show the height.)*

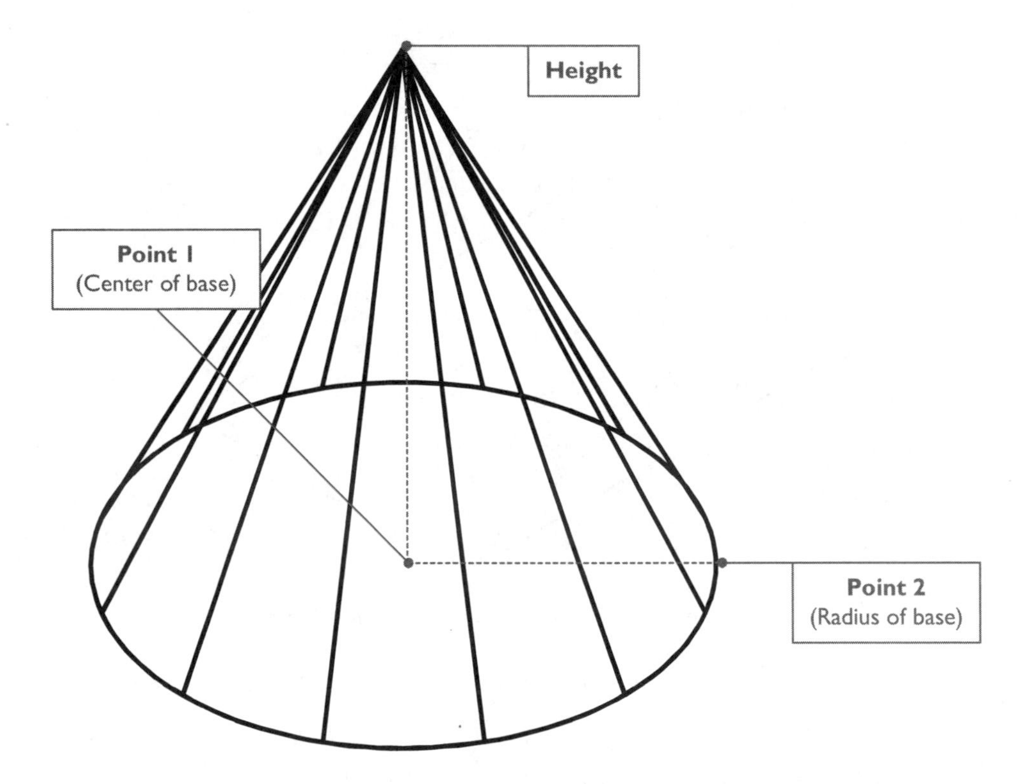

Cones defined by the radius of the base and the height.

DRAWING CONES: ADDITIONAL METHODS

The **CONE** command has a number of options that change the look of the cone:

- **Elliptical** option creates a cone with an elliptical base.
- **Diameter** option specifies the diameter of the cone's base.
- **Apex** option tilts the cone.

Elliptical

The **Elliptical** option creates a cone with an elliptical base in a manner identical to the CYLINDER command.

Specify center point for base of cone or [Elliptical] <0,0,0>: **e**

Specify axis endpoint of ellipse for base of cone or [Center]: *(Pick a point.)*

Specify second axis endpoint of ellipse for base of cone: *(Pick another point.)*

Specify length of other axis for base of cone: *(Pick a point.)*

Specify height of cone or [Apex]: *(Indicate the height.)*

Diameter

The **Diameter** option specifies the diameter of the cone.

Specify radius for base of cone or [Diameter]: **d**

Specify diameter for base of cone: *(Type a value, or pick a point.)*

Apex

The **Apex** option creates a tilted cone. This lets you rotate the cone's vertical axis.

Specify height of cone or [Apex]: **a**

Specify apex point: *(Pick a point, or enter x,y,z-coordinates.)*

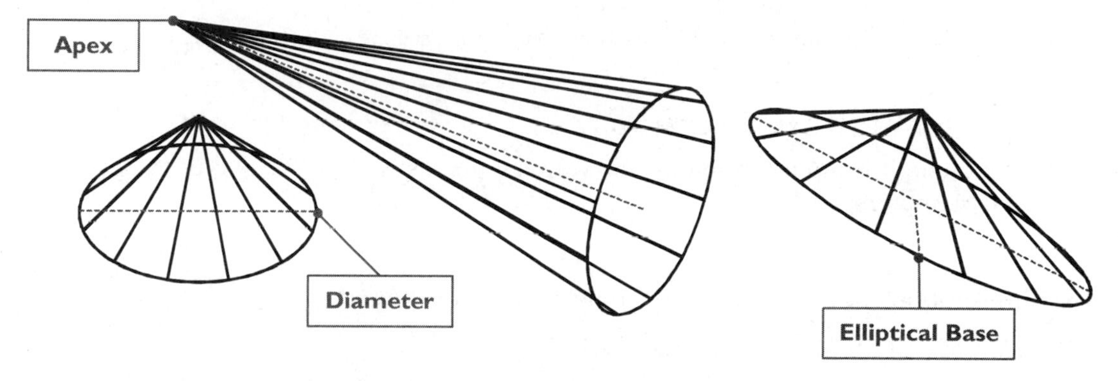

Cones drawn by diameter, apex, and with an elliptical base.

TORUS

The TORUS command creates solid model donuts.

BASIC TUTORIAL: DRAWING SOLID TORI

1. To draw 3D solid donuts, start the **TORUS** command:
   - From the **Draw** menu, choose **Solids**, and then **Torus**.
   - At the 'Command:' prompt, enter the **torus** command.

   Command: **torus** *(Press* ENTER.*)*
2. In all cases, AutoCAD asks for the center of the torus:

   Current wire frame density: ISOLINES=12

   Specify center of torus <0,0,0>: *(Pick a point, or enter x,y,z coordinates.)*
3. Indicate the radius of the torus:

   Specify radius of torus or [Diameter]: *(Pick a point, or enter a value.)*

4. And indicate the radius of the tube:

   Specify radius of tube or [Diameter]: *(Pick a point, or enter a value.)*

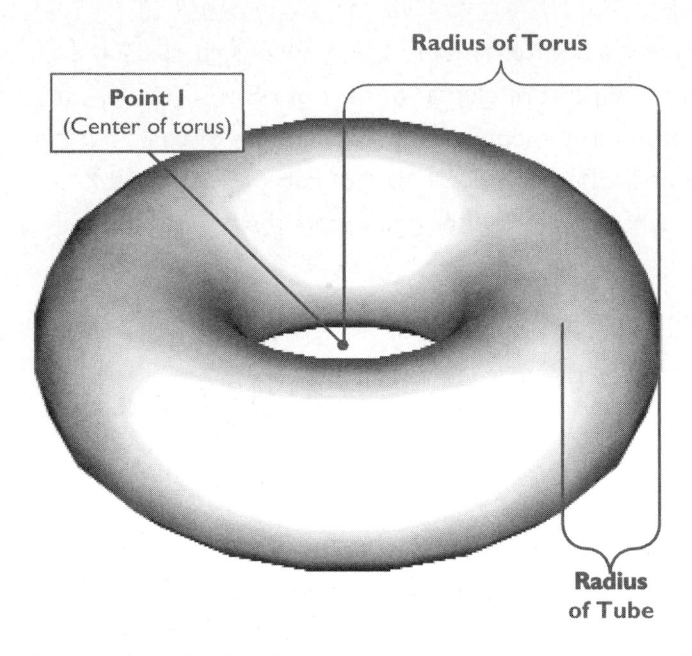

Torus (shown rendered) defined by its center, radius, and the radius of the tube.

DRAWING TORI: ADDITIONAL METHODS

The **TORUS** command has one option for drawing donuts, as well as the ability to accept unusual values for the diameters.

Diameter

The **Diameter** option specifies the diameter of the torus or the tube.

Specify radius of torus or [Diameter]: **d**

Specify diameter: *(Enter a value, or pick a point in the drawing.)*

Specify radius of tube or [Diameter]: **d**

Specify diameter: *(Enter a value, or pick a point in the drawing.)*

You can specify diameters and radii that "don't make sense." An example is when the tube radius exceeds the torus radius. Create a torus whose radius is 2, but with tube radius 5. Notice that the tube intersects the center point of the torus. Another example is giving a negative value to the torus radius. (The tube radius must, however, be a positive number of greater value.) Construct a torus with a torus radius of –2 and a tube radius of 5 (the tube radius must be greater than 2 in this case).

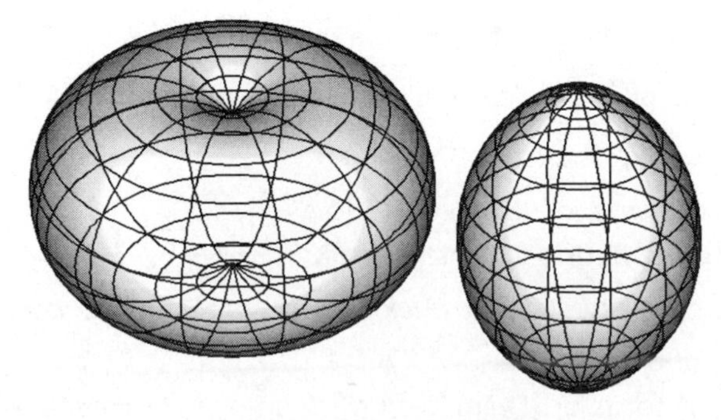

Torus makes pumpkin and football shapes.

REVOLVE

The **REVOLVE** command rotates 2D objects to create 3D solid models.

The **BOX**, **SPHERE**, and other commands are good for building basic objects, but are often too simplistic for creating complex solid models. The **REVOLVE** command and **EXTRUDE** command (covered next) are handy for easily creating complex 3D solid models:

- **REVOLVE** creates symmetrically round objects, like wineglasses and chair legs.
- **EXTRUDE** creates symmetrically lengthwise objects, like handrails and gears.

To revolve (convert) 2D objects into 3D solids, AutoCAD needs to know two things: (1) the objects to be revolved; and (2) the axis about which revolve them.

This command rotates circles, ellipses, closed polylines, polygons, donuts, closed splines, and regions. (AutoCAD cannot revolve objects within blocks, or polylines that cross over themselves.)

The revolution takes place around the x or y axes, or about an object (line or polyline segment). The length of the axis is unimportant; AutoCAD uses only its orientation in space. You can specify a full 360-degree revolution, or a partial revolution, with positive or negative angles.

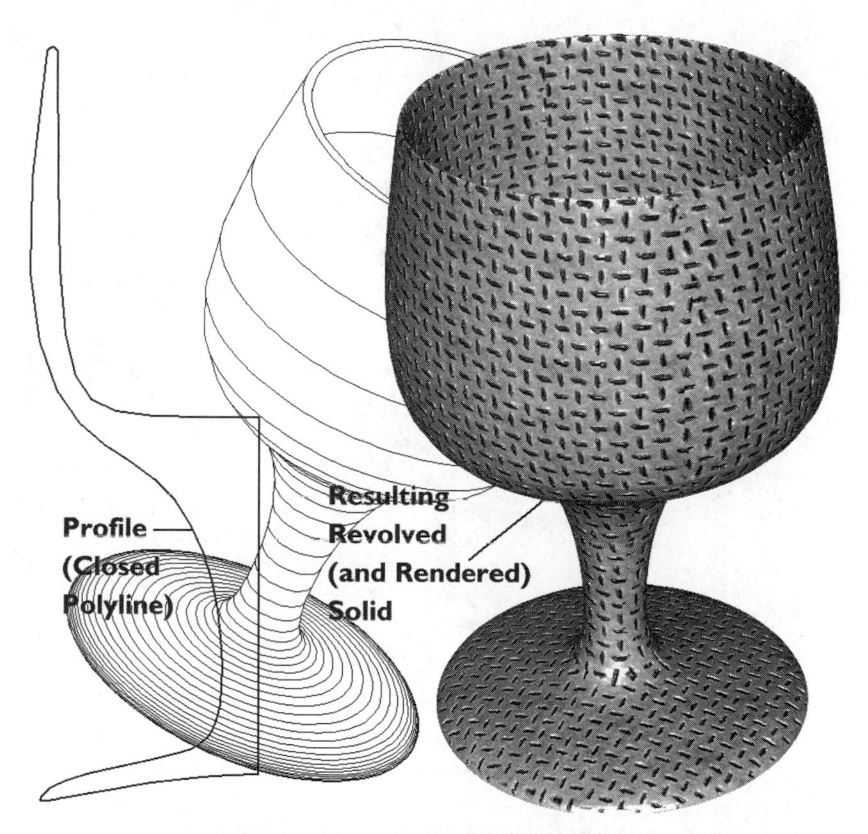

Revolve 2D profiles into 3D solid objects.

TUTORIAL: REVOLVING OBJECTS INTO SOLIDS

1. To revolve 2D objects into 3D solids, start the **REVOLVE** command:
   - From the **Draw** menu, choose **Solids**, and then **Revolve**.
   - At the 'Command:' prompt, enter the **revolve** command.
   - Alternatively, enter the alias **rev** at the keyboard.

   Command: **revolve** *(Press* ENTER.*)*

2. AutoCAD prompts you to select the objects to be revolved (converted into 3D solids), often called the "profile." Although Autodesk's documentation states that "You can revolve only one object at a time," AutoCAD can in fact revolve more that one; select as many as you want! Each object must form a single, closed profile: circle, ellipse, closed spline, or closed polyline.

   Current wire frame density: ISOLINES=12

   Select objects: *(Select one or more objects.)*

   Select objects: *(Press* **ENTER** *to end object selection.)*

Notes: To revolve an object about a central axis, such as a goblet, draw just one half of the profile. Because the profile must be closed, use the **PEDIT** command's **Join** and **Close** options to ensure the polyline is closed.

You can enter a positive or negative angle. AutoCAD uses the "right-hand rule" to determine the positive direction of rotation. When rotating an object, the positive direction of the axis is from the first (start) point to the second (end) point.

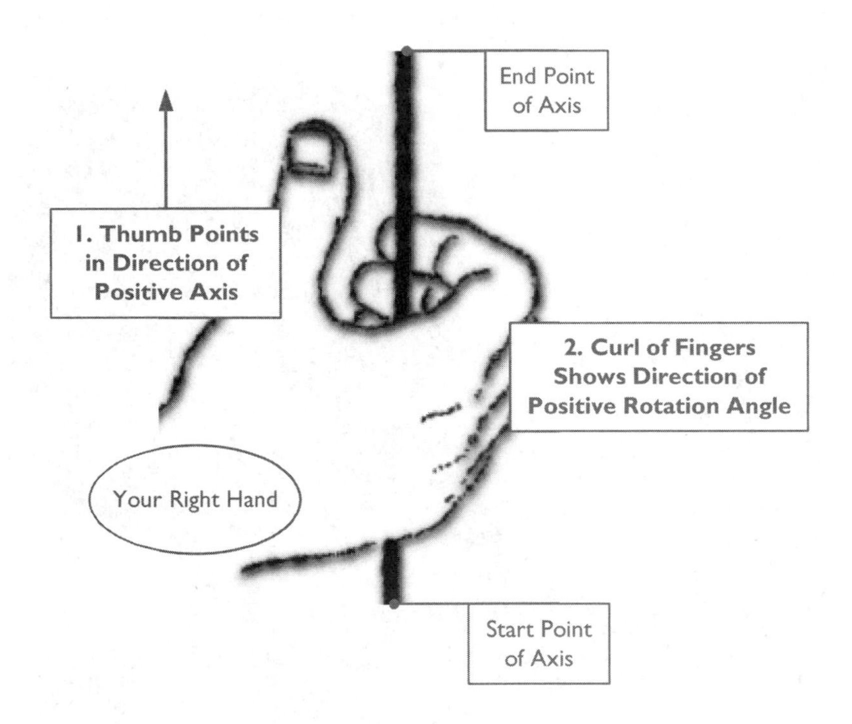

Using your right hand:
1. *Your thumb points in the direction of the positive axis.*
2. *The curl of your fingers shows the direction of positive rotation angle.*

3. Select the axis about which the object(s) will be revolved. You can choose:
   - Two points in the drawing that show AutoCAD the axis.
   - A line or polyline segment (object).
   - The x or y axis.
   - One segment of the profile

   The profile must be entirely on one side of the axis, or part of the axis.

   Specify start point for axis of revolution or define axis by [Object/X (axis)/Y (axis)]: *(Pick two points.)*

4. Specify the angle of revolution — how far around you want the profile revolved:
   Specify angle of revolution <360>: *(Type an angle.)*

Portion	*Angle of Revolution*
Full	360 degrees
3/4	270 degrees
2/3	240 degrees
1/2	180 degrees
1/3	120 degrees
1/4	90 degrees

EXTRUDE

The **EXTRUDE** command extrudes 2D objects along paths to create 3D solid models.

AutoCAD needs to know two things: (1) the objects to be extruded; and (2) the path along which to extrude.

This command extrudes circles, ellipses, closed polylines (including polygons and donuts), closed splines, regions, and 3D faces. AutoCAD cannot extrude objects within blocks, polylines that cross over themselves, or polylines with more than 500 vertices.

If you plan to extrude objects with holes in them, such as gears with axle holes, then you must convert the objects to a single region. Use the **SUBTRACT** command to remove holes.

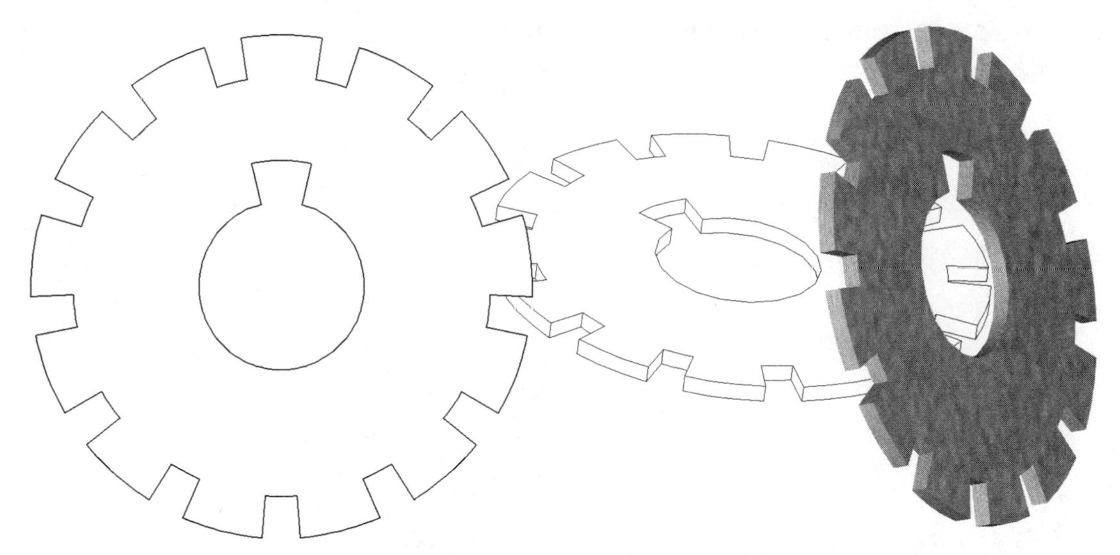

Objects with holes can be extruded, but must first be converted to a single region.

The path can be an object — lines, circles, arcs, ellipses, elliptical arcs, polylines, or splines — or a height. You can also specify an angle, which tapers the extrusion.

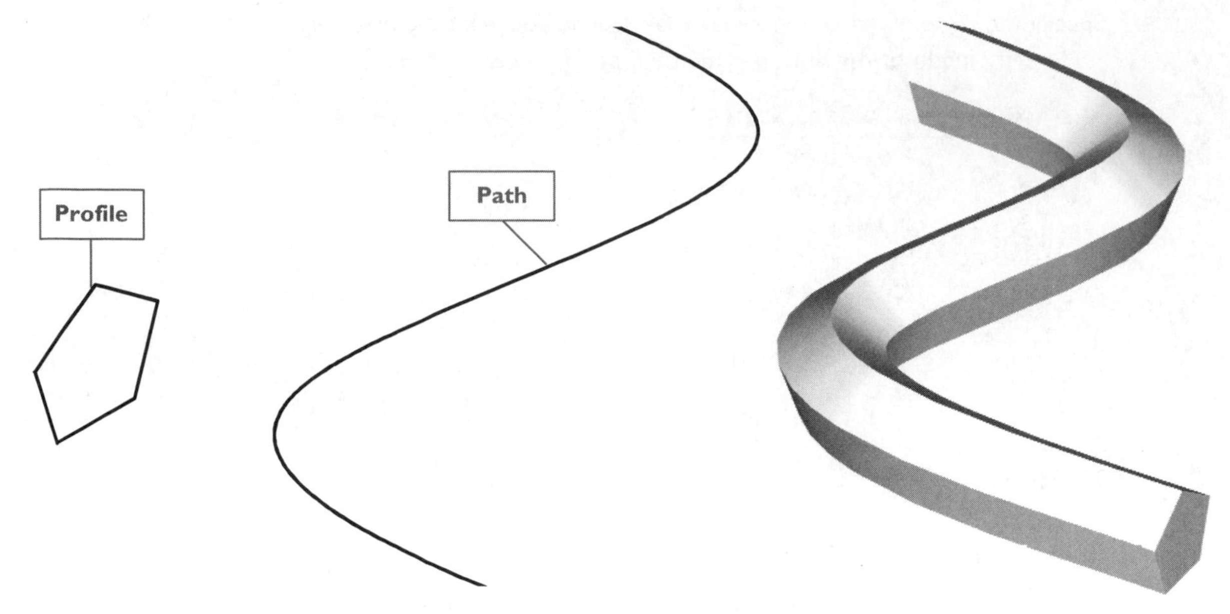

Extrusions are created from profiles and paths.

Caution! The extrusion process takes place at right angles to the centerline of the path. The profile can lie in another plane at any angle, providing it is not parallel to the path. If it is, AutoCAD complains, "Profile and path are tangential. Unable to extrude the selected object." The solution is to draw the path in one UCS and the profile in a second UCS at an angle to the first. AutoCAD reorients the profile to be perpendicular to the path.

TUTORIAL: EXTRUDING OBJECTS INTO SOLIDS

1. To extrude objects into 3D solids, start the **EXTRUDE** command:
   - From the **Draw** menu, choose **Solids**, and then **Extrude**.
   - At the 'Command:' prompt, enter the **extrude** command.
   - Alternatively, enter the alias **ext** at the keyboard.

   Command: **extrude** *(Press* ENTER.*)*

Applying a taper angle to an extrusion results in this bevel gear.

2. In all cases, AutoCAD prompts you select the object(s) to extrude.

   Current wire frame density: ISOLINES=12

   Select objects: *(Select one or more objects.)*

   Select objects: *(Press* **ENTER** *to end object selection.)*

3. Indicate the height to which the object(s) should be extruded through:
   - Two points in the drawing that show AutoCAD the height.
   - A distance.
   - A line or other object (Path).

   The path must be perpendicular to the object.

   Specify height of extrusion or [Path]: *(Enter a value.)*

4. Optionally, you can taper the extrusion. The angle you specify, however, must not allow the extruded solid to intersect itself.

   Specify angle of taper for extrusion <0>: *(Press* **ENTER**, *or enter an angle such as* **5**.*)*

UNION

The **UNION** command merges 2D regions and 3D solid models into a single body.

This command joins 2D regions into a single region, and 3D solid objects into a single solid model; regions and 3D solid models cannot, however, be mixed. The objects do not need to be touching or intersecting to be unioned.

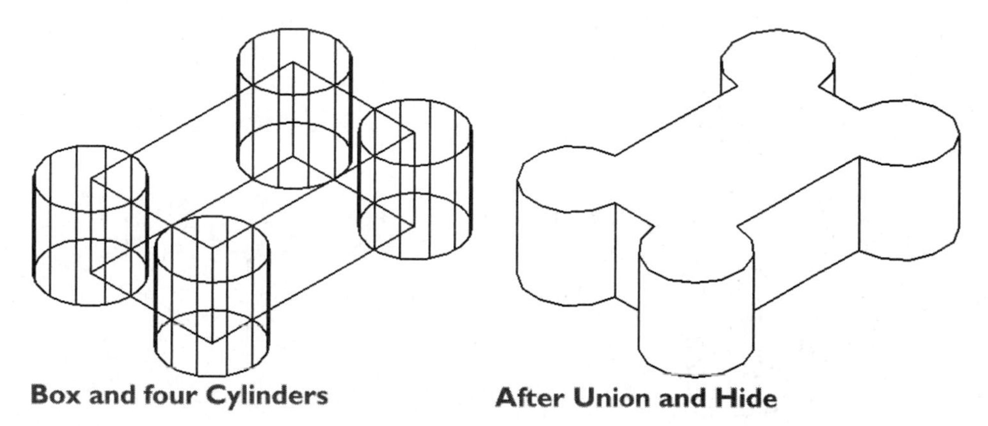

The Union command merges multiple solid objects into a single body.

Note: The **UNION, INTERSECT**, and **SUBTRACT** commands are collectively known as "Boolean operations," named after George Boole, a 19th century mathematician who developed Boolean logic. You may be familiar with using the terms **or** (union), **and** (intersect), and **not** (subtract) in narrowing down searches on the Internet. These are based on the work of Mr Boole. www.kerryr.net/pioneers/boole.htm

TUTORIAL: JOINING SOLIDS

1. To join two or more 3D solids, start the **UNION** command:
   - From the **Modify** menu, choose **Solids Editing**, and then **Union**.
   - At the 'Command:' prompt, enter the **union** command.
   - Alternatively, enter the alias **uni** at the keyboard.

   Command: **union** *(Press* **ENTER**.*)*

2. In all cases, AutoCAD prompts you to select the objects to union. Select at least two:

   Select objects: *(Select one or more objects.)*

   Select objects: *(Press ENTER to end object selection.)*

   The selected objects are merged into a single object. You will probably notice some faces or edges missing. This is normal.

SUBTRACT

The **SUBTRACT** command removes intersecting parts of 2D regions and 3D solid models. Regions and 3D solid models cannot be mixed. The objects must be intersecting to be subtracted.

BASIC TUTORIAL: SUBTRACTING SOLIDS

1. To remove 3D solids from each other, start the **SUBTRACT** command:
   - From the **Modify** menu, choose **Solids Editing**, and then **Subtract**.
   - At the 'Command:' prompt, enter the **subtract** command.
   - Alternatively, enter the alias **su** at the keyboard.

   Command: **subtract** *(Press ENTER.)*

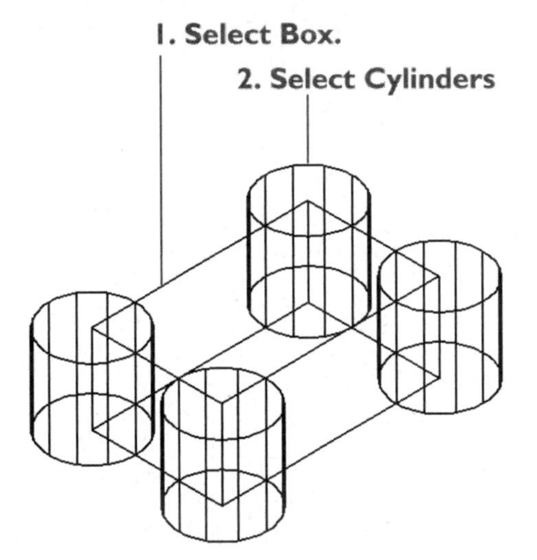
1. Select Box.
2. Select Cylinders

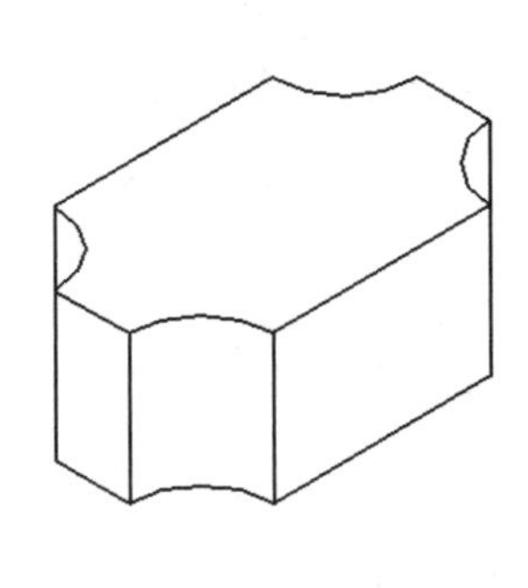
Result:
Cylinders removed from Box.

Subtracting the cylinders from the box gives one result.

2. In all cases, AutoCAD prompts you to select the objects that will be subtracted.

   Select solids and regions to subtract from ...

   Select objects: *(Select one or more objects.)*

   Select objects: *(Press ENTER to end object selection.)*

3. Select the objects that will do the subtracting.

   Select solids and regions to subtract ...

   Select objects: *(Select one or more objects.)*

   Select objects: *(Press ENTER to end object selection.)*

Note: The result may be unexpected, because the order in which you select objects is important:

1. First, select the objects from which the subtracting takes place.
2. Second, select that objects that do the subtracting.

If the result looks wrong, it is probably because you selected the objects in reverse order. Employ the **U** command to undo, and try again.

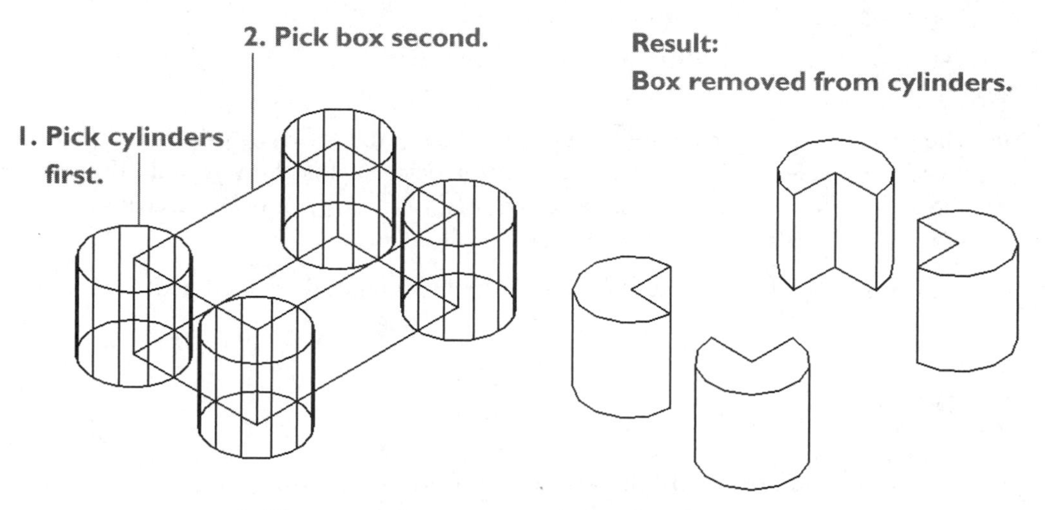

A different result from subtracting the box from the cylinders.

When two or more objects are selected to "subtract from," they are unioned into a single solid before others are subtracted.

 INTERSECT

The **INTERSECT** command removes all but intersecting parts of 2D regions and 3D solid models.

This command determines the overlapping area of regions or the common volume of 3D solids. Regions and solids cannot be mixed.

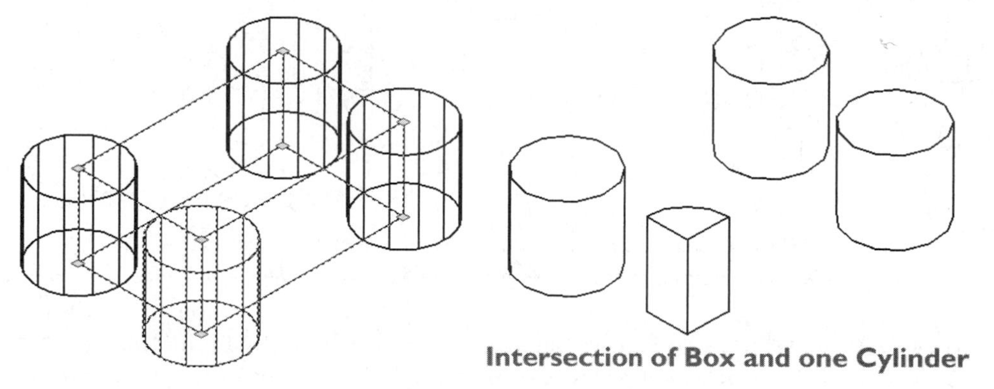

Intersection of the box and the left, front cylinder results in a pie-shaped solid.

TUTORIAL: INTERSECTING SOLIDS

1. To remove all but the intersections of 3D solids, start the **INTERSECT** command:
   - From the **Modify** menu, choose **Solids Editing**, and then **Intersect**.
   - At the 'Command:' prompt, enter the **insersect** command.

- Alternatively, enter the alias **in** at the keyboard.

  Command: **intersect** *(Press* **ENTER.***)*

2. In all cases, AutoCAD prompts you to select the objects from which to make the intersection. For this command to work correctly, all solids you select must intersect in a single volume.

   Select objects: *(Select one or more objects.)*

   Select objects: *(Press* **ENTER** *to end object selection.)*

Sometimes the result of this command is — nothing! AutoCAD reports, "Null solid created - deleted." This occurs when you select objects that do not intersect, such as the box and all four cylinders in the figure above. The four cylinders intersect the box, but not each other. Because this command finds the volume common to all five objects, the result is nothing.

The solution to this problem is to use the **INTERFERE** command, described next.

 INTERFERE

The **INTERFERE** command creates 3D bodies from intersecting solids.

This command has two features for dealing with interferences: (1) it highlights the solids that interfere (not the interference volumes themselves, unfortunately); and (2) it optionally turns the interference into independent solids.

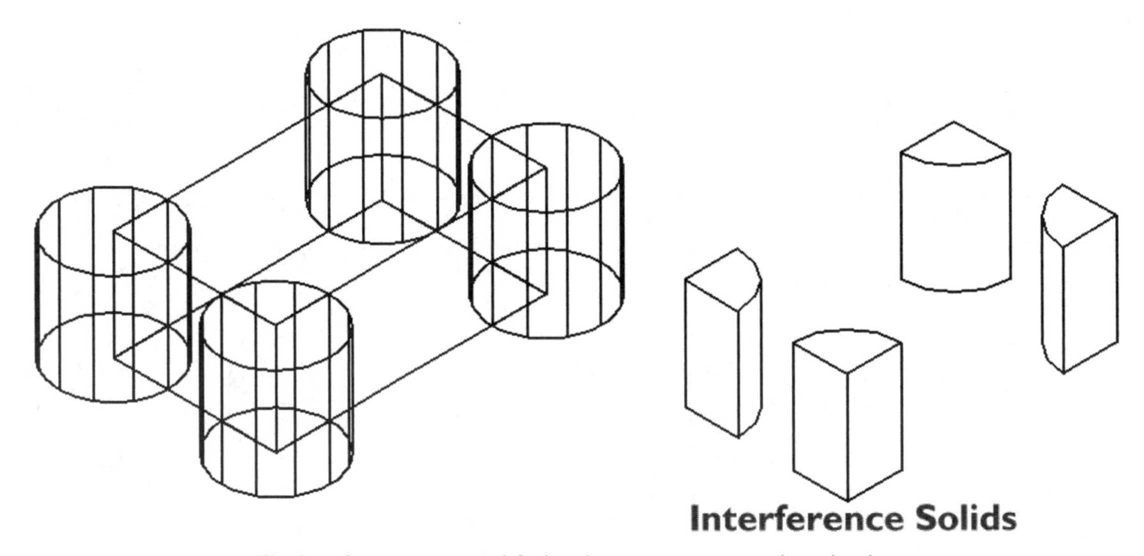

The Interference command finds volumes in common with each other.
Here, the interference solids have been moved away from the original solids.

I think you can see that these two features are immediately useful. By highlighting interfering pairs of solids, AutoCAD shows you which parts of a model collide or interfere — for example, pipes that go through other pipes when they ought not to, or heating ducts that penetrate walls.

This command is a more intelligent version of the **INTERSECTION** command, because it finds the intersection between *any* and all sets of solid models. The volumes where two or more solids intersect are converted into independent solids. You can, if you like, move them away from the rest of the solid model.

TUTORIAL: CHECKING INTERFERENCE BETWEEN SOLIDS

1. To determine if 3D solids interfere with each other, start the **INTERFERE** command:
   - From the **Draw** menu, choose **Solids**, and then **Interfere**.
   - At the 'Command:' prompt, enter the **interfere** command.
   - Alternatively, enter the alias **inf** at the keyboard.

   Command: **interfere** *(Press* ENTER.*)*

2. AutoCAD prompts you to select two sets of solids, but you don't have to. There are two ways of creating selection sets for this command:

   Select first set of solids: *(Select one or more objects.)*

   Select objects: *(Press* ENTER *to end object selection.)*

   - At the first prompt, select *all* objects, and press **ENTER** at the second prompt. AutoCAD checks all objects against each other for interference.
   - At the first prompt, select some objects, and then select other objects at the second prompt. AutoCAD checks for interference between the two selection sets.

3. Optionally, select a second set of objects.

   Select second set of solids: *(Press* ENTER *to skip, or select objects for the second selection set.)*

   AutoCAD compares the solids against each other, looking for interference.

   Comparing 1 solid against 4 solids.

   Interfering solids (first set): 1

   (second set): 4

   Interfering pairs : 4

4. Optionally, AutoCAD can create solids out of the volumes that interfere (intersect).

   Create interference solids? [Yes/No] <N>: **y**

5. If you want, AutoCAD can highlight pairs of interfering objects. This is useful when three or more objects interfere.

   Highlight pairs of interfering solids? [Yes/No] <N>: **y**

6. Press **N** to see the next pair of interfering solids.

   Enter an option [Next pair/eXit] <Next>: **n**

7. When finished checking, enter **x** to exit the command:

   Enter an option [Next pair/eXit] <Next>: **x**

At this point, the model may look no different. To isolate the newly created solids representing the interference volumes, use the MOVE command together with the **Previous** option. (When you selected the solids originally at the "Select first/second set of solids" prompt, AutoCAD placed them into the Previous selection set.)

Command: **move**

Select objects: **p**

Select objects: *(Press* ENTER.*)*

Specify base point or displacement: *(Pick a point.)*

Specify second point of displacement or <use first point as displacement>: *(Pick another point.)*

This moves the original solids away from the interference solids. The result is similar to the figure on the previous page.

SLICE

The **SLICE** command cuts 3D bodies in two.

This command uses a *cutting plane* to slice solid objects. You have the option of retaining both halves, or just one. No matter how complex the solid, this command never slices it into more than two parts.

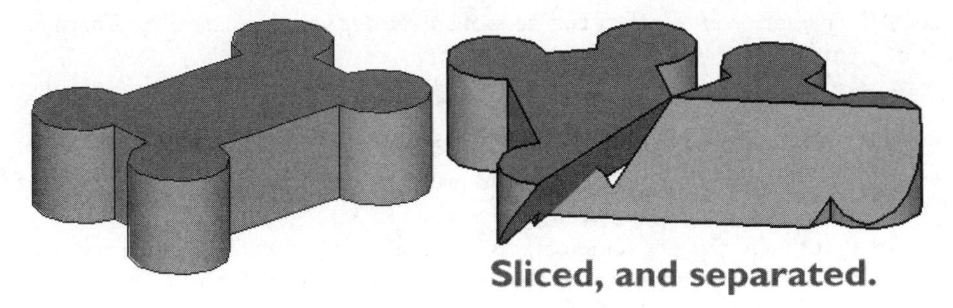

Slice command cuts solids in two.

The cutting plane can be defined in a number of ways: with another object, three picked points, the current view, and so on. The plane can, however, be difficult to place precisely. You indicate the cutting plane by a variety of methods:

- **Object**: the cutting plane is aligned in the plane of a 2D object (circle, ellipse, arc, spline, or 2D polyline segment).
- **3points**: the cutting plane is defined by three points.
- **View**: the cutting plane is aligned with the current viewport, along with a point.
- **XY, YZ, ZX**: the cutting plane is aligned with the x,y (or y,z or z,x) plane of the current UCS, along with a point.
- **Zaxis**: the cutting plane is defined by a point on the plane and a second point on the z axis to the plane.

BASIC TUTORIAL: SLICING SOLIDS

1. To slice 3D solids, start the **SLICE** command:
   - From the **Draw** menu, choose **Solids**, and then **Slice**.
   - At the 'Command:' prompt, enter the **slice** command.
   - Alternatively, enter the alias **sl** at the keyboard.

   Command: **slice** *(Press* ENTER.*)*
2. In all cases, AutoCAD prompts you to select the solid models to slice.

   Select objects: *(Select one or more objects.)*

   Select objects: *(Press* ENTER *to end object selection.)*
3. Here comes the hard part: placing the slicing plane.
   AutoCAD provides a number of options. An easy one is **View**: the slicing plane is parallel to the viewpoint; you just pick a point on the solid model to locate the plane.

   Specify first point on slicing plane by [Object/Zaxis/View/XY/YZ/ZX/3points]
   <3points>: **v**
4. AutoCAD asks for a point to locate the slicing plane "front to back."

   Specify a point on the current view plane <0,0,0>: **mid**

5. Pick more accurately with an object snap, such as ENDpoint or MIDpoint. (Keep in mind that INTersection osnap does not work with solid models.)

   of *(Pick point 1.)*

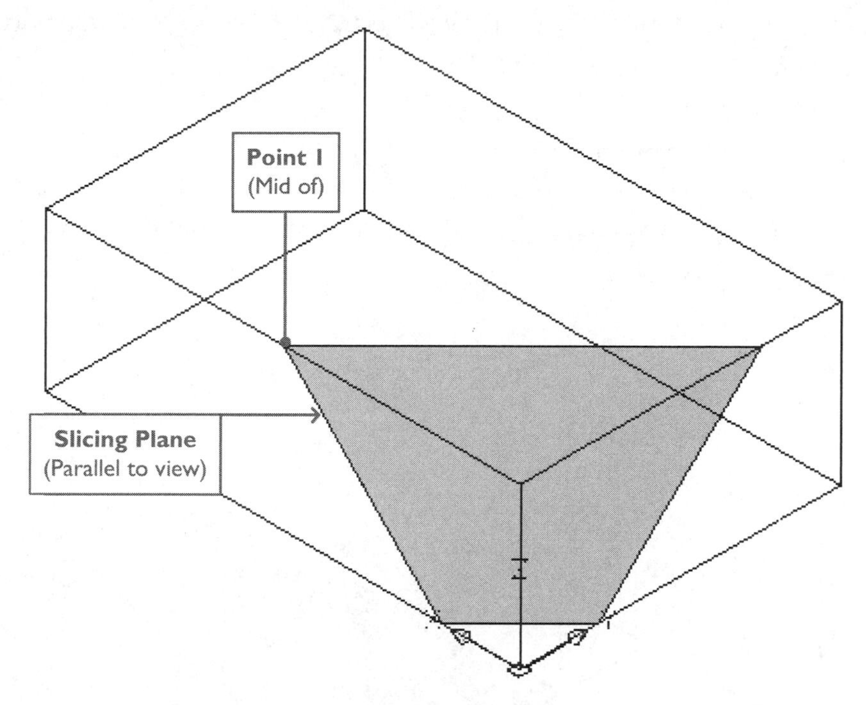

View option defines the cutting plane by the current view and a point.

6. Usually you want to keep both halves of the sliced objects; if not, pick a point on either side of the slicing plane to select the half to keep.

   Specify a point on desired side of the plane or [keep Both sides]: **b**

7. Use the **MOVE** command to move apart the two halves, if you so chose.

SLICING SOLIDS: ADDITIONAL METHODS

The **SLICE** command has a number of options for specifying the slicing plane, in addition to the **View** option.

- **3points** option defines the cutting plane by three points (default option).
- **XY**, **YZ,** and **ZX** options align the cutting plane with the x,y (or y,z or z,x) plane of the current UCS, along with a point.
- **Zaxis** option aligns the cutting plane is defined by a point on the plane, and a second point on the z axis to the plane.
- **Object** option aligns the cutting plane in the plane of a 2D object (circle, ellipse, arc, spline, or 2D polyline segment).

Let's look at them.

3Points

The **3points** option defines the cutting plane by three points. Remember to use object snaps, such as MIDpoint and ENDpoint.

Specify first point on slicing plane by [Object/Zaxis/View/XY/YZ/ZX/3points] <3points>: **3**

Specify first point on plane: **mid**

of *(Pick point 1.)*

Specify second point on plane: **mid**

of *(Pick point 2.)*

Specify third point on plane: **mid**

of *(Pick point 3.)*

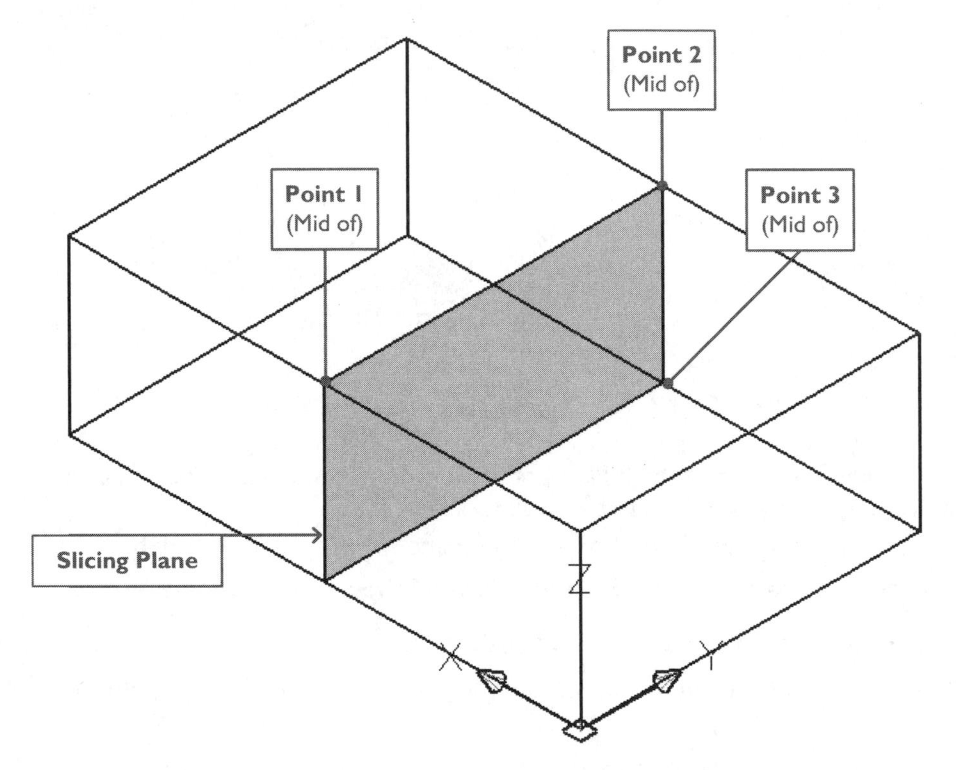

The 3point option defines the slicing plane by three points.

XY,YZ, ZX

The **XY**, **YZ**, and **ZX** options align the cutting plane with the x,y (or y,z or z,x) plane of the current UCS, along with a point.

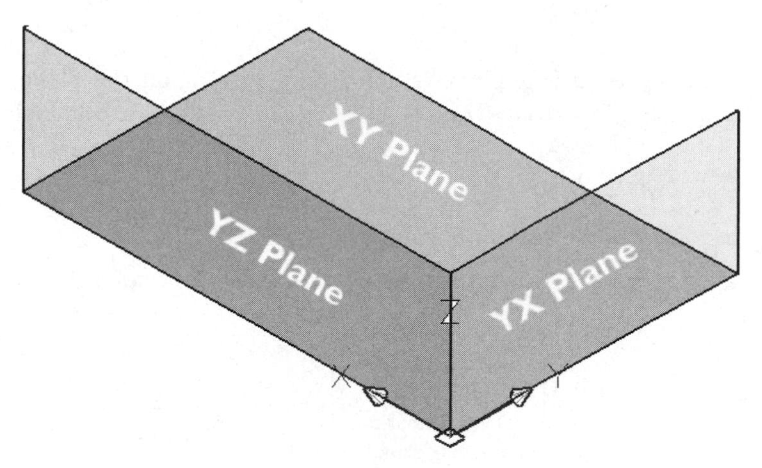

The three Cartesian planes are xy, yz, and yz.

These are easy but inflexible options — easy, because AutoCAD only needs only a single point from you; inflexible, because the slicing planes are restricted to being parallel to the three Cartesian planes:

Plane	*Point Needed*
x,y	z
y,z	x
z,x	y

Specify a plane, and then a point perpendicular to the plane. Here is an example for the x,y-plane:

Specify first point on slicing plane by [Object/Zaxis/View/XY/YZ/ZX/3points] <3points>: **xy**

Specify a point on the XY-plane <0,0,0>: *(Pick point 1, or coordinates, such as* **0,0,0.5***.)*

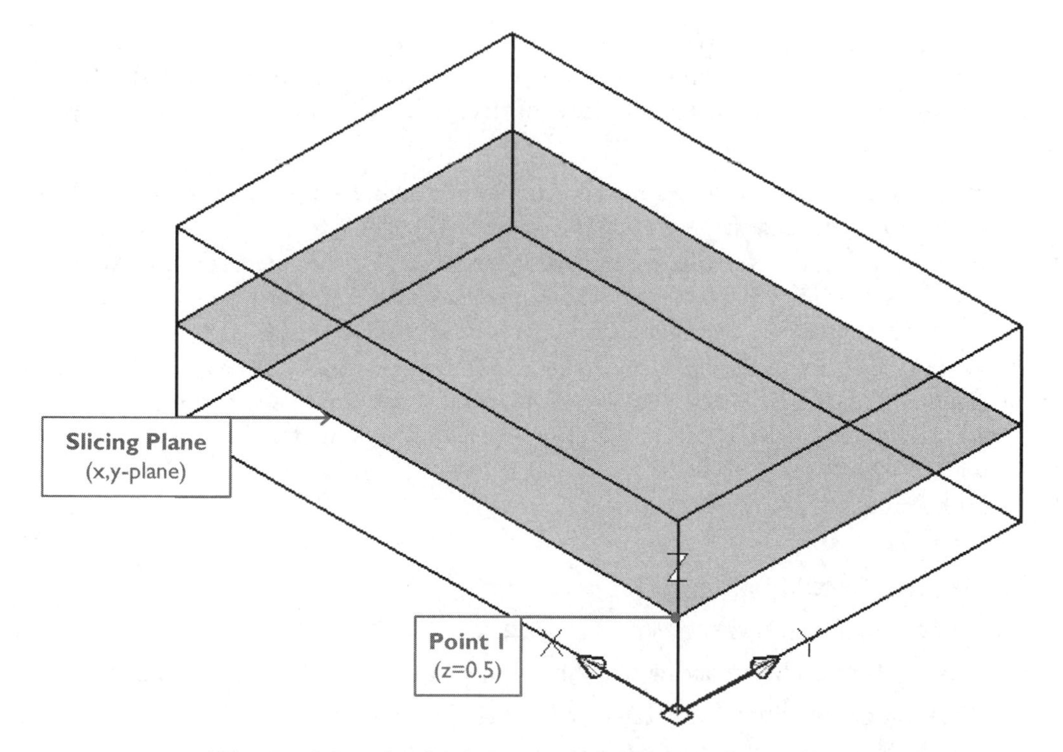

XY option defines the slicing plane by the height above the x,y-plane.

Zaxis

The **Zaxis** option defines the cutting plane is defined by a point on the plane, and a second point on the z axis to the plane. *Important*: This z-axis is perpendicular to the slicing plane, *not* to the UCS!

> [Object/Zaxis/View/XY/YZ/ZX/3points] <3points>: **z**
>
> Specify a point on the section plane: *(Pick point 1.)*
>
> Specify a point on the Z-axis (normal) of the plane: *(Pick point 2.)*

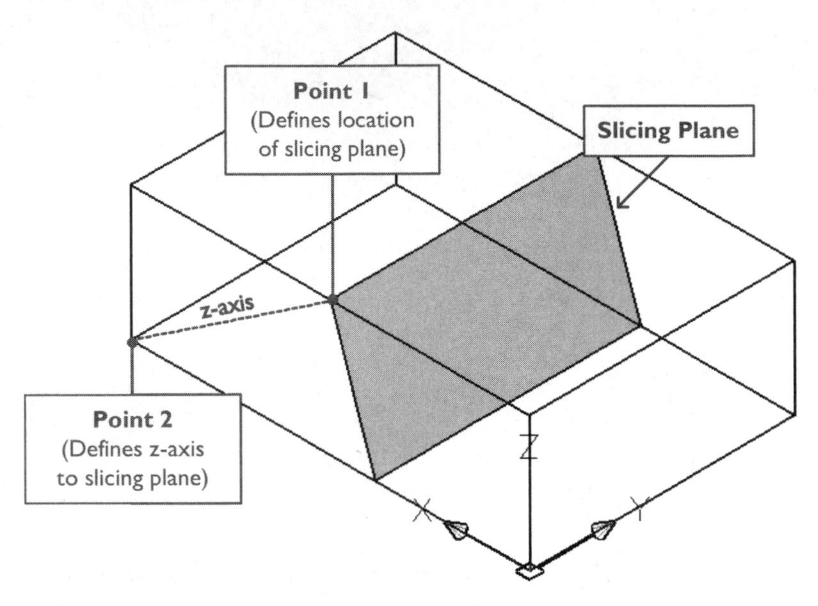

Two points define the slicing plane and its z-axis.

Object

The **Object** option aligns the cutting plane in the plane of a 2D object (circle, ellipse, arc, spline, or 2D polyline segment). Before drawing the object, you may need to reorient the UCS or change the elevation; otherwise the object might not intersect the solid model. Here's how, using a circle as the slicing object:

1. With the **BOX** command, draw a rectangular box with the dimensions of the monolith from *2001:A Space Odyssey* — length = 1, width = 4, height = 9.
2. With the **ELEV** system variable, change the elevation to 3 so that the circle is drawn through the box. (The value for the elevation doesn't matter, just as long it is somewhere within the height of the box (more than 0 and less than 9).
3. Draw the circle. The radius does not matter, because all AutoCAD needs is an object from which the slicing plane can be determined. In addition, the circle need not intersect or touch the solid, because it defines the plane. (Recall from geometry class that planes extend to infinity in all directions.)
4. Slice the monolith with the circle:

   Command: **slice**

   Select objects: *(Pick the box.)*

   Select objects: *(Press* **ENTER** *to end object selection.)*

   Specify first point on slicing plane by [Object/Zaxis/View/XY/YZ/ZX/3points] <3points>: **o**

   Select a circle, ellipse, arc, 2D-spline, or 2D-polyline: *(Pick the circle.)*

   Specify a point on desired side of the plane or [keep Both sides]: **b**

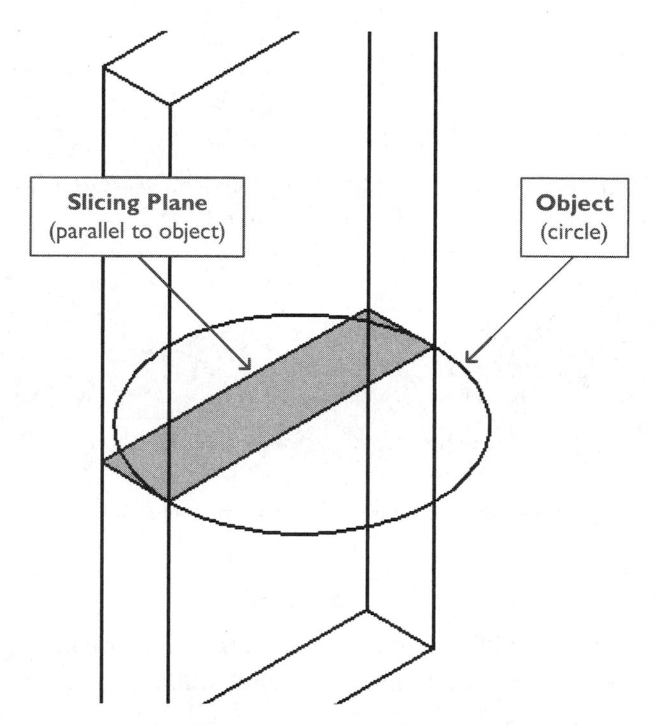

Plane defined by 2D object specifies the slice location.

Notice that the monolith has a rectangle on it. This is the slicing plane that separates the box into two pieces. If you like, move one part to prove to yourself the box is indeed sliced.

Note: One of the most common problems in 3D modeling is aligning the x,y-working plane to the angled face of a 3D solid. Here's how to do that using commands from the menu bar:

1. From the menu bar, select **Tools | New UCS | Face**.
2. Pick the edge of the face.
3. If AutoCAD highlights the wrong face, use the **Next** option to pick the other face abutting the edge. If the x and y axes of the UCS are not oriented correctly, use the **Xflip** and **Yflip** options to flip them

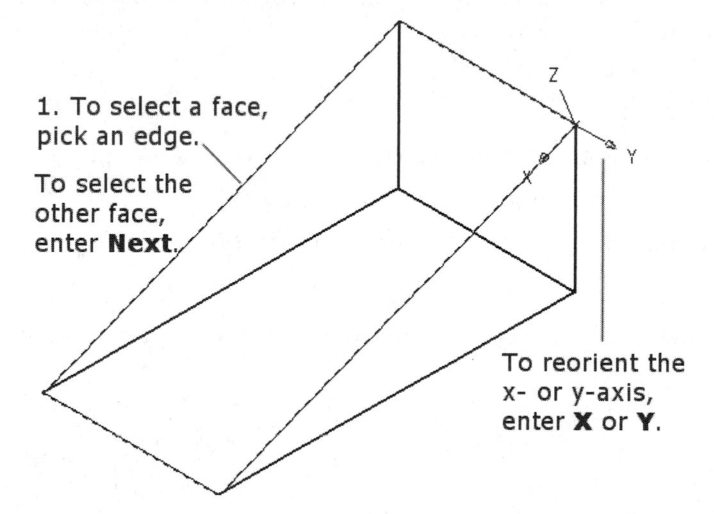

Orienting the working plane to angled faces.

4. When done, press **ENTER** to exit the **UCS** command.
5. Use the **PLAN Current** command to match the view to the new UCS.

SECTION

The **SECTION** command creates 2D regions from 3D solids.

This command uses the same technique as **SLICE** to specify the *section*, which is a 2D region. "Cross-section" might have been a better name for this command.

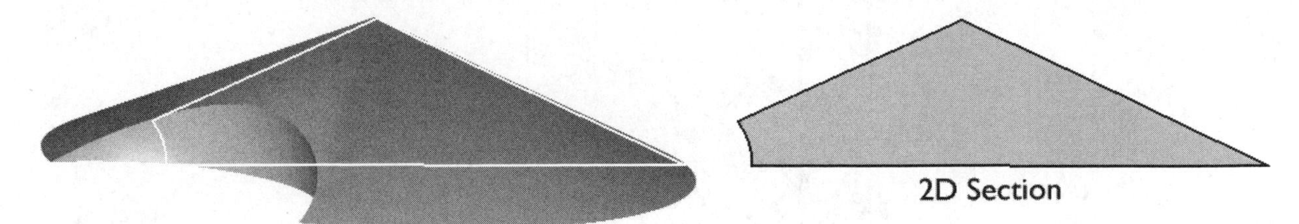

Section command creates 2D cross-sections of solid models.

BASIC TUTORIAL: 2D REGIONS FROM 3D SOLIDS

1. To create a 2D region from 3D solids, start the **SECTION** command:
   - From the **Draw** menu, choose **Solids**, and then **Section**.
   - At the 'Command:' prompt, enter the **section** command.
   - Alternatively, enter the alias **sec** at the keyboard.

   Command: **section** *(Press* **ENTER**.*)*
2. In all cases, AutoCAD prompts you to select the object from which to create the cross-section.

   Select objects: *(Select one or more solid models.)*

   Select objects: *(Press* **ENTER** *to end object selection.)*
3. Specify the section plane in the same manner as with the **SLICE** command.

   Specify first point on Section plane by [Object/Zaxis/View/XY/YZ/ZX/3points] <3points>: **v**

   Specify a point on the current view plane <0,0,0>: *(Pick a point.)*
4. Use the **MOVE** command to "pull" the section out of the solid model. AutoCAD stores the section in the Last selection set.

   Command: **move**

   Select objects: **l**

   Select objects: *(Press* **ENTER** *to end object selection.)*

   Specify base point or displacement: *(Pick a point.)*

   Specify second point of displacement: *(Pick another point.)*

You can use the **PROPERTIES** or **LIST** command to find the area and perimeter of the section.

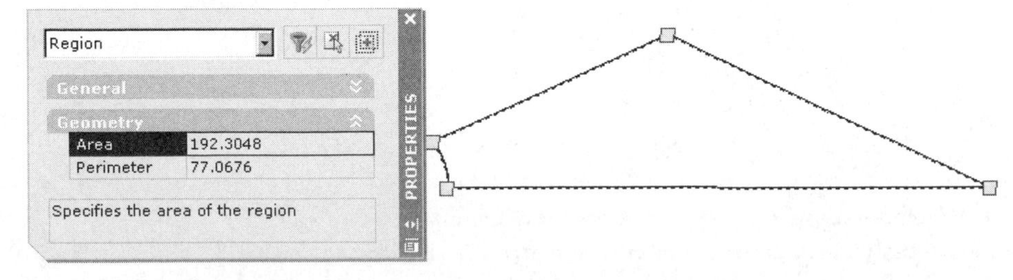

Properties command reports the section's area and perimeter.

MASSPROP

The **MASSPROP** command reports information about solid models, such as their area, perimeter, centroid, and so on (short for "mass properties").

TUTORIAL: REPORTING PROPERTIES

1. To find out information about solids, start the **MASSPROP** command:
   - From the **Tools** menu, choose **Inquiry**, and then **Region/Mass Properties**.
   - From the **Inquiry** toolbar, choose the **Mass Properties** button.
   - At the 'Command:' prompt, enter the **massprop** command.

   Command: **massprop** *(Press* ENTER.*)*
2. In all cases, AutoCAD prompts you to select objects. While you can select more than one object, it is better to pick just one, because of the amount of information generated.

   Select objects: *(Select solid object.)*

   Select objects: *(Press* ENTER *to end object selection.)*

   AutoCAD switches to the Text window, and displays the report:

```
----------------  SOLIDS  ----------------
Mass:                    4114.1195
Volume:                  4114.1195
Bounding box:            X: -20.4761  --  21.5163
                     Y: 11.0798  --  20.7801
                     Z: -60.3141  --  -18.3141
Centroid:                X: 1.5929
                     Y: 13.5965
                     Z: -39.9397
Moments of inertia:  X: 7608951.3091
                     Y: 7078292.1662
                     Z: 1020452.9262
Products of inertia: XY: 83775.7436
                     YZ: -2231040.1682
                     ZX: -230014.1591
Radii of gyration:   X: 43.0055
                     Y: 41.4788
                     Z: 15.7492
Principal moments and X-Y-Z directions about centroid:
                     I: 303884.0429 along [0.8643 -0.0306 -0.5021]
                     J: 505287.2270 along [0.0265 0.9995 -0.0154]
                     K: 231018.9604 along [0.5023 0.0000 0.8647]
```

   The data is relative to the current UCS.
3. AutoCAD asks if you wish to save the data to a file.

   Write analysis to a file? [Yes/No] <N>: **y**
4. If you answer **Y,** AutoCAD displays the Create Mass and Area Properties File dialog box. Enter a file name, and then click **Save**. AutoCAD creates the *.mpr* mass properties report file.
5. Press **F2** to return to the drawing window.

SOLID MODELING TUTORIAL 1

Let's draw a solid model using some of the commands from this chapter. You construct, edit, and display the model as a solid object. The figure below illustrates the finished model.

The completed model, rendered.

The figure below shows the dimensions of the 3D object. You may want to refer to it as you construct the base model from solid primitives.

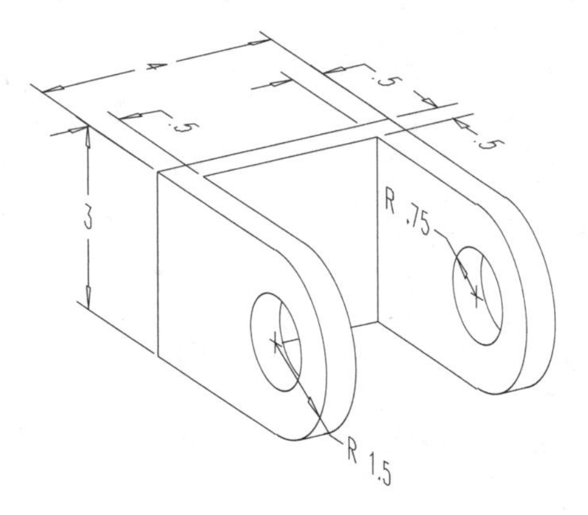

The dimensions of the model.

Before drawing any object, analyze it to determine the best primitives to use. In this tutorial, you use both solid modeling commands and some 3D principles to construct the model. The figure below illustrates the building block components used to "assemble" the model.

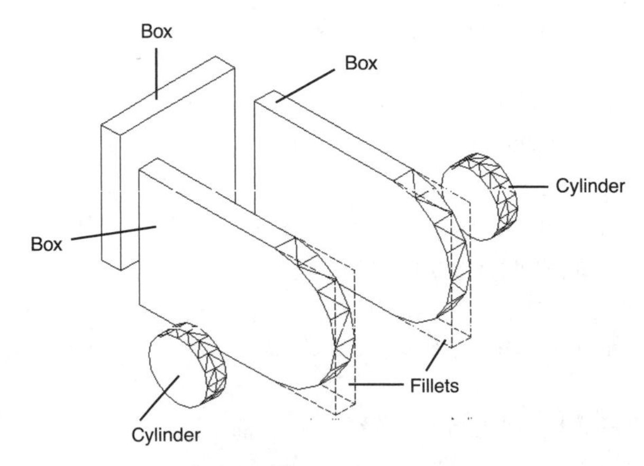

The solid model components.

1. Start AutoCAD with a new drawing.

   Set snap to 0.5.

2. Select **Box** from the **Draw | Solids** menu, and enter the points indicated in the following command sequence.

   Command: **box**

   Select the first corner of the box.

   Specify corner of box or [CEnter] <0,0,0>: **3,2**

   Define the location of the opposite corner of the box.

   Specify corner or [Cube/Length]: **@5,0.5**

   Now define the height of the box.

   Specify height: **3**

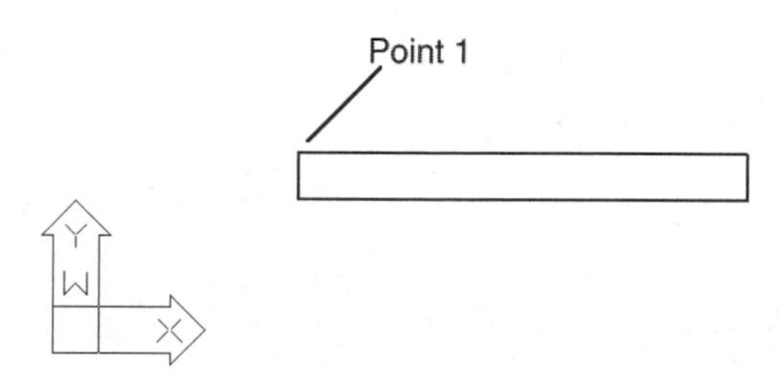

Constructing the first leg with a solid model box.

3. Repeat the **BOX** command. Refer to the figure for the points to enter.

   Command: *(Press spacebar to repeat the command.)*

   BOX Specify corner of box or [CEnter] <0,0,0>: *(Select point 1.)*

   Specify corner or [Cube/Length]: **@0.5,3**

   Specify height: **3**

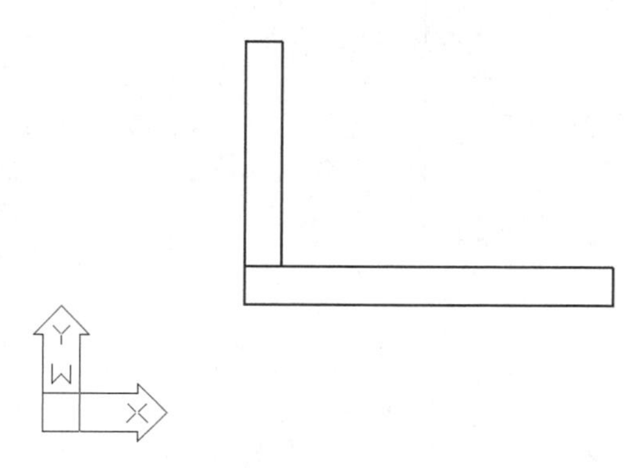

Adding a second box.

4. Use the **VPOINT** command to rotate the view so you can complete your work.

   Command: **vpoint**

   Current view direction: VIEWDIR=0.0,0.0,1.0

   Specify a view point or [Rotate] <display compass & tripod>: **r**

   Enter angle in XY plane from X axis <0>: **290**

   Enter angle from XY plane <0>: **23**

5. Use a zoom factor to "zoom out" the display of the drawing.

   Command: **zoom**

   Specify corner of window, enter a scale factor (nX or nXP), or [All/Center/Dynamic/Extents/Previous/Scale/Window] <real time>: **0.7x**

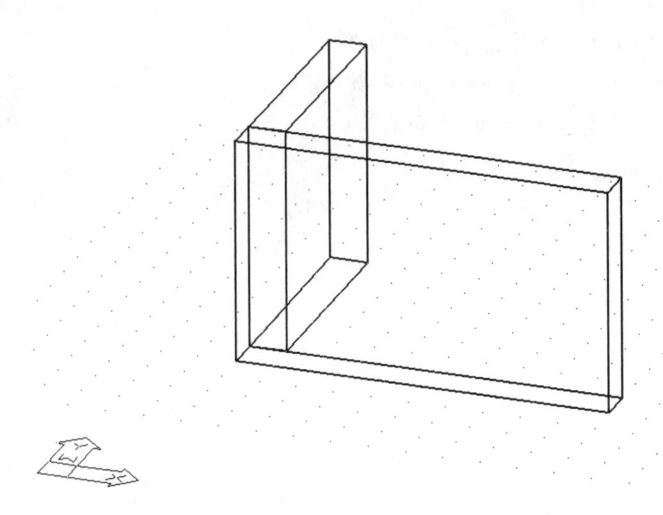

Rotating the viewpoint gives a better view of the model.

6. Add a drill hole to the side of the object. Begin by setting the UCS icon so you can see the origin of the UCS you will be working in.

   Command: **ucsicon**

   Enter an option [ON/OFF/All/Noorigin/ORigin] <ON>: **or**

7. Change the UCS to the lower-left corner of the model.

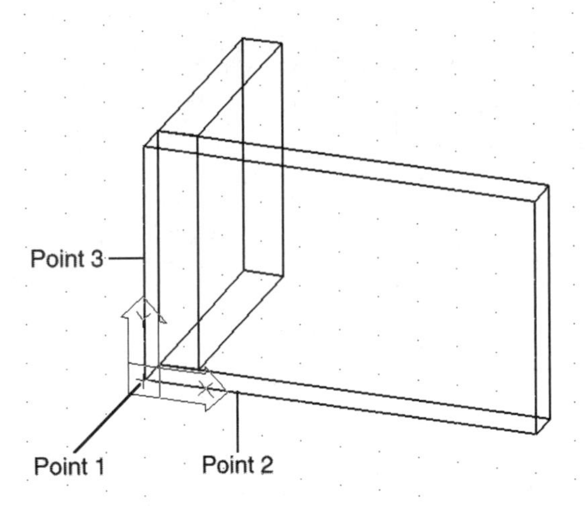

Changing the UCS with the 3-point method.

Command: **ucs**

Current ucs name: *WORLD*

Enter an option [New/Move/orthoGraphic/Prev/Restore/Save/Del/Apply/?/World] <World>: **n**

Specify origin of new UCS or [ZAxis/3point/ OBject/Face/View/X/Y/Z] <0,0,0>: **3**

Specify new origin point <0,0,0>: *(Select point 1.)*

Specify point on positive portion of X-axis <1.0,0.0,0.0>: **nea**

of *(Select point 2.)*

Specify point on positive-Y portion of the UCS XY plane <0.0,1.0,0.0>: **nea**

of *(Select point 3.)*

8. Now add the drill hole with the **CYLINDER** command.

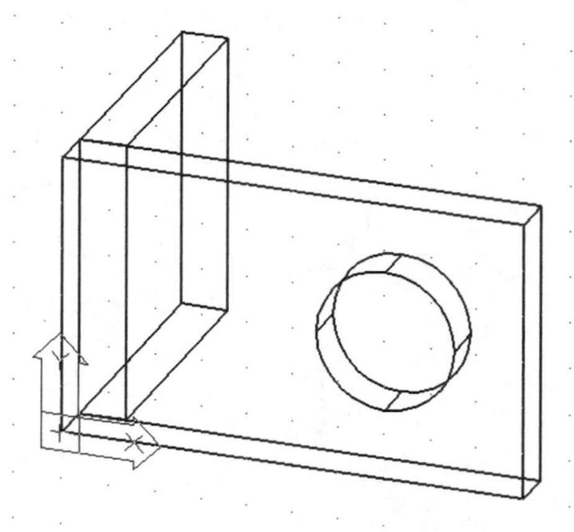

Holes start out as cylinders.

Command: **cylinder**

Specify the center point of the cylinder. Note that the absolute coordinates you use are relative to the origin of the new UCS.

Specify center point for base of cylinder or [Elliptical] <0,0,0>: **3.5,1.5**

Specify the radius of the cylinder.

Specify radius for base of cylinder or [Diameter]: **0.75**

Designate the extrusion height of the cylinder. Since you want the extrusion to extend in a negative direction, the value is negative.

Specify height of cylinder or [Center of other end]: **–0.5**

9. Rather than draw the second leg, make a copy of the first leg (with the cylinder):
   Set the UCS back to "World."
   Start the **COPY** command.
   Select the first box and its cylinder, and then specify a displacement of **@3.5<90**.

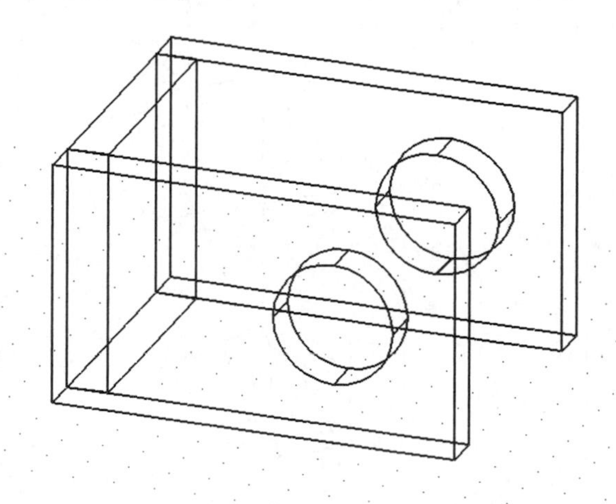

Copying the first leg to make the second leg.

10. Combine the solids as a single composite body. To do this, you must change the three boxes into a single solid object, and then subtract the cylinders that make the drill holes. You first use the **UNION** command to combine the boxes.

    Command: **union**

    Select objects: *(Select all three boxes and press* **ENTER**.*)*

    AutoCAD performs the necessary calculations to combine the boxes into a single solid object. You will notice that, when the drawing is redisplayed, the edge lines between the boxes are no longer a part of the object.

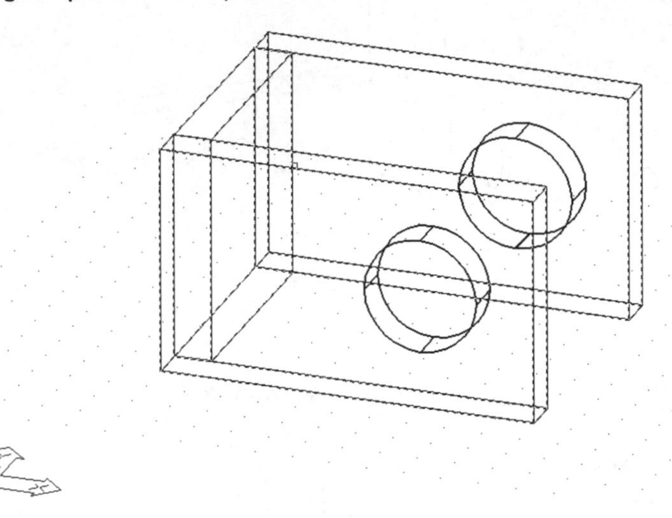

Union the three boxes into a single body.

11. Subtract the cylinders from the object with the **SUBTRACT** command.

    Command: **subtract**

    Select solids and regions to subtract from...

    Select objects: *(Select the unioned boxes.)*

    Select objects: *(Press* **ENTER**.*)*

    1 solid selected Select solids and regions to subtract...

    Select objects: (Select each of the cylinders.)

    Select objects: *(Press* **ENTER**.*)*

    AutoCAD removes the cylinders from the solid body created by the union of the three boxes. You will not see a difference.

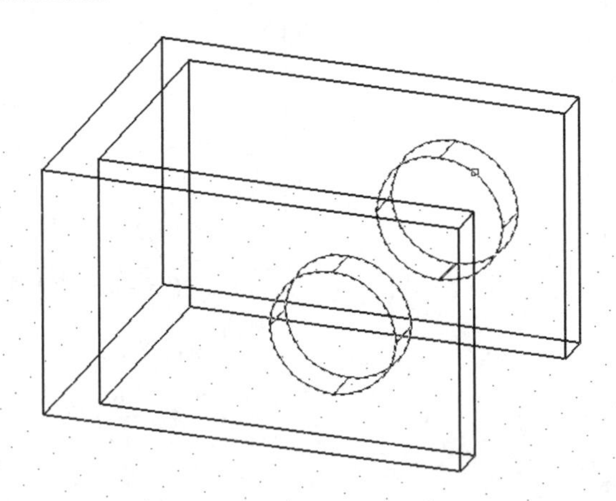

Subtracting cylinders from bodies creates round holes.

12. Next, fillet the corners.

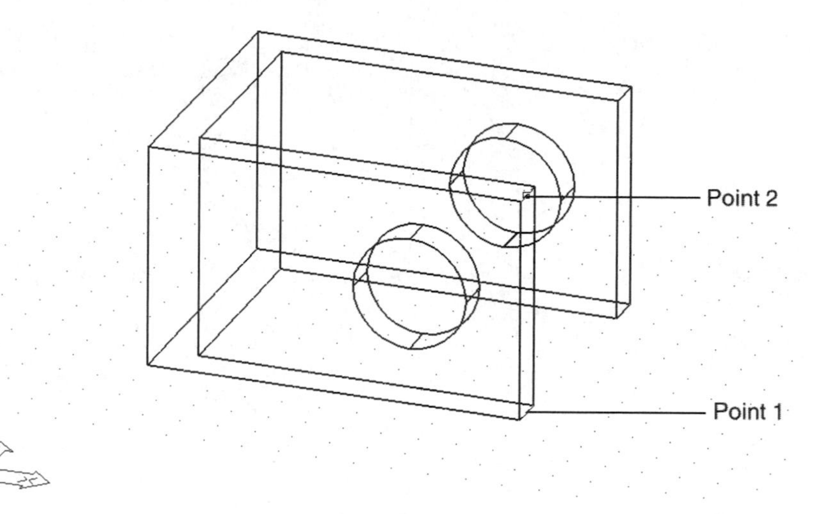

Fillet the ends of the legs.

It may be easier to select the edges with snap turned off, and to zoom in closer.

Command: **fillet**

Current settings: Mode = TRIM, Radius = 0.50

Select first object or [Polyline/Radius/Trim]: *(Select point 1.)*

Enter fillet radius <0.5000>: **1.5**

Select an edge or [Chain/Radius]: *(Select point 2.)*

Select an edge or [Chain/Radius]: *(Press* ENTER*.)*

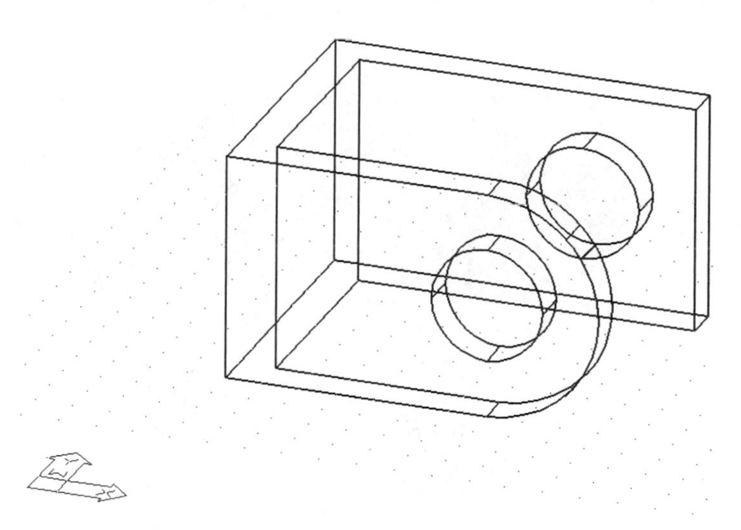

One leg filleted...

You may need to use the **REGEN** command to clean up the view.

13. Now use the **FILLET** command again to fillet the back leg of the object.

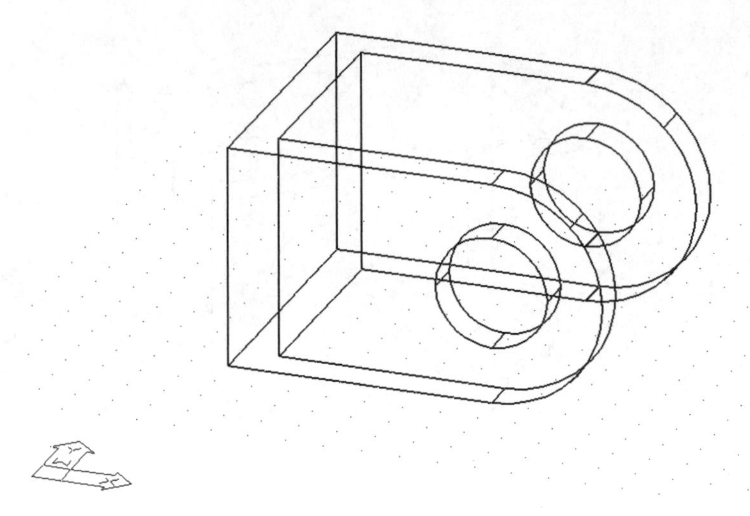

...and both legs filleted.

14. You can produce hidden-line and shaded models of your model.

    Command: **hide**

    From the menu bar, select **View | Shade | Gouraud**.

    Alternatively, you may want to use the **RENDER** command to produce a high-quality rendering of the model.

    Command: **render**

    ... and press **ENTER**.

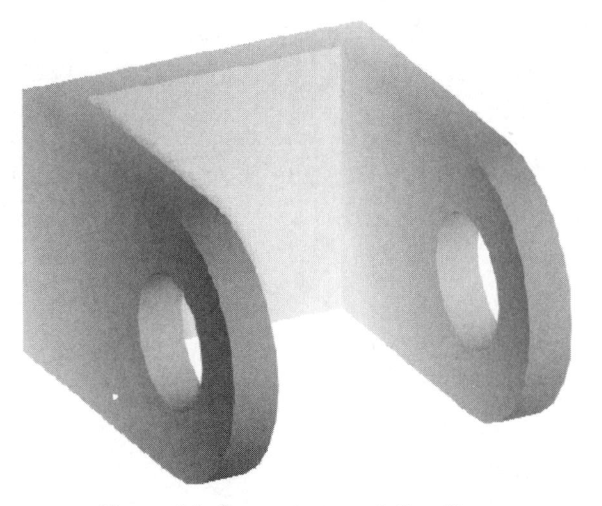

The model after rendering with fog effects.

15. Use the **MASSPROP** command to determine the properties of the model.
16. Save your work with the **SAVE** command, calling it "ch11model.dwg."

SOLID MODELING TUTORIAL II

The benefit to solid modeling is that you can create the same object in many ways. One approach may be preferable to another, because it requires fewer steps, or uses commands with which you are more familiar. The previous tutorial described one approach; here is a completely different approach developed by technical editor Bill Fane.

1. Start a new drawing.

   Set the grid and snap to 0.5
2. Draw the profile with the **PLINE** command (lines and arcs), and a circle.

   As an alternative, you can use the **LINE** command, and then create the rounded end with the **FILLET** command.

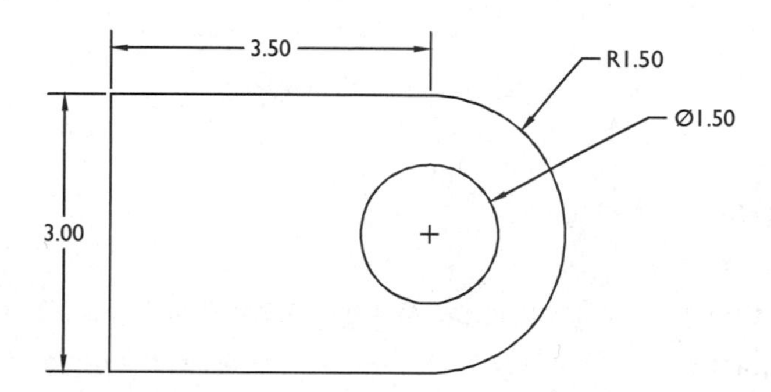

Profile made of polyline and circle.

2. Convert the profile into a region with the **REGION** command.

   Command: **region**

   Select all objects.

   Select objects: **all**

   Select objects: *(Press* **ENTER** *to end object selection.)*

   2 loops extracted.

   2 Regions created.

   The objects don't look any different, but now consist of two region objects.
3. Subtract the circle from the outside region with the **SUBTRACT** command.

   Command: **subtract**

   First select the outside region:

   Select solids and regions to subtract from ...

   Select objects: *(Pick the outside region.)*

   Select objects: *(Press* **ENTER** *to end object selection.)*

   And then the circle region:

   Select solids and regions to subtract ...

   Select objects: *(Pick the circle.)*

   Select objects: *(Press* **ENTER** *to end object selection.)*

   Again, the object looks no different, but now consist of a region with a hole (island).

4. Use the **VPOINT** command to rotate the view so you can see your work in 3D.

   Command: **vpoint**

   Current view direction: VIEWDIR=0.0,0.0,1.0

   Specify a view point or [Rotate] <display compass & tripod>: **r**

   Enter angle in XY plane from X axis <0>: **290**

   Enter angle from XY plane <0>: **23**

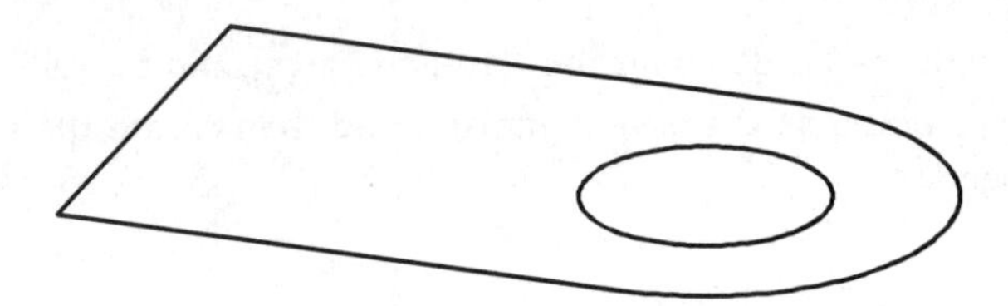

Region seen in 3D viewpoint.

5. To "tilt" the region up, rotate it about the x axis:

   Command: *(Select the region.)*

   Command: **rotate3d**

   Specify first point on axis or define axis by [Object/Last/View/Xaxis/Yaxis/Zaxis/2points]: **x**

   Specify a point on the X axis <0,0,0>: *(Pick a corner of the region.)*

   Specify rotation angle or [Reference]: **90**

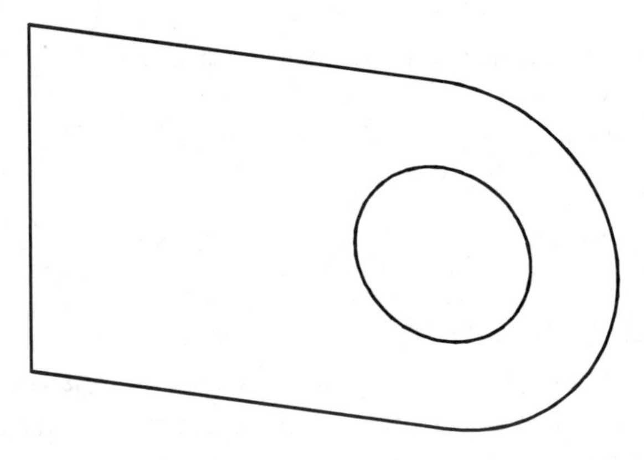

Region tilted about the x axis.

6. To convert the region into a solid model, use the **EXTRUDE** command. Give the model a thickness of 4.

   Command: **extrude**

   Current wire frame density: ISOLINES=4

   Select objects: *(Select region object.)*

   Select objects: *(Press **ENTER** to end object selection.)*

   Specify height of extrusion or [Path]: **4**

Specify angle of taper for extrusion <0>: *(Press **ENTER** for 0 degrees of taper.)*

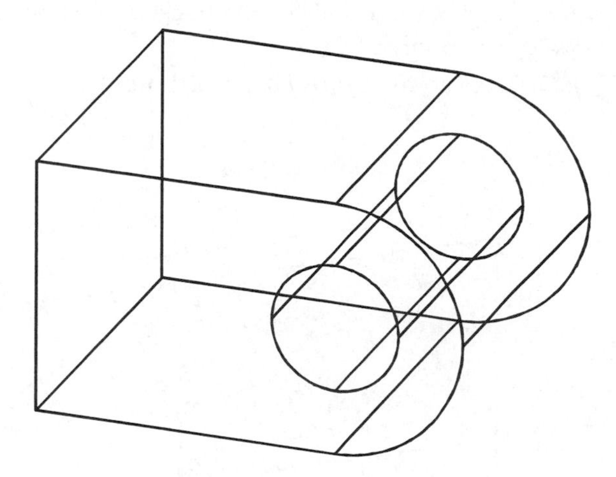

Model extruded to 4-inch width.

7. Before removing the "inside" of the part, set the UCS to the top face. That makes the model easier to work with.

   Command: **ucs**

   Enter an option [New/Move/orthoGraphic/Prev/Restore/Save/Del/Apply/?/World] <World>: **n**

   Specify origin of new UCS or [ZAxis/3point/OBject/Face/View/X/Y/Z] <0,0,0>: **f**

   Select face of solid object: *(Pick the top face, shown highlighted in the figure.)*

   Enter an option [Next/Xflip/Yflip] <accept>: *(Press **ENTER**.)*

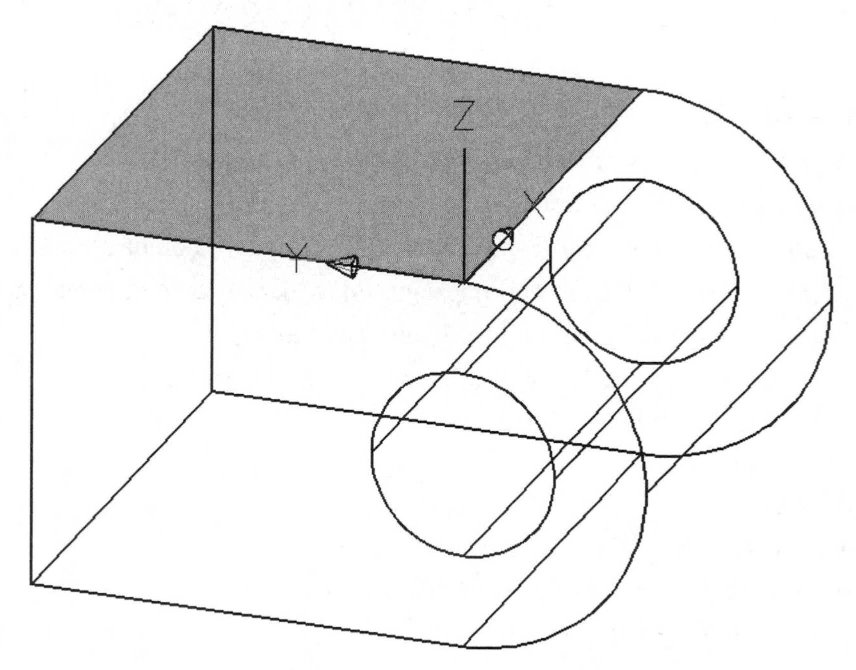

UCS set to the top face of the model.

8. To create the "removal volume," draw a 3D box on the top face with a negative depth. The box is offset by 0.5 units from three edges. Use tracking mode to locate the corner of the box. It's easier with ortho turned on.

   Command: *(Click* **ORTHO** *on the status bar.)* <Ortho on>

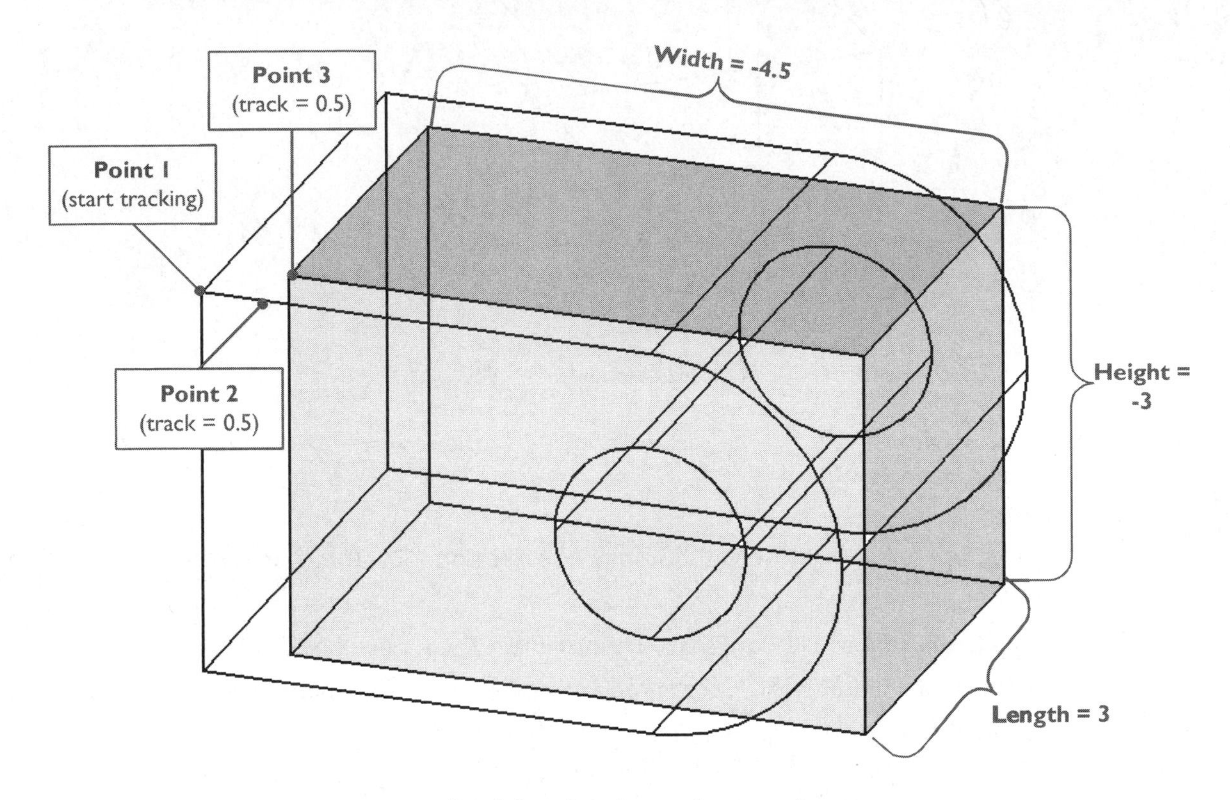

Box defines the volume to be removed.

   Command: **box**

   Specify corner of box or [CEnter] <0,0,0>: **tk**

   First tracking point: *(Pick point 1.)*

   Next point (Press ENTER to end tracking): *(Move cursor to point 2, and then type* **0.5.***)*

   Next point (Press ENTER to end tracking): *(Move cursor to point 3, and then type* **0.5.***)*

   Next point (Press ENTER to end tracking): *(Press* **ENTER.***)*

   Tracking has located the starting point of the box, which you now draw:

   Specify corner or [Cube/Length]: **l**

   Specify length: **3**

   Specify width: **-4.5**

   Specify height: **-3**

9. Use the **SUBTRACT** command to subtract the box from the model.

   Command: **subtract**

First select the model:

Select solids and regions to subtract from ...

Select objects: *(Pick the model.)*

Select objects: *(Press* **ENTER** *to end object selection.)*

And then the box:

Select solids and regions to subtract ...

Select objects: *(Pick the box.)*

Select objects: *(Press* **ENTER** *to end object selection.)*

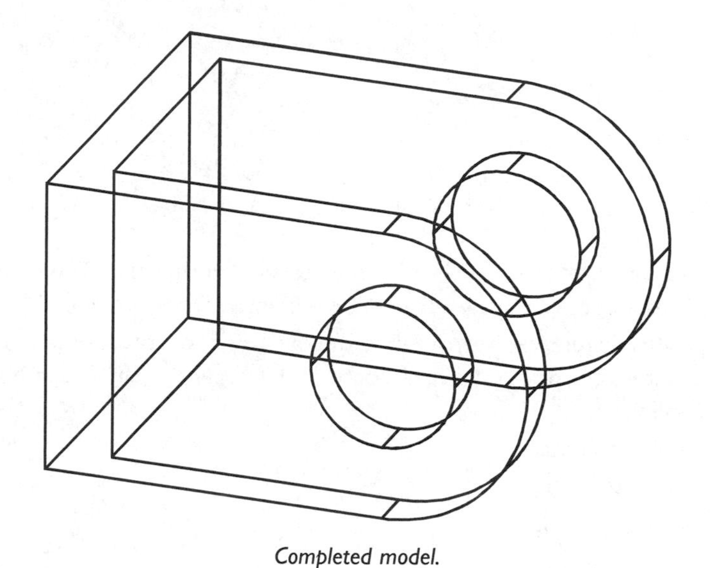

Completed model.

SOLID MODELING TUTORIAL III

In this tutorial, you learn how to extrude a closed object along a spline path. This is a way to model tubes, handrails, and piping with bends. The tricky part is drawing the profile at right angles to the path. Let's draw a tube.

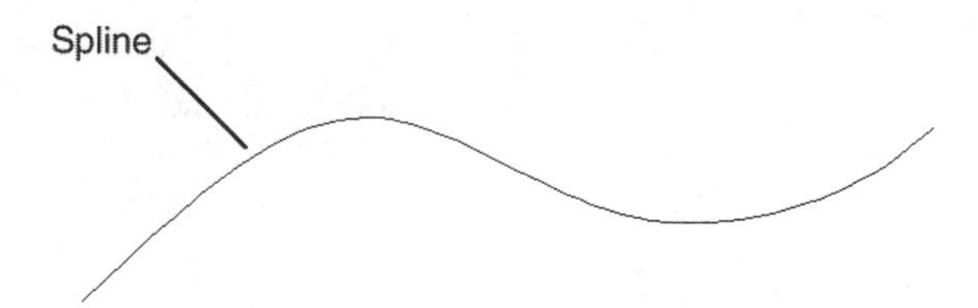

Spline defines extrusion path and direction.

1. Draw the path for the tube with the **SPLINE** command. Don't draw the spline with tight curves; otherwise the extrusion may fail.
2. Change the UCS so that you are looking at the end of the spline. This lets you draw the profile at a right angle. Rotate the UCS by 90 degrees about the y axis.

   Command: **ucs**

   Current ucs name: *WORLD*

   Enter an option [New/Move/orthoGraphic/Prev/ Restore/Save/Del/Apply/?/World]
   <World>: **y**

   Specify rotation angle about Y axis <90>: **90**

3. Set the view to match the UCS with the **PLAN** command:

   Command: **plan**

   Enter an option [Current ucs/Ucs/World] <Current>: *(Press* **ENTER**.*)*

   You may want to use the **ZOOM Window** command to see the spline better.

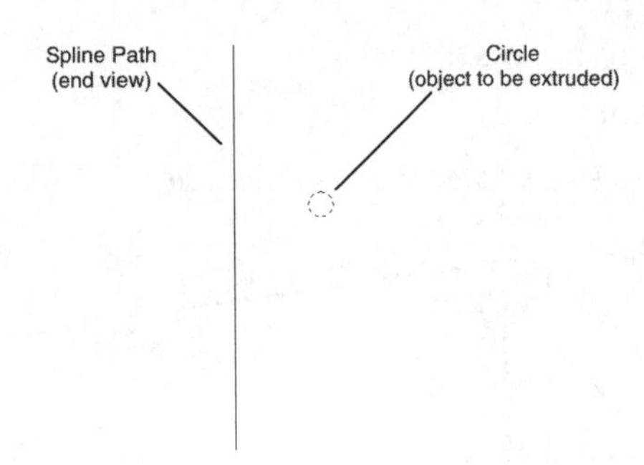

Profile drawn at right angles to the path.

4. Draw a shape to define the profile — the cross-section of the tube. The profile can be any closed shape, such as circle, ellipse, polygon, or rectangle. Draw a circle for this tutorial. You should draw the circle where you want the extrusion located, because the extrusion extends out of the circle, not along spline path. Don't draw the circle too big; otherwise the extrusion will fail.
5. Apply the **EXTRUDE** command:

   Command: **extrude**

   Select objects: *(Pick the circle.)*

   1 found Select objects: *(Press* **ENTER**.*)*

   Specify height of extrusion or [Path]: **p**

   Select extrusion path: *(Pick spline curve.)*

   Path was moved to the center of the profile.

   Profile was oriented to be perpendicular to the path.

   AutoCAD extrudes the circle along the path. Particularly complex paths and objects may take a long time to extrude on slow computers.

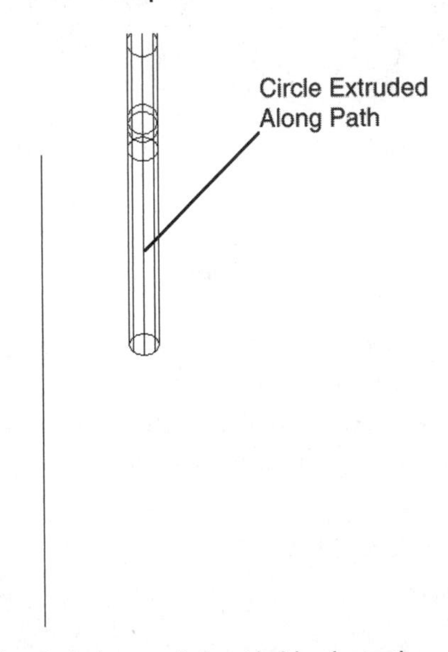

The circle is extruded, guided by the path.

6. Once AutoCAD is finished, change the viewpoint so that you can see the tube. Use the **3DORBIT** command. For the view in the figure below, I entered **VPOINT -1,1,0.7** .
7. If necessary, use the **ISOLINES** system variable to make the extrusion look better. Use the **HIDE** or **RENDER** commands to see the tube with hidden lines removed. The figure illustrates tubes drawn with a variety of profiles.

Tubes created with square, pentagon, closed polygon, triangle, and elliptical profiles.

EXERCISES

1. Draw these 3D solid objects with the parameters provided.

a. Sphere:	Radius = 2		
b. Wedge:	Cube = 2		
c. Cylinder:	Radius = 1.5	Height = 2.5	
d. Cone:	Diameter = 1	Height = 3	
e. Torus:	Torus Diameter = 4.8	Tube Diameter = 1.2	
f. Box:	Length = 3	Width = 2	Height = 1

2. From the CD-ROM, open the *Ch11Glass.dwg* drawing file, a profile of a goblet. Use the **REVOLVE** command to create the goblet.

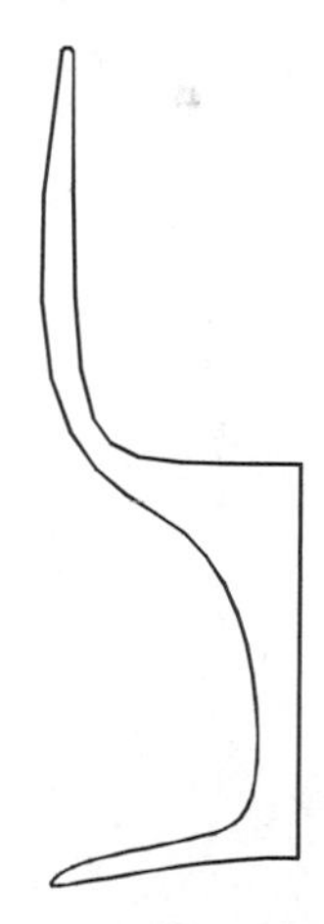

Ch11Glass.dwg: Profile of drinking goblet.

3. Create a profile of your own, and then apply the **REVOLVE** command.

4. From the CD-ROM, open the *Ch11Extrude.dwg* drawing file, a profile of a gear.
   a. Use the **EXTRUDE** command to create the gear 0.5 units thick.
   b. Repeat, but with a taper angle of 15 degrees.

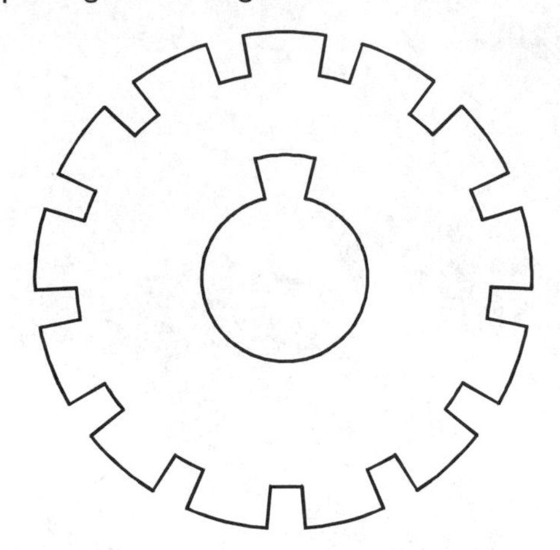

Ch11Path.dwg: Extrusion of a gear.

5. From the CD-ROM, open the *Ch11Path.dwg* drawing file, a pentagon profile and a path. Use the **EXTRUDE** command to extrude the pentagon along the spline path.

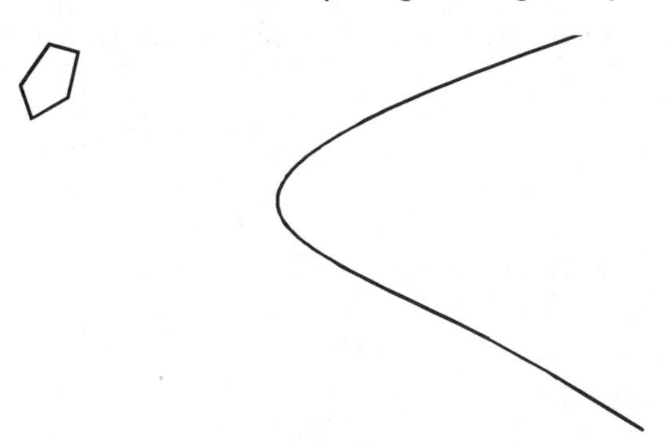

Ch11Path.dwg: Profile and path.

6. Create a profile of your own, and then apply the **EXTRUDE** command.

7. From the CD-ROM, open the *Ch11Pump2.dwg* file, a drawing of a pump. Use the **SLICE** command to show the inside of the pump.

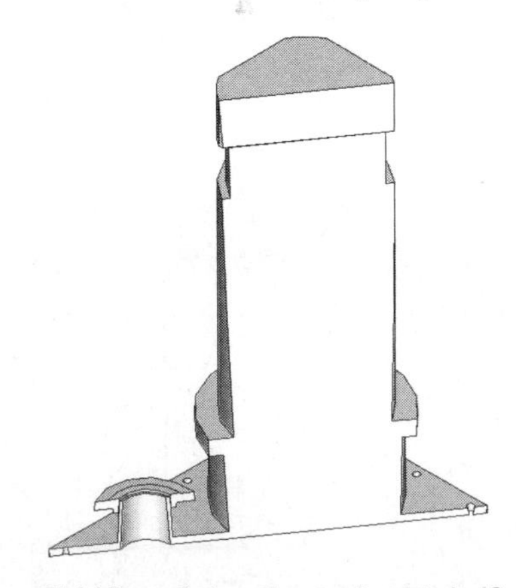

Ch11Pump2.dwg: Pump sliced in half.

8. From the CD-ROM, open the *Ch11Pump.dwg* file, a drawing of a pump.
   Use the **INTERFERE** and **MOVE** commands to isolate the interference solids.

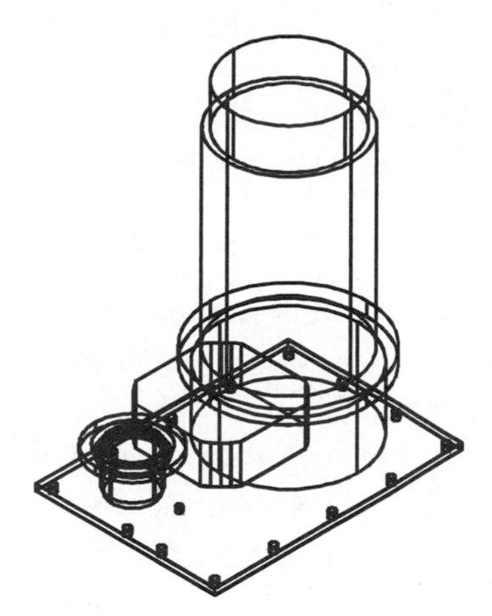

Ch11Pump.dwg: Pump

9. Use the **SECTION** command to obtain a cross-section profile of the bevel gear you extruded in exercise #4.

10. From the CD-ROM, open the *Ch11Sandpile.dwg* file, a drawing of a pile of sand, where a bulldozer took away a chunk of it.

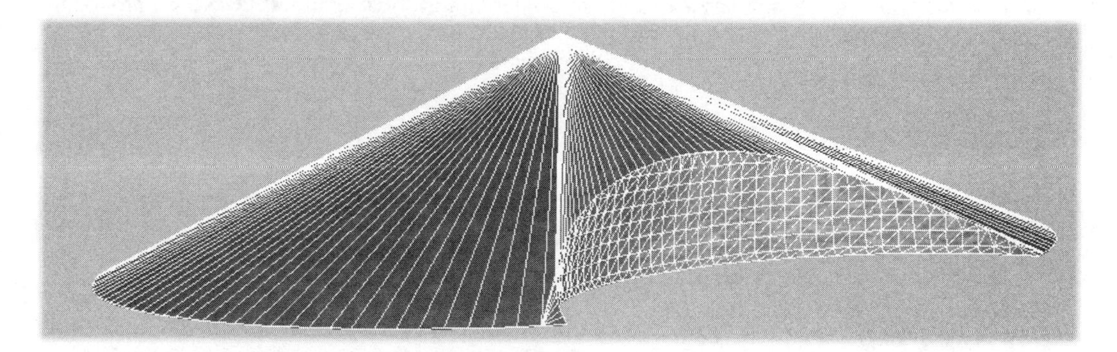

Pile'o sand with a bite taken outa it.

   a. Use the **MASSPROP** command to find the volume of sand (in cubic feet) in the pile.
   b. Assuming sand weights 90 pounds per cubic foot, how much does the pile weigh?

CHAPTER REVIEW

1. What is an *extrusion*?
2. What objects can be extruded?
3. What is created with the **REVOLVE** command?
4. What are some limitations to the **REVOLVE** command?
5. How can objects be revolved?
6. What kind of a shape is a torus?
7. What two types of solid cone bases can be constructed?
8. Does the **INTERSECT** command remove everything in common from two solid objects?
9. Does the **UNION** command combine two solid objects to create a region?
10. Why are the **UNION, INTERSECT**, and **SUBTRACT** commands called Booleans?
11. Describe how to create a bolt hole in a solid object.
12. Briefly explain the purpose of the **SLICE** command.
13. Does the **INTERFERE** command create solid objects from the intersection of two or more solid objects?
14. Match the command with its purpose:
    a. **INTERSECT** i. Cuts a solid model into two parts.
    b. **INTERFERE** ii. Removes one solid model from a second.
    c. **SUBTRACT** iii. Removes all but the solid volumes in common.
    d. **SLICE** iv. Creates a solid body defined by the intersection of two solid models.
15. With what does the **SLICE** command do its slicing?
16. Choose from the following list the kinds of object the **EXTRUDE** and **REVOLVE** commands convert into 3D solids:
    a. Self-intersecting polylines.
    b. Closed 2D objects.
    c. Open 2D objects.
    d. Lines.
17. Can the **REVOLVE** command revolve objects less than a full 360 degrees?
18. Which one of the following is the best definition of *planar*?
    a. All vertices lie in the same plane.
    b. The object is made of a plane.
    c. Part of a wing.
    d. Points in all directions at once.
19. What is the purpose of the **ISOLINES** system variable?

Editing Solid Models

To create more sophisticated designs with solid models, AutoCAD provides you with a number of commands that let you further shape and change solid models. In addition, a trio of commands generate 2D plans from 3D solid models.

In this chapter, you learn about:

CHAMFER chamfers the edges of solid models.

FILLET fillets the edges of solid models.

SOLIDEDIT changes edges, faces, and bodies.

SOLVIEW creates floating viewports with standard orthographic views: top, front, and sides.

SOLDRAW extracts the 2D images of the 3D solid model.

SOLPROF creates pictorial views of 3D solid models.

FINDING THE COMMANDS

On the **SOLIDS** and **SOLIDS EDITING** toolbars:

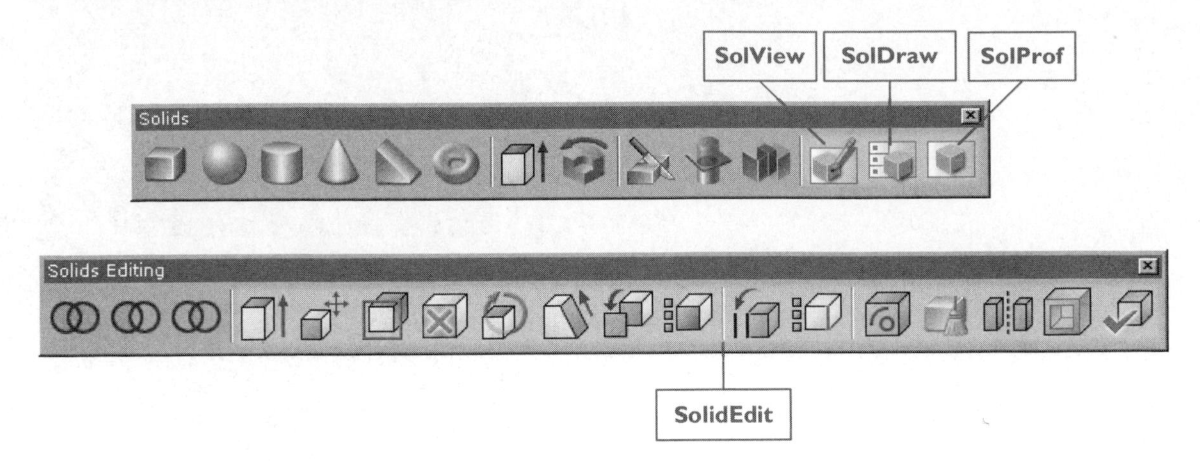

On the **DRAW** and **MODIFY** menus:

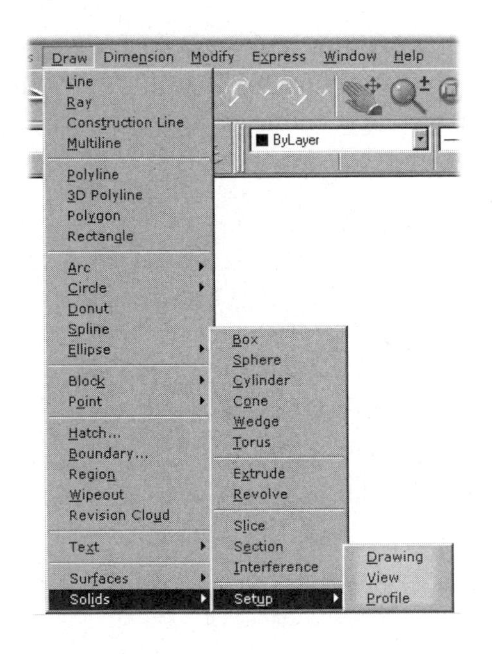

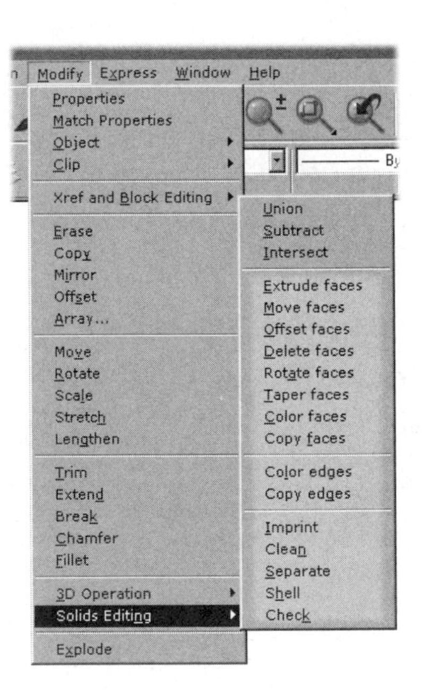

"2D" EDITING COMMANDS

AutoCAD has two groups of commands for editing solids. Some are "2D" editing commands that you are already familiar with from *Using AutoCAD: Basics*; other are 3D solids editing commands. The reason for the quotation marks around "2D" is that you've probably only used the commands for editing 2D objects, but some of them work equally well with 3D solids.

Specifically, these include (some commands are discussed in this or earlier chapters):

- **ALIGN** moves, resizes (scales), and/or rotates solid models in 3D space (Chapter 10).
- **CHAMFER** cuts the edges of faces (this chapter).
- **ERASE** erases the selected solid models from the drawing.
- **EXPLODE** converts a 3D solid model into 2D regions (this chapter).
- **FILLET** rounds the edges of faces (this chapter).
- **MIRROR** and **MIRROR3D** make mirror copies of the solid objects (Chapter 10).
- **MOVE** moves the solid model in 3D space (by any distance in the x, y, and/or z directions).
- **ROTATE** and **ROTATE3D** rotate the solid object in two and three dimensions (Chapter 10).
- **SCALE** changes the size of the solid object.
- **CHANGE**, **CHPROP**, and **PROPERTIES** commands change some properties of solid models, specifically color, layer, linetype, linetype scale, lineweight, and embedded hyperlink.

These 2D editing commands work as you would expect. Erasing and rotating 3D solid models are just like erasing and rotating 2D objects. Two commands, however, have been especially adapted for solid models: CHAMFER and FILLET. They affect the edges of solid objects, as shown by the figure.

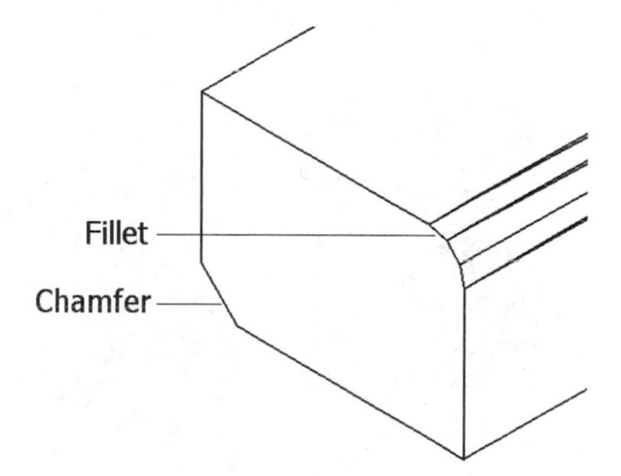

The Chamfer and Fillet commands operate on the edges of solid models.

These two commands work on *edges* only (the intersection between two faces); they do not work on round parts of solid models, such as spheres, cylinders, and curves. You can fillet and chamfer where edges meet curved surfaces.

EXPLODE

The EXPLODE command converts solid models into regions and bodies; flat surfaces become 2D regions, while curved surfaces become 3D bodies.

When you apply the command again to the resulting regions and bodies, the 2D regions become lines and arcs. Bodies become circles and arcs. (The isolines disappear.)

The figure below shows a solid model exploded into regions and bodies, and then pulled apart with the **MOVE** command.

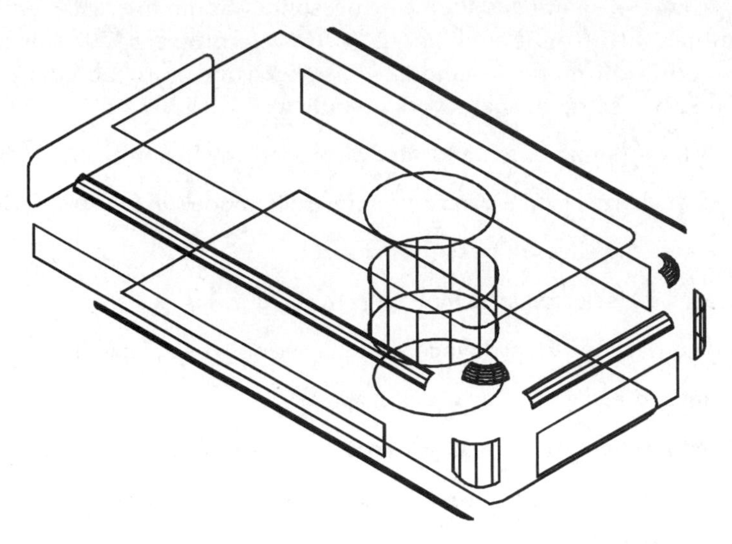

A sold model exploded into regions and bodies, then "pulled" apart.

LINETYPES

Applying linetypes to solid models results in funky effects, such as the Batting linetype applied to the sphere shown below, with isolines set to 6. The linetype effect, however, applies only to the isolines that define the surface and edges of the solid, and not to the solid itself.

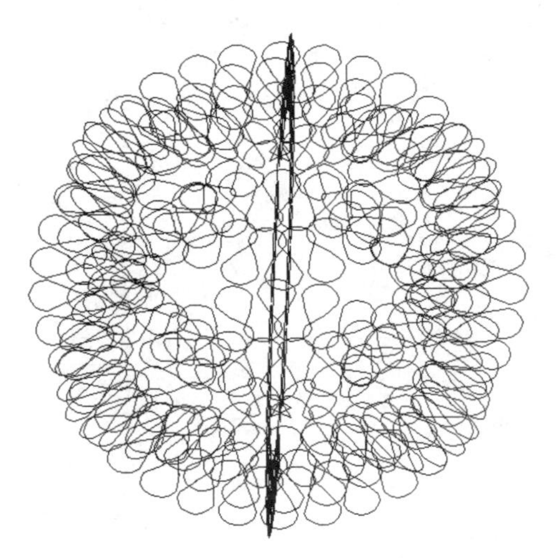

Applying the Batting linetype to the isolines defining a sphere.

2D COMMANDS THAT "DON'T WORK"

Not all 2D editing commands work with solid models. The **STRETCH** command moves solid models, but does not stretch them. **EXTEND** and **LENGTHEN** apply only to open objects, which solid models are not; the workaround is to use the **SOLIDEDIT** command, described later in this chapter.

BHATCH, **HATCH**, and **HATCHEDIT** don't work, because solid models cannot be hatched. The closest workaround is to change their color with the **COLOR** command; alternatively, use the **RMAT** command to apply a material to the solid model, which is then displayed by the **RENDER** command.

You cannot break solid models with the **BREAK** command. As a workaround, use the **SLICE** command to cut off a portion of the solid model.

DIVIDE and **MEASURE** do not apply to solid models. The **TRIM** command does not work with solid models; the workaround is to use the **INTERSECT** or **SLICE** commands. Solid models cannot be offset with the **OFFSET** command; the workaround is to use the **COPY** or **SOLIDEDIT** commands.

Note: AutoCAD LT can view, but not edit solid models.

SELECTING EDGES AND ISOLINES

Some commands ask you to select a *chain* (two or more edges). Others prompt you to select a *loop* (all the edges belonging to one face). Here is a reminder of solid modeling terminology:

Body is the entire solid model.

Faces are the flat and round sides of bodies.

Edges are the intersections between faces; rectangular faces have four edges.

Vertices are the intersections of edges.

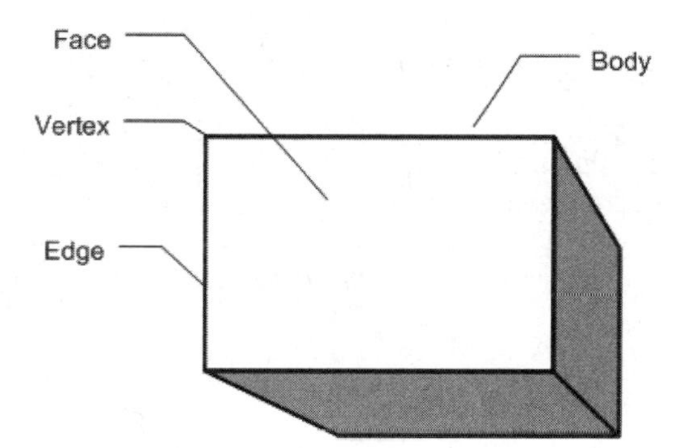

The parts of a 3D solid model.

When you pick an edge, AutoCAD highlights the adjacent faces, and reports the number of faces it found. The figure below shows the two faces highlighted: the side and the top.

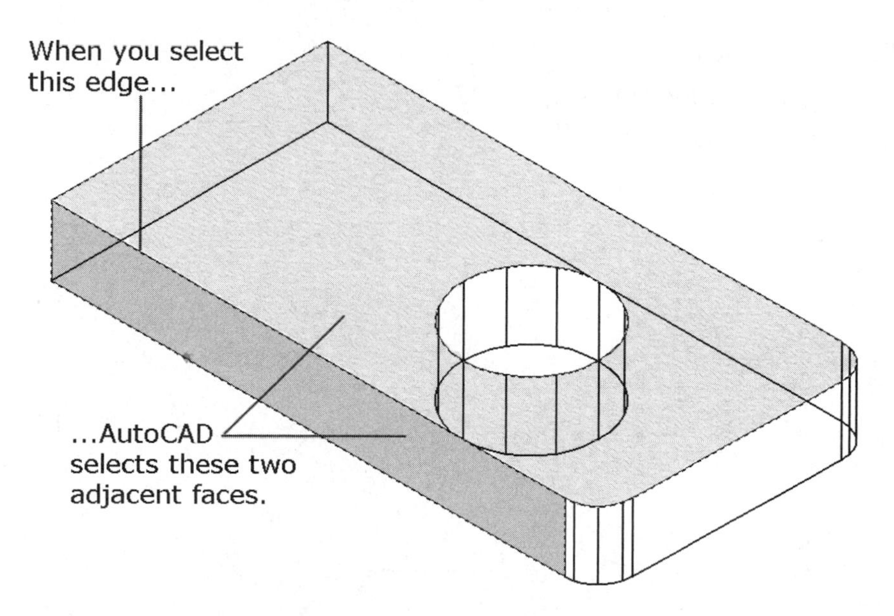

Picking one edge selects two faces.

To remove the faces you don't want to work with, use the **Remove** option to pick an *unique* edge (one that is not shared by the two selected faces).

Select faces or [Undo/Remove/ALL]: **r**

Remove faces or [Undo/Add/ALL]: *(Pick an unique edge).*

2 faces found, 1 removed.

Press **ENTER** when you are finished with the selection process:

Remove faces or [Undo/Add/ALL]: *(Press* ENTER *to end the selection process.)*

Notice that AutoCAD now highlights the single face (shown by dashed lines).

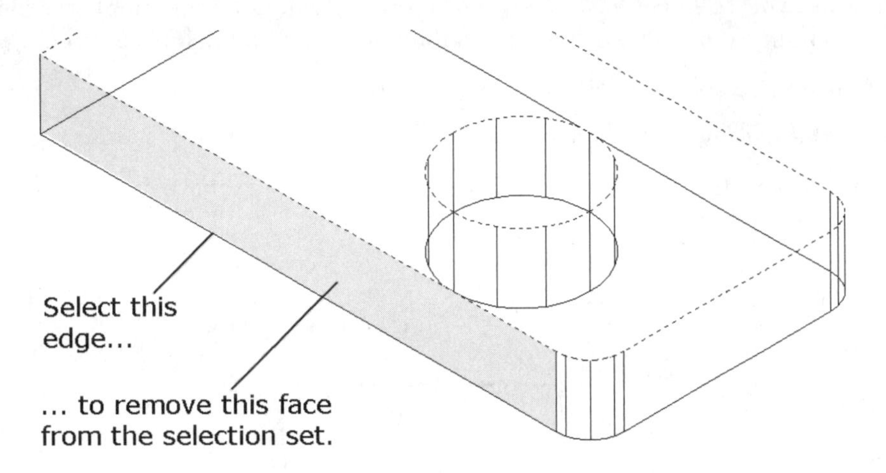
Select this
edge...
... to remove this face
from the selection set.

Pick a unique edge to de-select faces.

SELECTING ISOLINES

Curved surfaces have *isolines*, the vertical and curved lines that show you the curvature of the face. When you select a sphere or a fillet, you select its entire (single) face.

Isolines are a display aid, like grid dots. You can vary the number of isolines displayed on curved solids with the **ISOLINES** system variable, followed by the **REGEN** command. You can enter a value ranging from 0 to 2047, but a value of 12 or 16 looks best.

Command: **isolines**

Enter new value for ISOLINES <4>: **12**

Command: **regen**

Regenerating model.

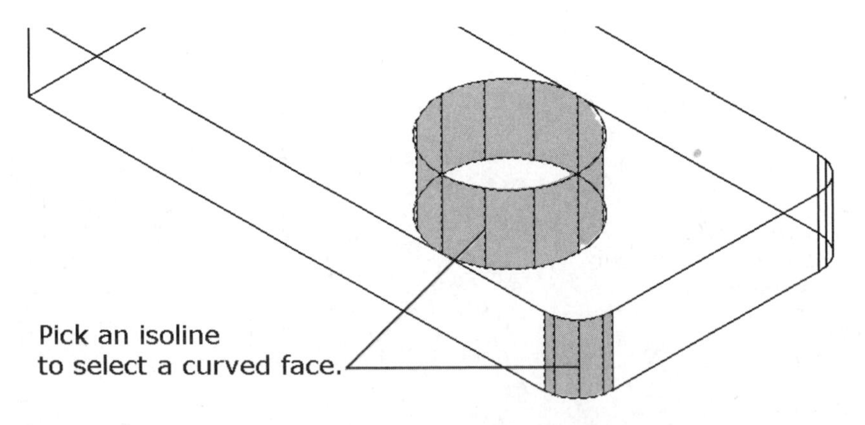
Pick an isoline
to select a curved face.

Select isolines to select curved faces.

CHAMFER

The **CHAMFER** command chamfers the edges of 3D solid objects.

When you select a solid model, this command automatically switches to solids-editing mode, and presents a set of prompts different from those you saw in 2D editing (see Chapter 10 in *Using AutoCAD: Basics*). This command operates on a single solid model at a time, but allows you to chamfer as many edges as you need during a single operation. (It does not work with surface models.)

BASIC TUTORIAL: CHAMFERING OBJECTS

1. To chamfer the intersection of two faces of a solid model, start the **CHAMFER** command:
   - From the menu bar, choose **Modify**, and then **Chamfer**.
   - From the **Modify** toolbar, choose the **Chamfer** button.
   - At the 'Command:' prompt, enter the **chamfer** command.
   - Alternatively, enter the **cha** alias at the 'Command:' prompt.

   Command: **chamfer** *(Press* ENTER.*)*
2. In all cases, AutoCAD displays prompts familiar to you from 2D editing:

   (TRIM mode) Current chamfer Dist1 = 0.0000, Dist2 = 0.0000

   Select first line or [Polyline/Distance/Angle/Trim/Method/mUltiple]: *(Pick a solid model.)*
3. In selecting the solid model, you pick an edge (the line between two faces) or an isoline (the lines that define curved surfaces). When you select an edge, AutoCAD creates a bevel between two faces; select an isoline to create a countersink.
   - **Select an edge,** and AutoCAD prompts you to select the *base surface* (a.k.a. the first face). Because the two chamfer distances (Dist1 and Dist2) can be unequal, AutoCAD needs to know which distance applies to which face. Dist1 is applied to the base surface; Dist2 to the other surface.
   - **Select an isoline**, and AutoCAD first asks for the chamfer distances (Dist1 and Dist2), and then asks for the other face.

   Notice that AutoCAD highlights one of the two surfaces adjoining the selected edge.

   Base surface selection...

   Enter surface selection option [Next/OK (current)] <OK>: *(Type* **N** *or* **OK**.*)*
   - Enter **OK** when the two chamfer distances are the same, or when AutoCAD correctly guesses the base surface.
   - Enter **N** (next) to select a different base surface.
4. Provide the chamfer distance(s); you cannot specify an angle, as with 2D chamfering. Notice the reference to the "base surface chamfer distance":

   Specify base surface chamfer distance: **0.5**

   Specify other surface chamfer distance <0.5000>: *(Press* ENTER, *or enter a value.)*

   If the chamfer distance is too large, AutoCAD complains, "Cannot blend edge with unselected adjacent tangent edge. Finding connected blend set failed. Failure while chamfering," and exits the command. Restart the command, and enter a smaller value.
5. AutoCAD can chamfer just the selected edge, or a *loop*. A "loop" is all edges that touch the base surface. A rectangular surface, for example, has a loop of four edges.

   Select an edge or [Loop]: *(Pick an edge, or type* L.*)*

   Select an edge or [Loop]: *(Press* ESC.*)*
6. The command repeats until you press ESC.

TUTORIAL: ADDING BEVELS TO SOLIDS' EDGES

Before applying the **CHAMFER** command, first create a simple solid model of a hole inside a cube.

1. Start AutoCAD with a new drawing.
2. Draw a cube, as follows:

   Command: **box**

   Specify corner of box or [CEnter] <0,0,0>: *(Press* ENTER.*)*

   Specify corner or [Cube/Length]: **c**

   Specify length: **1**
3. Add the cylinder, as follows:

   Command: **cylinder**

   Current wire frame density: ISOLINES=16

   Specify center point for base of cylinder or [Elliptical] <0,0,0>: **0.5,0.5,0**

   Specify radius for base of cylinder or [Diameter]: **0.125**

   Specify height of cylinder or [Center of other end]: **1**
4. To view the cube better, rotate the viewpoint, as follows:

   Command: **vpoint**

   Current view direction: VIEWDIR= 0,0,0

   Specify a view point or [Rotate] <display compass and tripod>: **3,-2,1**

   Regenerating model.

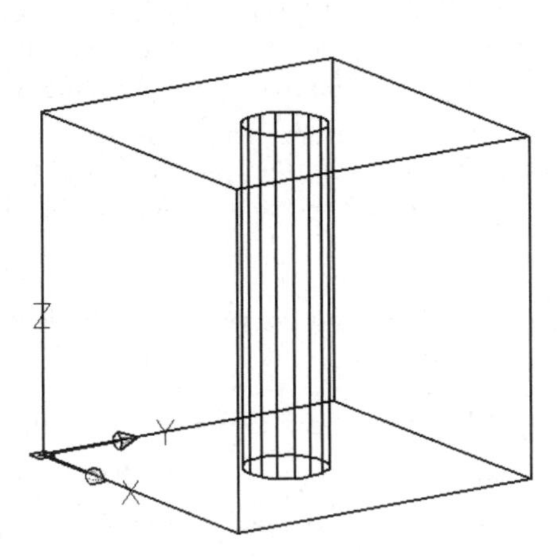

The cylinder inside the cube.

5. Subtract the cylinder from the box to create the hole:

   Command: **subtract**

   Select solids and regions to subtract from ..

   Select objects: *(Pick the cube.)*

   1 found Select objects: *(Press* ENTER.*)*

   Select solids and regions to subtract ..

   Select objects: *(Pick the cylinder.)*

   1 found Select objects: *(Press* ENTER.*)*

6. From the menu bar, select **Modify | Chamfer**.

   Command: **chamfer**

   (TRIM mode) Current chamfer Dist1 = 0.5000, Dist2 = 0.5000

   Select first line or [Polyline/Distance/Angle/Trim/Method]: *(Pick any edge of the cube.)*

   Base surface selection...

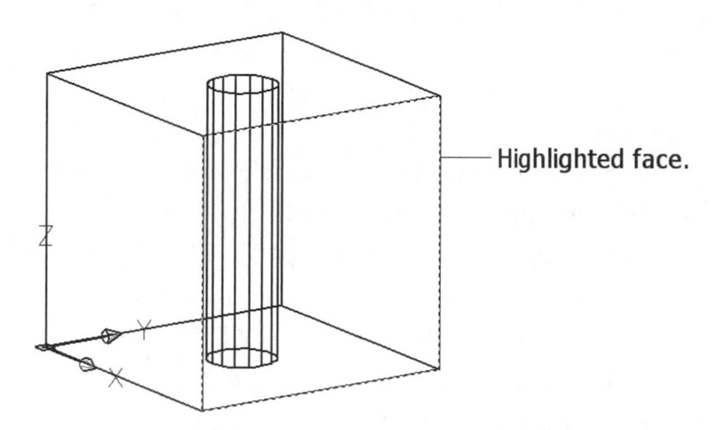

Pick an edge, and AutoCAD highlights one adjacent face.

   To highlight (select) a different face, enter **N** (short for "next") at the prompt; otherwise press **ENTER**:

   Enter surface selection option [Next/OK (current)] <OK>: *(Press* **ENTER**.*)*

7. Notice that you specify the chamfer distances here, and not with the **Distance** option earlier in this command:

   Specify base surface chamfer distance <0.5000>: **0.2**

   Specify other surface chamfer distance <0.5000>: **0.2**

8. Select the edges you want chamfered: pick two edges, and then press **ENTER** to exit the command.

   Select an edge or [Loop]: *(Pick edge 1.)*

   Select an edge or [Loop]: *(Pick edge 2.)*

   Select an edge or [Loop]: *(Press* **ENTER**.*)*

   Notice that AutoCAD performs a "double" chamfer: the intersection between the two chamfered edges is another 45-degree angle.

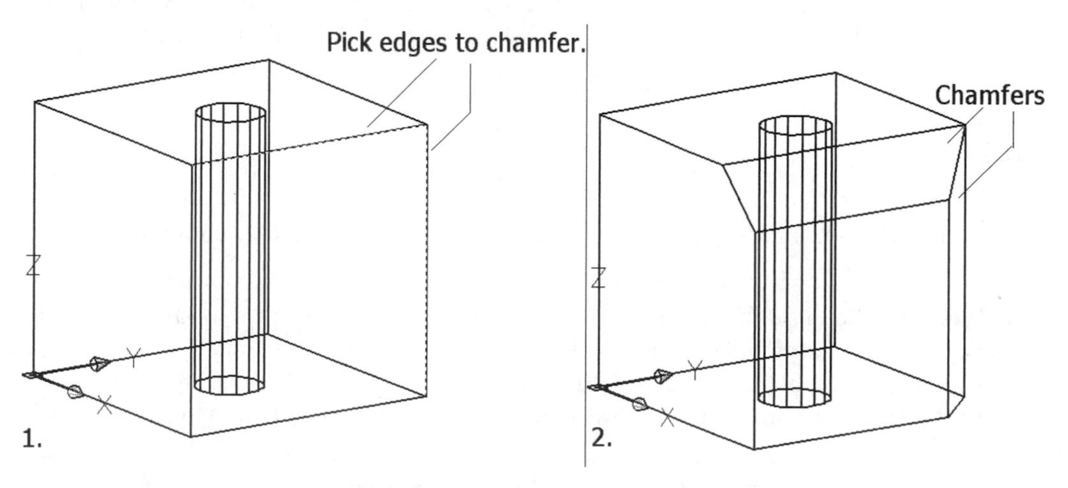

AutoCAD performs multiple chamfer on solid models.

TUTORIAL: ADDING COUNTERSINKS

In this tutorial, you create a second type of chamfer, one that adds a countersink to the hole. Continue with the drawing from the previous tutorial.

1. Start the **CHAMFER** command again:

   Command: **chamfer**

   (TRIM mode) Current chamfer Dist1 = 0.2000, Dist2 = 0.2000

   Select first line or [Polyline/Distance/Angle/Trim/Method]: *(Pick an edge on the top face of the cube, but not one of the chamfer edges.)*

   Base surface selection...

2. If the "wrong" face is selected, use the **Next** option to pick the top face:

   Enter surface selection option [Next/OK (current)] <OK>: **n**

   Enter surface selection option [Next/OK (current)] <OK>: *(Press* **ENTER.***)*

3. Use a larger chamfer distance:

   Specify base surface chamfer distance <0.2000>: **0.3**

   Specify other surface chamfer distance <0.2000>: **0.3**

4. Select the top of the cylindrical hole:

   Select an edge or [Loop]: *(Select the top of the cylindrical hole.)*

   Select an edge or [Loop]: *(Press* **ENTER.***)*

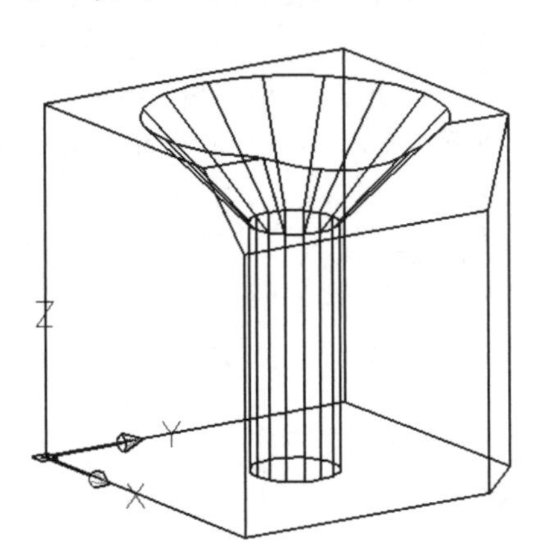

Chamfering holes creates countersinks.

Notice two things: the chamfer created a countersink, and the countersink extends into the first chamfer you created. To view the countersink better, select **View | Shade | Gouraud** from the menu.

FILLET

Applying the **FILLET** command to solids is similar to using the **CHAMFER** command, except that it creates rounded surfaces.

When you select a solid model, this command automatically switches to solids-editing mode. This command operates on a single solid model at a time, but allows you to fillet as many edges as you need during a single operation. It cannot fillet between two solid models. AutoCAD is limited to constant radius fillets; it cannot apply a variable radius to fillets. (It does not work with surface models.)

BASIC TUTORIAL: FILLETING OBJECTS

1. To fillet a solid model, start the **FILLET** command:
   - From the menu bar, choose **Modify**, and then **Fillet**.
   - From the **Modify** toolbar, choose the **Fillet** button.
   - At the 'Command:' prompt, enter the **fillet** command.
   - Alternatively, enter the **f** alias at the 'Command:' prompt.

   Command: **fillet** *(Press* ENTER.*)*

2. In all cases, AutoCAD first displays the current fillet settings, and then asks you to select objects:

   Current settings: Mode = TRIM, Radius = 0.0000

   Select first object or [Polyline/Radius/Trim/mUltiple]: *(Select a solid model.)*

3. Provide the fillet radius:

   Enter fillet radius: **0.1**

4. Select the edge to fillet:

   Select an edge or [Chain/Radius]: *(Pick an edge.)*

   - **Radius** option allows you to specify different radii for different edges.
   - **Chain** option allows you to pick one edge, and AutoCAD selects all connected edges.

   Select additional edges, or else press **ENTER** to exit the command:

   Select an edge or [Chain/Radius]: *(Press* ENTER.*)*

TUTORIAL: FILLETING SOLID EDGES

1. For this tutorial, draw a cylinder on top of a rectangular box, as follows:

   Command: **box**

   Specify corner of box or [CEnter] <0,0,0>: *(Press* ENTER.*)*

   Specify corner or [Cube/Length]: **l**

   Specify length: **9**

   Specify width: **4**

   Specify height: **1**

   Command: **cylinder**

   Current wire frame density: ISOLINES=16

   Specify center point for base of cylinder or [Elliptical] <0,0,0>: **4.5,2,1**

   Specify radius for base of cylinder or [Diameter]: **1**

   Specify height of cylinder or [Center of other end]: **3**

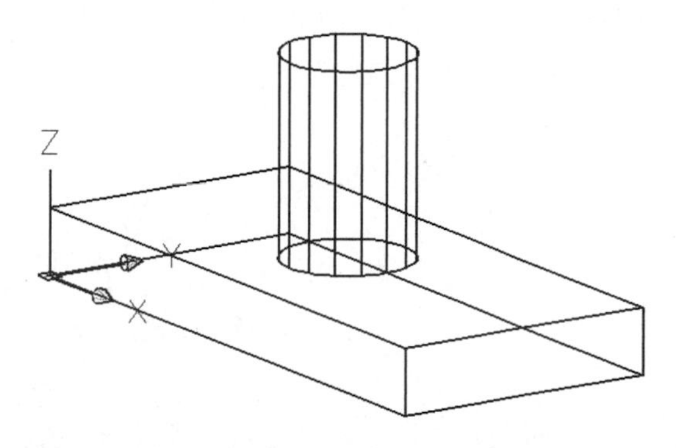

Creating a part from a box and a cylinder.

2. Start the **FILLET** command, and fillet the four edges of the end of the box:

   Command: FILLET

   Current settings: Mode = TRIM, Radius = 0.5000

   Select first object or [Polyline/Radius/Trim]: *(Pick an edge.)*

   Enter fillet radius <0.5000>: **0.2**

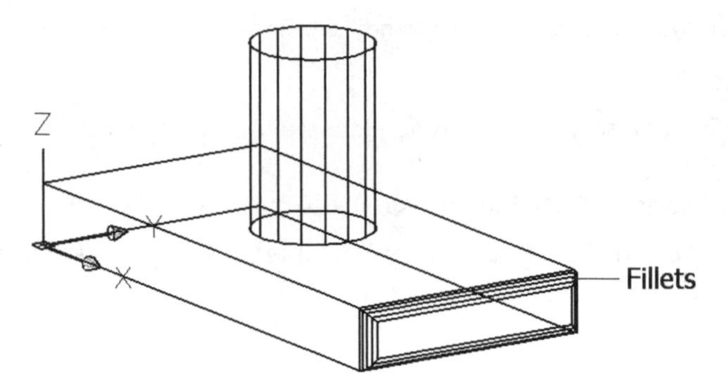

Adding fillets to the end of the part.

   Select an edge or [Chain/Radius]: *(Pick edge 1.)*

   Select an edge or [Chain/Radius]: *(Pick edge 2.)*

   Select an edge or [Chain/Radius]: *(Pick edge 3.)*

   Select an edge or [Chain/Radius]: *(Press* **ENTER.***)*

   4 edge(s) selected for fillet.

3. The other fillet is at the base of the cylinder. The **FILLET** command cannot fillet between two solid objects, so first apply the **UNION** command to join the box and the cylinder.

   Command: **union**

   Select objects: *(Pick the box.)*

   1 found Select objects: *(Pick the cylinder.)*

   1 found, 2 total Select objects: *(Press* **ENTER.***)*

   The objects won't look any different.

4. Apply the fillet between the box and the cylinder:

   Command: **fillet**

   Current settings: Mode = TRIM, Radius = 0.2000

   Select first object or [Polyline/Radius/Trim]: *(Pick the bottom edge of the cylinder.)*

   Enter fillet radius <0.5000>: **0.2**

   Select an edge or [Chain/Radius]: *(Press* **ENTER.***)*

   1 edge(s) selected for fillet.

Notice the curve at the base of the cylinder.

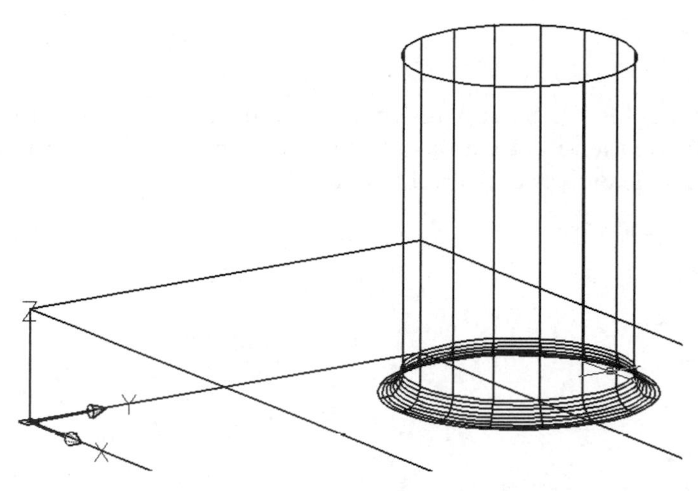

The fillet applied between the box and the cylinder.

5. Apply one last fillet, on top of the cylinder.

   Command: **fillet**

   Current settings: Mode = TRIM, Radius = 0.2000

   Select first object or [Polyline/Radius/Trim]: *(Pick the top edge of the cylinder.)*

   Enter fillet radius <0.5000>: **0.2**

   Select an edge or [Chain/Radius]: *(Press* **ENTER**.*)*

   1 edge(s) selected for fillet.

   Notice that the top of the cylinder is gracefully curved. Select **View | Shade | Gouraud** to see the shaded model.

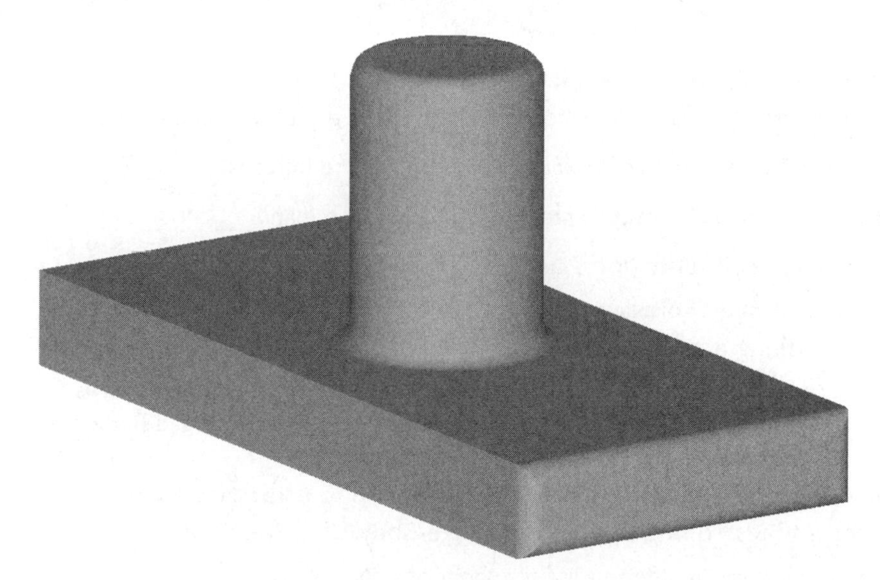

The completed model, shown shaded.

SOLIDEDIT

The **SOLIDEDIT** command is the Swiss Army Knife of solids model editing: 15 of the 18 items on the **Modify | Solids Editing** menu are options of this command. It handles the following tasks on the bodies, faces, edges, and vertices of 3D solid models:

- Copies faces and edges.
- Moves faces.
- Extrudes faces.
- Offsets faces.
- Rotates faces.
- Tapers faces.
- Changes the color of faces and edges.
- Shells bodies, making them hollow.
- Imprints geometry on faces.
- Deletes faces, fillets, and chamfers.
- Cleans up redundant edges, vertices, and removes imprints.
- Separates a disjointed body into independent objects.
- Checks the validity of the body.

The **SOLIDEDIT** command groups its options according to the aspect of the 3D solid: face, vertex, edge, or body. Some options occur in more than one category; for instance, faces can be copied, and edges can be copied. Thus, the following sections describe the options in the order listed above.

TUTORIAL: EDITING SOLID OBJECTS

1. To edit solid models, start the **SOLIDEDIT** command:
   - From the **Modify** menu bar, choose **Solids Editing**, and then an item from the submenu.
   - From the **Solids Editing** toolbar, choose one of the buttons.
   - At the 'Command:' prompt, enter the **solidedit** command.

     Command: **solidedit** *(Press* ENTER.*)*
2. In all cases, AutoCAD displays the following message:

   Solids editing automatic checking: SOLIDCHECK=1

   Each time you use this command, AutoCAD first checks to see if the body is a valid ShapeManager object — if **SOLIDCHECK** is turned on (set to 1). Invalid bodies are sometimes created during object creation and editing.
3. AutoCAD asks you to select which part of the solid to edit: a face, an edge, or the body.

   Enter a solids editing option [Face/Edge/Body/Undo/eXit] <eXit>:

   Each option displays a different list of suboptions:

   Face edits the faces of solid models.

   Enter a face editing option

   [Extrude/Move/Rotate/Offset/Taper/Delete/Copy/coLor/Undo/eXit] <eXit>:

   Edge edits the edges of solid models.

   Enter an edge editing option [Copy/coLor/Undo/eXit] <eXit>:

   Body edits the entire body.

   Enter a body editing option

[Imprint/seParate solids/Shell/cLean/Check/Undo/eXit] <eXit>:

Undo undoes the last action.

eXit ends the options or the command.

So, there you have it: a command with 15 options. (And you thought the **PEDIT** command was complex!) Let's work through all of them; I'm sure you'll find some quite handy.

 Copying Faces and  Copying Edges

The **Face / Copy** and **Edge / Copy** options copy selected faces and edges:

- By copying faces, you reuse existing features for new 3D solids; copied faces remain bodies.
- By copying edges, you create 2D plans of 3D objects; copied edges become lines or arcs.

Of all the options in the **SOLIDEDIT** command, only this one separates faces and edges from bodies.

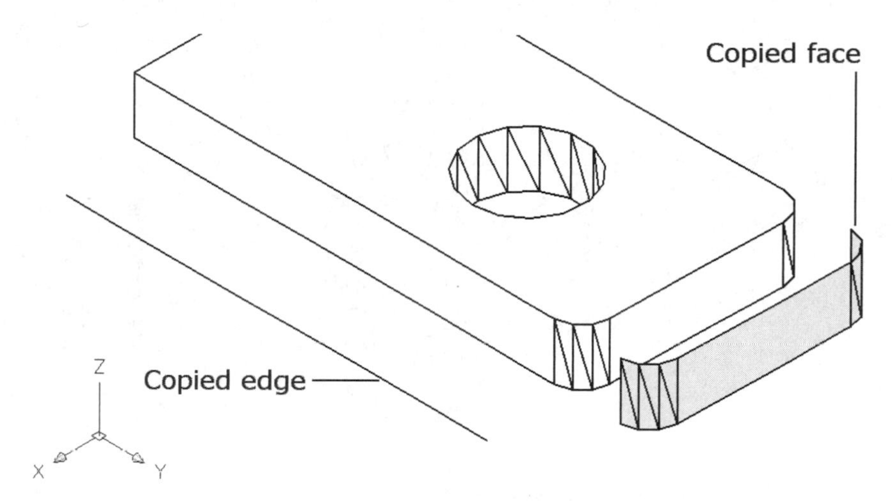

Making copies of faces and edges.

 Moving Faces

The **Face / Move** option moves selected faces. By moving faces, you relocate portions of the body, such as holes.

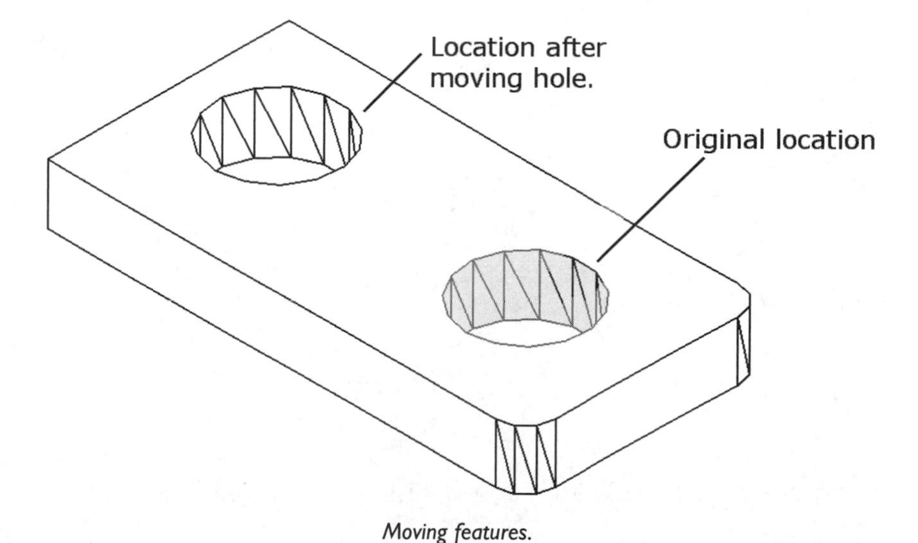

Moving features.

Extruding Faces

The **Face / Extrude** option extends a face. You can specify an extrusion distance and an angle of taper. A positive extrusion distance extends the face "outwards," while a negative distance extends the face inwards (see figure below). If the negative distance is equal to or larger than the thickness of the body, AutoCAD complains "This extrusion would have created an empty solid and has been rejected." Try a smaller distance.

As an alternative to specifying the extrusion distance, you can select an object that defines the extrusion "path." The object can be a line, arc, spline, circle, ellipse, and so on, and must not be tangential to the face (i.e., lie in the same plane). Be careful when you use a spline for the path; an unexpected solid may result!

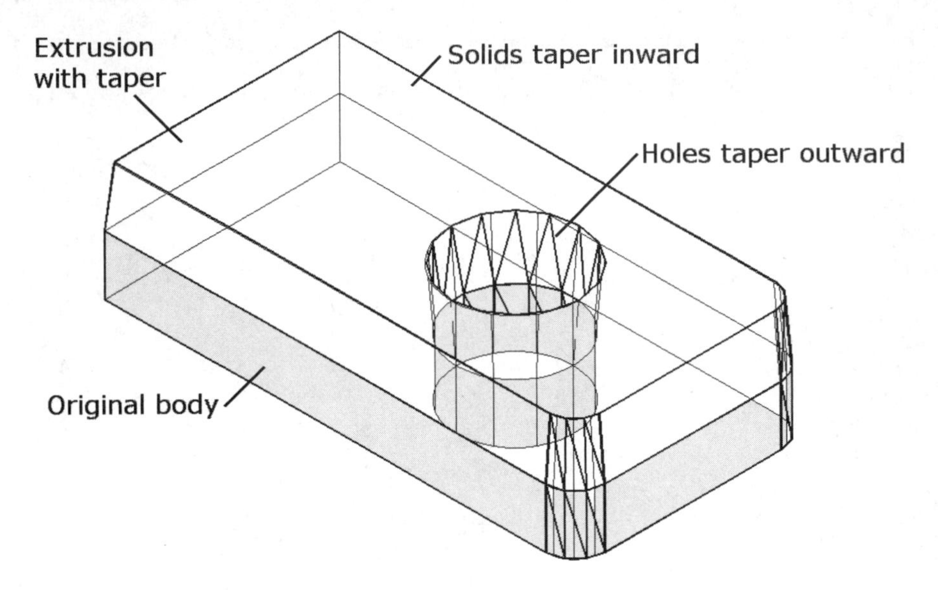

Extending a body outward with a positive distance and a taper.

With regard to positive taper angles, notice that the solid part of the body tapers "inward," but that holes taper "outward." That's because the body is tapering inward towards the hole. The opposite happens with a negative taper angle. Be careful with taper angles: if it results in a self-intersecting solid, AutoCAD refuses to execute your command.

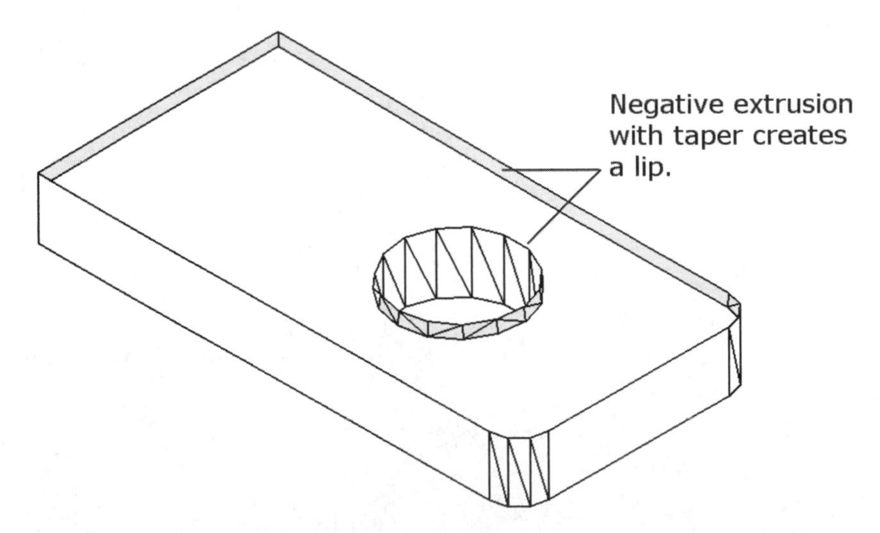

Extending a body inward with a negative distance and a taper creates a "lip."

Offsetting Faces

The **Face / Offset** option is somewhat misleading in its name. Whereas the 2D **OFFSET** command creates parallel copies, the **SOLIDEDIT Offset** option is more like the **Extrude** option, but with no tapers or paths. A negative offset distance makes the face smaller; positive values make it larger. (The opposite is true for holes: negative offset distance makes the hole larger, and a positive value makes the hole smaller.) *Careful*: When the offset distance equals the hole size, the hole disappears.

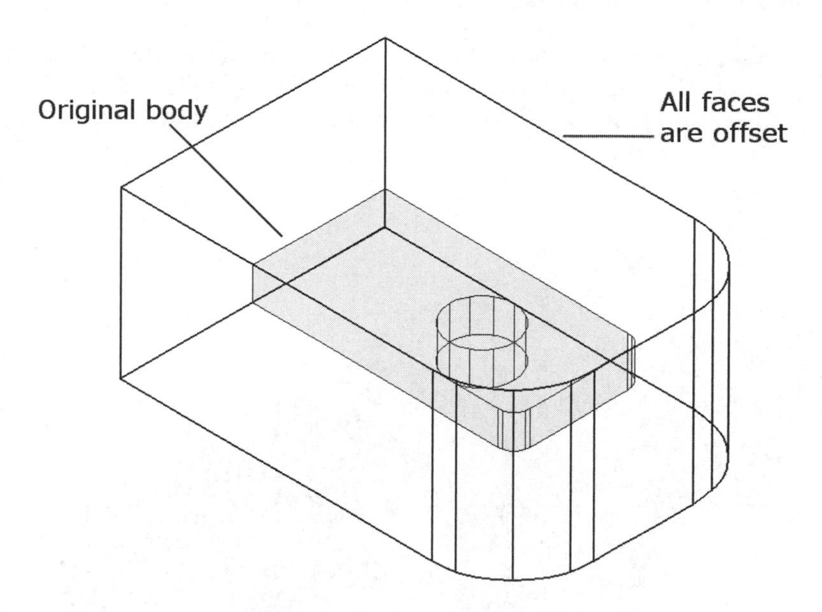

Offsetting faces can cause holes to disappear.

Rotating Faces

The **Face / Rotate** option rotates features in bodies. Below, the hole has been rotated by 45 degrees.

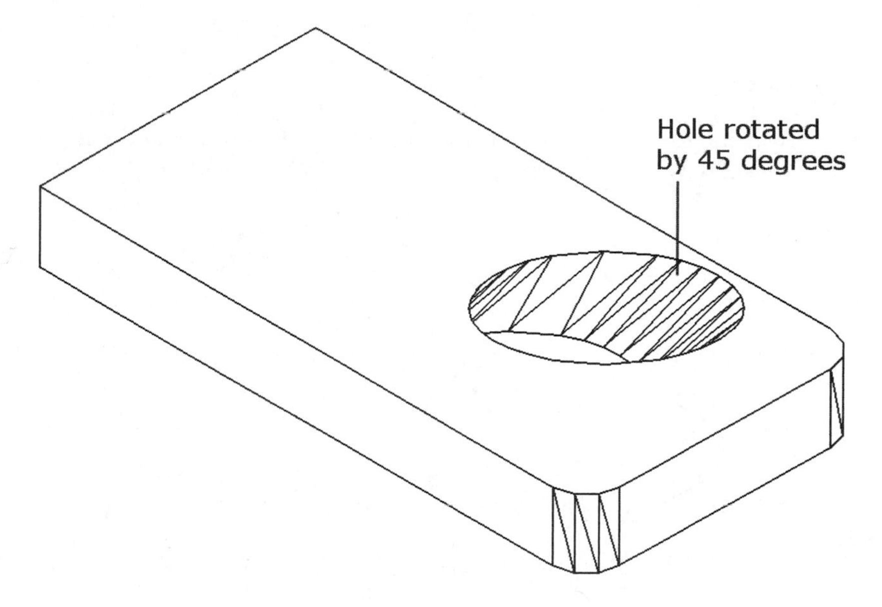

Rotating features.

Tapering Faces

The **Face / Taper** option is like the **Extrusion** option without the extrusion: **Taper** just tapers the body without extruding it. This lets you change a hole with straight sides, for example, to a tapered hole. (The same warnings apply with respect to specifying the taper angle.)

Coloring Faces and Coloring Edges

The **Face / coLor** and **Edge / coLor** options change the color of faces and edges. After you select the faces and edges, AutoCAD displays the standard Select Color dialog box. Select a color, and then click **OK**. The colors you select are displayed by the **SHADE** and **RENDER** commands. To change the color of the entire body, use the regular **COLOR** command.

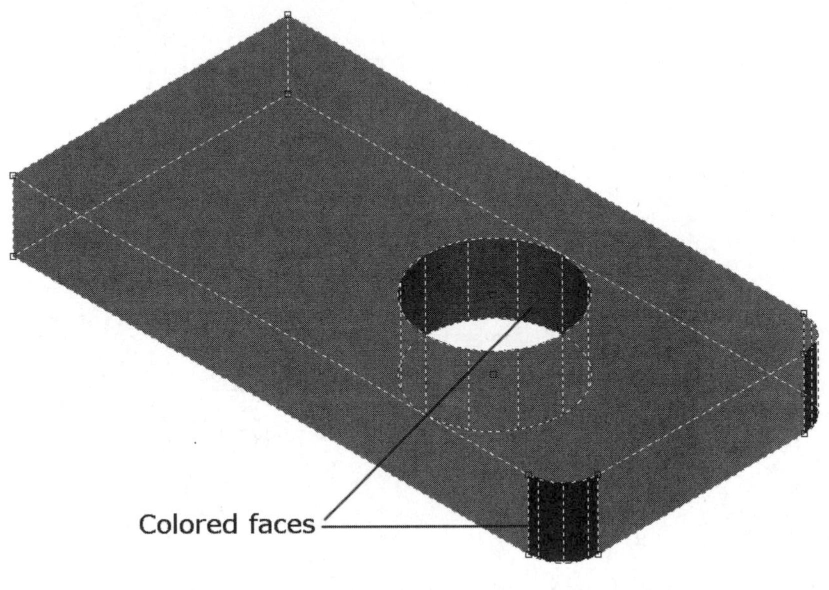

Coloring faces.

Shelling the Body

The **Body / Shell** option makes bodies hollow. In the figure below, I applied the **Shell** option, and then used the **SLICE** command to cut off the end to see the effect inside the body.

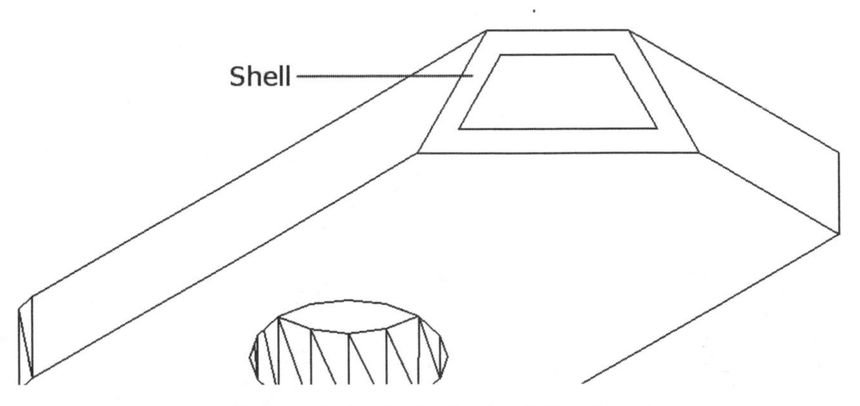

Making a hollow body with the shell option.

A solid can have only one shell; the shell distance is constant. The shell offset distance must be smaller than the smallest part of the body; using a negative distance shells the body outward.

Imprinting Bodies

The **Body / Imprint** option creates "imprints" at the intersects of two solids. As the figure shows, the result of imprinting the sphere on the body is two circles. The resulting imprints are part of the body and not separate objects. Imprints can be removed with the **cLean** option.

Note: There are limitations on what can be used to perform the imprinting. The imprinting object must be a region, body, another 3D solid, or a curve. Curves includes circles, arcs, splines, and so on. The selected object must intersect the faces of the solid.

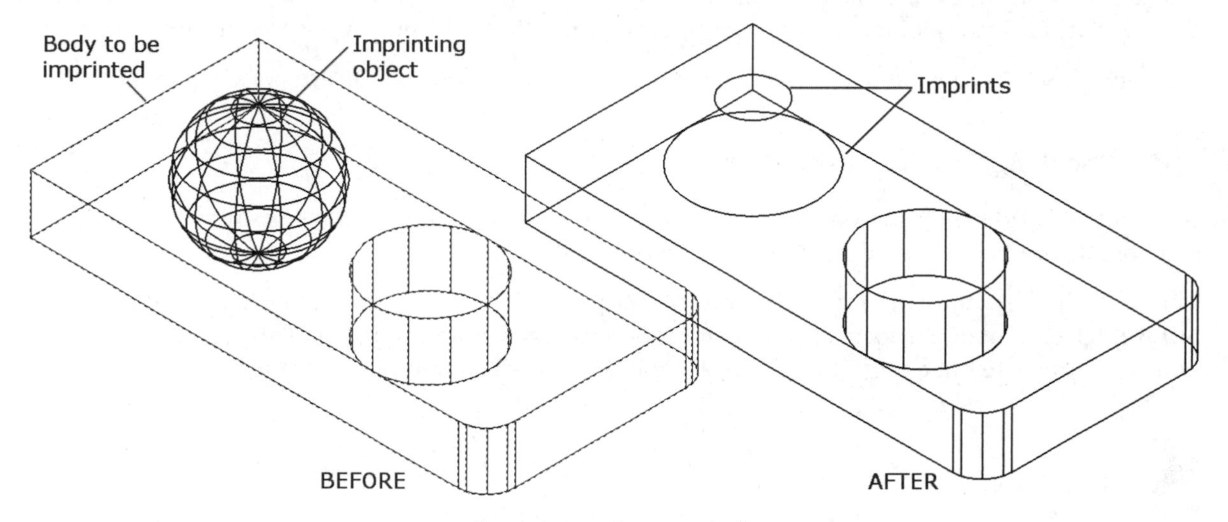

Imprinting a sphere on a body.

Deleting Faces, Fillets, and Chamfers

The **Face / Delete** option removes certain faces, as well as fillets and chamfers applied to bodies.

Not just any face is removable; AutoCAD must be able to "paper over" the face being removed. If you try to take the face off the end of the body, as in the figure, AutoCAD won't do it, complaining, "Gap cannot be filled."

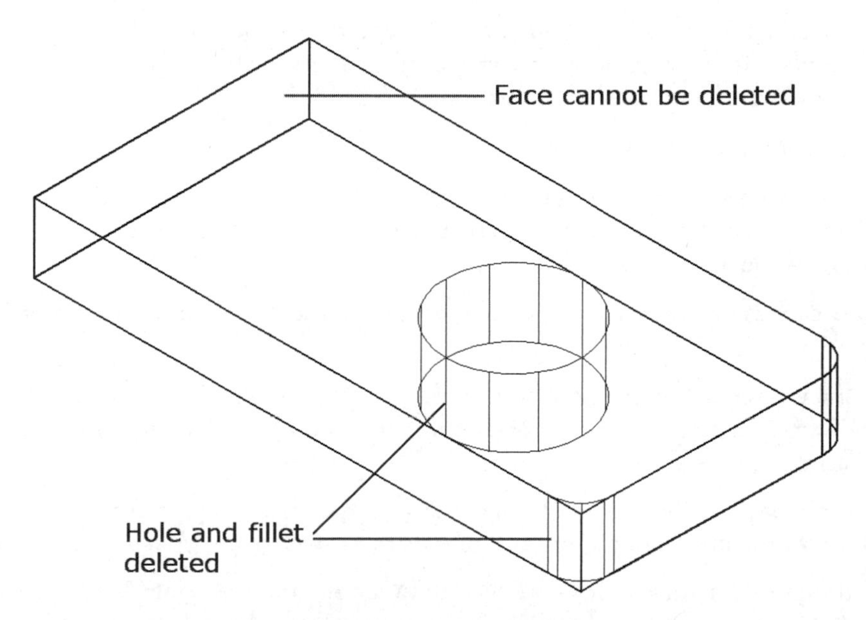

Removing faces and a fillet.

Cleaning Faces and Edges

The **Body / cLean** option removes redundant edges and vertices. The option also removes imprints created by the **Imprint** option; there is no control over the imprints; all are removed.

Separating the Body

The **Body / seParate solids** option separates bodies with disjointed volumes into independent bodies.

This option does not work on bodies created by Boolean operations (union, intersect, and subtract). You cannot, for example, use this option to remove a cylinder (hole) subtracted from a box.

This option works only on a "single" body created from two bodies that are physically separate.

Checking Validity of the Body

The **Body / Validate** option checks to see if the body is a valid ShapeManager solid. An invalid body can sometimes be created during object creation and editing.

This option is redundant: when the SOLIDCHECK system variable is on (as it is by default), the SOLIDEDIT command performs the validation process each time you use it. This explains the "Solids editing automatic checking: SOLIDCHECK=1" prompt displayed by the command.

SOLVIEW

AutoCAD has a set of solids-related commands that, frankly, I've had a hard time figuring out. Part of the problem is that the two must be used in a specific order. It didn't help me that AutoCAD's menu lists the commands in wrong order. The correct order is:

1. **SOLVIEW** creates floating viewports and sets the correct viewpoints for the standard orthographic views — top, front, and sides.
2. **SOLDRAW** extracts 2D images of the 3D solid model.
3. **SOLPROF** creates pictorial views of the 3D solid model, as described later in this chapter.

The purpose of these commands is to generate 2D views of 3D solid models. You may be thinking, "Big deal, I can fake it with paperspace viewports on my own." True, but these commands automatically create orthogonal views, add hatching to sections, and more.

The SOLVIEW command has four options:

- **Ucs** creates views relative to the current UCS. Where no viewports exist, this option creates the first viewport from which others are created. (The three options following require an existing viewport.)
- **Ortho** creates orthographic views from existing viewports — top, bottom, front, back, and sides.
- **Auxiliary** creates auxiliary views from existing viewports. Auxiliary views are drawn at an angle to show true lengths and areas of objects not drawn at right angles, such as the sloping surface of wedges.
- **Section** creates sectional views from existing viewports, and applies crosshatching. Section views show the insides of objects, as if they were cut with a saw to expose the innards.

Autodesk's documentation warns that the SOLDRAW command can only be used with viewports created by SOLVIEW. This means that your drawings cannot have clipped viewports, because SOLVIEW does not create them.

TUTORIAL: CREATING 2D VIEWS OF 3D SOLID MODELS

1. Start AutoCAD, and then open a 3D solid model. (For the purpose of this tutorial, open *truckmodel.dwg* from the CD-ROM.)

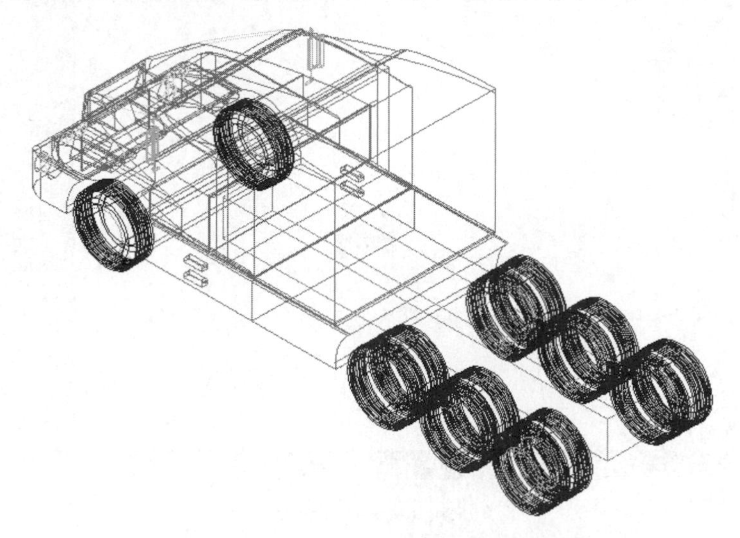

The truckmodel.dwg drawing in model space.

2. Before using **SOLVIEW**, tell AutoCAD which side of the 3D model is the *front view*. You do this in two steps:

   Step 1: Use the **VPOINT** command to view the front of the model. For this truck model:

   Command: **vpoint**

   Specify a view point or [Rotate] <display compass and tripod>: **0,-1,0**

   Step 2: Use the **UCS** command's **View** option to orient the user coordinate system to the view.

   Command: **ucs**

   Enter an option [New/Move/orthoGraphic/Prev/Restore/Save/Del/Apply/?/World]
   <World>: **v**

   The truck model should now look similar to the figure below.

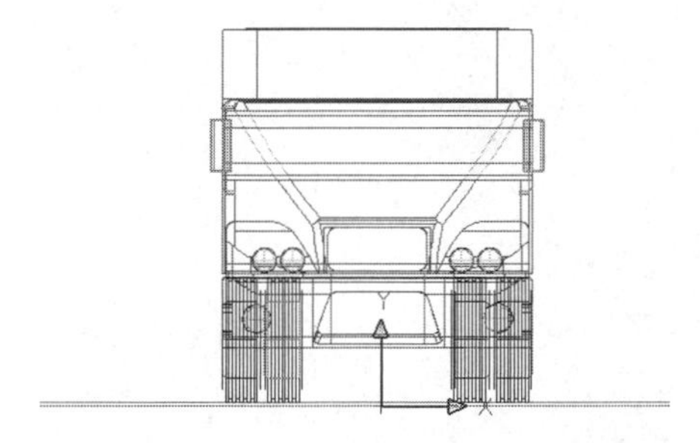

The 3D solid model oriented to display the front view.

3. **SOLVIEW** works in layout mode. To switch to layout mode, click the **Layout1** tab.
   When the Page Setup dialog box appears, dismiss it by clicking **OK**.
4. Notice that AutoCAD automatically creates one viewport that takes up most of the page. To make room for additional viewports, make this one smaller, and move it to the lower left corner. Follow these steps:
   a. Click the viewport's *frame* (the black rectangle that outlines the viewport).
   Notice it is becomes highlighted (dashed lines) with four blue squares at the corner. These are called "cold grips."

b. Select a cold grip (any will do). Notice that it turns red. This is called a "hot grip."
c. Drag the hot grip towards the truck model. Notice that the viewport becomes smaller.

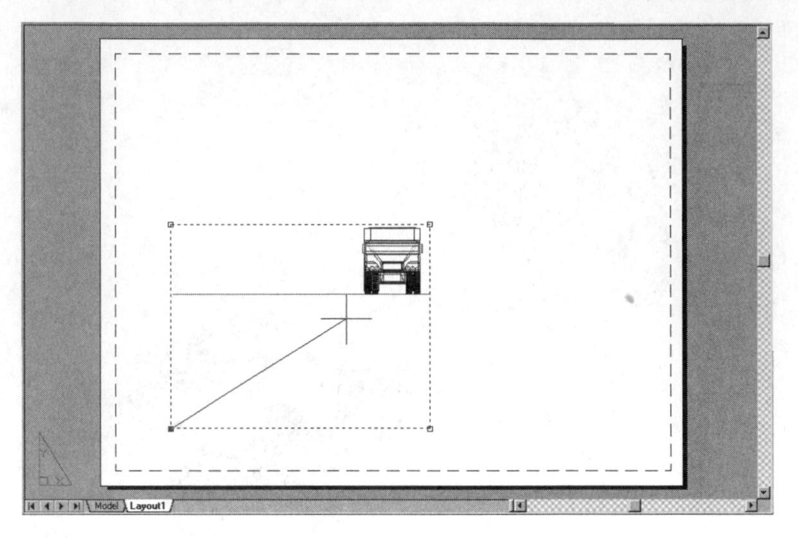

Making the viewport smaller.

d. Repeat for the opposite grip to center the truck in the viewport.
e. Finally, drag the viewport to the lower left corner. (Grab the viewport by its rectangular frame, and drag).

Note: Technical editor Bill Fane suggests an alternative method:

1. Drag the viewport's frame to the appropriate size and position.
2. On the status bar, click the **PAPER** button.
3. Execute the **ZOOM All** command. (For a specific scale factor, use instead **ZOOM nXP**.)
4. Click the **MODEL** button (not the Model tab).

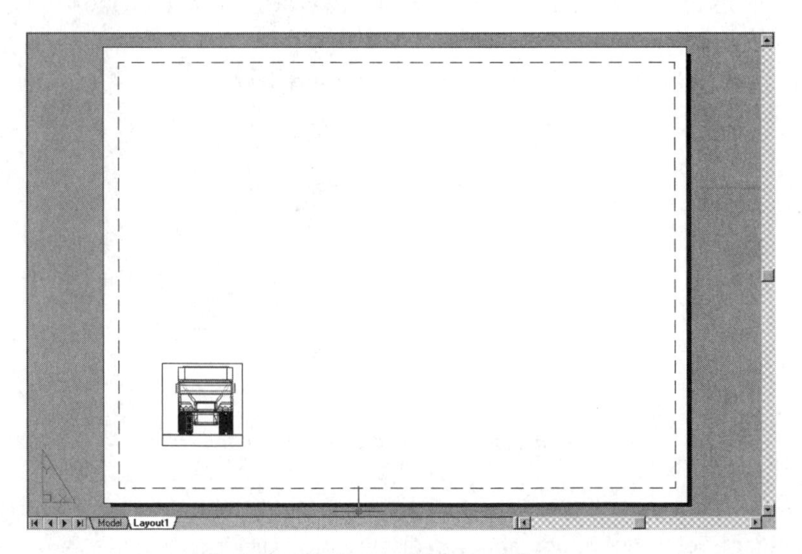

Drag the viewport to the lower left corner.

4. With the prep work behind us, we can finally start the **SOLVIEW** command. From the menu bar, select **Draw | Solids | Setup | View**.

   Command: **solview**

   Enter an option [Ucs/Ortho/Auxiliary/Section]: **o**

   Specify side of viewport to project: *(Pick the right edge of the viewport.)*

   This part is crucial: the *edge of the viewport you pick determines the orthographic view* that AutoCAD generates for you. For example, to get the right side view, pick the right edge of the viewport.

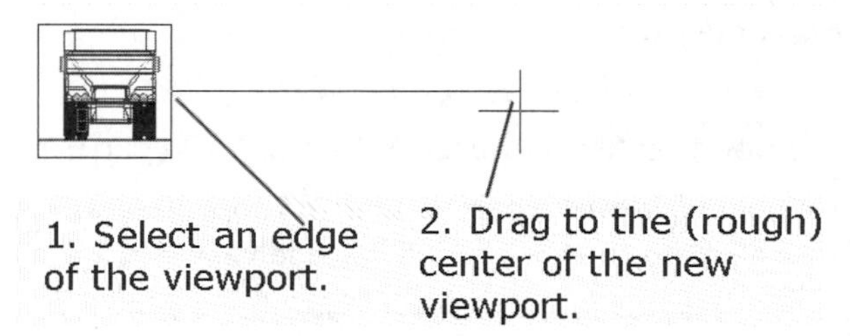
1. Select an edge of the viewport.
2. Drag to the (rough) center of the new viewport.

Creating the new viewport.

Note: To obtain a specific orthographic view with the **SOLVIEW** command's **Ortho** option, make these selections:

Top view: Pick the *top* edge of the viewport.
Right side view: Pick the *right* edge of the viewport.
Bottom view: Pick the *bottom* edge of the viewport.
Left side view: Pick the *left* edge of the viewport.

5. Give AutoCAD an idea of where to place the new viewport containing the side view:

   Specify view center: *(Pick a point for the center of the new viewport.)*

   The point you pick becomes the *centerpoint* of the view. It doesn't matter, however, where exactly you pick the point, because you can move the viewport later.

   Notice that AutoCAD draws the side view without a viewport; specifying the boundaries of the viewport is the next step.

6. Specify the size of the viewport by picking two corners.

   Specify view center <specify viewport>: *(Press* **ENTER** *to specify the viewport corners.)*

   Specify first corner of viewport: *(Pick a point for one corner.)*

   Specify opposite corner of viewport: *(Pick another point for the other corner.)*

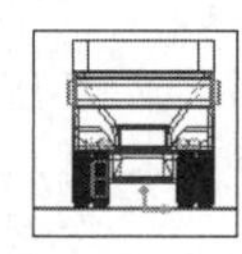
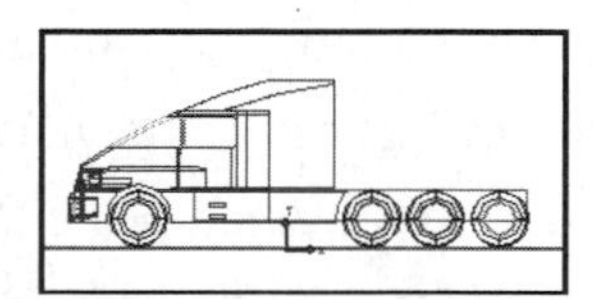

Pick two corners to specify the viewport showing the side view.

7. AutoCAD prompts you to give the viewport a name. Enter a name, such as "side":

   Enter view name: **side**

   UCSVIEW = 1 UCS will be saved with view

8. The **SOLVIEW** command returns to its first prompt, so that you can place other views. Repeat the steps to create the top view.

   Enter an option [Ucs/Ortho/Auxiliary/Section]: **o**

   Specify side of viewport to project: *(Pick the top edge of the viewport.)*

   Specify view center: *(Pick a point for the center of the new viewport.)*

   Specify view center <specify viewport>: *(Press* **ENTER** *to specify the viewport corners.)*

   Specify first corner of viewport: *(Pick a point for one corner.)*

   Specify opposite corner of viewport: *(Pick another point for the other corner.)*

   Enter view name: **top**

   UCSVIEW = 1 UCS will be saved with view

   Enter an option [Ucs/Ortho/Auxiliary/Section]: *(Press* **ENTER** *to exit the command.)*

   The drawing should now have the front, side, and top views, as illustrated by the figure. If you wish, you can create the bottom and left side view in the same manner.

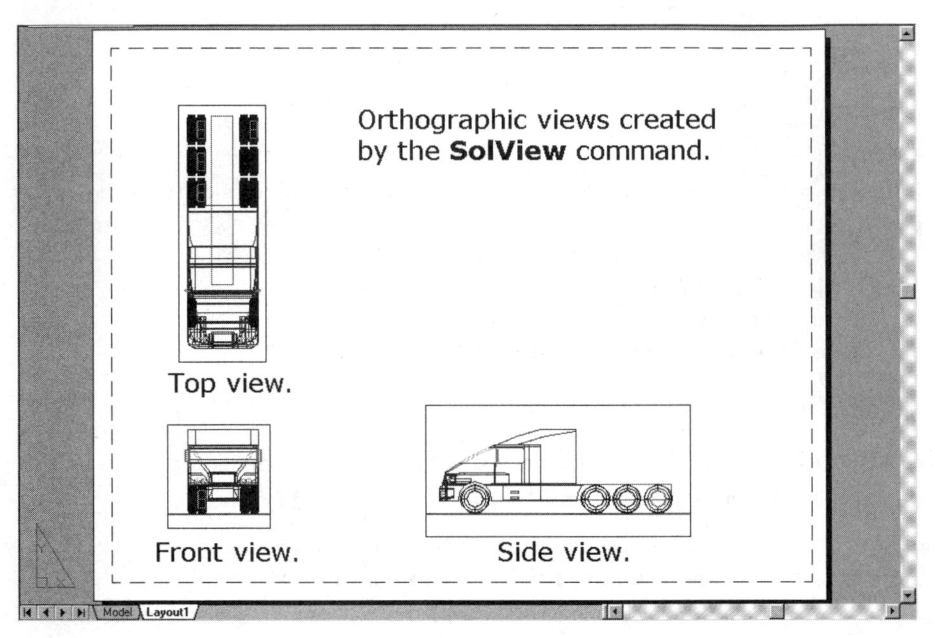

The front, side, and top views of the 3D solid model.

TUTORIAL: CREATING AUXILIARY AND SECTION VIEWS

The **SOLVIEW** command creates two other standard views used in drafting: auxiliary and section views.

1. Create an auxiliary view of the truck's windshield, because it is at an angle.

   Enter an option [Ucs/Ortho/Auxiliary/Section]: **a**

   Specify first point of inclined plane: *(Pick point 1 near windshield.)*

   Specify second point of inclined plane: *(Pick point 2 at other end of windshield.)*

   Specify side to view from: *(Pick point 3.)*

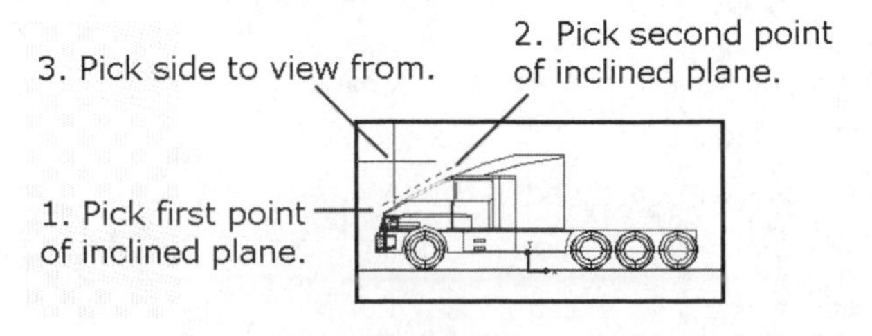

Defining the auxiliary view.

2. The remainder of the prompts are familiar:

   Specify view center: *(Pick a center point for the new viewport.)*

   Notice that the dragline is at right angles to the plane line:

   Specify view center <specify viewport>: *(Press* ENTER *to specify viewport corners.)*

   Specify first corner of viewport: *(Pick a point for one corner.)*

   Specify opposite corner of viewport: *(Pick another point for the other corner.)*

   The viewport does not need to encompass the model; you can pick points around the windshield.

3. Name the view "aux."

   Enter view name: **aux**

   UCSVIEW = 1 UCS will be saved with view

   Enter an option [Ucs/Ortho/Auxiliary/Section]: *(Press* ENTER *to exit the command.)*

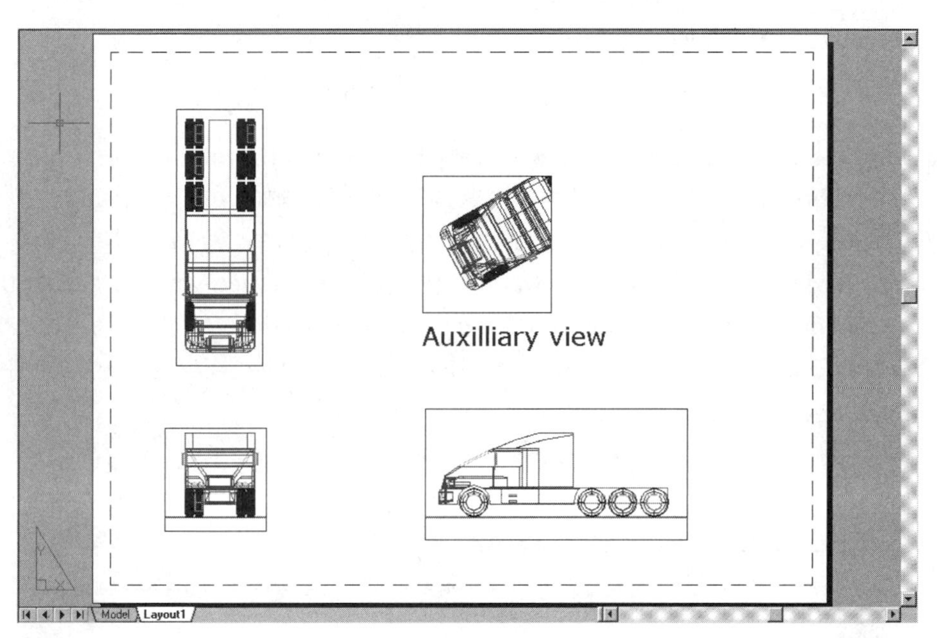

The completed auxiliary view.

4. Press **ENTER** to exit the command, and then move the "Aux" viewport into a better position.

5. Now create the section view, which will be straight through the center of the top view. It is helpful first to turn on ortho mode.

   Restart the **SOLVIEW** command, and then select the **Section** option, as follows:

   Command: **solview**

   Enter an option [Ucs/Ortho/Auxiliary/Section]: **s**

   Specify first point of cutting plane: *(Pick point 1 at one end of the truck.)*

   Specify second point of cutting plane: *(Pick point 2 at the other end of the truck.)*

   Specify side to view from: *(Pick point 3 to the right the truck.)*

The dotted line is the section line, as shown in the figure.

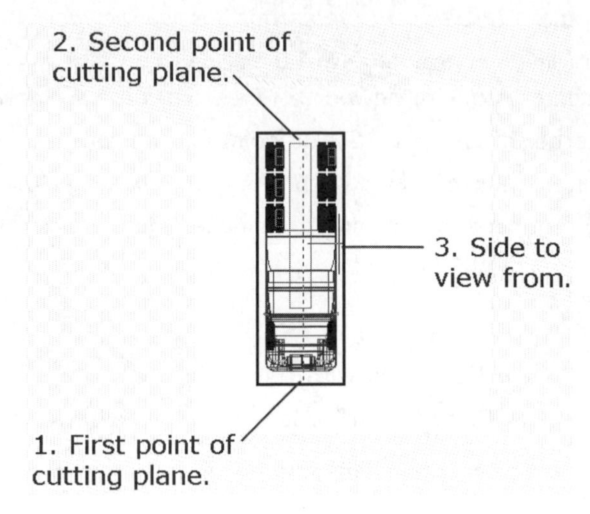

Specifying the section view.

AutoCAD automatically determines the scale factor that fits the other views:

Enter view scale <0.1172>: *(Press* **ENTER** *to accept the scale factor.)*

Specify view center: *(Pick a point for the center of the new viewport.)*

Specify view center <specify viewport>: *(Press* **ENTER** *to specify the corners of the viewport.)*

Specify first corner of viewport: *(Pick a point for one corner.)*

Specify opposite corner of viewport: *(Pick a point for the other corner.)*

Enter view name: **section**

UCSVIEW = 1 UCS will be saved with view

Enter an option [Ucs/Ortho/Auxiliary/Section]: *(Press* **ENTER** *to exit the command.)*

The complete section view is shown below.

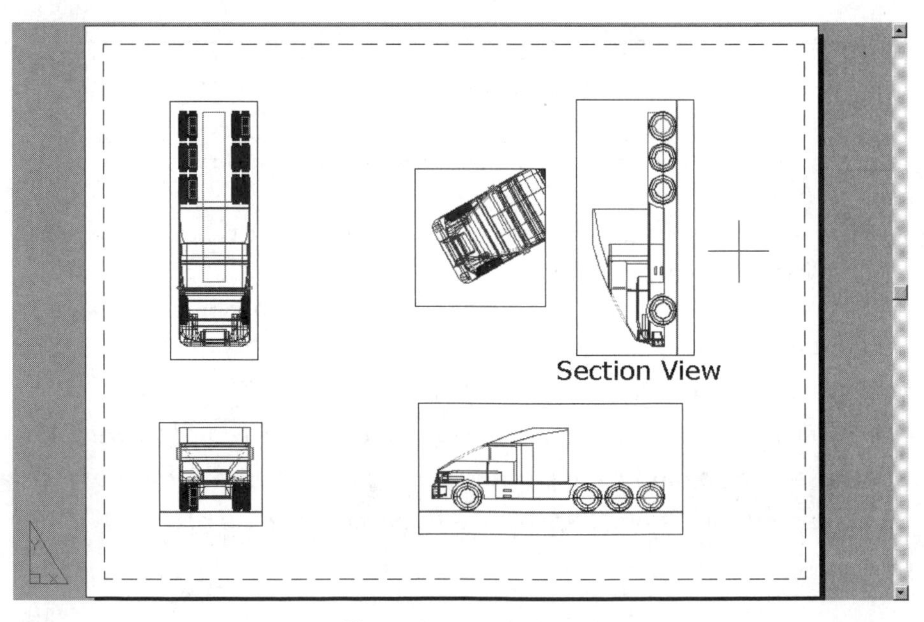

The section view in place.

SOLDRAW

The views are still fully 3D; never mind that they don't *look* 3D. The section view, for example, lacks hatching. This is where the curiously-name **SOLDRAW** command comes into play. It converts 3D-looking views into 2D drawings, and adds hatching appropriate for section views.

Take a moment to click the **Layer** droplist on the Layers toolbar. Notice that AutoCAD has added layer names: aux-DIM, aux-HID, aux-VIS, section-HAT, and so on. AutoCAD manipulates these layers to create the 2D look.

Compared to the **SOLVIEW** command, the **SOLDRAW** command is easy to use.

TUTORIAL: GENERATING 2D VIEWS

1. To generate 2D views from solid models, start the **SOLDRAW** command:
   - From the **Draw** menu, choose **Solids**, **Setup**, and then **Drawing**.
   - From the **Solids** toolbar, choose the **Setup Drawing** button.
   - At the 'Command:' prompt, enter the **solidedit** command.

   Command: **soldraw** *(Press* ENTER.*)*

   In all cases, AutoCAD displays the following prompt:

   Select viewports to draw..
2. It's easiest to select all objects, and let AutoCAD filter out those that aren't valid:

   Select objects: **all**

   62 found. 56 were not in current space. 1 was the paper space viewport.

   Select objects: *(Press* ENTER.*)*

   25 solids selected. 40 solids selected. Non-SOLVIEW viewport ignored.

   That's all it takes! It may take AutoCAD a few minutes to generate the views, even on a fast computer. Wait until the 'Command:' prompt returns. Trust me: waiting is faster than drawing the views yourself!

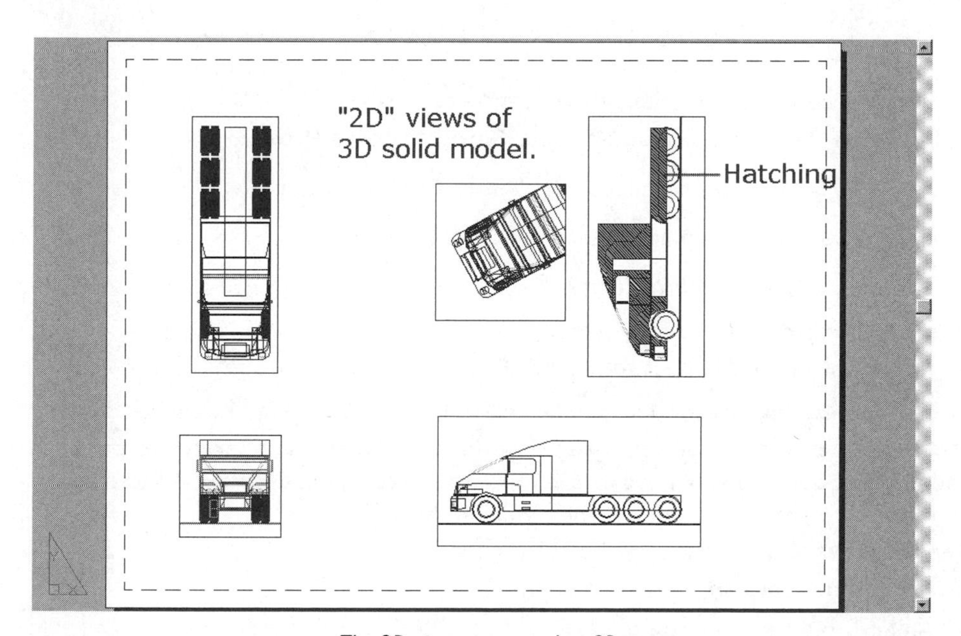

The 3D views converted to 2D.

Notice that the views now look 2D, and that the section view has hatching. Indeed, if you were to use the **LIST** command, you would find the viewports contain 2D entities, such as lines and arcs. Don't worry; the 3D solid model is not gone: its layers have been frozen, and the new 2D entities have been placed on the AutoCAD-generated layers I told you of earlier.

SOLPROF

The **SOLPROF** command (short for "solids profiling") creates pictorial views of 3D solid models.

BASIC TUTORIAL: CREATING 3D PICTORIAL VIEWS

1. In preparation for this tutorial, create a simple solid model with these commands:

   Box and **Cylinder** commands.

   Subtract cylinder from the box.

   Fillet two corners at one end.

   Set the viewpoint to 1,1,1.

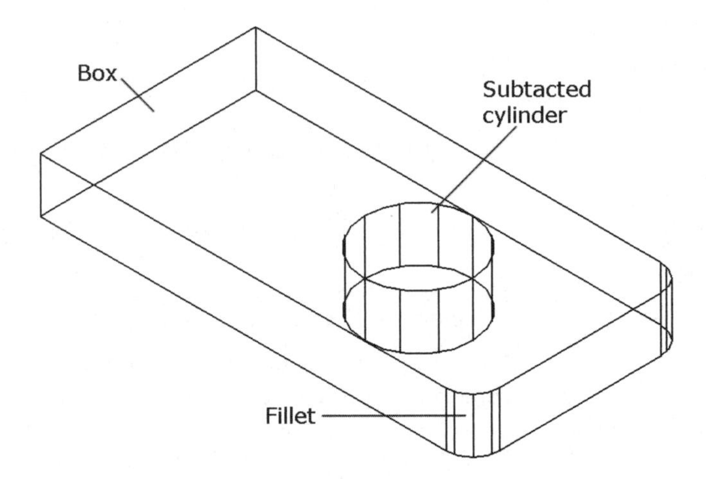

The 3D solid model oriented to display in the isometric view.

2. With this prep work done, start the **SOLPROF** command.
   - From the **Draw** menu, choose **Solids, Setup**, and then **Profile**.
   - From the **Solids** toolbar, choose the **Setup Profile** button.
   - At the 'Command:' prompt, enter the **solprof** command.

   Command: **solprof** *(Press* **ENTER***.)*

   AutoCAD may display the following prompt:

   You must be in a Layout to use SOLPROF.

3. Oops! **SOLPROF** expects you first to switch to layout mode. Why? I have no idea.
   To fix this problem, on the status bar, click the MODEL button to switch to paperspace and layout mode. You may need to use the **ZOOM Extents** and **ZOOM Window** commands to get a good view of the model in the viewport.
4. Repeat **SOLPROF**, and then pick the solid model.

   Command: **solprof**

   Select objects: *(Pick the solid.)*

   1 found Select objects: *(Press* **ENTER** *to end object selection.)*

5. The command asks three yes/no questions; first answer "Yes" to them all. (Later in the tutorial, answer "No" to see the difference.)

   Display hidden profile lines on separate layer? [Yes/No] <Y>: *(Press* **ENTER** *to accept the default, "Yes.")*

   Project profile lines onto a plane? [Yes/No] <Y>: *(Press* **ENTER.***)*

   Delete tangential edges? [Yes/No] <Y>: *(Press* **ENTER.***)*

   One solid selected.

   The solid may look no different. What has happened, however, is that AutoCAD has created a *second* model on top of the original, placing the new one on separate layers.

6. Click the **Layer** droplist to view the names of the new layers:

Layer Name	***Purpose***
PH-91	Profile hidden.
PV-91	Profile visible.

Note: The parts of the profile that are visible are placed on layer PV-91. (*PV* is short for "profile visible.") The parts of the profile that would be hidden are placed on layer PH-91. The number 91 has no significance; it happens to be the number AutoCAD assigns the paperspace viewport. In your drawing, the number is likely to be different.

7. Freeze layer 0 (or whichever layer you drew the original solid model on).

   If necessary, use the **LINETYPE** command to load the Hidden linetype, and then apply the Hidden linetype to layer PH-xx.

   The profile should look similar to the figure below.

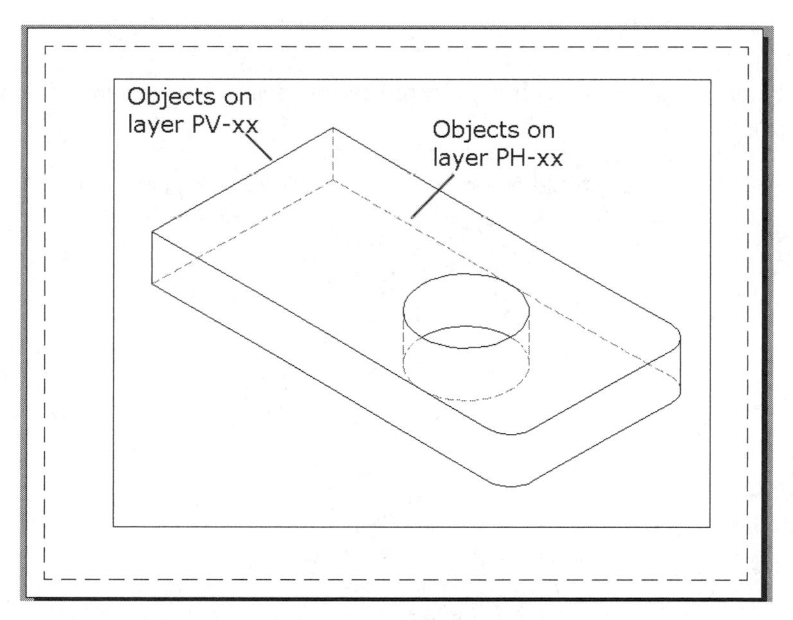

Visible and hidden lines are placed on separate layers.

8. To hide the hidden lines, simply freeze layer PH-xx.

To summarize: the default values of the **SOLPROF** command create visible and hidden lines on separate layers, overtop of the existing 3D solid model. Let's now return to the command's prompts to figure out what they mean.

9. Erase the objects on layer PH-xx and PV-xx.
   Thaw layer 0.
   Start the **SOLPROF** command:
      Command: **solprof**
      Select objects: *(Pick solid.)*
      1 found Select objects: *(Press* **enter** *to end object selection.)*
10. Earlier in this tutorial, you answered "Yes" to the following prompt, and AutoCAD created the PV-xx and PH-xx layers for visible and hidden objects.
    This time, answer "No." AutoCAD treats *all* profile lines as visible, including hidden ones.
       Display hidden profile lines on separate layer? [Yes/No] <Y>: **n**
11. When you earlier answered "Yes" to the following prompt, AutoCAD created the profile lines as *2D* drawings, as described by the phrase "*project* ... onto a plane."
    This time, answer "No." AutoCAD creates the profile as *3D* drawings.
       Project profile lines onto a plane? [Yes/No] <Y>: **n**

    In either case — "Yes" or "No" — the profile is collected together as a single block, with the name ***U**. (The * means that AutoCAD created the block, and the *U* means the block is "unnamed.")
12. Your answer earlier of "Yes" to the following prompt caused AutoCAD to leave off the tangential edges.
    This time, answer "No." AutoCAD includes the tangential edges.
       Delete tangential edges? [Yes/No] <Y>: **n**

    Tangental edges are lines that indicate the transition between flat and curved surfaces, such as fillets.
    Look at the figure to see the changes created by answering "No" instead of "Yes" to the three questions:

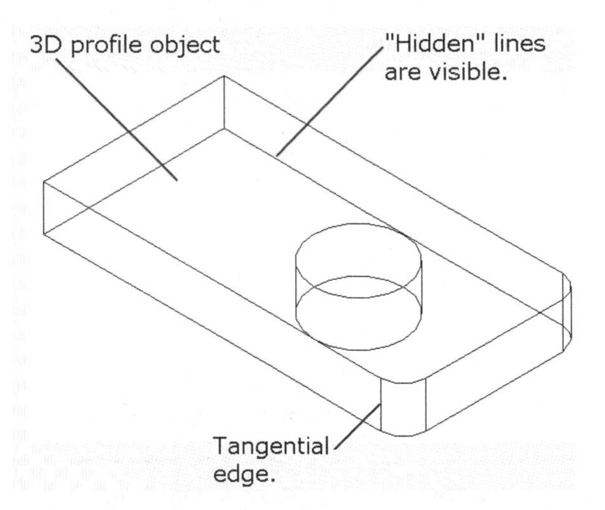

Changes made to the profile by answering "No" to the questions.

SOLIDS EDITING TUTORIAL

In this tutorial, you apply several of the commands from this chapter to design the plastic housing of a network router used by home offices. At the end of the tutorial, the solid model will look like the illustration below.

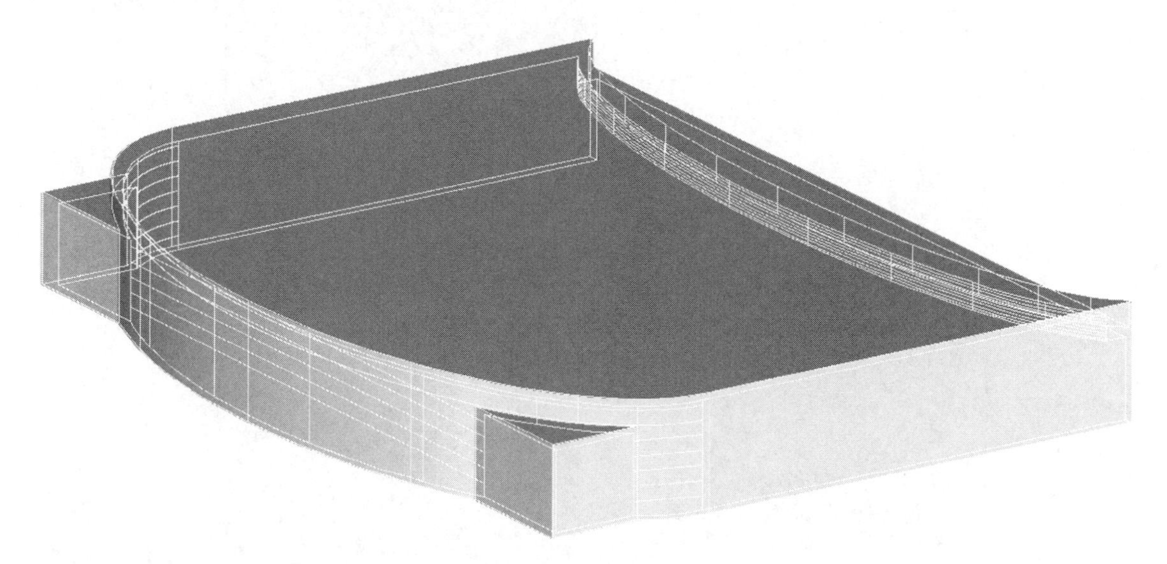

1. Start AutoCAD with a new drawing.
   Turn on snap and grid.
   Draw two boxes and a cylinder, as follows:

   Box #1:

   Command: **box**
   Specify corner of box or [CEnter] <0,0,0>: **0,0**
   Specify corner or [Cube/Length]: **8,6**
   Specify height: **1**

   Box #2:

   Command: *(Press* **ENTER** *to repeat the command.)*
   BOX Specify corner of box or [CEnter] <0,0,0>: **-0.2,1.5**
   Specify corner or [Cube/Length]: **8.2,6**
   Specify height: **1.25**

   Elliptical Cylinder #1

   Command: **cylinder**
   Current wire frame density: ISOLINES=12
   Specify center point for base of cylinder or [Elliptical] <0,0,0>: **e**
   Specify axis endpoint of ellipse for base of cylinder or [Center]: **-0.2,1.5**
   Specify second axis endpoint of ellipse for base of cylinder: **8.2,1.5**
   Specify length of other axis for base of cylinder: **-0.5,4**
   Specify height of cylinder or [Center of other end]: **1.25**

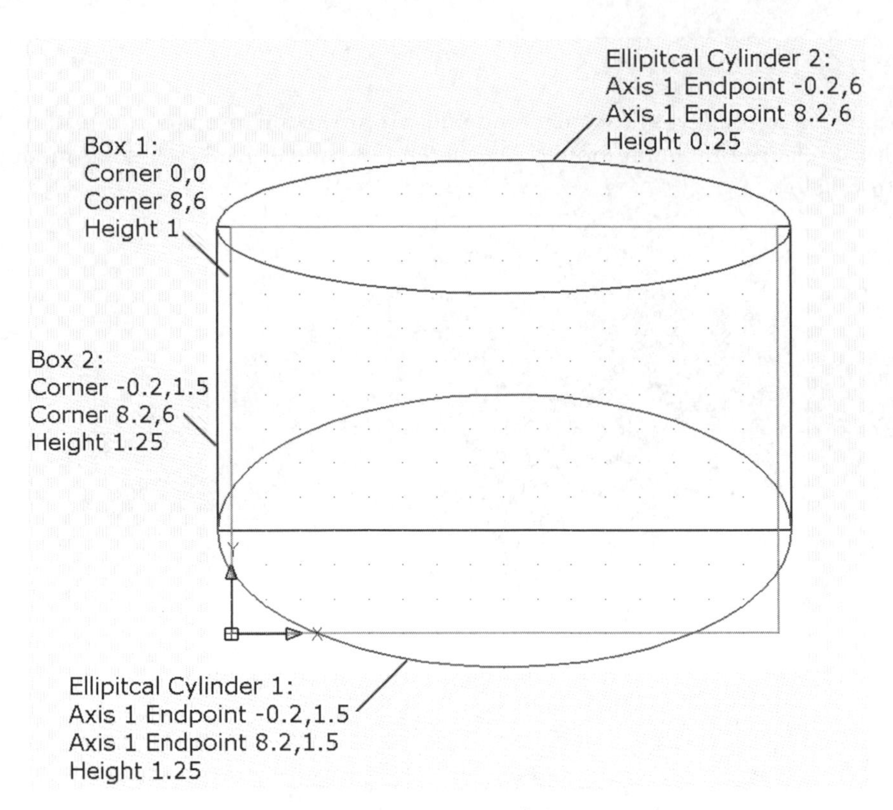

Plan view of the first four solids that make up the cover.

Elliptical Cylinder #2

Command: *(Press* **ENTER** *to repeat the command.)*

CYLINDER Current wire frame density: ISOLINES=12

Specify center point for base of cylinder or [Elliptical] <0,0,0>: **e**

Specify axis endpoint of ellipse for base of cylinder or [Center]: **-0.2,6**

Specify second axis endpoint of ellipse for base of cylinder: **8.2,6**

Specify length of other axis for base of cylinder: **4.5,4**

Specify height of cylinder or [Center of other end]: **0.25**

Use the **MOVE** command to move Cylinder #2 up by 1 unit:

Command: **move**

Select objects: **l** *(last)*

1 found Select objects: *(Press* **ENTER** *to end object selection.)*

Specify base point or displacement: **0,0,0**

Specify second point of displacement or <use first point as displacement>: **0,0,1**

You now have the basic building blocks in place.

2. It's time to switch to 3D viewing mode. From the menu bar, select **View | Viewports | New Viewports**.

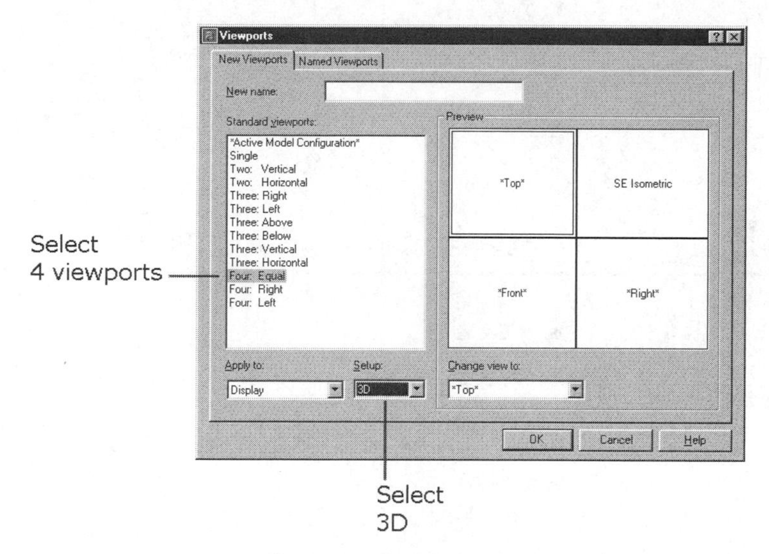

The Viewports dialog box.

In the **New Viewports** tab of the **Viewports** dialog box, select **Four: Equal** under **Standard Viewports**.

From the **Setup** listbox, select **3D**. Notice that the preview shows the four standard views: top, right, front, and isometric.

Click **OK**.

3. Mold the basic building blocks into the desired shape by manipulating them with Boolean operations, as follows:

Body #1: Union cylinder #1 with box #2.

Command: **union**

Select objects: *(Select cylinder #1.)*

1 found Select objects: *(Select box #2.)*

1 found, 2 total Select objects: *(Press* ENTER *to end the command.)*

Body #2: Union box #1 with body #1.

Command: **union**

Select objects: *(Select body #1.)*

1 found Select objects: *(Select box #1.)*

1 found, 2 total Select objects: *(Press* ENTER *to end the command.)*

Body #3: Subtract cylinder #2 from body #2.

Command: **subtract**

Select solids and regions to subtract from ..

Select objects: *(Select body #2.)*

1 found Select objects: *(Press* ENTER *to end object selection.)*

Select solids and regions to subtract ..

Select objects: *(Select cylinder #1.)*

1 found Select objects: *(Press* ENTER *to end the command.)*

The resulting body is shown by the figure.

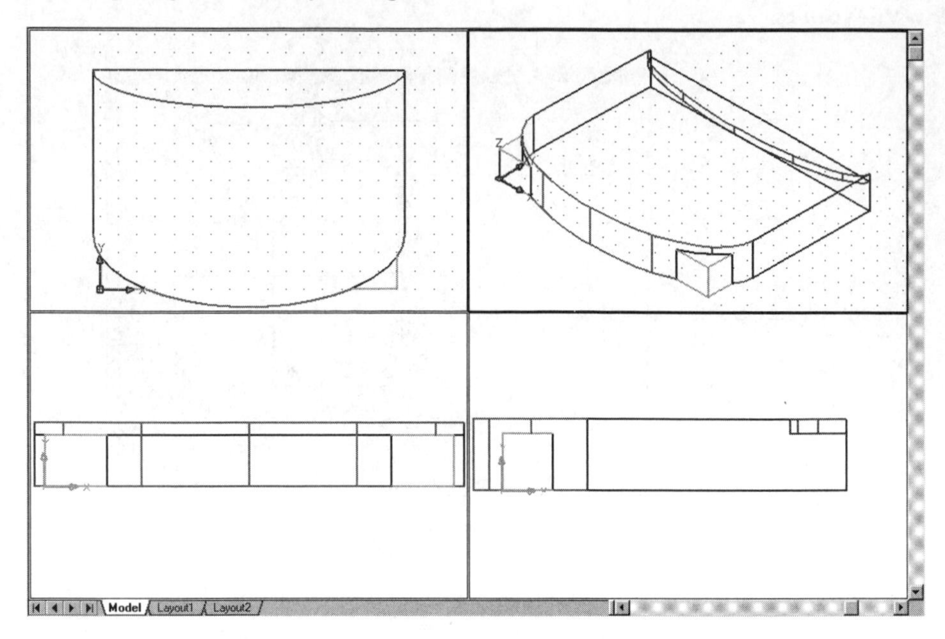

The body created from boxes and elliptical cylinders.

4. The next stage is to hollow out the body with the **Shell** option of the **SOLIDEDIT** command. The body changes to a shell just 0.2 units thick.

   Command: **solidedit**

   Solids editing automatic checking: SOLIDCHECK=1

   Enter a solids editing option [Face/Edge/Body/Undo/eXit] <eXit>: **b**

   Enter a body editing option

   [Imprint/seParate solids/Shell/cLean/Check/Undo/eXit] <eXit>: **s**

   Select a 3D solid: *(Select the body.)*

   Remove faces or [Undo/Add/ALL]: *(Press* ENTER *to end the option.)*

   Enter the shell offset distance: **0.2**

   Solid validation started. Solid validation completed. Enter a body editing option

   [Imprint/seParate solids/Shell/cLean/Check/Undo/eXit] <eXit>: *(Press* ENTER *to end the option.)*

   Solids editing automatic checking: SOLIDCHECK=1

   Enter a solids editing option [Face/Edge/Body/Undo/eXit] <eXit>: *(Press* ENTER *to end the command.)*

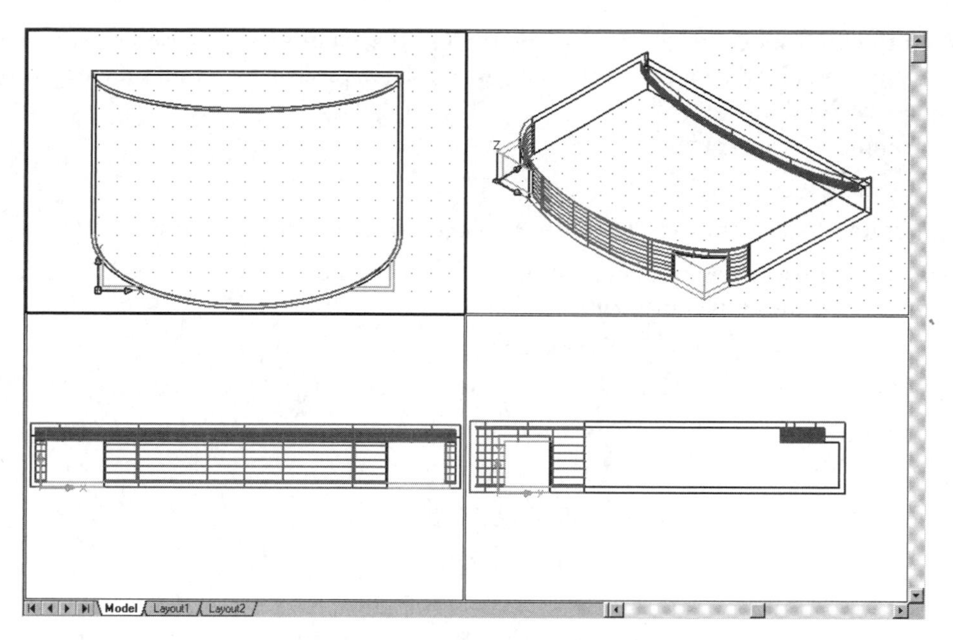

The body is hollowed out with the shell operation.

5. The final operation is to remove the bottom and the back of the body.
   Pick the edge between the two faces to remove them from the shell with the **SOLIDEDIT** command.
6. The result should look like the figure — after changing the viewpoint, and applying the **SHADE** command.

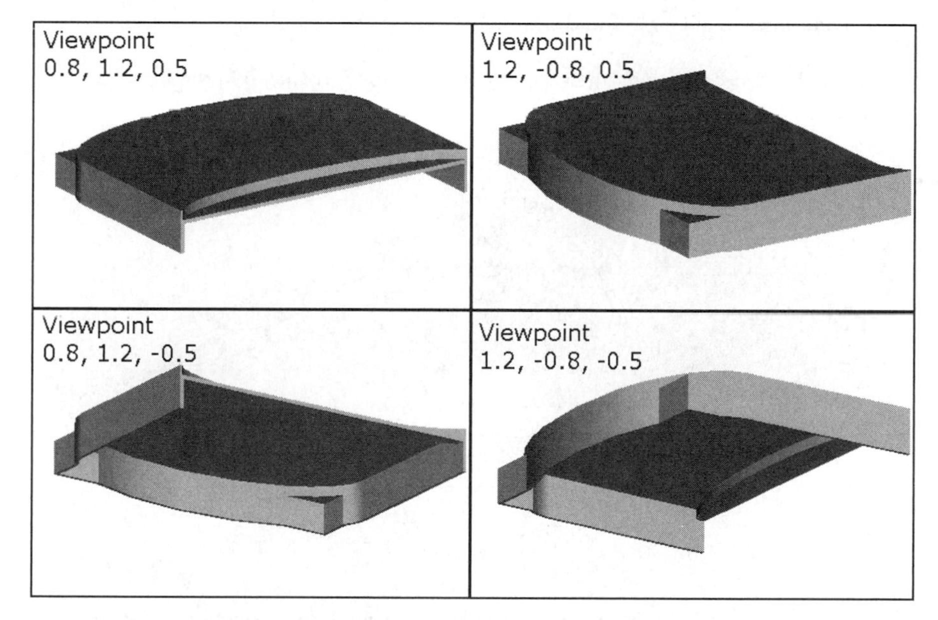

The hollowed out body, as seen from a variety of viewpoints.

I changed the color of the body to color 51 (pale yellow), a color that I find renders better than black or other dark colors.

7. With the body modelled to your satisfaction, use the **MASSPROP** command to determine its properties. (This command works only with solid models and regions, and not surface models or 2D drawings.)

   Command: **massprop**

   Select objects: *(Pick body.)*

   1 found Select objects: *(Press* **ENTER** *to end object selection.)*

   The command switches to the text window, and displays a report, a portion of which is shown below. You results may vary.

```
    ----------------  SOLIDS  ----------------
Mass:              7.2142
Volume:            7.2142
Bounding box:      X: -0.2000  --  8.2000
                   Y: -0.5365  --  5.8500
                   Z: -0.0086  --  1.2586
Centroid:          X: 4.0000
                   Y: 2.5896
                   Z: 1.0238
```

   ... and so on.

Note: If you need to convert a solid model to a surface model, there is no command for doing that in AutoCAD. There is a workaround, however: the **3DSOUT** and **3DSIN** commands operate as a solids-to-surface converter. These two **3DS**-commands are meant to let AutoCAD exchange drawings with 3D Studio rendering and animation software.

After using the two commands, the resulting surface model is shown in the figure, a mosaic that shows the surface lines and rendering.

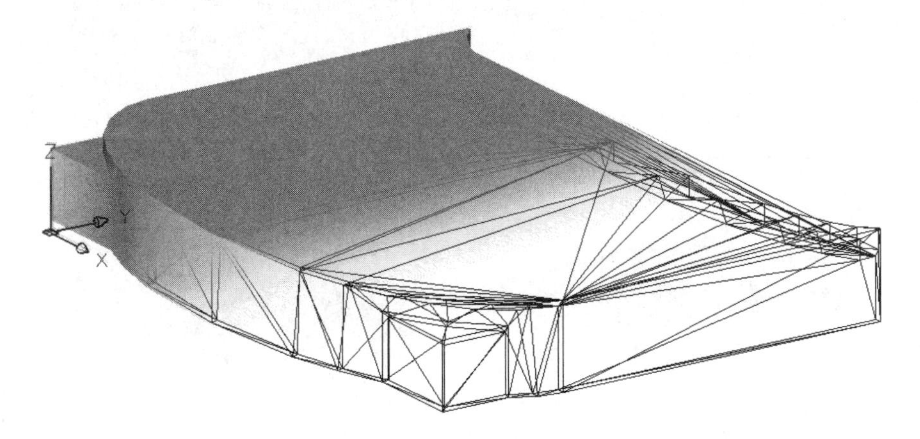

The body exported with 3DSOUT, and then imported with 3DSIN to create a surface model.

EXERCISES

1. From the CD-ROM, open the *model1.dwg* file. Change its color and linetype properties.
2. From the CD-ROM, open the *Ch12Fillet.dwg* file, a box with two holes.
   Apply a fillet radius of 0.1 to all edges. Save the result, which should be similar to the figure below.

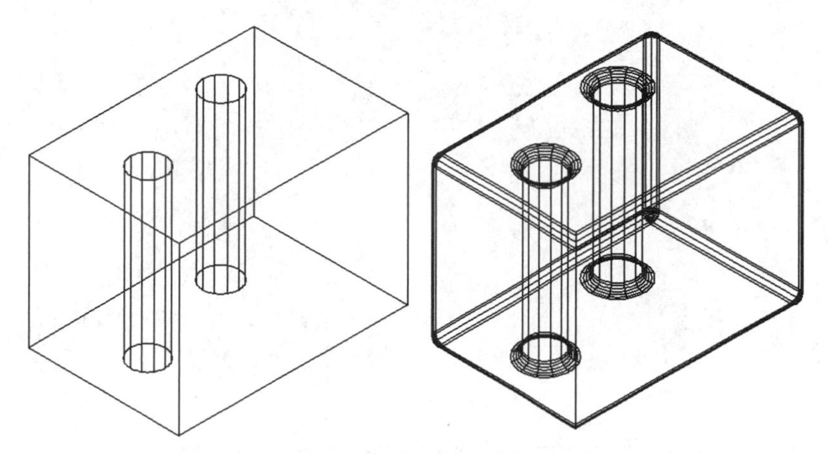

3. From the CD-ROM, open the *Ch12Chamfer.dwg* file, a wedge.
   Apply a chamfer of 0.1 to the two edges shown in the figure below.

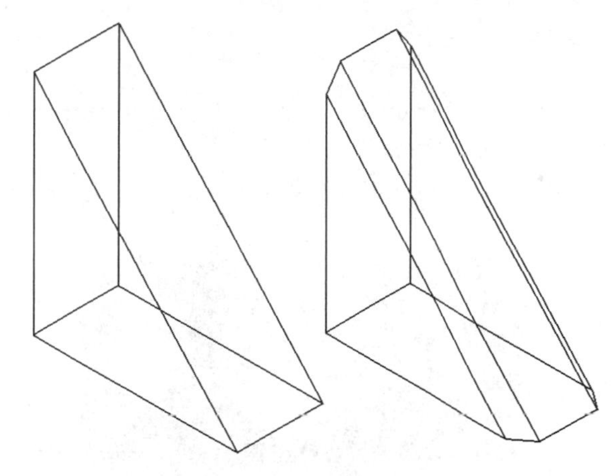

4. From the CD-ROM, open the *Ch12Solidedit.dwg* file. Use the **SOLIDEDIT** command to:
   a. Move the hole.
   b. Rotate the top face.
   c. Shell the body.

   The result may look similar to the figure below.

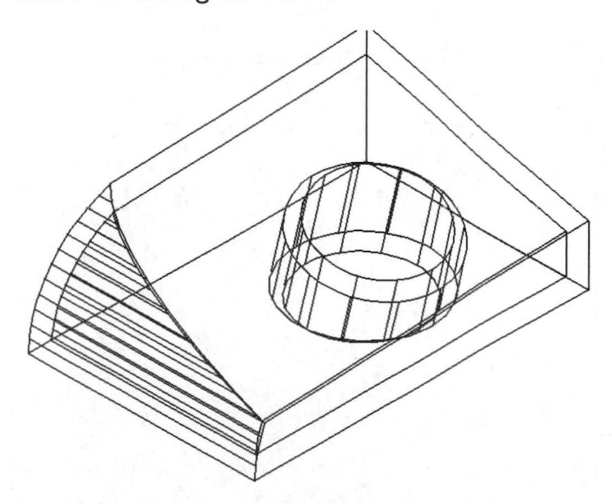

5. From the CD-ROM, open the *model1.dwg* file. Use the **SOLVIEW, SOLDRAW,** and **SOLPROF** commands to create the three standard orthographic views: front, top, and side.
6. Using the commands you have learned in this book, draw the dishwasher rack wheel, illustrated below, as a solid model.

The wheel is shown approximately full-size; some dimensions are:

Outer diameter = 1-3/4"

Wall thickness = 1/16"

Axle hole = 5/32"

Width of wheel = 3/4"

Use the **SOLVIEW, SOLDRAW,** and **SOLPROF** commands to create the three standard orthographic views: front, top, and side.

7. Draw the door latch cover, illustrated below, as a solid model.

The cover is shown approximately full-size; some dimensions are:

Overall length = 2-1/4"

Overall width = 1-1/8"

As an aid, the profile of the cover, shown below, is available on the CD-ROM as *Ch12Latch.dwg*.

Use the **SOLVIEW, SOLDRAW,** and **SOLPROF** commands to create the three standard orthographic views: front, top, and side.

CHAPTER REVIEW

1. Name two properties of solid objects that the **PROPERTIES** command can change.
2. Can the **FILLET** command create fillets between two solid objects?
3. Can the **CHAMFER** command create chamfers on more than one edge of a solid model at a time?
4. Briefly describe the purpose of the following commands:
   a. **ALIGN**
   b. **ERASE**
   c. **MOVE**
   d. **ROTATE3D**
5. What must you do before filleting or chamfering between two solid objects?
6. Briefly describe purpose of the **SOLIDEDIT** options:
   a. coLor
   b. Shell
   c. Imprint
   d. Delete
   e. cLean
7. When you select the edge of a body, what does AutoCAD select?
8. Under which condition can you select a face?
9. Describe the function of the following **SOLVIEW** options:
   a. Ucs
   b. Ortho
   c. Auxiliary
   d. Section
10. How do you place the hatching needed in a sectional view?
11. Does the **SOLDRAW** command work with any viewport?
12. What steps must you take to show hidden profile lines with the Hidden linetype?
13. Which command makes it easy to create the four standard views of a 3D model?
14. When chamfering a solid object, what is meant by the *base surface*?
15. Why do the **EXTEND** and **LENGTHEN** commands not work with solid models?
16. What happens when the **EXPLODE** command is applied to:
   a. Solid models
   b. Regions
17. What is a *chain*?
   A *loop*?
18. How many edges does a rectangular face have?
19. Can you select faces?
20. What is an *isoline*?
21. Briefly describe the result of moving a face with the **SOLIDEDIT** command.
   Rotating a face?
22. How would you create a lip around the top of a solid model?

UNIT IV

Rendering and Imaging

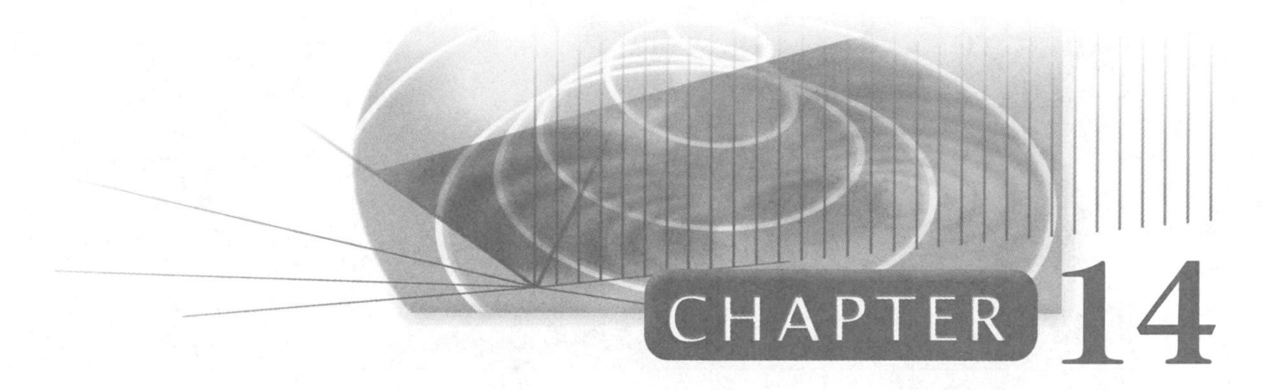

Hiding, Shading, and Rendering

Three-dimensional models are usually viewed as wireframe, as if they were transparent. But many faces and isolines can make the models hard to visualize, especially for non-drafters. For this reason, AutoCAD allows you to remove hidden lines, as well as realistically to render 3D models with lights, surface materials, and shadows.

In this chapter, you learn about the following commands:

HIDE removes hidden lines from 3D viewpoints.

HLSETTINGS displays hidden lines with color and linetype.

SHADEMODE renders 3D objects while editing.

RENDER generates photorealistic renderings of 3D views.

BACKGROUND places colors and images behind renderings.

FOG creates fog-like effects for depth cueing.

MATLIB and **RMAT** load and apply materials to 3D objects.

LIGHT inserts lighting into 3D scenes.

SCENE collects views and lights.

LSLIB, **LSNEW**, and **LSEDIT** insert and edit landscape objects.

FINDING THE COMMANDS

On the **RENDER** toolbar:

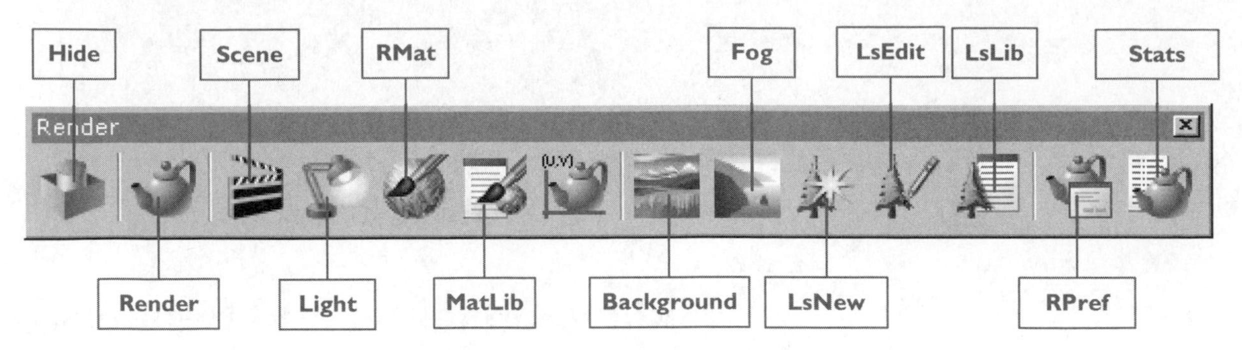

On the **VIEW** menu:

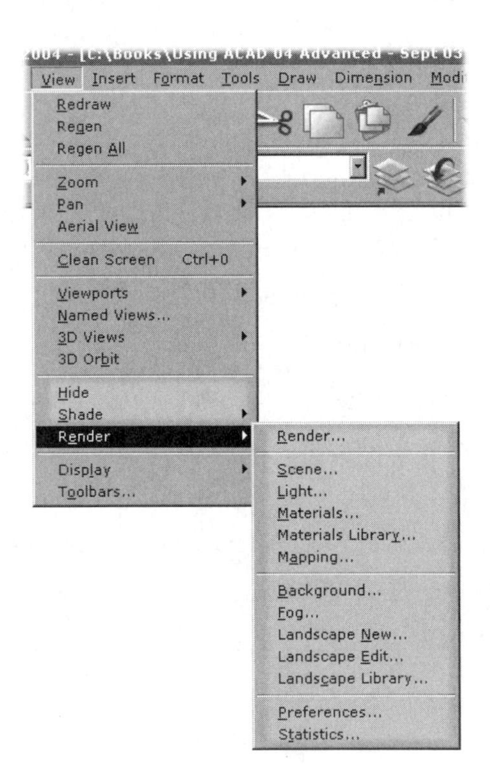

HIDE

The **HIDE** command removes hidden lines from 3D objects.

This command hides lines that are behind many kinds of objects: circles, 2D solids, traces, regions, wide polyline segments, 3D faces, and polygon meshes. In addition, the extruded edges of objects with thickness hide objects behind them: circles, 2D solids, traces, and wide polyline segments. These are seen as solid objects with top and bottom faces.

Text is displayed through objects; it is not hidden unless (1) it has a thickness, or (2) the **HIDETEXT** system variable is turned on. This applies to text created by the **TEXT** and **MTEXT** commands alike, as well as to attributes and dimension text, contrary to Autodesk's documentation. (Dimension lines are also hidden.)

To reduce display time, **HIDE** ignores objects on frozen layers, but does affect objects on layers turned off. (Historically, in the early days of desktop computers, AutoCAD could take an entire day to remove hidden lines from complex 3D drawings, a task usually left for overnight or weekends. With today's fast computers, hidden-line removal takes just seconds.)

BASIC TUTORIAL: REMOVING HIDDEN LINES

Before using this command, open a drawing containing 3D objects.

1. To remove hidden lines from 3D objects, start the **HIDE** command:
   - From the **View** menu, choose **Hide**.
   - At the 'Command:' prompt, enter the **hide** command.
   - Alternatively, enter the alias **hi** at the keyboard.

   Command: **hide** *(Press* **ENTER.***)*

   AutoCAD immediately processes the drawing, reporting "Removing hidden lines."
   When done, the hidden-line view is displayed.

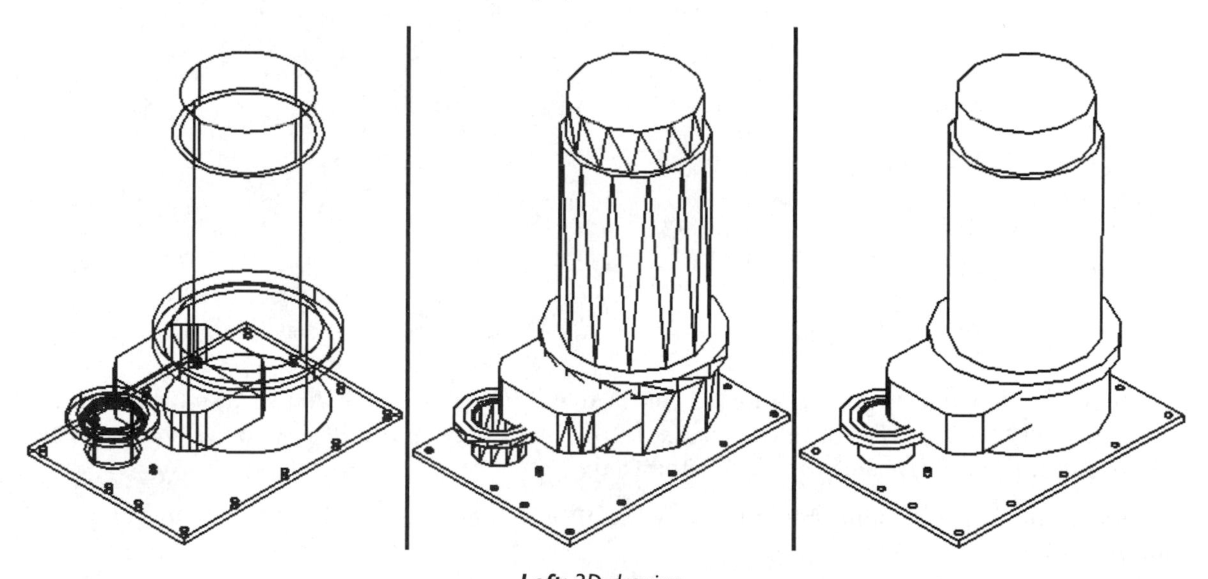

Left: *3D drawing.*
Center: *Hidden line removed.*
Right: *Hidden lines removed in silhouette mode.*

The image displayed by this command is temporary: as soon as you change the view, the hidden lines reappear. To edit in hidden-line mode, use the **SHADEMODE** command's **Hidden** option.

Use the **SAVEIMG** command to save the image as a raster file, or **WMFOUT** to save it as a vector file. You can also press **CTRL+C** to copy the image to Clipboard, and then paste the image into another document.

AutoCAD normally removes all lines that are hidden (or "obscured") by objects located in front of them. You can show the hidden lines, if you wish, in other colors and lineweights, as described below.

REMOVING HIDDEN LINES: ADDITIONAL METHODS

The **HIDE** command is controlled by a single command and a system variable.

- **HLSETTINGS** command controls hidden-line removal.
- **DISPSILH** system variable toggles the display of silhouette curves of solid models.

Let's look at them.

HlSettings

The **HLSETTINGS** command (short for "hidden line settings") controls the display of hidden objects. Lines that are normally invisible can be given color and linetype. This allows you to create drawings where the hidden lines take on a dashed linetype, as illustrated by the figure below.

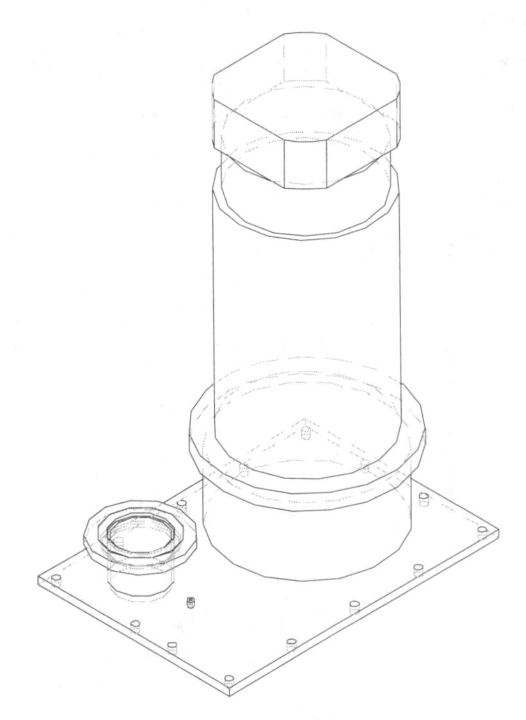

Hidden lines displayed with dashed linetype and gray color.

The linetypes provided by this command unfortunately do not match AutoCAD's other linetypes. Thus, the **HLSETTINGS** dialog box does not list a "Hidden" linetype; the closest equivalent is "Dashed." Furthermore, the linetypes cannot be customized.

The effects of this command apply to both the **HIDE** command and the **SHADEMODE** command's **Hidden** option.

(Historically, in older versions of AutoCAD, the **HIDE** command automatically placed hidden lines on a separate, frozen layer. The layer could be thawed, and then assigned color and linetype to show the hidden lines. About ten years ago, Autodesk changed the manner in which hidden-lines are calculated, making the process much faster, but in the process losing its ability to display hidden lines. With AutoCAD 2004, Autodesk added the **HLSETTINGS** command to replace the lost feature.)

Command: **hlsettings**

AutoCAD displays the Hidden Line Settings dialog box.

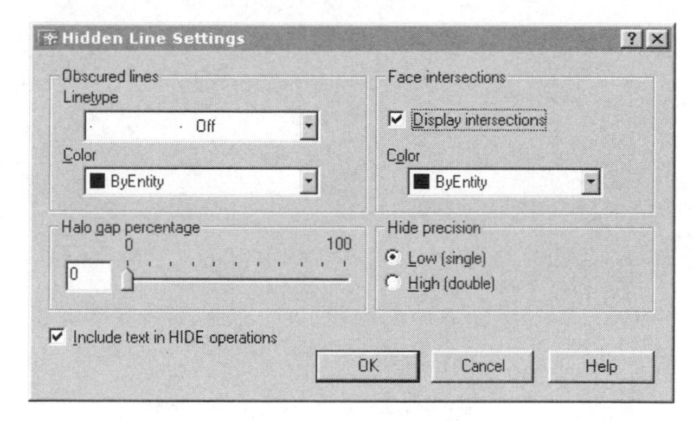

Hidden Line Settings dialog box.

Obscured Lines

The **Linetypes** option selects the linetype with which to display hidden lines. These linetypes are "intelligent": they keep the same dash size, no matter the zoom level. (The **LTSCALE** command has no effect on these linetypes.) "Off" displays no hidden lines, like the **HIDE** command's default setting. "Solid" shows continuous lines, which can be colored with the Color option. "Dashed" is most similar to AutoCAD's Hidden linetype.

The **Color** option specifies the color for hidden lines. I like using a shade of gray, such as color #254.

Halo Gap Percentage

The *halo gap* is an optional gap AutoCAD can add to the transition between visible and hidden lines. The slider bar changes the size of the gap from 0 to 100% of one unit; as with linetypes, the gap size is unaffected by zoom level.

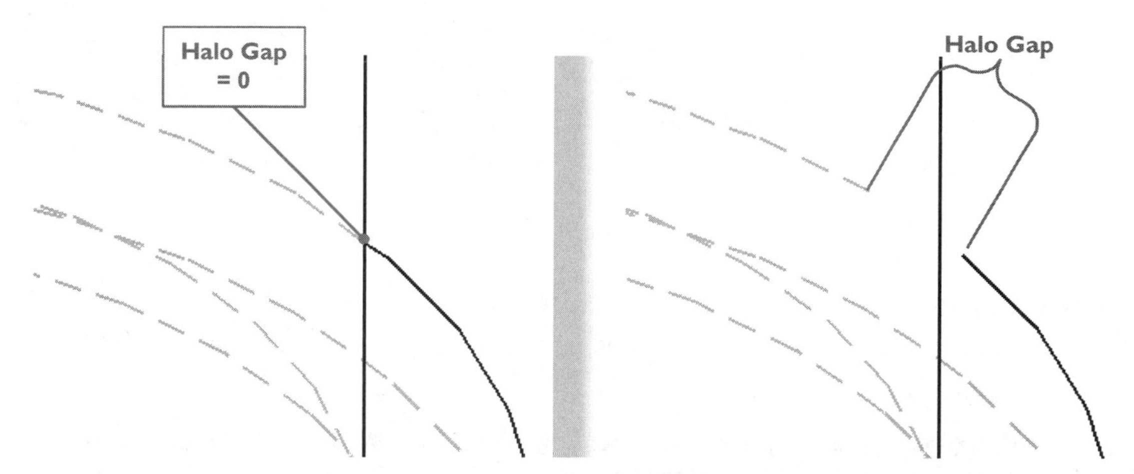

Halo gap between visible and hidden portion of a line.

Include Text in HIDE Operations

Normally, text is displayed through objects, ignoring the fact it may be obscured by an object — all text, including visible attributes and dimension text. (If you do not want text displayed in hidden-line views, place it on a frozen layer.) Text with thickness is always treated as a hidden object.

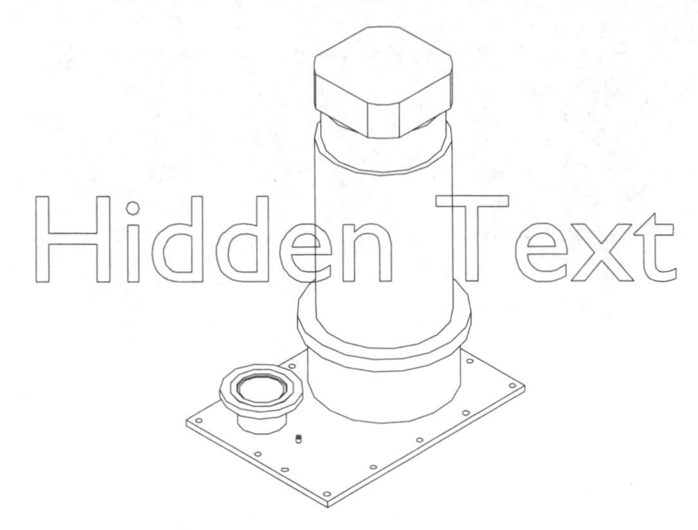

Text ignores other objects, and displays normally.

To make text obey the rules of hiddeness, turn on the **Include Text in HIDE Operations** option.

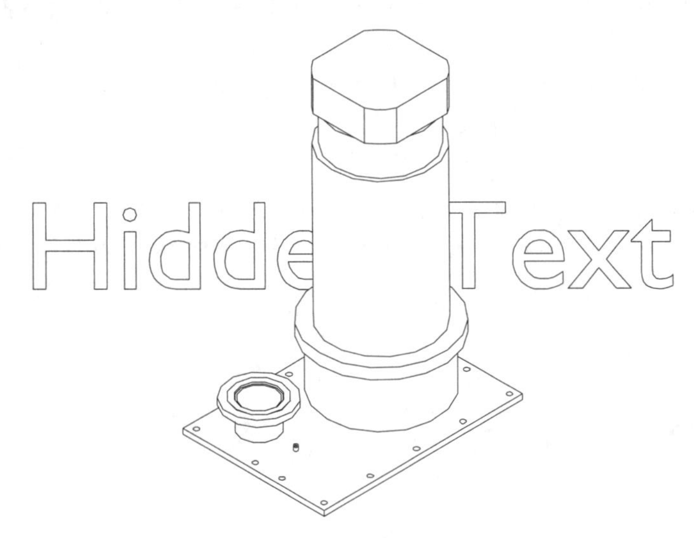

Text included in hide operation.

Hide Precision

The **Hide Precision** option toggles the accuracy of calculations for hide and shade operations.

- **Low** (single precision) uses less memory.
- **High** (double precision) uses more memory, but hidden-line removal is more accurate, particularly for those involving solid models.

Face Intersections

The **Face Intersections** option determines whether AutoCAD draws lines at the intersections of 3D surfaces. When turned on, it allows you to select a color for the intersection lines.

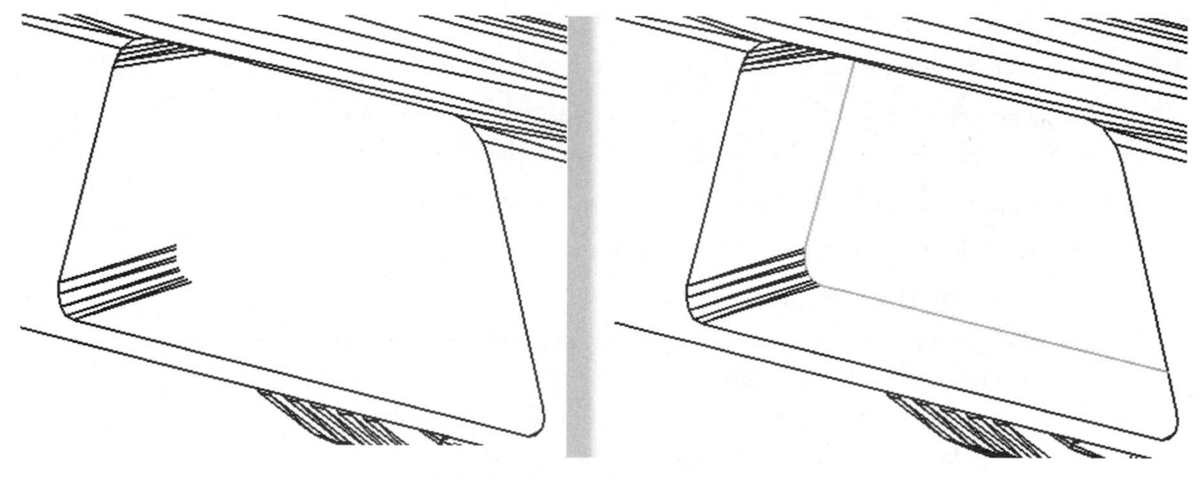

Left: *Face intersection not displayed.*
Right: *Face intersection displayed.*

Hidden-line removal is also available with the SHADEMODE command, which carries the added advantage of being able to edit in hidden-line mode.

SHADEMODE

The **SHADEMODE** command applies simple rendering to 3D objects.

This command allows you to edit objects that are shaded or have their hidden lines removed, unlike the RENDER and HIDE commands. 2D drawings cannot be shaded. The following forms of shading are available:

- **3D wireframe** generates wireframe models in 3D space.
- **Hidden** removes hidden faces similarly to the **HIDE** command.
- **Flat** flat-shades faces.
- **fLat+edges** flat-shades faces, and outlines them with the background color.
- **Gouraud** smooths faces.
- **gOuraud+edges** smooths shaded faces, and outlines them with the background color (illustrated at right).
- **2D wireframe** generates wireframe models in 2D space; returns shaded views to "normal" 2D display.

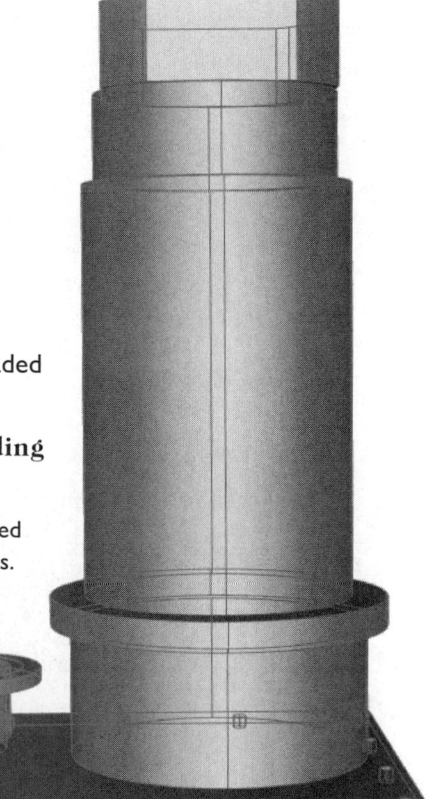

The same options are available through the 3DORBIT command's **Shading Modes** shortcut menu, though under slightly different names.

Note: Shading modes use some ambient light, plus light from a source located over your left shoulder. You can replace this light source with your own lights. Ambient, point, distant, and spot lights are used by SHADEMODE when (1) lights are defined in the drawing by the **LIGHT** command, (2) the modes are Flat, Gouraud, fLat+edges, or gOuraud+edges, and (3) the **Lights** option has been enabled in the **OPTION** command's **3D Graphics System Configuration** dialog box.

BASIC TUTORIAL: SHADING DRAWINGS

Before using this command, open a drawing containing 3D objects.

1. To shade drawings with 3D objects, start the **SHADEMODE** command:
   - From the **View** menu, choose **Shade**, and then one of the options.
   - At the 'Command:' prompt, enter the **shademode** command.
   - Alternatively, enter the alias **sha** at the keyboard.

   Command: **shademode** *(Press* ENTER.*)*
2. AutoCAD reports the current shading mode, and then prompts you to select a mode:

   Current mode: 2D wireframe

   Enter option [2D wireframe/3D wireframe/Hidden/Flat/Gouraud/fLat+edges/ gOuraud+edges] <Gouraud>: *(Enter an option.)*

SHADING DRAWINGS: ADDITIONAL METHODS

The SHADEDGE command has a number of options, and the quality of its display is controlled by several system variables.

2D Wireframe

The **2d wireframe** option returns the shaded mode to the "normal" wireframe display. All objects and properties are displayed normally. Even though this mode is called "2D," it is actually 3D.

3D Wireframe

The **3d wireframe** option displays objects in wireframe without lineweights or linetypes. The UCS icon is rendered. If materials have been applied, they are displayed (although barely discernible); raster and OLE objects are not displayed. Curved surfaces of solid models look denser when the system variable ISOLINES is set to higher values.

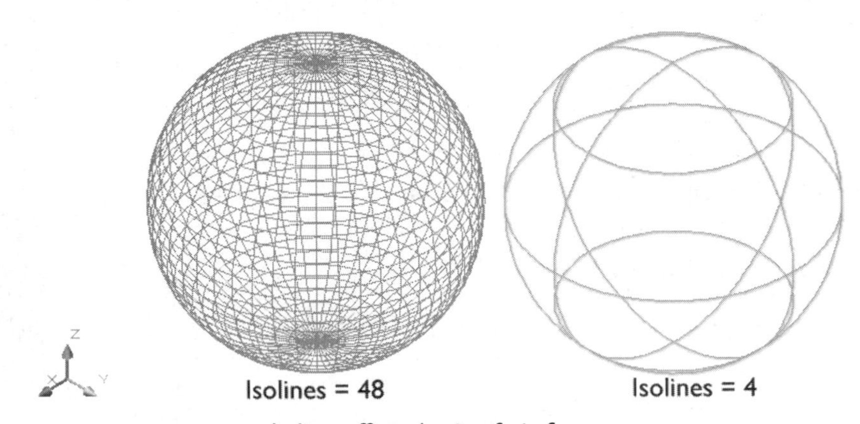

Isolines affects density of wireframes.

Hidden

The **Hidden** option removes hidden lines from the drawing.

Curved surfaces of solid models look smoother when the FACETRES system variable (short for "face tessellation resolution") is set to 10, its maximum value. You must use the REGEN command after FACETRES:

Command: **facetres**

Enter new value for FACETRES <0.5000>: **10**

Command: **regen**

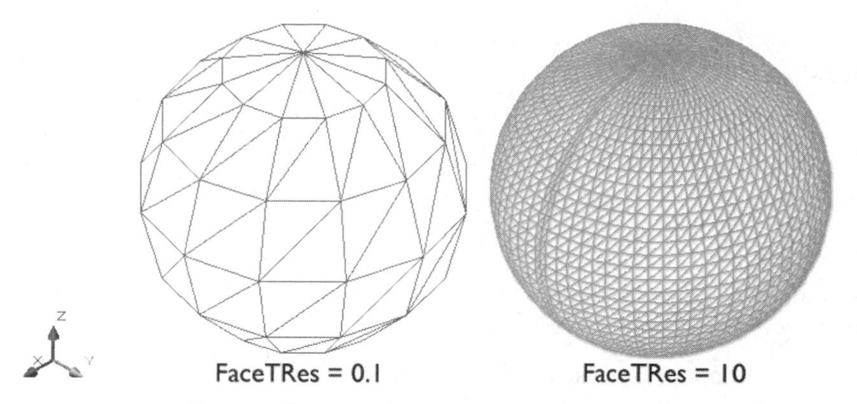

Facetres affects smoothness of hidden-line removal.

Flat

The **Flat** option shades the faces of 3D objects in a uniform manner, giving a faceted appearance to curved objects. (Technically, flat shading uses *normals* to determine the intensity of light falling on each face of the object.) Curved surfaces look smoother when **FACETRES** is set to higher values.

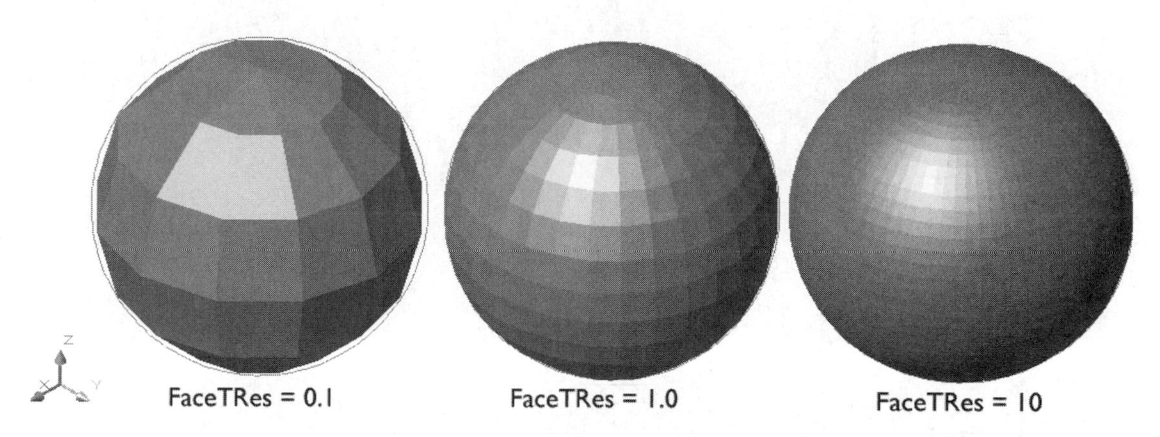

FaceTRes affects smoothness of flat shading.

Gouraud

The **Gouraud Shaded** option applies smoother shading to 3D objects, giving a somewhat more realistic appearance. Curved surfaces look smoother when **FACETRES** is set to higher values.

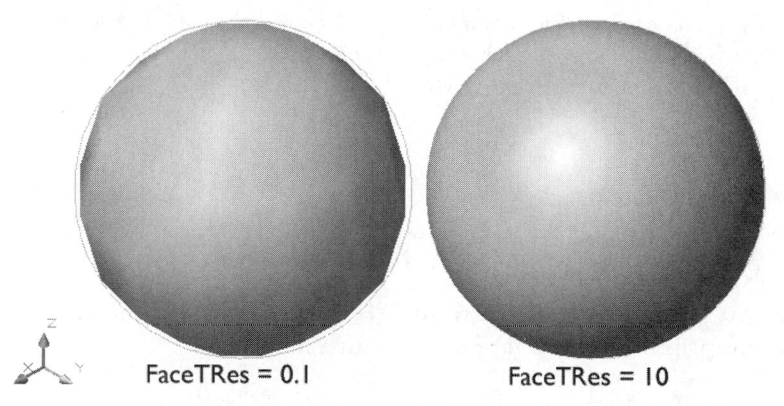

FaceTRes affects smoothness of Gouraud shading.

(Historically, this option is named after Henri Gouraud, who invented the shading algorithm in 1971 as a way for relatively-slow computers to "fake" the look of smoothly curved surfaces using relatively-fast integer mathematics. Technically, Gouraud shading averages the *normals* of the faces that surround each vertex to determine its normal. For more details, see freespace.virgin.net/hugo.elias/graphics/x_polygo.htm.

fLat+edges

The **fLat+Edges** option (flat shaded with edges) flat-shades 3D objects. It adds the hidden-line wireframe view, as well as hidden-line isolines on curved surfaces.

gOuraud+edges

The **gOuraud+edges** option (Gouraud shaded with edges) smooth shades 3D objects, and includes the hidden-line wireframes and isolines.

In both cases, turning on edges makes 2D objects appear that do not otherwise in shaded views. On curved surfaces, **ISOLINES** affects the number of edges, while **FACETRES** affects the smoothness of the shading.

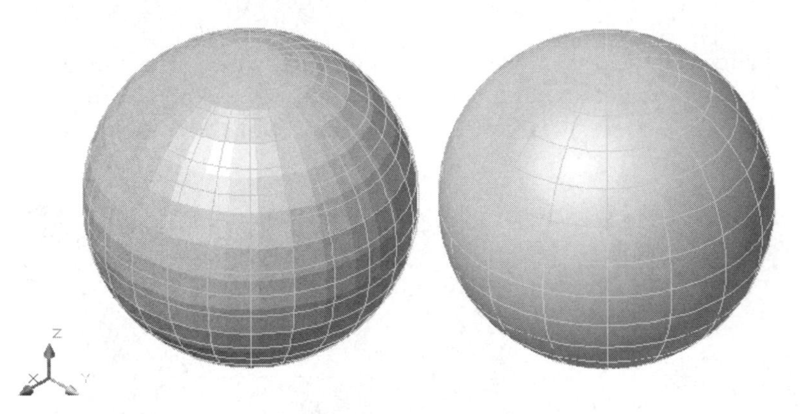

Left: *Flat shaded with edges on.*
Right: *Gouraud shaded with edges on.*

RENDER

The **RENDER** command generates photorealistic renderings of 3D views. (AutoCAD cannot render in paper space mode.)

Note: If you do not place lights (with **LIGHT**) or define scenes (with **SCENE**), the **RENDER** command uses the current view and ambient light. If you don't specify lights or a scene, then this command renders the drawing using *all* lights and the current view.

If you set up the **RENDER** command to skip its dialog box, you need to use the **RPREF** command to set rendering options. Uncheck the **Skip Render Dialog** option to bring back the **Render** dialog box.

The **RENDERUNLOAD** was removed from AutoCAD Release 14. To free up memory, use the **ARX** command's **Unload** option to unload *acrender.arx* instead.

BASIC TUTORIAL: RENDERING DRAWINGS

Before using this command, open a drawing containing 3D objects.

1. To render 3D drawings, start the **RENDER** command:
   - From the **View** menu, choose **Render**, and then **Render**.

- At the 'Command:' prompt, enter the **render** command.
- Alternatively, enter the alias **rr** at the keyboard.

Command: **render** *(Press* ENTER.*)*

AutoCAD displays the Render dialog box.
(When the **Skip Dialog Box** option is turned on, AutoCAD immediately renders the drawing.)

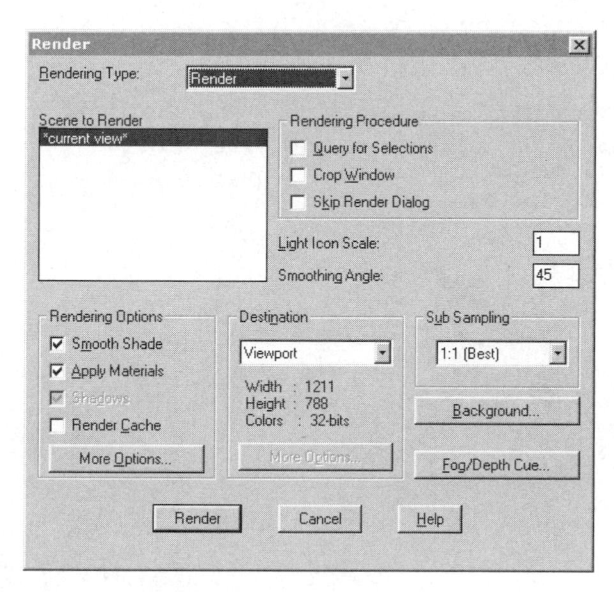

Render dialog box.

2. Click **Render**.
   AutoCAD reports, "Using current view. Default scene selected. 100% complete," and then displays the rendered drawing.
3. To return to the wireframe display, use the **REGEN** command:
   Command: **regen**

Note: There is no command-line version of the RENDER command. Instead, use the **(c:render)** AutoLISP routine to set the cropping window for rendering to the viewport, and the **(c:rfileopt)** routine to set options for rendering to files. For more information on these routines, search the online *AutoLISP Developer's Guide* provided with AutoCAD.

RENDERING DRAWINGS: ADDITIONAL METHODS

The RENDER command has many, many options. Most are accessed through its dialog box, but some are also accessed through separate commands:

- **RPREF** command presets rendering options to avoid displaying the Render dialog box.
- **BACKGROUND** command specifies the background to the rendering.
- **FOG** command variable specifies the color and intensity of the atmosphere.
- **STATS** reports statistics of the rendering.
- **REPLAY** command displays renderings that were saved as raster files.
- **SCENE** command creates scenes from lights and views.

We look at all these commands later in this chapter. The dialog box's other options are:

Rendering Type selects between three modes of rendering: Render, Photo Real, and Photo Raytrace. Each mode is progressively more realistic and requires more time to complete. If a third-party rendering program has been installed correctly, its name is also listed here.

Query for Selections renders only selected objects. After you click **Render**, AutoCAD prompts you to select objects to render.

Select objects: *(Select one or more objects.)*

Select objects: *(Press* ENTER *to start rendering.)*

This is useful when you want to see only part of the drawing rendered, either for speed or aesthetic reasons. To make the rest of the wireframe model appear with the rendering, use the **Background** button's **Merge** option, as described later in this chapter.

Rendering selected objects.

Crop Window renders a rectangular area of the drawing. (This option is available only when the **Viewport** option is selected.) After you click **Render**, AutoCAD prompts you to specify the rectangle:

Pick crop window to render: *(Pick two points.)*

The resulting crop, unfortunately, is surrounded by a black background, which cannot be altered by any of the **Background** button's options.

Rendering rectangular window.

Skip Render Dialog renders the drawing immediately after you enter the **RENDER** command. The Render dialog box is no longer displayed. To get back the Render dialog box, use the **RPREF** command to turn off this option.

Light Icon Scale controls the size of the Overhead, Direct, and Sh_Spot blocks that represent light sources in the drawing.

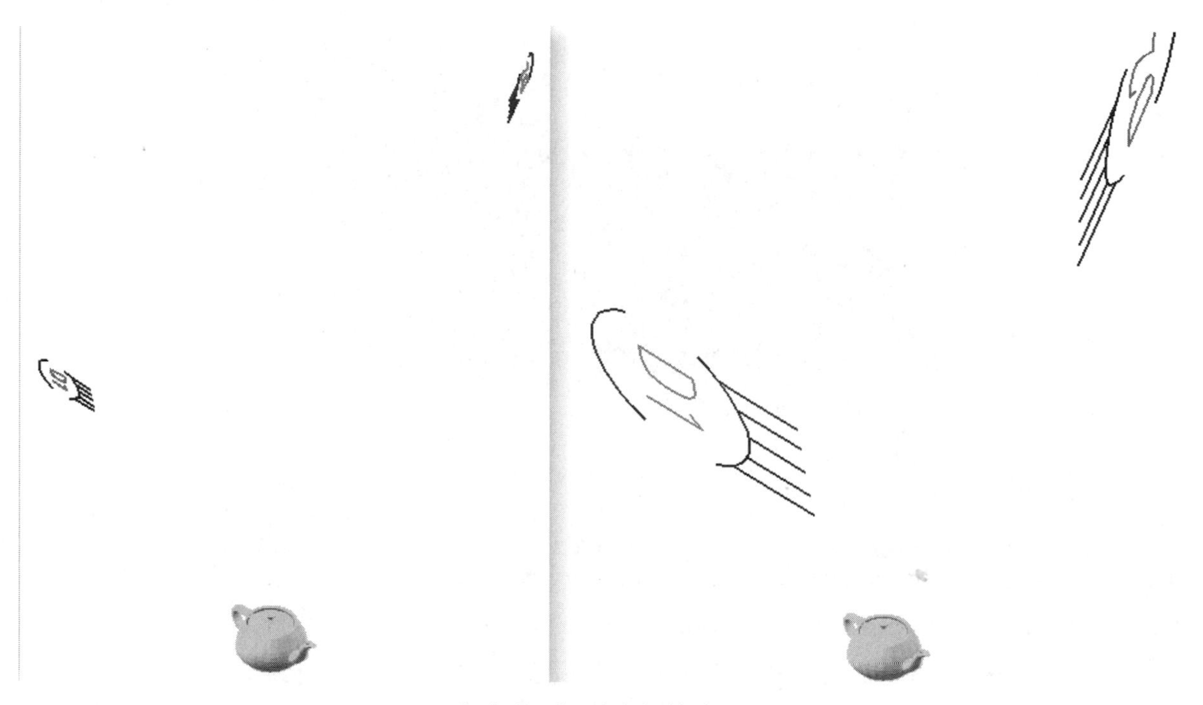

Left: *Small-scale light blocks.*
Right: *Large-scale light blocks.*

Smoothing Angle specifies the cutoff angle for determining the smoothing of edges. This option works together with the **Smooth Shade** option. Edges with larger angles are *not* smoothed; edges with smaller angles are smoothed.

Smooth Shade blends the colors across faces whose angle is less than that specified by the **Smoothing Angle** option. To smooth round surfaces of solid models, set the value of the **FACETRES** system variable to **10**. (This system variable has no effect on surface models.)

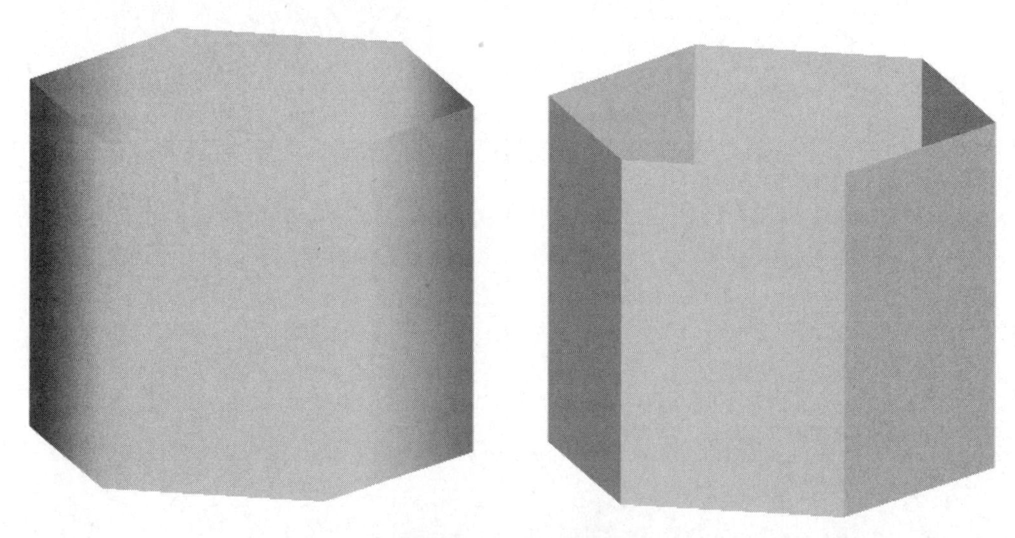

Left: *Edges are smoothed.*
Right: *Edges are not smoothed.*

Apply Materials applies surface materials defined by the RMAT command. When this option is turned off, all objects take on the parameters defined by the "Global" material: color (gray), ambient (0.70), reflection (0.2), roughness (0.5), transparency (0), refraction (1.0), and bump mapping (none).

Shadows casts shadows when the Photo Real and Photo Raytrace rendering modes are selected. The type of shadow is controlled by the **Shadow Options** option of the LIGHT command.

Shadow casting.

Render Cache speeds up the rendering by saving render data to a cache file. Unchanged objects are rendered the same way, reducing the number of required calculations.

Destination specifies where the rendering should be displayed:

- **Viewport** displays the rendering in the current viewport. New renderings replace previous ones.

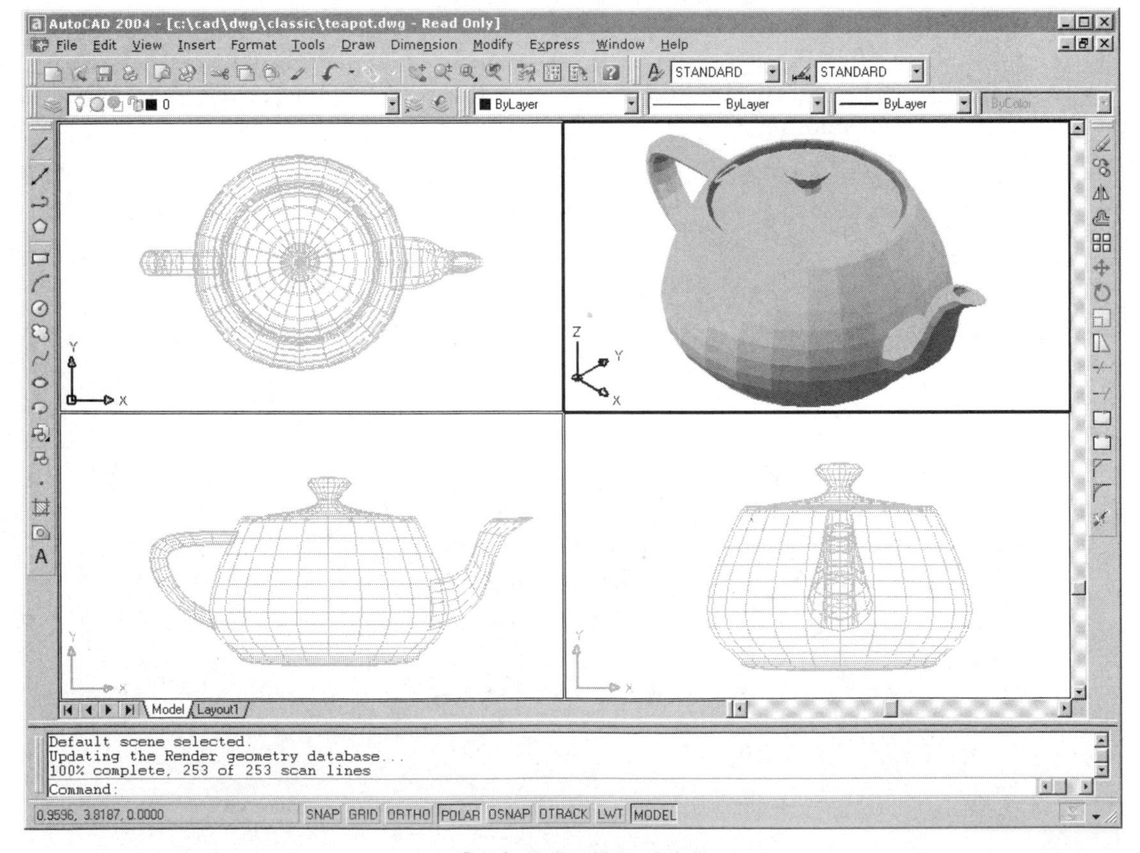

Rendering to a viewport.

- **Render Window** displays the renderings in an independent window. New renderings are opened in additional windows, allowing you to compare them.

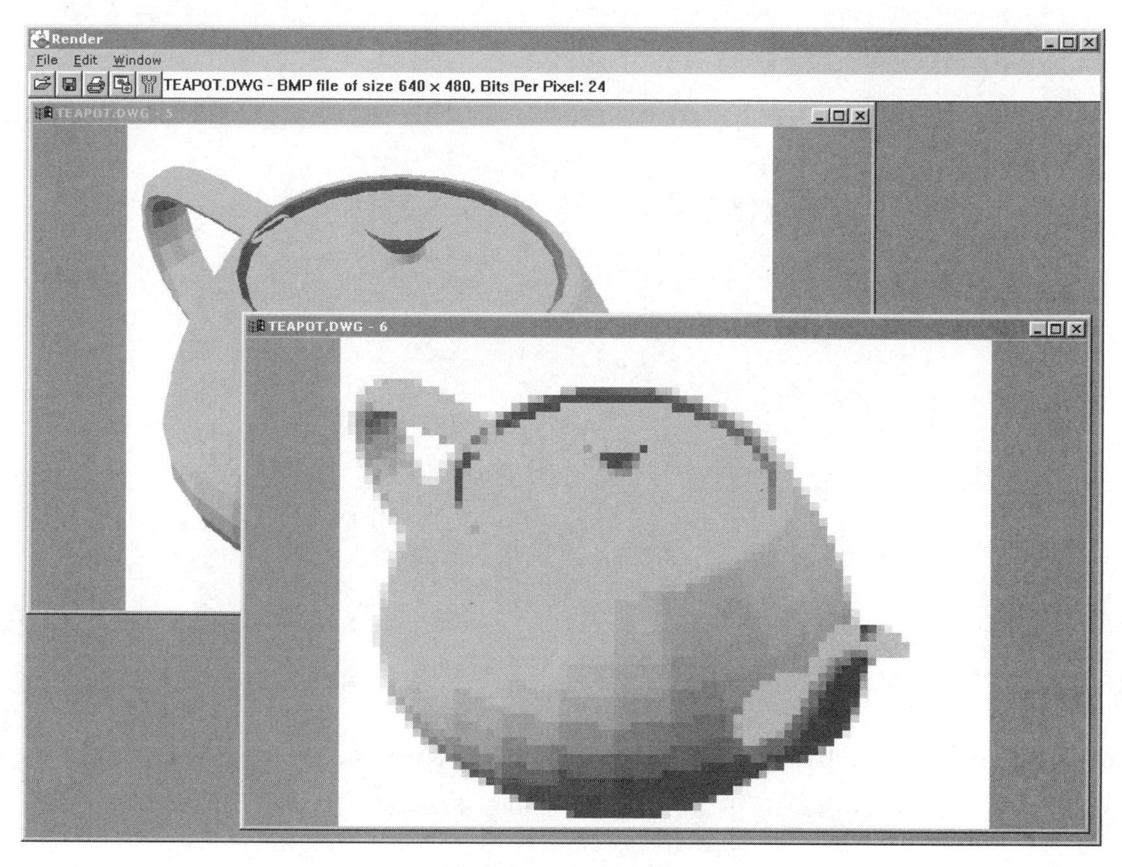

Rendering to the window.

- **File** outputs the renderings to raster files on disk; you do not see the rendering. Choose the **More Options** button to specify the file type and size, and other options.

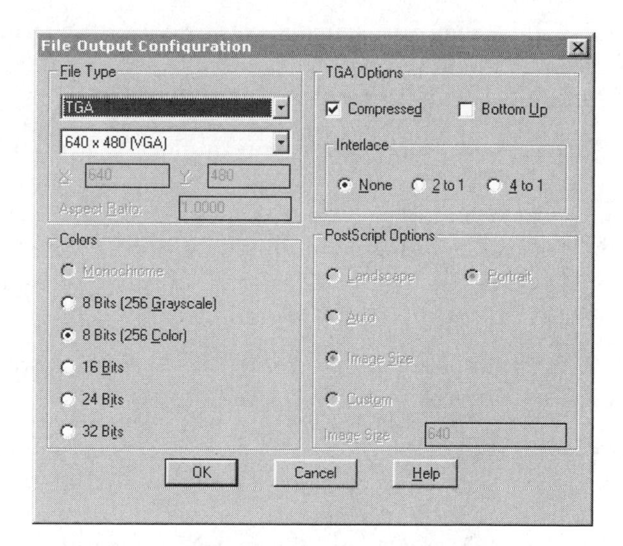

File Output Configuration dialog box.

Sub Sampling speeds up the display time by rendering fewer pixels at the cost of lower-quality renderings. You can choose from settings between 1:1 (best quality) to 8:1 (fastest).

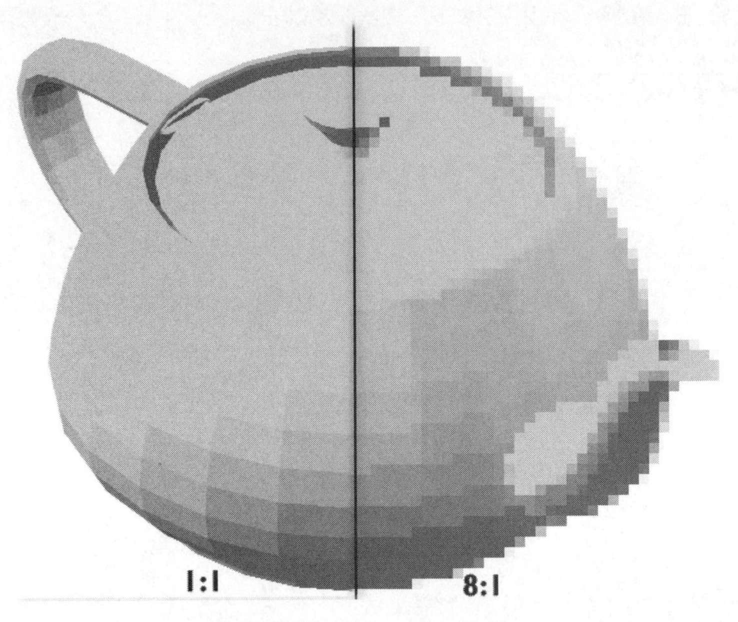

***Left:** No sub-sampling (1:1).*
***Right:** Largest amount of sub-sampling (8:1).*

Background displays the Background dialog box. See the BACKGROUND command.

Fog/Depth Cue displays the Fog/Depth Cue dialog box. See the FOG command.

ADVANCED OPTIONS

Choose the **More Options** button to display the Render Options dialog box.

Render Quality. Gouraud and Phong are the names of the algorithms used by RENDER to create its images. Phong is more realistic when your model has lights, but takes somewhat longer to render than does Gouraud.

Face Controls. AutoCAD uses face normals to determine which parts of the model are the front and the back. *Faces* are the 3D polygons that make up the surface of models; *normals* are the vectors that point at right angles from the faces. AutoCAD determines the direction of the normals by applying the right-hand rule to the order in which the face vertices were drawn. A positive normal points toward you; a negative normal points away from you. To save time in rendering, negative face normals are ignored, because they aren't seen in renderings.

Discard Back Faces. When this option is on, AutoCAD performs faster renderings, but may display visual mistakes.

Anti-Aliasing. This software technique reduces the stairstep-like edges of rendered images; it makes edges look smoother. The stairstep effect — or "jaggies"— are due to the monitor's limited resolution of about 96 dpi (dots per inch), much coarser than typical 600-dpi inkjet printers. AutoCAD adds subtle pixels of gray and other colors to make the edges look smoother. The higher the anti-aliasing setting, the smoother the look, but

Continued...

... continued.

slower the rendering using the techniques described by the table:

Anti-aliasing	***Meaning***
Minimal	Analytical horizontal anti-aliasing.
Low	Four shading samples per pixel.
Medium	Nine shading samples per pixel.
High	16 shading samples per pixel.

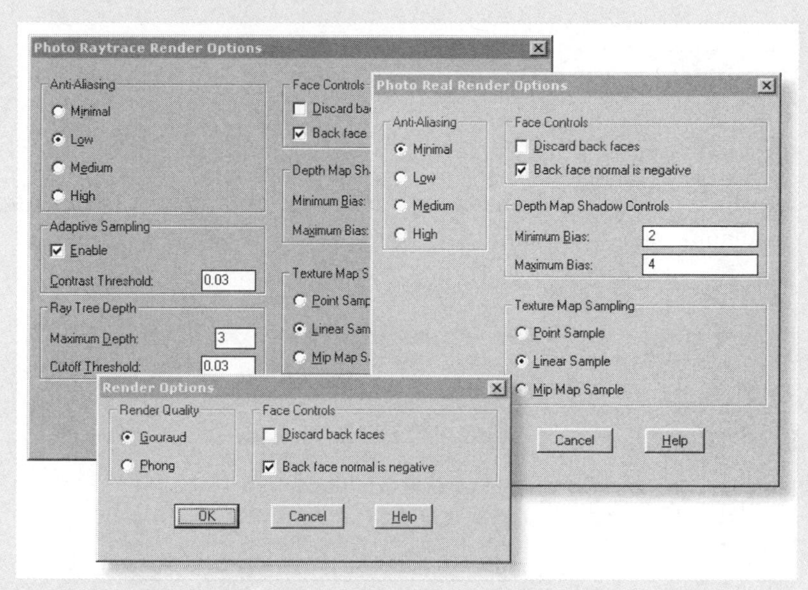

Render Options dialog boxes vary, depending on the rendering mode.

Depth Map Shadow Controls. This option helps prevent erroneous shadows, such as shadows that cast their own shadows, or shadows that don't connect with their objects (like Peter Pan's shadow). Autodesk recommends that the minimum bias should range between 2 and 20, while the maximum bias should range between 6 and 200. (Additional shadow controls are available when you define lights.)

Texture Map Sampling. Texture maps are raster images applied to 3D objects to make them look more realistic, such as pictures of bricks applied to fireplaces. This option controls how texture maps are handled when applied to objects smaller than the maps:

Sampling	***Meaning***
Point	The nearest pixel in the bitmap.
Linear	The average four pixel neighbors.
MIP Map	The pyramidal average of a square sample.

Adaptive Sampling. This technique of sampling allows for faster anti-aliased rendering by ignoring some pixels. For contrast threshold values closer to 0.0, AutoCAD takes more samples; for values closer to 1.0, AutoCAD takes fewer samples for faster rendering, but also possibly for lower quality.

Ray Tree Depth. In *ray tracing*, AutoCAD follows each beam of light as it reflects (bounces off opaque objects) and refracts (transmits through transparent objects) among the objects in the scene. This control allows you to speed up rendering by limiting the amount of reflecting and refracting. The maximum depth is the largest number of "tree branches" AutoCAD keeps track of; the range is 0 to 10. The cutoff threshold determines the percentage that a ray trace must affect a pixel before ray tracing stops; the range is 0.0 to 1.0.

RPREF

The **RPREF** command (short for "rendering preferences") presets rendering options to avoid displaying the Render dialog box.

This command displays a dialog box identical to that of the **RENDER** command, except that an **OK** button replaces the **Render** button. To avoid displaying the Render dialog box, turn on the **Skip Render Dialog** option.

BACKGROUND

The **BACKGROUND** command changes the background of the rendering.

AutoCAD normally displays nothing behind the rendering, except for the default background color (typically white or black). This command lets you choose from four different kinds of background:

- **Solid** specifies the color for the background.
- **Gradient** specifies two- and three-color gradients for the background.
- **Image** specifies a raster image for the background.
- **Merge** specifies the current AutoCAD drawing for the background.
- **Environment** adds reflection and refraction effects.

SOLID

The **Solid** option specifies a color for the background. (This option might be better named "Solid Color.")

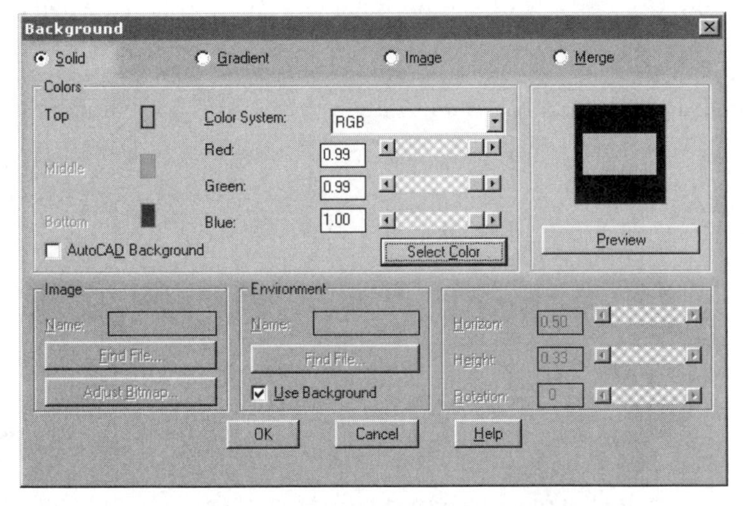

Background dialog box with Solid options.

1. Choose the **Solid** radio button.
2. To change the color of the background, turn off the **AutoCAD Background** option.
3. You can specify the color in three ways:
   - From the **Color System** drop list, select **RGB** (red, green, blue) or **HLS** (hue, lightness, saturation), and then move the sliders to create the color.
   - Alternatively, enter values in the **Red** (Hue), **Green** (Lightness), and **Blue** (Saturation) text entry boxes
   - Click the **Select Color** button, and then select an AutoCAD color from the dialog box.

Click **Preview** to see the color.

4. Click **OK** to exit the dialog box.

   Click **Render** to render the drawing with the background color.

Rendering with background color.

GRADIENT

The **Gradient** option specifies two- and three-color linear gradients for the background. This is a quick way to simulate backgrounds, such as grass (green), ground (brown), and sky (blue). On sunny days, the sky has a gradient that changes from light blue overhead to darker blue at the horizon.

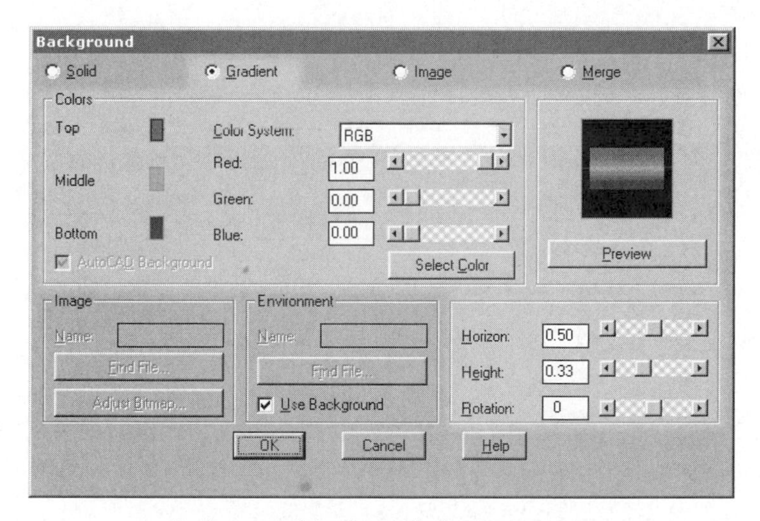

Background dialog box with Gradient options.

1. Choose the **Gradient** radio button.
2. Notice that three colors are represented as **Top**, **Middle**, and **Bottom**. To change the colors, click one of the rectangular color samples, and then specify the color — either by RGB or HLS, or by choosing the **Select Color** button.

To simulate a sky-grass background, use these colors:

Top	ACI color 4 (cyan).
Middle	ACI color 5 (blue).
Bottom	ACI color 92 (green).

3. To specify the gradient, change the settings:
   - **Horizon** moves the middle color up (closer to 0.9) and down (closer to 0.1).
   - **Height** varies the height of the middle color taller (closer to 0.9) and thinner (closer to 0.1). Set this to 0 for two-color gradient.
   - **Rotate** rotates the gradient from horizontal (0) to vertical (90).

   Click **Preview** to see the gradient.
4. Click **OK** to exit the dialog box.

   Click **Render** to render the drawing with the gradient colors.

Rendering with gradient colors in the background.

IMAGE

The **Image** option specifies a raster image for the background.

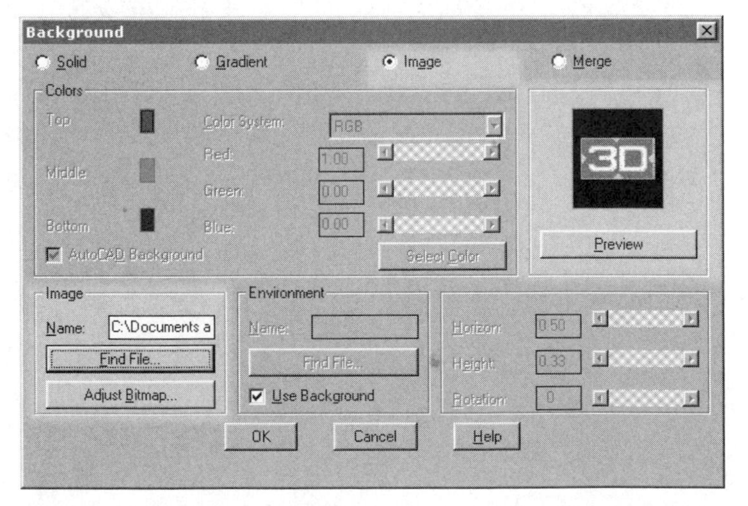

Background dialog box with Image option.

1. Choose the **Image** radio button.
2. To select the image for the background, choose the **Find File** button. The Background Image dialog box lacks a preview window; the work-around is to select **Thumbnails** from the **View Menu**. Render accepts these file types: *.bmp*, *.gif*, *.jpg*, *.pcx*, *.png*, *.tga*, and *.tif*.

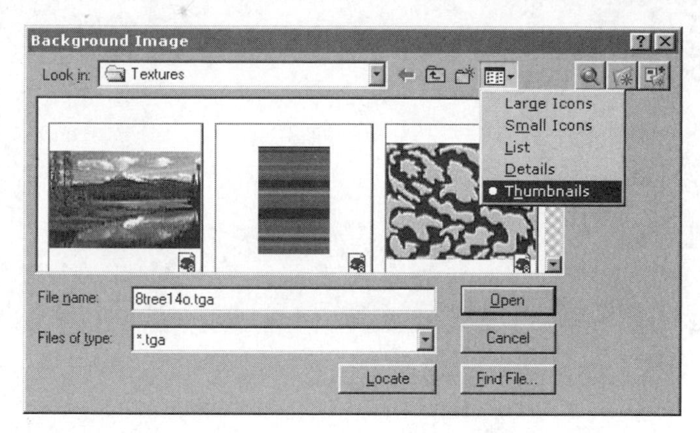

Background Image dialog box with thumbnail display.

3. Select a raster file, and then click **Open**.
4. AutoCAD normally makes the image fit the viewport. If you wish, you can change the size of the image by clicking **Adjust Image.**

   This dialog box lets you control the image in many ways:

   - **Fit to Screen** fits the image to the viewport.
   - **Use Image Aspect Ratio** forces the horizontal and vertical scales to stay in sync.
   - **Offset** and **Scale** position and size the image in the viewport.
   - **Tile** tiles (repeats) the image, when the image is smaller than the viewport.
   - **Crop** displays the image at its actual size, even when larger than the viewport.
   - **Center** centers the image in the viewport.

   Make changes to the background image's placement, and then click **OK**.

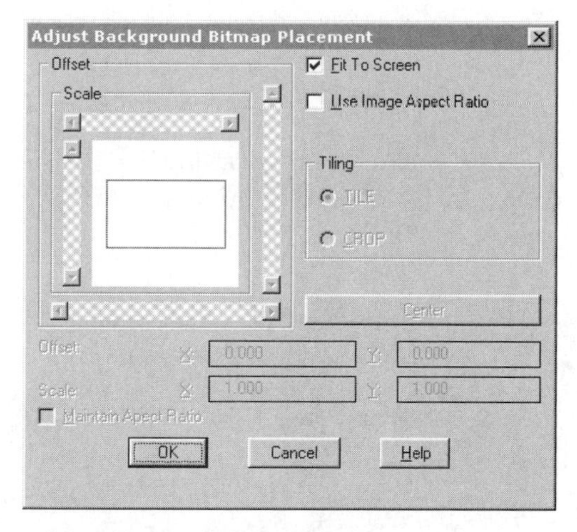

Adjust Background Bitmap Placement dialog box.

5. Click **OK** to exit the Background dialog box.

6. Click **Render** to render the drawing with the background image.

Rendering with background image, fitted to viewport.

Rendering with background image, tiled and centered.

MERGE

The **Merge** option specifies the current AutoCAD drawing for the background. (This option might be better named "Wireframe Drawing.") This option has no further options. It is unavailable when the destination is **Render Window** or **File**.

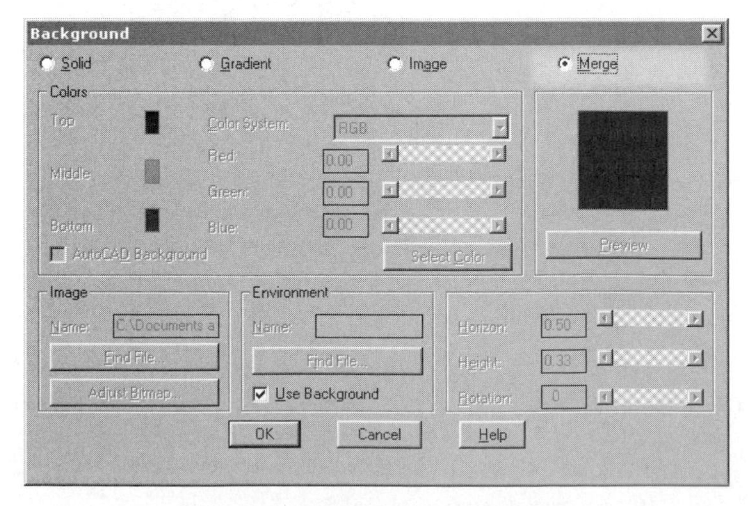

Background dialog box with Merge option.

1. Choose the **Merge** radio button.
2. Click **OK** to exit the dialog box.

   Click **Render** to render the drawing with the wireframe drawing. (Turn on the Render dialog box's **Query for Selections** option to create the hybrid image shown below.)

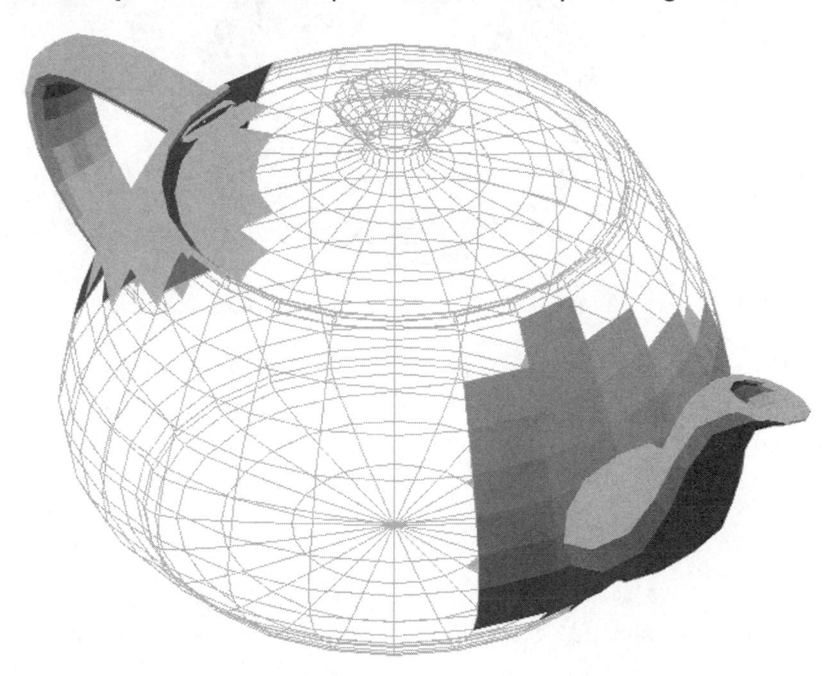

Partial rendering with merged background.

ENVIRONMENT

The **Environment** option is meant to add reflection effects. It is supposed to work with the other four options, solid, gradient, image, and merge.

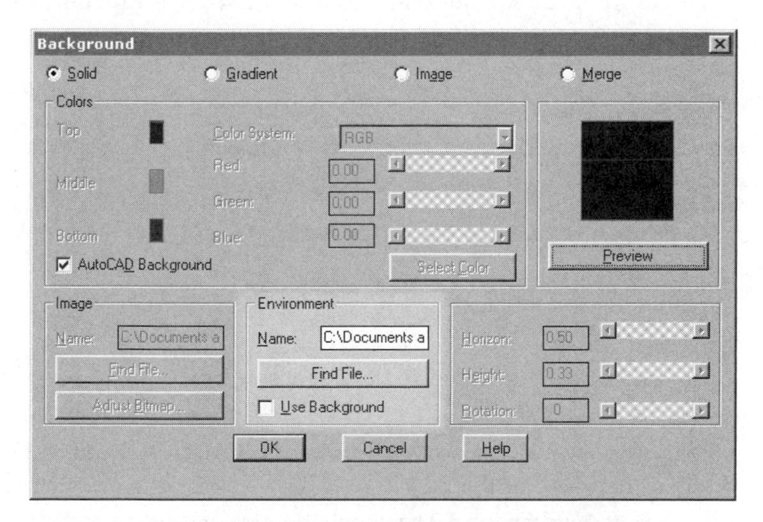

Environment options of the Background dialog box.

Shiny objects tend to reflect their surroundings. Realistic renderings should include 3D models of all objects in the "environment" — such as the glass in buildings reflecting the surrounding buildings, water reflecting mountains, and so on. Adding these objects is too much work, so the alternative is to place a suitable raster image of the surroundings. AutoCAD pastes the image inside an imaginary sphere that surrounds all objects to be rendered. Shiny objects then reflect this environment bitmap.

In the figure below, the shiny doughnut reflects the image of the water skier.

The environment reflects off shiny objects.

The technical editor and I spent considerable time attempting to make the environment option work — neither was successful. The image above was created in Autodesk Inventor, whose shading mode supports environment bitmaps. AutoCAD apparently does not.

 FOG

The **FOG** command variable specifies the color and intensity of the atmosphere.

AutoCAD's fog effect simulates fog by applying an increasing amount of white with distance. The farther away, the more dense the white. This is why AutoCAD also calls fog "depth cue."

The color need not be white. The subtle use of black, for example, enhances the illusion of depth, because objects farther away tend to be darker. A limited application of yellow creates the illusion of glowing lamps; the reckless use of green fog simulates Martian invasions.

You access the fog effect through the Render dialog box's **Fog/Depth Cue** button, or directly through the **FOG** command.

TUTORIAL: ADDING FOG

1. To add the fog effect to renderings, start the **FOG** command:
   - From the **View** menu, choose **Render**, and then **Fog**.
   - In the Render dialog box, click the **Fog/Depth Cue** button.
   - At the 'Command:' prompt, enter the **fog** command.

     Command: **fog** *(Press* ENTER.*)*
2. In all cases, AutoCAD displays the Fog/Depth Cue dialog box.
2. Click **Enable Fog** to turn on the fog effect.
   (To disable the fog effect temporarily, turn off this option; AutoCAD remembers all the fog settings the next time you turn it on.)
   Optionally, turn on **Fog Background**. This toggle determines whether the fog affects the background. For example, if the background color in your rendering is normally white but you choose black for the fog color, then the background becomes black.
3. Select a color for the fog. White is the default. Move all sliders to 0.0 for black, or choose a color from dialog boxes by clicking the **Select Color** and **Select Index** buttons — either button displays the same Select Color dialog box.

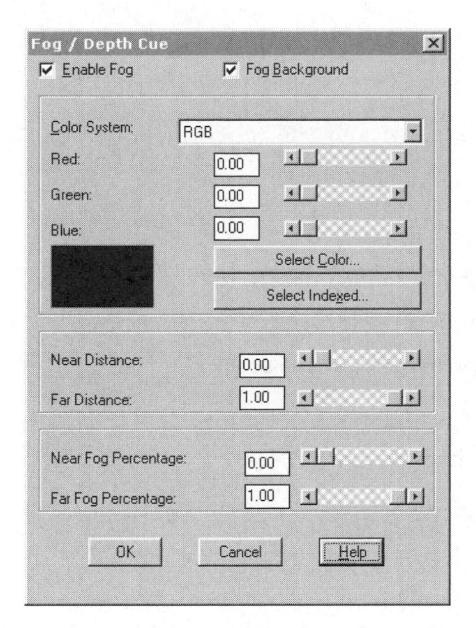

Fog/Depth Cue dialog box.

4. Set the extent of the fog. The **Near Distance** and **Far Distance** sliders determine where the fog begins and ends.
   - **Near Distance** positions the start of the fog effect. This slider can be tricky to understand, because it represents a relative distance from the "camera" (your eye) to the back clipping plane. Try starting with a value of **0.45**.
   - **Far Distance** positions the end of the fog effect. The value also represents the percentage distance from the camera to the back clipping plane. Try starting with a value of **0.55**.
5. Specify the strength of the fog. The **Near Fog Percentage** and **Far Fog Percentage** sliders determine the percentage of fog effect at the near and far distances. For a stronger fog effect, increase the value of **Near Fog**; for a weaker effect, reduce the value of **Far Fog**.
6. Click **OK**.
7. In the Render dialog box, click **Render** to see the fog effect.

   I find I have to adjust the fog distance and strength parameters a number of times, each time executing a render, until the effect looks right.

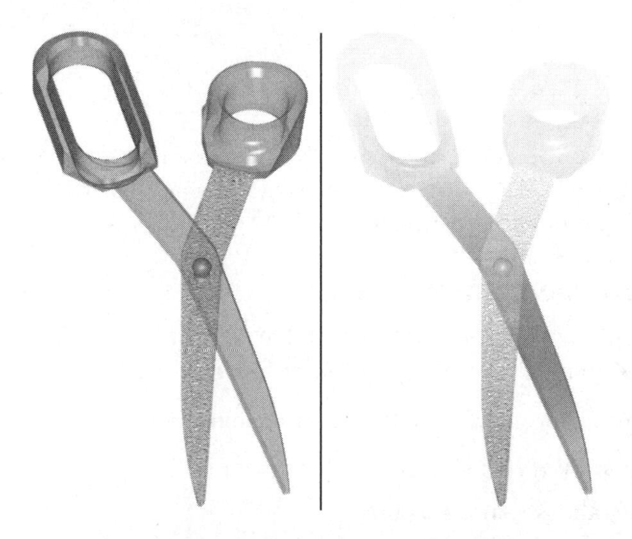

Left: *Rendering without fog effect.*
Right: *Rendering with white fog effect.*

STATS

The **STATS** command reports statistics of the rendering, such as the rendering mode, scene name, and total time.

TUTORIAL: REPORTING STATISTICS

1. To view statistics on the rendering process, start the **STATS** command:
   - From the **View** menu, choose **Render**, and then **Statistics**.
   - At the 'Command:' prompt, enter the **stats** command.

   Command: **stats** *(Press* ENTER.*)*

   AutoCAD displays the Statistics dialog box.

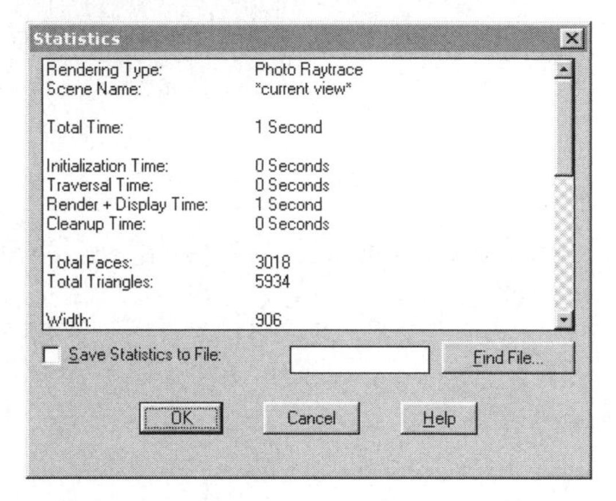

Statistics dialog box.

2. You can save the statistics to a text file. Turn on the **Save Statistics to File** option, and then enter a file name. Include the *.txt* extension with the file name, because AutoCAD does not add one.

   stats.txt

3. Click **OK** to exit the dialog box.

   If you wish, open the statistics file in Notepad or a word processor for viewing and printing.

REPLAY

The **REPLAY** command displays renderings saved as raster files, as well as raster files from other sources. This command works only if **RENDER** saved the images in BMP, TGA, or TIFF format. The images are displayed in the current viewport.

TUTORIAL: REPLAYING SAVED RENDERINGS

1. To view renderings saved to disk, start the **REPLAY** command:
   - From the **Tool** menu, choose **Display Image**, and then **View**.
   - At the 'Command:' prompt, enter the **replay** command.

   Command: **replay** *(Press* ENTER.*)*

2. AutoCAD displays the Replay dialog box.

   Select a file, and then click **Open**.

3. AutoCAD displays the Image Specifications dialog box. The purpose of this dialog box is to let you adjust the size of the image.

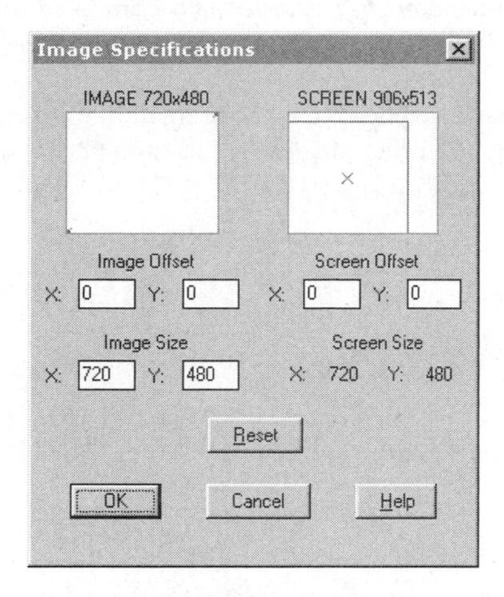

Image Specifications dialog box.

Your options are:

- Leave the image at its natural size, which may be smaller or larger than the viewport. When the image is smaller than the viewport, AutoCAD blacks out the margin area.
- Adjust the size of the image with the **Image Size** settings (in pixels), and move the position of the image with the **Image Offset** settings (from the lower left corner).

If you mess up, click **Reset**.

4. Click **OK** to exit the dialog box. AutoCAD displays the image in the current viewport.

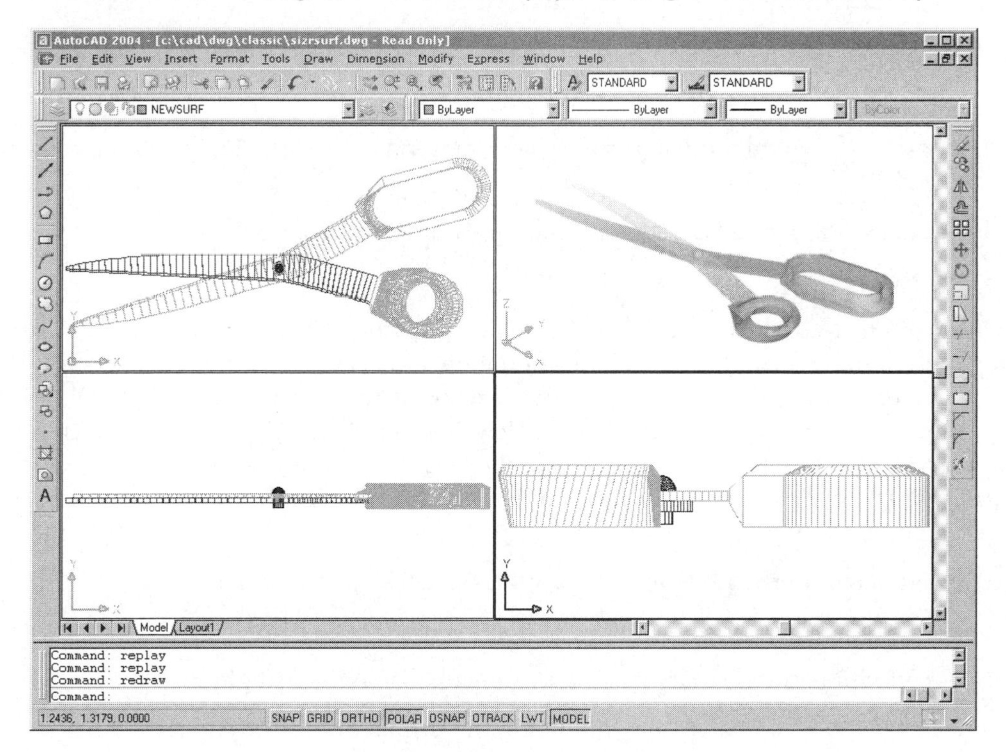

Replaying a rendering file in a viewport.

5. To remove the image from the viewport, use the **REDRAW** command.

Note: To save the renderings of 3D drawings, you have these two options:

After the scene has been rendered, use the **SAVEIMG** command to save the image to a file. Choose the file format — BMP, Targa, or TIFF. This command is limited by the resolution of the viewport.

Alternatively, in the Render dialog box, select **File** from the **Destination** area, and then choose the **More Options** button to specify the raster file format — BMP, PCX, PostScript, Targa, or TIFF. This option allows you to specify the image resolution.

MATLIB AND RMAT

The **MATLIB** and **RMAT** commands load and apply materials to 3D objects.

Materials define the look of rendered objects, They can consist of colors or bitmaps. By varying the amount shininess, roughness, and ambient reflection, colors can be made to look like physical materials, such as plastic, concrete, and wood. Materials do not define the density of 3D solids and bodies; they are strictly for looks.

Materials appear with the **RENDER** and **SHADEMODE** commands. For the renderings, turn on the **Apply Materials** option in the Render dialog box. For shadings, turn on all checkboxes in the Render Options section of the 3D Graphics System Configuration dialog box. (From the **Tools** menu, select **Options**. Click the **System** tab, and then select the **Properties** buttons in the Current 3D Graphics Display area.) While 2D wireframe shows object color, 3D wireframe shows the basic material color.

By default, new drawings contain a single material definition, called *GLOBAL*, which holds the default parameters for color, reflection, roughness, and ambience. You can use AutoCAD's extensive library of predefined materials, or create your own.

Applying materials to rendering takes these steps:

Step 1. Use the **MATLIB** command (short for "materials library") to load materials into drawings.

Step 2. Use the **RMAT** command (short for "render materials") to attach materials to objects, colors, or layers.

Step 3. In the **RENDER** dialog box, turn on the **Apply Materials** option.

Step 4. Click **Render**!

AutoCAD attempts to simulate real-world colors, which are never uniform. A blue automobile is not pure "blue": the hue of blue varies, depending on the deepness of shadows, the intensity of the Sun, the color of other light sources, and the texture of the paint. The Sun glancing off an edge turns the blue nearly white, while in deep shadow the blue is nearly black. A red light shining on the blue car results in areas that look purple. Object colors are also affected by colors reflected from surrounding objects; AutoCAD can simulate reflected color, but does not take into account the reflected color of surrounding objects.

AutoCAD simulates materials by varying the values of these parameters:

Diffuse color is the object's primary color, such as the blue of the car.

Ambient color modifies the primary color with the color of ambient light.

Reflection color (a.k.a. specular color) highlights the primary color with a second color.

Transparency varies the object's opaqueness, while *refraction* determines how light bends going through the object. (Refraction shows up in Photo Raytrace renderings only.)

Roughness creates the effect of shininess, the amount of which varies according to the material's roughness. Very shiny objects have small, bright highlights, while rough objects have large, dull highlights.

Bitmaps are raster files that simulate repeating surface textures, such as brick and tile.

First, second, third, and fourth color simulates surfaces with random textures, such as granite rock. The surfaces are further defined by sharpness and scale of texture.

Stone-vein color simulates sinewy two-color surfaces, such as marble rock. The surfaces are further defined by turbulence, sharpness, and scale of pattern.

Light-dark color simulates wavy line surfaces, such as wood. The surfaces are further defined by the ratio of light to dark color, ring density, width, and shape, and scale of pattern.

In general, new materials are very difficult to define from scratch, as the dialog boxes below illustrate. I recommend instead that you take an existing definition, and modify it.

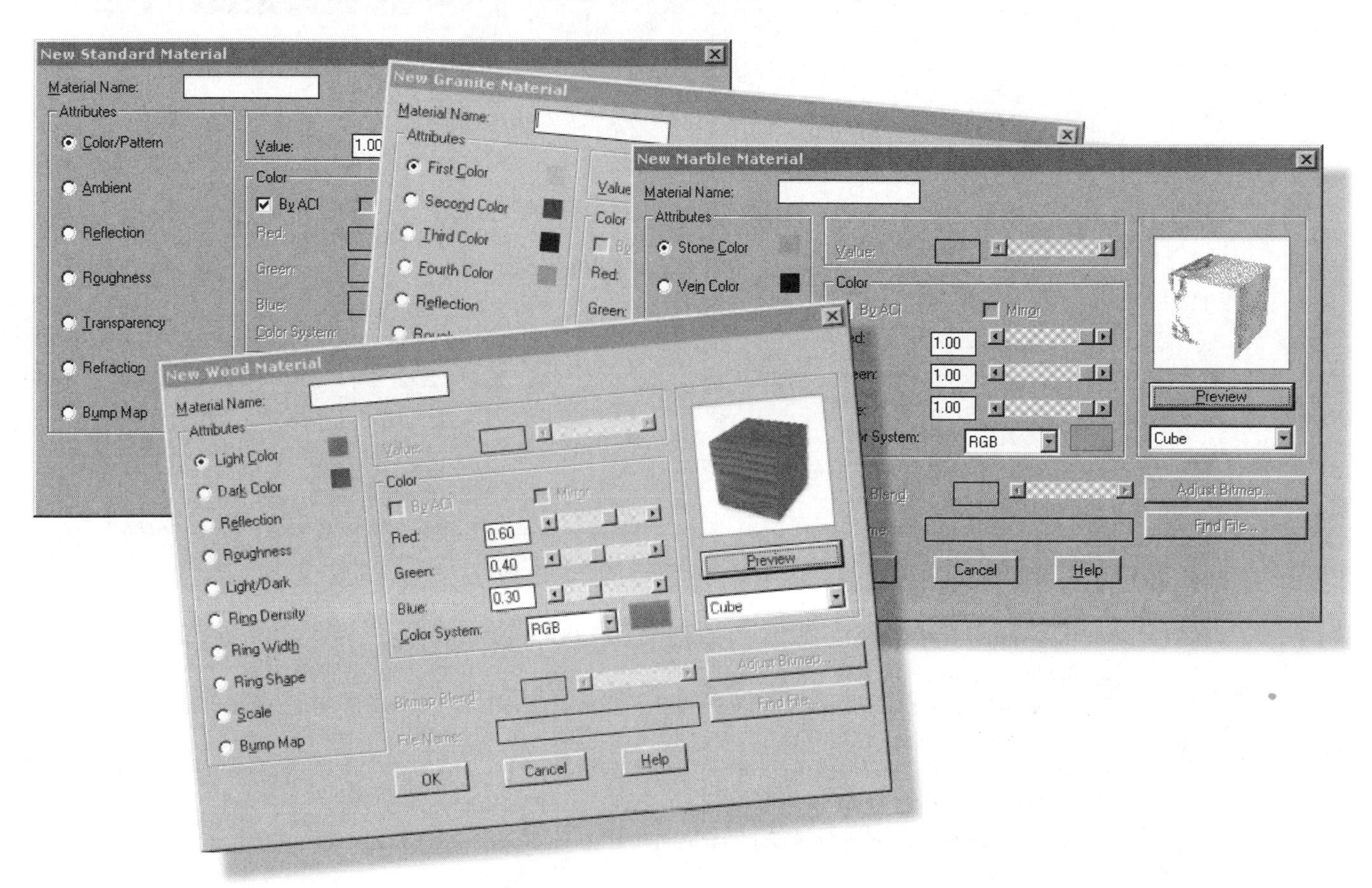

New Material dialog boxes.

TUTORIAL: ATTACHING MATERIALS

Before using this command, open a drawing containing 3D objects.

1. To attach materials to 3D objects, start the **RMAT** command:
   - From the **View** menu, choose **Render**, and then **Materials**
   - At the 'Command:' prompt, enter the **rmat** command.

     Command: **rmat** *(Press* ENTER.*)*

   AutoCAD displays the Materials dialog box.
2. Before you can apply materials, AutoCAD must load them into the drawing.
   Click the **Materials Library** button. (This is a shortcut to the **MATLIB** command.)
   AutoCAD displays the Materials Library dialog box, which is arranged backwards: you work from right to left.

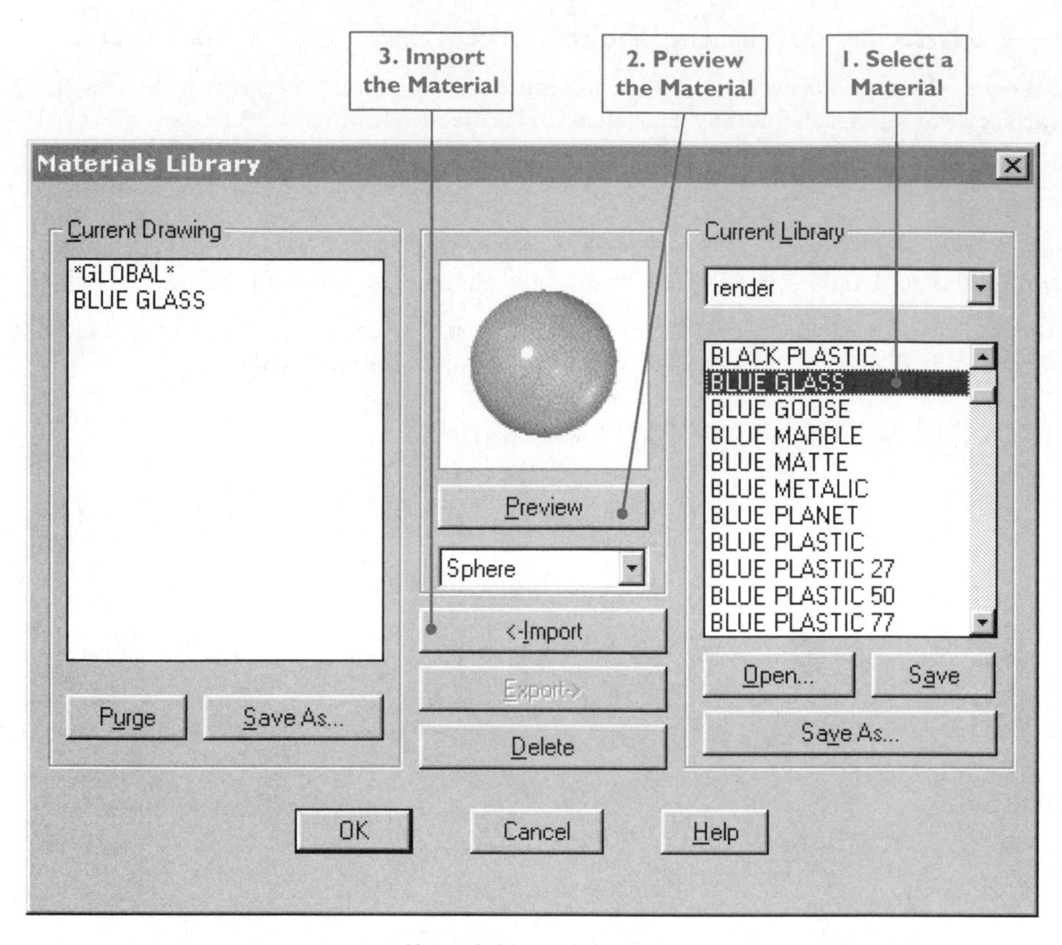

Materials Library dialog box.

3. Repeat these steps to add as many materials to the drawing as you wish:
   a. Select a material from the list under Current Library, such as "Blue Glass."
   b. Preview the look of the material by clicking the **Preview** button. The material is applied to a sphere or a cube.
   c. To add the material to the drawing, click **Import**.
4. Click **OK** to exit this dialog box, and return to the Materials dialog box.
5. Materials can be attached to objects, assigned to layers, or identified with colors. Follow these steps:
   a. Again select a material, this time from the list under Material.
   b. To attach the material to:
      - **Objects** in the drawing, click **Attach**. AutoCAD prompts you:
        Select objects: *(Select one or more objects.)*
        Select objects: *(Press* ENTER *to end object selection.)*
      - **Layers**, click **By Layer**, and select one or more layer names. (Hold down the CTRL key to select more than one layer name.)
      - **Colors**, click **By ACI**, and select one or more color numbers. (ACI is short for "AutoCAD color index.")
6. Click **OK** to exit the dialog box.
   The material assignments do not appear until the drawing is rendered.
7. Start the RENDER command.
   In the Render dialog box, ensure **Apply Materials** is turned on (check mark shows).
   Click **Render**.

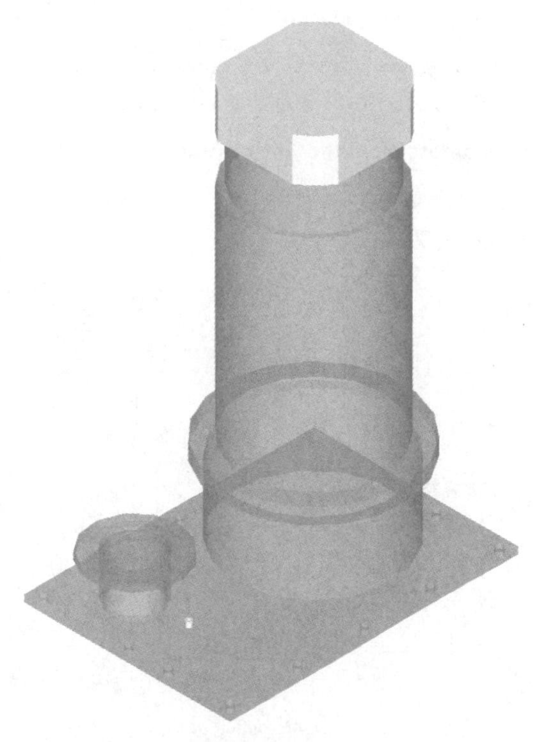

3D drawing rendered with "Blue Glass" material.

LIGHT

The **LIGHT** command inserts lights in 3D scenes.

The **SHADE** and **3DORBIT** commands are limited to a single light source, unless you turn on all render options for the current 3D graphics display. They and the **RENDER** command handle as many lights as you need. All lights can emit any color at any level of brightness.

AutoCAD supports the following types of light source:

- **Point lights** shine light in all directions, with inverse linear, inverse square, or constant intensity. These lights work well as light bulbs in lamps.
- **Distance lights** shine light in parallel beams with constant intensity. Typically, you place a single distant light to simulate the sun. To simulate a setting sun, change the color of the light to orange-red.
- **Spot lights** shine light in a specific direction, in a cone shape. These lights work well as spotlights to beam from light locations to targets, such as high-intensity desk lamps and vehicle-mounted spotlights. When you place spotlights in drawings, you specify the hotspot of the light (where the light is brightest) and the falloff, where the light diminishes in intensity.
- **Ambient light** shines an omnipresent light source to ensure every object is illuminated. There is a single ambient light in every rendering; turn off the ambient light to simulate nighttime scenes.

This command works in model space only.

TUTORIAL: PLACING LIGHTS

Before using this command, open a drawing containing 3D objects. Initially, no lights are defined other than a single default light source located at your eye.

1. To place lights in 3D drawings, start the **LIGHT** command:
   - From the **View** menu, choose **Render**, and then **Light**.
   - At the 'Command:' prompt, enter the **light** command.

     Command: **light** *(Press* **ENTER**.*)*

   AutoCAD displays the Lights dialog box.

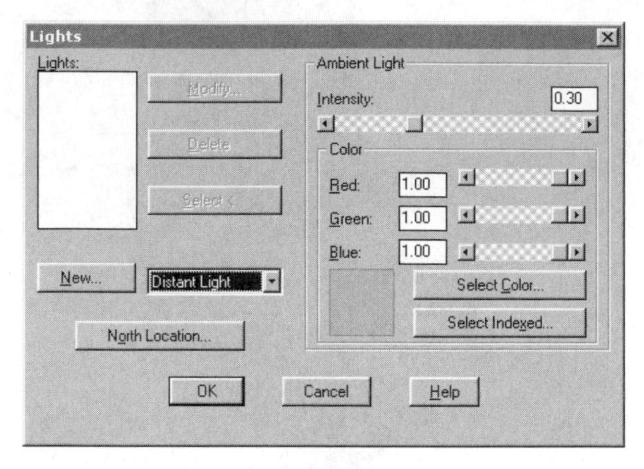

Lights dialog box.

2. Decide on the type of light: spot, point, or distant. (The ambient light is controlled by settings on the left half of the dialog box.)

   For this tutorial, select **Distant** light from the list box, because sun-like lights are the easiest to work with. (In contrast, spotlights are the hardest.)
3. To give the light a name and to specify its parameters, choose the **New** button.

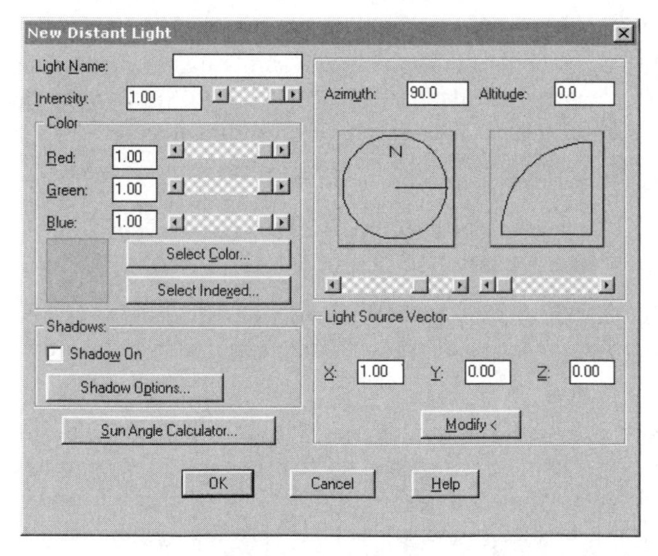

New Distant Light dialog box.

   AutoCAD displays the New Distant Light dialog box. The other light types have similar dialog boxes, with some variation for their innate differences.
4. Give the light a convenient name. AutoCAD distinguishes between the lights by name.

LIGHTING DEFINITIONS

Constant light — attenuation is 0; default intensity is 1.0.

Extents distance — distance from minimum lower-left coordinate to the maximum upper-right.

Inverse linear light— light strength decreases to ½-strength two units of distance away, and ¼-strength four units away; default intensity is ½ extents distance.

Inverse square light — light strength decreases to ¼-strength two units away, and 1/8-strength four units away; default intensity is ½ the square of the extents distance.

Falloff — angle of the full light cone; field angle ranges 0 to 160 degrees (default: 45 degrees).

HLS color — changes colors by hue (color), lightness, and saturation (less gray).

Hotspot — brightest cone of light; beam angle ranges from 0 to 160 degrees (default: 45 degrees).

RGB color — three primary colors — red, green, blue — shaded from black to white.

Light Name: **Sun**

5. The higher the intensity, the brighter the light. An intensity of zero turns off the light.
   For this tutorial, leave the **Intensity** setting at 1.0 (full).
6. You can specify colors for each light by choosing the **Select Color** button, which displays AutoCAD's Select Color dialog box. (The **Select Indexed** button displays the same dialog box.)
   For this tutorial, select the color yellow (#2) for the color of the light.
7. The New Distant Light dialog box has some settings specific to it:
   - **Azimuth** determines how far the Sun is around in the sky.
   - **Altitude** determines the Sun's height in the sky. Because it doesn't matter how far the Sun is from objects in the drawing, AutoCAD will place the distant light in drawings; you place the spot and point lights in drawings.
   - **Sun Angle Calculator** selects the sun's position by date, time and location.

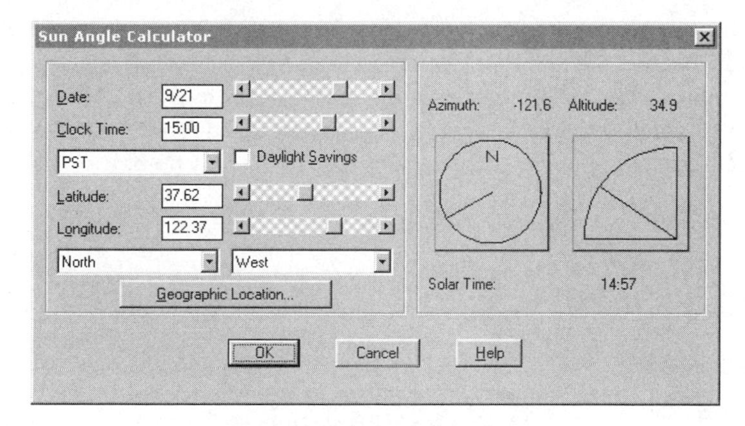

Sun Angle Calculator dialog box.

The fast way to set the sun's location is to click **Geographic Location**, and then select a location on the map.

(Don't rely on these maps, because they leave out major cities and it contains mistakes. Vancouver BC is missing. Victoria BC is apparently a suburb of Portland OR, and Cabo San Lucas, Mexico is about 300 miles offshore. Location data is stored in the *.map* files in AutoCAD's *\support* folder)

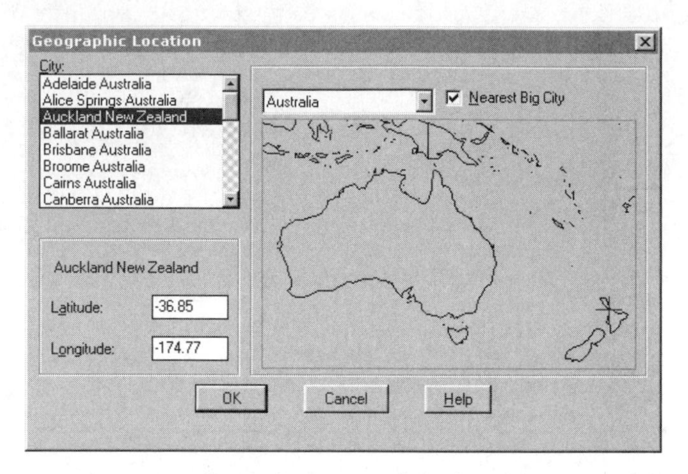

Geographic Location dialog box.

- **Shadows** toggles the display of shadows. Among the shadow options, **Volumes** is faster with sharp borders, while **Map** is slower but has soft borders.

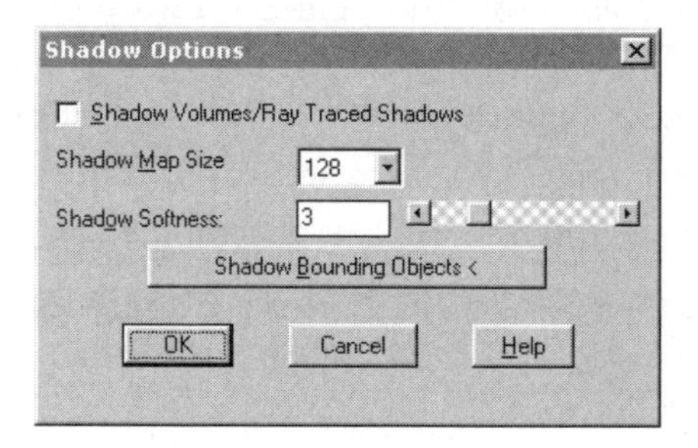

Shadow Options dialog box.

8. After you choose **OK** to exit the Lights dialog box, AutoCAD places a light block in the drawing. The block defines the light's parameters using attributes. It is placed on a locked layer called "AutoShade." (This layer, created automatically by AutoCAD, is named after Autodesk's first rending software, an add-on program called AutoShade.) Each type of light has a unique block shape, except for ambient light, which has no block.

AutoCAD blocks represent light sources in 3D drawings.
Left: *Point light.*
Center: *Distant light.*
Right: *Spot light.*

By default, AutoCAD uses all placed lights for renderings. If you wish to limit the rendering to some of the lights, use the SCENE command.

SCENE

The **SCENE** command collects views and lights.

Before using this command, you must create at least one named view (with the **VIEW** command) or place at least one light in the drawing (with the **LIGHT** command) — otherwise, there is no need for this command. If you select no lights for a scene, **RENDER** uses ambient light.

The scene parameters are stored as attribute definitions in a hidden block.

TUTORIAL: CREATING SCENES

Before using this command, open a drawing containing 3D objects, lights, and named views.

1. To create scenes, start the **SCENE** command:
   - From the **View** menu, choose **Render**, and then **Scene**.
   - At the 'Command:' prompt, enter the **scene** command.

     Command: **scene** *(Press* **ENTER**.*)*

   AutoCAD displays the Scenes dialog box.

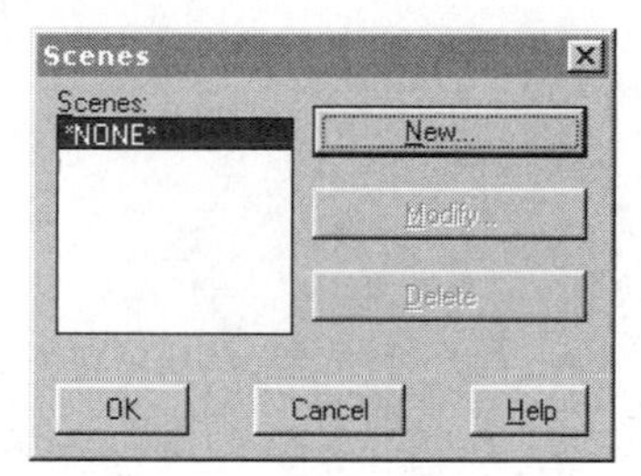

Scene dialog box.

   NONE is the name of the default scene, and despite its name, *is* a scene that consists of the current view and all lights placed in the drawing.

2. To create a scene, click **New**.
   AutoCAD displays the New Scene dialog box.

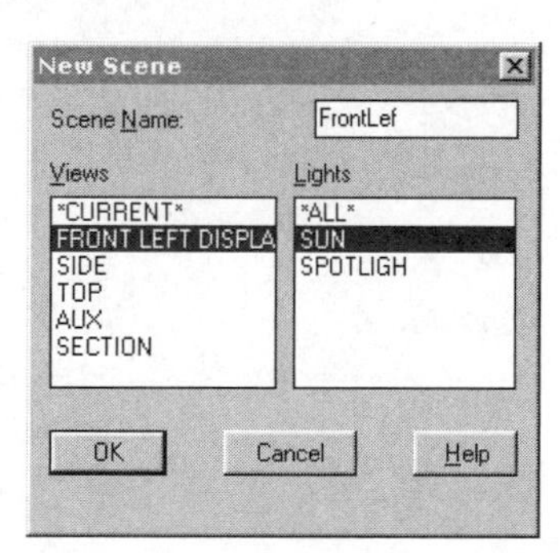

New Scene dialog box.

   CURRENT is the current viewpoint.

3. Select a view name from the **Views** column.
4. From the Lights column, select light names. To choose more than one light, hold down the **CTRL** key, or else select *ALL* to choose all lights.
5. To identify the scene, type a name of up to eight characters long.

6. Click **OK**.

   Notice that the scene name appears in the list.

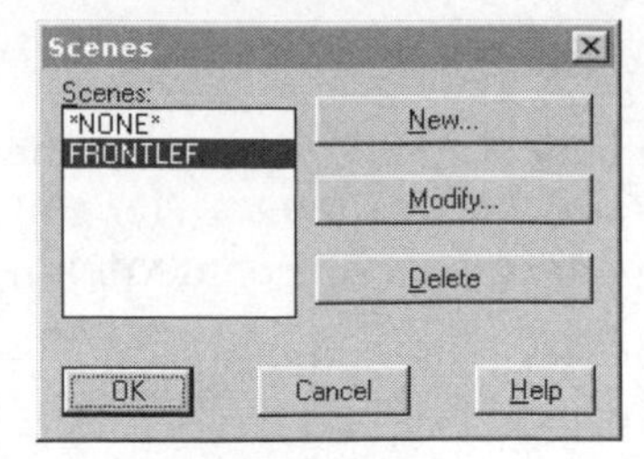

Scene names.

7. To use the scene, start the **RENDER** command.
8. Notice the scene name appears in the Scene to Render list. Select it, and then click **Render**.

   When the rendering is complete, the view changes to the one stipulated by the scene.

SPEEDING UP RENDERINGS

The time it takes for AutoCAD to create a rendering depends on the size of the 3D model and the options selected for the RENDER command.

Setting the following options generate the fastest renderings, at the expense of quality:

Rendering type:	**Render**
Smooth shading:	**Off**
Apply materials:	**Off**
Render quality:	**Gouraud**
Discard back faces:	**On**
Destination:	**Small viewport or 320x200 file**
Lights:	**None**
Render cache:	**On**
Shadows:	**Off**
Sub Sampling:	**8:1**
Enable Fog:	**Off**
Background:	**Solid**

Other factors include:

- The faster the computer's CPU and the larger the computer's RAM, the faster the rendering.
- The more complex the model, the longer the rendering. Rendering part of a model is faster than rendering the entire model. Rendering to smaller viewports greatly decreases the rendering time.

LSNEW, LSEDIT, AND LSLIB

The **LSNEW**, **LSEDIT**, **LSLIB** commands insert and edit landscape objects.

Landscape objects are meant for use in shaded and rendered views. They display bitmaps that populate renderings with trees, people, signs and other "landscape" objects.

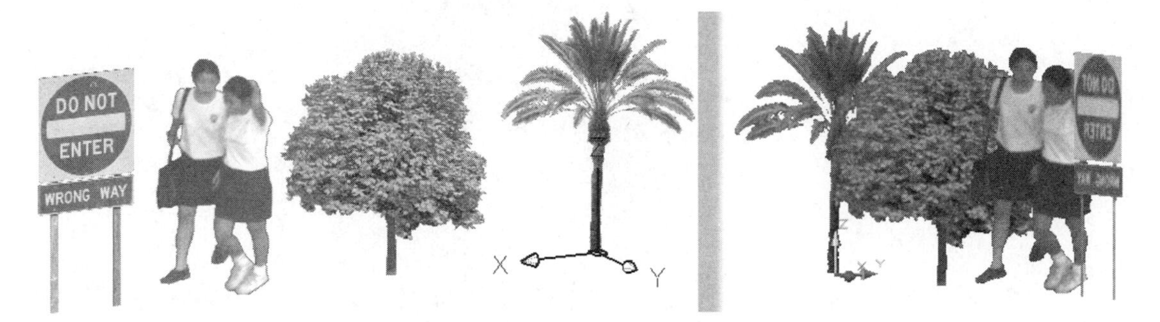

Left: *Landscape objects in shaded and rendered views.*
Right: *Rotated landscape objects still facing the viewpoint.*

They have a property that enables them always to face the viewpoint, no matter how it is rotated about the z axis. This feature works when the **View Aligned** option its turned on. It is off by default, so that landscape objects rotate like any other as the view is rotated.

In wireframe displays, landscape objects look like triangles and rectangles, as illustrated below.

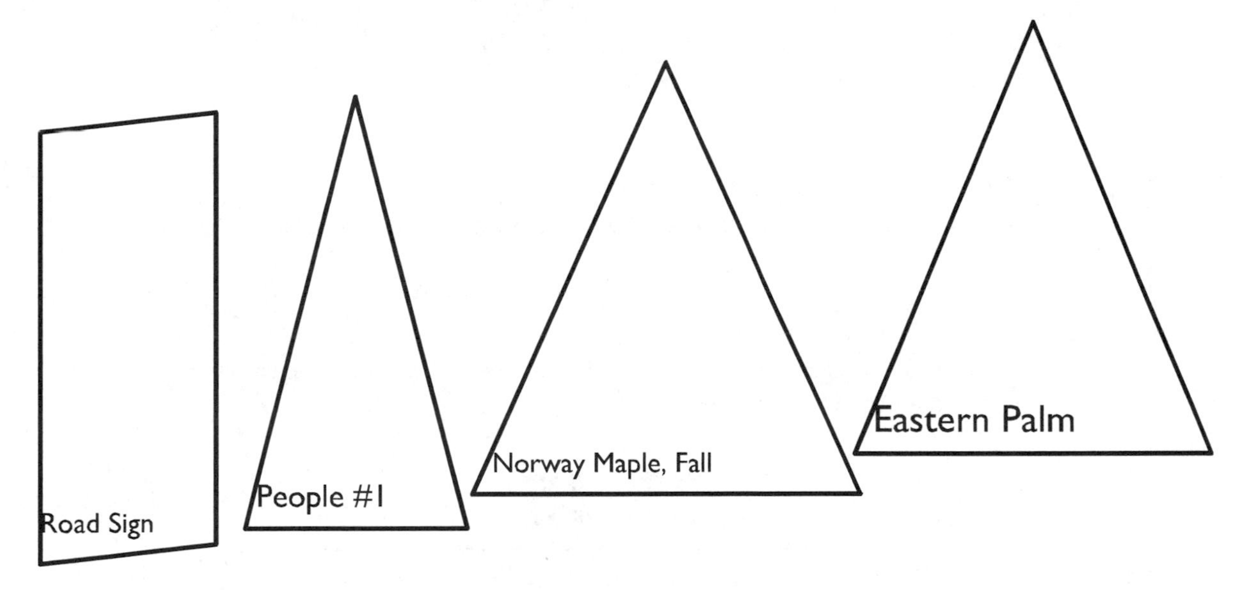

Landscape objects in wirefame display.

BASIC TUTORIAL: PLACING LANDSCAPE OBJECTS

Before using this command, ensure the drawing is in the WCS (with the UCS **World** command) and in plan view (with the PLAN **World** command).

1. To place landscape objects in drawings, start the **LSNEW** command:
   - From the **View** menu, choose **Render**, and then **Landscape New**.
   - At the 'Command:' prompt, enter the **lsnew** command.

Command: **lsnew** *(Press* **ENTER**.*)*

AutoCAD displays the Landscape New dialog box.

Landscape New dialog box.

2. From the Library list, select a landscape object, such as "Cactus."
   To see the image, click **Preview**.
3. The height is measured in current units. In drawings with default settings, "20" means 20 units — feet or meters. Change the height to match the object:

Landscape Object	*Height Ranges*	
Trees	20' to 100'	6m to 30m
People	5' to 6'	1.7m to 2.0m
Signs	10'	3m
Automobiles	4' to 7'	1.5m to 2.3m

4. AutoCAD normally places landscape objects at the origin (0,0,0).
   If you want the object placed elsewhere, click **Position**, and then pick a point in the drawing.
5. Click **View Aligned** to turn on the option (checkmark shows).
6. Click **OK** to exit the dialog box.
   Notice that AutoCAD places the landscape object in the drawing. You see just the line that makes up the edge of the triangle.

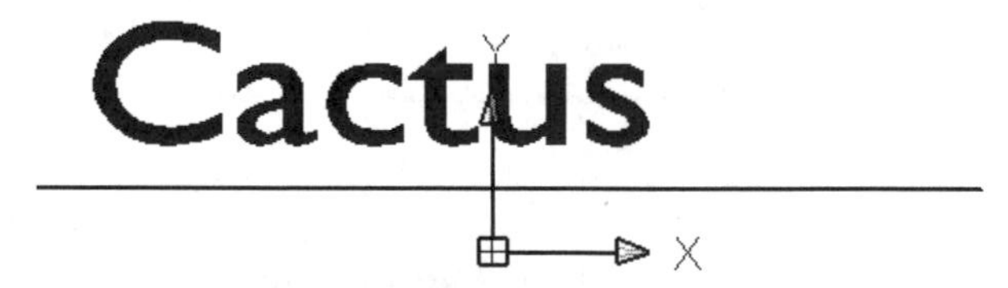

Landscape object in 2D plan view.

7. To see the landscape object, change the viewpoint (such as **VPOINT 1,1,1**), and then use the **SHADEMODE** or **RENDER** command.
   For interactive viewing, use **3DORBIT** set to flat or Gouraud shading. As you rotate the image horizontally (about the z-axis), notice that the cactus always faces you. (Rotating the image vertically, about the x-axis, tilts the cactus.)

Landscape object in 3D viewpoint.

PLACING LANDSCAPE OBJECTS: ADDITIONAL METHODS

The LSNEW dialog box has several options:

- **Height** option determines the size of landscape objects.
- **Single / Crossing Face** option creates landscape objects as one-sided or two-sided.
- **View Aligned** option toggles whether landscape objects face the current viewpoint.

Let's look at them.

Height

The **Height** option determines the size of landscape objects.

Single / Crossing Face

The **Single Face** option creates landscape objects as one-sided, and **Crossing Face** as two-sided.

View Aligned

The **View Aligned** option toggles whether landscape objects face the current viewpoint.

TUTORIAL: EDITING LANDSCAPE OBJECTS

The LSEDIT command is identical LSNEW, except that AutoCAD prompts you to select a single landscape object:

Command: **lsedit**

Select a landscape object: *(Pick a single landscape object.)*

AutoCAD displays the Landscape Edit dialog box, which allows you to change every aspect of the landscape object, except its picture. To change the picture, erase the landscape object, and then use the LSNEW command to insert the new one.

TUTORIAL: CREATING LANDSCAPE OBJECTS

Before starting this command, ensure that you have two raster files of the landscape object: (1) a cropped image, optionally in color, and (2) a matching monochrome opacity map file.

1. To create landscape objects, start the LSLIB command:
   - From the **View** menu, choose **Render**, and then **Landscape Library**.

- At the 'Command:' prompt, enter the **lslib** command.

  Command: **lslib** *(Press* ENTER.*)*

AutoCAD displays the Landscape Library dialog box.

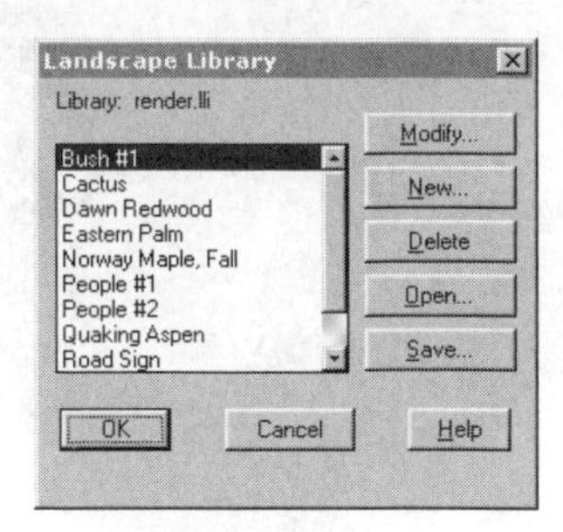

Landscape Library dialog box.

2. To create a new landscape object, click **New**.
   AutoCAD displays the Landscape Library New dialog box.

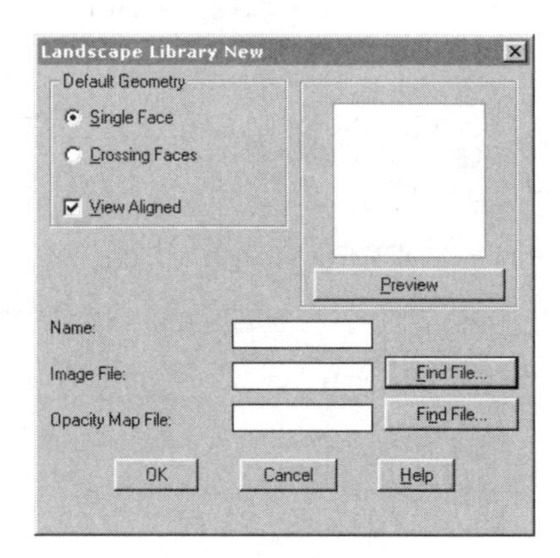

Landscape Library New dialog box.

3. Load the image file. Next to Image File, click **Find File**, and then select the file name from the dialog box, such as *ka.tif*.
4. You must also load the opacity map. Next to Opacity Map File, click **Find File**, and then select the file name from the dialog box, such as *kabmp.tif*.
5. Give the landscape object a name to identify it:

   Name: **Ka**

6. You've defined a new landscape object.
   Click **OK** to exit this dialog box.

New landscape object defined.

7. Save the landscape object to the master library file, which holds all the landscape object definitions.

   Click **Save**, and then save to the *render.lli* file. Click **Yes** when you are asked about replacing the file. (AutoCAD appends the new definition to the file; it does not replace it.)
8. Click **OK** to exit the dialog box.

   You can now place the landscape object in the drawing with the **LSNEW** command.

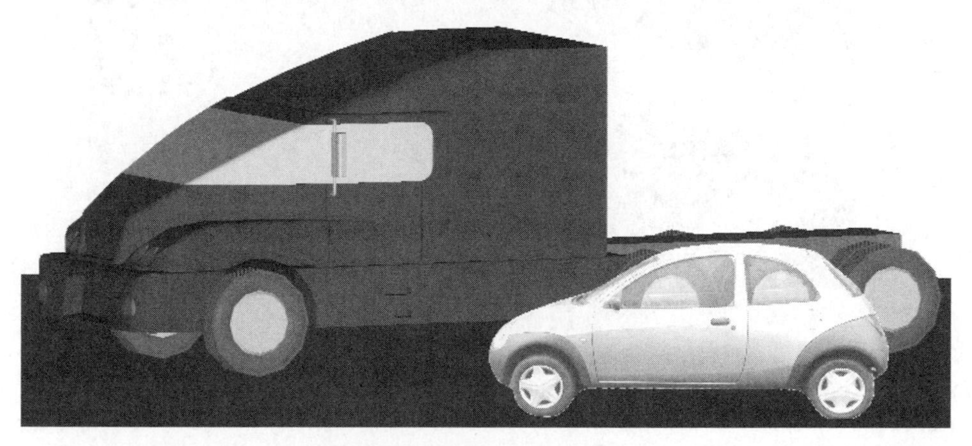

Landscape model in rendered view.

TUTORIAL: OPACITY MAPS FOR CROPPED IMAGES

If images are not rectangular ("cropped"), then you need to prepare a matching opacity maps. Unless the image is of a rectangular object, it should be cropped. *Cropping* isolates the object of interest from the rest of the picture. Below is a photograph of an automobile; we want to crop it so that it is isolated from the background.

Original digital photo of automobile.
(Photographed on Castle Hill in Budapest.)

Cropping Images

Images are available through online image banks, digital cameras, and scanning.

1. Open the file in an image editor, such as PaintShop Pro.
2. To select the outline of the object of interest, the car in this example, use the **SmartEdge** tool. I set smoothing to 20 and feathering to 0.

3. Before isolating the car from the rest of the photograph, ensure the background color is black.
4. To crop the picture to the selection, press **CTRL+R** . (Alternatively, select **Image | Crop to Selection** from the menu.)

Cropped image.

5. Save the cropped image with a meaningful name, such as "Ka" (the name of this automobile). AutoCAD accepts images in the following formats: *.tga*, *.bmp*, *.png*, *.jpg*, *.tif*, *gif*, and *.pcx*.

Creating the Opacity Map

The opacity map tells AutoCAD which parts of the rectangular image are opaque (not see-through) and which are transparent (where the rendering shows through). The "map" is a simple black-white image: white areas are opaque; black are transparent. In effect, you create a silhouette.

6. To create the opacity map, continue with the same image above. In PaintShop Pro, use the MagicWand tool to select the white (non car) areas.
7. Change the foreground color to black; ensure the background color is still white.
8. Press **CTRL+R** to crop to selection. Notice that PaintShop Pro creates a reverse silhouette.

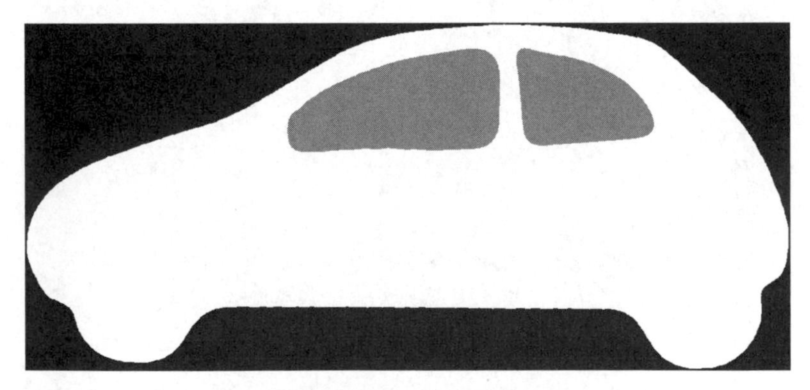

Monochrome opacity map.

9. If you wish areas to be translucent (semi-see-through), flood those areas with gray. The darker the shade of gray, the more the background shows through.

   For the car windows, I used 55% gray to show the inside of the car, yet allow the rendered image behind the car to show through the windows. (See figure two pages previous of the car in front of the truck.)

   Caution! The translucency feature works only with renderings created by the **RENDER** command; it does not work with any shadings created by the **SHADEMODE** command.
10. Save the opacity map with a related name, such as "KaBmp" in one the following formats: *.tga*, *.bmp*, *.png*, *.jpg*, *.tif*, *gif*, and *.pcx*.

EXERCISES

1. From the CD-ROM, open *Teapot.dwg* file, a 3D surface drawing of a teapot.
   a. Use the **HIDE** command. Do hidden lines disappear?
   b. Select the Dashed linetype from the **HLSETTINGS** command.
      Repeat the **HIDE** command. What happens to the hidden lines?
   c. Use the **REGEN** command. What happens to the hidden lines?

   (Historically, the teapot was used for testing purposes by 3D graphics programming pioneers as a complex 3D model. Hence, you see the teapot in AutoCAD's toolbar icons for rendering commands.)

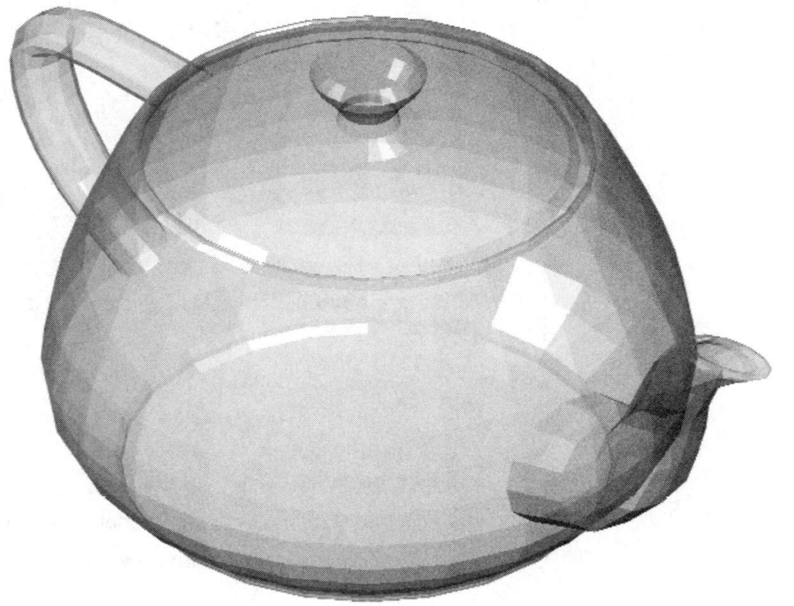

2. Continuing from exercise #1, use the **SHADEMODE** command, and specify the **Hidden** option. Does the image look different from that generated by the **HIDE** command?
   Use the **REGEN** command. What happens to the hidden lines?
3. Continuing from exercise #2, use the **RENDER** command on the teapot.
   What happens when you use the **REGEN** command?
4. Use the **RMAT** command to make the teapot look like glass.
5. From the CD-ROM, open the *Ch13TruckKa.dwg* file, a drawing of a 3D truck with a car landscape object.
   Start the **3DORBIT** command, and then select **Shading Modes | Gouraud Shade**.
   Use the **Orbit** feature to rotate the view. Does the landscape object always face you?

CHAPTER REVIEW

1. Briefly explain the purpose of the **HIDE** command.
2. Which command allows you to edit objects in hidden-line viewing mode?
3. How would you show hidden lines with the Hidden linetype?
4. What are *obscured lines*?
   A halo gap?
5. Explain the difference between Gouraud shading and Gouraud+Edges shading.
6. How do you turn off (return to wireframe display) the **SHADEMODE** command's shading modes?
7. What effect does **FACETRES** have on renderings of 3D solid models?
   On 3D surface models?
8. Name an advantage the **RENDER** command has over **SHADEDGE**.
   A disadvantage?

9. Can AutoCAD render just part of a drawing?
10. When would you render to a file?
11. What is a *scene*?
12. What are the two ways of defining 3D viewpoints for rendering?
    a.
    b.
13. When might you use *sub-sampling*?
14. What are the four types of lights available for rendering?
    a.
    b.
    c.
    d.
15. How would you place a picture in the background of a rendering?
16. What color creates the effect of "fog"?
    The illusion of increased depth?
17. What steps would you take to combine a rendering with a wirefame view of the same drawing?
18. How does anti-aliasing help a rendering?
    Hinder a rendering?
19. What type of light is the sun?
20. What is the difference between the **MATLIB** and **RMAT** commands?
21. How would you make a 3D model look like glass?
22. Explain the purpose of landscape objects.

Placing Raster Images

Drafters sometimes need to place raster images in drawings, such as scans of paper drawings, digital photographs, faxed images, and aerial photographs. The import formats discussed in Chapter 5 display raster images only temporarily; you cannot draw over them. That's the purpose of the **IMAGE** command, the subject of this chapter.

In this chapter, you learn about the following commands:

IMAGEATTACH attaches raster images to drawings.

IMAGE manages image files, much like xrefs.

IMAGEQUALITY and **IMAGEADJUST** adjust the quality of the images.

TRANSPARENCY turns designated pixels transparent.

IMAGECLIP clips images.

IMAGEFRAME toggles the display of the image's frame.

FINDING THE COMMANDS

On the **REFERENCE** toolbar:

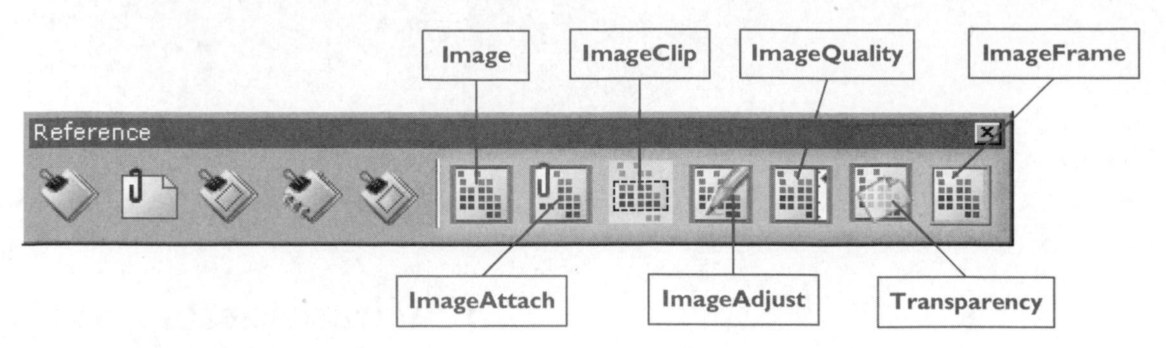

On the **INSERT** and **MODIFY** menus:

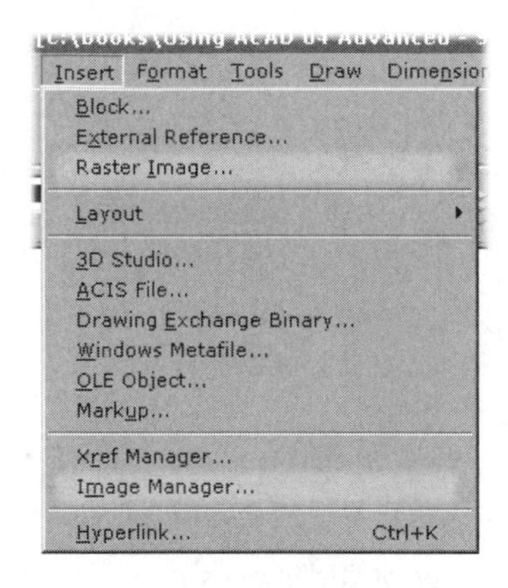

IMAGEATTACH

The **IMAGEATTACH** command attaches image files to drawings.

The advantage of this command over the raster display commands described in Chapter 5 is that the image is part of the drawing. You can move, copy, resize, and rotate it. You can draw over it, or use it as a background. You can perform certain modifications to the picture, such as change its contrast, or make certain areas transparent. You can clip the image, much like xrefs and viewports.

(Historically, in the mid-1980s Robert Godgart figured out how to get AutoCAD to display images in the backgrounds of drawings. He used the new ADI driver software, Autodesk's driver interface that made it easier for hardware vendors to make their graphics boards work with AutoCAD. His software, CAD Overlay, made it possible to trace over drawings inside AutoCAD, obviating the need for expensive, drawing-sized digitizing tablets. CAD Overlay was purchased by SoftDesk, which itself was later purchased by Autodesk. This CAD Overlay technology makes the **IMAGE** command possible.)

When AutoCAD attaches images, they are not made part of the drawing. Rather, AutoCAD remembers the link to the file location, and displays a copy of the image. The benefit is that the drawing size does not balloon, because image files can be many megabytes in size. The drawback is that the link to the image file, like xrefs and other support files, can be broken, resulting in no image being displayed.

The **IMAGEATTACH** command opens image files in the formats listed by the table. The images can be bitonal, 8-bit gray, 8-bit color, or 24-bit color. (This list is more accurate than that provided by Autodesk.) The most popular formats are highlighted in boldface.

Type	*Description*	*File Extensions*
BMP	Windows and OS/2 bitmap	.bmp .dib .rle
CALS-I	Mil-R-Raster I	.gp4 .mil .rst .cg4 .cal
FLIC	Autodesk Animator Animation	.flc .fli
GeoSPOT	Must be accompanied by HDR and PAL files	.bil
GIF	**Compuserve GIF**	**.gif**
IG4	Image Systems Group 4	.ig4
IGS	Image Systems Grayscale	.igs
JFIF, JPEG	**Joint Photographics Expert Group**	**.jpg .jpeg**
PCX	PC Paintbrush Picture	.pcx
PICT	Macintosh Picture	.pct
PNG	Portable Network Graphic	.png
RLC	Run-Length Compressed	.rlc
TARGA	True Vision	.tga
TIFF	**Tagged Image File Format**	**.tif .tiff**

BASIC TUTORIAL: ATTACHING IMAGES

1. To attach images to drawings, start the **IMAGEATTACH** command:
   - From the **Insert** menu, choose **Raster Image**.
   - At the 'Command:' prompt, enter the **imageattach** command.
   - Alternatively, enter the alias **iat** at the keyboard.

   Command: **imageattach** *(Press* ENTER.*)*

In all cases, AutoCAD displays the Select Image File dialog box, which lists only files AutoCAD is capable of opening.

The Preview area shows the content of the selected file.

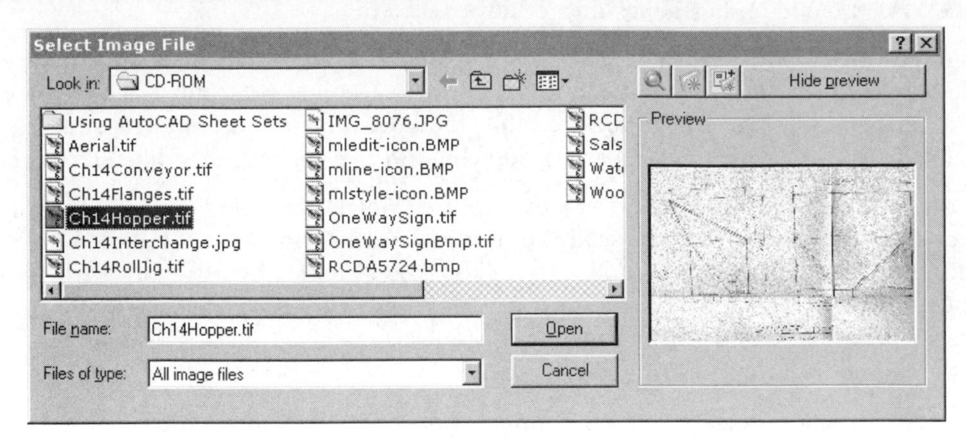

Select Image File dialog box.

2. For this tutorial, select the *Ch14Hopper.tif* file from the CD-ROM.
3. Click **Open**.

   AutoCAD displays the Image dialog box, which is similar to the Insert dialog box.

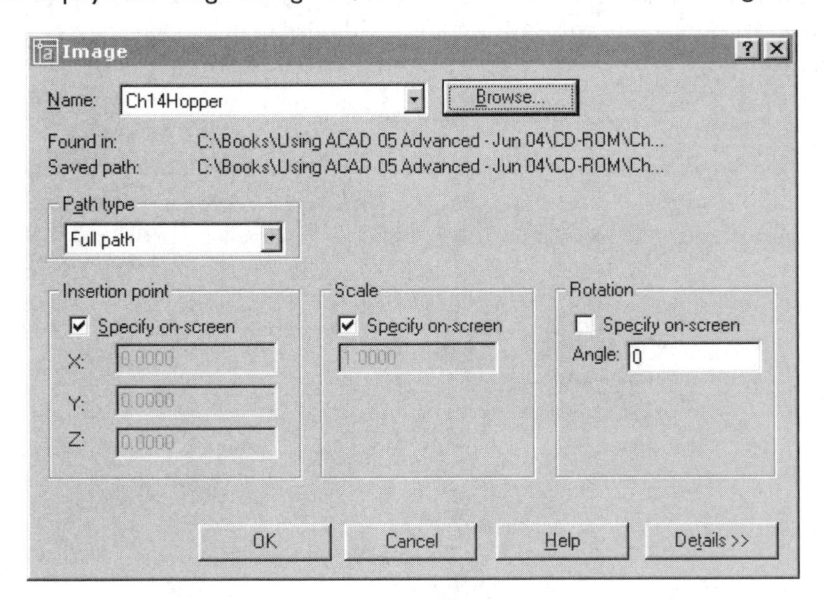

Image dialog box.

4. Attaching an image is like inserting a block. AutoCAD asks for the insertion point, scale, and rotation angle — which you specify in this dialog box or at the command-line.

   For this tutorial, turn off all **Specify On-screen** items (no check marks), because you adjust the scale and angle later.
5. Click **OK**.

   AutoCAD displays the image in all viewports of the current drawing. Scanned images are typically not straight or at the correct size. This is corrected with the **ROTATE** and **SCALE** commands.

Note: The rectangle surrounding the image is called the "frame." You cannot select the image directly; instead, you must select its frame.

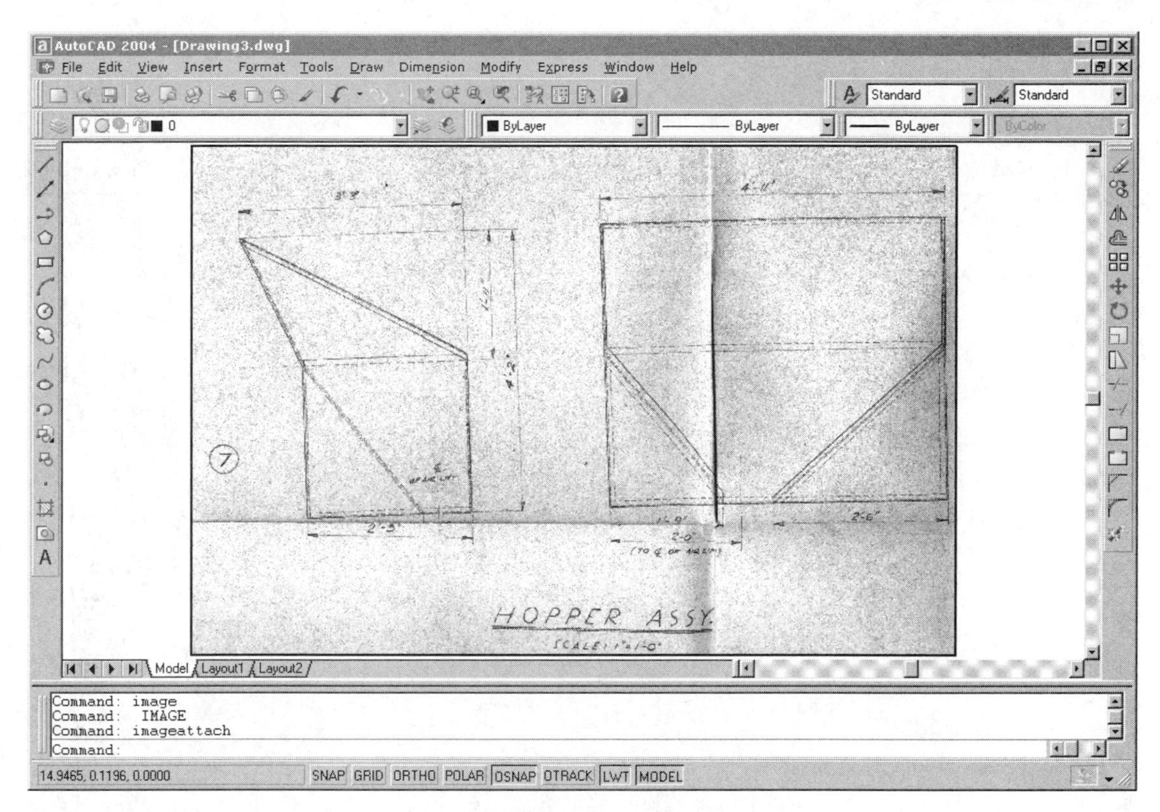

Image attached to drawing.

6. To straighten the image, use the **ROTATE** command's **Reference** option to rotate a "horizontal" line to 0 degrees.

   From the **Modify** menu, select **Rotate**.

   Command: **rotate**

   Select objects: *(Pick the image by its "frame," the rectangle surrounding the image.)*

   Select objects: *(Press* **ENTER** *to end object selection.)*

7. For the base point, use the image's insertion point, typically 0,0.

   Specify base point: **0,0**

8. Specify the **Reference** option:

   Specify rotation angle or [Reference]: **r**

9. For the reference angle, pick two points in the image. For best accuracy, the two points should be as far apart as possible. Ideally, the two points should be at the ends of a line that is horizontal (0 degrees) or vertical (90 degrees).

   Specify the reference angle <0>: *(Pick point 1.)*

   Specify second point: *(Pick point 2.)*

   See the figure on the next page for the pick points.

Note: The picking order is important. You must pick the left end of horizontal lines, or the lower end of vertical lines, before picking the right end or upper end. Reversing the pick sequence inverts the image.

10. Enter the angle the line should be.

    Specify the new angle: **0**

Notice that AutoCAD rotates the image. The frame no longer looks straight, but the scanned image is straight.

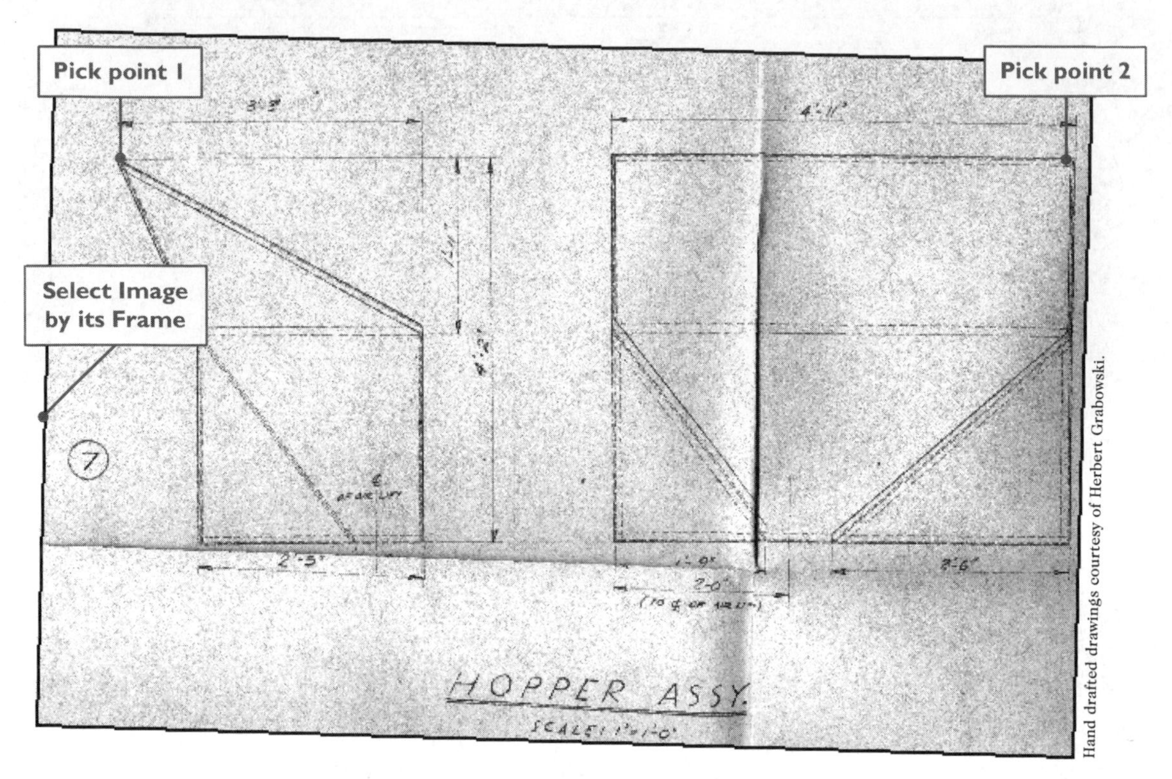

Scanned image straightened with Rotate command.

11. To resize the image, use the **SCALE** command's **Reference** option to change the scale length to true length.

    The best part of an image to resize is the longest dimension line. (It can be at any angle.) For the best accuracy, zoom into the dimension so that it fills the screen.

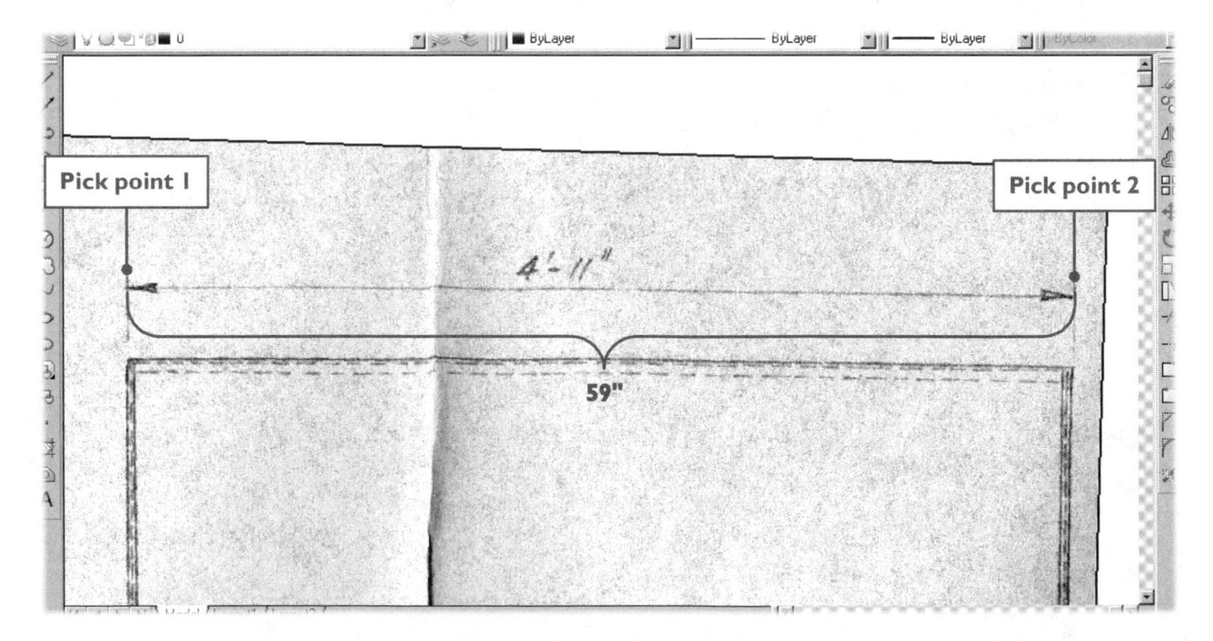

Scaling the image relative to a dimension line.

From the **Modify** menu, select **Scale**.

Command: **scale**

Select objects: **p** *(Previous selects the image again.)*

Select objects: *(Press* ENTER *to end object selection.)*

Specify base point: **0,0**

Specify scale factor or [Reference]: **r**

Specify reference length <1>: *(Pick one end of the dimension, point 1.)*

Specify second point: *(Pick one end of the dimension, point 2.)*

12. Enter the length of the dimension. In this example, 4'11" is the same as 59" (units).

    Specify new length: **59**

Notice that the image changes size. If necessary, use the ZOOM **Extents** command to see all of it.

Notes: To keep the image behind (or underneath) other objects in the drawing, use the DRAWORDER command's **Back** options:

Command: **draworder**

Select objects: *(Select the image by its border.)*

Select objects: *(Press* ENTER *to end object selection.)*

Enter object ordering option [Above object/Under object/Front/Back] <Back>: **back**

To use images as backgrounds for rendering, use the RENDER command's **Background** option, not the IMAGE command. This ensures the image always faces the rendering viewpoint. (Curiously, RENDER supports a smaller variety of raster files, and is much slower than IMAGE.)

1. In the Render dialog box, click the **Background** button.
2. Select the **Image** radio button, and then select the file with the **Find File** button.
3. If you wish, adjust its size with the **Adjust Bitmap** button.

The figure below illustrates an image placed in the background of a rendering with the RENDER command.

Image placed in background with Render command's Background option (Okanagan Lake, Kelowna, Canada).

ATTACHING IMAGES: ADDITIONAL METHOD

An alternative to the **IMAGEATTACH** command is to use the DesignCenter.

AdCenter

The **ADCENTER** command displays the DesignCenter window, and can attach images to drawings.

1. To display DesignCenter, press **CTRL+2** or select **Tools | DesignCenter** from the menu.
2. To find image files, click the **Folders** tab, and then navigate to the appropriate folder. For this tutorial, navigate to the CD-ROM.
3. To see a preview of the image, click its file name, such as *IMG_8076.jpg*.

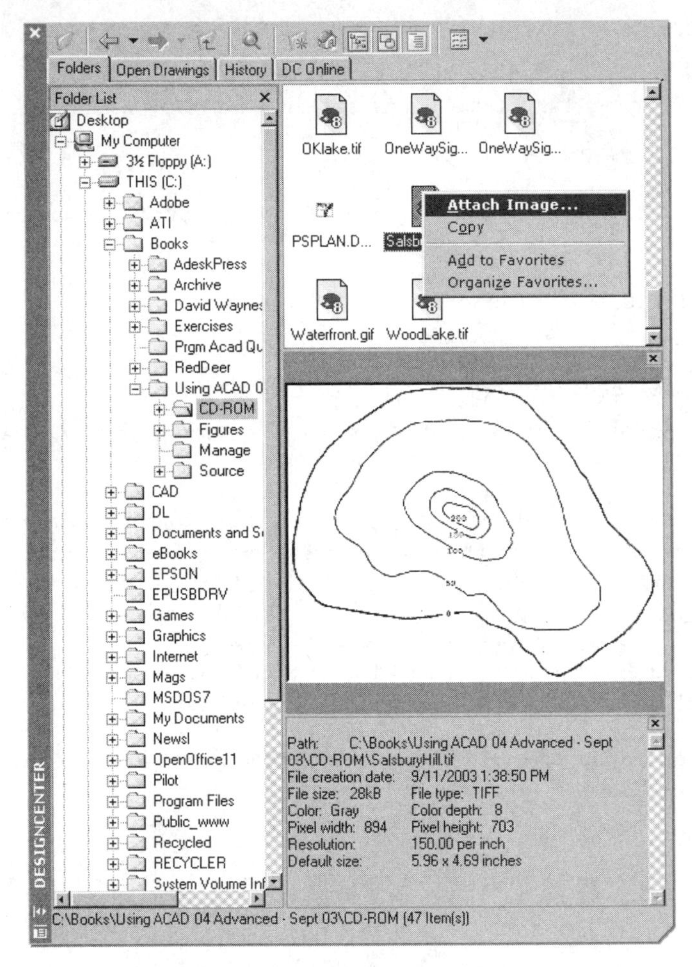

DesignCenter accesses image files.

4. Right-click the file name, and then from the shortcut menu, select **Attach Image**. This displays the same Image dialog box as does the **IMAGEATTACH** command.

The DesignCenter's **Copy** option copies the image to Clipboard, which can then be pasted into the drawing with the **PASTECLIP** command (press **CTRL+V**). The image can be pasted into only a very few applications other than AutoCAD, such as Office XP, but not Office 2000. AutoCAD displays prompts similar to inserting blocks:

Command: **pasteclip**

Specify insertion point <0,0>: *(Pick a point.)*

Base image size: Width: 4.333333, Height: 4.546667, Inches

Specify scale factor or [Unit] <1>: *(Enter a scale factor, or drag to show the size.)*

Specify rotation angle <0>: *(Press **ENTER**, or specify a rotation angle.)*

USING DIGITAL CAMERAS WITH AUTOCAD

AutoCAD reads the JPEG and TIFF image files generated by digital cameras, making the technology ideal for getting photos quickly into drawings. (AutoCAD cannot read the camera's native file format, often called RAW format.)

Digital cameras are useful for taking pictures of site views, as-completed projects, 3D "scans," and so on.

Panoramic. When you can't fit the entire object in a single photo, some digital cameras provide *panoramic* mode, which assists in taking a sequence of overlapping photos — vertically, horizontally, and 2x2. Later, with the help of stitching software, the sequence can be merged into a single large photo, and then attached to drawings.

1. *Four original digital photographs (Wellington, New Zealand).*
2. *Photos stitched together (black rectangles show overlappin seams).*
3. *Final image corrected for wide-angle distortion (black rectangle shows crop outline).*

Time Lapse. Some cameras take photos automatically, from once every 10 seconds to every 24 hours. This is handy for recording the progress of projects. The sequence of photos can be attached to drawings, or posted to project Web sites.

Time-lapse photographs show sequence of new office tower construction in Auckland.

Capabilities of cameras vary, and sometimes older ones are more capable. My old Epson PC800 digital camera takes as many time-lapse pictures as fit on a memory card (roughly 1,750 640x480 photos on a 96MB card), but my newer Canon G1 only does this when hooked up to a computer, and then is limited to a maximum of 100 pictures.

Movie Mode. Most cameras take short movies of 15 to 60 seconds. This is handy for recording the motion of mechanical devices, although AutoCAD cannot read the AVI format used by most cameras for recording movies.

Note: The **Units** option of this variation of the **PASTECLIP** command displays the following prompts:

Enter unit [MM/Centimeter/Meter/Kilometer/Inch/Foot/Yard/MILe/Unitless] <Inches>:

When you specify a unit other than unitless — such as feet or meters — *and* the image has resolution information, AutoCAD applies the scale factor after the image's width is determined.

When you specify "Unitless" (system variable **INSUNITS** = 0) *or* the image does not have resolution information, the scale factor is applied to the image's width in AutoCAD units.

IMAGE

The **IMAGE** command manages image files.

This command controls the attachment of image files in a manner identical to that of xrefs. For more information, refer to Chapter 2.

BASIC TUTORIAL: MANAGING IMAGES

1. To manage images attached to drawings, start the **IMAGE** command:
   - From the **Insert** menu, choose **Image Manager**.
   - At the 'Command:' prompt, enter the **image** command.
   - Alternatively, enter the alias **im** at the keyboard.

   Command: **image** *(Press* **ENTER**.*)*

   In all cases, AutoCAD displays the Image Manager dialog box.

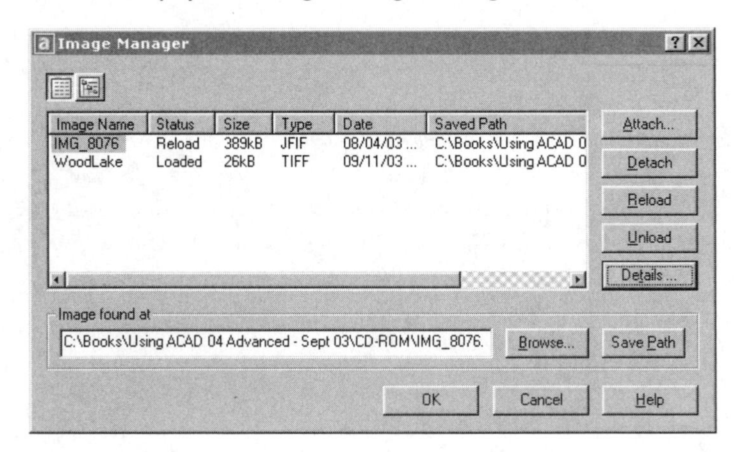

Image Manager dialog box.

2. You can now detach additional images, reload changed images, and search for images whose paths have changed.
3. Click **OK** to dismiss the dialog box.

IMAGEHLT

The **IMAGEHLT** system variable toggles highlighting of images. When on (set to 1), the entire image is highlighted when you select it; when off (set to 0, the default), only the frame is highlighted.

Command: **imagehlt**

Enter new value for IMAGEHLT <0>: *(Type* **0** *or* **1***.)*

Entire image is highlighted when ImageHlt is on.

IMAGEQUALITY

The **IMAGEQUALITY** command toggles the display quality of images. This command has two settings, draft and high.

TUTORIAL: SETTING IMAGE QUALITY

1. To change the image quality, start the **IMAGEQUALITY** command:
   - From the **Modify** menu, choose **Objects**, **Image**, and then **Quality**.
   - At the 'Command:' prompt, enter the **imagequality** command.

   Command: **imagequality**

2. AutoCAD prompts you to select a quality setting, which affects all images equally in the drawing.

   Enter image quality setting [High/Draft] <High>: **d**

The high setting is the default. Quite frankly, I find the draft setting displays a better quality image. It seems to me that the high setting slightly blurs the pixels, while draft keeps them sharp. The technical editor, however, says that high does a better job of reducing the "staircase" jaggies on nearly horizontal and vertical lines.

IMAGEADJUST

The **IMAGEADJUST** command adjusts brightness, contrast, and fade of images.

This command displays a dialog box with three sliders, one each for brightness, contrast, and fade control. The fade control is particularly useful for reducing the intensity of scanned images, over which you trace.

TUTORIAL: ADJUSTING IMAGES

1. To adjust images, start the IMAGEADJUST command:
   - From the **Modify** menu, choose **Objects**, **Image**, and then **Adjust**.
   - Double-click the image.
   - At the 'Command:' prompt, enter the **imageadjust** command.
   - Alternatively, enter the alias **iad** at the keyboard.

   Command: **imageadjust**

2. In all cases, AutoCAD prompts you to select one or more images:

   Select objects: *(Select the image by its border.)*

   Select objects: *(Press* ENTER *to end object selection.)*

   AutoCAD displays the Image Adjust dialog box.

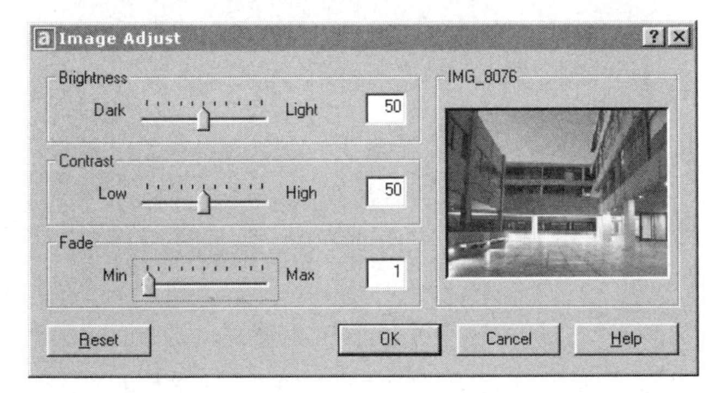

Image Adjust dialog box.

3. Move the sliders, or enter a value in the text boxes.
   - **Brightness** changes the image from dark (near 0) to very light (near 100).
   - **Contrast** changes the image from highly saturated colors (near 0) to grayish colors (near 100).
   - **Fade** changes the image from normal (near 0) to faint (near 100).

   Notice that the preview image changes.

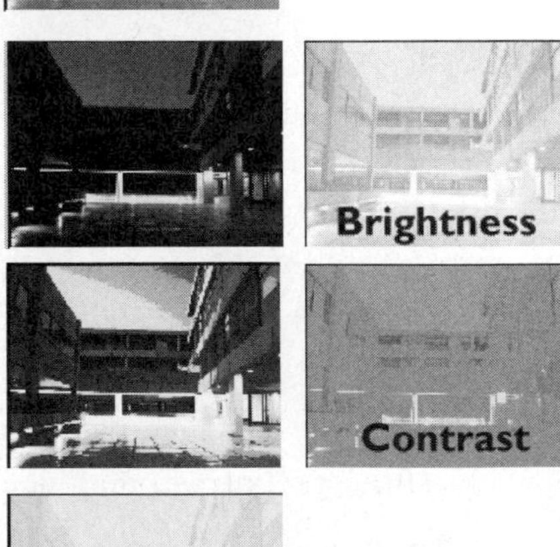

Effects of adjust images.

4. Once the image is adjusted to your satisfaction, click **OK**.
   AutoCAD changes the image to match the settings.

TRANSPARENCY

The TRANSPARENCY command toggles the transparency of background pixels.

This command has two settings: on and off. When on, "transparent" pixels become see-through; pixel transparency is set in the image, not in AutoCAD. Autodesk notes that "several image file formats allow images with transparent pixels," but fails to list those formats. They are GIF and PNG.

The figure below illustrates transparency. In the black-white image of the waterfront, the white color has been specified as transparent. On the right, transparency has been turned on.

TUTORIAL: SETTING IMAGE TRANSPARENCY

Before attaching the image to a drawing in AutoCAD, you must first designate a single color as the transparent one.

1. Open the image in an image editor, such as PaintShop Pro. For this tutorial, open *Waterfront.gif* from the CD-ROM.
   (If necessary, reduce the image to black-white (bi-tonal), 256 shades of gray, or alpha RGB.)
2. Set the background color to the color that should become transparent. For example, for white areas of the image to be transparent, set the background color to white.

Left: *Transparency off; white areas are opaque.*
Right: *Transparency on; white areas are see-through.*

3. While saving the image in GIF or PNG format, click the **Options** button in the Save As dialog box:
   - In PaintShop Pro 4, select the **Set the Transparency to the Background Color** option.
   - In PaintShop Pro 6, transparency is not available.
   - In PaintShop Pro 8, click the **Run Optimizer** button, and then select the **Areas That Match This Color** option. Use the dropper tool to select the transparency color (white).

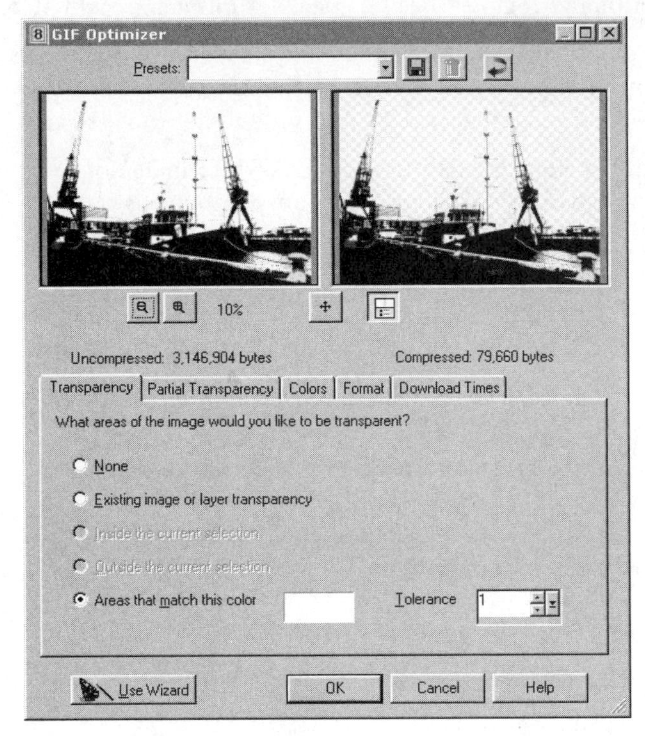

Selecting transparency color in PaintShop Pro 8.

4. Save the image.
5. In AutoCAD, attach the image with the IMAGEATTACH command.
6. Make the white areas transparent with the TRANSPARENCY command:
   - From the **Modify** menu, choose **Objects**, **Image**, and then **Transparency**.
   - At the 'Command:' prompt, enter the **transparency** command.

     Command: **transparency**
7. AutoCAD prompts you to select images:

   Select image(s): *(Select one or more images by their borders.)*

   Select image(s): *(Press* ENTER *to end object selection.)*
8. Turn on transparency by entering the **On** option:

   Enter transparency mode [ON/OFF] <OFF>: **on**

Notice that the white areas of the image become transparent.

IMAGECLIP

The **IMAGECLIP** command clips images.

This command draws a boundary on top of images. The parts of the image outside the boundary become invisible; inside the boundary, the image remains visible — when displayed and when plotted. The boundary can be rectangular or polygonal, but can consists only of straights lines, no curves.

This command also toggles the boundary on and off. When off, the entire image is displayed; clipping is turned off.

TUTORIAL: CLIPPING IMAGES

1. To clip images, start the **IMAGECLIP** command:
   - From the **Modify** menu, choose **Objects**, **Image**, and then **Clip**.
   - At the 'Command:' prompt, enter the **imageclip** command.
   - Alternatively, enter the alias **icl** at the keyboard.

   Command: **imageclip**
2. In all cases, AutoCAD prompts you to select a single image:
   Select image to clip: *(Select the image by its frame.)*
3. To create a new clipping boundary, enter the **New** option.
   Enter image clipping option [ON/OFF/Delete/New boundary] <New>: **n**
4. Select rectangular or polygonal clipping.
   Enter clipping type [Polygonal/Rectangular] <Rectangular>: *(Type* **r** *or* **p**.*)*

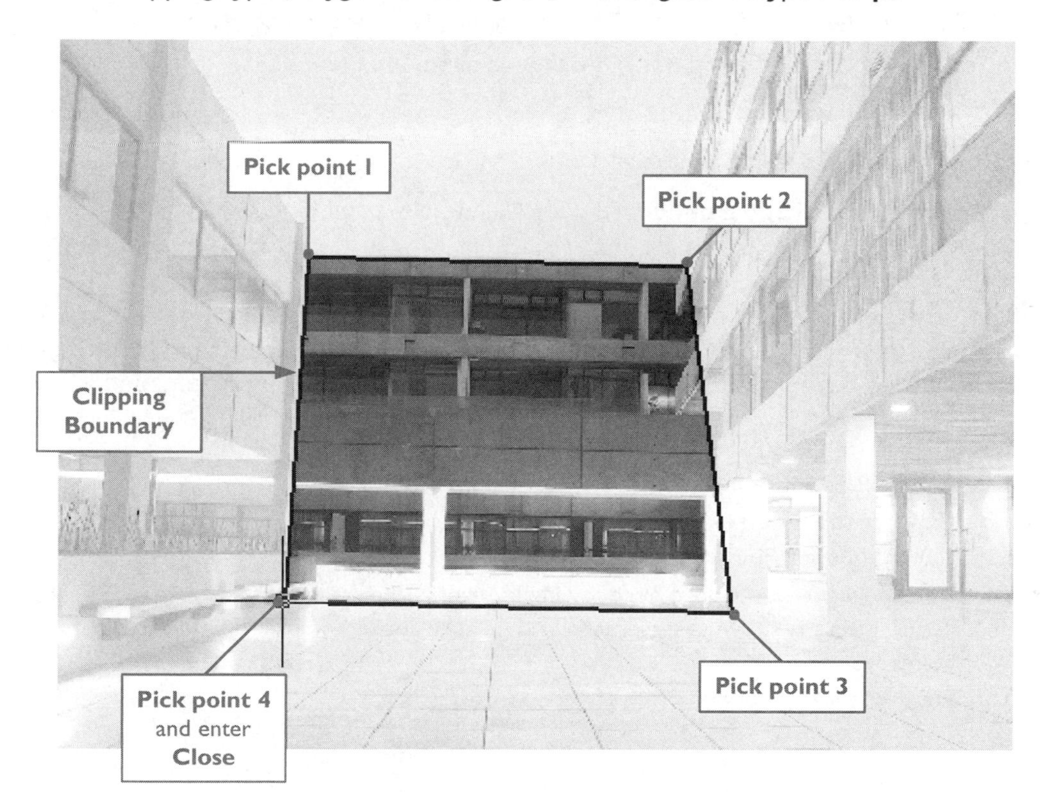

Applying clipping boundary to image.

5. Pick the points that specify the corners (vertices) of the rectangle/polygon.
   Specify first point: *(Pick point 1.)*

Specify next point or [Undo]: *(Pick point 2.)*

Specify next point or [Undo]: *(Pick point 3.)*

Specify next point or [Close/Undo]: *(Pick point 4.)*

6. Enter the **Close** option to end the command.

   Specify next point or [Close/Undo]: **c**

After clipping an image, its transparency is lost. If necessary, reapply the **TRANSPARENCY** command.

When you apply a new boundary to an image that already has a boundary, AutoCAD asks, "Delete old boundary? [No/Yes] <Yes>:". Answer "yes," and continue.

CLIPPING IMAGES: ADDITIONAL METHODS

The **IMAGECLIP** command has a number of options that control the look of the clipped images.

- **On** and **Off** options toggle the display of the boundary.
- **Delete** option removes the boundary.

In addition, with some careful picking, you can create holes in images.

On and Off

The **On** and **Off** options toggle the display of the boundary.

Enter image clipping option [ON/OFF/Delete/New boundary] <New>: **off**

Entering "off" turns off the boundary; the entire image is displayed.

Enter image clipping option [ON/OFF/Delete/New boundary] <New>: **on**

Entering "on" turns on the boundary; the image is again clipped.

Delete

The **Delete** option removes the boundary.

Enter image clipping option [ON/OFF/Delete/New boundary] <New>: **d**

Removing the boundary displays the entire image. All options can be reversed with the **U** command.

Clipping Holes

Clipping images is not as flexible as clipping xrefs and viewports, but it is possible to create a hole in an image. Follow back on the boundary path, as illustrated below by steps 2 and 4.

Use the **Polygonal** option. The tactic is as follows:

Step 1: Outline the entire image with the boundary.

Step 2: Move to the inside of the image.

Step 3: Outline the interior "hole" boundary.

Step 4: Return to the outer edge by following in reverse the same route as in step 2. The return route is shown exaggerated in the figure above; the pick points should coincide.

Step 5: Enter the **Close** option.

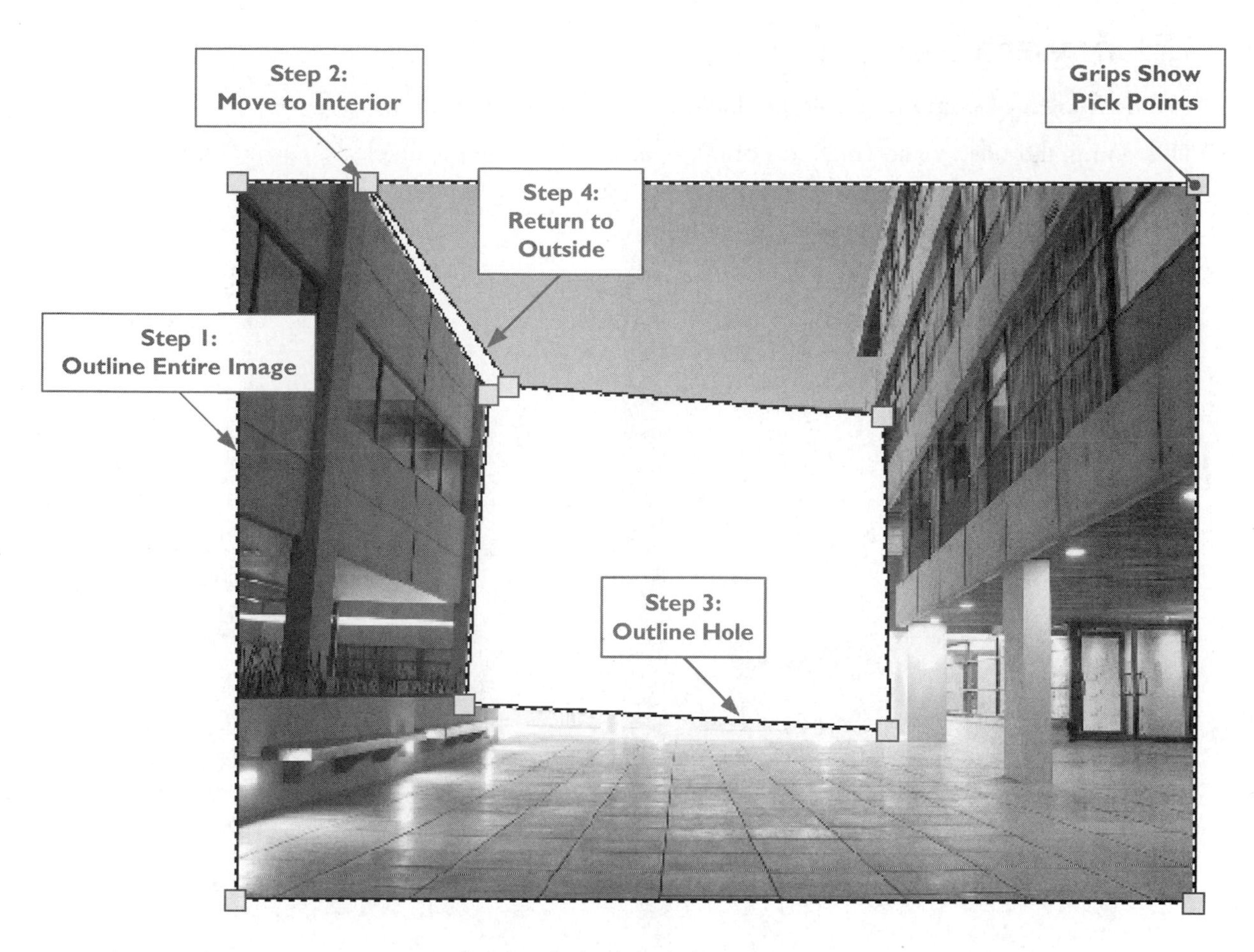

Creating clipping holes in images.

You can edit the boundary with grips; the clipping boundary cannot be edited with the **PEDIT** command.

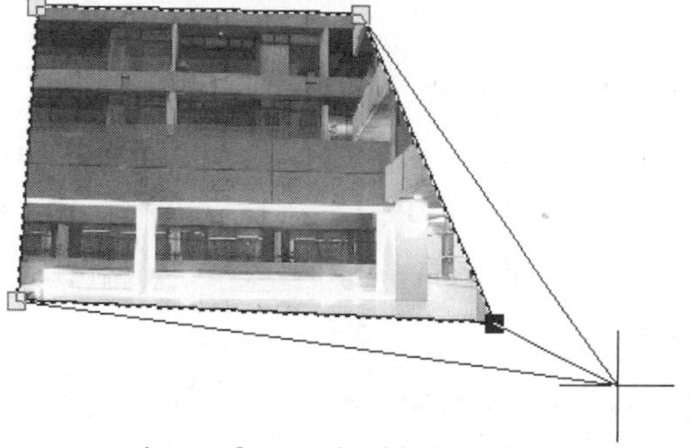

Image frame edited by its grips.

IMAGEFRAME

The **IMAGEFRAME** command toggles the display of image frames.

This command has two settings, on or off — either the frame is displayed, or it is not.

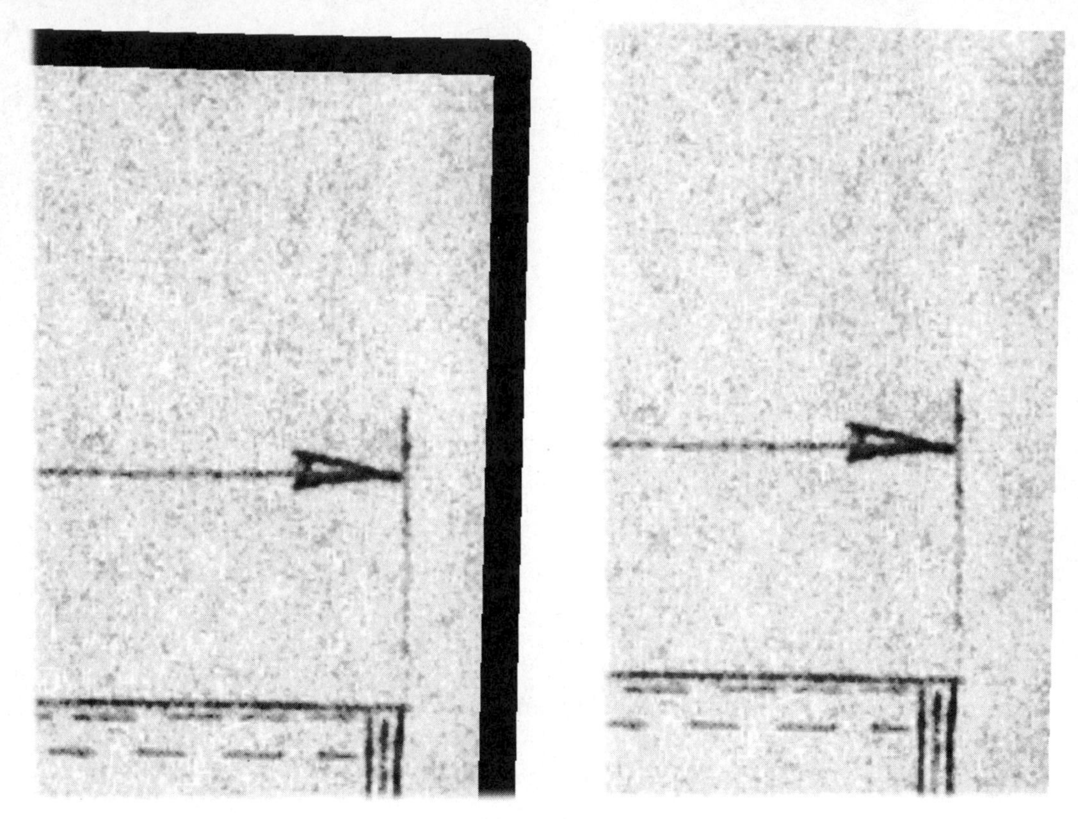

Left: *Image frame on.*
Right: *Image frame off.*

Normally, the frame is displayed. There are two conditions where you may want to turn off the frame: (1) before plotting so that the frame does not show, and (2) to make the image unselectable, because it can only be selected by its frame.

TUTORIAL: TOGGLING IMAGE FRAMES

1. To change the visibility of the images' frame, start the **IMAGEFRAME** command:
   - From the **Modify** menu, choose **Objects**, **Image**, and then **Frame**.
   - At the 'Command:' prompt, enter the **imageframe** command.

   Command: **imageframe**
2. AutoCAD prompts you to turn the frame on or off. All images in the drawing are affected equally.

   Enter image frame setting [ON/OFF] <ON>: **off**

Note: The thickness of the image frame is affected by lineweight.

ADJUSTING IMAGES: ADDITIONAL METHODS

Images can also be adjusted in other ways than through image-related commands:

- Right-click images to select commands from a shortcut menu.
- Double-click image frames to adjust the image.
- Edit the image's grips to resize, move, copy, and so on.
- Access image properties with the Properties window.

Right-click

Click on the image frame, and then right-click to display the shortcut menu. The **Image** item contains commands specific to images.

Right-click selected image for shortcut menu.

Adjust displays the Adjust Image dialog box. See the **IMAGEADJUST** command earlier in this chapter.

Clip displays the image clipping options on the command line. See **IMAGECLIP** earlier in this chapter.

Transparency displays the transparency options on the command line. See the **TRANSPARENCY** command earlier in this chapter.

Image Manager displays the Image Manager dialog box. See the **IMAGE** command earlier in this chapter.

Double-click

Double-clicking the image's frame displays the Image Adjust dialog box. See the **IMAGEADJUST** command earlier in this chapter.

Grips Editing

Selecting the image's frame displays four or more grips; more grips are displayed when the image has been clipped. The grips allow these editing functions:

Drag any one grip to resize the image.

Select a grip, and press the spacebar repeatedly to cycle through the grip editing options: move, rotate, scale, mirror, and stretch.

The **Copy** option works only when in move mode; the **Scale** and **Stretch** options are identical to resizing by repositioning a grip.

Properties

Select the image frame, right-click, and then select **Properties** from the shortcut menu. The Properties window combines several IMAGE-related commands into a single interface. In some cases, the options in the Property window are easier to use than the corresponding command, such as toggling the display of the clipping boundary. In most cases, however, the options are not as complete as the related command; it does not, for example, create the clipping boundary.

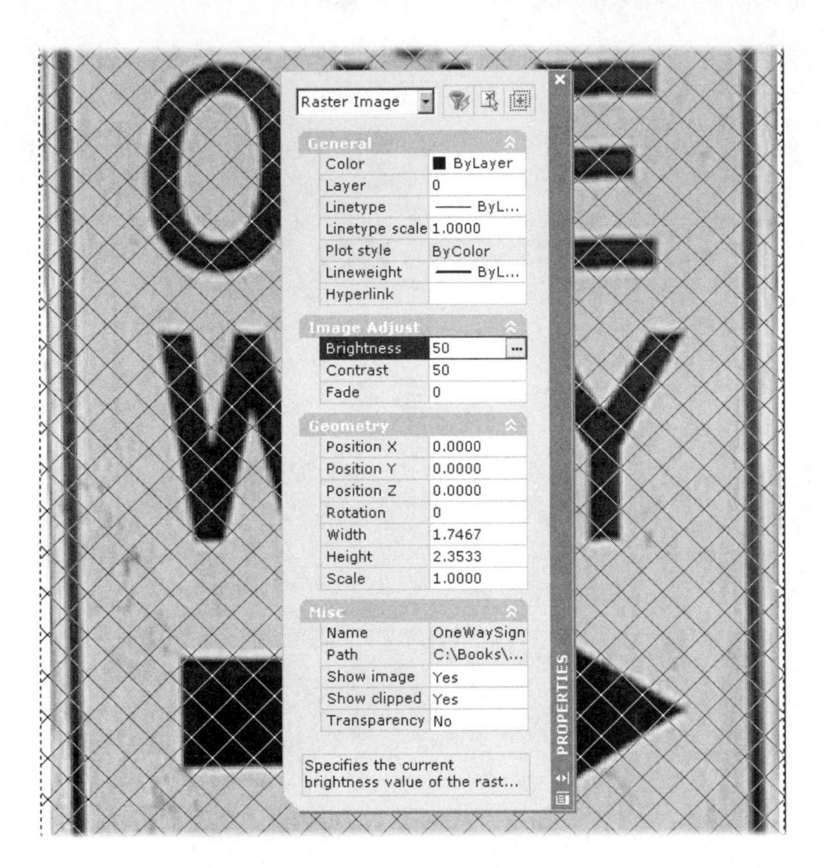

Properties window edits several image properties.

Lineweight changes the thickness of the image frame.

Brightness, **contrast**, **fade** adjust the image; click the **...** button to display the Adjust Image dialog box (IMAGEADJUST command).

Geometry modifies the image's insertion point, rotation angle, size, and scale (IMAGEATTACH command).

Name and **path** are displayed but cannot be edited; you need to use the IMAGE command.

Show image toggles the display of the image.

Show clipped toggles the display of clipped images (IMAGECLIP command's **On** and **Off** options).

Transparency toggles the transparency of pixels (TRANSPARENCY command).

EXERCISES

1. Use the **IMAGEATTACH** command to open the *Ch14Interchange.jpg* file, an aerial photograph of a freeway interchange.

Ch14Interchange.jpg

2. From the CD-ROM, attach the *Ch14Conveyor.tif* file. This scanned image is a drawing of a conveyor platform support. Use the **ROTATE** and **SCALE** commands to straighten the image, and size it correctly.

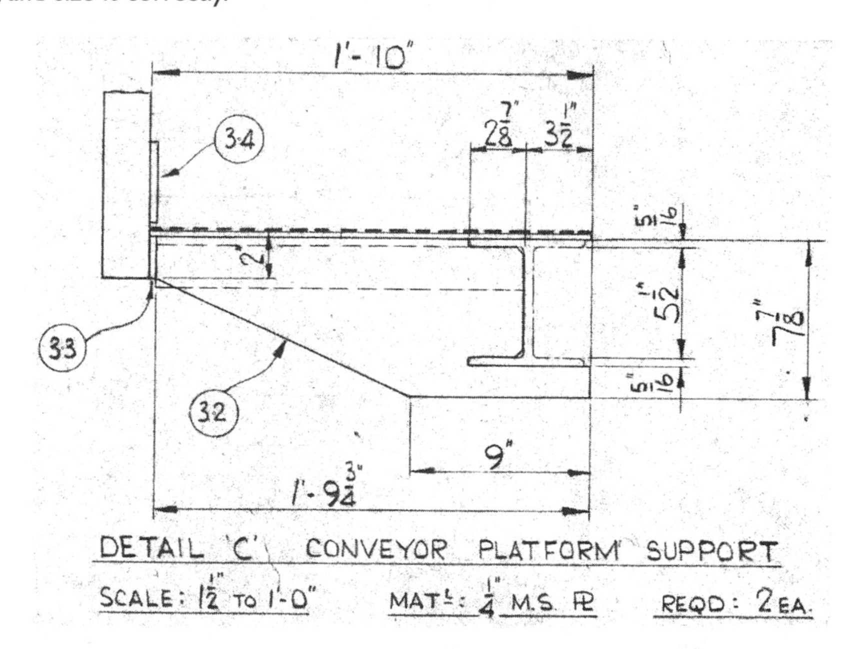

Ch14Conveyor.tif

3. Attach the *Ch14Flange.tif* file from the CD-ROM. This scanned image is a drawing of square and rectangular flanges. Straighten and scale the image. Create a new layer, and trace over with lines and text.

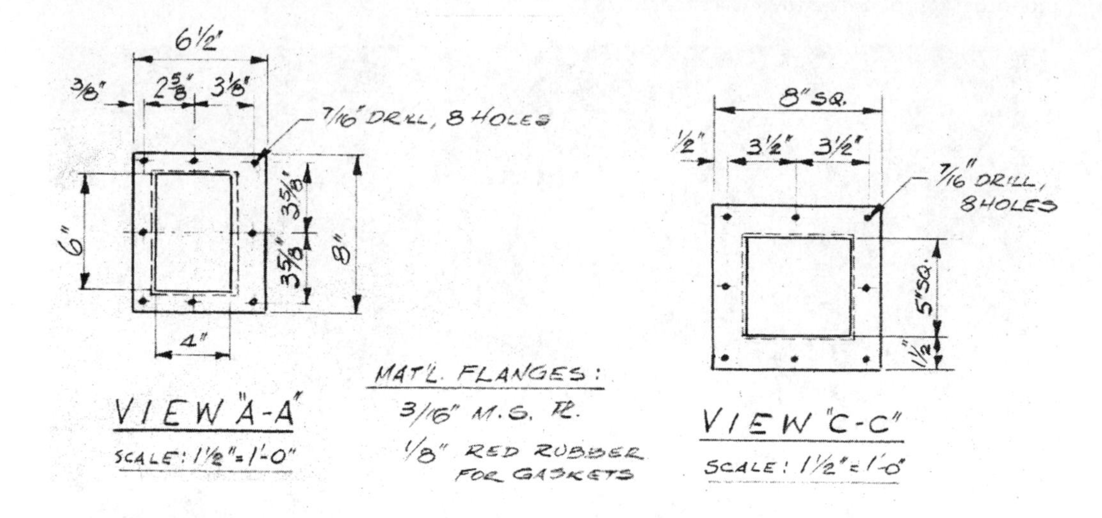

Ch14Flange.tif

(The technical editor notes that in mechanical design, the better practice is to redraw, rather than trace, for improved accuracy.)

4. Attach the *Ch14RollJig.tif* file from the CD-ROM. Straighten and scale the image. Create a new layer, and trace over with lines and text.

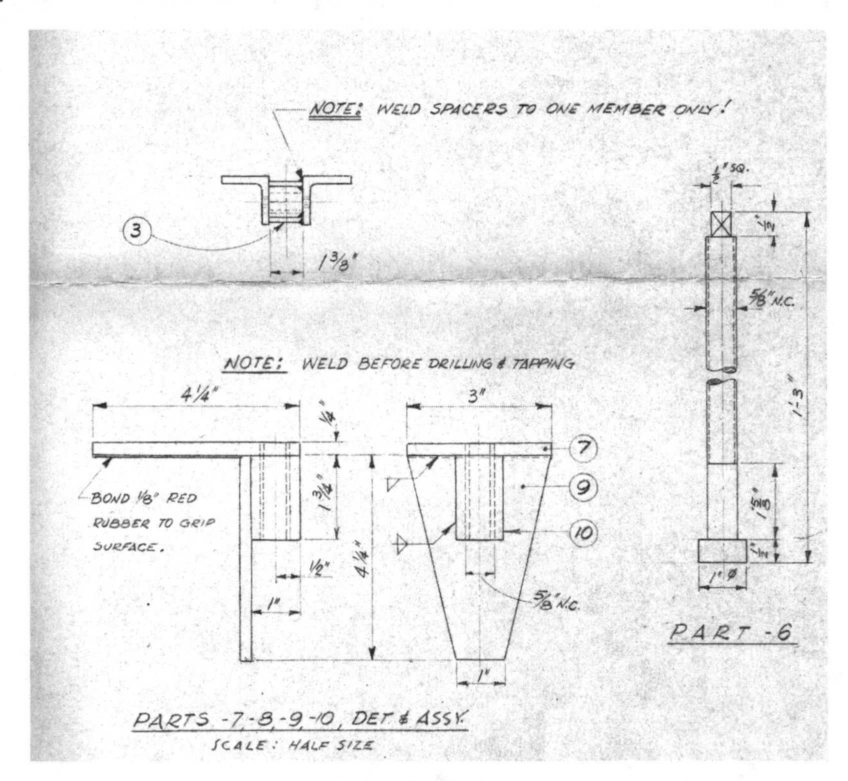

Ch14RollJig.tif

5. Attach the *Aerial.tif* file from the CD-ROM, an aerial photograph of a subdivision. Use the **IAMGECLIP** command to hide the nonresidential areas of the photo.

Aerial.tif

CHAPTER REVIEW

1. When AutoCAD attaches images, do they become part of the drawing?
2. When picking an image with the cursor, which part of the image must you select?
3. Which command keeps images underneath other objects in the drawing?
4. Can AutoCAD read the images created by digital cameras?
5. Briefly explain the difference between the **IMAGE** and **IMAGEATTACH** commands.
6. List a pro and a con to turning off the image's frame:
   Pro:
   Con:
7. What must you do to an image before using the **TRANSPARENCY** command?
8. How is the **IMAGEADJUST** command's **Fade** option useful?
9. What is the purpose of the clipping boundary?
10. Describe why you might want to clip an image.

Customizing and Programming

CHAPTER 16

Toolbar Customization

Toolbar buttons give you single-click access to nearly every command and group of commands. Instead of hunting through AutoCAD's maze of menus (is the HATCH command under **Draw** or **Construct**?) or trying to recall the exact syntax of typed command (was that "Viewpoint" or "Vpoint"?), you can use toolbars to collect the most-used commands in convenient strips.

Toolbars buttons operate by executing *macros* — one or more commands and options. Given the time it takes to write and debug macros, my rule-of-thumb is to write a macro any time the same action is repeated more than three times.

In addition, an obscure programming language included with AutoCAD, called Diesel, can be used to make toolbar macros more powerful.

In this chapter, you learn about the following commands:

TOOLBAR customizes toolbars.

-TOOLBAR toggles the display of toolbars.

TBCUSTOMIZE toggles the ability to customize toolbars (new to AutoCAD 2005).

MODEMACRO runs Diesel macros.

MACROTRACE debugs Diesel macros.

FINDING THE COMMANDS

Right-click any toolbar:

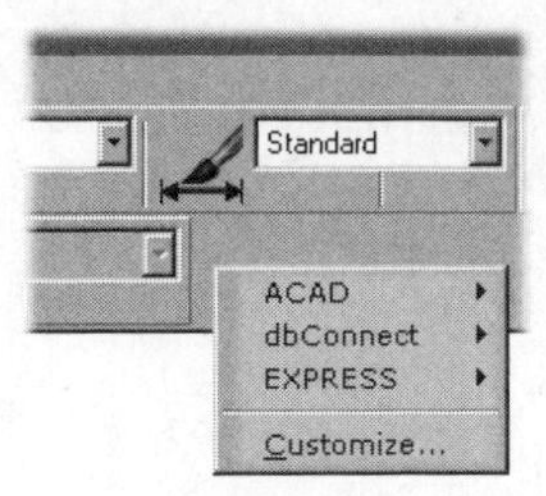

On the **TOOLS** menu:

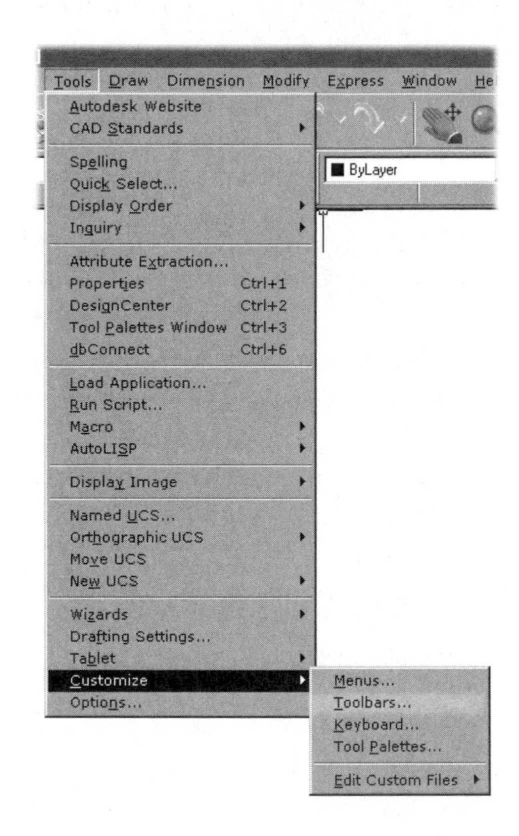

ABOUT TOOLBARS AND MACROS

AutoCAD allows you to customize many aspects of toolbars. You can change their position by dragging them, docking them at the edge of the AutoCAD window, or letting them float anywhere on your computer's screen. People fortunate enough to have two or three monitors sometimes move all toolbars to the second monitor, freeing up AutoCAD's window for more drawing area.

(Historically, toolbars were added to the first Windows release of AutoCAD, Release 11 "AutoCAD Windows Extension." AWE was intended by Autodesk as an experiment rather than a working release. Through Release 12, the toolbar was a single strip of buttons, and difficult to customize. The current iteration of toolbars was added with Release 13 for Windows.)

Toolbar macros are best suited for the quick customization of AutoCAD. They have some drawbacks, however, such as being limited to a maximum of 255 characters. Writing creative macros is limited compared to the possibilities offered by AutoCAD's programming languages, AutoLISP, Visual LISP, and ObjectARx. Despite that, programming the toolbar is the fastest and most convenient way to increase efficiency in AutoCAD by minimizing keystrokes and mouse clicks.

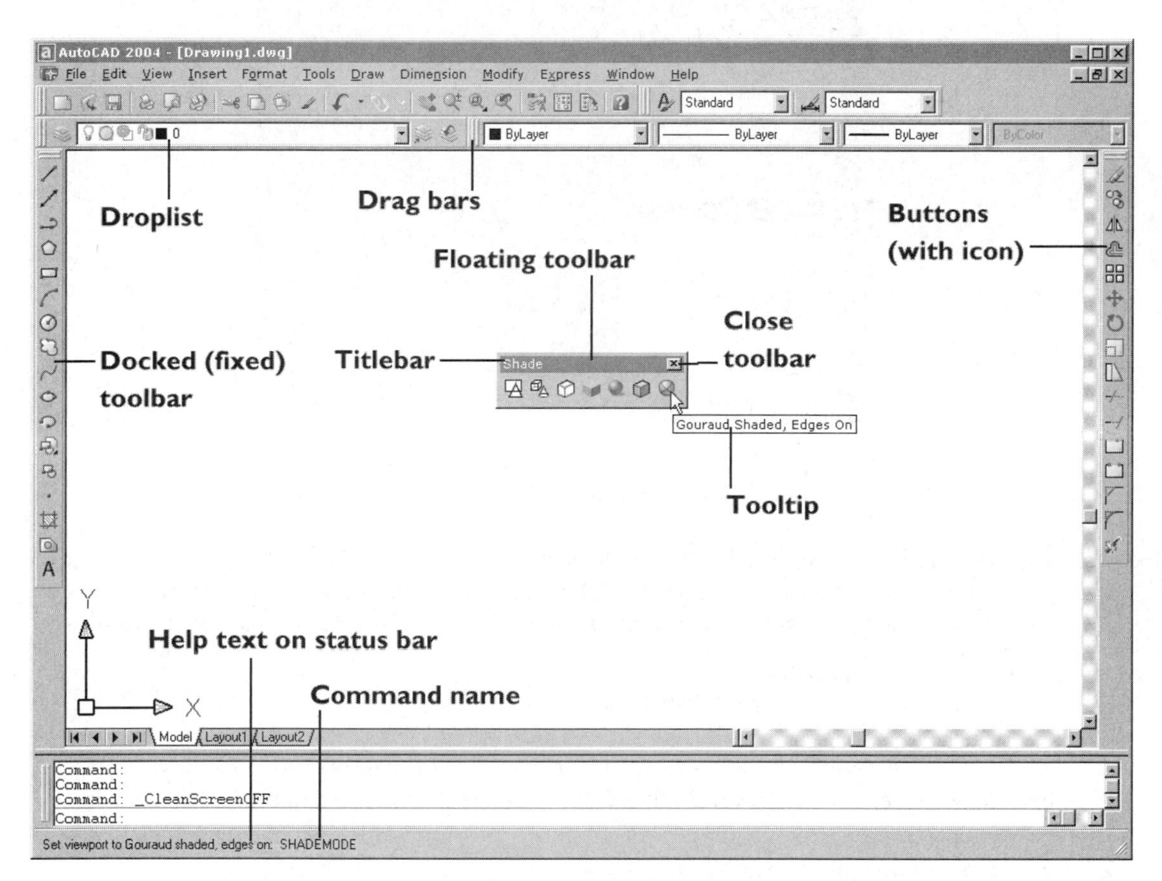

Anatomy of toolbars.

You can create new toolbars, and add and remove buttons. You can change the icons associated with buttons, and create new icons. You can write macros (short programming code) for each button, as well as change the help text that appears in tooltips and on the status line.

Note: If you find toolbar buttons too small, you can make them 50% larger, as follows:

1. Enter the **CUSTOMIZE** command, and then select the **Toolbars** tab.
2. Select **Large Buttons**. Notice that toolbar buttons immediately become larger.
3. Click **Close** to dismiss the dialog box.

To make the buttons smaller, uncheck the **Large Buttons** option.

TUTORIAL: CREATING NEW TOOLBARS

AutoCAD does not provide a toolbar for accessing multiline drawing and editing commands. In this tutorial, you create a toolbar for multilines.

1. To create new toolbars, start the **TOOLBAR** command:
   - From the **Tools** menu, choose **Customize**, and then **Toolbars**.
   - Right-click any toolbar, and from the shortcut menu, select **Customize**.
   - At the 'Command:' prompt, enter the **toolbar** command.
   - Alternatively, enter the alias **to** at the keyboard.

   Command: **toolbar** *(Press* ENTER.*)*

   In all cases, AutoCAD displays the Customize dialog box. (If the **Toolbars** tab is not showing, click it now.)

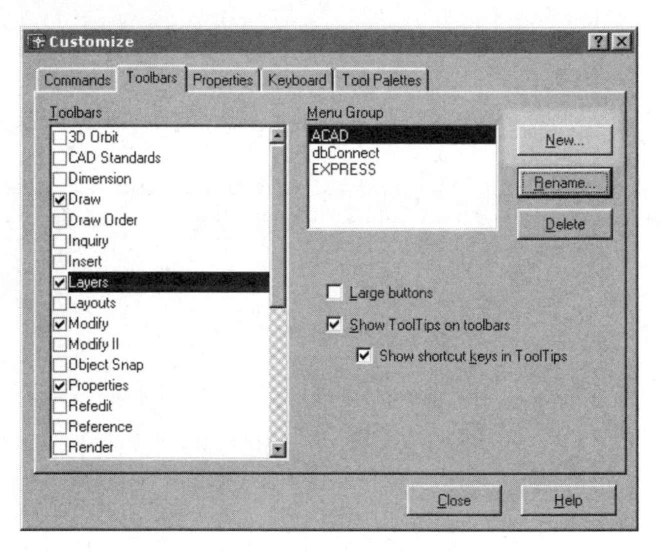

Customize dialog box's Toolbars tab.

Note: If you do not see the **Customize** item on any menu, the **TBCUSTOMIZE** system variable is turned off. This variable, introduced with AutoCAD 2005, prevents accidental tampering with toolbars by disabling the **CUSTOMIZE** and **TOOLBAR** commands. To reenable toolbar customization, turn on the variable:

Command: **tbcustomize**
Enter new value for TBCUSTOMIZE <0>: **1**

The Toolbars list names all toolbars available in AutoCAD, as well as those you create. A checkmark means the toolbar is displayed. To toggle the display of toolbars, click the box next to a name.

2. To create a new toolbar, click **New**.

   Notice the New Toolbar dialog box.

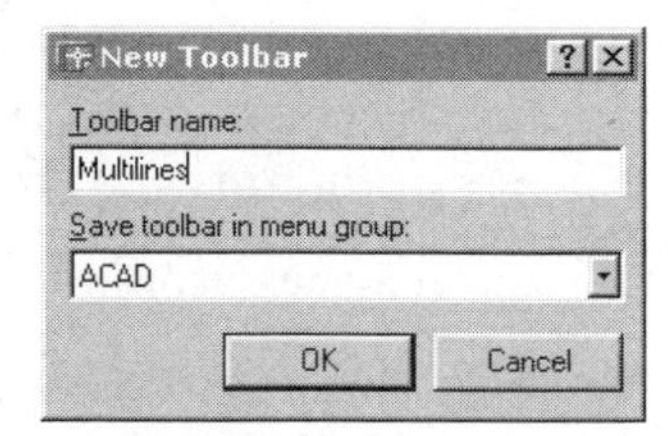

New Toolbar dialog box.

3. In the New Toolbar dialog box, change the default name, "Toolbar1," to the name your new toolbar will display, such as "Multilines." This name will appear on the new toolbar's titlebar.

    Toolbar name: **Multilines**

    Click **OK**.

    AutoCAD creates a new, empty toolbar. You may miss it, because it is tiny: hunt around the screen to find it.

Newly created toolbar.

4. Select the **Commands** tab of the Customize dialog box. You populate toolbars with buttons by dragging command names from this dialog box to the toolbars. (You can also drag buttons out of existing toolbars.)

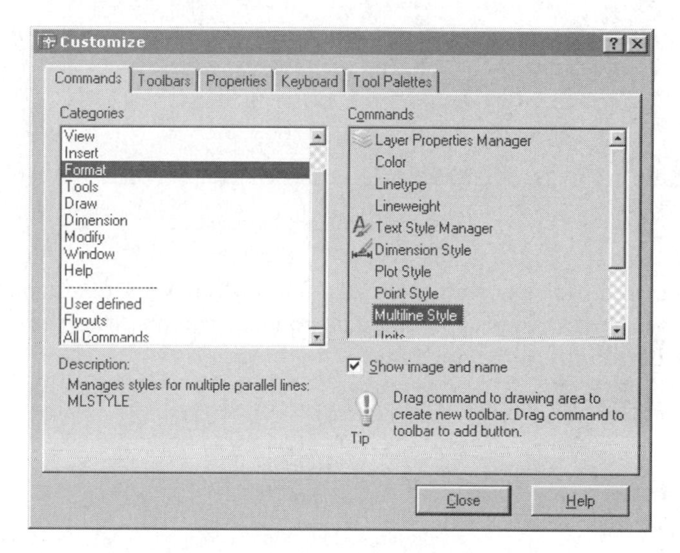

Customize dialog box's Commands tab.

This dialog box presents AutoCAD's commands in two ways: (1) categorized by function (View, Insert, and so on), and (2) alphabetized. The alphabetical listing is "hidden" at the end of the Categories list: scroll down to find **All Commands**.

(I have a couple of problems with the way Autodesk lists commands here. Not all commands are listed: **ACISIN** is there, but **ACISOUT** is not. And the listing doesn't use the *names* of commands. For example, the **VPCLIP** command is listed as "Clip Existing Viewport." Often, this means hunting through the entire list, guessing at what might be the descriptive equivalent to a specific command name.)

5. In the **Command** list, select a command description.

    For this tutorial, select the **Format** category, and then select the **Multiline Style** command.

    Look at the bottom half of the dialog box under Description. AutoCAD describes the purpose of the command, along with its name.

    Description:

    Manages styles for multiple parallel lines:

    MLSTYLE

6. Drag the command name to the new toolbar.

   As you drag it, the command name becomes a button. Notice the special cursor: the + sign indicates AutoCAD is adding the button to the toolbar.

   Once the cursor moves inside the new toolbar, the I-beam cursor helps you position the button between others.

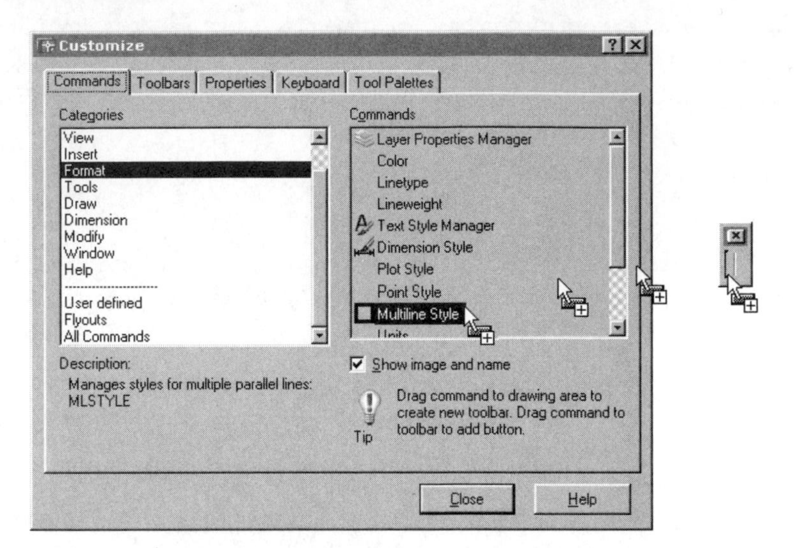

Dragging commands to toolbars.

7. Notice that the command's button appears in the toolbar. Well, maybe not. Autodesk did not provide icons for all commands, including none for the multiline commands. The Multiline Style button is there, but is blank.

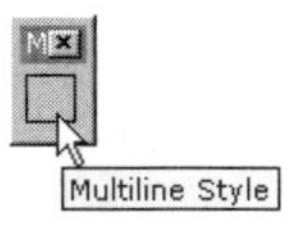

Button added to toolbar.

   Move the cursor to where the button should be on the Multilines toolbar. After a moment, the button's tooltip appears, confirming its existence.

Notes: While the Customize dialog box is open, *all* toolbars are customizable — not just the one you are working with. This means you can add and remove buttons from other toolbars, such as Draw, Standard, and so on.

Be careful about moving and deleting buttons from the standard toolbars. The **UNDO** command does not reverse your actions.

8. The blank button needs an icon to make it easier to identify. To do so, click the **Properties** tab of the Customize dialog box. (If AutoCAD displays the following message, click the Multiline Style button.)

Select a button to edit it.

   The **Button Properties** tab controls all aspects of buttons: their icons, macros, help strings, and tooltips. We leave most of these items for later tutorials.

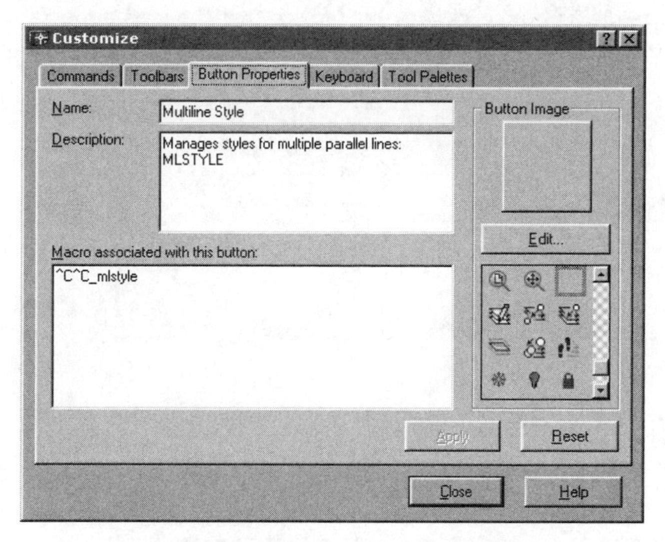

Button Properties tab of Customize dialog box.

9. Good-looking icons are hard to draw, so it's easier to adapt an existing icon, or to use icons created by others. For this tutorial, use the icon provided on the CD-ROM. Click **Edit**. Notice the Button Editor dialog box.

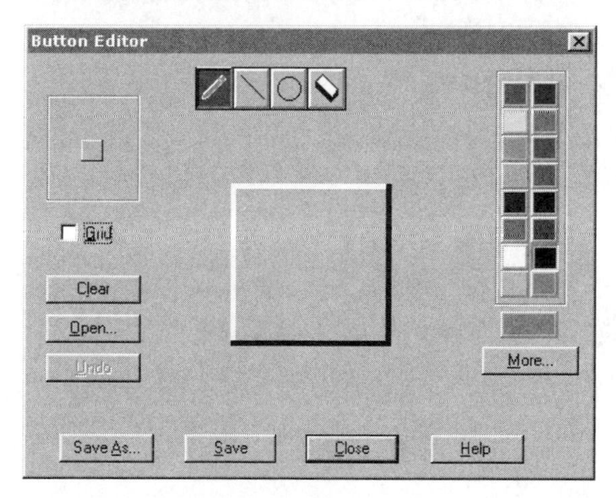

Button Editor dialog box.

In the Button Editor dialog box, click **Open**, and select *mlstyle-icon.bmp* from the CD-ROM. Click **Open**. The image appears in the dialog box.

10. Back in the Button Editor dialog box, click **Save**, and then click **Close**.

Notes: You can right-click any button in any toolbar, and then select **Copy Button Image** from the shortcut menu. The image is then pasted onto any other button with the **Paste Button Image** command. If you make a mistake, right-click, and then select **Reset Button Image**.

AutoCAD uses several conventions for designing icons. One is a paintbrush to indicate styles, and another is a pencil to indicate editing.

Left: *Paintbrush indicates "style."*
Right: *Pencil indicates "edit."*

11. In the Customize dialog box, click **Apply**. Notice that the icon appears on the button in the Multilines toolbar.

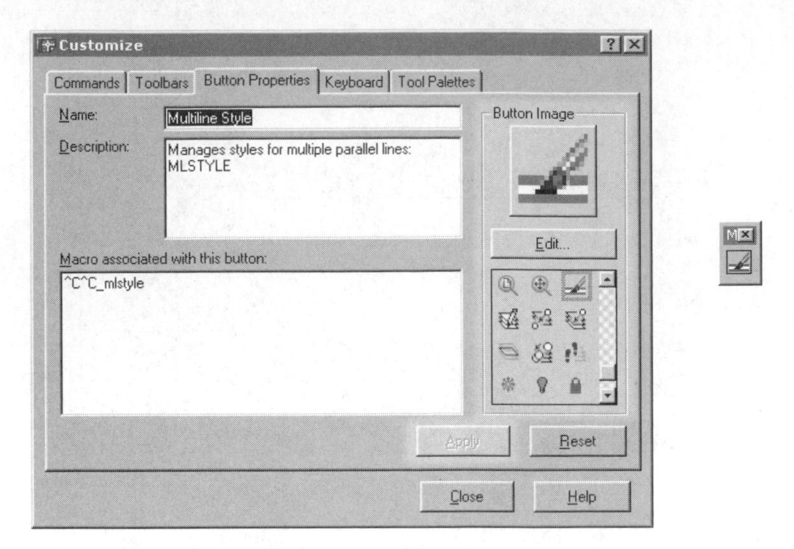

Applying icon to toolbar button.

12. Populate the toolbar with the other multiline commands. You find the icon files on the CD-ROM.

Command	*Command "Name"*	*Icon File*
MLINE	Multiline	*mline-icon.bmp*
MLEDIT	Object Multiline	*mledit-icon.bmp*

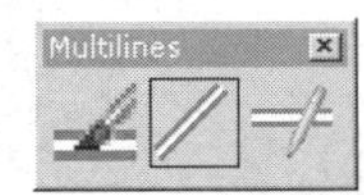

Completed Multiline toolbar.

If you intend to use Large Buttons, you have to create two icons for each button: one for small and one for large buttons.

13. When done, click **Close** to dismiss the Customize dialog box. There is no need to "save" the new toolbar; it is saved automatically.
14. Your new toolbar acts just like the any other toolbar in AutoCAD.

    It's always a good idea test after the customizing and programming tasks in AutoCAD. This ensures that the toolbar buttons operate as you expect them to. (This is sometimes called "debugging.")

    Click each button to ensure the correct command is executed.

In addition to creating new toolbars with customized icons, you can change the command(s) that lie behind each button. When you click a toolbar button, AutoCAD executes the *macro* (one or more commands) assigned to the button.

TUTORIAL: WRITING TOOLBAR MACROS

In this tutorial, you modify the macro for the Multiline button, so that multlines are drawn on a layer called "Multilines." This involves using the **-LAYER** command's **Make** option to set the current layer. This is a good option to use, because it also creates the layer, if it does not already exist.

1. Open the Customize dialog box by right-clicking the Multilines toolbar, and then selecting **Customize** from the shortcut menu.
2. Select the Multiline button on the Multiline toolbar. Notice that the Customize dialog box switches to the **Button Properties** tab.

   This tab has several areas that correspond to AutoCAD's user interface:

   - **Name** is the button's tooltip, a brief description of the button's function.
   - **Description** is status line help text, a longer description of the button's function.
   - **Macro** is one or more commands executed by clicking the button.

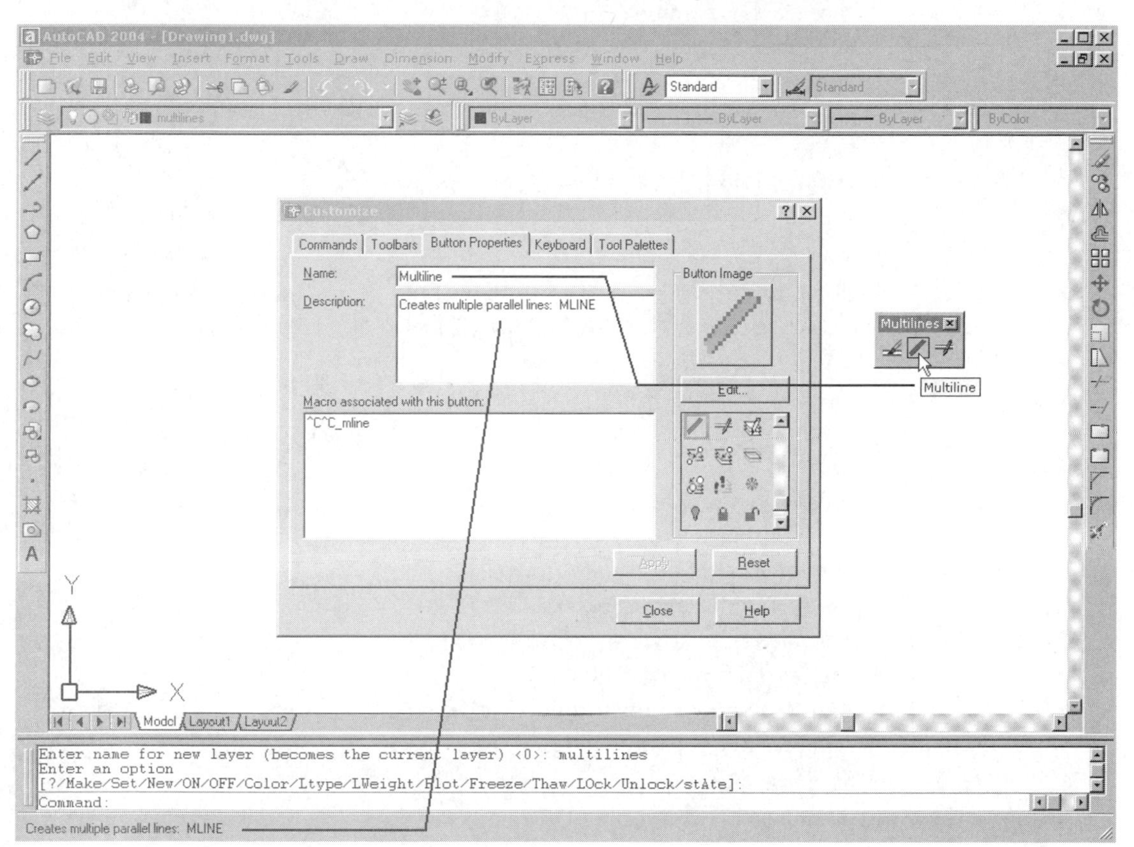

Name and description of buttons.

The simplest macros consist of a single AutoCAD command, prefixed by *control characters*. For the Multiline button, this is:

^C^C_**mline**

I have highlighted the command, "mline", in **boldface**. The ^C and _ (underscore) are control characters.

The ^C imitates pressing the **ESC** key on the keyboard to cancel any command in progress. The carat (^) alerts AutoCAD that C is a control character, and not a command. There are two ^Cs because it takes two presses of **ESC** to cancel some commands. Strictly speaking, there should be three, because some commands, like **PEDIT**, have options that go three levels deep.

(Historically, in versions of AutoCAD older than Release 13, you pressed **CTRL+C** to cancel commands; the C was short for "Cancel." Going back further in history to the days of telegraphs and teletypes, ASCII code 003 was used to indicated ETX — end-of-transmission. More than one hundred years later, ASCII codes are still in use, because programmers used teletype terminals to communicate with the earliest computers, and because some things never change. The letter C is the third letter of the alphabet, hence control-character 3 became **CTRL+C**. The use of control codes has lead to some clashes: ever since Windows was developed, **CTRL+C** came to mean "copy to Clipboard," but for compatibility reasons continues to mean "cancel" in macros used with AutoCAD .)

In addition to ^C, AutoCAD recognizes other control characters, including some not documented by Autodesk.

Control Character	*Meaning*
;	Same as pressing **ENTER**.
_	Internationalized commands.
'	Transparent command.
\	Pause for user input.
@	Read value stored in **LASTPOINT**.
$M=	Prefix for Diesel expressions.
Multiple	Repeats macros.
^[	Escape.
^A	Toggles group mode, like pressing **CTRL+A**.
^B	Toggles snap mode between on and off (**CTRL+B**).
^C	Cancels current command (**ESC**).
^D	Changes coordinate display (**CTRL+D**).
^E	Switches to next isometric plane (**CTRL+E**).
^G	Toggles grid display (**CTRL+G**).
^H	Backspaces (**CTRL+H**).
^I	Same as pressing **TAB**.
^M	Same as pressing **ENTER**.
^O	Toggles ortho mode (**CTRL+O**).
^P	Toggles **MENUECHO** system variable.
^Q[1]	Echoes prompts to the printer.
^T	Toggles tablet mode (**CTRL+T**).
^V	Switches to next viewport (**CTRL+V**).
^U[1]	Toggle polar mode (**CTRL+U**).
^W[1]	Toggle object snap tracking (**CTRL+W**).
^X[1]	Deletes, like pressing **DEL**.
^Z	Suppresses automatic **ENTER** at the end of a macro.

[1] Not documented by Autodesk.

When you want commands to operate transparently in macros, prefix them with the ' apostrophe, and *don't* prefix them with the ^C control characters.

The other control character in the **^C^C_mline** macro is the *underscore* (a.k.a. underline character). This is an Autodesk convention that *internationalizes* the command. You should always prefix command and option names with the underscore to ensure that your English-language macros works with the German, Japanese, Spanish, and other versions of AutoCAD.

And "mline" is the command name. In macros, you enter AutoCAD commands and their options in exactly the same way you type them on the keyboard at the 'Command:' prompt. You can use aliases instead of the full command name, but I discourage this practice, because aliases can change, leading to non-operational macros.

Nothing is needed to terminate the macro; AutoCAD automatically does the "press **ENTER**" for you. If, however, you want to suppress the automatic **ENTER**, add the ^Z control character to the end of the macro, like this:

```
^C^C_mline^Z
```

3. Before modifying the Multiline macro, think about it: you want the layer to change to one called "Multilines." What would the macro look like? One way to work out the macro code is to follow these steps:

   a. Enter the commands at the keyboard:

      Command: **-layer**

      Current layer: "0"

      Enter an option

      [?/Make/Set/New/ON/OFF/Color/Ltype/LWeight/Plot/Freeze/Thaw/LOck/Unlock/stAte]: **m**

      Enter name for new layer (becomes the current layer) <0>: **multilines**

      Enter an option

      [?/Make/Set/New/ON/OFF/Color/Ltype/LWeight/Plot/Freeze/Thaw/LOck/Unlock/stAte]:
      (Press ENTER.*)*

   b. Press **F2** to flip to the Text window to see all command prompts, and then write out your entries:

      -layer

      m

      multilines

      (Press ENTER.*)*

   c. Convert your entries into macro code. AutoCAD uses the semicolon (;) as the equivalent to pressing the **ENTER** key.

      ^C^C_-layer;m;multilines;;

      Two semicolons are needed at the end of the macro to exit the **-LAYER** command.

   d. Enter this code in the **Button Properties** tab of the Customize dialog box.

      Macro associated with this button: ^C^C_**-layer;m;multilines;;**_mline

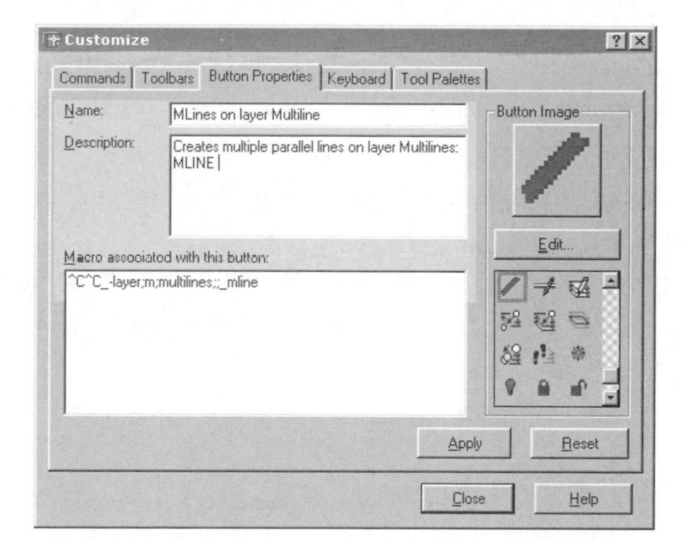

Editing the button's macro.

Notes: To make macros pause for user input, use the backslash symbol (\ ). In the following macro, the **BREAK** command waits for the user to select an object before continuing with the **First point** option:

^C^C_break _f \@

The at symbol (@) accesses the coordinates stored in the **LASTPOINT** system variable. To repeat macros, use the **MULTIPLE** command. (To exit the repeating macro, press **ESC**.) This macro also illustrates using the semicolon (;) as the **ENTER** or spacebar.

^C^C_**multiple**;_line

4. You may want to update the Name and Description to let users know that this button draws mlines on layer Multiline:

   Name: **MLines on layer Multiline**

   Destription: **Creates multiple parallel lines on layer Multilines: MLINE**

5. Click **Apply**, and then **Close**.
6. Test the modified button in two ways:

   a. Pause the cursor over the Multiline button. Take note of the wording in the tooltip and on the status line. Is it correct?

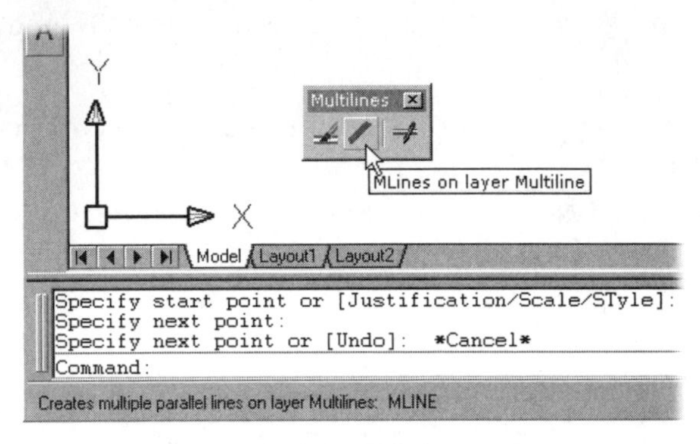

Testing the modified macro.

   b. Click the Multiline button. Does the macro execute correctly? The layer should change to "Multilines," and the **MLINE** command should start up. In the 'Command:' prompt area, you should see the following:

   Command: _-layer

   Current layer: "multilines"

   Enter an option

   [?/Make/Set/New/ON/OFF/Color/Ltype/LWeight/Plot/Freeze/Thaw/LOck/Unlock/stAte]: m

   Enter name for new layer (becomes the current layer) <multilines>: multilines

   Enter an option

   [?/Make/Set/New/ON/OFF/Color/Ltype/LWeight/Plot/Freeze/Thaw/LOck/Unlock/stAte]:

   Command: _mline

   Current settings: Justification = Top, Scale = 1.00, Style = STANDARD

   Specify start point or [Justification/Scale/STyle]: *(Pick a point.)*

   Specify next point: *(Pick another point.)*

   Specify next point or [Undo]: *(Press* ESC *to exit the command.)*

Notes: Press **CTRL+0** (zero, not O) to maximize AutoCAD's drawing area. This shortcut for the **CLEANSCREENON** command removes the title bar and all toolbars. Press **CTRL+0** a second time to return the AutoCAD window to normal.

To turn on (or off) all toolbars at once, use the undocumented **-TOOLBAR** command, as follows:

Command: **-toolbar**

Enter toolbar name or [ALL]: **all**

Enter an option [Show/Hide]: **s**

This command also turns on and off individual toolbars, which can be of use in a macro or AutoLISP routine.

DIESEL EXPRESSIONS

The purpose of Diesel, a simple programming language, is to customize the status line. Since its introduction more than a decade ago, Diesel has found its way into toolbar macros, and established itself as the only programming environment in AutoCAD LT.

Diesel has an unusual format for a macro language. Each function begins with the dollar sign and a bracket:

$(*function,varicble*)

The '$' alerts AutoCAD that a Diesel expression is on the way. The opening and closing parentheses signal the beginning and end of the function and its variables. Parentheses allow Diesel functions to be *nested*, where the result of one function is evaluated by the previous function.

Diesel consists of 28 functions names. All take at least one variable, and some as many as nine. A comma always separates function names from their variable(s). Diesel tolerates no spaces.

TUTORIAL: CUSTOMIZING THE STATUS LINE

Before showing how Diesel can be used in toolbar macros, here is an introduction to its intended purpose, changing the status line.

1. To customize the status line, enter the **MODEMACRO** system variable at the 'Command:' prompt, and then type a line of text:

   Command: **modemacro**

   New value for MODEMACRO, or . for none <"">: **Using AutoCAD 2004** *(Press* ENTER.*)*

   Notice that the words "Using AutoCAD 2004" appear next to the coordinate display on the status line.

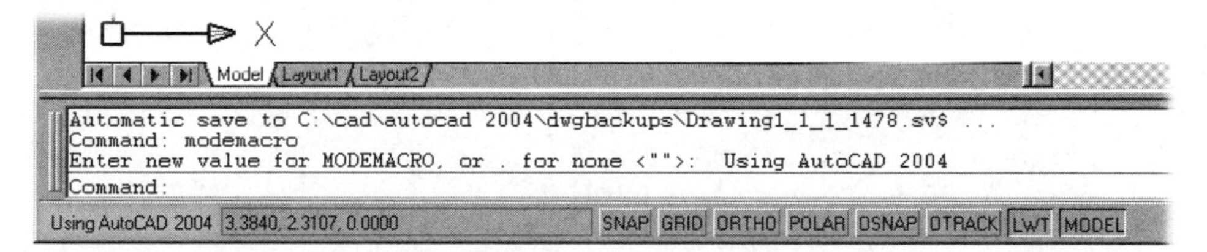

ModeMacro adds text to the status line.

> **HISTORY OF DIESEL**
>
> From the launch of AutoCAD v1.0 and for a dozen years following, AutoCAD's status line remained largely unchanged. Through to Release 11, it displayed **O** to indicate that ortho mode was on, **S** snap on, and **T** tablet on, as well as the x,y coordinates, and the layer name. Newer additions included **P** (paper space) and a square indicating the current color.
>
> The seven indicators were not enough; users wanted more information, such as the missing z coordinate. Autodesk's response was along the lines of, We wouldn't be able to add more without missing out on what some other user wants, so it's best we do nothing. (Some display drivers allowed users to customize the status line to a limited extent.)
>
> With AutoCAD Release 12, Autodesk made the status line fully customizable through Diesel, short for "direct interactively evaluated string expression language." Since then, Diesel has remained unchanged. Most users are unaware of Diesel, and hence fail to employ it.

Functions Supported by Diesel

Math	***Functions***
+	Addition.
-	Subtraction.
*	Multiplication.
/	Division.

Logic	***Function***
=	Equal.
<	Less than.
>	Greater than.
!=	Not equal.
<=	Less than or equal.
>=	Greater than or equal.
and	Logical bitwise AND.
eq	Determines if all items are equal.
if	If-then.
or	Logical bitwise OR.
xor	Logical bitwise XOR (exclusive OR).

Conversion	***Function***
angtos	Converts number to angle format.
fix	Converts real number to an integer.
rtos	Converts number to units format.

Strings	***Function***
index	Extracts an element from comma-separated series.
nth	Extracts *n*th element from one or more items.
strlen	Returns the number of characters of the string.
substr	Returns a portion of a string.
upper	Converts text string to uppercase characters.

System	***Function***
edtime	Displays time based on a specified format.
eval	Passes a string to Diesel.
getenv	Gets a variable from the system registry.
getvar	Gets a system variable.
linelen	Returns the length of the status area.

Error Message	***Meaning***	***Example***
$?	Right parentheses or left quotation is missing.	**$(+,1,2** *or* **$(eq,"To**
$(func)??	Incorrect function name entered.	**$(stringlenth, ...)**
$(func,??)	Incorrect number of arguments.	**$(if)**
$(++)	Output string too long.	...

2. To restore the status line, type the **MODEMACRO** system variable with a pair of double quotes: "" (a.k.a null string).

   Command: **modemacro**

   New value for MODEMACRO, or . for none <"Tailoring AutoCAD">: **""**

3. To display useful information on the status line, use system variables with the **$(getvar** function. This function gets the values of system variables, and displays their values on the status line.

   For example, to add the current thickness setting to the status line, enter the following:

   Command: **modemacro**

   New value for MODEMACRO, or . for none <"">: **$(getvar,thickness)**

   AutoCAD displays 0 (or another value) on the status line.

4. The 0 by itself is not informative. Other users would have no idea what it means. To add an explanation, prefix the macro with text:

   Command: **modemacro**

   New value for MODEMACRO, or . for none <"">: **Thickness = $(getvar,thickness)**

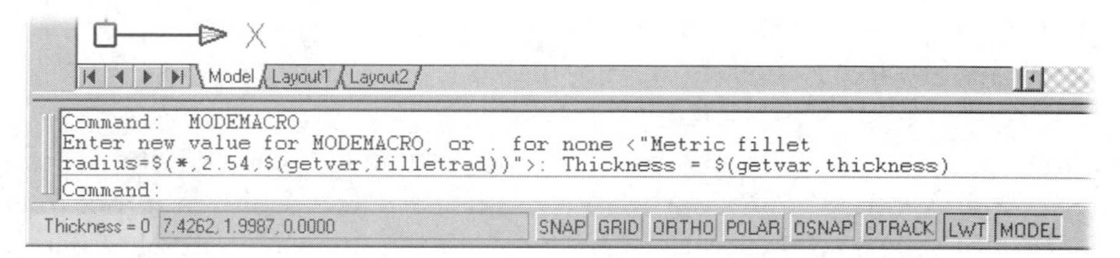

Thickness displayed on status line.

   AutoCAD updates the status line display whenever the system variable changes.

5. Change the thickness with the **ELEV** command.

   Command: **elev**

   Specify new default elevation <0.0000>: **5**

   Specify new default thickness <0.0000>: **10**

   Notice that the value of thickness is updated on the status line, while the z coordinate displays the value of the elevation.

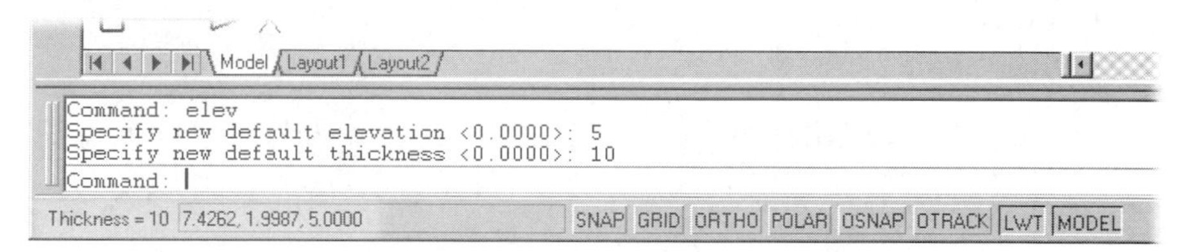

Thickness (and elevation) updated on status line.

6. Macros that perform calculations are possible with Diesel. For example, this one converts the value of the fillet radius to metric units:

   Command: **modemacro**

   New value for MODEMACRO, or . for none <"">: **Metric fillet radius = $(*,2.54, $(getvar,filletrad))**

AutoCAD initially displays 0 on the status line. Each time you change the value of the fillet radius — either through the **FILLET** command's **Radius** option or the **FILLETRAD** system variable — the value is updated on the status line.

Status line reports fillet radius in metric units.

Diesel in Toolbars

To use Diesel code in toolbar macros, prefix the code with the **$M=** control. This alerts AutoCAD that a Diesel function is coming. You can string together commands and Diesel expressions to create really complex toolbar macros. For example, Autodesk wrote the following macro for the RefSet button:

```
$M=$(if,$(eq,$(getvar,refeditname),""), ^C^C^P(ai_refedit_alert);
    ^P,$(if,$(and,$(=,$(getvar, tilemode),0),$(=,$(getvar,cvport),1)),
    ^C^C_refset;,^C^C_refset _add;))^Z
```

Complex looking, isn't it? Below, I've *parsed* the code, and added color to differentiate the Diesel functions. In addition, AutoCAD commands and system variables are shown in **boldface**. ("Parsing" shows the structure of programming code.) Each indentation indicates a level of parentheses.

Diesel Code	Comments
$M=	Start Diesel macro.
$(if,	If ...
$(eq,	... equals ....
$(getvar,**refeditname**),	... system variable REFEDITNAME ...
""),	... nothing (*not named*).
^C^C^P(**ai_refedit_alert**);^P,	Then run *ai_refedit_alert* routine.
$(if,	If ...
$(and,	... and ....
$(=,	... equals ....
$(getvar,**tilemode**),	... system variable TILEMODE ...
0),	... 0 (*AutoCAD is in paper space*) ...
$(=,	... equals ....
$(getvar,**cvport**),	... system variable CVPORT ...
1)	... 1 (*AutoCAD is in layout mode*)
),	(*End of* and *statement*).
^C^C_**refset**;,^C^C_**refset _add**;	Then run REFSET, wait for user to select a reference, and run REFSET **Add**.
)	(*End of if statement*).
)	(*End of first if statement*).
^Z	Suppress ENTER.

DEBUGGING DIESEL

The undocumented **MACROTRACE** system variable tracks down errors in Diesel macros. To start debugging, set it to 1, and then execute the **MODEMACRO** command.

```
Command: macrotrace
New value for MACROTRACE <0>: 1
```

AutoCAD displays the step-by-step evaluation of the macro in the Text window. Here is an example using the $(*,2.54, $(getvar,filletrad)) macro:

```
Command: modemacro
New value for MODEMACRO, or . for none <"">: $(*,2.54,$(getvar,filletrad))
Eval: $(*, 2.54, $(getvar,filletrad))
Eval: $(GETVAR, filletrad)
===> 0.5
===> 1.27
```

Turn off **MACROTRACE** when you no longer need it.

```
Command: macrotrace
New value for MACROTRACE <1>: 0
```

DIESEL PROGRAMMING TIPS

Each argument must be separated by a comma; there must be no spaces within the expression.

The maximum length of a Diesel macro is approximately 460 characters; the maximum display on the status line is roughly 32 characters.

To prevent evaluation of a Diesel macro, use quoted strings, such as "$(+,1)".

To display quotation marks on the status line, use double quotation marks, such as ""Test"".

The purpose of this routine is to prevent use of the RefEdit button when reference editing is already underway. The (ai_refedit_alert) routine displays the warning message "** Command not allowed unless a reference is checked out with REFEDIT command **" when you select an object. The value of system variable **CVPORT** = 1 when AutoCAD is in paper space mode.

(Curiously, buttons on the RefEdit toolbar cannot be edited, but if you edit the *acad.mns* menu source file, you can edit the toolbar.)

EXERCISES

1. Create a toolbar with the following import and export commands:

Export	*Import*
AicsOut	AcisIn
DxfOut	DxfIn
WmfOut	WmfIn
3dsOut	3dsIn

2. Create icons for the commands above. Hint: Think of common symbols that represent "export" and "import."

3. Write toolbar macros that for the following tasks:
   a. Draws center lines (lines with Center linetype) on layer "Center".
   b. Saves the drawing, and then plots it.
   c. Zooms to the selected object.

4. AutoCAD uses the following macro for the **New** button:

   ^C^C_**qnew**

   Analyze the meaning of each portion of the code, as parsed below.

   ^C^C ______________________________

   _ ______________________________

   qnew ______________________________

5. The following toolbar macro is used for the **Break at Point** button:

   ^C^C_**break** _**f** \@

   Analyze the meaning of the codes parsed below.

   _break ______________________________

   _f ______________________________

   \@ ______________________________

6. The following toolbar macro is used for the **Properties** button.

```
$M=$(if,$(and,$(>,$(getvar,opmstate),0)),^C^C_propertiesclose,^C^C_properties)
```

Analyze the meaning of each portion of the code, as parsed below.

```
$M=                             ______________________
$(if,                           ______________________
$(and,                          ______________________
$(>,                            ______________________
$(getvar,opmstate),0)),         ______________________
^C^C_propertiesclose,           ______________________
^C^C_properties)                ______________________
```

Hint: **OPMSTATE** is an undocumented system variable that reports the status of the Properties window:

OpmState	*Meaning*
0	Properties window closed.
1	Properties window open.

7. AutoCAD uses this macro for the Copy button on the Modify toolbar.

```
$M=$(if,$(eq,$(substr,$(getvar,cmdnames),1,4),grip),_copy,^C^C_copy)
```

a. Parse the macro.
b. Explain each line of the parse.
c. Briefly describe the purpose of this macro.

8. Using the **Button Properties** tab of the Customize dialog box, find which other buttons have macros similar to that in exercise #7.
Why are the macros are similar?

CHAPTER REVIEW

1. Briefly explain the purpose of *macros*.
2. What is the caret (^) symbol used for?
3. What was the original purpose of Diesel?
4. Why should you be careful when deleting toolbar buttons.
5. Can you create your own icons for toolbar buttons?
6. After creating a new toolbar macro, it is a good idea to test it out?
7. Explain the meaning of the following control characters:

   ;
   ^C
   ^G
   ^Z
8. What characters are at the start of every Diesel expression?
9. Can Diesel perform calculations?
10. Name the command that helps you debug Diesel expressions.

Accelerators, Aliases, and Scripts

AutoCAD provides several alternatives to specifying commands: menus, toolbars, the status bar, mouse buttons, tablets, and the keyboard. Some of these, such as menus and toolbars, group commands into logical structures, but at the expense of slower access (clicking through menus and puzzling over the meaning of button icons). Others limit the customization possibilities, such as mouse buttons and the status line.

In contrast, power users know that the keyboard can be one of the quickest ways to enter commands — once you have memorized the shortcuts. In this chapter, you learn these alternatives:

Accelerators execute commands in combination with function and control keys.

Aliases reduce the number of keystrokes needed to enter commands.

Scripts combine several commands and their options into a single operation.

FINDING THE COMMANDS

On the **TOOLS** and **EXPRESS** menus:

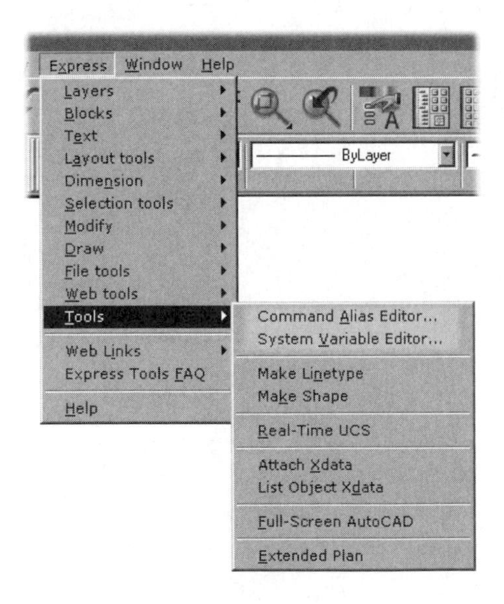

ACCELERATORS

To execute commands through shortcut keystrokes (a.k.a. accelerators), you press the assigned key combinations on the keyboard. For some users, this is faster than selecting the command from the menu or toolbar, or typing it completely at the keyboard. For instance, press **F1** to display the Help window, and press **CTRL+S** to save the drawing.

In addition to predefining numerous shortcut keystrokes (see Appendix A), AutoCAD permits you to assign additional ones. You can create and edit keystroke shortcuts for 188 key combinations, by my count.

- **Function keys** operate by pressing keys marked with the "F" prefix, such as **F2** and **F12**.
- **Control keys** operate by holding down the **CTRL** key, and then pressing a function key, a number, or a letter, such as **CTRL+F2**, **CTRL+6** and **CTRL+A**.
- **Shifted function keys** operate by holding down the **SHIFT** key, and then pressing a function key.
- **Control + Shift keys** operate by holding down both the **CTRL** and **SHIFT** keys, and then pressing a function, number, or letter key.
- **Control + Alternate keys** operate by holding down both the **CTRL** and **ALT** keys, and then pressing a number or letter key.
- **Control + Alternate + Shift** keys operate by holding down the **CTRL**, **ALT**, and **SHIFT** keys, and then pressing a number or letter key.

Only commands can be assigned, not macros (multiple commands), or the options of commands. For example, you can assign the **ZOOM** command, but not the **ZOOM Window** command-and-option.

Note: Do not redefine these keys used by Windows:

F1 displays help.
ALT+F4 exits AutoCAD.
CTRL+F4 closes the current window.
CTRL+F6 changes focus to the next window.
CTRL+C copies to the Clipboard.
CTRL+V pastes from the Clipboard.
CTRL+X cuts to the Clipboard.

AutoCAD 2005 makes the following additions to accelerators:

CTRL+4 toggles the display of the Sheetset Manager window; equivalent to the **SHEETSET** and **SHEETSETHIDE** commands.

CTRL+5 toggles the display of the Info Palette window; equivalent to the **ASSIST** and **ASSISTCLOSE** commands.

CTRL+7 toggles the display of the Markup Set Manager window; equivalent to the **MARKUP** and **MARKUPCLOSE** commands.

TUTORIAL: DEFINING SHORTCUT KEYSTROKES

AutoCAD has no shortcut for letting you get to the **Keyboard** tab of the **CUSTOMIZE** command quickly. In this tutorial, you assign it to CTRL+ALT+K:

1. To customize shortcut keystrokes, start the **CUSTOMIZE** command:
   - From the **Tools** menu, choose **Customize**, and then **Keyboard**.
   - At the 'Command:' prompt, enter the **customize** command.

     Command: **customize** *(Press* ENTER.*)*

   AutoCAD displays the Customize dialog box. If necessary, choose the **Keyboard** tab.

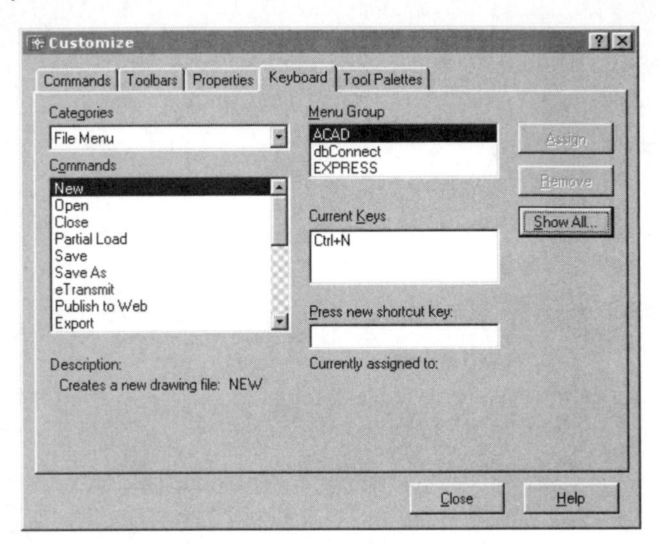

Keyboard tab of the Customize dialog box.

2. In the **Keyboard** tab, click the Categories list, and then select **Tools Menu**.

Notice that AutoCAD's commands are sorted three ways in the Categories list:

- **Alphabetical order**: select "AutoCAD Commands" (found at the end of the list).
- **Menu order**: select a menu name, such as "File Menu."
- **Toolbar order**: select a toolbar name, such as "Dimension Toolbar."

Notice that the list also includes the names of custom menus and toolbars you may have created.

As you select items from the Categories list, the Commands list shows the group of commands associated with the category. For example, the "File Menu" category shows **New**, **Open**, **Close**, and so on.

Note: In most cases, commands are listed with descriptions of their options. For example, the **UCS** command is listed with "3 Point UCS," "X Axis Rotate UCS," and so on. This is misleading, because AutoCAD does not assign options to accelerators — only commands — in most cases. (There are a few exceptions, as we see later.) If you were to assign the "3 Point UCS" command to **CTRL+ALT+U**, AutoCAD would execute only the **UCS** command, and not its **3point** option.

3. In the Commands list, select **Customize Keyboard**.
   In long lists, like this one, the fast way to get to a particular item is to press the key corresponding to the first letter. To get to **Customize Keyboard** quickly, press "C."

4. Look at the **Current Keys** text box to see if a shortcut has already been assigned.
   (If you wish to change an existing shortcut, select it in the **Current Keys** text box, and then click **Remove**. You cannot edit it directly, even though the text box is white, the sign of editable text.)
5. Click the cursor in the **Press new shortcut key** text box.
6. Press the key combination: hold down the **Ctrl** and **Alt** keys, and then press **K**.
7. Click **Assign**.
   Notice that CTRL+ALT+K appears in Current Keys list.
8. Click **Close** to dismiss the dialog box.
9. Test the keystroke shortcut by holding down the CTRL and ALT keys, and then pressing **K**.

DEFAULT ACCELERATORS

AutoCAD provides a handy list of all keyboard shortcuts that have been assigned to function and control keys — but it's not so handy accessing it. Here's how:

1. From the **Tools** menu, select **Customize** and then **Keyboard**.
2. In the dialog box, click **Show All**. AutoCAD displays the Shortcut Keys dialog box.

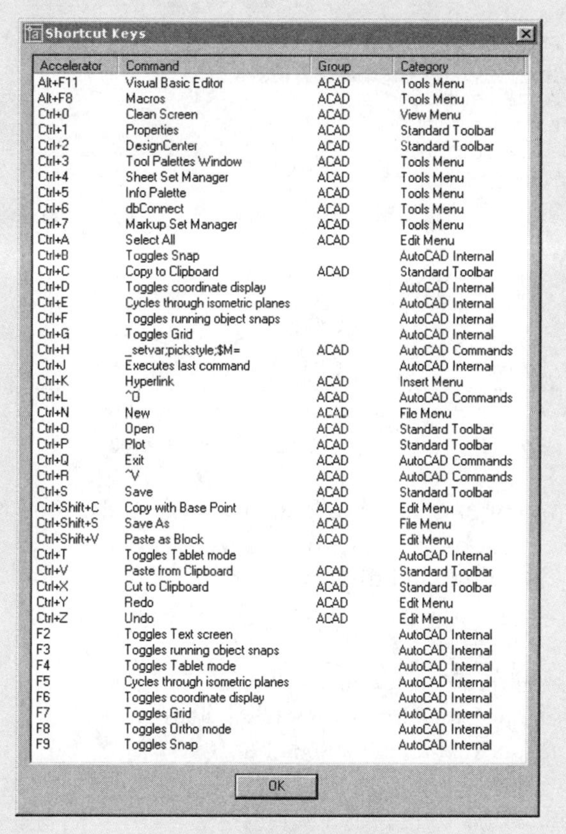

Accelerator	Command	Group	Category
Alt+F11	Visual Basic Editor	ACAD	Tools Menu
Alt+F8	Macros	ACAD	Tools Menu
Ctrl+0	Clean Screen	ACAD	View Menu
Ctrl+1	Properties	ACAD	Standard Toolbar
Ctrl+2	DesignCenter	ACAD	Standard Toolbar
Ctrl+3	Tool Palettes Window	ACAD	Tools Menu
Ctrl+4	Sheet Set Manager	ACAD	Tools Menu
Ctrl+5	Info Palette	ACAD	Tools Menu
Ctrl+6	dbConnect	ACAD	Tools Menu
Ctrl+7	Markup Set Manager	ACAD	Tools Menu
Ctrl+A	Select All	ACAD	Edit Menu
Ctrl+B	Toggles Snap		AutoCAD Internal
Ctrl+C	Copy to Clipboard	ACAD	Standard Toolbar
Ctrl+D	Toggles coordinate display		AutoCAD Internal
Ctrl+E	Cycles through isometric planes		AutoCAD Internal
Ctrl+F	Toggles running object snaps		AutoCAD Internal
Ctrl+G	Toggles Grid		AutoCAD Internal
Ctrl+H	_setvar;pickstyle;$M=	ACAD	AutoCAD Commands
Ctrl+J	Executes last command		AutoCAD Internal
Ctrl+K	Hyperlink	ACAD	Insert Menu
Ctrl+L	^O	ACAD	AutoCAD Commands
Ctrl+N	New	ACAD	File Menu
Ctrl+O	Open	ACAD	Standard Toolbar
Ctrl+P	Plot	ACAD	Standard Toolbar
Ctrl+Q	Exit	ACAD	AutoCAD Commands
Ctrl+R	^V	ACAD	AutoCAD Commands
Ctrl+S	Save	ACAD	Standard Toolbar
Ctrl+Shift+C	Copy with Base Point	ACAD	Edit Menu
Ctrl+Shift+S	Save As	ACAD	File Menu
Ctrl+Shift+V	Paste as Block	ACAD	Edit Menu
Ctrl+T	Toggles Tablet mode		AutoCAD Internal
Ctrl+V	Paste from Clipboard	ACAD	Standard Toolbar
Ctrl+X	Cut to Clipboard	ACAD	Standard Toolbar
Ctrl+Y	Redo	ACAD	Edit Menu
Ctrl+Z	Undo	ACAD	Edit Menu
F2	Toggles Text screen		AutoCAD Internal
F3	Toggles running object snaps		AutoCAD Internal
F4	Toggles Tablet mode		AutoCAD Internal
F5	Cycles through isometric planes		AutoCAD Internal
F6	Toggles coordinate display		AutoCAD Internal
F7	Toggles Grid		AutoCAD Internal
F8	Toggles Ortho mode		AutoCAD Internal
F9	Toggles Snap		AutoCAD Internal

Shortcut Keys dialog box.

Unfortunately, you cannot copy this list to the Clipboard. Instead, press ALT+PRTSCR to copy the dialog box to the Clipboard. Paste (with CTRL+V) the image into a paint program, and then print it.

If all goes well, AutoCAD executes the **CUSTOMIZE** command, and opens the dialog box at the **Keyboard** tab. The following is displayed on the command line:

```
Command: _+customize
Tab index <0>: 3
```

The + prefix can be used with commands that display tabbed dialog boxes. It forces the command to display a prompt asking which tab ("Tab index") to display. The first tab is #0; requesting tab #3 displays the fourth tab, Keyboard in this case.

It can be hard to know what to do in the Keyboard tab, because of the cluttered design. The figure below illustrates the order in which to carry out the steps for assigning keystroke shortcuts.

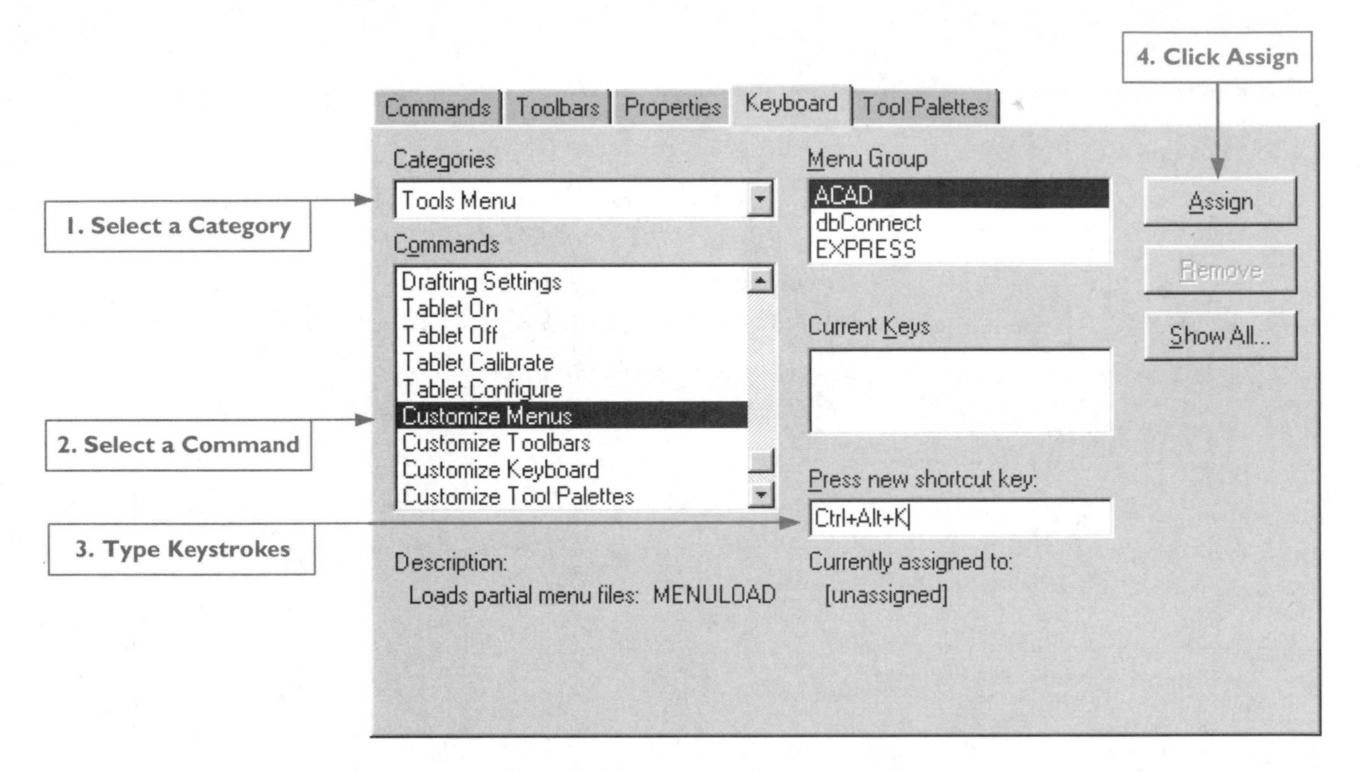

Sequence for keystroke assignments.

Although the tutorial above emphasized shortcuts for commands, the Command window has its own set of shortcut keystrokes, what are not customizable. Pressing the ↑ key repeatedly, for example, displays previous lines in the command history.

Keystroke	*Meaning*
Enter	Executes and repeats commands.
↑	Displays the previous line in command history.
↓	Displays the next line in command history.
←	Moves the cursor to the left.
→	Moves the cursor to the right.
Home	(Home) Places the cursor at beginning of the line.
End	(End) Places the cursor at end of the line.
Insert	(Insert) Toggles insertion mode.
Delete	(Delete) Erases characters to the right of cursor.
Ctrl+v	Pastes text from Clipboard.

COMMAND ALIASES

You may have become familiar with the reference "Alternatively, enter the alias at the keyboard" throughout Volumes 1 and 2 of this book. Aliases typically abbreviate command names, such as "C" for the CIRCLE command, and "CP" for COPY. You enter the one or two-letter abbreviation, and then press ENTER.

Command: **cp** (*Press* ENTER.)

This is a faster way to enter command names — provided you can remember the abbreviation. For instance, I usually enter **m** for "move," but always forget that **c** is short for "circle," not "copy."

Autodesk assigned 200+ commands and system variables to aliases in the *acad.pgp* file. (The complete list is provided in Appendix A.) AutoCAD allows you to create and edit aliases of your own. In addition, you can assign aliases to system variables. Like keyboard shortcuts, aliases work only with command names; they cannot include command options. You can assign an alias to ZOOM, but not to ZOOM **Window**. (The options of commands are already shortened; for example, you typically enter **l**, not "last.")

There are two ways to create and edit aliases: using a dialog box, and editing the *acad.pgp* file. To edit aliases through a dialog box, you must ensure that the Express Tools are installed with AutoCAD 2005.

AutoCAD 2005 adds these aliases:

msm — MARKUP **ssm** — SHEETSET

tb — TABLE **ts** — TABLESTYLE

TUTORIALS: CREATING NEW ALIASES

For this tutorial, you add the **cs** alias for the CUSORSIZE system variable. (This system variable changes the length of the crosshair cursor.)

1. To create command aliases, start the **ALIASEDIT** command:
   - From the **Express** menu, choose **Tools**, and then **Command Alias Editor**.
   - At the 'Command:' prompt, enter the **aliasedit** command.
   - Ironically, there is no alias for the alias editor.

   Command: **aliasedit** *(Press* ENTER.*)*

AutoCAD displays the AutoCAD Alias Editor dialog box.

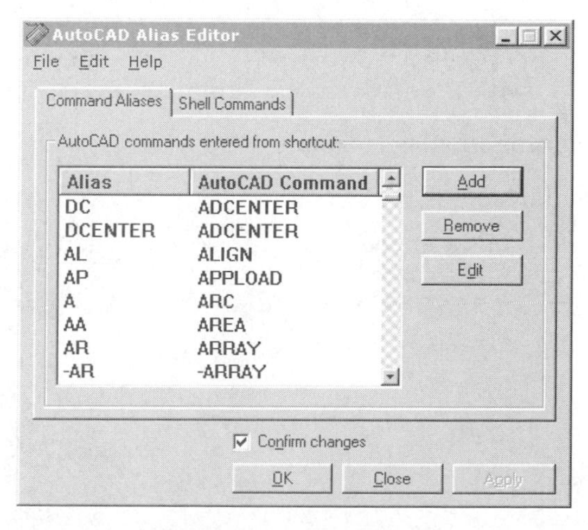

AutoCAD Alias Editor dialog box.

To sort the aliases and command names in alphabetical order, click the header, **Alias** or **AutoCAD Command**.

2. To create a new alias, click **Add**.

   Notice the New Command Alias dialog box.

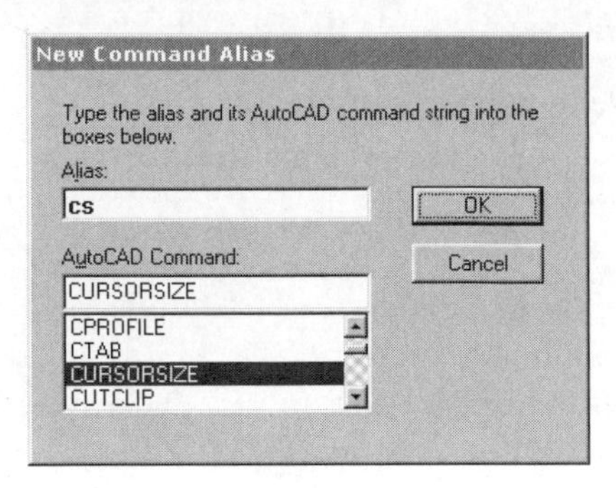

New Command Alias dialog box.

3. In the **Alias** text box, type an alias, such as **cs**.

   Alias: **cs**

   Under AutoCAD Command, you can type a command name:

   AutoCAD Command: **cusorsize**

   Or else select it from the list. The list does not include all of AutoCAD's commands, but does include many system variables.

4. Click **OK**. If you enter an alias identical to an existing one, AutoCAD warns that you are redefining it.

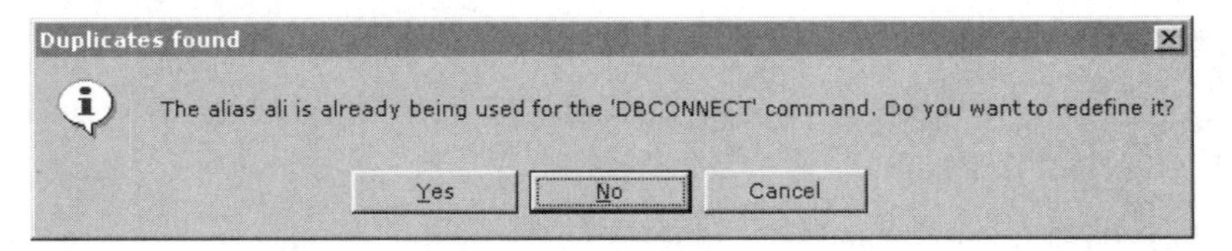

Duplicates Found dialog box.

5. Notice that the new alias is listed in the dialog box. Click **OK** to exit the editor.

   AutoCAD asks if you want to overwrite the *acad.pgp* file.

   Click **Yes**.

   AutoCAD confirms that the list of aliases has been updated.

If you prefer to edit the *acad.pgp* file directly, follow these steps. (PGP is short for "program parameters," but has been known informally as the "pig pen" file.)

The format for defining a command alias is simple:

```
alias, *fullCommandName
```

The alias is followed by a comma, a space, an asterisk, and then the complete command name, such as:

```
L, *LINE
```

For this tutorial, you add the **lin** alias for the **DIMLINEAR** command.

1. To open the *acad.pgp* file, select the **Tools** menu, choose **Customize**, **Edit Custom Files**, and then **Program Parameters (acad.pgp)**.
   Notice that the Notepad text editor opens with the file:

```
acad.pgp - Notepad
File  Edit  Format  Help
; This PGP file was created with the AutoCAD Alias Editor.
; Last modified 10/19/2003   3:24:50 PM

; Use the Alias Editor (EDITALIAS command) to edit this file.

; [Operating System Commands]

CATALOG,   DIR /W,           8,*File specification:,
DEL,       DEL,              8,*File to delete:,
DIR,       DIR,              8,*File specification:,
EDIT,      START EDIT,       9,*File to edit:,
EXPLORER,  START EXPLORER,   1,,
NOTEPAD,   START NOTEPAD,    1,*File to edit:,
PBRUSH,    START PBRUSH,     1,,
SH,        ,                 1,*OS Command:,
SHELL,     ,                 1,*OS Command:,
START,     START,            1,*Application to start:,
TYPE,      TYPE,             8,*File to list:,


; [AutoCAD Command Aliases]

3A,         *3DARRAY
3DO,        *3DORBIT
3F,         *3DFACE
3P,         *3DPOLY
A,          *ARC
```

Notepad text editor with acad.pgp file.

2. Scroll down to the alias section:

   ; [AutoCAD Command Aliases]

   And then scroll further until you reach the dimensioning commands:

   DIMHORIZONTAL, *DIMLINEAR

   DIMLIN, *DIMLINEAR

   DIMORD, *DIMORDINATE

3. Press **ENTER** to make room for another line of text.
4. Enter the following text:

   lin,* dimlinear

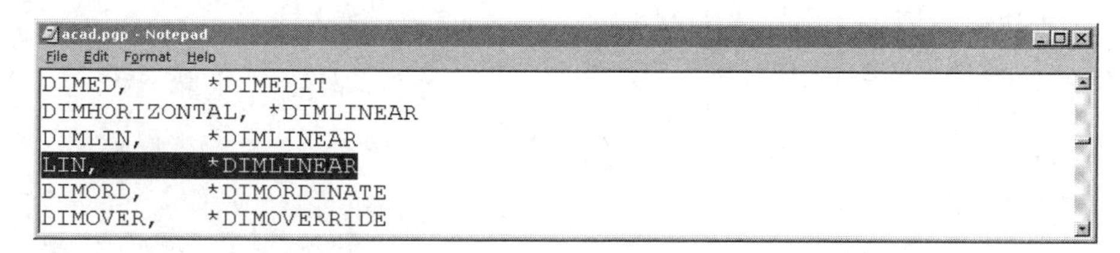

```
acad.pgp - Notepad
File  Edit  Format  Help
DIMED,       *DIMEDIT
DIMHORIZONTAL, *DIMLINEAR
DIMLIN,      *DIMLINEAR
LIN,         *DIMLINEAR
DIMORD,      *DIMORDINATE
DIMOVER,     *DIMOVERRIDE
```

Adding aliases.

5. From the **File** menu, select **Save**.

6. Return to AutoCAD.

   To make the new alias active, start the **REINIT** command, as follows:

   Command: **reinit**

   This reloads the *acad.pgp* file into AutoCAD.

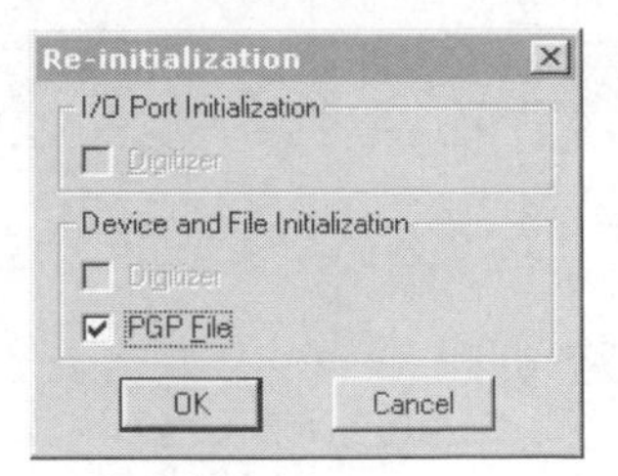

Re-initialization dialog box.

7. Select the **PGP File** option, and then click **OK**.

   AutoCAD reloads the *acad.pgp* file.

8. Test your change by entering "lin" at the 'Command:' prompt. AutoCAD should start the **DIMLINEAR** command:

   Command: **lin**

   DIMLINEAR

   Specify first extension line origin or <select object>:

   Notice that AutoCAD reports the full name of the command in uppercase letters, as in DIMLINEAR.

Autodesk makes the following suggestions for creating command aliases:

- Always try to use the first character of a command. If it is already taken by another command, then use the first two characters. For example, the alias **m** is assigned to the **MOVE** command, and **ml** is assigned to the **MLINE** command.
- For consistency, add suffix characters to related commands. For example, **h** is used for the **HATCH** command, so **he** is used for **HATCHEDIT**.
- You can (but don't have to) include the hyphen prefix for commands that display their prompts at the 'Command:' prompt. For example, use **-la** for the **-LAYER** command.
- Commands with a control key equivalents, status bar buttons, or function keys do not require an alias. Examples include the **NEW** command (already assigned to **CTRL+N**), **SNAP** (already on the status line), and **HELP** (already assigned to function key **F1**).
- Aliases should reduce the length of command names by at least two characters.

SCRIPTS

Scripts do one thing only: they replay commands, options, and responses that you type at the keyboard. Anything that shows at the 'Command:' prompt can be put into script files. That includes AutoCAD commands, their option abbreviations, your responses, and AutoLISP code. Anything you cannot do at the keyboard, including selecting buttons in dialog boxes and toolbars, cannot be included in script files. (Historically, scripts were introduced to AutoCAD with version 1.4 — w-w-a-a-y-y back in 1983.)

The purpose of the script is to reduce keystrokes by placing them in files — a simple form of "toolbar macro." For instance, a script file that saves the drawing and then starts the **PLOT** command looks like this:

```
qsave
plot
```

Script files have an extension of *.scr* and are in plain ASCII text. For that reason, don't use a word processor, such as WordPad, OpenOffice, or Word. Instead, use Notepad to write scripts.

Only one script file can be loaded in AutoCAD at a time. (The lone script can call other script files.) Scripts stall when they encounter invalid command syntax. Scripts cannot record mouse movements, and scripts cannot control dialog boxes.

An advantage of scripts is that they transcend windows. AutoLISP routines, for example, run only in the current drawing window, and terminate when the drawing is closed. A script, however, run at a "lower level" in AutoCAD. This allows scripts to open a series of drawings, plot each one, and then close them. The technical editor notes that this feature makes scripts particularly attractive, because you can take the afternoon off, and the completed plots will be waiting for you when you return to the office.

SCRIPT COMMANDS AND MODIFIERS

AutoCAD has four commands that load and control scripts.

Script

The **SCRIPT** command loads a script file and immediately runs it.

Command: **script**

Enter script file name <C:\AutoCAD\Drawing1.scr>: **filename**

Turning off the **FILEDIA** system variable (setting it to 0) displays prompts at the command line, instead of displaying dialog boxes. This allows scripts to access such commands.

Scripts can be run transparently during other commands. Prefix **SCRIPT** with the apostrophe, like this:

Command: **line**

Specify first point: **'script**

>>Script file: **filename**

AutoCAD reminds you with the double angle bracket (>>) that two commands are on the go. (All four of AutoCAD's script commands are transparent, even '**DELAY**.)

RScript

The **RSCRIPT** command (short for "repeat script") re-runs the script currently loaded in AutoCAD.

Command: **rscript**

Resume

The **RESUME** command resumes paused scripts. (You pause the script by pressing the **BACKSPACE** key.)

Command: **resume**

Delay

The **DELAY** "command" pauses scripts for a set amount of time. (While it is a command, it is never used at the 'Command:' prompt, only in scripts.)

delay 1000

The number specifies the pause in milliseconds, where 1,000 milliseconds equal one second. The minimum delay is 1 millisecond; the maximum 32767 (just under 33 seconds). **DELAY** is used in script files to wait while raster files are displayed, or to slow down portions of script files.

Special Characters

In addition to the script-specific commands, script files recognize some special characters and keys.

(space) Both the spacebar and the **ENTER** are represented by invisible spaces. Some commands required several blanks spaces in a row, representing repeated pressing of the **ENTER** key. For instance, using the **ATTEDIT** command:

```
; Edit the attributes one at a time:
attedit    1,2
```

It hard to count the number of spaces, so I recommend placing each on its own line in the script file, like this:

```
; Edit the attributes one at a time:
attedit





1,2
```

; *(semicolon)* allows comments in script files. AutoCAD ignores anything following the semicolon.

* *(asterisk)* is meant for use with the **VSLIDE** command in scripts. When you prefix it with the *, AutoCAD pre-loads slides for faster viewing:

```
*vslide
```

ESC stops script files. Use the **RSCRIPT** command to start it up again from the beginning.

EXERCISES

1. Assign the accelerators listed below to the following commands:

SHIFT+F1	**CIRCLE**
CTRL+ALT+L	**LINE**
CTRL+SHIFT+A	**ARC**

2. Assign accelerators to commands of your own choosing.

3. Assign the aliases listed below to the following commands:

bl	**BLOCK**
dimhor	**DIMHORIZONTAL**
cad	**CHECKSTANDARDS**

4. Assign aliases to commands of your own choosing.

5. Write a script that purges the drawing.
   Tip: Use the **-PURGE** command.

6. Write a script that draws lines for a title block in the corner of a vertical A-size drawing border.

CHAPTER REVIEW

1. List an advantage and a disadvantage of the following user interface elements:
   a. Accelerators (keyboard shortcuts)
   b. Menu bar
   c. Toolbar buttons
   d. Aliases
   e. Status bar
   f. Mouse buttons
2. What kind of keys do accelerators work with?
3. Describe at least two differences between accelerators and aliases.
4. Explain the purpose of the following accelerators:
   F1
   CTRL+0
   CTRL+S
   CTRL+SHIFT+S
   CTRL+F6
5. Do accelerators execute command options?
6. Does AutoCAD display a list of all accelerators?
7. Which keyboard key displays earlier commands from the command history?
8. Can you create new aliases?
9. Where are aliases stored?
10. What is the purpose of scripts?
11. Can scripts control dialog boxes?

AutoLISP Programming

Toolbar macros and scripts are easy to create, but they limit your ability to control AutoCAD. The more powerful alternative is AutoLISP. It lets you do things as simple as adding two numbers, and as complex as parametrically drawing staircases in 3D.

In this chapter, you learn how to write programs in AutoLISP, and how to use many AutoLISP functions, including:

Math functions perform mathematical calculations.

Conditional functions make decisions.

String and **conversion** functions manipulate text and convert numbers to text.

Command functions execute AutoCAD commands and load AutoLISP routines.

Get functions obtain data from the user.

Ss functions create and manipulate selection sets.

Ent functions manipulate objects.

ABOUT AUTOLISP

AutoLISP can be brief: you can write a line or two of code that automates your work. AutoLISP can be complex: it can integrate into toolbar macros, incorporate Diesel code, and manipulate data in the drawing database.

Some of AutoCAD's commands, such as **3DARRAY** and **EDGE**, are written as AutoLISP programs. AutoCAD automatically loads the programs when you enter the command names. A search of AutoCAD's folders finds 173 AutoLISP files. (Autodesk also uses the more powerful ObjectARX programming environment for new commands.)

BASIC TUTORIAL: ADDING TWO NUMBERS

AutoLISP can perform tasks as simple as adding two numbers. (In this chapter, AutoLISP functions are highlighted in **color** text.)

1. Start AutoCAD with a new drawing.
2. At the 'Command:' prompt, enter the following:

   Command: **(+ 2 3)** *(Press* ENTER.*)*

   5

   Notice that AutoCAD replies with the answer, 5.

The syntax (+ 2 3) may seem convoluted, because AutoLISP uses *prefix notation*: the *operator* + appears before the *operands*, 2 and 3. Think of "add 2 and 3."

In addition, parentheses surround every AutoLISP statement. (LISP is sometimes said to be short for "lost in stupid parentheses.") Every opening parenthesis — (— requires a closing parenthesis —). Balancing parentheses is the most frustrating aspect to AutoLISP. When you leave out closing parentheses, AutoCAD displays this prompt:

Command: **(+ 2 3** *(Press* ENTER.*)*

1>

The '1>' prompt tells you that a closing parenthesis is missing. Type) and AutoCAD completes the calculation:

1> **)**

5

The parentheses also alerts AutoCAD that you are entering an AutoLISP expression. Try typing the addition expression without parentheses. Notice how AutoCAD reacts:

Command: **+** *(Press spacebar.)*

Unknown command "+". Type ? for list of commands.

Command: **2** *(Press spacebar.)*

Unknown command "2". Type ? for list of commands.

Command: **3** *(Press spacebar.)*

Unknown command "3". Type ? for list of commands.

Without the parentheses, AutoCAD interprets each press of the spacebar as the end of a command. It doesn't know "commands" called **+**, **2**, and **3**.

3. AutoLISP provides the four basic arithmetic functions — addition, subtraction, multiplication, and division. Subtract the two numbers:

   Command: **(- 2 3)**

   -1

   AutoCAD correctly returns negative 1.

HISTORY OF AUTOLISP

LISP was one of the very first computer programming languages, invented by John McCarthy in 1958. The language was designed to work with lists of numbers and words; LISP is short for "list processing." You can read the history of LISP at www-formal.stanford.edu/jmc/history/lisp/lisp.html.

AutoLISP first appeared as an undocumented feature in AutoCAD v2.15, released during the summer of 1985. Autodesk programmers had taken XLISP, a public domain version of LISP written by David Betz, and adapted it for use in AutoCAD. Only third-party developers were told about the new feature.

Finally, in v2.17 the new programming language, initially called "Variables and Expressions," was documented by Autodesk, and made available to all users. Autodesk suggested it could be useful for creating "menu items that perform complex tasks." Early users created gimmicky routines, like repeatedly zooming into a drawing. The earliest releases of AutoLISP were weak, lacking even conditional statements. Clever users found a work-around: they used the **RSCRIPT** command to force the AutoLISP routine to repeat itself.

With AutoCAD v2.5, Autodesk got serious. They renamed the programming language "AutoLISP," and added the powerful **GET** and **SS** routines to directly access objects in the drawing database. Third-party developers wrote complex routines that manipulated entire drawings. Regular users found AutoLISP useful for writing short routines that automated everyday drafting activities.

The drawback to AutoLISP, however, is that it is an *interpreted* language, which means that it runs slowly. This is not an issue for short routines, but does affect large programs. To solve the problem, Autodesk introduced a new programming environment, called ADS, to AutoCAD Release 10 for OS/2. (ADS is short for "AutoCAD development system.") ADS allowed programmers to hook speedy C-language routines into AutoCAD. With Release 13, ADS was replaced by the ObjectARx programming interface.

Later, Autodesk purchased Visual LISP from Bitwise Solutions, which compiles LISP code to run faster. VLISP is now part of AutoCAD, but is not discussed in this book.

AUTOCAD LT

AutoCAD LT users are out of luck. AutoLISP is not available to them. Even though Autodesk had included AutoLISP in the beta versions of LT Release 1, the code was removed just days before the software first shipped to customers. Dealers were worried that the lower-cost LT would be too powerful with AutoLISP, and thus reduce sales of the more expensive AutoCAD.

When you try to use AutoLISP, LT responds: "AutoLISP command not available." This has not stopped third-party developers, however, from adding their own variant of AutoLISP to LT. In addition, several CAD software competitors to AutoCAD include a version of AutoLISP.

4. Multiplication is performed with the asterisk (*) symbol, with which you may be familiar from high school algebra.

   Command: **(* 2 3)**

   6

5. Division is performed with the slash (/) symbol:

   Command: **(/ 2 3)**

   0

Error! Dividing 2 by 3 is 0.666667, not 0. Until now, you have been working with *integer* numbers (also known as "whole numbers"). Integer numbers have no decimal points; notice that the 2 and the 3 have no decimal points. AutoLISP interprets the numbers as integers, and hence returns the result as an integer. (AutoLISP rounds down.)

To work with *real* numbers, add the decimal suffix. This forces AutoLISP to perform real-number division. It's sufficient just to turn one integer into a real number:

Command: **(/ 2.0 3)**

0.666667

AutoLISP returns the correct answer. Real numbers are displayed to six decimal places, even though calculations are performed to 32-bit accuracy.

6. The parentheses allow AutoLISP to *nest* calculations, just as you do in algebra:

   Command: **(+ (- (* (/ 9.0 7) 4) 3) 2)**

   4.14286

TUTORIAL: AUTOLISP IN COMMANDS

AutoLISP can perform calculations within commands. For example, suppose you need to fit thee circles into a 2" space. What diameter should the circles have? This example shows how you can use AutoLISP functions any time AutoCAD expects user input.

1. Start the **CIRCLE** command.

   Command: **circle**

   3P/2P/TTR/<Center point>: *(Pick a point.)*

   Diameter/<Radius>: **d**

2. To determine the diameter, enter the AutoLISP equation, as follows:

   Diameter: **(/ 2.0 3)**

   AutoCAD draws the circle with a diameter of 0.666667 inches.

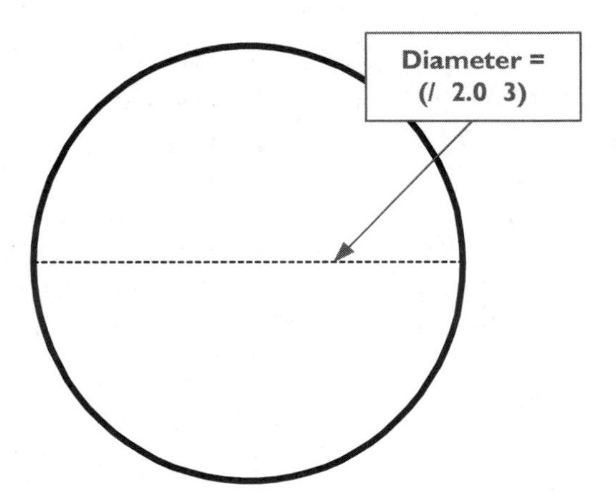

AutoCAD accepts AutoLISP expressions as values.

Note: AutoCAD has an alternative to AutoLISP for performing calculations at the command prompt and during commands. For example, to divide 2 by 3:

Command: **cal**

>> Expression: **2 / 3**

0.666667

The **CAL** command is capable of vector functions, which are lacking from AutoLISP.

SETQ AND **VARIABLES**

In the above tutorials, you enter the **(/ 2.0 3)** equation a couple of times. AutoLISP remembers the result of calculations, just like memory keys on calculators. Arguably the most commonly-used AutoLISP function is **SETQ**, which saves the results of calculations (short for SET eQual).

1. To save the results of calculations, use the **SETQ** function together with a *variable*, such as "x." It is called a "variable," because it can hold any value.

    Command: **(setq x (/ 2.0 3))**

    0.666667

    In this equation, x remembers the result of the (/ 2.0 3) calculation. Notice that an extra set of parentheses are needed to nest the calculation.

2. To prove to yourself that x does contains the value of 0.666667, use AutoCAD's ! (exclamation) prefix, as follows:

    Command: **!x**

    0.666667

    The ! prefix (sometimes called "bang") reports the values held by variables.

You can make up any combination of numbers and letters to create names for variable. Example could include Point2, Dia, and Xvalue — variable names that reflect their content. Names of variables are case insensitive. (You cannot use AutoLISP function names, such as **SETQ**, **T**, and **GETINT**.)

Unlike most other programming languages, AutoLISP does not require you first to "declare" the names and state of variables. Variables are not assigned a *type*; in AutoLISP any variable handles any type, whether integers, reals, strings, or point coordinates.

3. You make one variable equal another:

    Command: **(setq dia x)**

    0.666667

    Command: **!dia**

    0.666667

4. Re-enter the **CIRCLE** command, this time using variable **dia**, as follows:

    Command: **circle**

    3P/2P/TTR/<Center point>: *(Pick a point.)*

    Diameter/<Radius>: **!dia**

    AutoCAD draws the same size circle, precisely.

Note: **SETQ** has the ability to perform multiple "set equals." For example, to assign values to variables x, y, and z:

```
(setq x 1.0 y 2.0 z 3.0)
```

Or, parsing the code to make it more legible:

```
(setq  x 1.0
       y 2.0
       z 3.0)
```

AUTOLISP FUNCTIONS

Autodesk made AutoLISP so powerful that it can manipulate almost any aspect of AutoCAD's drawings. In this chapter, you get a taste of the many different kinds of functions AutoLISP offers you for manipulating numbers, words, and objects.

For a handy reference list, see Austin Community College's "AutoLISP Function Compendium" at www.austincc.edu/edg/fctnlist.html. In addition, you can refer to the on-line help. From the **Help** menu, select **Developer Help**. In the **Contents** tab, open AutoLISP Development Guide, and then open Appendices. Look for "AutoLISP Function Synopsis."

MATH FUNCTIONS

In addition to the four basic arithmetic functions, AutoLISP has many of the mathematical functions you might expect in a programming language. The list includes trigonometric, logarithmic, logical, and bit manipulation functions. (Missing are matrix manipulation functions.) For example, the **MIN** function returns the smallest (minimum) value of a list of numbers:

Command: **(min 7 3 5 11)**

3

To remember the result of this function, use the **SETQ** function:

Command: **(setq minnbr (min 7 3 5 11))**

3

Now each time you want to refer to the minimum value of that series of numbers, you can refer to variable **minnbr**. Here's an example of a trig function, sine:

Command: **(sin minnbr)**

0.14112

Returns the sine of the angle of 3 radians.

Here is the complete list of AutoLISP mathematical functions:

Mathematical Function	***Meaning***
+	Addition.
-	Subtraction.
*	Multiplication.
/	Division.
1+	Increment (adds 1).
1-	Decrement (subtract 1).
FIX	Truncates reals into integers.
FLOAT	Turns integers into reals.
REM	Returns the remainder after division.
EXPT	Exponent.
SIN	Sine of radians.
COS	Cosine of radians.
ATAN	Arctangent of two numbers.

Note: You must provide AutoLISP with angles measured in *radians*, not degrees. This is an inconvenience, because you probably work mostly with degrees. The solution is use AutoLISP to convert degrees to radians, keeping in mind that there are 2*pi radians in 360 degrees. For example, to get the sine of 45°, use this code:

Command: **(setq rad (* (/ 45 180.0) pi))**

0.707107

The 45 degrees are divided by 180 (half of 360), and then multiplied by **pi**. Either the 45 or the 180 needs the decimal (.0) to force division of real numbers. As a convenience, AutoLISP predefines the constant **pi** as 3.14159. To see the value, use the bang:

```
Command !pi
3.14159
```

GEOMETRIC FUNCTIONS

Because AutoCAD works with geometry, AutoLISP has a number of functions that deal with geometry. For example, the **DISTANCE** function is similar to the **DIST** command. First, assign p1 and p2 a pair of points:

```
Command: (setq p1 '(1.3 5.7))
(1.3 5.7)
Command: (setq p2 '(7.5 3.1 11))
(7.5 3.1 11)
```

You may have missed that single quote mark in front of the list of x, y coordinates, as in: '(1.3 5.7). It tells AutoLISP that you are specifying a pair (or triple) of coordinates, and that AutoLISP should not try to evaluate the numbers.

Apply the **DISTANCE** function to find the 3D distance between the two points.

```
Command: (distance p1 p2)
6.72309
```

Note: Don't use commas! Use spaces to separate the values of coordinates. When you leave out the z coordinate, AutoLISP assumes it equals 0.0.

Another geometric function of interest is **ANGLE**, which measures the angle from 0 degrees (usually pointing east) to the line defined by **p1** and **p2**:

```
Command: (angle p1 p2)
5.88611
```

The result is returned in radians.

The intersection of two lines is determined by the **INTERS** function. First, assign coordinates to four variables, and then apply the function:

```
Command: (inters pt1 pt2 pt3 pt4)
```

Object snaps are also available in AutoLISP. The code below shows the x, y, z coordinates that result from applying the MIDpoint object snap to a line defined by **p1** and **p2**.

```
Command: line
From point: !p1
To point: !p2
To point: (Press ENTER.)
Command: (osnap p1 "mid")
(4.4 4.4 5.5)
```

The "**mid**" refers to the MIDpoint object snap mode. Thus, in this function, you are finding the midpoint of the line that starts at **p1** (1.3, 5.7).

The important things to notice about the **(osnap p1 "mid")** statement are: (1) you need to specify only point **(p1)** laying on the line; and (2) the point **(p1)** must lie on an object; otherwise the function fails. AutoLISP cannot find the midpoint of imaginary lines.

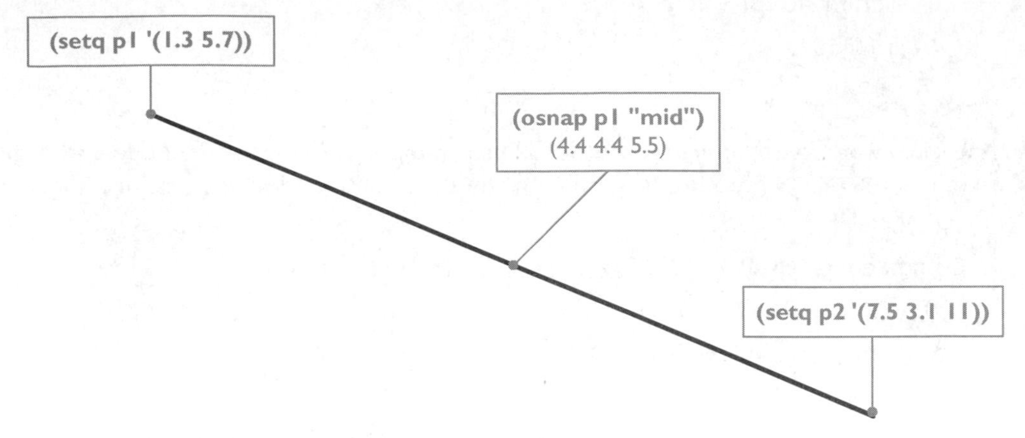

Determining the midpoint of a line using AutoLISP functions.

Other important geometric functions include **TEXTBOX** (finds the rectangular outline of a line of text) and **POLAR** (returns a 3D point of a specified distance and angle).

AutoLISP includes these geometric functions:

Geometric Function	***Meaning***
Angle	Angle (in radians) between the x axis and line defined by two points.
Distance	3D distance between the two points.
Polar	3D point at an angle (in radians).
OSnap	3D point after applying an object snap mode to a point.
Inters	3D point of an intersection defined by four points.

CONDITIONAL FUNCTIONS

Conditional functions could be considered the most important part of any programming language, because they allow computer programs to "think," and therefore make decisions. All conditionals ask a simple question: is the test true or false? As long as the test is true, do one thing; when false, do another thing. Conditional functions check if one value is less than or greater than another, or they repeat an action until something is false.

Consider an AutoLISP routine that draws a staircase. Part of the routine checks if the floor-to-ceiling distance is greater than eight feet. If so, then it draws 14 steps; if not, it draws 13 steps. A similar wording is used in condition functions:

```
Command: (if (> height 96) (setq steps 14) (setq steps 13))
13
```

Parsed, the code looks like this (easier to read):

```
(if  (> height 96)
     (setq steps 14)
     (setq steps 13)
)
```

Let's break this down to see how the function works:

(if	*If*
(>	*greater than*
height	*floor-to-ceiling distance*
96)	*of 8 feet:*
	Then
(setq steps 14)	*use 14 steps.*
	Else
(setq steps 13)	*use 13 steps.*
)	

Notice how the conditional function (>) skips statements:

- If the test is true, then execute the first statement, and skip the second:

  (if (> height 96)
  (setq steps 14)
  ~~(setq steps 13)~~
)

- If the test is false, then skip the first, and execute the second.

  (if (> height 96
  ~~(setq steps 14)~~
  (setq steps 13)
)

You must carefully construct conditional statements, so that the correct statement is executed, depending on whether the test is true or false.

Here is the complete list of conditional functions in AutoLISP:

Conditional Function	***Comment***
=	Equal to.
/=	Not equal to.
<	Less than.
<=	Less than or equal to.
>	Greater than.
>=	Greater than or equal to.
If	If-then-else.
Cond	Looks for true test expression.
Repeat	Repeats expression n number of times.
While	Repeats test expression until false.
ForEach	Evaluates expression for all members of the list.

STRING AND CONVERSION FUNCTIONS

You can manipulate *strings* (one or more characters) in AutoLISP, but to a lesser extent than numbers. For example, you can find the length of a string as follows:

```
Command: (strlen "Using AutoCAD")
13
```

The **STRLEN** (short for STRing LENgth) function reports that "Using AutoCAD" has 13 characters, counting the space.

Notice that "Using AutoCAD" is surrounded by quotation marks; this tells AutoLISP you are working with a string, and not variables called Using and AutoCAD. If you were to enter **(strlen using autocad)**, AutoLISP would attempt to find the length of the strings held by variables **using** and **autocad**, as in the following example:

```
Command: (setq autocad "A software package")
"A software package"
Command: (setq using "the planet Earth")
"The planet earth"
Command: (strlen using autocad)
34
```

Other string functions change all characters to upper or lower case, return part of a string, and find a pattern in a string. The **STRCAT** function (short for "string concantenation") lets you create reports by mixing variables and text, and joining strings together, as follows:

```
Command: (strcat autocad " used all over " using)
"A software package used all over the planet Earth"
```

The complete list of AutoLISP's string functions is:

String Function	***Meaning***
StrCase	Converts all text to upper or lowercase.
StrCat	Concatenates all strings.
StrLen	Counts the number of characters in strings.
SubStr	Returns a portion of a string.
WcMatch	Matches wildcards in strings.

Related to string functions are the *conversion* functions, because they convert numbers to and from strings. For example, recall the code converting degrees to radians, which works well for decimal degrees, such as 45.3711°. But how to convert 45 degrees, 37 minutes and 11 seconds, which AutoCAD represents as 45d37'11"? A conversion function, such as **ANGTOF** (short for "ANGle TO Floating-point") is useful. It converts strings containing formatted angles to radians (real numbers), as this example shows:

```
Command: (angtof "45d37'11\"" 1)
0.796214
```

Notice the use of **d** to indicate degrees, ' to indicate seconds, and \" to indicate seconds. The \" is used so that AutoLISP doesn't get confused with the closing quotation mark ("), which indicates the end of the string.

The function **ANGTOF** performs several kinds of several conversion — one for each type of unit supported by AutoCAD. To indicate which kind of conversion, use a *mode* number, 1 in this case. Several AutoLISP functions use the mode numbers listed here:

Mode	Meaning	Example
0	Decimal degrees	45.3711
1	Degrees-minutes-seconds	45d 37' 11"
2	Grad	100.1234
3	Radian	0.3964
4	Surveyor units	N 45d37'11" E

The mode numbers are identical to the value of the **AUNITS** system variable. This is no coincidence. When you don't know the setting of units, you use this system variable to provide the mode number:

```
Command: (angtof "45d37'11\"" (getvar "aunits"))
0.796214
```

The **GETVAR** function (short for "get variable") gets the value of system variables.

Other conversion functions convert one unit of measurement into another (via the **CVUNIT** function and the *acad.unt* file), number to and from strings, convert characters to and from ASCII values (for example, letter A into ASCII 65), and translates (i.e. moves) a point from one coordinate system to another. The complete list of conversion functions is:

Conversion Function	Meaning
String Conversions	
AtoF	Converts strings into real numbers.
AtoI	Converts strings into integers.
RtoS	Converts numbers into strings.
ItoA	Converts integers into strings.
Ascii	Converts the first character to its ASCII value.
Chr	Converts ASCII code to its character.
Numerical Conversion	
CvUnit	Converts units based on values found in *acad.unt*.

COMMANDS AND SYSTEM VARIABLES

"Powerful" often equates to "complicated," yet one of AutoLISP's most powerful functions is the easiest to understand: the **COMMAND** function. As the name suggests, this function executes AutoCAD commands.

This makes it easy to get AutoLISP to draw circles, place text, zoom into views — anything you *type* at the 'Command:' prompt is available with the **COMMAND** function. To see how this function works, first recall how the **CIRCLE** command works:

```
Command: circle
Specify center point for circle or [3P/2P/Ttr (tan tan radius)]: 2,2
Specify radius of circle or [Diameter]: d
Specify diameter of circle: 1.5
```

The **COMMAND** function mimics what you type at the 'Command:' prompt like scripts:

```
(command "circle" "2,2" "d" "1.5")
```

Notice that the typed text is in quotation marks; all are considered strings, even numbers.

After you type in the AutoLISP function, AutoCAD responds by drawing the circle:

Command: **(command "circle" "2,2" "D" "1.5")**

circle Specify center point for circle or [3P/2P/Ttr (tan tan radius)]: 2,2

Specify radius of circle or [Diameter] <0.2658>: D

Specify diameter of circle <0.5317>: 1.5

Command: nil

(Don't worry about that 'nil' at the end; that's just AutoLISP letting you know it has finished its work. At the end of this chapter, you learn how to suppress the nil.)

Let's look at another example, this time one of AutoCAD's more complex commands, **TEXT**. AutoCAD presents these prompts:

Command: **text**

Current text style: "Standard" Text height: 0.2000

Specify start point of text or [Justify/Style]: **5,10**

Specify height <0.2000>: **1.5**

Specify rotation angle of text <0>: *(Press* **ENTER***.)*

Enter text: **Using AutoCAD**

Enter text: *(Press* **ENTER***.)*

Converted to AutoLISP, this becomes:

Command: **(command "text" "5,10" "1.5" "" "Using AutoCAD")**

For the 'Rotation angle:' prompt, you often press the **ENTER** key to accept the default value. Notice how that was dealt with in the AutoLISP function: "" — a pair of empty quotation marks (shown in blue).

On the other hand, a final "" is missing that normally represents the **ENTER** that exits the **TEXT** command. The "" is missing because AutoLISP knows that the command should end with the closing parenthesis. The final "" *is* needed, however, with commands that automatically repeat themselves, such as the **LINE** command:

Command: **(command "line" "1,2" "3,4" "")**

When you leave out that final "", AutoCAD is left hanging with a 'To point:' prompt.

Entering the code at the 'Command:' prompt results in:

Command: **(command "text" "5,10" "1.5" "" "Using AutoCAD")**

text Current text style: "Standard" Text height: 1.5000

Specify start point of text or [Justify/Style]: 5,10

Specify height <1.5000>: 1.5

Specify rotation angle of text <0>:

Enter text: Using AutoCAD

To work effectively with the **COMMAND** function, you must know the prompt sequences of AutoCAD's 300 commands. One method is to purchase a "quick reference" book, such as my *The Illustrated AutoCAD 2005 Quick Reference* (Autodesk Press). This book lists AutoCAD's commands in alphabetical order, their prompt sequences, system variables, default values, and range of permissible values.

The **COMMAND** function has a failing. In making the statement, "Anything you *type* at the 'Command:' prompt is available with the Command function.' I placed the emphasis on the word *type*. This function fails to work with dialog boxes and toolbar macros — just like scripts. Commands that use dialog boxes don't work with **COMMAND** — or any other aspect of AutoCAD that involves the mouse. For this reason AutoCAD includes command line versions for most (but not all) commands — for example, the **-LAYER** command instead of **LAYER**, and **-ARRAY** instead of **ARRAY**. Sometimes it makes sense to use system variables instead of commands.

Accessing System Variables

While you could use the **COMMAND** function to access system variables, AutoLISP has a pair of more direct functions: **GETVAR** (gets the value of a system variables) and **SETVAR** (sets or changes the values).

For example, system variable **SPLFRAME** determines whether the frame of splined polyline is displayed. By default, its value is 0, meaning the frame is not displayed. Use **GETVAR** to confirm the setting of the variable:

```
Command: (getvar "splframe")
0
```

To display the frame, change the value to 1 with **SETVAR**:

```
Command: (setvar "splframe" 1)
1
```

GET FUNCTIONS

It's one thing to execute commands, such as **CIRCLE** and **TEXT**; it is trickier working with objects that already exist, such as moving circles and editing text. AutoLISP has a group of functions known collectively as "Getxxx" that get data from the screen and the keyboard.

To show how **GET** functions work, I'll recode the **TEXT** command to change the order of its prompts.

The **GETSTRING** function asks the user to enter text, and stores it in variable **txtstr**.

```
Command: (setq TxtStr (getstring T "What do you want to write? "))
What do you want to write? Using AutoCAD
"Using AutoCAD"
```

Notice that extra "**T**"; it's an AutoLISP workaround to allow **GETSTRING** to accept a string of text that contains spaces. Leave out the T, and **GETSTRING** accepts text only up to the first space, in the same manner of the 'Command:' prompt. (You would end up with just "Using" and no "AutoCAD.")

On to the next prompt:

```
Command: (setq TxtHt (getreal "How big do you want the letters? "))
How big do you want the letters? 2
2.0
```

The **GETREAL** function prompts the user to enter a height. It automatically converts the **2** (an integer) to a real number, **2.0**. Instead of entering numbers, you can also pick two points in the drawing, and AutoLISP reads the distance. If you incorrectly enter text, AutoCAD complains, "Requires numeric value," and reissues the prompt: "How big do you want the letters?"

Another prompt involves the rotation angle:

```
Command: (setq TxtAng (getangle "Tilt the text by how much? "))
Tilt the text by how much? 30
0.523599
```

The **GETANGLE** function prompts the user for an angle. It converts the degrees into radians. As with the previous function, you can pick two points, but not enter text.

And the final prompt:

```
Command: (setq TxtIns (getpoint "Where do you want the text to start? "))
Where do you want the text to start? (Pick a point.)
(27.8068 4.9825 0.0)
```

The **GETPOINT** function prompts for the coordinates of the text's starting point. It expects you to type in a single value, a 2D (x,y) coordinate, or a 3D (x,y,z) coordinate — or to pick a point in the drawing. Object snaps work. A single value is assumed to be the polar distance in the direction from 0,0 and the current cursor location. Entering text results in the error message, "Invalid point."

With the parameters stored in variables, you can execute the **TEXT** command:

```
Command: (command "text" TxtIns TxtHt TxtAng TxtStr)
text Justify/Style:
Height <1.5000>: 2.000000000000000
Rotation angle <0>: 0.523598775598299
Text: Using AutoCAD
Command: nil
```

That was an example of customizing the **TEXT** command. Not only were the prompts changed, but their order was also changed.

All of AutoLISP's Get functions are:

Get Function	***Meaning***
Strings, Numbers, and Angles	
GetAngle	Angle (in radians) picked by the user.
GetOrient	Angle (in radians) independent of ANGBASE and ANGDIR.
GetString	Text typed by the user.
GetInt	Value of the integer number typed by the user.
GetReal	Value of the real number typed by the user.
InitGet	Restricts the data input by user.
GetKWord	Get input based on INITGET restrictions.
Picked Points	
GetPoint	X,y,z coordinates of points picked or typed by user.
GetCorner	Second corner of a rectangle.
GetDist	Distance between two points.

SELECTION SET FUNCTIONS

To work with more than one object at a time, AutoLISP has a group of functions for creating selection sets. Whereas AutoCAD's **SELECT** command works with a single selection set at a time, AutoLISP's SSxxx commands work with up to 128 selection sets.

AutoLISP's functions for creating and examining selection sets are:

SS Function	***Meaning***
SsGet	Selects objects in the drawing.
SsAdd	Adds objects to the selection set.
SsDel	Deletes objects from the selection set.
SsLength	Reports the number of objects in the selection set.
SsMemb	Checks if an object is part of the selection set.
SsName	Reports the name of *n*th object in the selection set.

OBJECT MANIPULATION FUNCTIONS

The most powerful AutoLISP functions manipulate the drawing database. Unlike the **COMMAND** function, which is powerful but simple, the object manipulation functions are powerful and complex.

Most of these functions begin with "ent," short for *entity*, the old name for AutoCAD objects. Sometimes reference is made to the *symbol table*, which is the part of the drawing database that stores the names of layers, text styles, and all other named objects in the drawing.

To create and manipulate objects, the **ENT** functions work with a variant on the DXF format, called "dotted pairs." For example, to work with a layer named RightOfWay, you employ the following format:

```
"2 . RightOfWay"
```

The quotation marks indicate the start and end of the data. The dot (.) in the middle separates the two values: 2 is the DXF code for layer names, while RightOfWay is the name of the layer. To work with object manipulating functions, you need a good grasp of the DXF format, which is documented in AutoCAD's on-line Developer Help file (search for "DXF Reference").

AutoLISP's object manipulation functions are:

Ent Functions	Meaning
EntMake	Creates new objects.
EntGet	Gets data describing objects.
TblObjName	Gets names of objects in the symbol table.
HandEnt	Returns the handle (id) of objects.
EntNext	Gets the next object in the database.
EntLast	Gets the last-created object.
EntSel	Prompts user to select an object.
NEntSel	Prompts user to select an object, and returns coordinates of pick point.
EntMod	Modifies objects.
EntUpd	Updates screen with modified objects.
EntDel	Erases objects.

ADVANCED AUTOLISP FUNCTIONS

There is a whole host of AutoLISP functions that you may never use in your AutoCAD programming career. For example, Autodesk provides nearly four dozen functions for controlling the ASE (AutoCAD Structured query language Extension) link between objects in the AutoCAD drawing and records in external database files. These functions are not hard to spot: they all start with "Ase_", as in **ASE_LSUNITE** and **ASE_DOCMP**.

Another two dozen AutoLISP functions load and control dialog boxes, such as **LOAD_DIALOGUE** and **NEW_DIALOGUE**. A group of five functions are for memory management, whose use is so rarified that Autodesk recommends never using them.

Other functions, such as for such as saving data to files, are described in the following tutorials.

WRITING AUTOLISP PROGRAMS

If you are like many CAD users, you are too busy creating drawings to learn programming. You may wonder, "Why bother learning a programming language?" But learning some coding means letting AutoLISP perform tedious tasks for you, sometimes with just a couple of lines of code. The nice things about AutoLISP are that you don't need to know all of it; and you can program it on-the-fly.

In this tutorial, you learn how to use AutoLISP for a really simple but tedious task. AutoCAD's **ID** command reports the x, y, z coordinates of the points you pick in drawings. The problem is that the coordinates are reported in the command prompt area:

> Command: **id**
>
> Point: *(Pick a point.)*
>
> X = 8.9227 Y = 6.5907 Z = 0.0000

It could be useful to have **ID** place the coordinates in the drawing, next to the pick point. That would let you easily label x, y, z coordinates all over site plans. AutoLISP lets you do this.

Before writing any program, you need to think through the process: how are the coordinates going to get from the command prompt area to the drawings. There are two parts to solving the problem:

Part 1. Obtaining the coordinates from the drawing, by picking a point.

Part 2. Placing the coordinates in the drawing as text.

TUTORIAL: PART 1 — PICKING POINTS

AutoLISP provides several ways to get the coordinates of picked points. Browsing through AutoCAD's online *Customization Guide*, we find we could:

- Use the **ID** command with the **COMMAND** function, as in **(command "id")**.
- Use the **LASTPOINT** system variable with the **GETVAR** function, as in **(getvar "lastpoint")**.
- Use the **GETPOINT** function, as in **(getpoint "Pick a point: ")**.

It would be a useful lesson to use each of the three, and then see what happens. By experimenting, you make mistakes, but learn from the mistakes.

1. Start AutoCAD, load a drawing, and switch to the Text window with **F2**. At the 'Command:' prompt, type:

   Command: **(command "ID")**

2. Pick a point:

   Point: *(Pick a point.)*

   X = 8.9227 Y = 6.5907 Z = 0.0000

   AutoCAD stores the x,y,z- coordinates of the picked point in system variable **LASTPOINT**.

3. Before placing the coordinates as a line of text in the drawing, you need to store them in a variable with **SETQ**.

   Command: **(setq xyz (getvar "LastPoint"))**

   (8.9227 6.5907 0.0000)

Xyz is the name of the variable that stores the x, y, z coordinates.

GETVAR is the AutoLISP function that retrieves values stored in system variables.

"LastPoint" is the name of the system variable surrounded by quotation marks, because it is a string.

AutoLISP returns the value stored in variable **xyz**: (8.9227 6.5907 0.0000). Note that the coordinates are surrounded by parentheses: this is called a "list." Spaces separate the numbers, which are the x, y, and z coordinates, respectively:

x = 8.9227

y = 6.5907

z = 0.0000

AutoCAD always stores the values in the order of x, y, and z. You will never find the z coordinate first or the x coordinate last.

4. The third useful AutoLISP function is preferred by programmers, because it is more efficient than the **ID-LASTPOINT** combo used above. Type in the following to see how it works:

   Command: **(setq xyz (getpoint "Point: "))**

   Point: *(Pick a point.)*

   (8.9227 6.5907 0.0000)

   GETPOINT displays the prompt text, "Point:", and then waits for you to pick a point in the drawing.

   In this case, we mimicked the prompt of the **ID** command. But you could just as easily write:

   Command: **(setq xyz (getpoint "Press the mouse button: "))**

   Press the mouse button: *(Pick a point.)*

   (8.9227 6.5907 0.0000)

   Or, have no prompt at all,:

   Command: **(setq xyz (getpoint))**

   (Pick a point.)

   (8.9227 6.5907 0.0000)

You've now seen several approaches that solve the first problem. With the x,y,z coordinates stored in a variable, let's tackle the second problem.

TUTORIAL: PART 2 — PLACING TEXT

To place text in the drawing, there is one easy solution: the COMMAND function in conjunction with the TEXT command. (The advanced solution is to use ENTMAKE.)

1. Work out the **TEXT** command's prompts that need to be answered:
   - **Start point** needs a pair of numbers, x, y coordinates, to locate the text.
   - **Height** needs a number that makes the text tall enough to be legible.
   - **Rotation angle** needs a number that represents the angle, probably 0.
   - **Text** needs a string, the value of the x, y, z coordinates.
2. Construct the AutoLISP function for placing the coordinates as text:

(command	AutoLISP **COMMAND** function.
"text"	AutoCAD **TEXT** command.
xyz	Variable that stores the starting point for the text.
0.2	Height of the text.
	(Change this number to something convenient for your drawings.)

0	Rotation angle of the text.
xyz	TEXT command accepts numbers stored in variables as text.
)	Closing parenthesis — one for every opening parenthesis.

3. Try out the code by typing it at the 'Command:' prompt:

```
Command: (command "text" xyz 0.2 0 xyz)
text Justify/Style/<Start point>:
Height <1.0>: 200
Rotation angle <0>: 0
Text: 2958.348773815669,5740.821183398367
Command: nil
```

   AutoCAD runs through the **TEXT** command, inserts the stored responses for its prompts, and then places the coordinates as text. It worked!

4. For the complete AutoLISP program, put together the two lines of code:

```
(setq xyz (getpoint "Pick point: "))
(command "text" xyz 200 0 xyz)
```

You now have a full-fledged AutoLISP program.

Well, not quite: what you have is an *algorithm* — the core of every computer program that performs the work. What is lacking is the *user interface* — the part that makes it easier for the user to employ the program. All you have for a user interface is a rather plain prompt, 'Pick point:.'

SPRUCING UP THE CODE

There are many user interface problems with this short program. Can you think of any?

Here's are some ways to improve it:

- Find a way to avoid retyping the two lines of code each time you want to label a point.
- Save the program on disk so that it can easily be loaded and shared between AutoCAD sessions.
- Have a way to load the program automatically.
- Reduce the x, y, z coordinates from eight decimal places.
- Control the layer on which the text is placed.
- Specify the text style.
- Control the size and orientation of the text.
- Store the x, y, z coordinates to a file on disk, to reuse the data.

The problem with adding features is that they add a tremendous amount of code, which means more bugs and more debugging. A simple program quickly expands with *feature-bloat*. While all the features sound desirable, they may make the program less desirable. Image the irritation in answering the many questions about decimal places, text font, text size, text orientation, layer name, file name — each time you label a single point.

Therefore, take a second look at the wishlist, and cross off the features you could live without.

Some of the features are useful for showing how to write AutoLISP programs, as described next.

NAMING PROGRAMS

To name programs, surround the entire code with the **DEFUN** function, as follows:

```
(defun c:label ( / xyz)
    (setq xyz (getpoint "Pick point: "))
    (command "text" xyz 200 0 xyz)
)
```

Let's take a look at what's been added, piece by piece:

(**DEFUN** defines the name of the program. (In AutoLISP, the terms *function*, *program*, and *routine* are used interchangeably.)

label is the name we gave the program. You may use any name that does not conflict with the names of built-in AutoLISP functions, or global user-defined functions.

c: is a prefix that makes this AutoLISP routine appear like an AutoCAD command. This lets you simply type **label** at 'Command:' prompt, as follows:

Command: **label**

Pick point: *(Pick a point.)*

When the c: prefix is missing, then the program must be run as an AutoLISP function — with the parentheses.

Command: **(label)**

Pick point: *(Pick a point.)*

Note: You can use prefixes other than c: for your own purposes, such as to identify subroutines you write. An example is:

```
(defun rhg:label)
```

(/ **xyz**) declares the names of *input* and *local* variables; the slash separates the two types of variables.

- **Input** variables feed data to the AutoLISP routine; the names of input variables appear before the slash, such as **(name /)**.
- **Local** variable is used only within the program; the names of local variables appear after the slash, such as **(** name **/ xyz)**.

In this program, **xyz** is the name of the local variable. The benefit to declaring variables as local is that AutoCAD automatically frees up the memory used by the variable when the AutoLISP program ends. This is also a drawback, for the variable's value is lost, making debugging harder.

If variables are not declared as local, they are *global*. The advantage of global variables is that they can be accessed by *any* AutoLISP function loaded into AutoCAD. Again, this can be a disadvantage when two programs use the same variable names, because one will overwrite the other.

The solution is to keep local variables global until the program is debugged, then make them local.

) is the closing parenthesis that balances the opening parenthesis at the beginning of the program.

Notice that the two center lines of code are indented. This helps them stand out from the **DEFUN** line and closing parenthesis. This is standard practice among programmers, because the indents make the code easier to read. The indents are typically created by pressing the **TAB** key, but you can also use of spaces — AutoLISP cares not.

TUTORIALS: SAVING AND LOADING PROGRAMS

Save the AutoLISP program to a file on disk, so that you avoid retyping the code with each new AutoCAD session. To do this:

1. Start a text editor. (The Notepad supplied with Windows is a good choice).
2. Enter the code.:

```
(defun c:label ( / xyz)
     (setq xyz (getpoint "Pick point: "))
     (command "text" xyz 200 0 xyz)
)
```

3. Save the file with the name *label.lsp* in AutoCAD's *support* folder.

All AutoLISP programs use the *.lsp* extension to identify them.

To load AutoLISP programs into AutoCAD, use the **LOAD** function. (This AutoLISP function is different from AutoCAD's **LOAD** command, which loads shape files.)

1. At the command prompt, enter the **LOAD** function with the name of the program to load into AutoCAD:

   Command: **(load "label")**

2. If AutoCAD does not find the AutoLISP program, it complains, "; error: LOAD failed: "label""

   In that case, you may need to specify the path name. Assuming you saved *label.lsp* in another folder, such as *my documents*, you would type:

   Command: **(load "\\my documents\\label")**

   Notice the use of the double backward slashes (\\ ). You can also use a single forward slash: (load "/my documents/label")

3. With the program loaded, you can now run it:

   Command: **label**

   Pick point: *(Pick a point.)*

AutoCAD provides a way to load AutoLISP programs automatically. When AutoCAD starts up, it looks for a file called *acaddoc.lsp*. AutoCAD automatically loads the names of AutoLISP programs listed in the file.

To add *label.lsp* is not difficult:

1. Open the *acaddoc.lsp* file with a text editor.

   (If the *acaddoc.lsp* file does not exist, then start a new file, and store it in AutoCAD's *support* folder)

2. Add the name of the AutoLISP program.

   (load "label.lsp")

3. Save the *acaddoc.lsp* file.
4. Start AutoCAD and **label** program should load automatically.

USING CAR AND CDR

The x, y, z coordinates are printed in the drawing with eight decimal places — that's too many. There are two solutions. One is to ask the user for the number of decimal places, as shown by the following code fragment:

Command: **(setq uprec (getint "Label precision: "))**

Label precision: **1**

1

Or get the value stored in system variable **LUPREC**, which is the precision specified by the user via the **UNITS** command. (We are working under the not-necessarily-true assumption that the user wants consistent units.) The code for this is:

(setq uprec (getvar "LUPREC"))

Getting the precision is easy. It's tougher to apply the precision to the values of the x, y, z coordinates. This task requires three steps: (1) pick apart the coordinate triplet; (2) apply the precision factor; and (3) join the coordinates. Here's how:

1. Open *label.lsp* in Notepad or any other text editor.
2. Remove **/ xyz** from the code to make the variable "global," so that you can check its value at any time. The code should look like this:

```
(defun c:label ( )
    (setq xyz (getpoint "Pick point: "))
    (command "text" xyz 200 0 xyz)
)
```

3. Save the file, and then load *label.lsp* into AutoCAD.
4. Run **label**, and then pick a point in the drawing. This places a value in the **xyz** variable. (If you don't see the coordinates printed on the screen, use the **ZOOM Extents** command.)
5. Examine the contents of **xyz** variable at the 'Command:' prompt. Type the following:

   Command: **!xyz**

   (6.10049 8.14595 10.0)

   Recall that the exclamation mark forces AutoCAD to print the value of variables. (Your results will differ, depending on where you picked.)
6. AutoLISP has functions specific to picking apart lists. The **CAR** function extracts the first item from the list, the x coordinate in this case.

   Command: **(car xyz)**

   6.10049
7. The **CDR** function is **CAR**'s complement: it removes the first item from the list, and gives you what's left:

   Command: **(cdr xyz)**

   (8.14595 10.0)
8. AutoLISP allows you to combine the "a" and "d" in many ways to extract items in other positions of the list. To extract the y coordinate (the second item), use **CADR**:

   Command: **(cadr xyz)**

   8.14595
9. And to extract the z coordinate (the third item), use **CADDR**:

   Command: **(caddr xyz)**

   8.14595

   You now have a way to extract the x coordinate, the y coordinate, and the z coordinate from variable **xyz**.
10. Store each coordinate in its own variable, using the newly-learned code:

    Command: **(setq ptx (car xyz)**

```
1> pty (cadr xyz)
1> ptz (caddr xyz)
1> )
```

PtX stores the x coordinate, **PtY** the y coordinate, and **PtZ** the z.

Notice the use of AutoLISP shorthand that applies the **SETQ** function to all variables at once. Recall the reason for the '1>' prompt: it is a reminder that a closing parenthesis is missing.

11. With the three coordinates separated, you can now reduce the number of decimal places. The **RTOS** function does two things at once: changes the number of decimal places, and converts the real number into a string. (Why? You find out later.) Enter the **RTOS** function:

```
Command: (rtos ptx 2 uprec)
"6.1"
```

The **RTOS** function makes use of three parameters:

PtX is the name of the variable holding the real number.

2 is the mode of conversion, decimal in this case. The number 2 is based on system variable **LUNITS**, which defines five modes of units:

Mode	*Units*
1	Scientific.
2	Decimal.
3	Engineering.
4	Architectural.
5	Fractional.

UPrec is the name of the variable holding the precision. (Recall the code from earlier in this tutorial.) The meaning of the precision mode depends on the type of units. For example, mode 3 means "three decimal places" for decimal units, while mode 3 means "eighth-inch" for architectural units.

If the precision specified by **UPrec** is 1, then the **RTOS** function reduces the display of 6.10049 down to 6.1.

12. Truncate and save the values of x, y, and z, as follows:

```
Command: (setq ptx (rtos ptx 2 uprec)
1> pty (rtos pty 2 uprec)
1> ptz (rtos ptz 2 uprec)
1> )
```

Notice that you can set a variable equal to itself: the former **PtX** holds the new value of the x-coordinate after **RTOS** gets finished processing the later **PtX**. Reusing variable names helps conserve memory.

13. With the coordinates truncated, string them together (pardon the pun) with the **STRCAT** function:

```
Command: (strcat ptx pty ptz)
"6.18.110.0"
```

Hmmm... not quite the look we were hoping for. Something like

```
6.1, 8.1, 10.0
```

would look better. AutoLISP provides no spaces, so you have to.

14. Reuse **STRCAT** to add commas and spaces:

    Command **(setq xyz (strcat ptx ", " pty ", " ptz))**

    "6.1, 8.1, 10.0"

    That's more like it.

15. Back to the text editor. Add in the code we developed above, shown in **boldface**:

```
(defun c:label ( / xyz xyz1 uprec ptx pty ptz)
    (setq uprec (getint "Label precision: "))
    (setq xyz (getpoint "Pick point: "))
    (setq   ptx (car xyz)
            pty (cadr xyz)
            ptz (caddr xyz)
    )
    (setq   ptx (rtos ptx 2 uprec)
            pty (rtos pty 2 uprec)
            ptz (rtos ptz 2 uprec)
    )
    (setq xyz1 (strcat ptx ", " pty ", " ptz))
    (command "text" xyz 200 0 xyz1)
)
```

Notice that all the variables are local. Also, variable **xyz** is changed in the last few lines: I don't want the text placed at coordinates that are rounded-off, so employ **xyz1** as the variable holding the text string.

16. Finally, add comments to the code to remind you when you look at the code several months or years from now. The semicolon (;) indicates the start of a comment:

```
; Label.Lsp labels a picked point with its x,y,z coordinates.
; by Ralph Grabowski, 25 February, 1996.
(defun c:label ( / xyz xyz1 uprec ptx pty ptz)
 ; Ask user for the number of decimal places:
 (setq uprec (getint "Label precision: "))
 ; Ask the user to pick a point in the drawing:
 (setq xyz (getpoint "Pick point: "))
 ; Separate 3D point into individual x,y,z-values:
 (setq ptx (car xyz)
   pty (cadr xyz)
   ptz (caddr xyz)
 )
 ; Truncate values:
 (setq ptx (rtos ptx 2 uprec)
   pty (rtos pty 2 uprec)
   ptz (rtos ptz 2 uprec)
 )
 ; Recombine individual values into a 3D point:
 (setq xyz1 (strcat ptx ", " pty ", " ptz))
```

```
; Place text:
(command "text" xyz 200 0 xyz1)
)
```

17. Save the file as *label.lsp,* and then test the AutoLISP program by loading it into AutoCAD:

```
Command: (load "label")
"C:LABEL"
```

18. And run the routine:

```
Command: label
Label precision: 1
Pick point: (Pick a point.)
text Justify/Style/<Start point>:
Height <200.0000>: 200
Rotation angle <0>: 0
Text: 5012.3, 773.2, 0.0
Command: nil
```

SAVING DATA TO FILES

It would be handy to save the x, y, z coordinates as data in a file. This is performed with the **OPEN WRITE-LINE**, and **CLOSE** functions.

Working with files is simpler in AutoLISP than in most programming languages, because AutoLISP has relatively weak functions for accessing data in files. It is limited to reading and writing ASCII files in *sequential* order. AutoLISP cannot deal with *binary* files, nor does it access data in *random* order. (AutoCAD users have asked for those abilities, but Autodesk has chosen to not add them.)

There are three steps in writing data to a file:

Step 1: Open the data file.

Step 2: Write data to the file.

Step 3: Close the file.

Most other programming languages require you first to create the file. AutoLISP creates files if they do not already exist.

STEP 1: OPENING DATA FILES

The **OPEN** function opens a file for three purposes: (1) to *read* data from the file; (2) to *write* data to the file; or (3) to *append* data to the file. AutoLISP can do only one of these at a time, so you must specify the action with a mode indicator (*must* be in lower case):

- **"r" — read**: data is read from the file.
- **"w" — write**: all existing data is erased, and then new data is added.
- **"a" — append**: new data is added to the end of the existing data.

For our routine, we want to keep adding data to the file, so choose the append mode. The AutoLISP code looks like this:

```
Command: (setq FIL (open "xyzdata.txt" "a"))
#<file "xyzdata.txt">
```

Nearly all programming languages work with *file descriptors*, rather than file names. The descriptor is a name, some sequence of letters and numbers, to which the operating system assigns the actual file name. Notice that AutoCAD uses **#<file "xyzdata.txt">**.

The code stores the file descriptor in variable **fil**, and from now on, our program works with **fil**, not the file name, *xyzdata.txt*.

The "**a**" tells AutoLISP to open *xyzdata.txt* for appending data. It is important that the mode be lowercase; this is the only occasion where AutoLISP is case-sensitive. The modes for the **OPEN** function are:

Mode	Meaning
"a"	Append data to end of file.
"w"	Write data to file, after erasing existing data.
"r"	Read data from file.

STEP 2: WRITING DATA TO FILES

To write the data to the file, use the **WRITE-LINE** function. This function writes a single line of data to a file. (A related function, the **WRITE** function, writes a single character to files.) The code looks like this:

```
(write-line xyz1 fil)
```

Software, such as spreadsheets, database programs, and even some word processors, read data with commas separating numbers, called "comma-separated values" or CSV for short.

```
8.1548, 3.2752, 0.0000
```

These programs consider the comma as a separator, and not a literal comma. This allows the spreadsheet software to place each number in its own cell, following which you can manipulate the data.

Recall that we used the **STRCAT** function and the **CDR**, **CADR**, and **CADDR** functions to separate the x, y, and z component, and insert a comma between each.

STEP 3: CLOSING FILES

For good housekeeping purposes, close the file. (AutoCAD automatically closes the file for you if you forget, but good programmers clean up after themselves.) Closing the file is as simple as:

```
(close fil)
```

Using the Notepad text editor, make the additions shown on the next page in **boldface** to your copy of *label.lsp*.

```
(defun c:label ( / xyz xyz1 uprec ptx pty ptz)
 (setq uprec (getint "Label precision: "))
 (setq xyz (getpoint "Pick point: "))
 (setq ptx (car xyz)
   pty (cadr xyz)
   ptz (caddr xyz)
 )
 ; Format the x,y,z coordinates:
 (setq ptx (rtos ptx 2 uprec)
   pty (rtos pty 2 uprec)
   ptz (rtos ptz 2 uprec)
```

```
)
; Add commas between the three coordinates:
  (setq xyz1 (strcat ptx ", " pty ", " ptz))
  ; Write coordinates to the drawing:
  (command "text" xyz 200 0 xyz1)
  ; Open the data file for appending:
  (setq fil (open "xyzdata.txt" "a"))
  ; Write the line of data to the file:
  (write-line xyz1 fil)
  ; Close the file:
  (close fil)
)
```

Again, test the code by loading it into AutoCAD, and running it. As you pick points in the drawing, the routine labels the picked points and writes the 3D point data to file. After a while, the data file will look something like this:

```
8.1548, 3.2752, 0.0000
7.0856, 4.4883, 0.0000
6.4295, 5.6528, 0.0000
5.5303, 6.7688, 0.0000
5.4331, 8.3215, 0.0000
```

If you wish, import the *xyzdata.txt* file into a spreadsheet program.

ADDITIONAL PROGRAM ENHANCEMENTS

Recall that one wishlist item was to control the layer on which text is placed. There are two ways to approach this item:

- No-code method: set the layer before starting the AutoLISP function.
- Code method: ask the user for the name of the layer, then set it with system variable **CLAYER**:

```
(setq lname (getstring "Label layer: "))
(setvar "CLAYER" lname)
```

Add those two lines before the line with the 'Pick point' prompt.

Another enhancement was to specify the text style. The same two methods apply. The no-code method sets the text style before starting the routine. Otherwise, add the following AutoLISP code:

```
(setq tsname (getstring "Label text style: "))
(setvar "TEXTSTYLE" tsname)
```

Once again, add those two lines before the line with the 'Pick point' prompt.

By now, you might be noticing that the program is starting to look big. This is the "feature bloat" we spoke of earlier, and explains why twenty years ago AutoCAD fit on a single 360KB floppy disk, but now requires 190MB of hard disk space, a 540-fold increase. The biggest culprit is the user interface, which makes software balloon far beyond the size of the basic algorithms.

TIPS FOR USING AUTOLISP

To end off this chapter, here are some tips to help you write AutoLISP programs.

USE ASCII TEXT EDITORS

AutoLISP code must be written in plain ASCII text — no special characters of the kind that all word processors add. If you were to write AutoLISP code with, say, Word, and then save it as a DOC file (the default), AutoCAD would refuse to load the AutoLISP file, even though the file extension is *.lsp*.

The best thing is to use a pure ASCII text editor, such as Notepad, supplied free by Microsoft with all versions of Windows. Do not use the Write or WordPad applications supplied with Windows. While both of these have an option to save in ASCII, you're bound to forget sometimes and end up frustrated.

Almost any other word processor has an option to save text in plain ASCII, but not by default. Word processors have a number of different terms for what I mean by "pure ASCII format." OpenOffice calls it "Text"; WordPerfect calls it "DOS Text"; WordPad calls it "Text Document"; and Atlantis calls it "Text Files."

LOADING LSP CODE INTO AUTOCAD

To load AutoLISP code into AutoCAD, you use the **LOAD** command:

```
Command: (load "points")
```

After you've done this a few times for debugging purposes, you'll find it tedious. To solve the problem, write a one-line AutoLISP routine that loads other routines, and reduces keystrokes, like this:

```
Command: (defun c:x () (load "points"))
```

Now any time you need to load the *points.lsp* routine, just type "x" and press **ENTER**, as follows:

```
Command: x
```

Windows lets you use another shortcut method: drag the *.lsp* file from File Manager (or Windows Explorer) into AutoCAD. The code moves in just one direction: from the text editor to AutoCAD; you cannot drag code from AutoCAD back to the text editor.

TOGGLING SYSTEM VARIABLES

A problem in programming is this: how can you change values when you don't know what the values are? In AutoCAD, you come across this problem with system variables, many of which are toggles. A *toggle* system variable has a value of 0 or 1, indicating that the value is either off (0) or on (1). For example, system variable **GRIDMODE** is set to 0 by default: when off, the grid is not displayed.

But no computer programmer would assume that the value of **GRIDMODE** is zero just because that's its default value. In the case of toggle system variables, there are two solutions to the problem:

- Use the **IF** function to check if the value is 0 or 1.
- Subtract 1, and then take the absolute value.

When AutoLISP programs change the values of system variables, these should always set them back to the original state.

Many programmers write a set of generic functions that save the current settings at the beginning of the routine. After carrying out changes, restore the saved values at the end of the routine. Here's a code fragment that shows this:

```
(setq splvar (getvar "splframe"))
...
(setvar "splframe" splvar)
```

SUPPRESSING NIL

Any time you run an AutoLISP routine, there is that pesky **nil** that appears as the very last displayed item. There is a reason for nil appearing, but the reason isn't good enough for leaving it there. To prevent nil from appearing, end routines with a (PRINC) all by itself.

```
....
(write-line xyz1 fil)
  ; Close the file:
  (close fil)
)
(princ)
```

PUNCTUATION ISSUES

Escape codes used in text strings must remain lowercase. For example, \e is the escape character (equivalent to ASCII 27) and \t is the tab character. Note that they use backslashes; it is for this reason that you cannot use the backslash for separating folders names. AutoLISP would think you were typing an escape code.

AutoLISP uses quotation marks ("and") to identify strings. Thus, you cannot use a quotation mark to display quotation marks and inches, such as **25 inches** as **25**". The workaround is to use escape codes, specifically the *octal* code equivalent of the ASCII character for the quotation mark. Sound complicated? It is. But all you need to know is **042**, which is why it is the answer to life, the universe, and everything. Assign the strings to variables, as follows:

```
(setq disttxt "The length is ")
(setq distval 25)
(setq qumark "\042")
```

Notice that octal 042 is assigned variable **qumark**. The backslash tells AutoLISP the numbers following are in octal. (Octal is half of hexadecimal: 0 1 2 3 4 5 6 7 10 11 12 ... 16 17 20 21 ... and so on.) Then concatenate the three strings together with the STRCAT function:

```
(strcat distxt distval qumark)
```

This makes the text look correct:

```
The length is 25"
```

You can use the same in prompt statements:

```
(prompt "The diameter is 2.54\042")
```

Similarly, you can use \009 to space text with tab characters.

```
(prompt "The diameter is \009 2.54\042")
```

EXERCISES

1. Write AutoLISP code for performing the following mathematical functions:
   a. Adding 14 to 5.
   b. Finding how many 0.9 boxes fit linearly in a 12-inch space.
   c. Subtracting 1133.10 from 1169.
   d. Multiplying pi by 2.
2. Using AutoCAD, calculate the following expressions:
   a. (+ 15 4)
   b. (/ 16 0.3)
   c. (- 169 113)
   d. (* 99 2)
3. Write the AutoLISP code for:
   a. Calculating the circumference of a circle with radius **r**.
   b. Storing the answer in variable **rad**.
4. Write the code for drawing a line with AutoLISP.
5. Write the code for naming a program "Doorway" with input variable "Width" and local variable "Swing."
6. Write code that gets the current value of the **CLAYER** system variable, and stores it in variable layer1.

CHAPTER REVIEW

1. Why must AutoLISP expressions be surrounded by parentheses?
2. Why should at least one number be a real number when dividing?
3. Explain the purpose of *nested* expressions?
   Provide an example.
4. What does the **SETQ** function do?
5. Which function returns the intersection of two lines?
6. Why are *conditional* functions important?
7. Why are *mode* numbers necessary?
8. What symbol displays quotation marks in prompts?
9. What does the **ANGTOF** function do?
10. Can you use the **LAYER** command with the AutoLISP **COMMAND** function?
11. Which function extracts the x coordinate from a list of coordinates?
12. Explain the difference between *writing* and *appending* to files.

A

AutoCAD Commands, Aliases, and Keyboard Shortcuts

AUTOCAD COMMANDS

The following commands are documented by Autodesk as available in AutoCAD 2004. Commands prefixed with ' (apostrophe) are *transparent* — they can be executed during another command. Commands prefixed with - (hyphen) display their prompts at the command line only, and not through a dialog box. Shortcut keystokes are indicated as CTRL+F4.

(2005) This icon indicates the command is new in AutoCAD 2005.

A complete reference guide to all of AutoCAD's commands is found in *The Illustrated AutoCAD 2005 Quick Reference* by author Ralph Grabowski, available through Autodesk Press.

Command	*Description*
A	
'About	Displays an AutoCAD information dialog box that includes version and serial numbers.
AcisIn	Imports ASCII-format ACIS files into the drawing, and then creates 3D solids, 2D regions, or body objects.
AcisOut	Exports AutoCAD 3D solids, 2D regions, or bodies as ASCII-format ACIS files (file extension *.sat*).
AdcClose	CTRL+2: Closes the DesignCenter window.
AdCenter	CTRL+2: Opens DesignCenter; manages AutoCAD content.
AdcNavigate	Directs DesignCenter to the file name, folder, or network path you specify.
Align	Uses three pairs of 3D points to move and rotate (align) 3D objects.
AmeConvert	Converts drawings made with AME v2.0 and v2.1 into ACIS solid models.
'Aperture	Adjusts the size of the target box used with object snap.
AppLoad	Displays a dialog box that lets you list AutoLISP, Visual Basic, and ObjectARX program names for easy loading into AutoCAD.
Arc	Draws arcs by a variety of methods.
(2005) **Archive**	Archives sheet sets in DWG or DWF format.
Area	Computes the area and perimeter of polygonal shapes.
Array	Makes multiple copies of objects.

Command	Description
Arx	Loads and unloads ObjectARX programs. Also displays the names of ObjectARX program command names.
Assist	CTRL+5: Displays the real-time help window; user interface updated in AutoCAD 2005.
2005 **AssistClose**	CTRL+5: Closes the Info Palette window.
AttDef	Creates attribute definitions.
'AttDisp	Controls whether attributes are displayed.
AttEdit	Edits attributes.
AttExt	Extracts attribute data from drawings, and writes them to files for use with other programs.
AttReDef	Assigns existing attributes to new blocks, and new attributes to existing blocks.
AttSync	Synchronizes changed attributes with all blocks.
Audit	Diagnoses and corrects errors in drawing files.
B	
Background	Sets up backgrounds for rendered scenes; can be solid colors, gradient shades, images in BMP, PCX, Targa, JPEG, or TIFF formats, or the current AutoCAD wireframe views.
'Base	Specifies the origin for inserting one drawing into another.
BAttMan	Edits all aspects of attributes in a block; short for Block Attribute Manager.
BHatch	Fills an automatically-defined boundary with hatch patterns, solid colors, and gradient fills; previews and adjusts patterns without starting over.
'Blipmode	Toggles display of marker blips.
Block	Creates symbols from groups of objects.
BlockIcon	Generates preview images for blocks created with AutoCAD Release 14 and earlier.
BmpOut	Exports selected objects from the current viewport to raster *.bmp* files.
Boundary	Draws closed boundary polylines.
Box	Creates 3D solid boxes and cubes (solid modeling command).
Break	Erases parts of objects, breaks objects in two.
Browser	Launches your computer's default Web browser with the URL you specify.
C	
'Cal	Runs a geometry calculator that evaluates integer, real, and vector expressions.
Camera	Sets the camera and target locations.
Chamfer	Trims intersecting lines, connecting them a chamfer.
Change	Permits modification of an object's characteristics.
CheckStandards	Compares the settings of layers, linetypes, text styles, and dimension styles with those set in another drawing.
ChProp	Changes properties (linetype, color, and so on) of objects.
Circle	Draws circles by a variety of methods.
CleanScreenOn	CTRL+0: Maximizes the drawing area by turning off toolbars, title bar, and window borders.
CleanScreenOff	CTRL+0: Turns on toolbars, title bar, and window borders.
Close	CTRL+F4: Closes the current drawing.
CloseAll	Closes all open drawings; keeps AutoCAD open.
'Color	Sets new colors for subsequently-drawn objects.
Compile	Compiles shapes and *.shp* and *.pfb* font files.
Cone	Creates 3D cones (a solid modeling command).
Convert	Converts 2D polylines and associative hatches in pre-AutoCAD Release 14 drawings to the "lightweight" format to save on memory and disk space.
ConvertCTB	Converts drawings from plot styles to color-based tables.
ConvertPStyles	Converts drawings from color-based tables to plot styles.

Command	*Description*
Copy	Copies selected objects.
CopyBase	CTRL+SHIFT+C: Copies objects with a specified base point.
CopyClip	CTRL+C: Copies selected objects to the Clipboard in several formats.
CopyHist	Copies Text window text to the Clipboard.
CopyLink	Copies all objects in the current viewport to the Clipboard in several formats.
Customize	Customizes the toolbar and keyboard shortcuts.
CutClip	CTRL+X: Cuts selected objects from the drawing to the Clipboard in several formats.
Cylinder	Creates 3D cylinders (a solid modeling command).
D	
DbcClose	CTRL+6: Closes the dbConnect Manager.
DbConnect	CTRL+6: Connects objects in drawings with tables in external database files.
DblClkEdit	Toggles double-click editing.
DbList	Provides information about all objects in drawings.
DdEdit	Edits text, paragraph text, attribute text, and dimension text.
'DdPType	Specifies the style and size of points.
DdVPoint	Sets 3D viewpoints.
'Delay	Creates a delay between operations in a script file.
DetachURL	Removes hyperlinks from objects.
DIMENSIONS	
Dim	Specifies semi-automatic dimensioning capabilities.
Dim1	Executes a single AutoCAD Release 12-style dimension command.
DimAligned	Draws linear dimensions aligned to objects.
DimAngular	Draws angular dimensions.
DimBaseline	Draws linear, angular, and ordinate dimensions that continue from baselines.
DimCenter	Draws center marks on circles and arcs.
DimContinue	Draws linear, angular, and ordinate dimensions that continue from the last dimension.
DimDiameter	Draws diameter dimensions on circles and arcs.
DimDisassociate	Removes associativity from dimensions.
DimEdit	Edits the text and extension lines of associative dimensions.
DimLinear	Draws linear dimensions.
DimOrdinate	Draws ordinate dimensions in the x and y directions.
DimOverride	Overrides current dimension variables to change the look of selected dimensions.
DimRadius	Draws radial dimensions for circles and arcs.
DimReassociate	Associates dimensions with objects.
DimRegen	Updates associative dimensions.
DimStyle	Creates, names, modifies, and applies named dimension styles.
DimTEdit	Moves and rotates text in dimensions.
'Dist	Computes the distance between two points.
Divide	Divides objects into an equal number of parts, and places specified blocks or point objects at the division points.
Donut	Constructs solid filled circles and doughnuts.
'Dragmode	Toggles display of dragged objects.
DrawOrder	Changes the order in which objects are displayed: selected objects and images are placed above or below other objects.
'DSettings	Specifies drawing settings for snap, grid, polar, and object snap tracking.
'DsViewer	Opens the Aerial View window.

Command	Description
DView	Displays 3D views dynamically.
DwgProps	Sets and displays the properties of the current drawing.
DxbIn	Creates binary drawing interchange files.
E	
EAttEdit	Enhanced attribute editor.
EAttExt	Enhanced attribute extraction.
Edge	Changes the visibility of 3D face edges.
EdgeSurf	Draws edge-defined surfaces.
'Elev	Sets current elevation and thickness.
Ellipse	Constructs ellipses and elliptical arcs.
Erase	Removes objects from drawings.
eTransmit	Packages the drawing and related files for transmission by email or courier.
Explode	Breaks down blocks into individual objects; reduces polylines to lines and arcs.
Export	Exports drawings in several file formats.
Extend	Extends objects to meet boundary objects.
Extrude	Extrudes 2D closed objects into 3D solid objects (a solid modeling command).
F	
2005 **Field**	Inserts updatable field text in drawings.
'Fill	Toggles the display of solid fills.
Fillet	Connects two lines with an arc.
'Filter	Creates selection sets of objects based on their properties.
Find	Finds and replaces text.
Fog	Adds fog or depth effects to a rendering.
G	
GoToURL	Links to hyperlinks
'GraphScr	F2: Switches to the drawing window from the Text window.
'Grid	F7 and CTRL+G: Displays grid of specified spacing.
Group	Creates named selection sets of objects. (CTRL+SHIFT+A toggles group selection style).
H	
Hatch	Performs hatching at the command prompt.
HatchEdit	Edits associative hatch patterns.
'Help	? and F1: Displays a list of AutoCAD commands with detailed information.
Hide	Removes hidden lines from the currently-displayed view.
HlSettings	Adjusts the display settings for hidden-line removal.
HyperLink	CTRL+K: Attaches hyperlinks to objects, or modifies existing hyperlinks.
HyperLinkOptions	Controls the visibility of the hyperlink cursor and the display of hyperlink tooltips.
I	
'Id	Describes the position of a point in x,y,z coordinates.
Image	Controls the insertion of raster images with an xref-like dialog box.
ImageAdjust	Controls the brightness, contrast, and fading of raster images.
ImageAttach	Attaches raster images to the current drawing.
ImageClip	Places rectangular or irregular clipping boundaries around images.
ImageFrame	Toggles the display of the image frames.
ImageQuality	Controls the display quality of images.
Import	Imports a variety of file formats into drawings.

Command	*Description*
Insert	Inserts blocks and other drawings into the current drawing.
InsertObj	Inserts objects generated by another Windows application.
Interfere	Determines the interference of two or more 3D solids (a solid modeling command).
Intersect	Creates a 3D solid or 2D region from the intersection of two or more 3D solids or 2D regions (a solid modeling command).
'Isoplane	F5 and CTRL+E: Switches to the next isoplane.
J	
JpgOut	Exports views as JPEG files.
JustifyText	Changes the justification of text.
L	
'Layer	Creates and changes layers; toggles the state of layers; assigns linetypes, lineweights, plot styles, colors, and other properties to layers.
LayerP	Displays the previous layer state.
LayerPMode	Toggles the availability of the LayerP command.
Layout	Creates a new layout and renames, copies, saves, or deletes existing layouts.
LayoutWizard	Designates page and plot settings for new layouts.
LayTrans	Translates layer names from one space to another.
Leader	Draws leader dimensions.
Lengthen	Lengthens or shortens open objects.
Light	Creates, names, places, and deletes "lights" used by the Render command.
'Limits	Sets drawing boundaries.
Line	Draws straight line segments.
'Linetype	Lists, creates, and modifies linetype definitions; loads them for use in drawings.
List	Displays database information for selected objects.
Load	Loads shape files into drawings.
LogFileOff	Closes the *.log* keyboard logging file.
LogFileOn	Writes the text of the 'Command:' prompt area to *.log* log file.
LsEdit	Edits landscape objects.
LsLib	Accesses libraries of landscape objects.
LsNew	Places landscape items in drawings.
'LtScale	Specifies the scale for all linetypes in drawings.
LWeight	Sets the current lineweight, lineweight display options, and lineweight units.
M	
2005 **Markup**	CTRL+7: Reviews and changes the status of marked-up *.dwf* files.
2005 **MarkupClose**	CTRL+7: Closes the Markup Set Manager window.
MassProp	Calculates and displays the mass properties of 3D solids and 2D regions (a solid modeling command).
2005 **MatchCell**	Copies the properties from one table cell to other cells.
'MatchProp	Copies properties from one object to other objects.
MatLib	Imports material-look definitions; used by the Render command.
Measure	Places points or blocks at specified distances along objects.
Menu	Loads menus of AutoCAD commands into the menu area.
MenuLoad	Loads partial menu files.
MenuUnLoad	Unloads partial menu files.
MInsert	Makes arrays of inserted blocks.
Mirror	Creates mirror images of objects.
Mirror3D	Creates mirror images of objects rotated about a plane.

Command	Description
MlEdit	Edits multilines.
MLine	Draws multiple parallel lines (up to 16).
MlStyle	Defines named mline styles, including color, linetype, and endcapping.
Model	Switches from layout tabs to Model tab.
Move	Moves objects from one location to another.
MRedo	Redoes more than one undo operation.
MSlide	Creates *.sld* slide files of the current display.
MSpace	Switches to model space.
MText	Places formatted paragraph text inside a rectangular boundary.
Multiple	Repeats commands.
MView	Creates and manipulates viewports in paper space.
MvSetup	Sets up new drawings.
N	
New	CTRL+N: Creates new drawings.
NewSheetSet	Creates a new sheet set.
O	
Offset	Constructs parallel copies of objects.
OleConvert	Converts OLE (object linking and embedding) objects to other formats.
OleLinks	Controls objects linked to drawings.
OleOpen	Opens the source OLE file and application.
OleReset	Resets the OLE object to its original form.
OleScale	Displays the OLE Properties dialog box.
Oops	Restores objects accidentally erased by the previous command.
Open	CTRL+O: Opens existing drawings.
OpenDwfMarkup	Opens marked-up *.dwf* files.
OpenSheetset	Opens *.dst* sheet set files.
Options	Customizes AutoCAD's settings.
'Ortho	F8 and CTRL+L: Forces lines to be drawn orthogonally, or as set by the snap rotation angle.
'OSnap	F3 and CTRL+F: Locates geometric points of objects.
P	
PageSetup	Specifies the layout page, plotting device, paper size, and settings for new layouts.
'Pan	Moves the view.
PartiaLoad	Loads additional geometry into partially-opened drawings.
-PartialOpen	Loads geometry from a selected view or layer into drawings.
PasteBlock	CTRL+SHIFT+V: Pastes copied block into drawings.
PasteClip	CTRL+V: Pastes objects from the Clipboard into the upper left corner of drawings.
PasteOrig	Pastes copied objects from the Clipboard into new drawings using the coordinates from the original.
PasteSpec	Provides control over the format of objects pasted from the Clipboard.
PcInWizard	Imports *.pcp* and *.pc2* configuration files of plot settings.
PEdit	Edits polylines and polyface objects.
PFace	Constructs polygon meshes defined by the location of vertices.
Plan	Returns to the plan view of the current UCS.
PLine	Creates connected lines, arcs, and splines of specified width.
Plot	CTRL+P: Plots drawing to printers and plotters.

Command	*Description*
PlotStamp	Add information about the drawing to the edge of the plot
PlotStyle	Sets plot styles for new objects, or assigns plot styles to selected objects.
PlotterManager	Launches the Add-a-Plotter wizard and the Plotter Configuration Editor.
PngOut	Exports views as PNG files (portable network graphics format).
Point	Draws points.
Polygon	Draws regular polygons with a specified number of sides.
Preview	Provides a Windows-like plot preview.
Properties	CTRL+1: Displays and changes the properties of existing objects.
PropertiesClose	CTRL+1: Closes the Properties window.
PSetUpIn	Imports user-defined page setups into new drawing layouts.
PSpace	Switches to paper space (layout mode).
Publish	Plots one or more drawings as a drawing set, or exports them in DWF format,
PublishToWeb	Creates a Web page from one or more drawings in DWF, JPEG, or PNG formats.
Purge	Deletes unused blocks, layers, linetypes, and so on.

Q

QDim	Creates continuous dimensions quickly.
QLeader	Creates leaders and leader annotation quickly.
QNew	Starts new drawings based on template files.
QSave	CTRL+S: Saves drawings without requesting a file name.
QSelect	Creates selection sets based on filtering criteria.
QText	Redraws text as rectangles with the next regeneration.
Quit	ALT+F4 and CTRL+Q: Exits AutoCAD.

R

Ray	Draws semi-infinite construction lines.
Recover	Attempts recovery of corrupted or damaged files.
Rectang	Draws rectangles.
Redefine	Restores AutoCAD's definition of a command.
Redo	CTRL+Y: Restores the operations changed by the previous Undo command.
'Redraw	Cleans up the display of the current viewport.
'RedrawAll	Performs redraw in all viewports.
RefClose	Saves or discards changes made during in-place editing of xrefs and blocks.
RefEdit	Selects references for editing.
RefSet	Adds and removes objects from a working set during in-place editing of references.
Regen	Regenerates the drawing in the current viewport.
RegenAll	Regenerates all viewports.
'RegenAuto	Controls whether drawings are regenerated automatically.
Region	Creates 2D region objects from existing closed objects (a solid modeling command).
Reinit	Reinitializes the I/O ports, digitizer, display, plotter, and the *acad.pgp* file.
Rename	Changes the name of blocks, linetypes, layers, text styles, views, and so on.
Render	Creates renderings of 3D objects.
RendScr	Creates renderings on computers with a single, nonwindowing display configured for full-screen rendering.
Replay	Displays BMP, TGA, and TIFF raster files.
'Resume	Continues playing back a script file that had been interrupted by the ESC key.
RevCloud	Draws revision clouds.
Revolve	Creates 3D solids by revolving 2D closed objects around an axis (a solid modeling command).
RevSurf	Draws a revolved surface (a solid modeling command).

Command	*Description*
RMat	Defines, loads, creates, attaches, detaches, and modifies material-look definitions; used by the Render command.
RMLin	Imports *.rml* redline markup files created by Volo View.
Rotate	Rotates objects about specified center points.
Rotate3D	Rotates objects about a 3D axis.
RPref	Set preferences for renderings.
RScript	Restarts scripts.
RuleSurf	Draws ruled surfaces.

S

Command	*Description*
Save / SaveAs	CTRL+SHIFT+S: Saves the current drawing by a specified name.
SaveImg	Saves the current rendering in BMP, TGA, or TIFF formats.
Scale	Changes the size of objects equally in the x, y, z directions.
ScaleText	Resizes text.
Scene	Creates, modifies, and deletes named scenes; used by the Render command.
'Script	Runs script files in AutoCAD.
Section	Creates 2D regions from 3D solids by intersecting a plane through the solid (a solid modeling command).
SecurityOptions	Sets up passwords and digital signatures for drawings.
Select	Preselects objects to be edited; CTRL+A selects all objects.
SetiDropHandler	Specifies how to treat i-drop objects when dragged into drawings from Web sites.
SetUV	Controls how raster images are mapped onto objects.
'SetVar	Views and changes AutoCAD's system variables.
ShadeMode	Shades 3D objects in the current viewport.
Shape	Places shapes from shape files into drawings.
2005 **Sheetset**	CTRL+4: Opens the Sheet Set Manager window.
2005 **SheetsetHide**	CTRL+4: Closes the Sheet Set Manager window.
Shell	Runs other programs outside of AutoCAD.
ShowMat	Reports the material definition assigned to selected objects.
SigValidate	Displays digital signature information in drawings.
Sketch	Allows freehand sketching.
Slice	Slices 3D solids with a planes (a solid modeling command).
'Snap	F9 / CTRL+B: Toggles snap mode on or off, changes the snap resolution, sets spacing for the X- and Y-axis, rotates the grid, and sets isometric mode.
SolDraw	Creates 2D profiles and sections of 3D solid models in viewports created with the SolView command (a solid modeling command).
Solid	Draws filled triangles and rectangles.
SolidEdit	Edits faces and edges of 3D solid objects (a solid modeling command).
SolProf	Creates profile images of 3D solid models (a solid modeling command).
SolView	Creates viewports in paper space of orthogonal multi- and sectional view drawings of 3D solid model (a solid modeling command).
SpaceTrans	Converts length values between model and paper space.
Spell	Checks the spelling of text in the drawing.
Sphere	Draws 3D spheres (a solid modeling command).
Spline	Draws NURBS (spline) curve (a solid modeling command).
SplinEdit	Edits splines.
Standards	Compares CAD standards between two drawings.
Stats	Lists information about the current state of rendering.
'Status	Displays information about the current drawing.
StlOut	Exports 3D solids to a *.stl* files, in ASCII or binary format, for use with stereolithography (a solid modeling command).

Command	*Description*
Stretch	Moves selected objects while keeping connections to other objects unchanged.
'Style	Creates and modifies text styles.
StylesManager	Displays the Plot Style Manager.
Subtract	Creates new 3D solids and 2D regions by subtracting one object from a second object (a solid modeling command).
SysWindows	CTRL+TAB: Controls the size and position of windows.

T

Command	*Description*
Table	Inserts tables in drawings.
TablEdit	Edits the content of table cells.
TableExport	Exports tables as comma-separated text files.
TableStyle	Specifies table styles.
Tablet	F4: Aligns digitizers with existing drawing coordinates; operates only when a digitizer is connected.
TabSurf	Draws tabulated surfaces.
Text	Places text in the drawing.
TextToFront	Displays text and/or dimensions in front of all other objects.
'TextScr	F2: Displays the Text window.
TInsert	Inserts blocks and drawings in table cells.
TifOut	Exports views as TIFF files (tagged image file format).
'Time	Keeps track of time.
Tolerance	Selects tolerance symbols.
-Toolbar	Controls the display of toolboxes.
ToolPalettes	CTRL+3: Opens the Tool Palettes window.
ToolPalettesClose	CTRL+3: Closes the Tool Palettes window.
Torus	Draws doughnut-shaped 3D solids (a solid modeling command).
Trace	Draws lines with width.
Transparency	Toggles the background of a bilevel image transparent or opaque.
TraySettings	Specifies options for commands operating from the tray (right end of status bar).
'TreeStat	Displays information on the spatial index.
Trim	Trims objects by defining other objects as cutting edges.

U

Command	*Description*
U	CTRL+Z: Undoes the effect of commands.
UCS	Creates and manipulates user-defined coordinate systems.
UCSicon	Controls the display of the UCS icon.
UcsMan	Manages user-defined coordinate systems.
Undefine	Disables commands.
Undo	Undoes several commands in a single operation.
Union	Creates new 3D solids and 2D regions from two solids or regions (a solid modeling command).
'Units	Selects the display format and precision of units and angles.
UpdateField	Updates the contents of field text.
UpdateThumbsNow	Updates the preview images in sheet sets.

V

Command	*Description*
VbaIDE	ALT+F8: Launches the Visual Basic Editor.

Command	Description
VbaLoad	Loads VBA projects into AutoCAD.
VbaMan	Loads, unloads, saves, creates, embeds, and extracts VBA projects.
VbaRun	ALT+F11: Runs VBA macros.
VbaStmt	Executes VBA statements at the command prompt.
VbaUnload	Unloads global VBA projects.
View	Saves the display as a view; displays named views.
ViewPlotDetails	Displays a report on successful and unsuccessful plots.
ViewRes	Controls the fast zoom mode and resolution for circle and arc regenerations.
VLisp	Launches the Visual LISP interactive development environment.
VpClip	Clips viewport objects.
VpLayer	Controls the independent visibility of layers in viewports.
VpMin	Restores the maximized viewport.
VpMax	Maximizes the current viewport (in layout mode).
VPoint	Sets the viewpoint from which to view 3D drawings.
VPorts	CTRL+R: Sets the number and configuration of viewports.
VSlide	Displays *.sld* slide files created with MSlide.

W

Command	Description
WBlock	Writes objects to a drawing files.
Wedge	Creates 3D solid wedges (a solid modeling command).
WhoHas	Displays ownership information for opened drawing files.
WmfIn	Imports *.wmf* files into drawings as blocks.
WmfOpts	Controls how *.wmf* files are imported.
WmfOut	Exports drawings as *.wmf* files.

X

Command	Description
XAttach	Attaches externally-referenced drawing files to the drawing.
XBind	Binds externally referenced drawings; converts them to blocks.
XClip	Defines clipping boundaries; sets the front and back clipping planes.
XLine	Draws infinite construction lines.
XOpen	Opens externally-referenced drawings in independent windows.
Xplode	Breaks compound objects into component objects, with user control.
Xref	Places externally-referenced drawings into drawings.

Z

Command	Description
'Zoom	Increases and decreases the viewing size of drawings.

3

Command	Description
3D	Draws 3D surface objects of polygon meshes (boxes, cones, dishes, domes, meshes, pyramids, spheres, tori, and wedges).
3dArray	Creates 3D arrays.
3dClip	Switches to interactive 3D view, and opens the Adjust Clipping Planes window.
3dConfig	Configures the 3D graphics system from the command line.
3dCOrbit	Switches to interactive 3D view, and set objects into continuous motion.
3dDistance	Switches to interactive 3D view, and makes objects appear closer or farther away.
3dFace	Creates 3D faces.
3dMesh	Draws 3D meshes.
3dOrbit	Controls the interactive 3D viewing.
3dOrbitCtr	Centers the view.

Command	*Description*
3dPan	Invokes interactive 3D view to drag the view horizontally and vertically.
3dPoly	Draws 3D polylines.
3dsIn	Imports 3D Studio geometry and rendering data.
3dsOut	Exports AutoCAD geometry and rendering data in *.3ds* format.
3dSwivel	Switches to interactive 3D view, simulating the effect of turning the camera.
3dZoom	Switches to interactive 3D view to zoom in and out.

COMMAND ALIASES

Aliases are shortened versions of command names. They are defined in the *acad.ppg* file.

Command	*Aliases*
A	
AdCenter	Dc, Dcenter, Adc, Content
Align	Al
AppLoad	Ap
Arc	A
Area	Aa
Array	Ar
-Array	-Ar
AttDef	Att, Ddattdef
-AttDef	-Att
AttEdit	Ate, Ddatte
-AttEdit	-Ate, Atte
AttExt	Ddattext
B	
BHatch	H, Bh
Block	B, Bmake, Bmod, Acadblockdialog
-Block	-B
Boundary	Bo, Bpoly
-Boundary	-Bo
Break	Br
C	
Chamfer	Cha
Change	-Ch
CheckStandards	Chk
Circle	C
Color	Col, Colour, Ddcolor
Copy	Cp, Co
D	
Dbconnect	Ase, Aad, Dbc, Aex, Asq, Ali, Aro
Ddedit	Ed
Ddgrips	Gr
Ddvpoint	Vp
DIMENSIONS	
DimAligned	Dal, Dimali
DimAngular	Dan, Dimang
DimBaseline	Dba, Dimbase
DimCenter	Dce
DimContinue	Dco, Dimcont
DimDiameter	Ddi, Dimdia
DimDisassociate	Dda
DimEdit	Dimed, Ded
DimLinear	Dli, Dimlin
DimOrdinate	Dor, Dimord
DimOverride	Dov, Dimover
DimRadius	Dra, Dimrad
DimReassociate	Dre
DimStyle	D, Dimsty, Dst, Ddim
DimTEdit	Dimted
Dist	Di
Divide	Div
Donut	Doughnut
Draworder	Dr
DSettings	Se, Ds, Ddrmodes
DsViewer	Av
DView	Dv
E	
Ellipse	El
Erase	E
Explode	X
Export	Exp
Extend	Ex
Extrude	Ext
F	
Fillet	F
Filter	Fi
G	
Group	G
-Group	-G
H	
Hatch	-H
Hatchedit	He
Hide	Hi
I	
Image	Im
-Image	-Im
ImageAdjust	Iad
ImageAttach	Iat
ImageClip	Icl
Import	Imp
Insert	I, Inserturl, Ddinsert

Command	Aliases
-Insert	-I
Insertobj	Io
Interfere	Inf
Intersect	In
L	
Layer	Ddlmodes, La
-Layer	-La
-Layout	Lo
Leader	Lead
Lengthen	Len
Line	L
Linetype	Ltype, Lt, Ddltype
-Linetype	-Ltype, -Lt
List	Ls, Li
LtScale	Lts
Lweight	Lineweight, Lw
M	
Markup	Msm
MatchProp	Ma, Painter
Measure	Me
Mirror	Mi
MLine	Ml
Move	M
MSpace	Ms
MText	T, Mt
-MText	-T
MView	Mv
O	
Offset	O
Open	Openurl
Options	Op, Preferences
OSnap	Os, Ddosnap
-OSnap	-Os
P	
Pan	P
-Pan	-P
-PartialOpen	Partialopen
PasteSpec	Pa
PEdit	Pe
PLine	Pl
Plot	Print, Dwfout
PlotStamp	Ddplotstamp
Point	Po
Polygon	Pol
Preview	Pre
Properties	Props, Pr, Mo, Ch, Ddchprop, Ddmodify
PropertiesClose	Prclose

Command	Aliases
PSpace	Ps
PublishToWeb	Ptw
Purge	Pu
-Purge	-Pu
Q	
QLeader	Le
Quit	Exit
R	
Rectang	Rec, Rectangle
Redraw	R
RedrawAll	Ra
Regen	Re
RegenAll	Rea
Region	Reg
Rename	Ren
-Rename	-Ren
Render	Rr
Revolve	Rev
Rotate	Ro
RPref	Rpr
S	
Save	Saveurl
Scale	Sc
Script	Scr
Section	Sec
SetVar	Set
ShadeMode	Sha, Shade
Sheetset	Ssm
Slice	Sl
Snap	Sn
Solid	So
Spell	Sp
Spline	Spl
SplinEdit	Spe
Standards	Sta
Stretch	S
Style	St, Ddstyle
Subtract	Su
T	
Table	Tb
TableStyle	Ts
Tablet	Ta
Text	Dt, Dtext
Thickness	Th
Tilemode	Ti, Tm
Tolerance	Tol
Toolbar	To

Command	Aliases
ToolPalettes	Tp
Torus	Tor
Trim	Tr
U	
UcsMan	Uc, Dducs, Dducsp
Union	Uni
Units	Un, Ddunits
-Units	-Un
V	
View	V, Ddview
-View	-V
VPoint	-Vp
VPorts	Viewports
W	
Wblock	W, Acadwblockdialog
-Wblock	-W
Wedge	We

Command	Aliases
X	
XAttach	Xa
XBind	Xb
-XBind	-Xb
XClip	Xc
XLine	Xl
XRef	Xr
-XRef	-Xr
Z	
Zoom	Z
3	
3dArray	3a
3dFace	3f
3dOrbit	3do, Orbit
3dPoly	3p

KEYBOARD SHORTCUTS

CONTROL KEYS

Function keys can be customized with the **Tools | Customize | Keyboard** command.

Ctrl-key	*Meaning*
CTRL+0	Toggles clean screen.
CTRL+1	Displays the Properties window.
CTRL+2	Opens the AutoCAD DesignCenter window.
CTRL+4	2005 Toggles Sheetset Manager window.
CTRL+5	2005 Toggles Info Palette window (Assist).
CTRL+6	Launches dbConnect.
CTRL+7	2005 Toggles Markup Set Manager window.
CTRL+A	Selects all objects in the drawing.
CTRL+SHIFT+A	Toggles group selection mode.
CTRL+B	Turns snap mode on or off.
CTRL+C	Copies selected objects to the Clipboard.
CTRL+SH+C	Copies selected objects with a base point to the Clipboard.
CTRL+D	Changes the coordinate display mode.
CTRL+E	Switches to the next isoplane.
CTRL+F	Toggles object snap on and off.
CTRL+G	Turns the grid on and off.
CTRL+H	Toggles pickstyle mode).
CTRL+K	Creates a hyperlink.
CTRL+L	Turns ortho mode on and off.
CTRL+N	Starts a new drawing.
CTRL+O	Opens a drawing.
CTRL+P	Prints the drawing.
CTRL+Q	Quits AutoCAD.
CTRL+R	Switches to the next viewport.
CTRL+S	Saves the drawing.
CTRL+SH+S	Displays the Save Drawing As dialog box.
CTRL+T	Toggles tablet mode.
CTRL+V	Pastes from the Clipboard into the drawing or to the command prompt area.
CTRL+SH+V	Pastes with an insertion point.
CTRL+X	Cuts selected objects to the Clipboard.
CTRL+Y	Performs the REDO command.
CTRL+Z	Performs the U command.
CTRL+TAB	Switches to the next drawing.

COMMAND LINE KEYSTROKES

These keystrokes are used in the command-prompt area and the Text window.

Keystroke	*Meaning*
left arrow	Moves the cursor one character to the left.
right arrow	Moves the cursor to the right.
HOME	Moves the cursor to the beginning of the line of command text.
END	Moves the cursor to the end of the line.
DEL	Deletes the character to the right of the cursor.
BACKSPACE	Deletes the character to the left of the cursor.
INS	Switches between insert and typeover modes.
up arrow	Displays the previous line in the command history.
down arrow	Displays the next line in the command history.
PGUP	Displays the previous screen of command text.
PGDN	Displays the next screen of command text.
CTRL+V	Pastes text from the Clipboard into the command line.
ESC	Cancels the current command.

FUNCTION KEYS

Function keys can be customized with the **Tools | Customize | Keyboard** command.

Function Key	*Meaning*
F1	Calls up the help window.
F2	Toggles between the graphics and text windows.
F3	Toggles object snap on and off.
F4	Toggles tablet mode on and off; you must first calibrate the tablet before toggling tablet mode.
ALT+F4	Exits AutoCAD.
CTRL+F4	Closes the current drawing.
F5	Switches to the next isometric plane when in iso mode; the planes are displayed in order of left, top, right, and then repeated.
F6	Toggles the screen coordinate display on and off.
F7	Toggles grid display on and off.
F8	Toggles ortho mode on and off.
ALT+F8	Displays the Macros dialog box.
F9	Toggles snap mode on and off.
F10	Toggles polar tracking on and off.
F11	Toggles object snap tracking on and off.
ALT+F11	Starts Visual Basic for Applications editor.

ALTERNATE KEYS

The menu-related Alt keys can be customized in the *acad.mnu* file.

Alt-key	*Meaning*
ALT+TAB	Switches to the next application.
ALT+-	(dash) Accesses window control menu.
ALT+D	Accesses the Draw menu.
ALT+E	Accesses the Edit menu.
ALT+F	Accesses the File menu.
ALT+H	Accesses the Help menu.
ALT+I	Accesses the Insert menu.
ALT+M	Accesses the Modify menu.
ALT+N	Accesses the Dimension menu.
ALT+O	Accesses the Format menu.
ALT+T	Accesses the Tools menu.
ALT+V	Accesses the View menu.
ALT+W	Accesses the Window menu.
ALT+X	Accesses the Express menu.

MOUSE AND DIGITIZER BUTTONS

The mouse and digitizer buttons can be customized in the *acad.mnu* file.

Button #	*Mouse Button*	*Meaning*
...	**Wheel**	Zooms or pans.
1	**Left**	Selects objects.
3	**Center**	Displays object snap menu.
2	**Right**	Displays shortcut menus.
4		Cancels command.
5		Toggles snap mode.
6		Toggles orthographic mode.
7		Toggles grid display
8		Toggles coordinate display.
9		Switches isometric plane.
10		Toggles tablet mode.
11		Not defined.
12		Not defined.
13		Not defined.
14		Not defined.
15		Not defined.
SHIFT+1	SHIFT+Left	Toggles cycle mode.
SHIFT+2	SHIFT+Left	Displays object snap shortcut menu.
CTRL+2	CTRL+Right	Displays object snap shortcut men

MTEXT EDITOR SHORTCUT KEYS

Shortcut	Meaning
TAB	Moves cursor to the next tab stop.
CTRL+A	Selects all text.
CTRL+B	**Boldface** toggle.
CTRL+I	*Italicize* toggle.
CTRL+U	Underline toggle.
Case Conversion:	
CTRL+SHIFT+A	Toggles between all UPPERCASE and all lowercase.
CTRL+SHIFT+U	Converts selected text to all UPPERCASE.
CTRL+SHIFT+L	Converts selected text to all lowercase.
Justification:	
CTRL+L	Left-justifies selected text.
CTRL+E	Centers selected text.
CTRL+R	Right-justifies selected text.
Line Spacing:	
CTRL+1	Single-spaces lines.
CTRL+5	1.5-spaces lines.
CTRL+2	Double-spaces lines.
Font Sizing:	
CTRL+SHIFT+,	*(comma)* Reduces font size.
CTRL+SHIFT+.	*(period)* Increases font size.
CTRL+=	Superscript toggle.
CTRL+SHIFT+=	Subscript toggle.
Copy and Paste:	
CTRL+C	Copies selected text to the Clipboard.
CTRL+X	Cuts text from the editor and send it to the Clipboard.
CTRL+V	Pastes text from the Clipboard.
Undo and Redo:	
CTRL+Y	Redoes last undo.
CTRL+Z	Undoes last action.
Symbols:	
CTRL+ALT+E	Inserts Euro symbol.
ALT+SHIFT+X	Converts selected text to ASCII number (A becomes 41, and so on).

MTEXT CONTROL CODES

Control Code	Meaning
\~	Nonbreaking space.
\\	Backslash.
\{	Opening brac; encloses multiple controls codes.
\}	Closing brace.
\A*n*	Set vertical alignment of text in boundary box:
	0 = bottom alignment.
	1 = center alignment.
	2 = top alignment.
\C*n*	Sets color of text:
	1 = red.
	2 = yellow.
	3 = green.
	4 = cyan.
	5 = blue.
	6 = magenta.
	7 = white.
\F*x*;	Changes to font file name *x*.
\H*n*;	Changes text height to *n* units.
\L	Underline.
\l	Turns off underline.
\M+*nnn*	Multibyte shape number *nnn*.
\O	Turns on overline.
\o	Turns off overline.
\P	End of paragraph.
\Q*n*;	Changes obliquing angle to *n*.
\S*n*^*m*	Stacks character *n* over *m*.
\T*n*;	Changes tracking to *n*; for example, \T3; is wide spacing beween characters.
\U+*nnn*	Places Unicode character *nnn*.
\W*n*;	Changes width factor (width of characters) to *n*.

APPENDIX

B

AutoCAD System Variables

AutoCAD stores information about its current state, the drawing and the operating system in over 400 *system variables*. Those variables help programmers — who often work with menu macros and AutoLISP — to determine the state of the AutoCAD system.

CONVENTIONS

The following pages list all documented system variables, plus several more not documented by Autodesk. The listing uses the following conventions:

Bold	System variable is documented in AutoCAD.
Italicized	System variable is not listed by the **SETVAR** command or Autodesk documentation.
	System variable must be accessed via the **SETVAR** command.
NEW IN 2005	System variable is new to AutoCAD 2005.

COLUMN HEADINGS

Default	Default value, as set in the *acad.dwg* prototype drawing.
R/O	Read-only; cannot be changed by the user or by a program.
Loc	Location where the value of the system variable is saved:

Location	*Meaning*
ACAD	Set by AutoCAD.
DWG	Saved in current drawing.
REG	Saved globally in Windows registry.
...	Not saved.

Variable	Default	R/O	Loc	Meaning
_PkSer	*varies*	R/O	*ACAD*	*Software serial number, such as "117-69999999".*
_Server	*0*	R/O	REG	*Network authorization code.*
_VerNum	*varies*	R/O	REG	*Internal program build number, such as "N.41.101".*

A

Variable	Default	R/O	Loc	Meaning
AcadLspAsDoc	0	...	REG	*acad.lsp* is loaded into: **0** Just the first drawing. **1** Every drawing.
AcadPrefix	*varies*	R/O	...	Paths specified by the AutoCAD search path in Options | Files tab. May be controlled by the **ACAD** environment variable.
AcadVer	"16.1"	R/O	...	AutoCAD version number.
AcisOutVer	40	R/O	...	ACIS version number, such as 15, 16, 17, 18, 20, 21, 30, 40, or 70.
AcGiDumpMode	0	...	...	Value of 0 or 1.
AdcState	0	R/O	...	Specifies if **DesignCenter** is active.
AFlags	0	...	...	Attribute display code: **0** No mode specified. **1** Invisible. **2** Constant. **4** Verify. **8** Preset.
AngBase	0	...	DWG	Direction of zero degrees relative to UCS.
AngDir	0	...	DWG	Rotation of positive angles: **0** Clockwise. **1** Counterclockwise.
ApBox	0	...	REG	AutoSnap aperture box cursor: **0** Off. **1** On.
Aperture	10	...	REG	Object snap aperture in pixels: **1** Minimum size. **50** Maximum size.
AssistState	0	R/O	...	Specifies if **Info Palette** is active.
Area	0.0	R/O	...	Area measured by the last **Area**, **List**, or **Dblist** commands.
AttDia	0	...	DWG	Attribute entry interface: **0** Command-line prompts. **1** Dialog box.
AttMode	1	...	DWG	Display of attributes: **0** Off. **1** Normal. **2** On.
AttReq	1	...	REG	Attribute values during insertion are: **0** Default values. **1** Prompt for values.
AuditCtl	0	...	REG	Determines creation of *.adt* audit log file: **0** File not created. **1** *.adt* file created.
AUnits	0	...	DWG	Mode of angular units: **0** Decimal degrees. **1** Degrees-minutes-seconds. **2** Grads. **3** Radians. **4** Surveyor's units.

Variable	Default	R/O	Loc	Meaning
AUPrec	0	...	DWG	Decimal places displayed by angles.
AutoSnap	63	...	REG	Controls AutoSnap display: **0** Turns off all AutoSnap features. **1** Turns on marker. **2** Turns on SnapTip. **4** Turns on magnetic cursor. **8** Turns on polar tracking . **16** Turns on object snap tracking . **32** Turns on tooltips for polar tracking and object snap tracking .
AuxStat	*0*	...	DWG	*-32768 Minimum value.* *32767 Maximum value.*
~~AxisMode~~	*0*	...	DWG	*Removed from AutoCAD 2002.*
AxisUnit	*0.0*	...	DWG	*Obsolete system variable.*

B

Variable	Default	R/O	Loc	Meaning
BackZ	0.0	R/O	DWG	Back clipping plane offset.
BackgroundPlot	2	...	REG	Controls background plotting and publishing (ignored during scripts): **0** Plot foreground; publish foreground. **1** Plot background; publish foreground. **2** Plot foreground; publish background. **3** Plot background; publish background.
BgrdPlotTimeout	*20*	...	...	*Background plot timeout; ranges from 0 to 300 secs.*
BindType	0	...	...	When binding an xref or editing an xref, xref names are converted from: **0** **xref|name** to **xref0name**. **1** **xref|name** to **name**.
BlipMode	0	...	DWG	Display of blip marks: **0** Off. **1** On.

C

Variable	Default	R/O	Loc	Meaning
CDate	*varies*	R/O	...	Current date and time in the format YyyyMmDd.HhMmSsDd, such as 20010503.18082328
CeColor	"BYLAYER"	...	DWG	Current color.
CeLtScale	1.0	...	DWG	Current linetype scaling factor.
CeLType	"BYLAYER"	...	DWG	Current linetype.
CeLWeight	-1	...	DWG	Current lineweight in millimeters; valid values are 0, 5, 9, 13, 15, 18, 20, 25, 30, 35, 40, 50, 53, 60, 70, 80, 90, 100, 106, 120, 140, 158, 200, and 211, plus the following: **-1** BYLAYER. **-2** BYBLOCK. **-3** DEFAULT as defined by **LwDdefault**.
ChamferA	0.5	...	DWG	First chamfer distance.
ChamferB	0.5	...	DWG	Second chamfer distance.
ChamferC	1.0	...	DWG	Chamfer length.
ChamferD	0	...	DWG	Chamfer angle.

Variable	*Default*	*R/O*	*Loc*	*Meaning*
ChamMode	0	...	...	Chamfer input mode: **0** Chamfer by two lengths. **1** Chamfer by length and angle.
CircleRad	0.0	...	...	Most-recent circle radius.
CLayer	"0"	...	DWG	Current layer name.
CleanScreenState	0	R/O	...	Specifies if cleanscreen mode is active.
CmdActive	1	R/O	...	Type of current command: **1** Regular command. **2** Transparent command. **4** Script file. **8** Dialog box. **16** AutoLISP is active .
CmdEcho	1	...	...	AutoLISP command display: **0** No command echoing. **1** Command echoing.
CmdNames	*varies*	R/O	...	Current command, such as "SETVAR".
CMLJust	0	...	DWG	Multiline justification mode: **0** Top. **1** Middle. **2** Bottom.
CMLScale	1.0	...	DWG	Scales width of multiline: **-*n*** Flips offsets of multiline. **0** Collapses to single line. *n* Scales by a factor of *n*.
CMLStyle	"STANDARD"	...	DWG	Current multiline style name.
Compass	0	...	...	Toggles display of the 3D compass: **0** Off. **1** On.
Coords	1	...	DWG	Coordinate display style: **0** Updated by screen picks. **1** Continuous display. **2** Polar display upon request.
CPlotStyle	"ByColor"	...	DWG	Current plot style; values defined by AutoCAD are: "ByLayer" "ByBlock" "Normal" "User Defined"
CProfile	"<<Unnamed Profile>>"	R/O	REG	Current profile.
CTab	"Model"	R/O	DWG	Current tab.
CursorSize	5	...	REG	Cursor size, in percent of viewport: **1** Minimum size. **100** Full viewport.
CVPort	2	...	DWG	Current viewport number.

Variable	*Default*	*R/O*	*Loc*	*Meaning*
D				
Date	*varies*	R/O	...	Current date in Julian format, such as 2448860.54043252
DBMod	4	R/O	...	Drawing modified, as follows: **0** No modification since last save. **1** Object database modified. **2** Symbol table modified. **4** Database variable modified. **8** Window modified. **16** View modified.
DbcState	0	R/O	DWG	Specifies if dbConnect Manager is active.
DctCust	"d:\acad 2005\support\sample.cus"	...	REG	Name of custom spelling dictionary.
DctMain	"enu"	...	REG	Code for spelling dictionary: **ca** Catalan. **cs** Czech. **da** Danish. **de** German; sharp 's'. **ded** German; double 's'. **ena** English; Australian. **ens** English; British 'ise'. **enu** English; American. **enz** English; British 'ize'. **es** Spanish; unaccented capitals. **esa** Spanish; accented capitals. **fi** Finish. **fr** French; unaccented capitals. **fra** French; accented capitals. **it** Italian. **nl** Dutch; primary. **nls** Dutch; secondary. **no** Norwegian; Bokmal. **non** Norwegian; Nynorsk. **pt** Portuguese; Iberian. **ptb** Portuguese; Brazilian. **ru** Russian; infrequent 'io'. **rui** Russian; frequent 'io'. **sv** Swedish.
2005 ***DefaultViewCategory***	""	...	...	Default name for View Category in the **View** command's New View dialog box
DefLPlStyle	"ByColor"	R/O	REG	Default plot style for new layers.
DefPlStyle	"ByColor"	R/O	REG	Default plot style for new objects.
DelObj	1	...	REG	Toggle source objects deletion: **0** Objects deleted. **1** Objects retained.
DemandLoad	3	...	REG	When drawing contains proxy objects: **0** Demand loading turned off. **1** Load app when drawing opened. **2** Load app at first command. **3** Load app when drawing opened or at first command.

Variable	*Default*	*R/O*	*Loc*	*Meaning*
DiaStat	1	R/O	...	User exited dialog box by clicking: **0** **Cancel** button. **1** **OK** button.

Dimension Variables

Variable	*Default*	*R/O*	*Loc*	*Meaning*
DimADec	0	...	DWG	Angular dimension precision: **-1** Use **DimDec** setting (default). **0** Zero decimal places (minimum). **8** Eight decimal places (maximum).
DimAlt	Off	...	DWG	Alternate units: **On** Enabled. **Off** Disabled.
DimAltD	2	...	DWG	Alternate unit decimal places.
DimAltF	25.4	...	DWG	Alternate unit scale factor.
DimAltRnd	0.0	...	DWG	Rounding factor of alternate units.
DimAltTD	2	...	DWG	Tolerance alternate unit decimal places.
DimAltTZ	0	...	DWG	Alternate tolerance units zeros: **0** Zeros not suppressed. **1** All zeros suppressed. **2** Include 0 feet, but suppress 0 inches . **3** Includes 0 inches, but suppress 0 feet. **4** Suppresses leading zeros. **8** Suppresses trailing zeros.
DimAltU	2	...	DWG	Alternate units: **1** Scientific. **2** Decimal. **3** Engineering. **4** Architectural; stacked. **5** Fractional; stacked. **6** Architectural. **7** Fractional. **8** Windows desktop units setting.
DimAltZ	0		DWG	Zero suppression for alternate units: **0** Suppress 0 ft and 0 in. **1** Include 0 ft and 0 in. **2** Include 0 ft; suppress 0 in. **3** Suppress 0 ft; include 0 in. **4** Suppress leading 0 in dec dims. **8** Suppress trailing 0 in dec dims. **12** Suppress leading and trailing zeroes.
DimAPost	""	...	DWG	Prefix and suffix for alternate text.
DimAso	On	...	DWG	Toggle associative dimensions: **On** Dimensions are created associative. **Off** Dimensions are not associative.
DimAssoc	2	...	DWG	Controls creation of dimensions: **0** Dimension elements are exploded. **1** Single dimension object, attached to defpoints. **2** Single dimension object, attached to geometric objects.
DimASz	0.18	...	DWG	Arrowhead length.

Variable	Default	R/O	Loc	Meaning
DimAtFit	3	...	DWG	When insufficient space between extension lines, dimension text and arrows are fitted: **0** Text and arrows outside extension lines. **1** Arrows first outside, then text. **2** Text first outside, then arrows. **3** Either text or arrows, whichever fits better.
DimAUnit	0	...	DWG	Angular dimension format: **0** Decimal degrees. **1** Degrees.Minutes.Seconds. **2** Grad. **3** Radian. **4** Surveyor units.
DimAZin	0	...	DWG	Supress zeros in angular dimensions: **0** Display all leading and trailing zeros. **1** Suppress 0 in front of decimal. **2** Suppress trailing zeros behind decimal. **3** Suppress zeros in front and behind the decimal.
DimBlk Closed filled Closed blank Closed Dot Architectural tick Oblique Open Origin indicator Origin indicator 2 Right angle Open 30 Dot small Dot blank Dot small blank Box Box filled Datum triangle Datum triangle filled Integral None	""	R/O	DWG	Arrowhead block name: Architectural tick: "Archtick" Box filled: "Boxfilled" Box: "Boxblank" Closed blank: "Closedblank" Closed filled: "" (default) Closed: "Closed" Datum triangle filled: "Datumfilled" Datum triangle: "Datumblank" Dot blanked: "Dotblank" Dot small: "Dotsmall" Dot: "Dot" Integral: "Integral" None: "None" Oblique: "Oblique" Open 30: "Open30" Open: "Open" Origin indication: "Origin" Right-angle: "Open90"
DimBlk1	""	R/O	DWG	Name of first arrowhead's block; uses same list of names as under **DimBlk**. **.** No arrowhead.
DimBlk2	""	R/O	DWG	Name of second arrowhead's block; uses same list of names as under **DimBlk**. **.** No arrowhead.
DimCen	0.09	...	DWG	Center mark size: **-*n*** Draws center lines. **0** No center mark or lines drawn. **+*n*** Draws center marks of length *n*.
DimClrD	0	...	DWG	Dimension line color: **0** BYBLOCK (default) **1** Red. ... **255** Dark gray. **256** BYLAYER.
DimClrE	0	...	DWG	Extension line and leader color.

Variable	*Default*	*R/O*	*Loc*	*Meaning*
DimClrT	0	...	DWG	Dimension text color.
DimDec	4	...	DWG	Primary tolerance decimal places.
DimDLE	0.0	...	DWG	Dimension line extension.
DimDLI	0.38	...	DWG	Dimension line continuation increment.
DimDSep	"."	...	DWG	Decimal separator (must be a single char.)
DimExe	0.18	...	DWG	Extension above dimension line.
DimExO	0.0625	...	DWG	Extension line origin offset.
DimFrac	0	...	DWG	Fraction format when **DimLUnit** set to 4 or 5: **0** Horizontal. **1** Diagonal. **2** Not stacked.
DimGap	0.09	...	DWG	Gap from dimension line to text.
DimJust	0	...	DWG	Horizontal text positioning: **0** Center justify. **1** Next to first extension line. **2** Next to second extension line. **3** Above first extension line. **4** Above second extension line.
DimLdrBlk	""	...	DWG	Block name for leader arrowhead; uses same name as **DimBlock**. **.** Supresses display of arrowhead.
DimLFac	1.0	...	DWG	Linear unit scale factor.
DimLim	Off	...	DWG	Generate dimension limits.
DimLUnit	2	...	DWG	Dimension units (except angular); replaces **DimUnit**: **1** Scientific. **2** Decimal. **3** Engineering. **4** Architectural. **5** Fractional. **6** Windows desktop.
DimLwD	-2	...	DWG	Dimension line lineweight; valid values are BYLAYER, BYBLOCK, or an integer multiple of 0.01mm.
DimLwE	-2	...	DWG	Extension lineweight; valid values are BYLAYER, BYBLOCK, or an integer multiple of 0.01mm.
DimPost	""	...	DWG	Default prefix or suffix for dimension text (maximum 13 characters): **""** No suffix. **<>mm** Millimeter suffix. **<>Å** Angstrom suffix.
DimRnd	0.0	...	DWG	Rounding value for dimension distances.
DimSAh	Off	...	DWG	Separate arrowhead blocks: **Off** Use arrowhead defined by **DimBlk**. **On** Use arrowheads defined by **DimBlk1** and **DimBlk2**.
DimScale	1.0	...	DWG	Overall scale factor for dimensions: **0** Value is computed from the scale between current modelspace viewport and paperspace. **>0** Scales text and arrowheads.
DimSD1	Off	...	DWG	Suppress first dimension line: **On** First dimension line is suppressed. **Off** Not suppressed.

Variable	*Default*	*R/O*	*Loc*	*Meaning*
DimSD2	Off	...	DWG	Suppress second dimension line: **On** Second dimension line is suppressed. **Off** Not suppressed.
DimSE1	Off	...	DWG	Suppress the first extension line: **On** First extension line is suppressed. **Off** Not suppressed.
DimSE2	Off	...	DWG	Suppress the second extension line: **On** Second extension line is suppressed. **Off** Not suppressed.
DimSho	On	...	DWG	Update dimensions while dragging: **On** Dimensions are updated during drag. **Off** Dimensions are updated after drag.
DimSOXD	Off	...	DWG	Suppress dimension lines outside extension lines: **On** Dimension lines not drawn outside extension lines. **Off** Are drawn outside extension lines.
DimStyle	"STANDARD"	R/O	DWG	Current dimension style.
DimTAD	0	...	DWG	Vertical position of dimension text: **0** Centered between extension lines. **1** Above dimension line, except when dimension line not horizontal and **DimTIH** = 1. **2** On side of dimension line farthest from the defining points. **3** Conforms to JIS.
DimTDec	4	...	DWG	Primary tolerance decimal places.
DimTFac	1.0	...	DWG	Tolerance text height scaling factor.
DimTIH	On	...	DWG	Text inside extensions is horizontal: **Off** Text aligned with dimension line. **On** Text is horizontal.
DimTIX	Off	...	DWG	Place text inside extensions: **Off** Text placed inside extension lines, if room. **On** Force text between the extension lines.
DimTM	0.0	...	DWG	Minus tolerance.
DimTMove	0	...	DWG	Determines how dimension text is moved: **0** Dimension line moves with text. **1** Adds a leader when text is moved. **2** Text moves anywhere; no leader.
DimTOFL	Off	...	DWG	Force line inside extension lines: **Off** Dimension lines not drawn when arrowheads are outside. **On** Dimension lines drawn, even when arrowheads are outside.
DimTOH	On	...	DWG	Text outside extension lines: **Off** Text aligned with dimension line. **On** Text is horizontal.
DimTol	Off	...	DWG	Generate dimension tolerances: **Off** Tolerances not drawn. **On** Tolerances are drawn.
DimTolJ	1	...	DWG	Tolerance vertical justification: **0** Bottom. **1** Middle. **2** Top.
DimTP	0.0	...	DWG	Plus tolerance.
DimTSz	0.0	...	DWG	Size of oblique tick strokes: **0** Arrowheads. **>0** Oblique strokes.

Variable	*Default*	*R/O*	*Loc*	*Meaning*
DimTVP	0.0	...	DWG	Text vertical position when **DimTAD**=0: **1** Turns **DimTAD** on. **>-0.7** *or* **<0.7** Dimension line is split for text.
DimTxSty	"STANDARD"	...	DWG	Dimension text style.
DimTxt	0.18	...	DWG	Text height.
DimTZin	0	...	DWG	Tolerance zero suppression: **0** Suppress 0 ft and 0 in. **1** Include 0 ft and 0 in. **2** Include 0 ft; suppress 0 in. **3** Suppress 0 ft; include 0 in. **4** Suppress leading 0 in decimal dim. **8** Suppress trailing 0 in decimal dim. **12** Suppress leading and trailing zeroes.
DimUPT	Off	...	DWG	User-positioned text: **Off** Cursor positions dimension line. **On** Cursor also positions text.
DimZIN	0	...	DWG	Suppression of 0 in feet-inches units: **0** Suppress 0 ft and 0 in. **1** Include 0 ft and 0 in. **2** Include 0 ft; suppress 0 in. **3** Suppress 0 ft; include 0 in. **4** Suppress leading 0 in decimal dim. **8** Suppress trailing 0 in decimal dim. **12** Suppress leading and trailing zeroes.
DispSilh	0	...	DWG	Silhouette display of 3D solids: **0** Off. **1** On.
Distance	0.0	R/O	...	Distance measured by last **Dist** command.
DonutId	0.5	...	...	Inside radius of donut.
DonutOd	1.0	...	...	Outside radius of donut.
DragMode	2	...	REG	Drag mode: **0** No drag. **1** On if requested. **2** Automatic.
DragP1	10	...	REG	Regen drag display.
DragP2	25	...	REG	Fast drag display.
DrawOrderCtrl	3	...	DWG	Determines behavior of draw order: **0** Draw order not restored until next regen or drawing reopened. **1** Normal draw order behavior. **2** Draw order inheritance. **3** Combines options 1 and 2.
DwgCheck	0	...	REG	Toggles checking if drawing was edited by software other than AutoCAD: **0** Supresses dialog box. **1** Displays warning dialog box.
DwgCodePage	*varies*	R/O	DWG	Drawing code page, such as "ANSI_1252".
DwgName	*varies*	R/O	...	Current drawing filename, such as "drawing1.dwg".
DwgPrefix	*varies*	R/O	...	Drawing's drive and folder, such as "d:\acad 2005\".

Variable	*Default*	*R/O*	*Loc*	*Meaning*
DwgTitled	0	R/O	...	Drawing filename is: **0** "drawing1.dwg". **1** User-assigned name.

E

Variable	*Default*	*R/O*	*Loc*	*Meaning*
EdgeMode	0	...	REG	Toggle edge mode for **Trim** and **Extend** commands: **0** No extension. **1** Extends cutting edge.
Elevation	0.0	...	DWG	Current elevation, relative to current UCS.
EntExts	*1*	...	...	*Controls how drawing extents are calculated:* *0 Extents calculated every time; slows down AutoCAD but uses less memory.* *1 Extents of every object are cached as a two-byte value (default).* *2 Extents of every object are cached as a four-byte value (fastest but uses more memory).*
EntMods	*0*	R/O	...	*Increments by one each time an object is modified to indicate that an object has been modified since the drawing was opened; value ranges from 0 to 4.29497E9.*
ErrNo	0	...	...	Error number from AutoLISP, ADS, & Arx.
Expert	0	...	...	Suppresses the displays of prompts: **0** Normal prompts. **1** "About to regen, proceed?" and "Really want to turn the current layer off?" **2** "Block already defined. Redefine it?" and "A drawing with this name already exists. Overwrite it?" **3** **Linetype** command messages. **4** **UCS Save** and **VPorts Save**. **5** **DimStyle Save** and **DimOverride.**
ExplMode	1	...	...	Toggles whether **Explode** and **Xplode** commands explode non-uniformly scaled blocks: **0** Does not explode. **1** Explodes.
ExtMax	-1.0E+20, -1.0E+20, -1.0E+20	R/O	DWG	Upper-right coordinate of drawing extents.
ExtMin	1.0E+20, 1.0E+20, 1.0E+20	R/O	DWG	Lower-left coordinate of drawing extents.
ExtNames	1	...	DWG	Format of named objects: **0** Names are limited to 31 characters, and can include A - Z, 0 - 9, dollar ($), underscore (_), and hyphen (-). **1** Names are limited to 255 characters, and can include A - Z, 0 - 9, spaces, and any characters not used by Windows or AutoCAD for special purposes.

F

Variable	*Default*	*R/O*	*Loc*	*Meaning*
FaceTRatio	0	...	...	Controls the aspect ratio of facets on rouunded 3D bodies: **0** Creates an *n* by 1 mesh. **1** Creates an *n* by *m* mesh.
FaceTRres	0.5000	...	DWG	Adjusts smoothness of shaded and hidden-line objects: **0.01** Minimum value. **2** Recommended value. **10** Maximum value.
FieldDisplay	1	...	REG	Toggles gray background to field text.

Variable	*Default*	*R/O*	*Loc*	*Meaning*
FieldEval	31	...	DWG	Determines how fields are updated: **0** Not updated **1** Updated with **Open**. **2** Updated with **Save**. **4** Updated with **Plot**. **8** Updated with **eTransmit**. **16** Updated with regeneration.
FileDia	1	...	REG	User interface for file-accessing commands: **0** Command-line prompts. **1** File dialog boxes.
FilletRad	0.5	...	DWG	Current fillet radius.
FillMode	1	...	DWG	Fill of solid objects and hatches: **0** Off. **1** On.
Flatland	*0*	R/O	...	*Obsolete system variable.*
FontAlt	"simplex.shx"	...	REG	Font name that substitutes for missing fonts.
FontMap	"acad.fmp"	...	REG	Name of font mapping file.
Force_Paging	*0*	...	...	***0*** *Minimum (default).* ***4.29497E9*** *Maximum.*
FrontZ	0.0	R/O	DWG	Front clipping plane offset.
FullOpen	1	R/O	...	Drawing is: **0** Partially loaded. **1** Fully open.

G

Variable	*Default*	*R/O*	*Loc*	*Meaning*
GfAng	0	...	...	Angle of gradient fill; 0 to 360 degrees.
GfClr1	"RGB 000,000,255"	...	...	First gradient color in RGB format.
GfClr2	"RGB 255,255,153"	...	...	Second gradient color in RGB format.
GfClrLum	1.0	...	...	Level of gray in one-color gradients: **0** Black. **1** White.
GfClrState	1	...	...	Specifies type of gradient fill: **0** Two-color. **1** One-color.
GfName	1	...	...	Specifies style of gradient fill: **1** Linear. **2** Cylindrical. **3** Inverted cylindrical. **4** Spherical. **5** Inverted spherical. **6** Hemispherical. **7** Inverted hemispherical. **8** Curved. **9** Inverted curved.
GfAShift	0	...	...	Specifies the origin of the gradient fill: **0** Centered. **1** Shifted up and left.

Variable	*Default*	*R/O*	*Loc*	*Meaning*
GlobCheck	*0*	...	...	*Reports statistics on dialog boxes:* ***-1*** *Turn off local language.* ***0*** *Turn off.* ***1*** *Warns if larger than 640x400.* ***2*** *Also reports size in pixels.* ***3*** *Additional information.*
GridMode	0	...	DWG	Display of grid: **0** Off. **1** On.
GridUnit	0.5,0.5	...	DWG	X,y-spacing of grid.
GripBlock	0	...	REG	Display of grips in blocks: **0** At block insertion point. **1** Of all objects within block.
GripColor	160	...	REG	ACI color of unselected grips: **1** Minimum color number; red. **160** Default color; blue. **255** Maximum color number.
GripHot	1	...	REG	ACI color of selected grips: **1** Default color, red. **255** Maximum color number.
GripHover	3	...	REG	ACI grip fill color when cursor hovers.
GripObjLimit	100	...	REG	Grips not displayed when more than this number: **1** Minimum. **32767**Maximum.
Grips	1	...	REG	Display of grips: **0** Off. **1** On
GripSize	3	...	REG	Size of grip box, in pixels: **1** Minimum size. **255** Maximum size.
GripTips	1	...	REG	Determines if grip tips are displayed when the cursor hovers over custom objects: **0** Off. **1** On.

H

Variable	*Default*	*R/O*	*Loc*	*Meaning*
HaloGap	0	...	DWG	Distance to shorten a haloed line; specified as the percentage of 1".
HidePrecision	0	...	DWG	Controls the precision of hide calculations: **0** Single precision, less accurate, faster. **1** Double precision, more accurate, but slower (recommended).
HideText	0	...	...	Determines whether text is hidden during the **Hide** command: **0** Text is not hidden nor hides other objects, unless text object has thickness. **1** Text is hidden and hides other objects.
Highlight	1	...	...	Object selection highlighting: **0** Disabled. **1** Enabled.
HPAng	0	...	...	Current hatch pattern angle.
HpAssoc	1	...	...	Determines if hatches are associative: **0** Not associative. **1** Associative.

Variable	Default	R/O	Loc	Meaning
HpBound	1	...	REG	Object created by **BHatch** and **Boundary**: **0** Region. **1** Polyline.
HpDouble	0	...	...	Double hatching: **0** Disabled. **1** Enabled.
HpDrawOrder	3	...	...	Draw order of hatch patterns and fills: **0** None. **1** Behind all other objects. **2** In front of all other objects. **3** Behind the hatch boundary. **4** In front of the hatch boundary.
HpGapTol	0	...	REG	Largest gap allowed in hatch boundary; ranges from 0 to 5000 units.
HpName	"ANSI31"	...	...	Current hatch pattern name **""** No default. **.** Set no default.
HpScale	1.0	...	...	Current hatch scale factor; cannot be zero.
HpSpace	1.0	...	...	Current spacing of user-defined hatching; cannot be zero.
HyperlinkBase	""	...	DWG	Path for relative hyperlinks.

I

Variable	Default	R/O	Loc	Meaning
ImageHlt	0	...	REG	When a raster image is selected: **0** Image frame is highlighted. **1** Entire image is highlighted.
IndexCtl	0	...	DWG	Creates layer and spatial indices: **0** No indices created. **1** Layer index created. **2** Spatial index created. **3** Both indices created.
InetLocation	"www.autodesk.com"	...	REG	Default browser URL.
InsBase	0.0,0.0,0.0	...	DWG	Insertion base point relative to the current UCS for **Insert** and **XRef** commands.
InsName	""	...	...	Current block name: **.** Set to no default.
InsUnits	1	...	...	Drawing units when a block is dragged into drawing from DesignCenter: **0** Unitless. **1** Inches. **2** Feet. **3** Miles. **4** Millimeters. **5** Centimeters. **6** Meters. **7** Kilometers. **8** Microinches. **9** Mils. **10** Yards. **11** Angstroms. **12** Nanometers. **13** Microns. **14** Decimeters. **15** Decameters.

Variable	Default	R/O	Loc	Meaning
				16 Hectometers. **17** Gigameters. **18** Astronomical Units. **19** Light Years. **20** Parsecs.
InsUnitsDefSource	1	...	REG	Source drawing units value; ranges from 0 to 20; see above.
InsUnitsDefTarget	1	...	REG	Target drawing units value; ranges from 0 to 20.
IntersectionColor	257	...	DWG	Color of intersection polylines: **0** Color is byblock. **1-255** AutoCAD color index. **256** Color is bylayer. **257** Color is byentity.
IntersectionDisplay	0	...	DWG	Determines 3D surface intersections during **Hide** command: **0** Does not draw intersections. **1** Draws polylines at intersections.
ISaveBak	1	...	REG	Controls whether *.bak* file is created: **0** No file created. **1** *.bak* backup file created.
ISavePercent	50	...	REG	Percentage of waste in saved *.dwg* file before cleanup occurs: **0** Every save is a full save. **>0** Faster partial saves.
IsoLines	4	...	DWG	Isolines on 3D solids: **0** No isolines; minimum. **16** Good-looking. **2,047** Maximum.

L

Variable	Default	R/O	Loc	Meaning
LastAngle	0	R/O	...	Ending angle of last-drawn arc.
LastPoint	*varies*	...	...	Last-entered point, such as 15,9,56.
LastPrompt	""	R/O	...	Last string on the command line; includes user input.
LayoutRegenCtl	2	...	REG	Controls display list for layouts: **0** Display list regen'ed with each tab change. **1** Display list is saved for model tab and last layout tab. **2** Display list is saved for all tabs.
LensLength	50.0	R/O	DWG	Perspective view lens length, in mm.
LimCheck	0	...	DWG	Drawing limits checking: **0** Disabled. **1** Enabled.
LimMax	12.0,9.0	...	DWG	Upper right drawing limits.
LimMin	0.0,0.0	...	DWG	Lower left drawing limits.
LispInit	1	...	REG	AutoLISP functions and variables are: **0** Preserved from drawing to drawing. **1** Valid in current drawing only.
Locale	"enu"	R/O	...	ISO language code; see DctMain.

Variable	*Default*	*R/O*	*Loc*	*Meaning*
LocalRootPrefix	"d:\documents and Settings*username*\local settings\appli..."	R/O	REG	Path to folder holding local customizable files.
LogFileMode	0	...	REG	Text window written to *.log* file: **0** No. **1** Yes.
LogFileName	"d:\acad 2005\Drawing1_1_1_0000.log"	R/O	DWG	Filename and path for *.log* file.
LogFilePath	"d:\acad 2005\"	...	REG	Path for the *.log* file.
LogInName	"*username*"	R/O	...	User's login name; max = 30 chars.
~~LongFName~~				*Removed from AutoCAD Release 14.*
LTScale	1.0	...	DWG	Current linetype scale factor; cannot be 0.
LUnits	2	...	DWG	Linear units mode: **1** Scientific. **2** Decimal. **3** Engineering. **4** Architectural. **5** Fractional.
LUPrec	4	...	DWG	Decimal places (or inverse of smallest fraction) of linear units.
LwDefault	25	...	REG	Default lineweight, in millimeters; must be one of the following values: 0, 5, 9, 13, 15, 18, 20, 25, 30, 35, 40, 50, 53, 60, 70, 80, 90, 100, 106, 120, 140, 158, 200, or 211.
LwDisplay	0	...	DWG	Toggles whether lineweight is displayed; setting is saved separately for Model space and each layout tab. **0** Not displayed. **1** Displayed.
LwUnits	1	...	REG	Determines units for lineweight: **0** Inches. **1** Millimeters.

M

Variable	*Default*	*R/O*	*Loc*	*Meaning*
MacroTrace	*0*	...	...	*Diesel debug mode:* ***0*** *Off.* ***1*** *On.*
MaxActVP	64	...	...	Maximum viewports to regenerate: **2** Minimum. **64** Maximum (increased from 48 in R14).
MaxObjMem	*0*	...	...	*Maximum number of objects in memory; object pager is turned off when value = 0, <0, or 2,147,483,647.*
MaxSort	200	...	REG	Maximum names sorted alphabetically.
MButtonPan	1	...	REG	Determines behavior of wheelmouse: **0** As defined by AutoCAD menu file. **1** Pans when dragging with wheel.
MeasureInit	0	...	REG	Drawing units for default drawings: **0** English. **1** Metric.
Measurement	0	...	DWG	Current drawing units: **0** English. **1** Metric.

Variable	Default	R/O	Loc	Meaning
MenuCtl	1	...	REG	Submenu display: **0** Only with menu picks. **1** Also with keyboard entry.
MenuEcho	0	...		Menu and prompt echoing: **0** Display all prompts. **1** Suppress menu echoing. **2** Suppress system prompts. **4** Disable **^P** toggle. **8** Display all input-output strings.
MenuName	"acad"	R/O	REG	Current menu filename.
MirrText	0	...	DWG	Text handling during **Mirror** command: **0** Retain text orientation. **1** Mirror text.
ModeMacro	""	...	...	Invoke Diesel macro.
MsmState	0	R/O	...	Specifies if **Markup Set Manager** is active.
MsOleScale	1.0	...	DWG	Determines the size of text-containing OLE objects when pasted in model space: **-1** Scaled by value of **PlotScale**. **0** Scale by value of **DimScale**. **>0** Scale factor.
MTextEd	"Internal"	...	REG	Name of the **MText** editor: **.** Use default editor. **0** Cancel the editing operation. **-1** Use the secondary editor. **"blank"** MTEXT internal editor. **"Internal"** MTEXT internal editor. **"Notepad"** Windows Notepad editor. **":lisped"** Built in AutoLISP function. *string* Name of editor fewer than 256 characters long using this syntax: *:AutoLISPtextEditorFunction#TextEditor.*
MTextFixed	0	...	REG	Specifies mtext editor appearence: **0** Display mtext editor and text at same size and position as object being edited. **1** Display mtext editor at the same location as last used; fixed height text.
MTJigStrings	"abc"	...	REG	Sample text displayed by mtext editor; maximum 10 letters; enter . for no text.
MyDocumentsPrefix	"C:\Documents and Settings*username*\My Documents"	R/O	REG	Path to the *my documents* folder of the currently logged-in user.

N

Variable	Default	R/O	Loc	Meaning
NodeName	*"AC$"*	R/O	REG	*Name of network node; range is one to three characters.*
NoMutt	0	...	...	Suppresses the display of message (a.k.a. muttering) during scripts, LISP, macros: **0** Display prompt, as normal. **1** Suppress muttering.
NwfState	*1*	...	...	*Reports whether New Features Workshop displays when AutoCAD starts.*

Variable	*Default*	*R/O*	*Loc*	*Meaning*
O				
ObscureColor	0	...	DWG	Color of objects obscured by **Hide** command: **0** Invisible. **1 - 255** Color number.
ObscureLtype	0	...	DWG	Linetype of objects obscured by **Hide** command: **0** Invisible. **1** Solid. **2** Dashed. **3** Dotted. **4** Short dash. **5** Medium dash. **6** Long dash. **7** Double short dash. **8** Double medium dash. **9** Double long dash. **10** Medium long dash. **11** Sparse dot.
OffsetDist	1.0	...	...	Current offset distance: **<0** Offsets through a specified point. **>0** Default offset distance.
OffsetGapType	0	...	REG	Determines how to reconnect polyline when individual segments are offset: **0** Extend segments to fill gap. **1** Fill gap with fillet (arc segment). **2** Fill gap with chamfer (line segment).
OleFrame	2	...	DWG	Controls the visibility of the frame around OLE objects: **0** Frame is not displayed and not plotted. **1** Frame is displayed and is plotted. **2** Frame is displayed but is not plotted.
OleHide	0	...	REG	Display and plotting of OLE objects: **0** All OLE objects visible. **1** Visible in paper space only. **2** Visible in model space only. **3** Not visible.
OleQuality	1	...	REG	Quality of display and plotting of embedded OLE objects: **0** Line art quality. **1** Text quality. **2** Graphics quality. **3** Photograph quality. **4** High quality photograph.
OleStartup	0	...	DWG	Loading OLE source application improves plot quality: **0** Do not load OLE source app. **1** Load OLE source app when plotting.
OpmState	*0*	...	...	*Specifies if **Properties** window is active.*
OrthoMode	0	...	DWG	Orthographic mode: **0** Off. **1** On.
OsMode	4133	...	REG	Current object snap mode: **0** NONe. **1** ENDpoint. **2** MIDpoint. **4** CENter. **8** NODe. **16** QUAdrant.

Variable	Default	R/O	Loc	Meaning
				32 INTersection. **64** INSertion. **128** PERpendicular. **256** TANgent. **512** NEARest. **1024** QUIck. **2048** APPint. **4096** EXTension. **8192** PARallel. **16383** All modes on. **16384** Object snap turned off via **OSNAP** on the status bar.
OSnapCoord	2	...	REG	Keyboard overrides object snap: **0** Object snap overrides keyboard. **1** Keyboard overrides object snap. **2** Keyboard overrides object snap, except in scripts.
OSnapHatch	0	...	...	Toggles whether hatches are snapped.
OSnapNodeLegacy	1	...	...	Toggles whether osnap snaps to text insertion points.

P

Variable	Default	R/O	Loc	Meaning
PaletteOpaque	0	...	REG	Determines if palettes can be made transparent: **0** Turned off by user. **1** Turned on by user. **2** Unavailable, but turned on by user. **3** Unavailable, and turned off by user.
PaperUpdate	0	...	REG	Determines how AutoCAD plots a layout with paper size different from plotter's default: **0** Displays a warning dialog box. **1** Changes paper size to that of the plotter configuration file.
PDMode	0	...	DWG	Point display mode: **0** Dot. **1** No display. **2** +-symbol. **3** x-symbol. **4** Short line. **32** Circle. **64** Square.
PDSize	0.0	...	DWG	Point display size, in pixels: **>0** Absolute size. **0** 5% of drawing area height. **<0** Percentage of viewport size.
PEdit Accept	0	...	REG	Suppresses display of the **PEdit** command's "Object selected is not a polyline. Do you want to turn it into one? <Y>" prompt.
PEllipse	0	...	DWG	Toggle ellipse creation: **0** True ellipse. **1** Polyline arcs.
Perimeter	0.0	R/O	...	Perimeter calculated by the last **Area**, **DbList**, and **List** commands.
PFaceVMax	4	R/O	...	Maximum vertices per 3D face.
PHandle	*0*	...	*ACAD*	*Ranges from 0 to 4.29497E9.*

Variable	Default	R/O	Loc	Meaning
PickAdd	1	...	REG	Effect of SHIFT key on selection set: **0** Adds to selection set. **1** Removes from selection set.
PickAuto	1	...	REG	Selection set mode: **0** Single pick mode. **1** Automatic windowing and crossing.
PickBox	3	...	REG	Object selection pickbox size, in pixels: **0** Minimum size. **50** Maximum size.
PickDrag	0	...	REG	Selection window mode: **0** Pick two corners. **1** Pick a corner; drag to second corner.
PickFirst	1	...	REG	Command-selection mode: **0** Enter command first. **1** Select objects first.
PickStyle	1	...	REG	Groups and associative hatches in selection sets: **0** Neither included. **1** Include groups. **2** Include associative hatches. **3** Include both.
Platform	"Microsoft Windows NT Version 5.0 (x86)"	R/O	...	Name of the operating system.
PLineGen	0	...	DWG	Polyline linetype generation: **0** From vertex to vertex. **1** From end to end.
PLineType	2	...	REG	Automatic conversion and creation of 2D polylines by **PLine**: **0** Not converted; creates old-format polylines. **1** Not converted; creates optimized lwpolylines. **2** Polylines in older drawings are converted on open; **PLine** creates optimized lwpolyline objects.
PLineWid	0.0	...	DWG	Current polyline width.
2005 **PlotOffset**	0	...	REG	Sets the plot offset: **0** Relative to edge of margins. **1** Relative to edge of paper.
PlotRotMode	1	...	DWG	Orientation of plots: **0** Lower left = 0,0. **1** Lower left plotter area = lower left of media. **2** X, y-origin offsets calculated relative to the rotated origin position.
Plotter	*0*	...	*REG*	*Obsolete; has no effect in AutoCAD.*
PlQuiet	0	...	REG	Toggles display during batch plotting and scripts (replaces **CmdDia**): **0** Plot dialog boxes and nonfatal errors are displayed. **1** Nonfatal errors are logged; plot dialog boxes are not displayed.
PolarAddAng	""	...	REG	Contains a list of up to 10 user-defined polar angles; each angle can be up to 25 characters long, each separated with a semicolon (;). For example: 0;15;22.5;45.
PolarAng	90	...	REG	Specifies the increment of polar angle; contrary to Autodesk documentation, you may specify any angle.
PolarDist	0.0	...	REG	The polar snap increment when **SnapStyl** is set to 1 (isometric).

Variable	*Default*	*R/O*	*Loc*	*Meaning*
PolarMode	0	...	REG	Settings for polar and object snap tracking: **0** Measure polar angles based on current UCS (absolute), track orthogonally; don't use additional polar tracking angles; and acquire object tracking points automatically. **1** Measure polar angles from selected objects (relative). **2** Use polar tracking settings in object snap tracking. **4** Use additional polar tracking angles (via **PolarAng**). **8** Press SHIFT to acquire object snap tracking points.
PolySides	4	...	...	Current number of polygon sides: **3** Minimum sides. **1024** Maximum sides.
Popups	1	R/O	...	Display driver support of AUI: **0** Not available. **1** Available.
Product	"AutoCAD"	R/O	ACAD	Name of the software.
Program	"acad"	R/O	ACAD	Name of the software's executable file.
ProjectName	""	...	DWG	Project name of the current drawing; searches for xref and image files.
ProjMode	1	...	REG	Projection mode for **Trim** and **Extend** commands: **0** No projection. **1** Project to x,y-plane of current UCS. **2** Project to view plane.
ProxyGraphics	1	...	REG	Proxy image saved in the drawing: **0** Not saved; displays bounding box. **1** Image saved with drawing.
ProxyNotice	1	...	REG	Display warning message: **0** No. **1** Yes.
ProxyShow	1	...	REG	Display of proxy objects: **0** Not displayed. **1** All displayed. **2** Bounding box displayed.
ProxyWebSearch	0	...	REG	Object enablers are checked: **0** Do not check for object enablers. **1** Check for object enablers if an Internet connection is present.
PsLtScale	1	...	DWG	Paper space linetype scaling: **0** Use model space scale factor. **1** Use viewport scale factor.
PsProlog	*""*	...	*REG*	*PostScript prologue filename.*
PsQuality	*75*	...	*REG*	*Resolution of PostScript display, in pixels:* *<0 Display as outlines; no fill.* *0 Not displayed.* *>0 Display filled.*
PStyleMode	1	...	DWG	Toggles the plot color matching mode of the drawing: **0** Use named plot style tables. **1** Use color-dependent plot style tables.
PStylePolicy	1	...	REG	Determines whether the object color is associated with its plot style: **0** Not associated. **1** Associated.

Variable	Default	R/O	Loc	Meaning
PsVpScale	0	...	...	Sets the view scale factor (the ratio of units in paper space to the units in newly-created model space viewports) for all newly-created viewports: **0** Scaled to fit.
PUcsBase	""	...	DWG	Name of UCS defining the origin and orientation of orthographic UCS settings in paper space only.

Q

Variable	Default	R/O	Loc	Meaning
QAFlags	*0*	...	...	*Quality assurance flags:* ***0*** *Turned off.* ***1*** *The ^C metacharacters in a menu macro cancels grips, just as if user pressed* ***ESC.*** ***2*** *Long text screen listings do not pause.* ***4*** *Error and warning messages are displayed at the command line, instead of in dialog boxes.* ***128*** *Screen picks are accepted via the AutoLISP (command) function.*
QaUcsLock	*0*	...	...	*Either 0 or 1.*
QTextMode	0	...	DWG	Quick text mode: **0** Off. **1** On.
QueuedRegenMax	*2147483647*	...	...	*Ranges between very large and very small numbers.*

R

Variable	Default	R/O	Loc	Meaning
R14RasterPlot	*0*	...	...	*Either 1 or 0.*
RasterPreview	1	R/O	REG	Preview image: **0** None saved. **1** Saved in BMP format.
RefEditName	""	...	...	The reference filename when it is in reference-editing mode.
RegenMode	1	...	DWG	Regeneration mode: **0** Regen with each view change. **1** Regen only when required.
Re-Init	0	...	...	Reinitialize I/O devices: **1** Digitizer port. *2 Plotter port.* **4** Digitizer. *8 Plotter.* **16** Reload PGP file.
RememberFolders	1	...	REG	Controls path search method: **0** Path specified in desktop AutoCAD icon is default for file dialog boxes. **1** Last path specified by each file dialog box is remembered.
ReportError	1	...	REG	Determines if AutoCAD sends an error report to Autodesk: **0** No error report created. **1** Error report is generated and sent to Autodesk.
RoamableRootPrefix	"d:\documents and settings*username*\application aata\aut..."	R/O	REG	Path to root folder where roamable customized files are located.
RTDisplay	1	...	REG	Raster display during real-time zoom and pan: **0** Display the entire raster image. **1** Display raster outline only.

Variable	*Default*	*R/O*	*Loc*	*Meaning*
S				
SaveFile	"auto.sv$"	R/O	REG	Automatic save filename.
SaveFilePath	"d:\temp\"	...	REG	Path for automatic save files.
SaveName	""	R/O	...	Drawing save-as filename.
SaveTime	10	...	REG	Automatic save interval, in minutes: **0** Disable auto save.
ScreenBoxes	0	R/O	ACAD	Maximum number of menu items **0** Screen menu turned off.
ScreenMode	3	R/O	...	State of AutoCAD display screen: **0** Text screen. **1** Graphics screen. **2** Dual-screen display.
ScreenSize	*varies*	R/O	...	Current viewport size, in pixels, such as 719.0000,381.0000.
SDI	0	...	REG	Toggles multiple-document interface (SDI is "single document interface"): **0** Turns on MDI. **1** Turns off MDI (only one drawing may be loaded into AutoCAD). **2** MDI disabled for apps that cannot support MDI; read-only. **3** (R/O) MDI disabled for apps that cannot support MDI, even when **SDI**= 1.
ShadEdge	3	...	DWG	**Shade** style: **0** Shade faces; 256-color shading. **1** Shade faces; edges background color. **2** Hidden-line removal. **3** 16-color shading.
ShadeDif	70	...	DWG	Percent of diffuse to ambient light: **0** Minimum. **100** Maximum.
ShortcutMenu	11	...	REG	Toggles availability of shortcut menus: **0** Disables all default, edit, and command shortcut menus. **1** Enables default shortcut menus. **2** Enables edit shortcut menus. **4** Enables command shortcut menus whenever a command is active. **8** Enables command shortcut menus only when command options are available at the command line.
ShpName	""	...	...	Current shape name: **.** Set to no default.
SigWarn	1	...	REG	Determines whether a warning is displayed when a file is opened with a digital signature.
SketchInc	0.1	...	DWG	**Sketch** command's recording increment.
SkPoly	0	...	DWG	Sketch line mode: **0** Record as lines. **1** Record as a polyline.
SnapAng	0	...	DWG	Current rotation angle for snap and grid.
SnapBase	0.0,0.0	...	DWG	Current origin for snap and grid.
SnapIsoPair	0	...	DWG	Current isometric drawing plane: **0** Left isoplane. **1** Top isoplane. **2** Right isoplane.

Variable	*Default*	*R/O*	*Loc*	*Meaning*
SnapMode	0	...	DWG	Snap mode: **0** Off. **1** On.
SnapStyl	0	...	DWG	Snap style: **0** Normal. **1** Isometric.
SnapType	0	...	REG	Toggles between standard or polar snap for the current viewport: **0** Standard snap. **1** Polar snap.
SnapUnit	0.5,0.5	...	DWG	X,y-spacing for snap.
SolidCheck	1	...	...	Toggles solid validation: **0** Off. **1** On.
SortEnts	96	...	DWG	Object display sort order: **0** Off. **1** Object selection. **2** Object snap. **4** Redraw. **8** Slide generation. **16** Regeneration. **32** Plot. **64** PostScript output.
SpaceSwitch	*1*	...	...	*Either 1 or 9.*
SplFrame	0	...	DWG	Polyline and mesh display: **0** Polyline control frame not displayed; display polygon fit mesh; 3D faces invisible edges not displayed. **1** Polyline control frame displayed; display polygon defining mesh; 3D faces invisible edges displayed.
SplineSegs	8	...	DWG	Number of line segments that define a splined polyline.
SplineType	6	...	DWG	Spline curve type: **5** Quadratic Bezier spline. **6** Cubic Bezier spline.
SsFound	""	...	...	Path and file name of sheet set.
SsLocate	1	...	USER	Determine whether sheet set files are opened with drawing.
SsmAutoOpen	1	...	USER	Determines whether the **Sheet Set Manager** is opened with drawing.
SsmState	0	R/O	...	Reports whether **Sheet Set Manager** is open.
StandardsViolation	2	...	REG	Determines whether alerts are displayed when CAD standards are violated: **0** No alerts. **1** Alert displayed when CAD standard violated. **2** Displays icon on status bar when file is opened with CAD standards, and when non-standard objects are created.
Startup	0	...	REG	Determines which dialog box is displayed by the **New** and **QNew** commands: **0** Displays **Select Template** dialog box. **1** Displays **Startup** and **Create New Drawing** dialog box.

Variable	Default	R/O	Loc	Meaning
SurfTab1	6	...	DWG	Density of surfaces and meshes: **5** Minimum. **32766**Maximum.
SurfTab2	6	...	DWG	Density of surfaces and meshes: **2** Minimum. **32766**Maximum.
SurfType	6	...	DWG	Pedit surface smoothing: **5** Quadratic Bezier spline. **6** Cubic Bezier spline. **8** Bezier surface.
SurfU	6	...	DWG	Surface density in m-direction: **2** Minimum. **200** Maximum.
SurfV	6	...	DWG	Surface density in n-direction: **2** Minimum. **200** Maximum.
SysCodePage	"ANSI_1252"	R/O	...	System code page.

T

Variable	Default	R/O	Loc	Meaning
TabMode	0	...	...	Tablet mode: **0** Off. **1** On.
Target	0.0,0.0,0.0	R/O	DWG	Target in current viewport.
Tbaskbar	*1*	...	...	*Determines whether each drawing appears as a button on the Windows taskbar.*
TbCustomize	*1*	...	...	*Determines whether toolbars can be customized.*
TDCreate	*varies*	R/O	DWG	Time and date drawing created, such as 2448860.54014699.
TDInDwg	*varies*	R/O	DWG	Duration drawing loaded, such as 0.00040625.
TDuCreate	*varies*	R/O	DWG	The universal time and date the drawing was created, such as 2451318.67772165.
TDUpdate	*varies*	R/O	DWG	Time and date of last update, such as 2448860.54014699.
TDUsrTimer	*varies*	R/O	DWG	Decimal time elapsed by user-timer, such as 0.00040694.
TDuUpdate	*varies*	R/O	DWG	The universal time and date of the last save, such as 2451318.67772165.
TempPrefix	"d:\temp"	R/O	...	Path for temporary files set by **Temp** envar.
TextEval	0	...	...	Interpretation of text input: **0** Literal text. **1** Read **(** and **!** as AutoLISP code.
TextFill	1	...	REG	Toggle fill of TrueType fonts: **0** Outline text. **1** Filled text.
TextQlty	50	...	DWG	Resolution of TrueType fonts: **0** Minimum resolution. **100** Maximum resolution (preferred).
TextSize	0.2000	...	DWG	Default height of text (2.5 in metric units).
TextStyle	"Standard"	...	DWG	Default name of text style.
Thickness	0.0000	...	DWG	Default object thickness.

Variable	Default	R/O	Loc	Meaning
TileMode	1	...	DWG	View mode: **0** Display layout tab. **1** Display model tab.
ToolTips	1	...	REG	Display tooltips: **0** Off. **1** On.
TpState	0	R/O	...	Determines if Tool Palettes is open.
TraceWid	0.0500	...	DWG	Current width of traces.
TrackPath	0	...	REG	Determines the display of polar and object snap tracking alignment paths: **0** Display object snap tracking path across the entire viewport. **1** Display object snap tracking path between the alignment point and "From point" to cursor location. **2** Turn off polar tracking path. **3** Turn off polar and object snap tracking paths.
TrayIcons	1	...	REG	Determines if the tray is displayed on the status bar.
TrayNotify	1	...	REG	Determines whether service notifications are displayed by the tray.
TrayTimeout	5	...	REG	Specifies the length of time (in seconds) that tray notificaitons are displayed: **0** Minimium. **10** Maximum.
TreeDepth	3020	...	DWG	Maximum branch depth in *xxyy* format: *xx* Model-space nodes. *yy* Paper-space nodes. *>0* 3D drawing. *<0* 2D drawing.
TreeMax	10000000	...	REG	Limits memory consumption during drawing regeneration.
TrimMode	1	...	REG	Trim toggle for **Chamfer** and **Fillet** commands: **0** Leave selected edges in place. **1** Trim selected edges.
TSpaceFac	1.0	...	...	Mtext line spacing distance; measured as a factor of "normal" text spacing; valid values range from 0.25 to 4.0.
TSpaceType	1	...	...	Type of mtext line spacing: **1** At Least: adjust line spacing based on the height of the tallest character in a line of mtext. **2** Exactly: use the specified line spacing; ignores character height.
TStackAlign	1	...	DWG	Vertical alignment of stacked text (fractions): **0** Bottom aligned. **1** Center aligned. **2** Top aligned.
TStackSize	70	...	DWG	Sizes stacked text as a percentage of the current text height: **1** Minimum %. **127** Maximum %.

U

Variable	Default	R/O	Loc	Meaning
UcsAxisAng	90	...	REG	Default angle for rotating the UCS around an axes (via the **UCS** command using the **X**, **Y**, or **Z** options; valid values are limited to: 5, 10, 15, 18, 22.5, 30, 45, 90, or 180.
UcsBase	""	...	DWG	Name of the UCS that defines the origin and orientation of orthographic UCS settings.

Variable	Default	R/O	Loc	Meaning
UcsFollow	0	...	DWG	New UCS views: **0** No change. **1** Automatically align UCS with new view.
UcsIcon	3	...	DWG	Display of UCS icon: **0** Off. **1** On. **2** Display at UCS origin, if possible. **3** On, and displayed at origin.
UcsName	"World"	R/O	DWG	Name of current UCS view: "" Current UCS is unnamed.
UcsOrg	0.0,0.0,0.0	R/O	DWG	Origin of current UCS relative to WCS.
UcsOrtho	1	...	REG	Determines whether the related orthographic UCS setting is restored automatically: **0** UCS setting remains unchanged when orthographic view is restored. **1** Related orthographic UCS is restored automatically when an orthographic view is restored.
UcsView	1	...	REG	Determines whether the current UCS is saved with a named view: **0** Not saved. **1** Saved.
UcsVp	1	...	DWG	Determines whether the UCS in active viewports remains fixed (locked) or changes (unlocked) to match the UCS of the current viewport: **0** Unlocked. **1** Locked.
UcsXDir	1.0,0.0,0.0	R/O	DWG	X-direction of current UCS relative to WCS.
UcsYDir	0.0,1.0,0.0	R/O	DWG	Y-direction of current UCS relative to WCS.
UndoCtl	5	R/O	...	State of undo: **0** Undo disabled. **1** Undo enabled. **2** Undo limited to one command. **4** Auto-group mode. **8** Group currently active.
UndoMarks	0	R/O	...	Current number of undo marks.
UnitMode	0	...	DWG	Units display: **0** As set by **Units** command. **1** As entered by user.
UpdateThumbnail	7	...	DWG	Determines whether thumbnails are created when drawing is saved: **0** Thumbnail previews not updated. **1** Sheet views updated. **2** Model views updated. **4** Sheets updated.
UserI1 *thru* **UserI5**	0	...	DWG	Five user-definable integer variables.
UserR1 *thru* **UserR5**	0.0	...	DWG	Five user-definable real variables.
UserS1 *thru* **UserS5**	""	...	...	Five user-definable string variables.

Variable	Default	R/O	Loc	Meaning
V				
ViewCtr	*varies*	R/O	DWG	X,y,z-coordinate of center of current view, such as 6.2,4.5,0.0.
ViewDir	*varies*	R/O	DWG	Current view direction relative to UCS, such as 0.0,0.0,1.0 for plan view.
ViewMode	0	R/O	DWG	Current view mode: **0** Normal view. **1** Perspective mode on. **2** Front clipping on. **4** Back clipping on. **8** UCS-follow on. **16** Front clip not at eye.
ViewSize	9.0	R/O	DWG	Height of current view in drawing units.
ViewTwist	0	R/O	DWG	Twist angle of current view.
VisRetain	1	...	DWG	Determines xref drawing's layer settings — on-off, freeze-thaw, color, and linetype: **0** Xref-dependent layer settings are not saved in the current drawing. **1** Xref-dependent layer settings are saved in current drawing, and take precedence over settings in xrefed drawing next time the current drawing is loaded.
VpMaximizedState	0	R/O	...	Specifies whether viewport is maximized by **VpMax** command.
VSMax	*varies*	R/O	DWG	Upper-right corner of virtual screen, such as 37.46,27.00,0.00.
VSMin	*varies*	R/O	DWG	Lower-left corner of virtual screen, such as -24.97,-18.00,0.0.
W				
WhipArc	0	...	REG	Display of circlular objects: **0** Displayed as connected vectors. **1** Displayed as true circles and arcs.
WhipThread	3	...	REG	Controls multithreaded processing on two CPUs (if present) during drawing redraw and regeneration: **0** Single-threaded calculations. **1** Regenerations multi-threaded. **2** Redraws multi-threaded. **3** Regens and redraws multi-threaded.
WmfBkgnd	1	...	...	Controls background of *.wmf* files: **0** Background is transparent. **1** Background is same as AutoCAD's background color.
WmfForegnd	0	...	...	Controls foreground colors of exported WMFimages: **0** Foreground is darker than background. **1** Foreground is lighter than background.
WorldUcs	1	R/O	...	Matching of WCS with UCS: **0** Current UCS does not match WCS. **1** UCS matches WCS.
WorldView	1	...	DWG	Display during **3dOrbit**, **DView**, and **VPoint** commands: **0** Current UCS. **1** WCS.
WriteStat	1	R/O	...	Indicates whether drawing file is read-only: **0** Drawing file cannot be written to. **1** Drawing file can be written to.
X				

Variable	Default	R/O	Loc	Meaning
XClipFrame	0	...	DWG	Visibility of xref clipping boundary: **0** Not visible. **1** Visible.
XEdit	1	...	DWG	Toggles whether drawing can be edited in-place when referenced by another drawing: **0** Cannot in-place refedit. **1** Can in-place refedit.
XFadeCtl	50	...	REG	Fades objects not being edited in-place: **0** No fading; minimum value. **90** 90% fading; maximum value.
XLoadCtl	2	...	REG	Controls demand loading: **0** Demand loading turned off; entire drawing is loaded. **1** Demand loading turned on; xref file opened. **2** Demand loading turned on; a *copy* of the xref file is opened.
XLoadPath	"d:\temp"	...	REG	Path for storing temporary copies of demand-loaded xref files.
XRefCtl	0	...	REG	Determines creation of *.xlg* xref log files: **0** File not written. **1** Log file written.
XrefNotify	2	...	REG	Determines if alert is displayed for updated and missing xrefs: **0** No alert displayed. **1** Icon indicates xrefs are attached; a yellow alert indicates missing xrefs. **2** Also displays balloon messages when an xref is modified.
XrefType	*0*	...	...	*Determines whether xrefs are attached or overlaid.*

Z

Variable	Default	R/O	Loc	Meaning
ZoomFactor	60	...	REG	Controls the zoom level via mouse wheel; valid values range between 3 and 100; 15 is recommended.

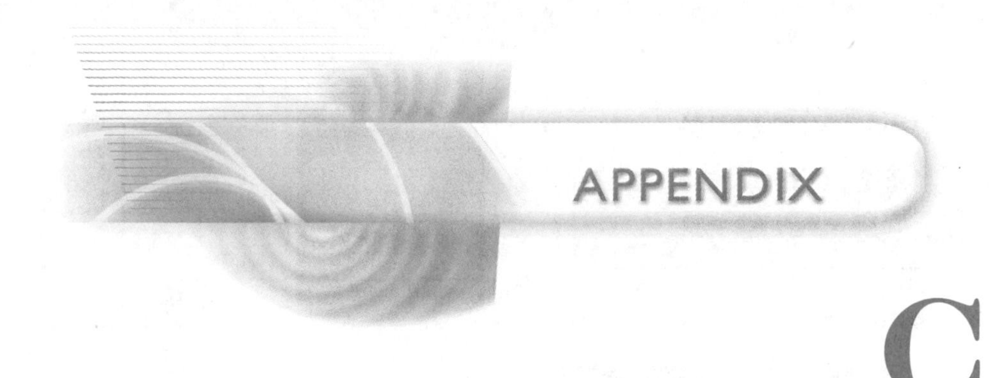

C

AutoCAD Toolbars and Menus

AUTOCAD TOOLBARS

NEW IN 2005 The indicates the toolbar is changed in AutoCAD 2005.

CAD STANDARDS TOOLBAR

DIMENSION TOOLBAR

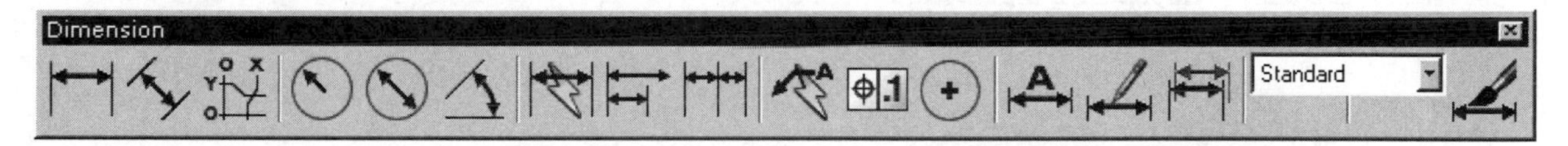

 DRAW TOOLBAR

DRAWORDER TOOLBAR

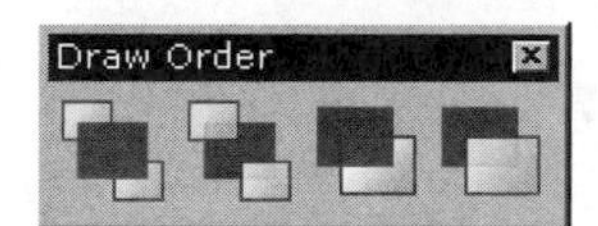

INQUIRY TOOLBAR

INSERT TOOLBAR

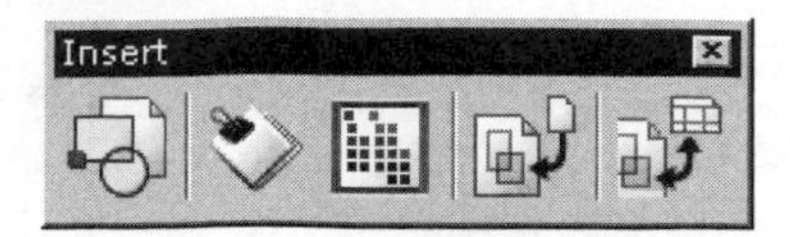

LAYERS TOOLBAR

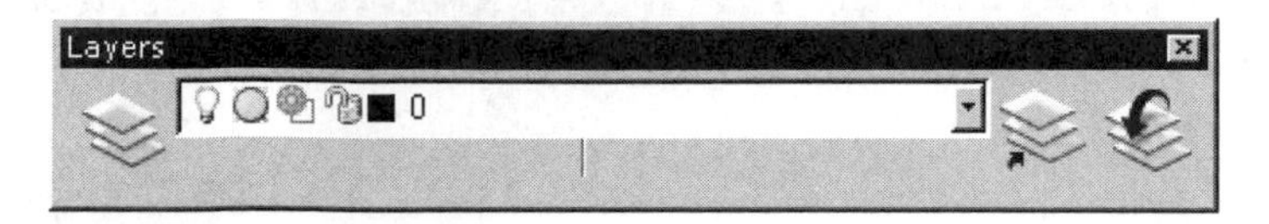

LAYOUTS TOOLBAR

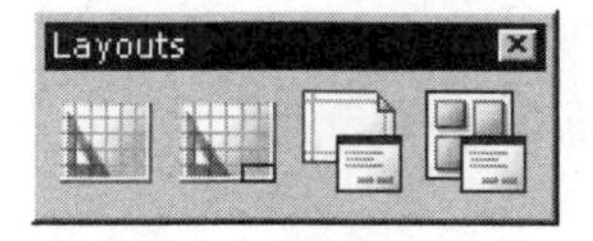

MODIFY TOOLBAR

MODIFY II TOOLBAR

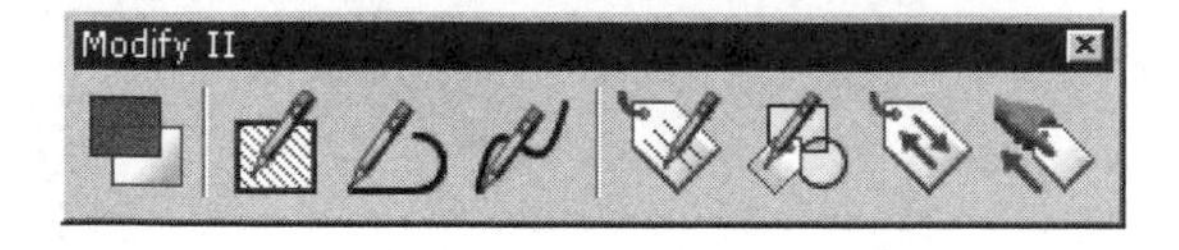

OBJECT SNAP TOOLBAR

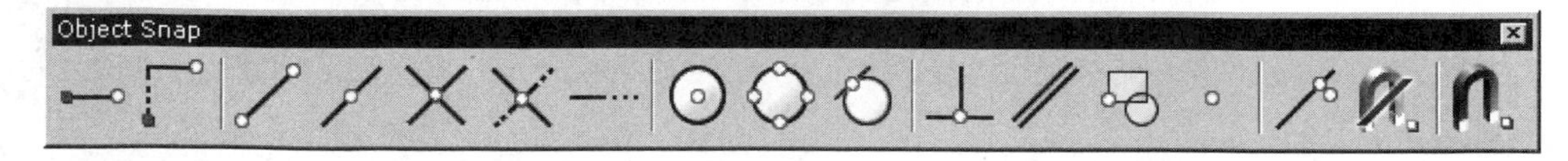

PROPERTIES TOOLBAR

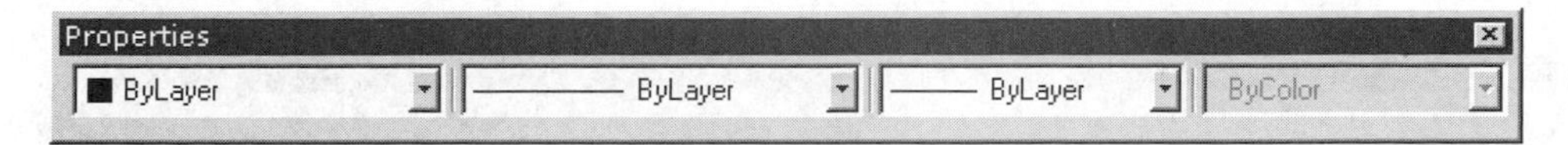

REFEDIT TOOLBAR

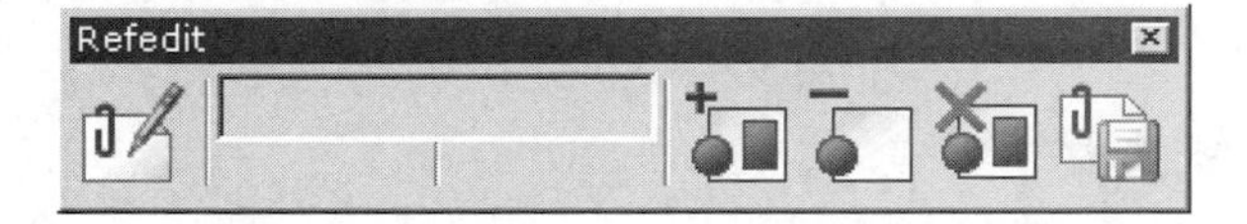

REFERENCE TOOLBAR

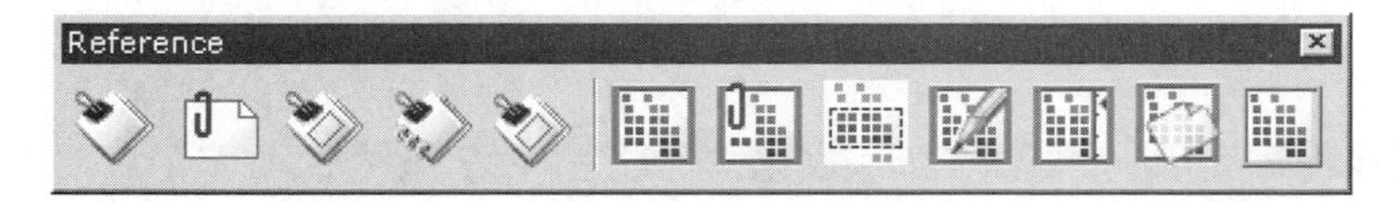

RENDER TOOLBAR

SHADE TOOLBAR

SOLIDS TOOLBAR

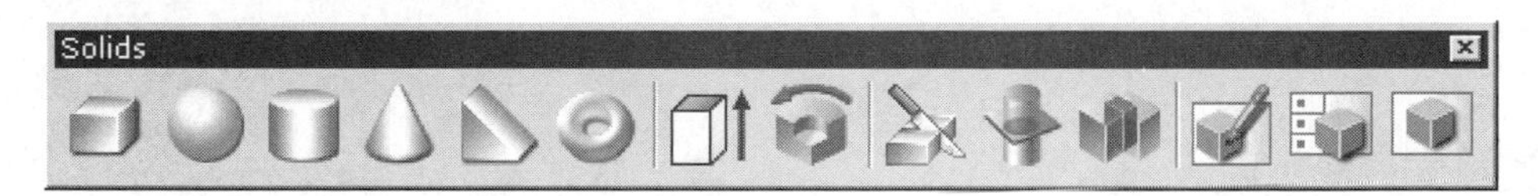

SOLIDS EDITING TOOLBAR

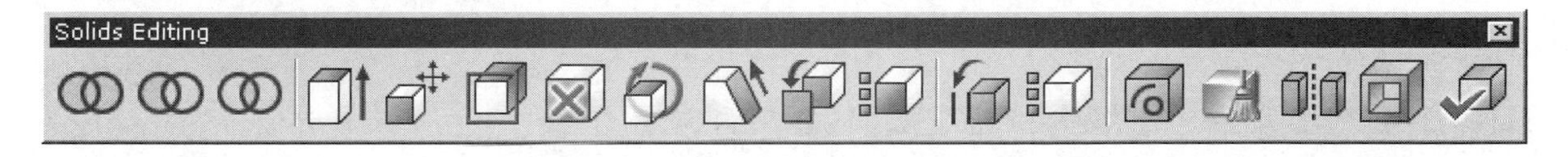

STANDARD TOOLBAR

STYLES TOOLBAR

SURFACES TOOLBAR

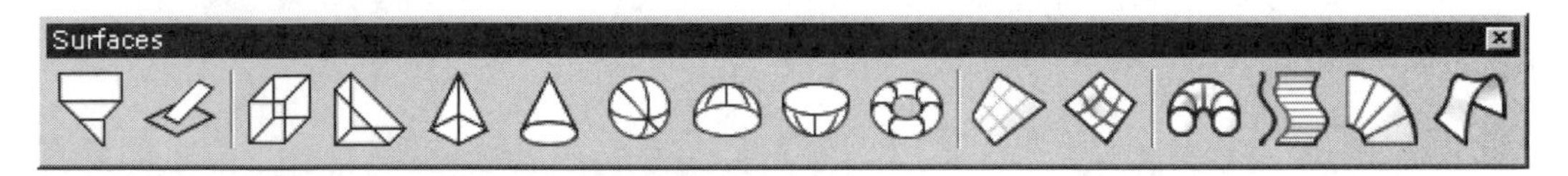

TEXT TOOLBAR

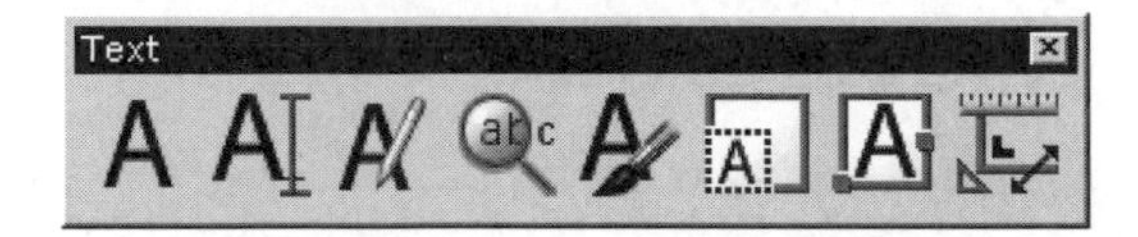

UCS TOOLBAR

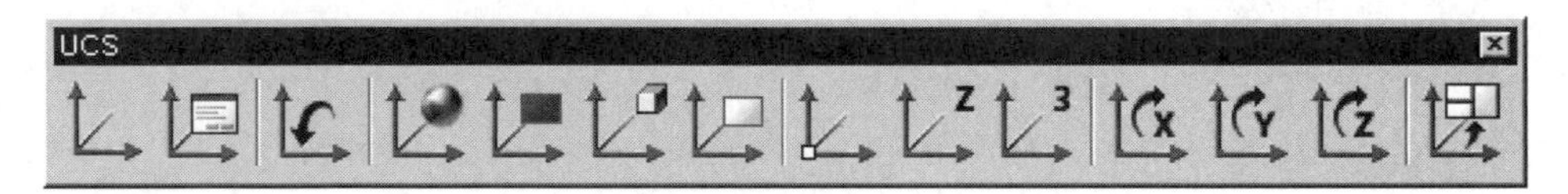

UCS II TOOLBAR

VIEW TOOLBAR

VIEWPORTS TOOLBAR

WEB TOOLBAR

ZOOM TOOLBAR

3D ORBIT TOOLBAR

AUTOCAD MENUS

 The icon indicates the menu is changed in AutoCAD 2005.

 FILE MENU

File

New...	Ctrl+N
New Sheet Set...	
Open...	Ctrl+O
Open Sheet Set...	
Load Markup Set...	
Close	
Partial Load	
Save	Ctrl+S
Save As...	Ctrl+Shift+S
eTransmit...	
Publish to Web...	
Export...	
Page Setup Manager...	
Plotter Manager...	
Plot Style Manager...	
Plot Preview	
Plot...	Ctrl+P
Publish...	
View Plot and Publish Details...	
Drawing Utilities ▸	
Send...	
Drawing Properties...	
1 C:\Books\...\ordinate.dwg	
2 C:\Books\...\CDROM\17_34.DWG	
8 Welding Fixture Model.dwg	
9 Taisei Door Window Sheet.dwg	
Exit	Ctrl+Q

Drawing Utilities submenu:

- Audit
- Recover...
- Update Block Icons
- Purge...

EDIT MENU

Edit

Undo Copy to clipboard	Ctrl+Z
Redo Group of commands	Ctrl+Y
Cut	Ctrl+X
Copy	Ctrl+C
Copy with Base Point	Ctrl+Shift+C
Copy Link	
Paste	Ctrl+V
Paste as Block	Ctrl+Shift+V
Paste as Hyperlink	
Paste to Original Coordinates	
Paste Special...	
Clear	Del
Select All	Ctrl+A
OLE Links...	
Find...	

VIEW MENU

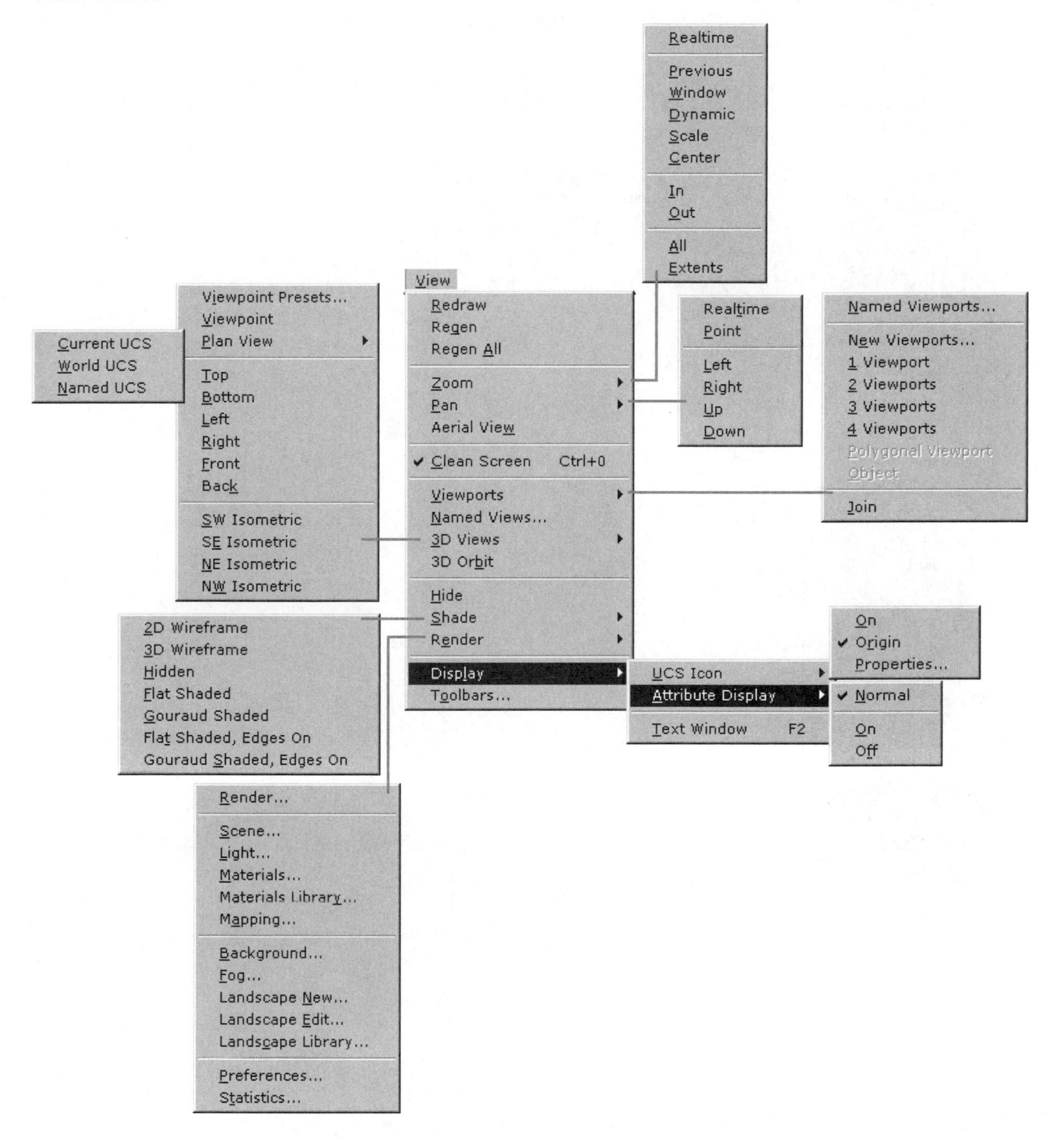
View
Redraw
Regen
Regen All
Zoom
Pan
Aerial View
Clean Screen Ctrl+0
Viewports
Named Views...
3D Views
3D Orbit
Hide
Shade
Render
Display
Toolbars...
Realtime
Previous
Window
Dynamic
Scale
Center
In
Out
All
Extents
Realtime
Point
Left
Right
Up
Down
Named Viewports...
New Viewports...
1 Viewport
2 Viewports
3 Viewports
4 Viewports
Polygonal Viewport
Object
Join
Viewpoint Presets...
Viewpoint
Plan View
Top
Bottom
Left
Right
Front
Back
SW Isometric
SE Isometric
NE Isometric
NW Isometric
Current UCS
World UCS
Named UCS
2D Wireframe
3D Wireframe
Hidden
Flat Shaded
Gouraud Shaded
Flat Shaded, Edges On
Gouraud Shaded, Edges On
UCS Icon
Attribute Display
Text Window F2
On
Origin
Properties...
Normal
On
Off
Render...
Scene...
Light...
Materials...
Materials Library...
Mapping...
Background...
Fog...
Landscape New...
Landscape Edit...
Landscape Library...
Preferences...
Statistics...

 INSERT MENU

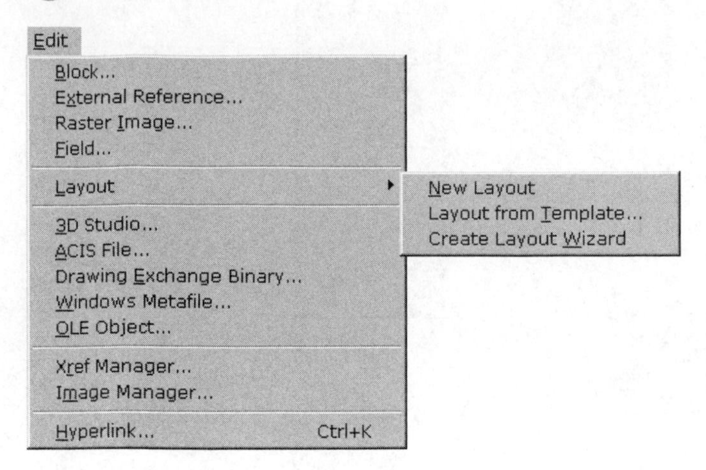

FORMAT MENU

Format

- Layer...
- Color...
- Linetype...
- Lineweight...
- Text Style...
- Dimension Style...
- Table Style...
- Plot Style...
- Point Style...
- Multiline Style...
- Units...
- Thickness
- Drawing Limits
- Rename...

TOOLS MENU

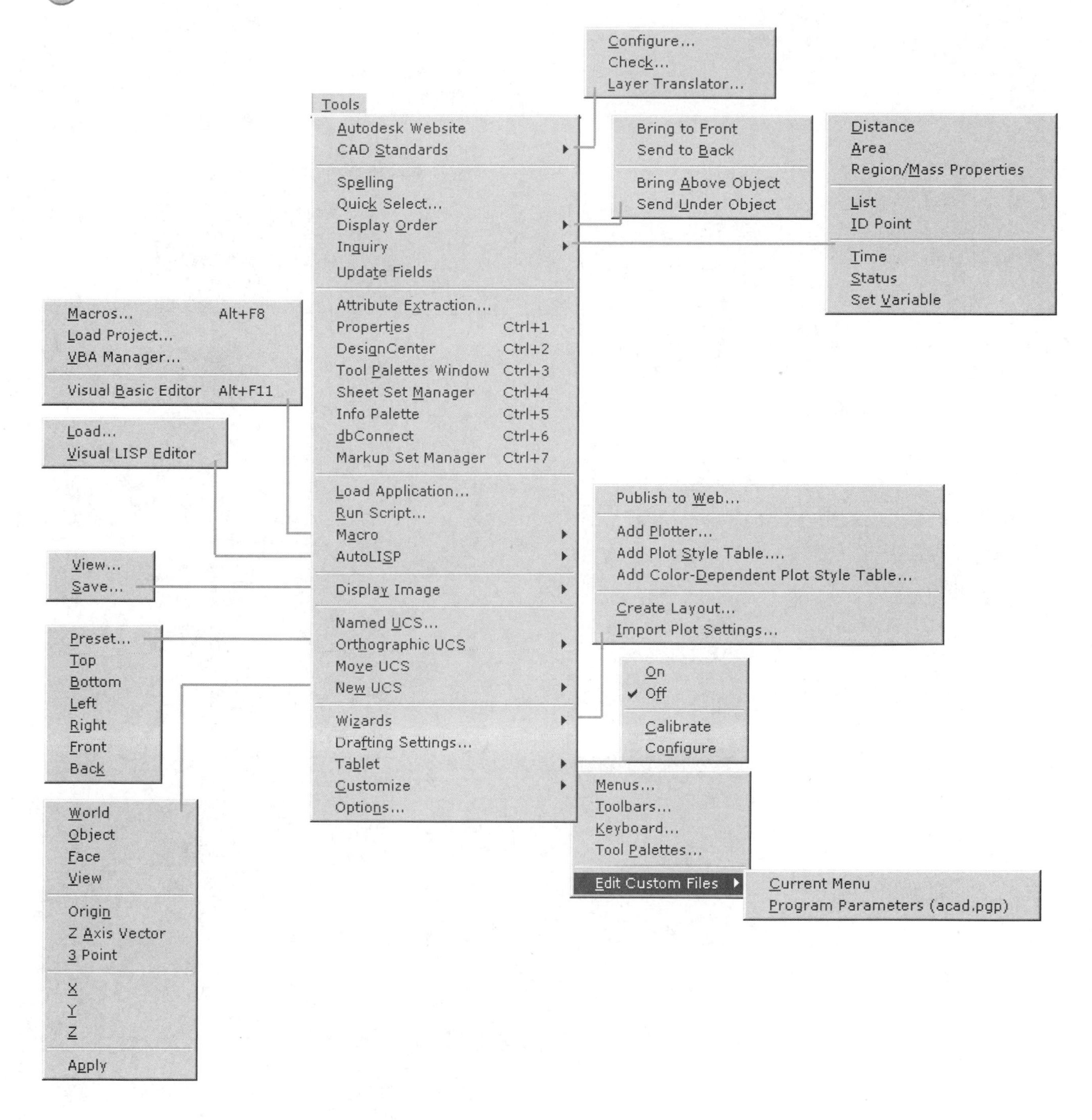

NEW IN 2005
Tools
Autodesk Website
CAD Standards
Spelling
Quick Select...
Display Order
Inquiry
Update Fields
Attribute Extraction...
Properties Ctrl+1
DesignCenter Ctrl+2
Tool Palettes Window Ctrl+3
Sheet Set Manager Ctrl+4
Info Palette Ctrl+5
dbConnect Ctrl+6
Markup Set Manager Ctrl+7
Load Application...
Run Script...
Macro
AutoLISP
Display Image
Named UCS...
Orthographic UCS
Move UCS
New UCS
Wizards
Drafting Settings...
Tablet
Customize
Options...
Configure...
Check...
Layer Translator...
Bring to Front
Send to Back
Bring Above Object
Send Under Object
Distance
Area
Region/Mass Properties
List
ID Point
Time
Status
Set Variable
Macros... Alt+F8
Load Project...
VBA Manager...
Visual Basic Editor Alt+F11
Load...
Visual LISP Editor
View...
Save...
Preset...
Top
Bottom
Left
Right
Front
Back
World
Object
Face
View
Origin
Z Axis Vector
3 Point
X
Y
Z
Apply
Publish to Web...
Add Plotter...
Add Plot Style Table....
Add Color-Dependent Plot Style Table...
Create Layout...
Import Plot Settings...
On
Off
Calibrate
Configure
Menus...
Toolbars...
Keyboard...
Tool Palettes...
Edit Custom Files
Current Menu
Program Parameters (acad.pgp)

DRAW MENU

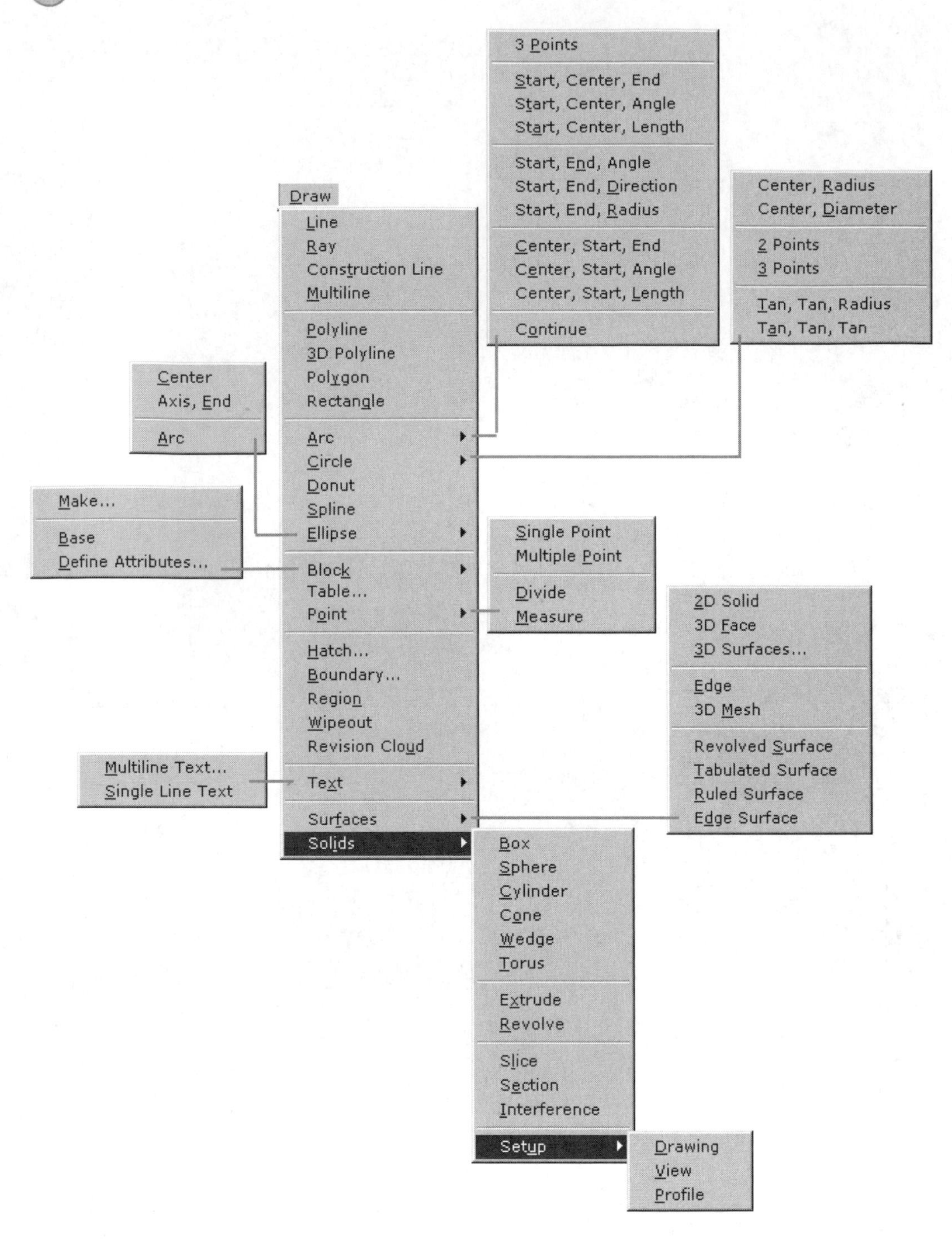

Draw
Line
Ray
Construction Line
Multiline
Polyline
3D Polyline
Polygon
Rectangle
Arc
Circle
Donut
Spline
Ellipse
Block
Table...
Point
Hatch...
Boundary...
Region
Wipeout
Revision Cloud
Text
Surfaces
Solids
3 Points
Start, Center, End
Start, Center, Angle
Start, Center, Length
Start, End, Angle
Start, End, Direction
Start, End, Radius
Center, Start, End
Center, Start, Angle
Center, Start, Length
Continue
Center, Radius
Center, Diameter
2 Points
3 Points
Tan, Tan, Radius
Tan, Tan, Tan
Center
Axis, End
Arc
Make...
Base
Define Attributes...
Single Point
Multiple Point
Divide
Measure
2D Solid
3D Face
3D Surfaces...
Edge
3D Mesh
Revolved Surface
Tabulated Surface
Ruled Surface
Edge Surface
Multiline Text...
Single Line Text
Box
Sphere
Cylinder
Cone
Wedge
Torus
Extrude
Revolve
Slice
Section
Interference
Setup
Drawing
View
Profile

DIMENSION MENU

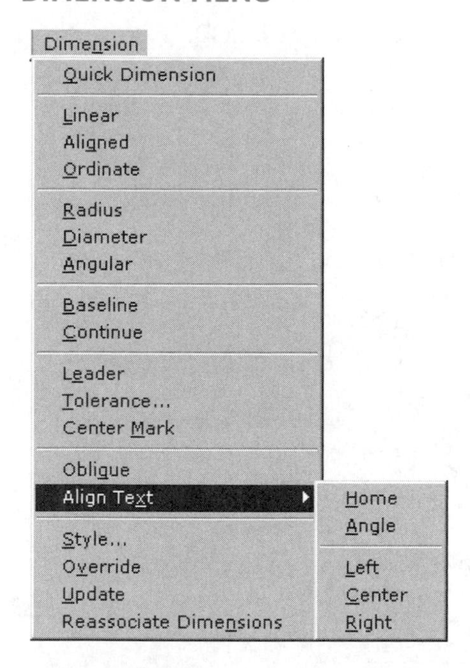

MODIFY MENU

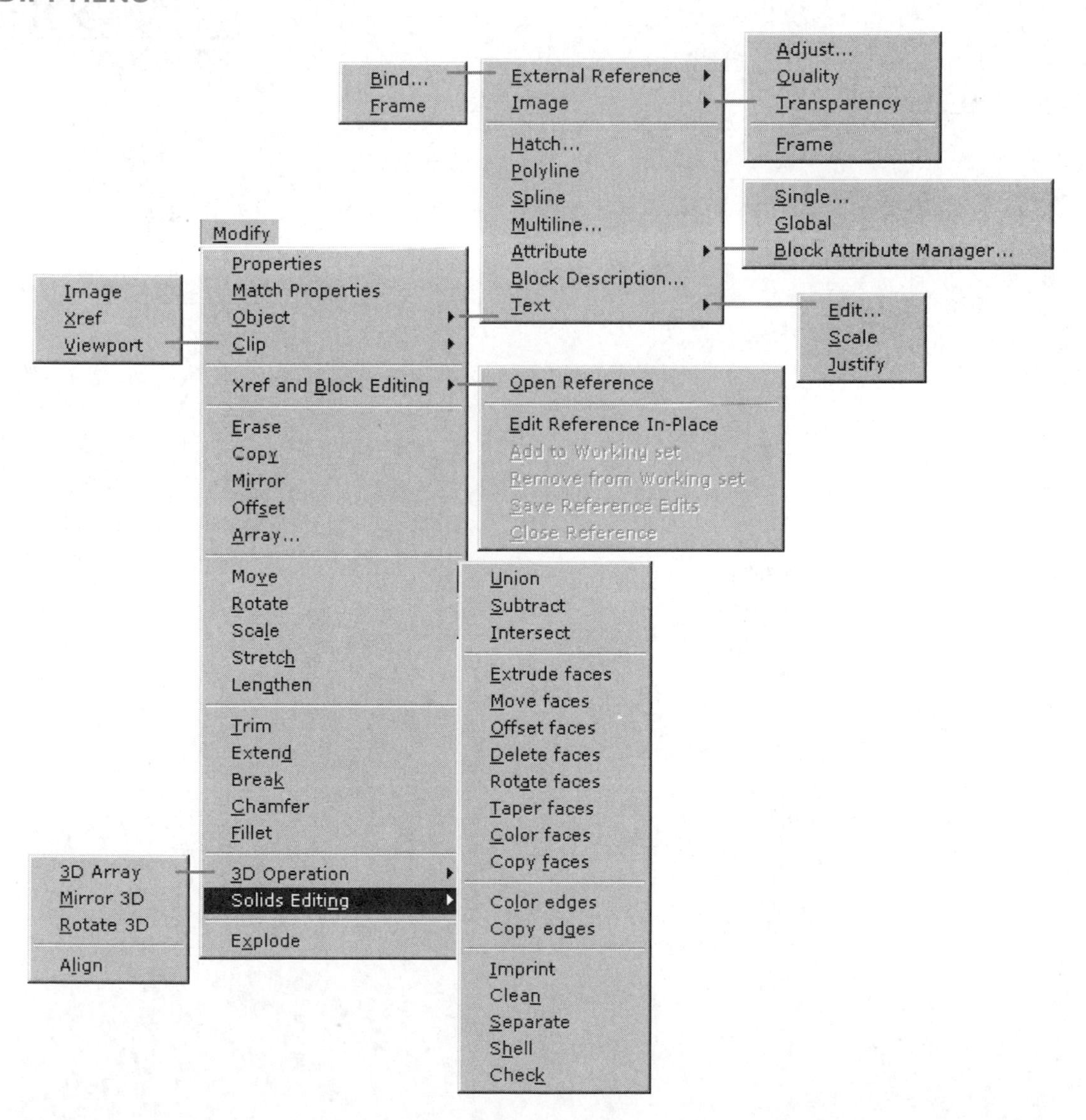

EXPRESS MENU

This menu is available only after being installed separately from the distribution CD.

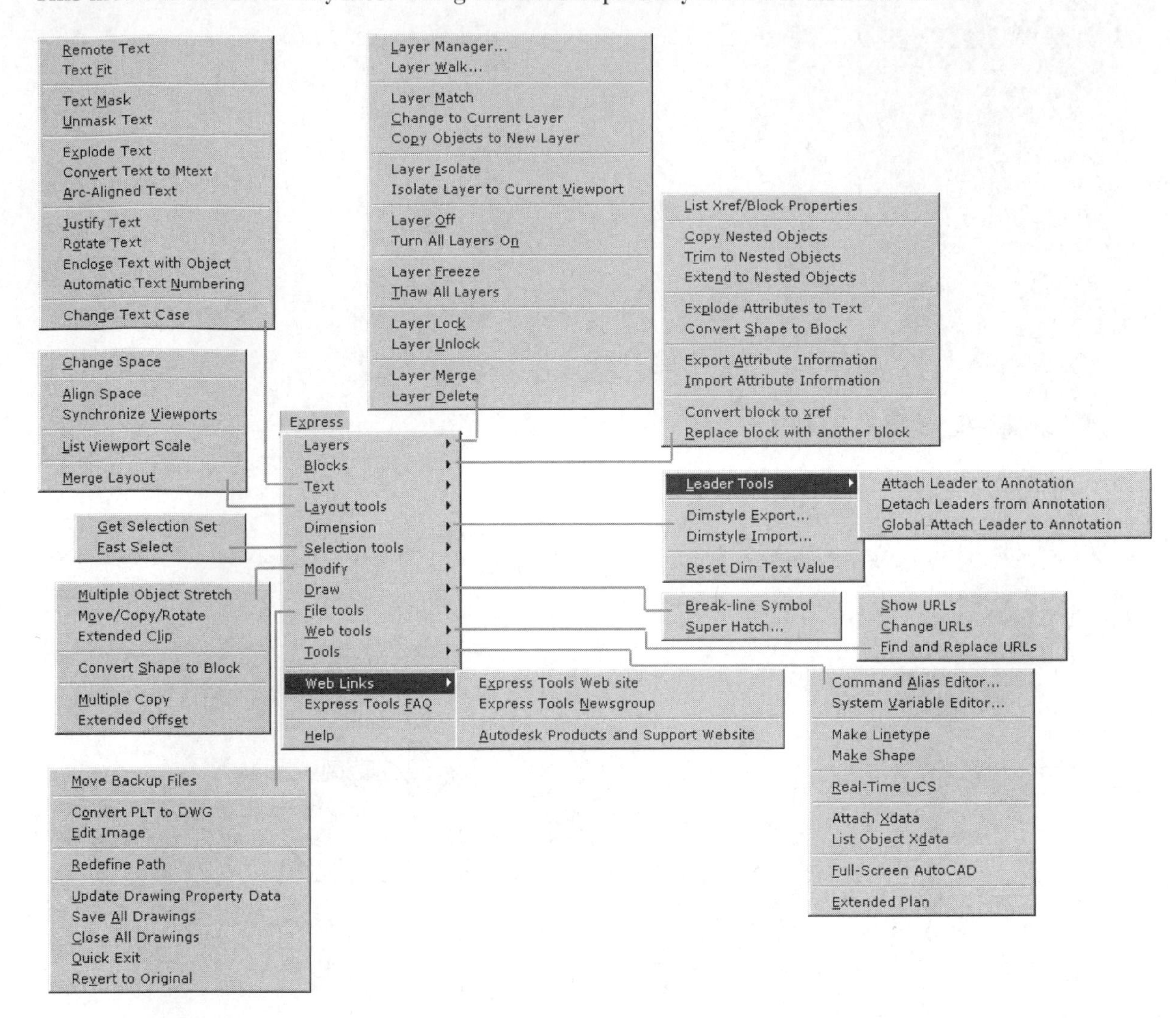

WINDOW MENU

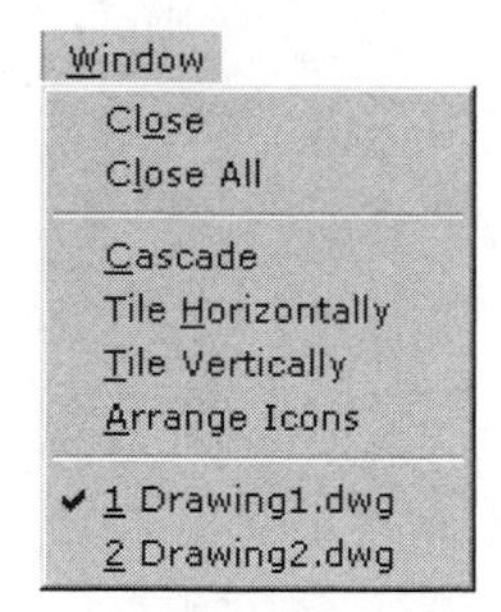

HELP MENU

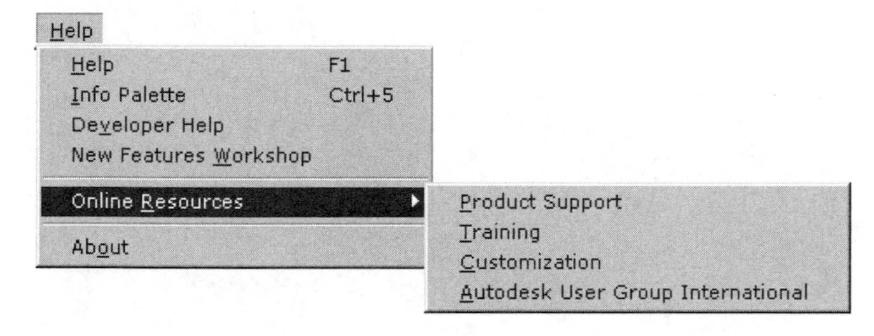

WINDOW CONTROL MENU

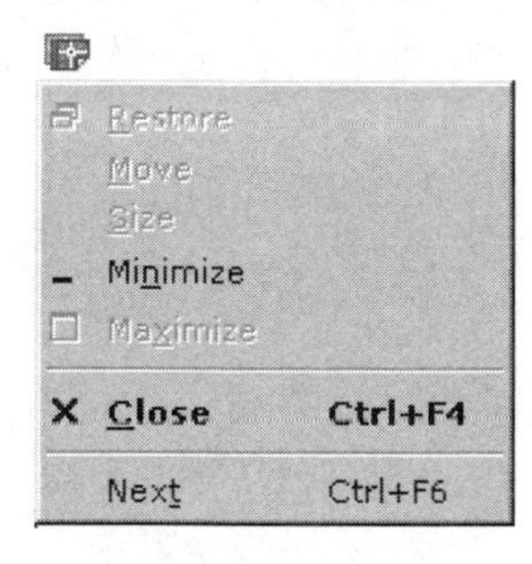

D

AutoLISP Functions

A

ACTION_TILE	Assigns action to be evaluated when user selects dialog box tile.
ADD_LIST	Adds and modifies strings in current dialog box listbox.
AND	Returns T if all expressions evaluate to non-nil.
ANGLE	Returns the angle in radians measured from the x axis.
APPEND	Combines many lists into a single list.
ASCII	Converts first character of string to its ASCII code.
ASSOC	Searches lists.
ATAN	Arctangent of a number, or the division of two numbers.
ATOF	Converts angle string into a real number.
ATOI	Converts angle string into an integer.
ATOM	Returns T if not a list.

C

CADDR	Returns the third item of a list (z).
CADR	Returns the second item of a list (y).
CAR	Returns the first item of a list (x).
CDR	Remove the first item from a list.
CHR	Converts ASCII code to a single character.
CLOSE	Closes open files.
COMMAND	Executes AutoCAD commands.
COND	Finds non-nil values.
CONS	Adds items to the beginning of a lists, and constructs dotted pairs.
COS	Cosine of angle in radians.
CVUNIT	Converts units.

D

DEFUN	Defines functions.
DIMX_TILE	X-dimension of the dialog box image tile.
DIMY_TILE	Y-dimension of the dialog box image tile.
DISTANCE	3D distance between two points.
DONE_DIALOG	Terminates dialog box.

E

END_IMAGE	Ends creation of dialog box image tile.
END_LIST	Ends processing of dialog box list box.
ENTDEL	Deletes and undeletes objects.
ENTGET	Returns the records of objects.
ENTLAST	Returns the name of the last non-deleted object in a drawing.
ENTMAKE	Creates new objects in drawings.
ENTMOD	Modifies the records of objects.
ENTNEXT	Returns the name of the next non-deleted object in the drawing database.
ENTSEL	Prompts the user to select a single object in the drawing.
ENTUPD	Updates the screen display of objects modified by ENTMOD.
EQ	Determines whether two expressions point to the same memory space.
EQUAL	Determines whether two expressions are equal using an optional fuzz factor.
EXPT	Exponent.

F

FILL_IMAGE	Draws filled rectangles in dialog boxes.
FINDFILE	Searches for files.
FIX	Converts reals into integers (rounds down).
FLOAT	Converts integers into real numbers.
FOREACH	Evaluates expressions for all items in a list.

G

GET_ATTR	Retrieves the DCL value of dialog box tile attributes.
GET_TILE	Retrieves the value of dialog box tiles.
GETANGLE	Prompts user for an angle.
GETCORNER	Prompts user for a rectangle's second corner.
GETDIST	Prompts user for a distance.
GETFILED	Prompts user to select a file name from a dialog box.
GETINT	Prompts user for an integer.
GETKWORD	Prompts user for a key word specified by INITGET.
GETORIENT	Prompts user for an angle.
GETPOINT	Prompts user for a point.
GETREAL	Prompts user for a real number.
GETSTRING	Prompts user for a string.
GETVAR	Returns the value of system variables.

H

HANDENT	Returns the object name specified by its handle.

I

IF	Tests which of two expressions is true.
INITGET	Specifies allowable input options for GETKWORD.
INTERS	Finds the intersection of two lines.
ITOA	Converts an integer to a string.

L

LAST	Returns the last element of a list.
LENGTH	Reports the number of elements in the list.

LIST	Combines expressions into a list.
LISTP	Determines if an item is a list.
LOAD	Reads *.lsp* program files.
LOAD_DIALOG	Loads *.dcl* files that define dialog boxes.

M

MINUSP	Determines if a number if negative.
MODE_TILE	Sets the mode of dialog box tiles.

N

NENTSEL	Prompts the user to select an object.
NEW_DIALOG	Activates dialog boxes.
NOT	Determines if an item is nil.
NTH	Returns the *n*th element of a list.
NULL	Determines if an item is nil.
NUMBERP	Determines if an item is a number.

O

OPEN	Opens ASCII files for reading, writing, or appending.
OR	Determines if an expression is non-nil.
OSNAP	Returns a 3D point after object snapping to an object.

P

POLAR	Returns the 3D point of an angle and distance.
PRIN1	Prints a string to the command line, and suppresses nil.
PRINC	Prints a string to the command line or open file.
PRINT	Prints a return, the string, and a blank space to the command line or open file.
PROGN	Evaluates expressions sequentially.
PROMPT	Prints string to the command line and the Text window.

Q

QUOTE	Does not evaluate expressions (also can use ').

R

READ-CHAR	Reads ASCII character from keyboard or from file.
READ-LINE	Reads a string from keyboard or from file.
REM	Divides one number by another, and returns the remainder.
REPEAT	Repeats an expression *n* times.
RTOS	Converts numbers to strings.

S

SET_TILE	Sets the value of dialog box tiles.
SETQ	Assigns values to variables.
SETVAR	Sets system variables.
SIN	Sine of an angle in radians.
SLIDE_IMAGE	Displays slides in dialog box image tiles.
SSADD	Adds entities to selection sets.
SSDEL	Deletes entities from selection sets.
SSGET	Prompts the user to select objects.
SSLENGTH	Determines the number of entities in a selection set.

SSMEMB — Determines whether an object is a member of a selection set.
SSNAME — Returns the name the *n*th object in a selection set.
START_DIALOG — Displays the current dialog box.
START_IMAGE — Starts creating images in dialog boxes.
START_LIST — Starts processing lists in dialog boxes.
STRCASE — Changes strings to upper or lower case.
STRCAT — Concatenates strings.
STRLEN — Determines the length of a string (in characters).
SUBST — Substitutes portions of strings.
SUBSTR — Returns the substring of a string.

T

TBLNEXT — Finds the next item in the specified symbol tables.
TBLOBJNAME — Returns the entity name of specified symbol table entries.
TBLSEARCH — Searches symbol tables.
TERM_DIALOG — Terminates dialog boxes.
TYPE — Determines the type of items.

U

UNLOAD_DIALOG — Unloads *.dcl* files from memory.

V

VECTOR_IMAGE — Draws vectors in dialog box image tiles.

W

WCMATCH — Wild card matching.
WHILE — Repeats an expression until nil.
WRITE-CHAR — Writes a character to the command line or to a file.
WRITE-LINE — Writes a string to the command line or to a file.

Z

ZEROP — Determines if a number is zero.

MISC.

1- — Decrements by 1.
1+ — Increments by 1.
+ — Sum.
- — Subtraction.
* — Multiplication.
/ — Division.
/= — Not equal.
< — Less than.
<= — Less than or equal to.
= — Equal.
> — Greater than.
>= — Greater than or equal to.

PI — 3.14159

INDEX

A

AutoCAD commands and system variables are shown in uppercase boldface, such as **ADCENTER**.

D

E

F

G

H

I

N

O

P

Q

R

2

3

LICENSE AGREEMENT FOR AUTODESK PRESS

A Thomson Learning Company

Educational Software/Data

You the customer, and Autodesk Press incur certain benefits, rights, and obligations to each other when you open this package and use the software/data it contains. BE SURE YOU READ THE LICENSE AGREEMENT CAREFULLY, SINCE BY USING THE SOFTWARE/DATA YOU INDICATE YOU HAVE READ, UNDERSTOOD, AND ACCEPTED THE TERMS OF THIS AGREEMENT.

Your rights:

1. You enjoy a non-exclusive license to use the enclosed software/data on a single microcomputer that is not part of a network or multi-machine system in consideration for payment of the required license fee, (which may be included in the purchase price of an accompanying print component), or receipt of this software/ data, and your acceptance of the terms and conditions of this agreement.

2. You own the media on which the software/data is recorded, but you acknowledge that you do not own the software/data recorded on them. You also acknowledge that the software/data is furnished "as is," and contains copyrighted and/or proprietary and confidential information of Autodesk Press or its licensors.

3. If you do not accept the terms of this license agreement you may return the media within 30 days. However, you may not use the software during this period.

There are limitations on your rights:

1. You may not copy or print the software/data for any reason whatsoever, except to install it on a hard drive on a single microcomputer and to make one archival copy, unless copying or printing is expressly permitted in writing or statements recorded on the diskette(s).

2. You may not revise, translate, convert, disassemble or otherwise reverse engineer the software/data except that you may add to or rearrange any data recorded on the media as part of the normal use of the software/ data. 3. You may not sell, license, lease, rent, loan, or otherwise distribute or network the software/data except that you may give the software/data to a student or and instructor for use at school or, temporarily at home. Should you fail to abide by the Copyright Law of the United States as it applies to this software/data your license to use it will become invalid. You agree to erase or otherwise destroy the software/data immediately after receiving note of Autodesk Press' termination of this agreement for violation of its provisions. Autodesk Press gives you a LIMITED WARRANTY covering the enclosed software/data. The LIMITED WARRANTY can be found in this product and/or the instructor's manual that accompanies it. This license is the entire agreement between you and Autodesk Press interpreted and enforced under New York law.

Limited Warranty

Autodesk Press warrants to the original licensee/ purchaser of this copy of microcomputer software/ data and the media on which it is recorded that the media will be free from defects in material and workmanship for ninety (90) days from the date of original purchase. All implied warranties are limited in duration to this ninety (90) day period. THEREAFTER, ANY IMPLIED WARRANTIES, INCLUDING IMPLIED WARRANTIES OF MERCHANTABILITY AND FITNESS FOR A PARTICULAR PURPOSE ARE EXCLUDED. THIS WARRANTY IS IN LIEU OF ALL OTHER WARRANTIES, WHETHER ORAL OR WRITTEN, EXPRESSED OR IMPLIED.

If you believe the media is defective, please return it during the ninety day period to the address shown below. A defective diskette will be replaced without charge provided that it has not been subjected to misuse or damage. This warranty does not extend to the software or information recorded on the media. The software and information are provided "AS IS." Any statements made about the utility of the software or information are not to be considered as express or implied warranties. Delmar will not be liable for incidental or consequential damages of any kind incurred by you, the consumer, or any other user. Some states do not allow the exclusion or limitation of incidental or consequential damages, or limitations on the duration of implied warranties, so the above limitation or exclusion may not apply to you. This warranty gives you specific legal rights, and you may also have other rights which vary from state to state. Address all correspondence to:

AutodeskPress
Executive Woods
5 Maxwell Drive
Clifton Park, NY 12065-2919